Socio-ecological Studies in Natural Protected Areas

Alfredo Ortega-Rubio
Editor

Socio-ecological Studies in Natural Protected Areas

Linking Community Development and Conservation in Mexico

Springer

Editor
Alfredo Ortega-Rubio
Red Nacional Áreas Naturales Protegidas RENANP-CONACYT
Centro de Investigaciones Biológicas del Noroeste CIBNOR-CONACYT
La Paz, Baja California Sur, Mexico

ISBN 978-3-030-47263-4 ISBN 978-3-030-47264-1 (eBook)
https://doi.org/10.1007/978-3-030-47264-1

This Springer imprint is published by the registered company Springer Nature Switzerland AG
The registered company address is: Gewerbestrasse 11, 6330 Cham, Switzerland

Foreword

In the early 1970s a silent revolution was taking place in UNESCO's central offices at Place de Fontenoy in Paris. An Italian ecologist, Francesco di Castri, had been hired by UNESCO to implement the goals and further the results of the International Biological Program (IBP), which he was a part of in the 1960s. Di Castri, who spent several years as a researcher in Chile and knew South America extremely well, teamed up with a French ecologist, Michel Batisse, and started discussing the accomplishments and shortcomings of the IBP. As a French scientist, Batisse had worked extensively, doing research in Francophone Africa, and had, like di Castri, acquired a serious concern about the social conditions of the underprivileged populations in poor, undeveloped countries. "How are we going to establish nature conservation programs," they wondered, "among people that have been exploited for centuries, have seen their natural resources being stolen from them under colonial regimes, and have very little means to survive and prosper?" Conservation, they realized, is a long-term commitment that can seem contradictory in the face of abject poverty and deprivation.

In order to tackle this challenge they contacted a number of scientists around the World, including two extraordinary Mexican ecologists: Gonzalo Halffter and Arturo Gómez-Pompa. Together, they developed a critique of the "isolated wilderness" paradigm in conservation. In a globalized planet, they argued, no ecosystem is entirely devoid of human influence. The idea of pristine ecosystems for conservation is a myth, argued Arturo Gómez-Pompa in a now famous paper. Conservation must be done with the human populations that live in those diversity-rich ecosystems, and it must be of benefit for them. Otherwise, they argued, it will keep on failing in the poorest nations, as the pressures to extract resources from the often impoverished land keep mounting. In the wake of these discussions, UNESCO's Man and the Biosphere (MAB) program was born, and, soon after, in 1974, the first Biosphere Reserves were created.

The legacy of the creators of the MAB program is remarkable. At a time in which almost no one spoke of global environmental change, the subject was a top priority for their program. They understood that if a significant amount of the planet's biodiversity was to be saved, an international network of reserves should be designed to cover all the biomes around the planet, the whole of the global biosphere. Lastly, they realized that if conservation in the poorest countries in the world was to be successful, it necessarily had to incorporate a social dimension. Ecological criteria were not enough, Biosphere Reserves had to include the collaborative participation of the communities that lived in, or around, the new model of protected areas. Visionaries, pioneering scientists, the concepts that were discussed and adopted in the UN World Summit at Rio in 1992 were already the driving ideas of the founders of the MAB program 20 years earlier.

Fifty years after the MAB program was established, it is now time to look back and evaluate achievements and disappointments of MAB's socio-ecological approach. In Mexico, we have had undeniable progress, but also some serious obstacles to implement MAB's vision. Many of these obstacles are related to the recalcitrant nature of governmental bureaucracies and the difficulties embedded in trying to implement what essentially should be a grassroots model into a complex and centralized government that has—by its very nature and organization—serious difficulties working with self-governing communities and grassroots organizations. But the other obstacle, I believe, is within our own selves, within the conservation organizations and the mind of conservation scientists. Truly collaborative multi- or transdisciplinary work implies for individual scientists to be able to venture into fields of scientific enquiry that are often not our own. Social scientists need to learn more about ecology and evolution, about the complexities of the flow of life and energy in the biosphere, or the tragic dimension of biological extinction and the risks involved in biodiversity loss. Natural scientists, on the other hand, need a much deeper understanding of how social sciences work, and the patience necessary for the challenging processes involved in recording social data, interacting with local communities, and developing the intellectual flexibility to understand the view of life and cosmos in other cultures.

This book is a very important contribution from a group of Mexican and international scientists to this goal, namely, to build bridges between the social and the ecological sciences, and to promote a new paradigm, that of a new scientific discipline which we can call "Socioecology." In that sense, it is of great and emblematic significance that the primary leader behind this project, Alfredo Ortega-Rubio, is a disciple of one of the founders of the MAB program, Gonzalo Halffter, and was formed as a scientist by Halffter himself under the tenets of the Biosphere Reserve initiative. Contemplating back to the early 1970s, when a handful of researchers got together in UNESCO's Place de Fontenoy headquarters thinking how to change the World, one cannot but feel a certain optimism seeing how much their ideas have taken root in a whole country like Mexico. The social dimension has now taken strong roots in the science of conservation and in the dream for a viable biosphere and a sustainable future.

It is now part of the geography of hope.

University of California, Riverside
Riverside, CA, USA

Exequiel Ezcurra

Preface

Globally, the most useful tool to make biodiversity conservation compatible with promoting sustainable development of rural communities is Protected Natural Areas (PNA). The PNA are terrestrial, coastal, or aquatic areas of a given territory, which are subjected to special protection, conservation, restoration, and development regimes that are representative of the different ecosystems and their biodiversity. To be considered as PNA, such territory must contain a high biodiversity and where the original environment has not been significantly altered by man. There is also the case when the territory to be considered as PNA contains a biodiversity that is not necessarily high but unique, in other words, it contains species that inhabit exclusively in this territory, that is, endemic species.

In either case, whether it is high diversity or high uniqueness, it is very important to keep in mind that because of their special legal regimes, PNA have their own jurisdiction. Likewise, in all of them, either within their polygons or in their areas of influence, human communities have settled there. Most of these communities were there prior to the PNA decree.

In countries with high population growth rates and urgent socio-economic needs, such as in those countries called "In Developing Process," it is simply not feasible to think that biodiversity conservation cannot be aligned to meeting the enormous needs of local inhabitants. Thus, PNA should be seen as strategic mechanisms, spaces that generate a transcendent stream of socio-economic benefits, including the conservation of environmental services and also of economic benefits, jointly with increasing life quality of human inhabitants, specially their ancestral communities.

The PNA are not only very valuable geographical spaces and invaluable germplasm reservoirs but also deposits of ancestral knowledge of their native inhabitants, who have had a tradition for generations of the sustainable use of their valuable natural resources in such important aspects as medicine and food. In this context, the increasingly generalized and advanced concepts in sustainable development and environmental protection necessarily imply both for local socio-economic development and conservation per se; they are essential to conceptualize the indissoluble articulation between the natural and social universe. In other words, it should arrive at the conceptualization of Socioecology.

In this manner, Socioecology should be considered as a focus that has not been competently developed in the Natural Protected Areas yet. This framework requires incorporating both the contextualization of environmental knowledge in the socio-economic matrix and properly integrating social considerations into the study of ecological aspects. We must understand that social and environmental aspects are intrinsically embedded in the PNA. Thus, in this context, the expansion of ecological knowledge has finally articulated, as never before in the history of this science, the knowledge of the social world with those of the environment. In fact, the PNA are a clear example that the current challenge is to decipher the co-evolution existing in the protected natural areas between local culture and the use, management and conservation of biodiversity.

The management of biodiversity from this logical point of view must be conceived indissolubly framed with the cultural, social, and economic aspects of local people, so the issues should be addressed from a socio-ecological perspective. It is essential to determine this tetra-dimensionality between environmental, cultural, social, and economic aspects to establish holistic proposals that really positively impact the sustainability of PNA.

From the Socio-ecological approach, sustainable social development is impossible to achieve without biodiversity sustainability. Likewise, the ecological balance can never be achieved if it is not through the in-depth study of the socioeconomic and sociological aspects of the human communities inserted in that ecological environment. Undoubtedly Socioecology is a young science although in full development. Thus, this aspect of science is surely essential because it could reconcile the philosophical contradictions that exist between ultra-developmentalism, which seeks to ignore how ecological systems actually work, and fundamentalist ecology, which does not take into account the aspirations of the social groups that have been in dynamic equilibrium in such ecosystems for generations. Undoubtedly, for the purpose of achieving such compatibility, it should be based on socio-ecological considerations.

All authors contributing to this work consider that although this new science is not fully developed, very valuable contributions have been developed with a holistic approach to study the real cultural, ecological, social, and economic interactions in the PNA that deserve to be summarized and disseminated, which are included in this context. This book contributes with very valuable approaches and experiences of those who, in the PNA, have worked with this holistic approach; the result of the synthesis of knowledge and experiences, which in the light of some basic information provided by the ecological sciences, aspires to give meaning and find explanation for situations that have been addressed independently by the social sciences.

Undoubtedly the contributions in the different sections of this book, all of them with a socio-ecological approach, will contribute to highlight the sustainability results with explanations based on the knowledge of the different sector agents, usually fragmentary because of the recognized methodological inadequacies of mono-disciplinary sciences.

Hopefully, this book will not only be a paradigmatic contribution to a young science, still in development, but also enjoyable reading of each of the chapters contained herein.

We acknowledge the economic support of CONACYT to the Thematic Research Network of Natural Protected Areas RENANP-CONACYT, the CONACYT Basic Science project number 251919, and the support of CIBNOR-CONACYT.

La Paz, Baja California Sur, Mexico
Spring 2020

Alfredo Ortega-Rubio

Contents

Part II Methodological Aspects

Part III Social Groups and Ecological Knowledge

Contributors

Melany Aguilar-López ECOYDES A. C., Mineral de la Reforma, Hidalgo, México

Marco Antonio Almendarez-Hernández Office of Intellectual Property and Technology Marketing and the Patent Center, Centro de Investigaciones Biológicas Noroeste (CIBNOR), La Paz, Baja California Sur, México

Romina Alzugaray-Martínez Centro de Investigaciones Pesqueras (CIP), La Habana, Cuba

José Alfredo Arreola-Lizárraga Centro de Investigaciones Biológicas del Noroeste SC, Guaymas, Sonora, México

Victor Javier Arriola-Padilla National Institute of Forestry, Agricultural and Livestock Research, National Centre for Disciplinary Research on Conservation and Improvement of Forest Ecosystems (INIFAP), Mexico City, Mexico

David Aurioles-Gamboa Facultad de Ingeniería Ambiental, Universidad Popular Autónoma del Estado de Puebla, Puebla, México

Cielo María Ávila-López Universidad Autónoma de Yucatán, Mérida, Yucatán, Mexico

Gerzaín Avilés-Polanco CONACYT-Centro de Investigaciones Biológicas del Noroeste, La Paz, Baja California Sur, Mexico

Araceli Avilés-Quevedo Freelance Services for R & D in Marine Aquaculture, La Paz, Baja California Sur, Mexico

Isaac Azuz-Adeath Centro de Excelencia en Innovación y Diseño, Colegio de Ingeniería, Centro de Enseñanza Técnica y Superior (CETYS Universidad), Ensenada, Baja California, Mexico

Isabel Banos-Gonzalez Departamento de Didáctica de las Ciencias Experimentales. Facultad de Educación, Universidad de Murcia, Murcia, Spain

Everardo Barba-Macías Sustainable Management of Basin and Coastal Zones, Sustainability Sciences Department, El Colegio de la Frontera Sur, Villahermosa, Tabasco, Mexico

Juan Manuel Barragán Muñoz Research Goup on Integrated Coastal Zone Management, University of Cadiz, Cádiz, Spain

Luis Felipe Beltrán-Morales Centro de Investigaciones Biológicas del Noroeste S.C. (CIBNOR), La Paz, Baja California Sur, Mexico

Vianney Beraud-Macías Universidad Autónoma de Aguascalientes, Aguascalientes, Mexico

Abel Betanzos-Vega Centro de Investigaciones Pesqueras (CIP), La Habana, Cuba

María Carmen Blázquez Behavioural Ecology Lab. Environmental Planning and Conservation Department, Centro de Investigaciones Biológicas Noroeste (CIBNOR), La Paz, Baja California Sur, México

Aurora Breceda Centro de Investigaciones Biológicas del Noroeste, La Paz, Baja California Sur, Mexico

Micheline Cariño-Olvera Universidad Autónoma de Baja California Sur, La Paz, Mexico

Jorge Carrión-Tacuri Dirección del Parque Nacional Galápagos, Santa Cruz, Ecuador

Pamela Castro-Figueroa Maestría en Innovación en Ingeniería, Colegio de Ingeniería, Centro de Enseñanza Técnica y Superior (CETYS Universidad), Ensenada, Baja California, Mexico

Luz María Cruz-García Universidad Autónoma de Baja California Sur, La Paz, Baja California Sur, México

María de Andrés García Research Goup on Integrated Coastal Zone Management, University of Cadiz, Cádiz, Spain

Yven Echeverría-Ayala Independent Consultant on Climate Change Adaptation based on Ecosystems (AbE), Mexico City, Mexico

Julio César Errejón-Gómez Environmental Planning and Conservation Program, Centro Investigaciones Biológicas Noroeste (CIBNOR), La Paz, Baja California Sur, Mexico

Juan Pablo Esparza-Carlos Departamento de Ecología y Recursos Naturales, Universidad de Guadalajara, Autlán, Jalisco, Mexico

Miguel Ángel Esteve-Selma Departamento de Ecología e Hidrología, Universidad de Murcia, Murcia, Spain

Exequiel Ezcurra University of California, Riverside, CA, USA

Macarena Fernández Génova Centro de Estudios del Cuaternario, de Fuego – Patagonia y Antártica (Fundación CEQUA), Punta Arenas, Región de Magallanes, y de La Antártica Chilena, Chile

Alba E. Gámez Department of Economics, Universidad Autónoma de Baja California Sur, La Paz, Baja California Sur, Mexico

Blanca C. Garcia Colegio de la Frontera Norte, COLEF, (Northern Borderlands Research College), Monterrey, Mexico

Milagro García-Bernal Centro de Investigaciones Biológicas del Noroeste, La Paz, Baja California Sur, Mexico

Javier García Sanabria Research Goup on Integrated Coastal Zone Management, University of Cadiz, Cádiz, Spain

Andrés Giménez-Casalduero Área de Ecología, Departamento de Biología Aplicada, Universidad Miguel Hernández, Elche, Spain

Maria Eugenia González-Ávila Departamento de Desarrollo Urbano y Medio Ambiente, El Colegio de la Frontera Norte, Monterey Nuevo León, Mexico

Department of Urban Development and Environment, The North Border College, Tijuana, Mexico

Inti González Ruiz Centro de Estudios del Cuaternario de Fuego – Patagonia y Antártica (Fundación CEQUA), Punta Arenas, Región de Magallanes y de La Antártica Chilena, Chile

Cambio Climático y Variaciones Recientes, Centro de Investigación Fundación CEQUA (CEQUA), Punta Arenas, Magallanes y Antártica Chilena, Chile

Irma González-López Comisión Nacional de Áreas Naturales Protegidas, Secretaría de Medio Ambiente y Recursos Naturales, La Paz, Baja California Sur, México

Juan Carlos Graciano Department of Economics, Universidad Autónoma de Baja California Sur, La Paz, Baja California Sur, Mexico

Eduardo Gutiérrez-Rivero Ministerio de Ciencia, Tecnología y Medio Ambiente, Centro Meteorológico Provincial, Bayamo, Cuba

Claudia J. Hernández-Camacho Centro Interdisciplinario de Ciencias Marinas, Instituto Politécnico Nacional, La Paz, Baja California Sur, México

Sergio Hernández-Solorzano Independent Consultor, Monterrey, Nuevo León, Mexico

Edgar Mauricio Hoyos-Padilla Pelagios Kakunjá, A.C, La Paz, Baja California Sur, México

Juan Pedro Ibarra-Michel Departamento Académico de Ciencias Económico Administrativas, Universidad Autónoma de Occidente, Mazatlán, Sinaloa, Mexico

Luis Ignacio Iñiguez-Dávalos Departamento de Ecología y Recursos Naturales, Universidad de Guadalajara, Autlán, Jalisco, Mexico

Mercedes Isla-Molleda Centro de Investigaciones Pesqueras (CIP), La Habana, Cuba

Cecilia L. Jiménez-Sierra Universidad Autónoma Metropolitana – Unidad Iztapalapa, Ciudad de México, México

Juan Juárez-Flores Sustainable Management of Basin and Coastal Zones, Sustainability Sciences Department, El Colegio de la Frontera Sur, Villahermosa, Tabasco, Mexico

James Thomas Ketchum Pelagios Kakunjá, A.C, La Paz, B.C.S, México

Centro de Investigaciones Biológicas del Noroeste, La Paz, Baja California Sur, México

Magdalena Lagunas-Vázques CONACYT-CCGS Centro de Cambio Global y La Sustentabilidad, Villahermosa, Tabasco, Mexico

Walter López-Báez Centro de Investigación Regional Pacífico Sur, Instituto Nacional de Investigaciones Forestales Agrícolas y Pecuarias, Inifap, Chiapas, Mexico

Elia López-Greene Comisión Nacional de Áreas Naturales Protegidas, Secretaría de Medio Ambiente y Recursos Naturales, La Paz, Baja California Sur, México

Romina López Márquez Centro de Estudios del Cuaternario de Fuego – Patagonia y Antártica (Fundación CEQUA), Punta Arenas, Región de Magallanes y de La Antártica Chilena, Chile

Divulgación Científica para la educación, Centro de Investigación Fundación CEQUA (CEQUA), Punta Arenas, Magallanes y Antártica Chilena, Chile

Julia Martínez-Fernández Fundación Nueva Cultura del Agua, Zaragoza, Spain

María Loraine Matias-Palafox Comisión Nacional para el Conocimiento y Uso de la Biodiversidad, Ciudad de México, México

José Manuel Mazón-Suástegui Centro de Investigaciones Biológicas del Noroeste, La Paz, Baja California Sur, Mexico

Carlos Mena Instituto de Geografía, Diego de Robles y Pampite (Cumbaya), Quito, and Universidad San Francisco de Quito, Quito, Ecuador

Manuel Mendoza-Carranza Sustainable Management of Basin and Coastal Zones, Sustainability Sciences Department, El Colegio de la Frontera Sur, Villahermosa, Tabasco, Mexico

Gonzalo Merediz-Alonso Amigos de Sian Ka'an A.C., Cancún, Quintana Roo, Mexico

Ananda Monteforte-Cariño Universidad Nacional Autónoma de México, Mexico City, Mexico

Mario Monteforte-Sánchez Centro de Investigaciones Biológicas del Noroeste (CIBNOR), La Paz, Baja California Sur, México

Efrén Moreno-Arzate Doctorado en Biosistemática, Ecología y Manejo de Recursos Naturales y Agrícolas, Departamento de Ecología y Recursos Naturales, Universidad de Guadalajara, Autlán, Jalisco, Mexico

René Moreno-Terrazas-Troyo Universidad Autónoma de Baja California Sur, La Paz, Mexico

Elizabeth Olmos-Martínez Department of Economic-Administrative Sciences, Autonomous University of the West, Cali, Mexico

Departamento Académico de Ciencias Económico Administrativa, Universidad Autónoma de Occidente, Mazatlán, Sinaloa, Mexico

Alfredo Ortega-Rubio Red Nacional Áreas Naturales Protegidas RENANP-CONACYT, Centro de Investigaciones Biológicas del Noroeste CIBNOR-CONACYT, La Paz, Baja California Sur, México

Gustavo Padilla-Arredondo Centro de Investigaciones Biológicas del Noroeste SC, Guaymas, Sonora, México

Miguel Palmeros Marine Conservation Program, Sociedad de Historia Natural Niparajá, La Paz, Baja California Sur, México

Óscar Orlando Parra-Barrientos University of Concepción, Concepción, Chile

Carmen Pasos-Acuña Pelagios Kakunjá, A.C, La Paz, Baja California Sur, México

Lili Pelayo-González Centro Interdisciplinario de Ciencias Marinas, Instituto Politécnico Nacional, La Paz, Baja California Sur, México

Mario del Roble Pensado-Leglise Instituto Politécnico Nacional, CIIEMAD, Ciudad de México, México

Irene Pérez-Ibarra Agencia Aragonesa de Investigación y Desarrollo (ARAID), Zaragoza, Spain

Raúl Francisco Pineda-López Autonomous University of Querétaro, Santiago de Querétaro, Mexico

Manuel Jesús Pinkus-Rendón Universidad Autónoma de Yucatán, Mérida, Yucatán, Mexico

Miguel Pinkus-Rendón Centro Peninsular en Humanidades y Ciencias Sociales, Universidad Nacional Autónoma de México, Mérida, Yucatán, Mexico

Juan Fernando Pío-León Universidad Politécnica del Mar y la Sierra, La Cruz de Elota, Mexico

Elia Inés Polanco-Mizquez Centro Intercultural de los Desiertos y Océanos, Puerto Peñasco, Sonora, México

María Magdalena Ramírez-Martínez Departamento de Ciencias de la Salud y Ecología Humana, Universidad de Guadalajara, Autlán, Jalisco, Mexico

Nicolás Recabarren Traub Ridepatagonia, Punta Arenas, Región de Magallanes y de La Antártica Chilena, Chile

Fiorella Repetto-Giavelli Centro de Estudios del Cuaternario, de Fuego – Patagonia y Antártica (Fundación CEQUA), Punta Arenas, Región de Magallanes y de La Antártica Chilena, Chile

Evelia Rivera-Arriaga Instituto de Ecología, Pesquerías y Oceanografía del Golfo de México, Universidad Autónoma de Campeche, Campeche, Mexico

Enrique Rodríguez-Balam Centro Peninsular en Humanidades y Ciencias Sociales, Universidad Nacional Autónoma de México, Mérida, Yucatán, Mexico

Gerardo Rodríguez Quiroz Instituto Politécnico Nacional, CIIDIR Sinaloa, Guasave, Sinaloa, México

Martín Enrique Romero-Sánchez National Institute of Forestry, Agricultural and Livestock Research, National Centre for Disciplinary Research on Conservation and Improvement of Forest Ecosystems (INIFAP), Mexico City, Mexico

Heidi L. Romero-Schmidt Centro de Investigaciones Biológicas del Noroeste, La Paz, México

Martha P. Rosas-Hernández Centro de Investigación Científica y de Educación Superior de Ensenada, Unidad La Paz, La Paz, Baja California Sur, México

Thelma Michelle Ruiz-Ruiz Centro de Investigaciones Biológicas del Noroeste SC, Guaymas, Sonora, México

Carolina Sampedro Instituto de Geografía, Diego de Robles y Pampite (Cumbaya), Quito, and Universidad San Francisco de Quito, Quito, Ecuador

Alberto de Jesús Sánchez-Martínez Diagnostic and Management of Tropical wetlands laboratory, División Académica de Ciencias Biológicas, Universidad Juárez Autónoma de Tabasco, Villahermosa, Tabasco, Mexico

Ricardo V. Santes-Álvarez Department of Urban Studies and Environment, El Colegio de la Frontera Norte, Tijuana, Baja California, Mexico

Joaquín Sosa-Ramírez Universidad Autónoma de Aguascalientes, Aguascalientes, México

Alicia Tenza-Peral Área de Ecología, Departamento de Biología Aplicada, Universidad Miguel Hernández, Elche, Spain

Center for Behavior, Institutions and the Environment, Arizona State University, Tempe, AZ, USA

Clara Margarita Tinoco-Navarro University of Concepción, Concepción, Chile

Juliana Torres Mendoza Estudios Territoriales, Centro de Investigación Fundación CEQUA (CEQUA), Punta Arenas, Magallanes y Antártica Chilena, Chile

Centro de Estudios del Cuaternario de Fuego – Patagonia y Antártica (Fundación CEQUA), Punta Arenas, Región de Magallanes y de La Antártica Chilena, Chile

Daniel Torres-Orozco Jiménez Universidad Nacional Autónoma de México, Facultad de Ciencias, Mexico City, Mexico

Hitandehui Tovar Marine Conservation Program, Sociedad de Historia Natural Niparajá, La Paz, Baja California Sur, México

Cinthia Trinidad-Ocaña Sustainable Management of Basin and Coastal Zones, Sustainability Sciences Department, El Colegio de la Frontera Sur, Villahermosa, Tabasco, Mexico

Roberto Urrutia-Pérez Water Research Center for Agriculture and Mining (CRHIAM), University of Concepción, Concepción, Chile

Guiyer Eulselmar Vázquez-González Independent Consultor, Monterrey, Nuevo León, Mexico

Benito Vázquez-Quesada Universidad Nacional Autónoma de México, Facultad de Ciencias, Mexico City, Mexico

Mónica Velarde-Valdez Departamento Académico de Ciencias Económico Administrativas, Universidad Autónoma de Occidente, Mazatlán, Sinaloa, Mexico

Germaynee Vela-Ruiz Figueroa Centro de Estudios del, Cuaternario de Fuego - Patagonia y Antártica (Fundación CEQUA), Punta Arenas, Región de Magallanes y de La Antártica Chilena, Chile

Juana Isabel Vera López Department of Public Administration Studies, The North Border College, Tijuana, Mexico

Lorraine A. Williams-Beck Centro de Investigaciones Históricas y Sociales, Universidad Autónoma de Campeche, Campeche, Mexico

About the Editor

Alfredo Ortega-Rubio, Ph.D. in Ecology, has been granted with three national awards by the President of Mexico for his career in scientific research applied to biodiversity conservation and local communities' improvement in Natural Protected Areas. He has been author of 212 original research articles, published 26 books and 115 book chapters, and led 167 strategic projects of resource management, ecological restoration, and environmental impact assessment, including those required to create the Sierra de La Laguna, the Vizcaíno Desert, and the Revillagigedo Archipelago Biosphere Reserves. He has directed 73 theses, which include 40 Ph.D. and 28 Master of Science. Since 2004 he is Researcher "E," top Researcher level at the CONACYT Research Center System, and has been ratified for the third consecutive time as National Researcher Level III, the top category.

Part I
Theoretical Aspects

Chapter 1
Socioecology

Elizabeth Olmos-Martínez and Alfredo Ortega-Rubio

Abstract This chapter shows the importance and need of a socioecological approach in different disciplines to generate new knowledge of the nature-society relationship to deal holistically with different systems (social and ecological) and toward sustainability, starting from adaptive resilience. The importance of ethnology stands out in managing natural areas where consumptive and nonconsumptive uses of the provided environmental services are made. The concepts and interactions converging in a dynamism, each time stronger, are described. Some studies with a socioecological approach have been rescued, for example, multi- and trans-disciplines that favor conservation, use, and management of natural resources, as well as territory and ecosystem management.

Keywords Sustainability · Resilience · Social · Ecologic

1.1 Introduction

The socioecological system concept has generated a growing interest among scientific disciplines given its usefulness in the framework of sustainability and knowledge of the relationship between social and natural systems integrated in search for development. Such concept has modified the theoretical frameworks of different scientific disciplines and their fields of study when they refer to the human-environment relationship since it incorporates both not only as external actors that alter the ecosystems by exogenous pressures but also as one more of its components (integral, inseparable, and dependent) that intervenes integrally in its evolution (Gallopín et al. 1989; Gallopín 2003, 2004; Maass 2012).

E. Olmos-Martínez (✉)
Departamento Académico de Ciencias Económico Administrativas, Universidad Autónoma de Occidente, Mazatlán, Sinaloa, Mexico

A. Ortega-Rubio
Red Nacional Áreas Naturales Protegidas RENANP-CONACYT, Centro de Investigaciones Biológicas del Noroeste CIBNOR-CONACYT, La Paz, Baja California Sur, Mexico
e-mail: aortega@cibnor.mx

A. Ortega-Rubio (ed.), *Socio-ecological Studies in Natural Protected Areas*,
https://doi.org/10.1007/978-3-030-47264-1_1

In this sense, socioecology has gained great importance in different disciplines, such as environmental and conservation management, among others, because it promotes diverse advantages; however, it is also subjected to challenges where all types of social, political, and economic actors should be involved. For example, within the public policies directed to instruments of environmental management in Mexico, they do not take into account the socioecological system; nevertheless, the legal framework of the hydrographic basins gets a little closer to this combined system in terms of management stages (Challenger et al. 2014).

In this manner, socioecological systems show a wide approach where one of them interacts with the other one from the theoretical and practical points of view. Currently, research studies, in which the sustainability concept stands out, frequently only deal with the relationship of a system with a particular interest (Barton 2006; Sarandón and Flores 2009; Sabatini and Arenas 2000; Badii and Landeros 2007; Dieleman and Juárez-Nájera 2008; Izazola 2001; Plaza 2006), giving place to contradiction in the way of understanding it. Therefore, a socioecological analysis and its application in science is essential to deal with this concept from this point of view.

1.2 The Socioecological System

The socioecological system deals with interactions between society and nature, specifically, trying to understand the dynamic character of their interactions (Kates et al. 2001). It can also be understood as a social system (its subsystems and elements) integrated by an ecological system (its subsystems and elements) forming an inseparable set, in which the reciprocal relationships among the components and subsystems lead to the evolution of this type of system as a whole (Challenger et al. 2014).

According to Salas-Zapata et al. (2011), the interactions among this combined system emerge for different reasons, and the array of systems is so wide that any system is susceptible to be defined as a socioecological system. These authors argue that systems exist as diverse as enterprises, industries, economies, cities, and regions, among others, and all of them are object of interest in sustainability.

On the other hand, sustainability could be understood differently when diverse disciplines deal with it. In other words, the theoretical construction on the society-nature relationship turns difficult when it is dealt with in a parallel manner instead of a multidiscipline, which is when the importance of socioecology emerges.

The complex interaction of the different socioecological systems is shown in Fig. 1.1 where according to Salas-Zapata et al. (2011) they converge in a stronger dynamism since sustainability does not assume the objects of study as isolated elements but coupled to both society and ecology, as mentioned by Gallopín (2003, 2006). In this sense, system couplings are interactions among the domains that cause impacts and disturbances. For example, in the social domain, the subsystems are culture, politics, economy, and social organization; on the other hand, those in the ecological domain are nature and environment (created by humans). Thus, all systems may be considered as such as long as couplings or interactions between them are delimited.

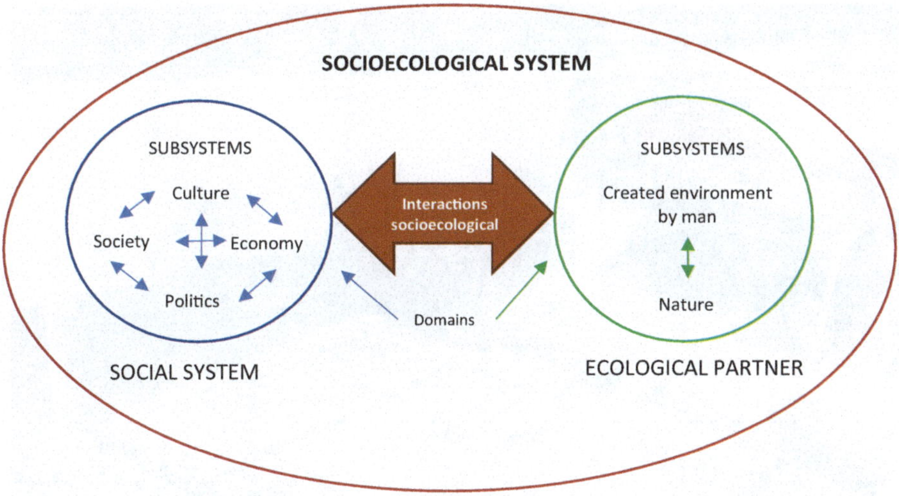

Fig. 1.1 Socioecological system. (Source: Authors' own production, adapted from Salas-Zapata et al. 2011)

According to Fiksel (2006) and Cumming and Norberg (2008), the content of the interactions may be of different nature; for example, if material, they are a flux of natural resources, money, raw matter, food waste, and persons; if nonmaterial, they may be a flux of information and knowledge, power, trust, norms, values, decision, public actions, and so on. Thus, a socioecological system usually mixes ethical, political, and environmental considerations, among others. As an example of such interactions, Fig. 1.2 shows the use of regional flora in a local nursery where women and youth work. Similarly, Fig. 1.3 shows another group of women gathered in a workshop for compost production, taking advantage of organic waste from housing and local decomposing flora; both groups work to increase their family income and belong to the protected natural area (PNA) of the Sierra de la Laguna Biosphere Reserve (REBIOSLA for its abbreviation in Spanish) in Baja California Sur, Mexico.

Additionally, Maass (2012) and Galán et al. (2013) stated that the concept of socioecological systems has gained a growing consensus with respect to its usefulness as an interdisciplinary research framework to understand the relationships between social and natural systems integrated as a potential management model. It also represents an important evolution in the underlying concepts of public policies in environmental management that have passed from a resource management approach in the last decades and, currently, toward a socioecological system management.

Moreover, Ostrom and Cox (2010) also mentioned that the socioecological system approach has increased as a proposal trying to explain the complexity of the system and its environment better because of the interrelation, scales, functions, and spatial heterogeneity that emerge.

Fig. 1.2 Nursery in San Jorge, REBIOSLA, Mexico. (Source: Elizabeth Olmos-Martínez)

1.3 Sustainability of the Socioecological Systems

According to Gallopín (2006) and Böhringer and Jochem (2007), understanding the sustainability of these systems may be starting from developing index and marker frameworks, such as the model Pressure, State, Response (PSR), Daly triangle, ecological footprint, Bossel model, environmental sustainability index, and sustainable economic welfare index (SEWI), among others. However, these markers describe sustainability, but they do no explain it since they allow observing how a system is sustainable but without understanding the organization models, processes, or the elements that make it sustainable or not (Salas-Zapata et al. 2012).

In the same manner, Salas-Zapata et al. (2012) mentioned that under the resilience approach, system sustainability may be dealt with under three different perspectives: (1) centered in equilibrium (recovering after a disturbance), (2) multiple equilibrium status (absorbing disturbances while maintaining its essential functions and relationships), and (3) adaptive change (having the capacity of adaptively organizing itself to preserve its essential attributes after a disturbance); however, according to these authors, not all of them are appropriate to understand the sustainability of a socioecological system.

Based on the previous information, according to Perrings (2006) and Fiksel (2003, 2006) within the adaptive change, resilience is considered the property and

Fig. 1.3 Women in the compost workshop, REBIOSLA, Mexico. (Source: Elizabeth Olmos-Martínez)

foundation of sustainable socioecological systems because the system must find the different possible status to have the capacity of experimenting adaptive changes and achieving them. In summary, socioecological resilience is the capacity of a system to adaptively organize the arrangement of socioecological interaction by itself to confront and mitigate disturbances and maintain essential attributes (Fiksel 2006; Holling 1996; Norberg and Cumming 2008).

At the same time, the interactions between social and ecological systems are two-way relationships: (a) social, cultural, political, and economic interventions and activities produce changes and transformation in the environment and nature (for example, economic activities); (b) ecosystem dynamics influences the relationships of power, culture, and economic activities of human beings (for example, natural disasters, intense precipitation, and climate change) (Salas-Zapata et al. 2012). As an example of the double interaction, Fig. 1.4 shows the sociocultural and economic intervention on nature transformation for the development of ecotourism activity in the Santuario Tortuguero El Verde Camacho PNA in Sinaloa, México, because part of the mangrove had to be cut down to provide access to boats that offer tours in the estuary of the same name. In the same site, Fig. 1.5 shows an example of the mangrove ecosystem dynamics interacting with the hydric resource; in this case, a dock had to be improvised for boarding the boats providing tours in the estuary because water level was greater than usual in the area due to intense precipitation.

Fig. 1.4 Transforming the natural environment with socioeconomic purposes, Santuario Tortuguero El Verde Camacho PNA, Mexico. (Source: Elizabeth Olmos-Martínez)

The importance of the adaptive capacity of a socioecological system stands out when social activities are adjusted to the special features and dynamics of the ecosystems they are related with because they do not produce transformations that lead to prolonged states of human stress (Anderies et al. 2004). Thus, one must understand the adaptive process in which disturbances are involved to know the sustainability or unsustainability in a system, characteristics that respond to these disturbances and changes and those produced in such system.

Based on the previous information, Berkes et al. (2003) mentioned that it was essential to conceive system sustainability as the socioecological resilience of such, given that a high resilience in this type of complex system is a synonym of sustainability and vice versa. It is worth to mention that different authors have supported such stance (Fiksel 2003, 2006; Folke 2006; Folke et al. 2011; Norberg and Cumming 2008; Salas-Zapata et al. 2012). In this sense, a sustainable system is not the one that lasts forever but that which carries out adaptive changes to respond to different disturbances and thus maintain its essential attributes.

Considering the wide array of socioecological systems, the characteristics that determine their adaptive capacity facing disturbances are variable. Salas-Zapata et al. (2012) highlighted four characteristics that have been related with socioecological resilience: (1) modular connectivity forms an adaptive behavior in two ways in which all the system is constituted by a set of interrelated elements, which may belong to social (persons) or ecological (nature) resources where connectivity and

Fig. 1.5 Influence of ecosystem dynamics in socioeconomic activities, Santuario Tortuguero El Verde Camacho PNA, Mexico. (Source: Elizabeth Olmos-Martínez)

its role in resilience is given by qualitative and quantitative criteria (Norberg and Cumming 2008, in Salas-Zapata et al. 2012); (2) diversity refers to the range of options that a system has to respond to a disturbance and continue with its crucial processes; it is expressed starting from redundancy (existence of multiple similar units but not identical and replaceable among them) and heterogeneity (understood as variety of behaviors or modes of acting that the system has to respond to the disturbance) (Walker et al. 2006 in Salas-Zapata et al. 2012); (3) feedback mechanisms refer to the coupling stimulus-response that allows a system to respond to disturbances and effects of its own behavior; it refers to the internal control of the system and how it responds to signals received from the environment (Norberg and Cumming 2008 in Salas-Zapata et al. 2012); and (4) efficiency is understood as the capacity a system has to perform its essential processes without exhausting the resources (economic, ecological, and social) on which it depends (Fiksel 2003 in Salas-Zapata et al. 2012).

1.4 Background

Several disciplines and fields of knowledge that deal with socioecological approach studies have emerged; this chapter includes a brief review of some of them.

Martín-López et al. (2010) dealt with Doñana socioecological system (southwest Spain) since it provides diverse benefits for human welfare through the multiple services it generates to society; they are concentrated in supply, regulation, and cultural services. By means of a service provider analysis and its benefits, starting from tendency markers of the last years, the results have shown that the services that have a growing tendency are those related with food provision and certain forms of knowledge (cultural and environmental education). Likewise, they found that regulation services have been degrading, mainly those related with the hydrological cycle.

On the other hand, Albarracín et al. (2019) dealt with ecosystem services in protected natural areas of the Colombian Amazonia, starting from payment for environmental services as a tool for conservation and supporting indigenous communities, favoring intercultural dialogue concerning caring for nature. The socioecological risks of the scheme design were analyzed and agreed among the actors; they established reciprocal agreements for maintaining biodiversity as a result of shared management to comply with the conservation objectives of the protected natural area and the life plan of the indigenous community of the region.

Likewise, Galicia et al. (2018) proposed the integration of a socioecological approach in conservation, use, and environmental service pay in the temperate forests of Mexico. The authors mentioned the need to understand the influence of pine and oak tree species on primary productivity besides the impact that land-use change and regulated and nonregulated biomass extraction have on the structure and function of the ecosystem. Moreover, other important factors were social and economic aspects such as protecting biological diversity and maintaining the provision of ecosystem services for human welfare, both related with the different conservation tools that represent interaction schemes of nature-society to reduce the high deficit in wood production that persists in Mexico.

Henao (2013) made a methodological proposal of agroecological resilience measurement in socioecological systems starting from a case study in the Colombian Andes centered in farming systems by comparing cultural management practices with agroecological and conventional approaches from resilience markers. The selected markers dealt with characterizing climate events (as threat), estimating the vulnerability level and response capacity by means of sustainable agriculture practices. The results showed differences in resilience between agroecological and conventional farms where practices of agroecological management had a tendency to show greater resilience capacity facing climate risk; moreover, identifying these capacities showed the risk management potential dealing with response and adaptation capacity; thus, the farmers with alternative management were capable of facing, resisting, and recovering from extreme meteorological events.

On the other hand, Otero and Boada (2007) performed a research study in Spain from the socioecological perspective of integrating interdisciplinary natural and social sciences. The authors analyzed the environmental history of Matadepera municipality in the socioecological patrimony of Valle de Olzinelles, linked with management and conservation of protected natural areas of the metropolitan region of Barcelona; they concluded that the socioecological focus was appropriate for assessing efficiency and proposing protection and conservation strategies for the patrimony they contained.

Similarly, Braasch et al. (2018) performed a study in La Sepultura Biosphere Reserve located north of the Sierra Madre in Chiapas, where since its decree (1995), the use of natural resources has been restricted to farmers. Currently, the way of exploiting trees without compromising conservation interests of the National Commission of Protected Natural Areas (CONANP for its abbreviation in Spanish) is "resin-action," known as everlasting forest use. Under a scheme of good practices, it does not affect forest function and generates an important supplementary income for family economy where the use of resin from the *Pinus oocarpa* species generates income for the local population of California and Tres Picos common land areas. The authors performed a participative socioecological research to study the recruitment problem of pine trees and know the interests and views of the actors involved in the resin-action project. Experiments, interviews, forums, and meetings with common land owners were carried out. Contact was established with key actors (forest technicians, common land owners, and resin buyers). The actors involved built a decision tree, where all of them agreed that the resin-action project was a viable productive activity in the long term. However, each group perceived the needs of the project differently; for example, one group considered the need of reforesting with plants; another one mentioned that natural regeneration was better. All the participants commented that raising cattle was the best method to control exotic grass although they did not rule out the prescribed fire. The farmers highlighted the need of technical and financial support, concluding that the resin-action project was an example of use and conservation of natural resources in the protected natural areas. However, it depended on highly dynamic ecological and social processes that could affect its viability, which is why a flexible operation integrating actors in decision-making was needed.

García-Frapolli and Toledo (2008) made a conceptual appropriation of nature from ecological economy to assess natural resource management in Mayan communities located in a protected natural area in Yucatan. The model offered the basis to make an instrument that allowed assessing and monitoring the protected areas with human presence. Their study demonstrated how this instrument could be useful for a greater understanding of the protected natural areas contributing in the design and policy conservation practices of biodiversity in a cultural, social, and productive context.

Ravera et al. (2009) mentioned that integrated and participative assessment processes were a good methodological-operational framework for decision-making facing complex environmental matters, which concern socio-agroecosystems of high and unpredictable dynamisms to changes with conflictive interests of the actors implicated in its management. The authors carried out a study in a protected natural area in the dry Nicaraguan tropic and showed the potential for applying an intercultural and interdisciplinary approach to develop a multiscale and multi-objective process of critical assessment for environmental degradation. They concluded that integrating discipline and knowledge could explain doubts on understanding the system and processes of local socioecological changes, planning and validating new working hypotheses, and obtaining accuracy and relevance of degradation assessment. Likewise, they assured assessment quality from the participative process because of the diversity of the actors, their interests, and the values involved.

Similarly, Segrado et al. (2015) used the socioecological analysis in a sustainable tourism model for the use of protected natural areas where tourism load capacity was determined in Chankanaab Natural Park, Cozumel, Mexico, and analyzed physical, ecological, and administrative aspects. They concluded that social and economic dimensions of the host community should be incorporated, as well as the perception of visitors that allow designing and implementing conservation, mitigation, and adaptation strategies in protected natural areas for tourism. According to the authors, an integral approach should be considered from a socioecological system; it should be conceived as a dynamic and organized totality to connect society with nature and achieve a more realistic understanding, acknowledging the interaction between humans and nature and pertinence of incorporating human, cultural, and economic needs in the interaction analysis.

On the other hand, Rescia et al. (2010) mentioned that the socioecological system resilience of Picos in Europe was threatened by changes in the traditional way of livestock exploitation, which aggravated the economic vulnerability of local population and migration toward urban areas together with landscape homogenization. The authors applied surveys and interviews to analyze the population capacity to influence and be influenced by the dynamics of change of land use and their perception to the changes that had already occurred. A conflict of interests was considered among the socioecological system components, and probable evolution tendencies were set up, integrating historical, cultural, environmental parameters under different management policies. The results showed that recovery and persistence of the socioecological services with acceptable levels of productivity and spatial diversity would be viable with an adaptive management and active participation of the population in decision-making.

Roca and Villares (2014) carried out a study to reinforce socioecological resilience of tourism destinations in Costa Brava (Spain), characterized by a great natural diversity and tourist models. The authors mentioned the concern for the crisis and pressures of climate change, which is why they had discussed strategies to confront these threats with the least cost possible. Interviews in depth were carried out with key actors to know their perception and proposals set out from local reality. Strategies were suggested to reinforce resilience in these coastal destinations, strengthen local identity, increase perception to change, and promote network collaboration and citizen participation.

On the other hand, Uribe (2018) dealt with socioecology and gender equality. The study was performed in Hidalgo, México, where the problem referred to the evolution of Hidalgo society since it had experienced a series of difficulties proper to sustainable development, among which vulnerability and gender inequality had been observed. The author made a reflection between ecological advance and gender equality as aspects that should be taken into account to implement an integral development. The result has shown that it is possible to aim for social harmony integrating the greatest number of dimensions of the urban work agenda where gender equality should be present. Furthermore, it includes the topic of socioecology as a way of reducing vulnerability that results from social evolution when integrating people to the urban and rural contexts.

Another important socioecological topic dealt with is the governance strategy. Rojas (2014) mentioned that the Colombian National System of Protected Natural Areas (SINANP for its abbreviation in Spanish) was a socioecological system that intended to coordinate its governmental norms by making adjustments to its structure and function according to the analyses and follow-up mechanisms it had. The intention was to identify the reasons for the socioecological crisis, which according to Kaufman (2008, in Rojas 2014) were judgment problems due to the subjectivity of reality reading caused by the actors' mental models. Thus, to consider SINAP as an eco-friendly alternative to the environment that contributes to human development integrally, it should achieve its anticipation and learning capacities to allow building efficient responses or adaptations to changes.

On the other hand, Anderson et al. (2010) analyzed the Chilean Network for socioecological studies in the long term, concentrating on advances, approaches, and relevance; formal ecological research programs started in 1980, which served as a successful framework to create research agendas and financing mechanisms to deal with significant ecological phenomena at the scale they occurred. The authors mentioned that such research programs had expanded their geographical and disciplinary focus; nonetheless, temperate and subantarctic biomass of the southern part of South America have lacked formal research networks and sites in the long term. However, numerous long-term research efforts have been found in Chile and Argentina but without coordination between them, which recommended to value efforts integrating research and education with results and social processes for decision-making and widening the array of participations by inviting different actors.

Escalera and Ruiz (2011) dealt with sociological resilience though anthropology from research studies in Ecuador and Costa Rica; they mentioned that environmental sustainability had started to reformulate itself from the natural sciences through socioecological resilience, claiming a transdisciplinarity capable of operationally articulating natural and cultural dimensions of the environment. Thus, anthropology provided a solid perspective surrounding the theoretical-methodological development of socioecosystem notions and socioecological resilience; it also contributed to the debate between culture and nature and ethnographic commitment; therefore, it would contribute to a more accurate design of intervention strategies.

Finally, Challenger et al. (2014) performed an analysis applying the socioecological system concept in Mexico to see the scope, possibilities, and limitations in environmental management. The authors detected the concepts and approaches that guided the instruments of public policies in the country from a literature review that comprised the legal and regulatory framework. The objective was to identify the documents that contained accurate information on the different management units, currently in force in the country, clarifying that the search was centered in the management units officially established by the federal government.

First, the authors selected eleven documents: (1) National Water Law (Reglamento de la Ley de Aguas Nacionales, 1994); (2) General Law of Ecological Equilibrium and Environmental Protection (LGEEPA, 2000); (3) Protected Natural Areas

Regulation (LGEEPA, 2000); (4) Ecological Regulation (LGEEPA, 2003); (5) General Law of Sustainable Forest Development (Reglamento de la Ley General de Desarrollo Forestal Sustentable, 2005); (6) General Law of Wildlife Regulation (Reglamento de la Ley General de Vida Silvestre, 2006); (7) Ecological Process Guide (Manual del Proceso de Ordenamiento Ecológico, 2006); (8) Specific Organization Guide of the General Directorate of Ecological Regulation and Ecosystem Conservation Research (Manual de Organización Específica de la Dirección General de Investigación de Ordenamiento Ecológico y Conservación de los Ecosistemas, 2008); (9) Territorial Ecological Regulation for Municipal Authorities (Guía de Ordenamiento Ecológico del Territorio para Autoridades Municipals, 2009); (10) Interior Regulation of the Ministry of the Environment and Natural Resources (Reglamento Interior de la Secretaría del Medio Ambiente y Recursos Naturales, 2012), and (11) Land Law in Rural Property (Reglamento de la Ley Agraria en Materia de Ordenamiento de la Propiedad Rural, 2012).

In the first analysis, document 1 was eliminated because it was mainly educational work; the rest of the documents were subjected to a software content analysis (Atlas.ti) to find keywords or phrases, such as ecosystem, habitat, basin, and those related with management as the objective of public policies, such as environmental management and administration.

Additionally, phrases that described different spatial units established under federal law for environmental management were also searched, for example, Environmental Management Unit (UGA for its abbreviation in Spanish), Wildlife Conservation Management Unit (UMA for its abbreviation in Spanish), critical habitat, zonification, and management. At the end of the review, 1358 coded citations were found among twelve concepts for ten of the previous documents. The phrase "environmental management of natural resources" and the word "ecosystem" were mentioned the most in those documents; "basin," "critical habitat," and "environmental management" were the least frequent.

The authors concluded that the exercise was a preliminary approximation to some key legal documents that supported the conceptual bases that regulate environmental management in Mexico by means of the most important policy instruments for this purpose. They emphasized it did not analyze or evaluate the results of such policies. Likewise, they suggested adjusting the regulation framework to include the proper attributes of socioecological system management, such as systemic approach, delimitation by basins or sub-basins, transversality, cotransversality, co-management, and adaptive monitoring. Finally, they mentioned that forest management and wildlife conservation units included very little of the necessary attributes for managing based on socioecological systems.

1.5 Conclusions

The socioecological approach allows showing wider and integrated vision of the relationship between human societies and natural resources. The appropriation of nature has to be analyzed integrally, with all its segments, and not starting from its fragments. Moreover, it should allow identifying timely and spatially the key variables and processes to be measured and analyzed.

The concept of socioecological resilience, which is the foundation of the sustainable systems and processes involved including the theoretical perspective that goes with it, constitutes a very appropriate analytical framework to study the situation between humans and the environment, thus enriching environmental anthropology and the resilience theory. Therefore, to enrich environmental anthropology, the resilience theory suggests new and more holistic perspectives capable of overcoming doubts in the relationship nature-culture-society, providing the opportunity of transdisciplinary research with the value of ethnography and microstudy for a general and deeper understanding of socioenvironmental phenomena.

Implementing public policy instruments, such as protected natural areas, requires establishing interactions among biophysical, social, economic, and political instruments that conform a socioecological system. To achieve this purpose, governing mechanisms are necessary for an inclusive decision-making process, as well as schemes for developing local capacities. This combined approach favors the transition of an exclusion scheme by the social part to one of participation and inclusion of the local communities and other multiple actors where defining and identifying the adaptive local managements are key strategies for the construction of socioecological resilience.

The application of a socioecological approach in protected natural areas is key for their maintenance, sustainable use, and resilience and to secure the use of tangible and intangible goods. In Mexico, the PNAs and people that live in them are closely linked and function within the complex adaptive systems since the relationships and feedback between the ecological structures and processes and those of society and institutions are intrinsically linked. Social needs through consumptive use, as provision of ecosystem services, and nonconsumptive, as cultural ecosystem services and choices in management practices, affect the ecosystem processes and its services, flux, and distribution; thus, they should be understood and incorporated in governmental decisions in both socioecosystem components. The previous information highlights the need of developing new alternatives in governing and managing decision-making to promote human action that allow the sustainable use of these ecosystems.

Understanding how the collective action emerges, importance of a new conservation scheme, and need of solving common alternatives should allow the communities and institutions (governmental and social) to be better prepared for a joint land and ecosystem management.

References

Albarracín Álvarez OL, Novoa Mahecha D, Rodríguez Peña SM (2019) Elementos de enfoque y estudios de caso para abordar los servicios ecosistémicos en áreas protegidas de la Amazonía colombiana. *Biodiversidad práctica*, Edición cambio climático 4(1):30–50

Anderies J, Janssen M, Ostrom E (2004) A framework to analyze the robustness of social-ecological systems from an institutional perspective. Ecol Soc 9(1):18

Anderson C, Rozzi R, Armesto JJ, Gutiérrez JR (2010) Construyendo la red chilena para estudios socioecológicos a largo plazo: avances, enfoques y relevancia. Revista chilena de historia natural 83:1–11

Badii MH, Landeros J (2007) Plaguicidas que afectan a la salud humana y la sustentabilidad. CULCyT//Toxicología de Plaguicidas 4(19):21–34

Barton JR (2006) Sustentabilidad urbana como planificación estratégica. EURE (Santiago) 32(96):27–45

Berkes F, Colding J, Folke C (2003) Introduction. In: Berkes F, Colding J, Folke C (eds) Navigating social-ecological systems: building resilience for complexity and change. Cambridge University Press, Cambridge

Böhringer C, Jochem P (2007) Measuring the immeasurable – a survey of sustainability indices. Ecol Econ 63(1):1–8

Braasch M, García Barrios LE, Ramírez Marcial N, Cortina Villar HS, Huber-Sannwald E, García Marmolejo G (2018) ¿Resinar, pastorear y conservar pinares en una reserva de la biósfera? Exploración socioecológica participativa. El Colegio de la Frontera Sur, Chiapas, p 45

Challenger A, Bocco G, Equihua M, Lazos Chavero E, Maass M (2014) La aplicación del concepto del sistema socio-ecológico: alcances, posibilidades y limitaciones en la gestión ambiental de México. Revista Investigación ambiental 6(2):1–21

Cumming G Norberg J (2008) Scale and complex systems. En: Norberg J Cumming G (eds) Complexity theory for a sustainable future. Columbia University Press, New York, pp 246–276

Dieleman H, Juárez-Nájera M (2008) ¿Cómo se puede diseñar educación para la sustentabilidad? Revista internacional de contaminación ambiental 24(3):131–147

Escalera Reyes J, Ruiz Ballesteros E (2011) Resiliencia Socioecológica: aportaciones y retos desde la Antropología. Revista de Antropología Social 20:109–135

Fiksel J (2003) Designing resilient, sustainable systems. Environ Sci Technol 37(23):5330–5339

Fiksel J (2006) Sustainability and resilience: toward a systems approach. En: Sustain Sci Pract Policy 2:14–21

Folke C (2006) Resilience: the emergence of a perspective for social-ecological systems. Glob Environ Chang 16(3):253–267

Folke C, Jansson A, Rockstrom J et al (2011) Reconnecting to the biosphere. AMBIO 40:719–738. Springer

Galán C, Balvanera P Castellarini F (2013) Políticas públicas hacia la sustentabilidad: integrando la visión ecosistémica. Centro de Investigaciones en Ecosistemas (CIE-co), UNAM y Conabio, México

Galicia L, Chávez-Vergara BM, Kolb M, Jasso-Flores RI, Rodríguez-Bustos LA et al (2018) Perspectivas del enfoque socioecológico en la conservación, el aprovechamiento y pago por servicios ambientales de los bosques templados de México. Madera y bosques 24(2):1–18

Gallopín G (2003) Sostenibilidad y desarrollo sostenible: un enfoque sistémico. En: Cepal Serie Medioambiente y desarrollo 64:1–44

Gallopín G (2004) Sustainable development: epistemological challenges to science and technology. Background paper prepared for the Workshop. In: Sustainable development: Epistemological Challenges to Science and Technology, ECLAC, Santiago de Chile. October 13–15

Gallopín G (2006) Los indicadores de desarrollo sostenible: aspectos conceptuales y metodológicos. Seminario de expertos sobre indicadores de sustentabilidad en la formulación y seguimiento de políticas. Memorias. Santiago de Chile. 4, 5 y 6 de octubre

Gallopin GC, Gutman P, Maletta H (1989) Global impoverishment, sustainable development and the environment: a conceptual approach. Int Soc Sci J XLI(121):375–397

García-Frapolli E, Toledo VM (2008) Evaluación de sistemas socioecológicos en áreas naturales protegidas: un instrumento desde la economía ecológica. Argumentos 21(56):103–116

Henao Salazar A (2013) Propuesta metodológica de medición de la resiliencia agroecológica en sistemas socio-ecológicos: un estudio de caso en los Andes colombianos. Agroecología 8(1):85–91

Holling C (1996) Surprise for science, resilience for ecosystem, and incentives for people. Ecol Appl 3:733–735

Izazola H (2001) Agua y sustentabilidad en la Ciudad de México. Estudios Demográficos Y Urbanos 47:285–320

Kates R et al (2001) Sustainability science. Science 292(5517):641–642

Kaufman S (2008) Complex systems, anticipation, and collaborative planning. Universidad de Ohio. Virginia Tech Symposium on Enhancing Resilience to Catastrophic Events through Communicative Planning

Maass JM (2012) El Manejo Sostenible de Socio-Ecosistemas. En: Calva JL (coord) Agenda para el Desarrollo 2012–2018, Volumen 14: Sustentabilidad y Desarrollo Ambiental. UNAM y Cámara de Diputados, Editorial Porrúa, México, D.F

Martín-López B, García-Llorente M, Gómez-Baggethun E, Montes C (2010) Evaluación de los servicios de los ecosistemas del sistema socio-ecológico de Doñana. Forum de sostenibilidad 4:91–111

Norberg J, Cumming G (2008) Introduction. In: Norberg J, Cumming G (eds) Complexity theory for a sustainable future. Columbia University Press, New York

Ostrom E, Cox M (2010) Moving beyond panaceas: a multi-tiered diagnostic approach for social-ecological análisis. Environ Conserv 37(4):451–463

Otero Armengol I, Boada Junca M (2007) Perspectiva socioecológica en el análisis de espacios naturales protegidos de la Región Metropolitana de Barcelona (España). Investigación 10(4):125–130

Perrings C (2006) Resilience and sustainable development. Environ Dev Econ 11:417–427

Plaza Gutiérrez JI (2006) Territorio, geografía y políticas públicas. Desarrollo y sustentabilidad en las áreas rurales. Boletín de la A.G.E. 41:69–95

Ravera F, Tarrasón D, Pastor PA, Grasa R (2009) Proceso y métodos de evaluación integrada participativa de degradación en agroecosistemas semiáridos. Un caso de estudio en un área protegida en el trópico seco nicaragüense. Revista Iberoamericana de Economía Ecológica 13:79–99

Rescia AJ, Fungairiño SG, Dover JW (2010) Reactivación del sistema socioecológico ganadero de Picos de Europa (norte de España). Ecosistemas 19(2):137–145

Roca E, Villares M (2014) Reforzar la resiliencia socioecológica de los destinos turísticos: el caso de la Badia de Roses (Costa Brava). Arquitectura, Ciudad y Entorno 9(25):493–524

Rojas Lenis Y (2014) La historia de las áreas protegidas en Colombia, sus firmas de gobierno y las alternativas para ala gobernanza. Revista sociedad y economía 27:155–175

Sabatini F, Arenas F (2000) Entre el Estado y el mercado: resonancias geográficas y sustentabilidad social en Santiago de Chile. EURE (Santiago) 26(79):95–113

Salas-Zapata WA, Ríos-Osorio LA, Álvares del Catillo J (2011) Bases conceptuales para una clasificación de los sistemas sociecológicos de la investigación en sostenibilidad. Revista Lasallista de investigación 8(2):136–142

Salas-Zapata WA, Ríos-Osorio LA, Álvarez-Del Castillo J (2012) Marco conceptual para entender la sustentabilidad de los sistemas socioecológicos. Ecol Austral 22:000–000

Sarandón SJ Flores CC (2009) Evaluación de la sustentabilidad en agroecosistemas: Una propuesta metodológica. Agroecología 4:19–28

Segrado Pavón RG, Arroyo Arcos L, Amador Soriano K, Palma Polanco M, RDC SB (2015) Hacia un modelo de aprovechamiento turístico sustentable en áreas naturales protegidas: Estudio de caso del Parque Natural Chankanaab de Cozumel, México. Pasos 13(1):25–42

Uribe J (2018) Socio-ecología y equidad de género en Hidalgo. Edición especial Buzón de Pacioli, I Congreso Internacional de Género, Año XVIII, 105:65–69

Walker B, Gunderson L, Kinzig A, Folke C, Carpenter S et al (2006) A handful of heuristics and some propositions for understanding resilience in social-ecological systems. Ecol Soc 11(1):13

Chapter 2
Beyond Paradigms: Socio-ecology's Heritage and Prospective

Blanca C. Garcia

Abstract This chapter aims to review the ecological, political, and social foundations for sustainable socio-ecological systems. Specifically, it will explore the assumptions and elements behind ecological systems and services. To that respect, Milbrath (1989) was one of the first researchers to link the "social learning" concept to sustainable development through the "self-educating community" expression to define situations of mutual learning, where actors learn from each other and from nature. From the socio-ecological perspective, sustainable development is linked to the resilience and capacity building of actors who negotiate and reach collective decisions (Pahl-Wostl et al. 2007b; Schusler and Pfeffer 2003; Woodhill 2004; Muro and Jeffrey 2008). These perspectives help considering a better view of the communities' decision-making practices, their capacity to endure adaptation and change, as well as their ability to collectively learn how to adjust toward new scenarios of governance policies about resource, access, and sustainability. Hence, this chapter could elicit a glimpse into these and other components of the human-natural complex adaptive systems that are the interest of socio-ecology as a field of study.

Keywords Socio-ecology · Social learning · Knowledge-based development
Adaptive social ecosystems · Anthropocene

2.1 Introduction

In order to bring a meaningful contribution to this book, this opening chapter had first the temptation to make a review of the epistemic and ontological assumptions underlying the major development paradigms. This in order to support the theoretical underpinnings of sustainable development, which is seemingly one of the core

B. C. Garcia (✉)
Colegio de la Frontera Norte, COLEF, (Northern Borderlands Research College), Monterrey, Mexico
e-mail: bgarcia@colef.mx

A. Ortega-Rubio (ed.), *Socio-ecological Studies in Natural Protected Areas*,
https://doi.org/10.1007/978-3-030-47264-1_2

values of ***social ecology***. From Thomas Kuhn's *Structure of Scientific Revolutions* (1962, 2012) to the Actor-Network Theory or the Object-Oriented Ontology, we can easily end up in the aisle of the philosophy and methodology of ***sustainability***. However, if a multilevel outlook is proposed by a "systems perspective," it is nowadays arguably accomplished by socio-ecological models, which as a final view become rather diluted when confronted with the overwhelming evidence of the Anthropocene. This latter paradigm highlights how we are going through a dramatic and rather catastrophic climate change with all its implications, questioning "the world's ecosystems and their ability to sustain human life" (Göpel 2019: 29). However, since this book aims to provide a stimulating and honest account of community experiences into sustainable development, the very idea of knowledge-based development (KBD) and adaptive models is worth revisiting from this stance.

Following this line of thought, it is also worth to mention alternative economic models, in the sense that mainstream as well as alternative economic paradigms had each one an impact upon sustainability thinking and practical applications. These include the Economy for the Common Good, Transition Towns, the Commoning Movement, and Bhutan's Gross National Happiness framework. However, critics are finding that the Economy of the Common Good is mostly a prescriptive business model with no contingencies attached, Transition Towns is a relatively unknown network, and Bhutan's GNH has practically phased out. Nonetheless, the Commoning Movement exhibits a more mature conceptual and institutional design. Above all, it has developed a set of contingencies based on empirical research, in which "a radical revision of technological means begins the transition. A reconsideration of human goals completes it" (Göpel 2019: 52). This is as close as it gets between Strong Sustainability and Anthropocene Perspective – only the other way around: A reconsideration of human goals is first, followed by technological means enabling path dependencies. Alternative economic models can include elements such as integrated value systems (happiness, subjective well-being, human development index, quality of life, discretionary time, etc.). More particularly relevant to knowledge-based development (which includes *capital systems* and the reference to *values research,* etc.), these approaches lead us to questions such as "What types of capital exist and where do they come from?". This is how "Strong Sustainability" emerges once an integrated value perspective is taken into account. Hence, for Göpel (2019: 113), "understanding humans in nature is a pre-requisite to survival." Moreover, perhaps the most clear convergence between the whole spectrum of both sustainability and Anthropocene studies is Elinor Oström and her discussion on the commons as an economic practice. Most important is her call for building "institutions that bring out the best in humans." Even if survival is the ultimate purpose, at present, it can become our best sustainability model proposal.

Hence, as the interconnected environmental, energy, economic, and equity crises of our twenty-first century are posing complex and often unpredictable challenges to our communities around the world, we are located right at the core of the sustainability and Anthropocene studies discussion. Clearly, conventional forms of urban planning, design, and governance – often centralized, hierarchical, and inflexible – are ill-suited to these new realities. Capacity development for governance in our communities is therefore setting the pace for building and strengthening our communities' resilience. Moreover, resilience thinking is about understanding and engaging with a changing world. By understanding how and why the system as a whole is changing, we are better placed to build a capacity to work with social and environmental crisis, as opposed to being a victim of it (Walker and Salt 2006: 9). Among those crises, resource scarcity is characterized by connectedness, complexity, uncertainty, conflict, multiple stakeholders, and, thus, multiple perspectives. Resources are thus unknowable in objective terms although this understanding does not currently conform the dominant paradigm for sustainable development. Clearly, it is no longer possible to rely only on scientific knowledge for management and policy prescriptions. *Social learning*, as a core part of some socio-ecological models, which is built on different paradigmatic and epistemological assumptions, seems to offer managers and policy-makers some alternative and complementary possibilities for a number of contemporary ecological crises.

Hence, this chapter aims to explore the interrelationships between environmental sustainability, social learning, and resilience through the economic and social challenges and opportunities that sustainable socio-ecological systems bring. The uniquely complex nature of these interdependent issues opens up opportunities for social ecology to make groundbreaking contributions to address the challenges that community development spaces pose. This chapter will therefore intend to explore the convergence between ***resilience***, social ***learning,*** and ***governance***, which are concepts embedded in development models such as ***knowledge-based*** and ***sustainable development*** for specific communities. In that sense, this chapter aims to develop a better understanding of development processes through the lens of social ecology and other frameworks that follow parallel principles and thereby observe communities' ability to develop and support them to exploit their resources, their access, and their communities' knowledge base.

2.2 Social Ecology

Social ecology is a study approach that embraces an ecological, reconstructive, and communitarian view on society. This theory looks to reconstruct and transform current outlooks on both social issues and environmental factors while promoting direct democracy (Clark 1988). Social ecologists make emphasis on small-scale economic structures and the social dimensions of the ecological crisis. Social ecology traces the causes of environmental degradation to the existence of unjust, hierarchical relationships in human society, which is seen as endemic to the large-scale social structures of modern capitalist states. Social ecologists advance models of decentralized small-scale communities and systems of production. They provide guidance for developing successful programs through social environments and the idea that behaviors both shape and are shaped by the social environment. The principles of social ecological models are consistent with social cognitive theory concepts which suggest that creating an environment conducive to change is important to making it easier to adopt healthy behaviors.

A major proponent of social ecology was the American environmental anarchist Murray Bookchin (1921–2006). Bookchin made a number of crucial contributions to further development of a Social Ecology Theory. Most significantly, he broadened the theoretical basis of the communitarian, organicist, and regionalist tradition developed by Reclus, Geddes, and Mumford by making dialectical analysis a central focus (ibidem). He opened the possibility for more critical and theoretically sophisticated discussions of concepts like holism, unity-in-diversity, development, and relatedness (Bookchin 2007). He also develops Mumford's defense of an organic world view into an account of diverse forms of domination and of the rise of hierarchical society. Of particular importance is Bookchin's emphasis on the central role of the developing global capitalist economy into ecological crises, which is a highly contemporary issue. Back in the 1960s, Bookchin felt that "by the very logic of its grow-or-die imperative, capitalism may well be producing ecological crises that gravely imperil the integrity of life on this planet" (Bookchin et al. 2015). Since then, social ecology developed in the deepest sense into a radical ecology. It actually stemmed into other theoretical variations such as *deep ecology*, *ecofeminism*, and *eco-socialism*, all of them encompassing a look into social problems (first) in order to discover the roots of ecological crises (Clark 2000). After Bookchin, a number of authors further developed the social ecology concept, of which the working group at the University of California, in Irvine, is worth mentioning in the following paragraphs.

2.2.1 Conceptual Social Ecology

Social ecology as a concept was conceived after social ecology programs at UCI had celebrated their first 25 years of existence. *Conceptual Social Ecology* presents a number of facets of social ecology: its current definition and basic assumptions; its founding scientific presentations; its evolution as an organizational unit within the University of California, Irvine (UCI); and its approach to research on contemporary problems of the social and physical environments. Key conceptual authors of social ecology's original intellectual foundations in UCI are Arnold Binder, Daniel Stokols, and Ray Catalano. Arnold Binder founded the social ecology undergraduate interdisciplinary program in 1970 which was accorded status as a formal academic school at UCI in 1992. Authors of distinctive definitions of social ecology are Daniel Stokols, Thomas Crawford, Dave Taylor, and Valerie Jenness.[1]

It was Daniel Stokols who identified four assumptions of the social ecology perspective and core principles of Social Ecological Theory (see Table 2.1). He has

[1] UCI. School of Social Ecology https://socialecology.uci.edu/pages/conceptual-social-ecology

Table 2.1 Elements of social ecology paradigm

SE	Four assumptions	Six underlying recommendations	Five principles
	Assumption 1: Multiple facets of both the physical environment (for example, geography, architecture, and technology) and the social environment are integral to a social ecological analysis.	1. Identify a phenomenon as a social problem.	1. Principle one: Multiple dimensional analysis. Environmental settings have multiple dimensions which influence the person-environment interaction.
		2. View the problem from multiple levels and methods of analysis.	
	Assumption 2: The relative scale and complexity of environments may be characterized in terms of a number of components such as:	3. Utilize and apply diverse theoretical perspectives.	2. Principle two: Differential dynamic interplay (the emphasis is on interrelationships between personal and situational factors).
	(a) Physical and social components	4. Recognize human-environment interactions as dynamic and active processes.	
	(b) Objective (actual) or subjective (perceived) qualities		
	(c) Scale or immediacy to individuals and groups		
	Assumption 3: The social ecological perspective incorporates multiple levels of analysis and diverse methodologies.	5. Consider the social, historical, cultural and institutional contexts of people-environment relations.	3. Principle three: Relevance of systems theory.
			4. Principle four: Interdependence of environmental conditions.
	Assumption 4: The social ecological perspective incorporates concepts from systems theory to take into account both the interdependencies that exist among immediate and more distant environments and the dynamic interrelations between people and their environments.	6. Understand people's lives in an everyday sense.	5. Principle five: Inherent interdisciplinarity. Social ecology analyses emphasize the integration of multiple levels of analysis with diverse methodologies.

Source: Adapted from Stokols (1996, p. 7)

described the development of the ecological paradigm and applied the social ecological perspective to problems of health promotion.[2]

Last but not least, Eleanor Finlay describes Social Ecology Theory as "a coherent leftist vision that underscores the potential for human beings to play a mutualistic and creative role in natural evolution." This could be possible by uprooting the irrational, hierarchical, and ecologically destructive society we currently live under and by replacing it with a "socially enlightened and ecological society" (Finley 2017). An essential element of such a society would be the Aristotelian notion of politics, that is, the direct management of towns, cities, and villages by the people who live in them. According to Finley, social ecology maintains that we can

[2] https://socialecology.uci.edu/pages/conceptual-social-ecology

supplant capitalism and the state with a global federation of directly governed democratic municipalities.[3] In this sense, social ecology intersects directly with resilience, social learning, and governance as it will be depicted in the next paragraphs.

2.3 Social Ecology Paradigms, Social Learning, and Governance

Relentlessly, one of the most important indicators of a city's progress and development is its sustainability. The characteristic connectedness, complexity, uncertainty, conflict, multiple stakeholders, and thus multi-perspectives of resource management makes it unsuitable to depend only on scientific knowledge for planning and strategies. Social learning, a process-based theoretical model, seems to offer professionals and policy-makers new options and harmonizing potentials (Ison et al. 2007). The following development models and definitions aim to get a better perspective of where principles of social ecology, such as resource management and sustainable community development, converge.

2.3.1 Social Learning Processes

Bandura's (1977) *Social Learning Theory* explains that individuals learn from each other through observation, imitation, and modelling. It is often described as bridging the behaviorist and cognitive learning theories mainly because it includes attention, memory, and motivation concepts. However, Reed et al. (2006) argue that social learning has more than bridging functions; it extends to (1) demonstrating that a cognitive change has occurred among the participating persons, (2) demonstrating that this new awareness is not limited to the participants but extends to the wider communities of practice, and (3) showing that it occurs through social exchanges between and among networked social actors. The meaning of social learning could improve our capacity to critically assess outcomes and to better appreciate the processes through which social learning takes place. In this way, it may be possible to better facilitate the desired outcomes of social learning processes (Reed et al. 2006).

With this in mind, our rural and urban communities, through social capital, are likely to adapt to contextual changes due to a systemic coordination and cooperation through their learning processes. In order to examine possible outcomes in terms of adaptive responses to change in city systems, we will need to understand the links

[3] http://www.kurdishquestion.com/index.php/insight-research/social-ecology-kurdistan-and-the-origins-of-freedom.html

between resources management, resilience, social learning, and governance within the rural and urban systems (Ruiu 2017).

Milbrath (1989) was among the first researchers to link "social learning" concept to sustainable development through the "self-educating community" expression to define situations of mutual learning where actors learn from each other and from nature. Sustainable development is viewed as a wicked problem (Rittel and Webber 1973), whose solution is function of the capacity of diverse actors to discuss, negotiate, and reach collective decisions (Pahl-Wostl (2006), Pahl-Wostl et al. (2007a), Pahl-Wostl et al. (2007b), Pahl-Wostl et al. (2007c), Pahl-Wostl (2009), Pahl-Wostl (2015), Pahl-Wostl (2017)). These perspectives help considering a better view of the communities' decision-making practices, their capacity to endure change, as well as their ability to learn how to adjust toward new scenarios of governance policies about resource, energy, and sustainability. This chapter elicits a glimpse into these factors as possible examples of complex adaptive systems problems.

2.3.2 Social Learning and Governance

Surely one of the successful definitions of the state in terms of public management has been that of the "enabling state" (Gilbert 1989, 2002, 2005). In theorizing a potential relational state, Gilbert has advanced that the enabling state seemingly offers an approach oriented to the market that focuses on benefits that promote participation in the workplace and overall individual responsibility (Gilbert 2002). Gilbert's approach emphasizes a state whose role is to provide social protection through public support, thus generating private responsibilities (Gilbert 2005). On the other hand, Pierpaolo Donati, in his seminal work on the *Relational Sociology* (2010), assumes that knowledge creation is dependent on relational/social capital – it is at least "processed" in social interactions – but the dynamics of knowledge production and associated increasing competition undermines the traditional social capital (families, workplaces, local communities).

These ideas were previously embraced by Etzioni (1988) and Giddens himself (Giddens 1994, 1998), but in the last decade, emerging notions of public management have also included the iconic concept of governance. Kooiman (2003) defines governance as the totality of public and private interactions dedicated to solve problems and create social opportunities. He calls it *interactive governance*. There are relevant aspects in this perspective that relate to the definition of relational state, where a strong state is not something that is derived from a Constitution, but it happens to be something contextual and entrepreneurial (Pierre and Peters 2000). Governance approaches that would be relevant to the notion of relational state are those in which the institutions are strengthened and those in which the state is analyzed not only from the inside out but also from the "outside-in" (Mendoza and Vernis 2008). All these elements and changes clearly portray the need for public managers' relational leadership and network entrepreneurship skills. Indeed, today's context is already requiring public managers to exercise a "humble and facilitating" leadership style, based on openness, dialogue, and participation, with the ability to think at the partnership/network level, give strategic direction, and encourage experimentation and diversity (Mendoza and Vernis 2008: 25).

2.3.3 Dimensions of Governance

With the new millennium, the concept of governance has been extended with some "adjectives." Among others, we are talking about evolutionary governance, cooperative governance, and associative governance. Some authors affirm that governance is expressed in "modes" or "ways": through hierarchies, persuasion, markets, and community involvement and by association (Bell and Hindmoor 2009). For the purposes of this chapter, we have concentrated on evolutionary governance and associative governance, which have been raised from the theoretical explanation of social processes and the relationship between the state (government) and the actors around it.

For example, the theoretical position of the Evolutionary Governance Theory (EGT) is based on the explanation, from various theoretical bases, of the different stages in which the **dynamic** capacity of governability of social groups is developed as they are organized. From this perspective, governance is not something given or fixed but something that is changing according to the elements and actors of the context. This vision is complemented by the theoretical vision of the associative governance, with which we try to emphasize the relationships that arise between the actors and the bonds of trust that arise from a regional context in the use of the resource (Storker 1998).

Thus, the contemporary concept of governance refers to an empirical phenomenon that has been impacting the forms of state intervention and how the state has been transformed (Dancause and Morin 2013). It also comes from a new theoretical perspective that helps us to understand the processes of integration and management of the social system, as well as the role of the state in these processes (Enjolras 2005). Indeed, the concept of governance is used to better understand "the introduction of new methods for the development of public policies based on negotiation, as well as new ways of implementing these policies, especially through partnerships" (Canet 2004). On the other hand, the concept of associative governance allows us to rethink the relationships between the different economic, political, and social actors, as well as forms of public intervention. Its starting point is that the state no longer plays the same role it had been doing in the past. But above all, it notes that its fundamental characteristic is the nature of the links between the actors (Stoker 1998). For their part, Hamel and Jouve (2006) have also rightly pointed out the ideological role played by governance, which has served to justify and legitimize the transition of governments from Keynesianism to neoliberalism (Bell and Hindmoor 2009).

On the other hand, it is possible to link the increasingly widespread use of the concept of governance with the fact that the state has lost its central role in the formulation of public policies. Experts have pointed out that this has happened on three levels:

international relations, economic regulation, and the relations of the state with local powers (Canet 2004). Some see the state as one of several actors, acting in concert with other representatives of the private sector and civil society (Duchastel and Canet 2004). In terms of public policy and state regulation, and due to the complexity of social structures, it is observed that hierarchical, vertical, and coercive relationships have gradually been replaced by more horizontal relationships, with incentives, negotiation, and cooperation, which they involve a wider variety of economic and social actors (Canet 2004; Levesque 2001). It is therefore clear that in the areas of economic development and social development (where the social economy is at the intersection of these two), the state no longer plays the leading role. It has become, rather, a "partner" with multiple normative, human, and financial resources.

The interest in governance arises from finding ourselves in a period of constant change and climate crisis, in which social ecology has something to say. In it, new problems are faced and new solutions are also proposed; it is no coincidence that governance is such a mentioned topic. Governance is one of the ways in which academia has found a mechanism to analyze change (Levi-Faur 2014: 7). These changes arise and are institutionalized in different levels or spheres. They can be conceptualized in three directions: upward (regional, transnational, intergovernmental, and global), downward (local, regional, and metropolitan, as recommended by social ecology perspectives), and horizontally (private and civil authority spheres) (Levi-Faur 2014: 7). For the purposes of this chapter, we will understand governance as a complex social process, and it is advanced that "governance is an ongoing process of steering or enhancing the institutional capacity to steer and coordinate" (Pierre and Peters 2000; Kooiman 2003). It is a "norm-generating process" (Humrich and Zangl 2010: 343) as well as a group of "practices of governing" (Bevir 2011:1) and the exercise of public authority (Heinrich 2011: 256, in Levi-Faur 2014). Most importantly, "the notion of governance, **when applied to resources management** refers to the capability of a social system to mobilize energies, in a coherent manner, for the sustainable development of resource resources" (Rogers 2002). Clearly, resource governance refers to the range of political, social, economic, and administrative systems that are in place to develop and manage resource resources, and the delivery of resource services, at different levels of society (Rogers and Hall 2003). As mentioned above, relevant to this chapter is to define resource management and governance processes as social learning processes mainly because they can be linked to concepts that are embedded in collective ways of learning. One of those embedded concepts, in which resource management, governance, and learning converge, is resilience.

2.3.4 Social Learning, Governance, and Resilience

Increased resource demand and the consequences of climate change clearly pose serious risks to the provision of sustainable urban resource services, for example, drinking resource, sanitation, and safe drainage, especially in cities. These challenges call for a transition toward improved resource management, including considerations of "resilience." However, because the resilience concept has multidisciplinary origins, it is open to multiple interpretations, which poses a challenge to understanding and operationalizing the concept. It is thought that building resilience needs to take three distinctive dimensions into consideration: socioeconomic, external hazard considerations, and larger social-ecological (regional) systems to be sustainable. Indeed, resilience as a notion adds value to understanding and addressing the complex urban resource management within its three dimensions (Fath et al. 2015).

Resilience is a rich and complex concept. It has roots in systems theory (a key concept for social ecology), and it has a variety of interpretations and applications every time human agency and/or networked, collective action is involved. The concept has been defined by a number of authors (see Table 2.2), but for the purposes of this chapter, we follow upon the work of the *Resilience Alliance*, one of the leading scholarly bodies working on the resilience of social-ecological systems.

They have defined *resilience* as "the capacity of a social-ecological system to absorb or withstand perturbations and other stressors such that the system remains within the same regime, essentially maintaining its structure and functions" (Holling 1973; Gunderson and Holling 2002; Walker et al. 2004). Moreover, it describes the degree to which the system is capable of self-organization, learning, and adaptation, as illustrated in Fig. 2.1.

The concept of resilience started in Systems Ecology, framing the resilience concept explicitly in the adaptive cycle with the observation that resilient systems are the "ones that successfully navigate all stages of growth, development, collapse, and reorientation of this cycle" (Johannessen and Wamsler 2017). Resilience refers to the capacities, competencies, and cultures needed by social organizations to bounce back, requiring consideration of the entire life cycle for success (Johannessen and Wamsler 2017). *Resilience* is also a key concept in terms of social capital, which is relevant for the purposes for this chapter, but has the following challenges:

Table 2.2 Definitions of resilience – sampler

First author, year	Domain	Definition
Kimhi, 2004	Community	Individuals' judgment of the ability of their community to successfully solve the ongoing political violence
Allenby, 2005	Community	The system's capacity to retain its function and structure in the face of internal and external transformations and to degrade graciously if it must
Gunderson, 2003	Community	The recovery time of a social-ecological system determined by (1) that system's ability for regeneration in an active environment and (2) group's ability to learn and change (which, in turn, is mainly a function of the institutional context for knowledge sharing, learning, and management and partly by the group's social capital)

Source: CARRI Report (2013)

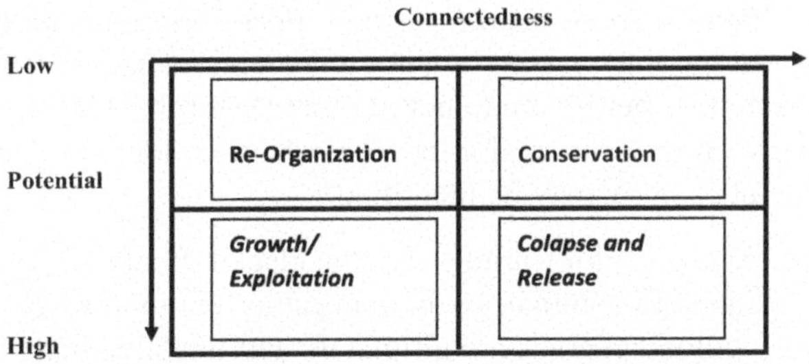

Fig. 2.1 The adaptive cycle. (Source: Adapted from Gunderson and Holling 2002)

(a) Transferring and integrating different forms of resilience across diverse communities and social settings
(b) The evolution of resilience as an adaptive capacity for the institutions and practices that regulate the formation and utilization of sustainable *development* strategies
(c) The complex combination of system forces, including policy, technology, and city capitals, with the creation, measurement, and representation of *resource* initiatives
(d) The new role of government, which advances learning adaptive capacities and *sustainable development* schemes (Fath et al. 2015)

Moreover, a first definition of Social Ecology System (SES) dates back to 1988. In it, Berkes and Folke found that it implied to analyze resilience in local resource management systems (Berkes and Folke 1998). Since then, studies of interlinked human and natural systems have emerged as a field on its own right, promoting interdisciplinary dialogue and collaboration in a wide set of fields and practices. As the SES concept celebrates its 20-year existence, an overview emerged on how authors use the concept in relation to research that deals with social and ecological linkages (Colding and Barthel 2019). Resilience is still a sought-after variable, but it is also true that a number of resilient systems have become less effective under the new ecological conditions they are facing.

2.4 Social Ecology Prospective

Hence, an analysis from a social ecology (SE) standpoint is seemingly a very robust social construction. However, it is contended that resilience and sustainability indicators for SE are becoming less meaningful under the emerging Antropocenic views. First, the sense of purpose in socioeconomic techno-systems makes it incompatible with current Anthropocene thinking. Perhaps the first and most common ground from major philosophers of the Anthropocene (Timothy Morton, Donna Haraway, Slavoj Zizek, Jean Dupuy) is that by definition, this complex reality we call the Anthropocene is now by far beyond human control. The key result is the decentralization of human agency. In that sense, social planning without empowerment, decision-making, and agency becomes quite challenging.

Even if *social ecology* and *deep ecology* reconcile their differences, the Anthropocene process remains an emerging, major challenge, not only for these two disciplines but for a number of others less robust in their social constructs. Based on the writings of its major theorists, deep ecology's basic areas of disagreement with social ecology may be identified.

(a) Social ecology argues that the idea of dominating nature resulted from the domination of human by human rather than the reverse. That is, the causes of the ecological crisis are ultimately and fundamentally social in nature. The historical emergence of hierarchies, classes, states, and finally the market economy and capitalism itself are the social forces that have, both ideologically and materially, produced the present plundering of the biosphere.
(b) Deep ecology, by contrast, locates the origin of the ecological crisis in belief systems, be they religions or philosophies. Most particularly, deep ecologists identify ancient near eastern religions and the scientific worldview as fostering a mindset that seeks to dominate nature. It is by asking deeper questions, as Arne Naess puts it, that these origins are identified so that the social causes of the ecological crisis are somehow relegated to the category shallow.

But social ecology at large views the natural world as a process – and not just any process but a development toward increasing complexity and subjectivity. Unlike sociobiology, which reduces the social to the biological, social ecology emphasizes the gradations between first and second nature: Second nature emerged out of first nature. Yet the boundary between human and nonhuman nature is real and articulated.

This is why it is important to draw parallels and comparisons between conceptual models and practices that are shedding some light into the sustainable development phenomena. Table 2.3 depicts a first-approach comparison between the principles of the commons (Ostrom 1990), the principles of social ecology, and the social processes of knowledge-based development (Carrillo 2004). With different views and assumptions, they are bringing to our postmodern world different voices that echo that of Bookchin in his critique to the capital system we are still in: "It is gravely imperil the integrity of life on this planet" (Bookchin 1988, 2007; Bookchin et al. 2015). Some of the parallel views between these three contemporary schools of thought converge not only in concepts such as sustainable resource management, governance, collective learning, or resilience. Their basic assumptions seemingly lie on ethics, inclusivity, diversity, creativity, etc. And they also lie on justice and freedom, which no matter how old they are, they still constitute the basic aspirations of most human groups and organizations. In this sense, Table 2.3 can shed some light on contemporary concepts and tools that could trigger a necessary *third way* to the multiple environments and human crises right on the present and immediate scenarios we are facing.

Indeed, while the meaning of development in Anthropocenic times gets so fuzzy, the realities of climate change are at our door. In terms of development, one can ask: Has *sustainability* become *unsustainable*? If that is the case, can social ecology report accurate results on multidisciplinary research? Can it generate new indicators

Table 2.3 Analogy between the principles of the commons (Ostrom 1990), social ecology (Stokols 1996), and the social processes of KBD (Carrillo 2004; Fachinelli et al. 2014)

Principles of the commons (Ostrom 1990)	Principles of social ecology (Stokols 1996)	Social processes of knowledge-based development (Carrillo 2004)
1. Clearly defined limits	1. Identify a phenomenon as a social problem.	Every city's capital system is unique, and its assets can be leveraged in a number of ways, depending on the social processes and the social groups involved. KBD seeks development from city's assets within the capital system and on the distinctive aspects of any given city.
	2. View the problem from multiple levels and methods of analysis.	
	3. Utilize and apply diverse theoretical perspectives.	
	(P1) Multiple dimensional analysis. Environmental settings have multiple dimensions which influence the person-environment interaction.	
	(P5) Interdisciplinarity. Social ecology analyses emphasize the integration of multiple levels of analysis with diverse methodologies.	
2. Congruence between appropriation and provision rules and local conditions	4. Recognize human-environment interactions as dynamic and active processes.	Development is a conscious and deliberate effort through citizenship interactions in a dynamic and active knowledge-based processes.
		It involves the constant recreation of its entire capital system and its identity, intelligence, and financial, relational, human (individual and collective), and tangible and intangible capitals.
3. Collective choice arrangements	(P2) Differential dynamic interplay (the emphasis is on interrelationships between personal and situational factors).	The social processes aim to leverage a city's assets in order to trigger development. The emphasis is on interrelationships and networks in which citizens interact and generate *knowledge moments*, or *insights*, in which they are able to create and innovate the existing knowledge structures. This is intrinsically a collective process and takes place in public places.
4. Monitoring	(P3) Relevance of systems theory and (P4) interdependence of environmental conditions evaluate the whole interdependent system. Resilience is considered a paramount value, although it does not show clear indicators.	Citizens are accountable for the way they are able to provide and manage knowledge, in different contexts within the city. Their accountability is key to find a way forward and is mostly self-monitored.

(continued)

Table 2.3 (continued)

Principles of the commons (Ostrom 1990)	Principles of social ecology (Stokols 1996)	Social processes of knowledge-based development (Carrillo 2004)
5. Mechanisms for sanctions		The city's capital system generates a unique set of indicators that set the path forward. With time, the city is able to offer a profile through which it has been able to generate its own definition and characteristics of its KBD model and its regulating mechanisms.
6. Mechanisms for conflict resolutions	5. Understand people's lives in an everyday sense.	Citizens go through a number of social processes through which a decision-making path generates a set of regulating mechanisms, which can then be applied to conflict resolutions.
7. Recognition of organizational rights	6. Consider the social, historical, cultural, and institutional contexts of people-environment relations.	Citizenship is thought to go through a number of social learning experiences and come out of them with the technical capacity to articulate and develop highly organized social capital systems (i.e., networks), in which relational capital is paramount.
8. Self-governance	Governance with the participation of diverse groups.	Multi-governance with the participation of City Champions and a critical mass of change agents with sufficient understanding of KBD.

that describe the state of the human struggle and the ecology loss? Can it be combined with other schemes and models as its foundational aspirations suggested?

One can advance that if social ecology combines its powerful tools with other disciplines such as knowledge-based development (KBD), it could be asking and answering questions triggered by the Anthropocene events about sustainable development, social learning, resilience, and governance. Or even those devoted to KBD: What knowledge? For whom? Also, who knows? Who decides? (Carrillo 2019). These are clearly questions that only the multidisciplinary work could attempt to answer.

References

Allenby B (2005) Toward Inherently Secure and Resilient Societies. Science 309 (5737):1034–1036

Bandura A (1977) Social learning theory. General Learning Press, New York

Bell SR, Hindmoor AM (2009) Rethinking governance. Earthscan, London

Berkes F, Folke C (1998) Linking Social and Ecological Systems: Management Practices and Social Mechanisms for Building Resilience. Cambridge University Press, New York

Bevir M (2011) The Sage handbook of governance. Los Angeles: SAGE
Bookchin M (1988) Social ecology versus deep ecology. In: Kick it over (Special Supplement, Winter 1988)
Bookchin M (2007) What is social ecology? From Social Ecology and Communalism, AK Press, first printing, 2007
Bookchin M, Bookchin D, Taylor B (2015) The next revolution: popular assemblies and the promise of direct democracy (with a foreword by Ursula K. Le Guin). Verso, London. ISBN 978-1-78168-581-5
Canet R (2004) "Qu'est-ce que la gouvernance?". Conférence prononcée dans le cadre du Séminaire Les nouveaux modes de gouvernance et la place de la société civile, organisé par le Service aux collectivités de l'UQAM, –Montréal, Écomusée du fier monde –16 mars 2004. Retrieved from: http://er.uqam.ca/nobel//ieim/IMG/pdf/canet-mars-2004.pdf
CARRI Report (2013) http://www.resilientus.org/wp-content/uploads/2013/08/definitions-of-community-resilience.pdf
Carrillo FJ (2004) Capital cities: a taxonomy of capital accounts for knowledge cities. J Knowl Manag 8(5):28–46
Carrillo J (2019) The Anthropocene turn in Kowledge based development J. Knowledge-Based Development, 10(4): 293–295
Clark J (1988) What is social ecology? Telos 82(22, 4):52–70
Clark J (2000) The matter of freedom: ecofeminist lesson for social ecology. Capital Nat Social 11(3):62–80
Colding J, Barthel S (2019) Exploring the social-ecological systems discourse 20 years later. Ecol Soc 24(1):2
Dancause L, Morin R (2013) Governance and the Associative Sector of the Social Economy: The Partnership of the State and the Civil Society in question. In M. Bouchard. Innovation and the Social Economy: The Quebec Experience. Toronto, Canada: University of Toronto Press
Duchastel J, Canet R (2004) Du local au global –Citoyenneté et transformation de la démocratie. En Jouve y P. Both. Démocraties métropolitaines, Collection Géographie contemporaine. (20–41). Sainte-Foy, Québec
Enjolras B (2005) Économies sociale et solidaire et régimes de gouvernance. Revue Internationale De L'économie Sociale: Recma, (296):56–69
Etzioni A (1988) The moral dimension: Toward a new economics. Free Press
Fachinelli AC, Carrillo FJ, D'Arisbo A (2014) Capital system, creative economy and knowledge city transformation: insights from Bento Gonçalves, Brazil. Expert Syst Appl 41(12):5614–5624
Fath BD, Dean CA, Katzmair H (2015) Navigating the adaptive cycle: an approach to managing the resilience of social systems. Ecol Soc 20(2):24. https://doi.org/10.5751/ES-07467-200224. Accessed on 25 Oct 2019
Finley E (2017) New review of "ecology or catastrophe". Anarch Stud 25(1):103–105
Giddens A. (1994) Politics, Sociology and Social Theory: Encounters with Classical and Contemporary Social Thought. Cambridge: Polity
Giddens A. (1998) The Third Way. The Renewal of Social Democracy. Cambridge: Polity
Gilbert N, Gilbert B (1989) The Enabling State: Modern Welfare Capitalism in America, Oxford University Press, New York
Gilbert N (2002) Transformation of the Welfare State: the Silent Surrender of Public Responsibility, Oxford University Press, New York
Gilbert N (2005) The 'Enabling State?' from Public to Private Responsibility for Social Protection: Pathways and Pitfalls, OECD (OECD Social, Employment and Migration Working Papers; N° 26), Paris
Göpel M (2019) The great mindshift –how a new economic paradigm and sustainability transformations go hand in hand. Springer open, 2019, 184 pp. Volume 2 in the Springer Series The Anthropocene: Politik – Economics – Society – Science (Hans Günter Brauch, series editor)
Gunderson LH, Holling CS (2002) Panarchy: Understanding transformations in human and natural systems. Island Press, Washington, D.C., USA

Gunderson LH (2003) Adaptive dancing: interactions between social resilience and ecological crises. (33–52) in F. Berkes, J. Colding, and C. Folke editors. Navigating social-ecological systems: building resilience for complexity and change. Cambridge University Press, Cambridge, UK
Hamel P, Jouve B (2006) Un modèle québécois? Montréal: Presses de l'Université de Montréal.
Heinrich A (2011) Network Journalism: Journalistic Practice in Interactive Spheres. New York City: Routledge
Holling CS (1973) Resilience and stability of ecological systems. Annu Rev Ecol Syst 4(1):1–23
Humrich C, Zangl B (2010) Global governance through legislation. In Handbook of Multilevel Governance, (ed) Enderlein H., Wälti S., Zürn M. Cheltenham: Edward Elgar, 343–357
Ison R, Röling N, Watson D (2007) Challenges to science and society in the sustainable management and use of resource: investigating the role of social learning. Environ Sci Policy 10(6):499–511
Johannessen Å, Wamsler C (2017) What does resilience mean for urban resource services? Ecol Soc 22(1):1. https://doi.org/10.5751/ES-08870-220101
Kimhi S, Shamai M (2004) Community resilience and the impact of stress. Wiley Periodicals, Inc. J Comm Psychol 32(4):439–451
Kooiman J (2003) Governing as Governance. Parts I, II, III. Sage Publications Ltd, London
Kuhn TS (1962) The Structure of Scientific Revolutions (1st ed.). University of Chicago Press (172)
Kuhn TS (2012) The Structure of Scientific Revolutions. 50th anniversary. Ian Hacking (intro) (4th edn). University of Chicago Press
Levesque B (2001) Le partenariat: une tendance lourde de la nouvelle gouvernance a l'ere de la mondialisation. Enjeux et defis pour les entreprises publiques et d'economie sociale. Annals Of Public And Cooperative Economics 72(3):323–338
Levi-Faur D (2014) From "Big Government" to "Big Governance", in D. Levi-Faur, The Oxford Handbook of Governance (2nd ed:7–13). New York: Oxford University Press
Mendoza X, Vernis A (2008) The changing role of governments and the emergence of the relational state. Corporate Governance: The international journal of business in society 8(4):389–396
Milbrath LW (1989) Envisioning a sustainable society: Learning our way out. SUNY series in environmental public policy. State University of New York Press: NY
Muro M, Jeffrey P (2008) A critical review of the theory and application of social learning in participatory natural resource management processes. Journal of Environmental Planning and Management 51(3):325–344
Ostrom E (1990) Governing the commons: the evolution of institutions for collective action. Cambridge University Press, Cambridge. ISBN 9780521405997
Pahl-Wostl C (2006) The importance of social learning in restoring the multifunctionality of rivers and floodplains. Ecol Soc 11(1):10. http://www.ecologyandsociety.org/vol11/iss1/art10/ (online)
Pahl-Wostl C (2009) A conceptual framework for analysing adaptive capacity and multi-level learning processes in resource governance regimes. Glob Environ Chang 19(3):354–365. http://www.gwsp.org/fileadmin/downloads/Publications/Pahl-Wostl__2009_-_A_conceptual_framework_for_analysing_adaptive_capacity_and_multi-level_learning_processes_in_resource_governance_regimes.pdf
Pahl-Wostl C (2015) Resource governance in the face of global change: from understanding to transformation. Springer, London
Pahl-Wostl C (2017) An evolutionary perspective on resource governance: from understanding to transformation. Water Resour Manag 31:2917–2932
Pahl-Wostl C, Mo ltgen J, Ebenhoh E, Holtz G (2007a) The new resource management and transition framework—state and development process. In: Pahl-Wostl C, Kabat P, Mo¨ltgen J (eds) Adaptive and integrated resource management. Coping with complexity and uncertainty. Springer, Heidelberg, pp 75–96
Pahl-Wostl C, Craps M, Dewulf A, Mostert E, Tabara D, Taillieu T (2007b) Social learning and resource resources management. Ecol Soc 12(2):5. http://www.ecologyandsociety.org/vol12/iss2/art5/ [online]

Pahl-Wostl C, Sendzimir J, Jeffrey P, Aerts J, Berkamp G, Cross K (2007c) Managing change toward adaptive resource management through social learning. Ecol Soc 12(2):30. [online] http://www.ecologyandsociety.org/vol12/iss2/art30/

Pierre J, Peters G (2000) Governance, Politics and the State. Macmillan: London

Reed MS, Evely AC, Cundill G, Fazey I, Glass J, Laing A, Newig J, Parrish B, Prell C, Raymond C, Stringer LC (2006) What is social learning? Ecol Soc 15(4):r1. [online] URL: http://www.ecologyandsociety.org/vol15/iss4/resp1/

Rittel H, Webber MM (1973) Dilemmas in a General Theory of Planning. Policy Sciences 4(2):155–169. Elsevier Scientific Publishing Company, Inc., Amsterdam, 1973. [Reprinted in N. Cross (ed.), Developments in Design Methodology, J. Wiley & Sons, Chichester, 1984:135–144]

Rockström J, Falkenmark M, Folke C, Lannerstad M, Barron J, Enfors E, Gordon L, Heinke J, Hoff H, Pahl-Wostl C (2014) Water resilience for human prosperity. Cambridge University Press, Cambridge

Rogers P (2002) Resource governance in Latin America and the Caribbean. Inter-American Development Bank (IDB), Washington, DC. (http://www.iadb.org, https://www.ircwash.org/sites/default/files/202.2-02WA-17802.pdf)

Rogers P, Hall A (2003) Effective resource governance. GWP-TEC, The Background Papers, vol 7. Elanders Novum, Sweden

Ruiu M (2017) Participatory processes in designing cohousing communities: the case of the community project. Housing and Society 43:1–14

Schusler T, Pfeffer M (2003) Social Learning for Collaborative Natural Resource Management. Society & Natural Resources 16:309–326

Stoker G (1998) Governance as theory: five propositions. Int Soc Sci J 50(1):17–28

Stokols D (1996) Translating social ecological theory into guidelines for community health promotion. Am J Health Promot 10(4):282–293

Walker B, Salt D (2006) Resilience thinking. Sustaining ecosystems and people in a changing world. NE. BE. LED. Island Press, Washington

Walker B, Holling CS, Carpenter SR, Kinzig A (2004) Resilience, adaptability and transformability in social-ecological systems. Ecol Soc 9(2)

Woodhill J (2004) Monitoring and Evaluation as Learning: Rethinking the Dominant Paradigm, in de Graaff J. (ed.), Monitoring and Evaluation of Soil Conservation and Watershed Development Projects. Taylor and Francis Group: London

Chapter 3
Effective Ecological Sustainability as a Complex System from a Social Dimension

Juan Pedro Ibarra-Michel and Elizabeth Olmos-Martínez

Abstract The ecosystem of the planet has many inputs from the natural environment as well as inputs from human activity that has in the past and will in the future effect long-term sustainability. The interdependence of the ecosystem, and its interactions with social and economic systems, translates into competition for resources that predetermine their viability, availability, renewal, and permanence. Taking into account some of the characteristics of the complex systems, the authors argue that sustainability of the ecosystem can be best modeled as such, since it meets most of the characteristic defined by such a system, such as having a large number of elements, and is rich in non-linear interactions between these elements, with interacting feedback loops that are recurrent, open, and operate in conditions that tend to unbalance the system. The authors conclude indicating that for "effective sustainability," it is necessary to take into account the complexity of the ecological environment and to accept its interdisciplinary nature in order to find the required tools for its accurate understanding.

Keywords Sustainability · Complexity · Modeling

3.1 Introduction

In order to understand the multiple dimensions of ecological sustainability, it is necessary to put into perspective the complexity implied by the concept itself. Sustainability as such is not an isolated phenomena, but encompasses various areas of social activity and its relationship with nature, its relationship with its congeners, and the various power structures that make up any society.

One of the ways to understand the complexity of sustainability is by the scheme proposed by Sala et al. (2013), including a series of elements (that need to be

---

J. P. Ibarra-Michel (✉) · E. Olmos-Martínez
Departamento Académico de Ciencias Económico Administrativas, Universidad Autónoma de Occidente, Mazatlán, Sinaloa, Mexico

A. Ortega-Rubio (ed.), *Socio-ecological Studies in Natural Protected Areas*,
https://doi.org/10.1007/978-3-030-47264-1_3

sustainable, for how long and that needs to be developed) linked together and which imply a basic limitation regarding the way in which a scientific and valid approach to the phenomena can be given.

Ecological sustainability is the scientific field where the interpretation and recognition of these complex relationships in the study of sustainability and make a clear delimitation of the limits and scope of its study necessary, amen of the use of diverse methodologies that allow for more accessible information availability and validates what you want to elucidate with the investigative work.

Theories of complexity transcend the limitations of traditional paradigmatic approaches, whether positivist and phenomenological, as Hudson (2000) points out, provides more precise and organic models for the study of this type of phenomena. Another valid approach to the problem of the complexity of the study of sustainability may be the critical realism proposed by Bhaskar (2013) that proposes the creation of creative and innovative methodologies for the study of observed phenomena.

The previous paradigms of science have been the result of a break with traditional approaches to the problems perceived in reality, especially that of sustainability, which is now conceived with multiple interconnections and interdependencies that lead to seek a more complete and integrative methodology than allows to obtain a more strengthened knowledge with epistemological bases aligned with the problems contemplated.

This work makes a theoretical review of multiple approaches and arguments against the complexity and the phenomena of ecological sustainability in the field of the relationship of organized social groups that seeks to determine how to achieve a higher state of development without breaking the balance with nature. A graphic scheme is also proposed where the main elements of ecological sustainability and their mechanisms are visualized, which, from their own point of view, provides a more comprehensive model for the researcher to visualize the limits and scope of their work.

3.2 The Complexity and the Study of Sustainability as a System

With the advent of sustainability came the generalized idea of including many stakeholders and groups of resource users in participatory and local-rooted approaches, which in the end becomes an expression of a knowledge dilemma. The excess of information to be managed in the strategies of the sustainable management of the resources makes adequate selection of management strategy a difficult problem. There is not a single discipline or thematic area that provides the necessary knowledge in its entirety for the achievement of sustainability, hence requiring that continuously different components of knowledge need to be negotiated (Bruckmeier 2009). On the other hand, the evaluation of the effectiveness of participatory research and development strategies in sustainability (Kasenmir et al. 2003) is not an easy solution either, because the given situations vary so strongly in spatial parameters and temporary, as in the social contexts of the action that it seems impossible to verify or falsify standardized solutions.

A general vision of what sustainability implies can be observed in the following scheme proposed by Sala et al. (2013), which shows the relationships and interdependencies of the different dimensions of sustainability (Fig. 3.1).

Starting from the above, a basic difficulty in the study of sustainability as a whole is presented, although it seeks to validate the results through a variety of methodological techniques its linkage with a wide variety of indicators does not allow for achieving a complete understanding with this obvious limitation.

For Bruckmeier (2009) "the very idea of sustainability implies too many unresolved problems of the past such as poverty, population growth, human well-being, economic growth, industrialization and its unintended consequences," all within the not promising scenario where the threat of climate change looms over the planet, in addition to the problems of that has brought globalization and an economic neoliberalism that has opened an increasingly widening gap between social classes because it prioritizes the immediate and growing utility for investors more than equality (Freeman and Gilbert Jr. 1992). All these problems do not favor a unique and unidirectional cognitive process; thus the great challenge for researchers has been to combine the disciplines that transcend and encompass the seemingly incommensurable into something that approaches a valid and generalizable knowledge of sustainability as a whole.

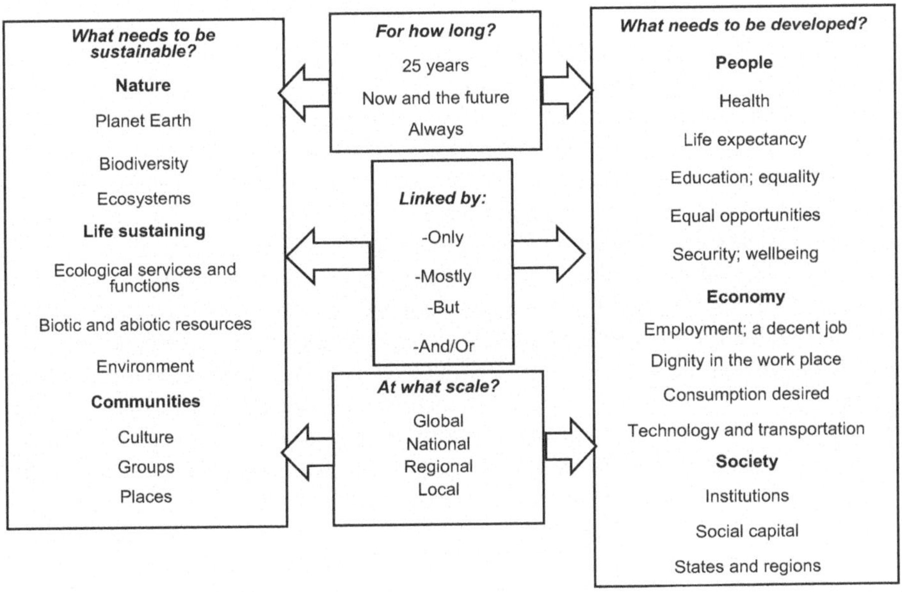

Fig. 3.1 Key principles of sustainable development. (Sala et al. 2013)

One of the ideas that can be very useful for understanding and visualizing the scope of sustainability is to understand sustainability as a system. Manderson (2006, p. 85) proposes the concept of sustainability as "the ability of one or more systems to sustain one or more systems over time"; this refers to the characteristics of the systems. Clayton and Radcliffe (1996, cited by Manderson 2006) where, in addition to their components, resources, relationships, their limits, and the mechanisms of control and regulation, there are what is called "emergent properties" that represent "something extra" that cannot be explained only by examining the sum of the parts of the system and that ideas such as holism and synergy are generally used to explain them.

Under the previous premises, the environment surrounding the phenomena of sustainability becomes very ambiguous and difficult to predict due to the changing dynamics of the systems (Manderson 2006). Clayton and Radcliffe (1996) alludes to the need to understand that the study of sustainability must be done under the perspective of an "open system" that receives multiple influences from other external systems, so the idea that this phenomena occurs in "closed" systems that do not experience this dynamic homeostasis with the world is beyond its limits of analysis.

For other authors such as Batanović et al. (2011), the study of sustainability refers to the use of various tools that could provide an adequate response to the problem of sustainability; these tools are commonly applied to complex systems among which they find the multicriteria analysis, the predictive methods, the simulations, and the unconventional treatment of the uncertainty. These tools constitute an acceptance of the underlying difficulty represented by the study of sustainability framed in a set of interconnected and interdependent systems and that hints at the mixture of some paradigmatic concepts in the obtaining of knowledge related to complexity.

For others like Levin (2006), the interdependence of the ecological system and its interactions with social and economic systems translates into competition for resources that predetermine their viability and permanence. Levin states that the ecological and socioeconomic systems are "complex adaptive systems in which patterns at the macroscopic level emerge from interactions and selection mechanisms mediated by the many levels of organization, from individuals to collective agents to whole systems and even above them" (Levin 2006, p. 25).

These selection mechanisms, Levin continues, and their non-linear nature can cause unpredictable changes whose consequences can be very important for the systems involved. This translates into a growing difficulty in predicting the behavior of the elements of the system, especially in the social and economic systems where the human being intervenes. Given this, defining some mechanism that allows a clearer approach to the study of sustainability is seen as an urgent need not only in the scientific field but in all areas and levels involved in the task of successfully completing the task of achieving a real sustainability.

Due to the ideas outlined above, it could be concluded that sustainability is impossible to know in its totality because it inherently brings too many problems of a very different nature and that range from cultural to ecological through economic and social issues (Milbrath 1995; Tibbs 2011). These problems have led to seek new ways of confronting them from a new position in the study of social, symbolic, and material reality, as well as in the analysis of complex adaptive processes (Urry 2005).

According to Urry (2005), for better policies or strategies of resource management, a new social paradigm is needed to better use the "knowledge base" available. The result should provide an effective model for evaluating interacting social and natural systems while also dealing with conflicting objectives and requirements. This brings with it the idea of "complexity" as an element that allows a critical analysis that could allow finding solutions to problems derived from the interdisciplinarity of the phenomena of sustainability with its many inputs and feedback loops. According to Urry (2005) "complexity investigates emerging, dynamic and self-organized systems that interact in a way that greatly influences the probabilities of subsequent events. Systems are irreducible to elementary laws or simple processes" (p. 3).

This goes against the reductionist ideas of the traditional positivist analysis of science and opens the door to the proposal of much more adequate evaluation of the heterogeneous and multidimensional system of sustainability. This has occurred because "panacea" solutions have failed to solve the complexity of the reality of sustainability (Ibarra-Michel 2018).

Considering certain characteristics of the complex systems of Cilliers (2002), sustainability can be placed within this category since it meets most of them as having a large number of elements, a rich non-linear interaction between these elements and at a short distance, and the interactions are recurrent, open, and operate in conditions that tend to unbalance the system. Cilliers also includes two qualities necessary for a system to be classified as complex; one of them is the representation or ability to "gather information about that environment and store it for future use", this does not imply that it be stored as a random collection of elements, but preferably must have some meaning that is important for its existence. Another important concept is "self-organization"; this is understood as the ability of the system to adapt according to changes in the environment.

The point mentioned by Cilliers (2002) moves to some of the philosophical discussions that have taken place since the emergence of postmodernity and post-structuralism, especially the writings of Saussure (1974) and Derrida (1976), who see the problem of science as a problem of deconstruction of reality in which the representations are marked by the social context; therefore we turn to a study of the qualities (syntax) and meanings of language (semantics) as a method to know the intentions and hidden meanings.

Cilliers (2002) proposes a model for the study of complex systems based on what he calls "connectionism" which, according to his words: "consists of a large number of units, richly interconnected with feedback circuits, but responding only to information local"(p. 5); the foregoing is related to the functioning of neural networks making the analogy of a human brain. Sustainability can then be seen as a complex system that can also be represented as a place of multiple connections in different interposed dimensions whose elements influence one another in a recurrent manner which limits its study under the traditional paradigms of science.

One way of understanding sustainability in a more holistic and systemic way would be the conceptualization proposed by Ibarra, Ibarra et al. (2016), which takes into account various elements that act as mechanisms within a complex system and at the same time make it dynamic and interconnected; in this manner the term "sustainability" should be linked to:

(a) A series of principles that determine its dynamic and lasting character over time;
(b) Must be understood as an active system that feeds itself through immovable interconnections that determine the health and survival of natural ecosystems and human societies;
(c) In such a manner that its continuity is determined by actions that are taken in the present but that have an exponential impact in the future; and.
(d) This requires a thorough review of ecological, social and economic assets to achieve their intelligent use to ensure a long-term balanced development (Ibarra et al. 2016, p. 97).

These mechanisms and concurrent elements in the complicated system of sustainability set the tone to discern the need for a different paradigmatic and epistemological approach in order to obtain more complete and comprehensive responses that take into account the different edges that can make a difference in perception and consistency in the events that the senses perceive. This refers to the point made by Kuhn (1996) in the sense that the paradigms of science must change, modify, or destroy if they do not meet their original purpose of providing reliable knowledge.

Reinforcing the above, Heylighen et al. (2007) emphasize that the paradigm of mechanistic science and Newtonian science has only brought incomplete answers to the complex problems of reality. Furthermore, their simplistic and unifying character does not correspond with the essence of many of the phenomena that are observed in it; and this has not only brought inconsistencies, paradoxes and multiple interpretations, but has not providing adequate or complete understanding for accurate modeling.

For Heylighen et al. (2007), the study of complexity must be linked to two characteristics of the systems, holism and emergence, where holism is defined as the tendency of a whole to be more than the sum of its parts and the emergence that are qualities of the system that cannot be reduced to the qualities of its parts. This direct allusion to Bertalanffy's systems theory (1969) leads to Morin's (1996) assertion that reality and its phenomena cannot be viewed in isolation and that its approach requires a much more inclusive perspective and therefore it is more complex.

Morin (1996) reaffirms the limitations of the simplifying scientific paradigm that uses the principles of "reduction and disjunction," which only impoverish knowledge, and advocates a new paradigm that contains as principles the "distinction, conjunction and implication." This vision of Morin moves to the awareness that there are different dimensions in each of the phenomena that you want to study from which you cannot subtract and therefore simplify. Human knowledge, says Morin, is condemned to be incomplete and partial because it is very difficult to cover all the spheres where the phenomena studied occur since there are a multitude of relationships that are kept with a multiplicity of concurrent elements which are unpredictable in a high grade. Uncertainty is, according to Morin, the fundamental characteristic of reality and that trying to frame it in its total magnitude is impossible, "the total is not the truth," he concludes.

To understand complexity Morin (1996) proposes three principles that aim to facilitate knowledge of a multidimensional reality. The first calls it "dialogical" that include the contradiction, complementarity, and antagonism tending to maintain the "duality within the unit" that is, the phenomena that occur in reality cannot be explained only with dissociated and excluding elements because that would limit your understanding. The second is that of organizational recursion according to which "a recursive process is one in which products and effects are, at the same time, causes and producers of what produces them." This is understood as constant feedback inside and outside of the phenomena that determine a constant dual causality. The third and last is the "hologrammatic" principle, which states that not only that the parts make up the whole system, but the whole system can also be found inside its component parts.

The proposal by Morin leads to the conclusion that the study of reality implies the inclusion of various disciplines that have a constant communication under a systemic perspective of the occurrence of the phenomena that occur within it. Sustainability as a complex and inclusive phenomena is therefore a tangible example of what Morin expresses in his approach, and this refers to what the author expressed as a sense of the need to create or find a new scientific paradigm that provides the basis for a deeper and more comprehensive understanding of this type of phenomena.

The problems to face in the concept of sustainability require new ways of understanding complex problems in science, policies, and technologies. The emergence of a new strategy for knowledge management allows the interaction of social, economic, and natural systems whose objectives are generally in conflict when establishing plans and policies that allow the advent of a real sustainability and not apparent.

Even understanding that the complexity of the problems of sustainability is difficult to reduce to simple and generalizable answers given the amount of knowledge and strategies that could be generated in a complex situation, although there have been attempts to include ideas from different disciplines like biology, sociology, economy and others. The large number of disciplines involved give as a result that it remains to be discovered a generalization understanding, and it is still a pending task to find a complete model that encompasses the large number of particular situations that each context implies and where sustainability is also applied. Due to this impossibility to define and operationalize sustainability in a universal way, the ideas that are presented as solutions should take into account temporal and spatial scales, as well as the conflicts that may arise at a certain moment.

3.3 The Study of Sustainability as a Complex System

Sustainability can be thought of as having a set of substrates or levels that interact through complex mechanisms that give it its characteristic systemic complexity. From this conceptualization, it is appropriate to understand how critical realism is presented as an acceptable alternative for its study, taking into account, as already mentioned, and the unique context in which this phenomenon occurs.

Given the impossibility of obtaining a single and generalizable explanation of sustainability, addressing the problem by trying to elucidate the ontological reality of events through the mechanisms that generate them is, from a particular point of view, the correct way to proceed. The perception of the human being is limited, so that his interpretation of what happens in the world will always be limited to his imperfect senses and his historical-cultural constructions, which leaves a very large margin of fallibility in what he obtains from his investigative work.

It is necessary to use different instruments and methods to describe a complex, holistic reality that is interconnected and interdependent, which does not justify the reductionist or partial intentions of traditional paradigms such as positivism and interpretative. The mix of methodologies and the important role that triangulation plays make it plausible to minimize the margins of error in research. Taking into account the breadth and diverse composition of the various elements of sustainability, it is necessary to put into perspective its operation that finally is the reason for this reflection. Figure 3.2 represents a more precise representation in which this phenomenon is conceived as a multilevel network of relationships and interdependencies, as well as the areas that should be explored for a partial or total understanding of it.

Faced with this multilevel network that involves the search for sustainability, trying to reduce its measurement only to econometric indicators is to minimize the influence that cultural value systems, for example, have on the actions of individuals that will precisely determine the progress toward a balanced development. It is necessary to look for another type of approach to know in a reliable way the occurrence of complex phenomena like this; sustainability is a social and economic phenomena that in turn has repercussions on environmental and natural processes, as a consequence the proposal to study it through alternative methodologies that give the researcher some freedom is not only plausible but desirable.

In the scheme, the ontological base, the environment, and human intervention are identified, from which the whole process is given toward sustainability; the argument that is proposed is that the natural environment within the geographical space is constituted as the architect and promoter of life without which the development of human societies would be impossible and, through their intervention of individuals in ecosystems, it gives rise to the efforts directed toward the equilibrium that seeks the lasting welfare as a desirable event for societies and the planet.

Regarding the context where the causal mechanisms are perceived, it is argued that society and culture as well as the economic environment are the ones that determine the form and background of the efforts toward sustainability. It is inherent to

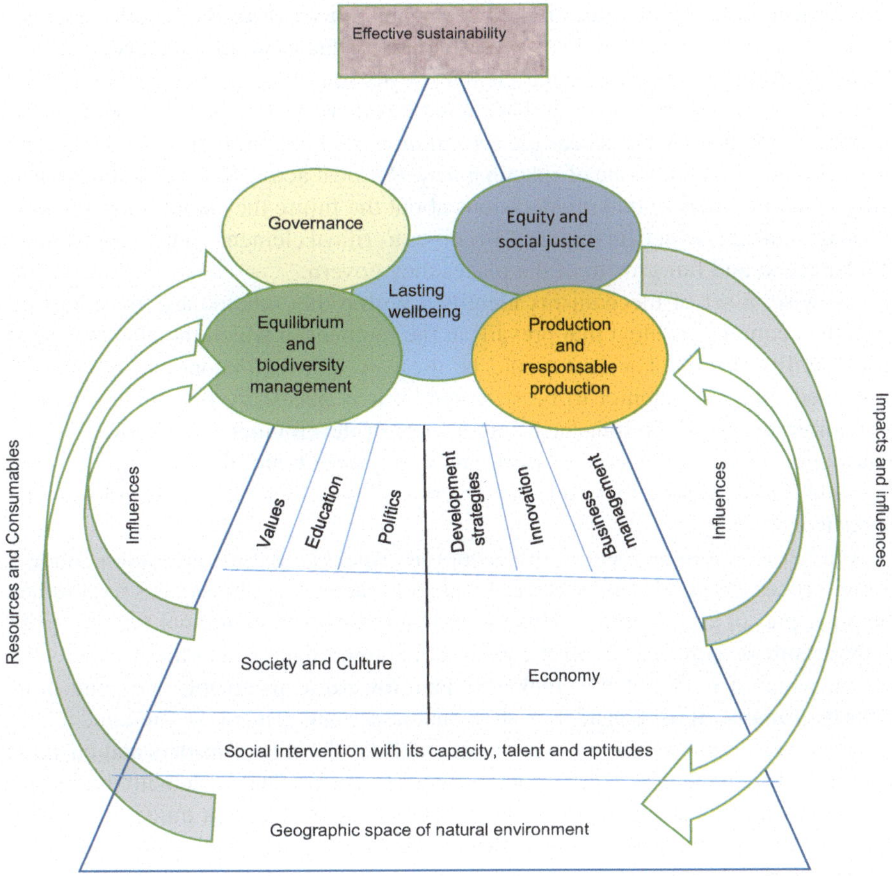

Fig. 3.2 Ecosystem sustainability as a complex system. (Source: elaborated by authors)

each society a series of unique characteristics that will distinguish it from others and that will lead to how the challenges of a development that pretends to be sustainable are addressed. The economy, on the other hand, determines the scope of the efforts both in development and research for change or those that are aimed at industrial reconversion, market regulations, or the flow of capital destined for this purpose.

The mechanisms that are indicated in the scheme are derived from the previous context that was already pointed out and are divided into economic, social, and cultural ones. In the first instance, values, education, and policies are proposed as the most important to be considered within this process of sustainability derived from the sociocultural context; values as convictions or guiding principles of behavior are those that set the tone in the way in which individuals accept or not the premises of sustainability in their daily lives, which translates into the moral and ethical satisfaction they may feel when doing what is considered correct.

This accumulation of values linked to sustainability is linked to another mechanism that has been considered key in this context: education. In effect, education is the pillar of any permanent social change, so its relationship with the success or failure of sustainability is a factor that cannot be put aside; to this is added the political aspect that directly influences the formulation of plans, strategies, and long-term projects that affect the scope of sustainability. Political actors have a great responsibility when it comes to making decisions about the future they want for communities and countries, which, in this case, becomes a critical element for the achievement of a balanced and fair growth of the places they govern.

Likewise, a set of mechanisms identified within this scheme are those derived from the economic context that prevails at the moment in which the phenomena of sustainability occurs. The first would be the way in which economic progress is conceived. For any attempt to be sustainable, it is necessary that the concept of "economic progress" be translated into a form of development that allows the fair distribution of wealth, access to work and entrepreneurship, the reduction of poverty and basic access to goods, and services that guarantee a healthy social coexistence.

Another essential element in the economic aspect is that of innovation, since it allows processes, goods, and services to adapt to the new circumstances required by the principles of sustainability. This process of innovation should not only be given in the business sector but in all the macroeconomic management that redirects the way in which it is intended to make the leap toward a sustainable economy in its entirety. Finally, it is considered that business management is fundamental to achieve a real transformation toward sustainability; the role of leaders and business leaders in this change is conceived in two aspects, the first has to do with the formulation of strategies and policies in the production processes that guide the desired goal and the second with the necessary work for a change lasting cultural in the organization and society.

Understanding the mechanisms that lead to a permanent well-being and a balance in the human activities of production and survival implies an implicit importance that should not be left aside within the academy or among those seeking its implementation.

Regarding the events inherent in the change toward sustainability, there is a distinction in four fundamental aspects to achieve lasting well-being in societies. Beginning with the governance and the accumulation of policies, rules, and regulations that it engenders, it is extremely important that the individuals and institutions in charge commit themselves to carry out a work in favor and without hindrances for the fulfillment of the strategies that facilitate the achievement of sustainability and that will only be achieved with harmony and joint work between society and its representatives as proposed by Stiftung (2011).

In reference to the necessary balance in the management of natural resources, it refers to one of the fundamental pillars of sustainability that depends to a large extent on what civil society, government, and business entities do or do not do. This point requires the joint work of a large number of actors or stakeholders to assert their opinions as to what is most convenient for the present and future generations, this does not leave a very wide margin for mistakenly thinking in the short term, which implies a mutual commitment without restrictions between them for the benefit of all (Manetti and Toccafondi 2012).

On the other hand, the scope of equity and social justice goes through the work and commitment of all social sectors to make affordable a higher stage of development where no individual is deprived of opportunities and access to opportunities that allow him to meet their needs basic and self-realization. If this objective is achieved, it is undoubted that the members of societies will no longer only seek to survive but will be more attentive to other fundamental issues such as sustainability (Inglehart 1990).

The last element that promotes lasting well-being and sustainability, responsible production, and consumption has to do with cultural change in organizations and society in general, which can occur in two ways, both by the pressure exerted by consumers and authorities on companies and as a business strategy that allows organizations to achieve objectives that, in addition to increasing their competitiveness, meet the intentions and expectations of customers and employees (Ibarra-Michel et al. 2019).

All the previous events converge in a desirable consequence for any social group, the lasting well-being that has to do with both objective and subjective factors (Eckersley 2006) that have to do with, not only economic and political elements, but with social elements and cultural aspects that are related to the perception of life itself and how members of any social group enjoy it and that in turn allows physical, mental, social, and spiritual well-being.

The above is consequently transferred to sustainability whose impacts and influences are perceived at the base of the pyramid, both in the natural environment and in the way in which human beings interact with it: how we interpret their needs and ours, how balance is sought, how we adapt, how we take advantage of their resources, and, of course, how we seek mutual survival.

3.4 Conclusions

Postmodernism has been a milestone in the change of not only cultural paradigms but in all the ways in which relations with power, science, nature, and other aspects of life in society are understood. The human being has been aware for the first time of the negative consequences that his actions are having on the environment; he has also realized that current societies are still far from being as fair and just as they would like. Thus, ideas such as sustainability, as a postmodern concept that seeks to lay the foundations for a world closer to the ideal, are seen as a very desirable alternative.

But the above is not easy, there are too many factors that influence the final results of the sustainability phenomena. Specifically, this is a problem with two aspects: first, how to implement it and, second, how to evaluate and study the results in a way that allows for a more complete understanding of it. The latter has brought with it a dilemma among academics who intend to study it as any phenomena using a traditional methodology and others that seek the integration of new paradigmatic approaches from the philosophy of science that leads to the creation of more innovative methodologies that conform to the complex reality and not vice versa trying to adjust the phenomena to the method.

The proposed scheme, as already mentioned at the beginning of this work, has the purpose of facilitating the visualization of the various elements, their influence, and interrelations that allow to dimension the way in which the phenomena of sustainability could be addressed both in its investigative aspects and didactic facilitating scientific and educational work at a higher level.

It is the opinion of the authors that for effective ecosystem sustainability research requires an approach that is interdisciplinary and holistic and cannot be simply modeled or solved by using only one narrow discipline of science. The complexity of sustainability is such that it needs to involve multiple scientific areas of study. Thus, the future of this science requires more understanding of the interactions to better generate models that can more effectively evaluate the complex phenomena of sustainability.

3.4.1 Lessons Learned

Due to the rapid evolution of science, it is important to provide better and accurate tools to capture the small glimpse of reality that we are allowed to perceive. This work is, of course, a tiny effort in the pursuit of a better understanding of the complex, and sometimes confusing, phenomena of sustainability. The proposed model tried to link the most important factors interfering in the reach of a balanced development and well-being of societies. Researchers must be aware that it is close to impossible trying to embrace all elements of a such holistic event as sustainability but, as a rule of thumb, it is very important to stablished links and relations to have a wider perspective resulting in a better understanding and possible solutions to this matter. That is the sole purpose of this proposal, to widening and clarifying the perspective of sustainability as a complex phenomenon.

References

Batanović V et al (2011) System theoretic approach to sustainable development problems. Yugosl J Oper Res 21:1–10

Bertalanffy LV (1969) General system theory: foundations, development, applications, vol 1. Braziller, New York, pp 1–10

Bhaskar R (2013) A realist theory of science. Routledge, London

Bruckmeier K (2009) Sustainability between necessity, contingency and impossibility. Sustainability 1:29–1411

Cilliers P (2002) Complexity and postmodernism understanding complex systems. Taylor & Francis e-Library, London
Clayton AM, Radcliffe NJ (1996) Sustainability: a systems approach. Earthscan, London
Derrida J (1976) Of grammatology. Johns Hopkins University Press, Baltimore
Eckersley R (2006) Progress, sustainability and human wellbeing: is a new worldview emerging? Int J Innov Sust Dev 1(4):306–317
Freeman RE, Gilbert DR Jr (1992) Business, ethics and society: a critical agenda. Bus Soc 31:9
Heylighen F et al (2007) Complexity and philosophy. In: Bogg J, Geyer R (eds) Complexity, science and society. Radcliffe Publishing, Oxford/New York, pp 117–134. Retrieved from: http://arxiv.org/abs/cs.CC/0604072
Hudson CG (2000) At the edge of Chaos: a new paradigm for social work? J Soc Work Educ 36:215–230
Ibarra JP et al (2016) Sustainability leadership for competitiveness in the hospitality industry in Mazatlán Mexico. J Intercult Manag 8(1):91–120
Ibarra-Michel JP (2018) El estudio de la sustentabilidad en su dimensión social desde el realismo crítico. Revista Metropolitana de Sustentabilidade (ISSN 2318–3233) 8(3):142–158
Ibarra-Michel JP et al (2019) Evaluación de la cultura organizacional en empresas hoteleras sostenibles de Mazatlán. Investigaciones Turísticas17:71–102. Retrieved from: https://doi.org/10.14198/INTURI2019.17.04
Inglehart R (1990) Culture shift in advanced industrial society. Princeton University Press, Princeton
Kasenmir J et al (2003) Public participation in sustainability science. A handbook. Cambridge University Press, Cambridge
Kuhn TS (1996) The structure of scientific revolutions. University of Chicago Press, Chicago
Levin KA (2006) Study design III: cross-sectional studies. Evid Based Dent 7(1):24
Manderson AK (2006) A systems-based framework to examine the multi-contextual application of the sustainability concept. Environ Dev Sustain 8(1):85–97
Manetti G, Toccafondi S (2012) The role of stakeholders in sustainability reporting assurance. J Bus Ethics 107(3):363–377
Milbrath LW (1995) Psychological, cultural, and informational barriers to sustainability. J Soc Issues 51(4):101–120
Morin E (1996) Introducción al pensamiento complejo. Editorial Gedesa, Barcelona
Sala S et al (2013) Progress in sustainability science: lessons learnt from current methodologies for sustainability assessment: part 1. Int J Life Cycle Assess 18(9):1653–1672
Saussure F et al (1974) Course in general linguistics. Fontana, London
Stiftung B (2011) Sustainable governance indicators 2011. Bertelsmann/OECD, Gütersloh
Tibbs H (2011) Changing cultural values and the transition to sustainability. J Futures Stud 15(3):13–32
Urry J (2005) The complexity turn. Theory, culture & society 22(5):1–14

Chapter 4
In Search of Long-Term Conservation: Objectives, Effectiveness, and Participation Schemes in Protected Areas

Benito Vázquez-Quesada and Daniel Torres-Orozco Jiménez

Abstract Protected areas (PAs) are instruments designed to ensure in situ long-term conservation. However, the selection of their objectives, design, and implementation strategies has changed over time according to different environmental value systems. In this sense, the Mexican environmental value system has changed from preservation to conservation, contributing to the development of different categories such as National Parks, Biosphere Reserves, Voluntary Conservation Areas, and Biocultural Landscapes. This chapter aims to historically describe these environmental value systems and their effect in the pursuit of long-term conservation in protected areas. To achieve this, we relate the environmental value system with each stage of Mexican conservation and we describe the objective of PAs and we address the effectiveness, participation, and access to power mechanisms in decision-making of each stage.

Keywords Governance type · Management regimes · Environmental value system

4.1 Introduction

4.1.1 Protected Areas

Protected areas (PAs) are considered as the major public environmental policy to maintain habitat integrity and species diversity against the environmental crisis (Naidoo et al. 2006; Rodrigues 2006). Currently, most of the nations have pledge support to the designation of PAs. As a result, 15% of the Earth's surface (an area similar to the surface of Europe and Antarctica together) and 17% of the marine realm (a surface similar to all North America) are being formally protected (UNEP-WCMC, IUCN, and NGS 2018).

B. Vázquez-Quesada · D. Torres-Orozco Jiménez (✉)
Universidad Nacional Autónoma de México, Facultad de Ciencias, Mexico City, Mexico
e-mail: benitovq@ciencias.unam.mx; daniel.toj@ciencias.unam.mx

A. Ortega-Rubio (ed.), *Socio-ecological Studies in Natural Protected Areas*,
https://doi.org/10.1007/978-3-030-47264-1_4

Despite the rapid success on the expansion of the marine and terrestrial protected areas worldwide, the loss of biodiversity remains alarming (Barnosky et al. 2012; Beresford et al. 2011; Butchart et al. 2010). In fact, one third of protected land is under a high rate of human pressure (e.g., construction roads, agriculture, urbanization), compromising the ecosystem integrity and its ability to provide environmental services (Jones et al. 2018; Vitousek et al. 1999) and its role in achieving the Sustainable Development Goals (SDGs; Dudley et al. 2017).

The perception of the role of PAs in the conservation of landscape has changed since their origin due to the continuous shift of the environmental value system. Conservation areas have different objectives and purposes, ranging from strict conservation areas (Category Ia, Ib, and II of IUCN) to regions permitting human sustainable use (Category IV) (see Table 4.1). Presently, IUCN recognizes four different categories of PA governance: government, shared, private, and indigenous people and local communities. Currently, 83% of the PAs worldwide rely on the government of its governance; however, Latin America is the region with more percentage under indigenous and local community governance (7.1%). This puts Latin America in a unique position, since comparisons in the mechanisms and forms of action of PA should be different from other regions (e.g., Europe and North America) where there is no indigenous/local governance (UNEP-WCMC IUCN and NGS 2020).

4.1.2 Environmental Value System

Conservation means human behavior (Saunders 2003). Despite the fact that human behavior is not always rational, it is predictable (Ajzen 1991; Clayton and Brook 2005). In this sense, behavior can be analyzed through the individual or collective values, social norms, and attitudes toward the natural world.

Values are fundamental aspects to understand individual's behaviors; however, they are largely neglected within conservation (Jones et al. 2016). Collectively, societies form an *environmental value system* (EVS) that shapes the manner in which individuals and societies perceive and evaluate nature. It strongly influences their views as to how natural resources should be protected or managed (Jones et al. 2016; Reser and Bentrupperba 2005). Therefore, values are some of the more stable guidelines that underpin our behavior (Stern et al. 1999), and EVS is being continuously shaped by cultural, economic, and socio-political events in all societies and changes over generations. These shifts on EVS explain how the relationships between people and nature are viewed and have deep consequences on the conservation goals, policies, and institutions that are pushed forward (Mace 2014).

Thus, all behavior, divisions of power, planning, and execution of conservation actions vary according to the different EVS and result in distinct management regimes and conservation categories of PAs. Currently, the Mexican law recognizes seven different protected areas according to their usage and management: National Park (NP), Flora and Fauna Protection Area (FFPA), Natural Resources Protection Area (NRPA), Natural Monument (NM), Sanctuary and Biosphere Reserves (BR),

Table 4.1 Protected areas categories according to IUCN and CONANP type of management

IUCN			CONANP					
	Category	Description	Category	Description	Year of creation	Amount by 2019	Area by 2019 (ha)	Management scheme
Ia	Strict reserve nature	Protected to all but light human use (e.g., science, education)		Core areas of PAs				Strict
Ib	Wilderness area	Protected from human perturbation		Core areas of PAs				
II	Natural Park	Protection of ecosystems with educational, spiritual, and recreational purpose	NP	Matching description	1917	67	16,218,589	Strict/flexible
III	Natural Monument	Protection of a specific Natural Monument (e.g., cave, landform, ancient grove)	NM	Matching description	1991	5	16,269	Strict
IV	Habitat/species management area	Protection of species/ habitat through active interventions	FFPA[b]	Protection of species/habitat within large areas where a small portion allows sustainable development	1936	40	6,996,864	Strict/flexible
			S	Protection of species/habitat within small areas	1986	18	150,193	Strict/flexible
V	Protected landscape/seascape	Protection of areas where the interaction of humans and nature has produced an area of novel character	BL	Protection of landscape promoting horizontal participation of all actors, sense of belonging, and an integral vision of the territory	2014	1	245,000	Flexible/ anti-statist
VI	Protected areas with sustainable use of natural resources	Protection of large areas where a small portion is under sustainable management	BR[b]	Matching description	1972	44	62,952,750	Flexible/ anti-statist

(continued)

Table 4.1 (continued)

IUCN			CONANP					
	Category	Description	Category	Description	Year of creation	Amount by 2019	Area by 2019 (ha)	Management scheme
	No equivalence		NRPA	Protection of soils, basins, and natural resources that are not included in another category	1938	8	4,503,345	Strict/flexible
	Other Effective Area-Based Conservation	Protection of ecosystems with biological, cultural, and spiritual relevance outside formal protected areas	ADVC	Areas Designated Voluntarily for Conservation Areas voluntarily assigned for its conservation by indigenous peoples, social organizations, and public or private entities	2000	352	545,067	Flexible/ anti-statist

NP Natural Park, *NM* Natural Monument, *FFPA* Flora and Fauna Protection Area, *S* Sanctuary, *BL* Biocultural Landscape, *BR* Biosphere Reserve, *RNPA* Natural Resources Protection Area, *ADVC* Voluntarily Designated Protection Area

[a]Data retrieved form UNEP-Protected Planet (https://www.protectedplanet.net/country/MX)

[b]CONANP categories that have more than one IUCN category equivalence

and Areas Designated Voluntarily for Conservation (ADVC) (see Table 4.1). These categories could be classified according to their management regimes as *hard, flexible*, or *anti-statist* (CNDH 2019). Here we aim to review how these different management regimes are the result of the change in the environmental value system, resulting in novel PA categories, models of participation, and conceptions about their effectiveness.

4.2 Hard Statism

Hard statism proposes that regions with greater biodiversity should become *public entities and seeks to relocate human populations* while compensating them for conservation (CNDH 2019).

4.2.1 Environmental Value System, 1917–1970s

Even though the protection of Mexican landscapes has been done since Pre-Columbian times in areas like Chapultepec and Oaxtepec (De la Maza 1999), the modern protected area movement started in Mexico using hard statism as a form of governance in the form of the National Parks.

In this sense, the *Desierto de los Leones*, an area near Mexico City, was designated as the first Mexican PAs in 1917 by President Venustiano Carranza under the Natural Park category with the objective of preserving the landscape's beauty and ensuring a future reservoir of water for the capital (De la Maza 1999). It would not be until almost two decades later that another protected area was declared as President Lazaro Cardenas (1934–1940) promoted the creation of Natural Parks. In fact, 49% of the current National Parks (n = 67) were created during this period of government.

The instauration of these PAs relies principally on the influence of Miguel Angel de Quevedo (1862–1946) and his efforts transmitting the relevance of these relatively untouched areas to the sitting Mexican Presidents in this period. From the institutional point of view, the appreciation of physical appeal and beauty of nature (e.g., aesthetic) and structure, function, and relationships of this ecosystems (e.g., ecologistic-scientific) were the main values that explain the creation of these Natural Parks (Kellert 1996). In the statism regime, nature is appreciated on its own (Mace 2014). For that reason, Natural Parks were conceived as areas set up to protect beautiful landscapes and wildlife, usually in areas of little economic potential (Watson et al. 2014). However, these areas did have a large potential for tourism. In fact, the current aim of Natural Parks is to designate and preserve areas that are relevant for their scenic beauty and scientific, educational, recreational, and historical value and where *only* activities related to the protection of natural resources as well as research and tourism are allowed (LGEEPA 2018).

The creation of two other PA categories during the 1930s, Flora and Fauna Protection Area (FFPA) and Natural Resources Protection Area (NRPA), also followed the statism regime. In both categories, the government aims to ensure the preservation of the resources from potential threats (see Fig. 4.1., focus on the period between 1917 and 1970).

4.2.2 *Mechanisms of Participation and Governance*

It would seem that conservation institutions formed by the government following the statism regime regard local communities as the enemy to defeat. The National Park model is characterized for ownership of resources in the hands of the government, which functions as administration agency and is funded based on public resources and usually has an expropriatory character. The FFPA and NRPA categories allow different zoning which is determined by the government in a Management Plan (LGEEPA 2018).

4.2.3 *Effectiveness*

During this first stage, management of effectiveness was not considered. There was an increment on the extension of PAs; however, during this initial instauration, neither the effectiveness on the biological outcomes nor its management was assessed.

By the end of the 1960s, only three PA categories were recognized: Natural Parks, Flora and Fauna Protection Area, and Natural Resources Protection Area, all managed using a hard statism. However, hard statism is difficult to maintain. During the end of the 1960s and early 1970s, the Mexican government put aside the protection of the environment. Instead, the State focused on promoting economic development by creating institutions such as the *Comisión Nacional de Desmonte* that aimed to transform the rainforest into agricultural and cattle lands. These new State policies caused great commotion among the scientific community. As such, scientist began to look for new governance schemes and new forms of participation where ecosystem protection does not exclusively depend on the will of the State (De la Maza 1999). Additionally, they emphasized the need of systematically evaluating the effectiveness of PAs.

4.3 Flexible Statism

In flexible statism, the State maintains its leadership in the creation of PAs, but it is accompanied by multiple initiatives to seek the sustainable development of the area and to transform the local communities into potential conservation allies (Carabias et al. 2015). However, in this case, the State keeps monopoly on the legitimate use of force (CNDH 2019).

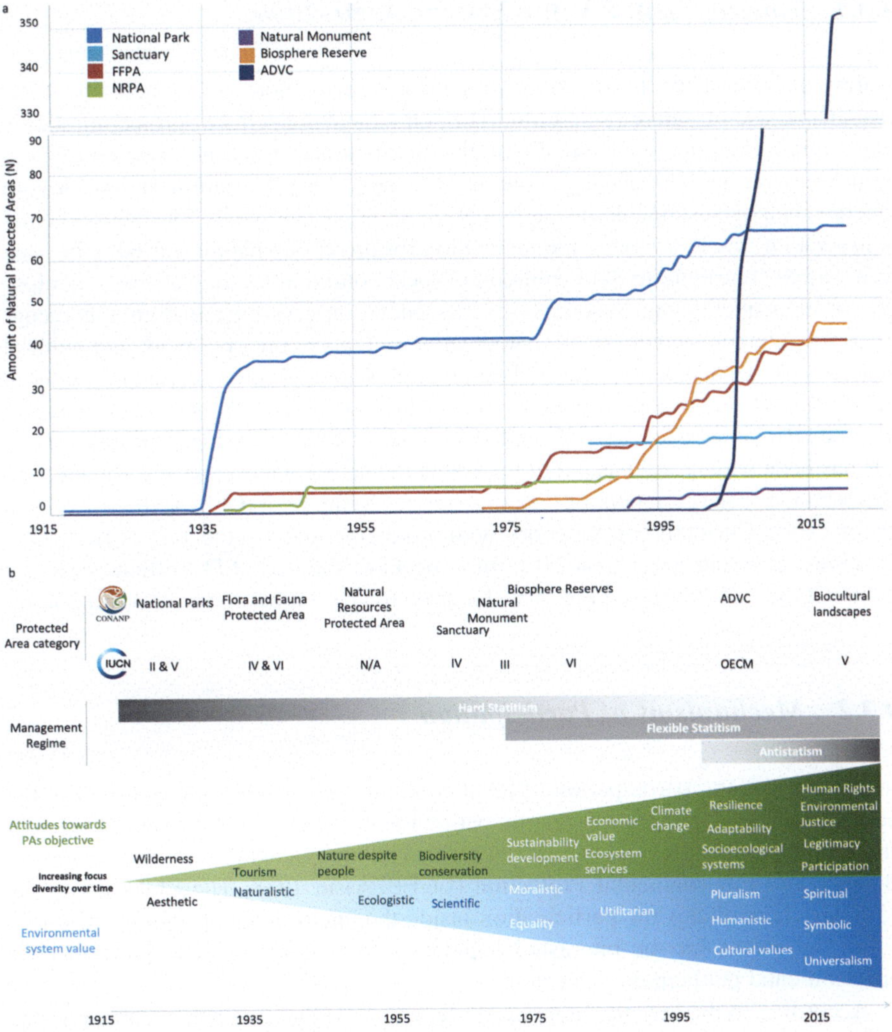

Fig. 4.1 Mexican natural protected areas time trend shows how (**a**) the category of protected areas (PA) and its amount have changed since 1917 (data from CONANP 2019, 2020). (**b**) The design and implementation of these categories have been influenced by several factors such as the environmental value system (EVS; blue) that determined the attitudes toward the PA's goal (green). The institutional EVS and attitude toward nature determine a specific management regime (e.g., hard, flexible or anti-statism) that has been translated in different international (e.g., IUCN) or national (e.g., CONANP) management categories. This time trend shows how the diversity of values, attitudes, management regimes, and PA categories underpinning conservation has shift and diversified through time

4.3.1 Environmental Value System, 1980–2000

During the end of the 1970s, the EVS began to focus more on the benefits of the sustainable use of nature (e.g., utilitarian), on ethical concerns of protecting nature (e.g., moralistic), and in the equality in the distribution of the benefits received from nature between individuals (e.g., equality) (Kellert 1996). This allowed new ideas to emerge such as sustainable development, to serve as new goals for environmental conservation. In other words, the motivation for protecting nature started to be seen as a way to contribute to the livelihood of local communities and as a way to close the social inequality gap. Sustainability sets a new conservation paradigm that recognizes the interconnectedness of ecological, social, economic, political, and cultural aspects. A new category of PA, the Biosphere Reserve (BR), was created in order to match this new conservation goal. *Laguna Ojo de Liebre* and *Montes Azules* became the firsts BRs in Mexico, in 1972 and 1978, respectively. Since then, the category of BR has been widely used as a tool to protect the natural landscape and enhance the socioeconomic development of the communities living on it (CONANP 2020; De la Maza 1999). Currently, the category with more area under protection is Biosphere Reserves, covering more than 70 million hectares (approx. 173 million acres) by 2020 (CONANP 2020) (see Fig. 4.1., focus on the period between 1980 and 2000).

4.3.2 Mechanisms of Participation

In flexible statism, participation of local communities is more relevant than in the hard statism. In order to ensure participation inside the PAs and to reach the goal of sustainability, the Mexican government created the General Law of Ecological Balance and Environmental Protection (LGEEPA) in 1988. The LGEEPA in the 15th Article regulates the participation inside the environmental policy and establishes the means for the interested population to be informed and participate in environmental public policy decisions.

According to the law, the State must recognize (1) that ecosystems are common heritage of society; (2) that the government must take responsibility for protecting the ecological balance; (3) that the coordination between agencies and entities of the federal public administration of the different levels of government and the agreement of society are indispensable for the effectiveness of ecological actions; (4) that individuals and societies are the main actors in the ecological consultation and the mechanism to redirect the relationship between nature and society; and finally, (5) that the eradication of poverty is necessary for sustainable development.

Moreover LGEEPA, in its Article 157, also establishes that the Federal Government must promote the *co-responsible participation* of society in the planning, execution, evaluation, and monitoring of environmental policy and also include mechanisms of public consultation (Art. 159) in which entities and agencies of public administration, academic institutions, and physical and moral individuals (e.g., social, business, organizations) participate.

Since the creation of the LGEEPA, all PAs, despite their category, must involve local participation to a certain degree. Still, the state remains the most powerful entity in the decision-making process. In this sense, PA participation is done by three types of *advisory councils* to ensure participation: (1) National PA Council, (2) Advisory PA Council, and (3) other councils to address specific issues.

4.3.2.1 National PA Council

The National Council is made up of representatives of the secretariat, other agencies and entities of the Federal Public Administration, academic institutions, research centers, producer and business groups, civil society organizations, and other social or private organizations. This council serves as a *consulting body* of the Ministry of Environment in the formulation, execution, monitoring, and evaluation of the policy for the establishment, management, and monitoring of the ANP of its competence.

4.3.2.2 Advisory PA Council

The Advisory Council, on the other hand, is the instrument that promotes and guarantees the organized participation of society inside a specific PAs. It aims to support and advise the governmental directors of the PA in the formulation of the Management Plan (MP), in the management and evaluation of the actions undertaken, as well as in the negotiation for decision-making, problem solving, search for sources of financing, and project development (Art. 18). This congress is made up of an Honorary President (Governor or Representative), an Executive President elected by the members, a Technical Secretary (Director of the PAs), Municipal Presidents, representatives of *ejidos* and communities, owners and holders, social organizations, and academic institutions. It consists of a maximum of 21 members. The decisions are taken by the vote of the majority of the members present, and the information and evaluations that are generated must be public and open for consultation and evaluation according to the National Transparency Law. The council must meet at least once a year.

4.3.3 Effectiveness

The term "effectiveness" as it relates to PAs is ambiguous because it could be assessed in several terms, such as ecological (e.g., animal population trends, changes in land cover), social (e.g., poverty relive, education level), or management (e.g., inputs in terms of staff or budget and the actions done) outcomes of creating PAs (Eklund and Cabeza 2017). The Mexican government did not assess the effectiveness of the PAs on any of these terms during their first eight decades of implementation. However, since the 1990s, the amount of research that justifies the existence of PAs has steadily increased.

4.3.3.1 Biological and Social Outcomes

Literature focusing on the biological effectiveness or outcomes of PAs is common. In general, global and Mexican, terrestrial PAs have been effective on reducing habitat cover change (Figueroa and Sánchez-Cordero 2008; Jiménez-Sierra et al. 2017) but have been inconclusive on their effect to halt species population declines (Geldmann et al. 2013). On the other hand, the evaluation of the effectiveness of PAs to influence socioeconomic characteristics is scarce but is conclusive on the increasing benefits of empowerment and co-management of the local communities (Berkes 2004).

4.3.3.2 Management Effectiveness

Probably the most common type of effectiveness measurement in PAs is the "management" measure. Currently, there are several systems for assessing effectiveness of management in PAs (see Hockings 2003 for review). The IUCN World Commission on Protected Areas (WCPA) has developed an evaluation framework for management effectiveness assessment (PAME).

PAME evaluations assess how well management inside the PAs aids to achieve the conservation goals (Hockings et al. 2006). PAME exposes areas of improvement on PA's management, accountability, and communication with public and assists in prioritization between conservation actions and equivalent PAs (Hockings et al. 2006; Leverington et al. 2010). Nowadays, 169 countries assess the management effectiveness of PAs using around 95 different methodologies as reported in the Global Database on PAME (UNEP-WCMC 2020).

Management Effectiveness Tracking Tool (METT) is the most common PAME approach to assess management effectiveness around the globe (Geldmann et al. 2015). METT is a questionnaire usually completed by a park manager and stakeholders that collect info on the objectives, threats (e.g., human settlements, farming, mining, transportation, tourism, natural system modifications, pollution, invasive species), budgets, staffing, size, and designations (e.g., legal status, law enforcement, management plan) (Stolton et al. 2007).

In Mexico, it wasn't until 2001 that CONANP founded the System of Information, Monitoring and Evaluation for Conservation (SIMEC) that focuses on (a) giving information to managers and general public, (b) monitoring population trends, and (c) evaluating the management effectiveness of PAs. By then, more than 21,448,190 hectares (approx. 53 million acres) were under a PA status without knowing if PAs were effective or not. Currently, less than 10% of the Mexican PAs have been assessed for its management effectiveness (UNEP-WCMC 2020), and globally only 20–50% of protected areas are effectively managed (Leverington et al. 2010).

Coad et al. (2019) analyzed the Global Datasets of PAME to understand the development of PA worldwide. Their study shows that around 60% of the Neotropical PA have inadequate funding and staff, becoming the most inadequately managed region of the world. Moreover, the global METT scores showed an improvement over time in the biological terms and in the planning, design, and establishment of formal user rights. However, local communities and indigenous people involvement in the decision-making process and the actual conservation outcomes showed the least improvement (Geldmann et al. 2015).

Despite the fact that flexible statism governance follows the ideas of sustainability, there is still a poor integration between social and biological outcomes. Management effectiveness evaluations show that involvement of local communities is rare (Geldmann et al. 2015). This could be related to the notion that the State should regulate the relationship between society and the PAs. Other regimes of governance, such as anti-statism, could be more effective in achieving a positive long-term outcome. Anti-statism policies emerge from the relationship between the environment and the people that live, own, depend on, and manage it. This scheme of governance is more prone to acceptance of the property rights of people to their land and consequently, increasing participation and legitimacy of PAs.

4.4 Anti-statism, 2000–Current

Some authors point out that the communities that inhabit the PAs are frequently associated with indigenous people identities (Boege 2008; Garnett et al. 2018), and these communities should be those who carry out the necessary actions to achieve environmental sustainability (Toledo 2005), with little to non-state intervention other than the guarantee of property rights over land and its natural resources (CNDH 2019). Anti-statism PAs should also be accompanied by recognition of self-determination and autonomy of indigenous peoples and similar communities under equal conditions (Borrini-Feyerabend et al. 2004; Boege 2017).

4.4.1 Environmental Value System

Since the beginning of the new century, conservationists started to question the flexible statism environmental governance. Now, the underpinning values determining the goals, forms of participation, and effectiveness of PAs rely more on the humanistic, spiritual, symbolic, and pluralistic values (Kellert 1996). Current conservation actions are rooted on the idea that there is not a central view of the environmental issues and that we should embrace the plurality of viewpoints in different socio-ecosystems. In this way, each society should be able to manage their resources as they best see fit.

Under these new EVS, the conservation goal is to create a community by enriching bottom-up (participatory) decision-making and recognizing Other Effective Area-Based Conservation Measures (OECM). In spite of the great complexity of involving several actors in the process, several authors agree that without the involvement and participation of the local population in the planning, management, evaluation, and decision-making, it will not be possible to achieve the social and ecological goals that are expected from a PA in the long term (Brenner 2010; Andrade and Rhodes 2012; Borrini-Feyerabend et al. 2013; Oldekop et al. 2016). It would also prevent the PA socio-ecological resilience or adaptation to change (Olsson et al. 2004).

In this way, the management of PAs *must incorporate a strong participatory, co-management, and representation component*. To address the inequalities and omissions by state institutions, a new value system associated with the conservation and sustainability in the management of PAs is necessary and is based on three fundamental changes (Merçon et al. 2019): (i) *ontological* that recognizes diverse ways of conceiving and experiencing nature and its cultural roots; (ii) *epistemological* that recognizes the interdependence between nature and culture as well as that the construction of knowledge and decision-making must be collaborative; and (iii) *ethical-political* that explicitly considers the plurality of values, governance systems, and power relations.

In this way, PAs will overcome the socio-structural conditions of injustice that have prevailed over indigenous people and local communities since (a) they have historically been deprived of interference in the public policies that affect them (Ruiz-Mallén and Corbera 2013; Garnett et al. 2018) and (b) they suffer from the worst conditions of poverty and marginalization (Paz-Salinas 2005; West et al. 2006; Adams and Hutton 2007; Brockington and Wilkie 2015). Paradoxically, indigenous and similar communities have control of the best preserved territories (Boege 2008; Garnett et al. 2018).

Since 2018, OECM were formally defined as "a geographically area other than a Protected Area, which is governed and managed in ways that achieve positive and sustained long-term outcomes for the *in situ* conservation of biodiversity, with associated ecosystem functions and services and, where applicable, cultural, spiritual, socioeconomic, and other locally relevant values" (UNEP-WCMC 2020). This is a great step on understanding the relevance of indigenous and similar communities worldwide.

Anti-statism governance in Mexico is exemplified by better bottom-up PAs, such as the Areas Designated Voluntarily for Conservation (ADVC) and the implementation of Biocultural Landscape (BL) PAs. In ADVC, local people decide to include their land on a formal protection scheme for a determined period of time (ranging from 15 years to perpetuity). On the other hand, Biocultural Landscape is a widely unexplored category that searches to conserve an area where "the interaction of people and nature over time has produced an area of distinct character... where safeguarding the integrity of this interaction is vital to protecting and sustaining the area and its associated nature conservation" (Dudley 2008). In this sense, this category seeks to preserve the environment, meanwhile maintaining the cultural and spiritual values and fomenting social cohesion and economic improvements to allow the long-term conservation of the area (Borrini-Feyerabend et al. 2004; Mitchell 2006).

These new types of governance have had a positive reception in society. Today, ADVC are the most common PA in Mexico (n = 353) and cover an area of 544, 106 ha (i.e., an area equivalent to the surface of the Alcaldies of Tlalpan and Milpa Alta together) (see Fig. 4.1, focus on the period between 2000 and 2019).

4.4.2 Mechanisms of Participation

In the anti-statism conservation, the participation of the government is being reduced, and the local people increase in power of the decision-making processes. During the last decade, it has begun to be used by indigenous peoples and similar communities as a legal strategy to claim a different relationship with the territory and commons (Borrini-Feyerabend et al. 2004). Through the defense of human rights, they seek to create and maintain identities, local symbolic resources (festivities, rituals, spiritualities, values), their own fields of action, links, and self-management to face common needs (Boege 2017).

This situation establishes a fundamental challenge to comply with the SDGs through PA since it involves breaking with the top-down protection schemes (hard statism) and establishing a bottom-up relationship based on respect for the human rights of the indigenous and similar communities as well as their forms of organization and particular livelihoods. Today, we see an increase in novel views on conservation, focusing on protecting indigenous rights to traditionally manage their common goods, self-determination, and autonomy (Borrini-Feyerabend et al. 2004; CNDH 2019). This focus might increase the effectiveness of PAs through the co-management of the territory (Berkes 2004) while reducing the social, economic, and political inequalities that indigenous and similar communities have historically suffered and allow biodiversity and agrobiodiversity conservation over the long term (Oldekop et al. 2016; Martínez-Esponda et al. 2019).

4.4.3 Effectiveness

The novelty and complexity of anti-statism conservation initiatives, such as OECM, ADVC, or Biocultural Landscapes, have limited the amount of studies focusing on their effectiveness. Currently, further research is being developed to understand how to determine the extent, governance type, and measurement of effectiveness.

4.5 Discussion

Historically, the Mexican PA's decree has been the main source of conflict for its implementation (CNDH 2019) because PAs have often been created without consulting the local population. Normally, the PAs decree imposes a new legal classification on the territory that consolidates its public interest and a new regulatory framework (CNDH 2019). This commonly overlaps previous classifications and activities that were carried out in the territory and establishes new regulations that affect directly the relationship that its inhabitants have established with the commons (e.g., water, forests, fisheries, food systems, seeds) mainly related to restrictions on the use of their common goods, access to sacred sites, as well as the ability to make decisions about their territories (Alcorn and Toledo 1998; Anaya and Espírito-Santo 2018; CNDH 2019).

The current EVS promotes changes that have also resulted in the emergence and consolidation of *new forms of physical and symbolic appropriation of territories, management, and governance frameworks* (Walkid 2011). To achieve the modern social, economic, political, and environmental goals through PAs, it is necessary to recognize the relevant role that *indigenous people and similar communities* (i.e., peasants and afro-descendants) have in the conservation of nature (Boege 2008; Garnett et al. 2018). Indigenous communities own or manage over 40% of the PA's territory worldwide, placing them as the group with the most territory inside these institutions (Garnett et al. 2018). This is not a coincidence, since in many cases, the collective practices of the indigenous and similar communities have given rise to multifunctional landscapes and particular forms of land management that increase the diversity of habitats, biodiversity, and agrobiodiversity that is necessary to address climate change (Borrini-Feyerabend et al. 2004; Robson 2007; Toledo and Barrera-Bassols 2008; Berkes 2009; Martínez-Esponda et al. 2019) and the current environmental crisis. Even when ecosystems are subject to a high degree of human pressure, indigenous territories perform better than PAs in protecting biodiversity and environmental services (Nolte et al. 2013). Paradoxically, indigenous and similar communities are also the group with the greatest social lag and greater vulnerability to climate change (Swiderska and Palmer 2015; IPCC 2018) and in many cases low participation or neglected participation inside the PA management.

Therefore, it is necessary to adapt the governance schemes, legal frameworks, and governmental structures that are involved in the management of PAs (Oldekop et al. 2016; Geldmann et al. 2015) because, on occasion, their implementation has come into conflict with local governance practice groups and institutions (Alcorn and Toledo 1998; Berkes 2004, 2009; Anaya and Espírito-Santo 2018). In order to achieve sustainable development, it is essential to recognize the socio-political context at local, regional, and national levels, as well as to establish mechanisms of participation that consider the inequalities and the underlying power structures in the decision-making and agency capacity of the actors involved in the management of environmental public policy instruments (Holmes 2007; Oldekop et al. 2016; Merçon et al. 2019).

4.6 Conclusion

4.6.1 Lessons Learned

4.6.1.1 Environmental Value System

Conservation and management of natural resources have undergone a change in recent decades and have moved away from *centralism of state* (hard statism) and regulation schemes *top-down* in environmental matters toward *bottom-up participatory schemes* (flexible statism and anti-statism). Novel schemes value the different cosmovisions of local population and believe that the power of decision-making and in the benefits of natural resource management should be shifted toward them (Shackleton et al. 2002; Dovers et al. 2015).

In addition, the objectives of the PAs have been expanding to *include economic, social, spiritual, and cultural goals* by using transdisciplinary knowledge and always aiming to reach a *more just and equitable society* that advances toward sustainability (Merçon et al. 2019).

Current conservation strategies must take into consideration the local institutions, collective practices, and biocultural heritage to provide culturally appropriate mechanisms for decision-making (Borrini-Feyerabend et al. 2013; Toledo 2015; Boege 2017). If considered, new conservation policies will allow the right to autonomy and self-determination to enhance sustainability (Martínez-Esponda et al. 2019).

4.6.2 Participation

Biosphere Reserves, FFPA, and other categories have participation mechanisms that are designed under a hard or flexible statism regime. Even though there is participation, the state maintains the control over strategic resources, financing, and development plans in detriment of the traditional land management systems and natural resources (Shackleton et al. 2002; Durand et al. 2014; Monterrubio-Solís 2019; Brenner 2010). And in some cases, an effective participation of society is not established as decision-making does not consider inequality in the exercise of power of the different actors involved (Brenner 2010; Durand et al. 2014; Toledo 2015). For example, in the advisory councils, some members (e.g., international and national agencies and civil society organization) have the monopoly of the information and decision-making and are usually not culturally appropriate (Durand et al. 2014; Monterrubio-Solís 2019).

Novel conservation strategies, ADVC and Biocultural Landscape, shift the participation mechanisms and give more autonomy to the local people. Now, it is important to analyze the underlying power structure and the potential conflicts in the management of natural resources between actors to develop effective conservation strategies (Durand et al. 2014).

4.6.3 Effectiveness

Historically, most Mexican PAs follow the hard or flexible statism. As discussed before, these schemes focus on the biological-ecological realm and neglect, at least in some degree, people's cultural practices and political organization. As a result, most of these PAs have shown more effectiveness on the biological or management fields than on the social ones. New effectiveness methods to assess flexible or antistatism schemes should evaluate the *power relationships* between the different actors inside the PAs (Merçon et al. 2019) and be guided by ethical principles to achieve "good governance," based on the principles of efficiency, social justice, and legitimacy (Brenner 2010).

References

Adams WM, Hutton J (2007) People, parks and poverty: political ecology and biodiversity conservation. Conserv Soc 5(2):147–183

Ajzen I (1991) The theory of planned behavior. Org Behav Human Decision Proc 50. https://doi.org/10.1016/0749-5978(91)90020-T

Alcorn J, Toledo VM (1998) Resilient resource management in Mexico's forest ecosystems: the contribution of property rights. In: Berkes F, Folke C (eds) Linking social and ecological systems. Cambridge University Press, Cambridge, pp 216–249

Anaya FC, Espírito-Santo MM (2018) Protected areas and territorial exclusion of traditional communities: analyzing the social impacts of environmental compensation strategies in Brazil. Ecol Soc 23(1). https://doi.org/10.5751/ES-09850-230108

Andrade GSM, Rhodes JR (2012) Protected areas and local communities: an inevitable partnership toward successful conservation strategies? Ecol Soc 17(4). https://doi.org/10.5751/ES-05216-170414

Barnosky AD, Hadly EA, Bascompte J, Berlow EL, Brown JH, Fortelius M, … others (2012) Approaching a state shift in Earth's biosphere. Nature 486(7401): 52–58. https://doi.org/10.1038/nature11018

Beresford AE, Buchanan GM, Donald PF, Butchart SHM, Fishpool LDC, Rondinini C (2011) Poor overlap between the distribution of protected areas and globally threatened birds in Africa. Anim Conserv 14(2):99–107. https://doi.org/10.1111/j.1469-1795.2010.00398.x

Berkes F (2004) Rethinking community-based conservation. Conserv Biol 18(3):621–630. https://doi.org/10.1111/j.1523-1739.2004.00077.x

Berkes F (2009) Community conserved areas: policy issues in historic and contemporary context. Conserv Lett 2(1):20–25. https://doi.org/10.1111/j.1755-263x.2008.00040.x

Boege E (2008) El patrimonio biocultural de los pueblos indígenas de México: hacia la conservación in situ de la biodiversidad y agrodiversidad en los territorios indígenas. Instituto Nacional de Antropología e Historia (INAH) Comisión Nacional para el Desarrollo de los Pueblos Indígenas (CDI) Mexico City: Mexico

Boege E (2017) El patrimonio biocultural y los derechos culturales de los pueblos indígenas, comunidades locales y equiparables. Diario de Campo 0(1):39–70

Borrini-Feyerabend G, Kothari A & Oviedo G (2004) Indigenous and Local Communities and Protected Areas: Towards Equity and Enhanced Conservation IUCN Gland, Switzerland and Cambridge, UK. xviii + 111pp. 2–8317–0675-0

Borrini-Feyerabend G, Dudley N, Jaeger T, Lassen B, Broome NP, Phillips A, Sandwith T (2013) Governance of protected areas: from understanding to action. IUCN, Gland

Brenner L (2010) Gobernanza ambiental, actores sociales y conflictos en las Áreas Naturales Protegidas mexicanas. Rev Mex Sociol 2(55):283–310

Brockington D, Wilkie D (2015) Protected areas and poverty. Philos Trans R Soc B: Biol Sci 370(1681). https://doi.org/10.1098/rstb.2014.0271

Butchart SHM, Walpole M, Collen B, Strien A, Bastian B, Brown C, Watson R (2010) Global biodiversity: indicators of recent declines. Science 328(May):1164–1168

Carabias J, De la Maza J, Cadena R (2015) Conservación y desarrollo sustentable en la Selva Lacandona. Natura y Ecosistemas Mexicanos AC, México

Clayton S, Brook A (2005) Can psychology help save the world? A model for conservation psychology. Anal Soc Issues Public Policy 5(1):87–102. https://doi.org/10.1111/j.1530-2415.2005.00057.x

CNDH (2019) Estudios sobre el cumplimiento e impacto de las recomendaciones generales, informes especiales y pronunciamientos de la CNDH (2001–2017) TOMO VI. Áreas naturales protegidas y derechos humanos Ciudad de México: Universidad Nacional Autónoma de México; Cordinación de Humanidades, UNAM; Programa Universitario de Derechos Humanos, UNAM; Comisión Nacional de Derechos Humanos, México City: México

Coad L, Watson JEM, Geldmann J, Leverington F, Hockings M, Knights K, Di Marco M (2019) Widespread shortfalls in protected area resourcing. Front Ecol Environ 17(5):259–264

CONANP (2019) Áreas Destinadas Voluntariamente a la Conservación. Retrieved from the web https://advc.conanp.gob.mx/listado-de-advc/

CONANP (2020) Listado de las Áreas Naturales Protegidas de México (LISTANP) Retrieved form the web site: http://sig.conanp.gob.mx/website/pagsig/listanp/

De la Maza ER (1999) Una historia de las áreas naturales protegidas en México. Gaceta Ecológica 51:15–34

Dovers S, Feary S, Martin A, McMillan L, Morgan D Tollefson M (2015) Engagement and participation in protected area management: who, why, how and when. Protected Area Governance and Management 413–440 Retrieved from https://www.jstor.org/stable/pdf/j.ctt1657v5d.21.pdf

Dudley N (2008) Guidelines for applying protected area management categories International Union for Conservation of Nature, Gland, Switzerland

Dudley N, Ali N, Ketiunen M, MacKinnon K (2017) Editorial essay: protected areas and the sustainable development goals. Parks 23(2):10–12. https://doi.org/10.2305/IUCN.CH.2017.PARKS-23-2ND.en

Durand L, Figueroa F, Trench T (2014) Inclusion and exclusion in participation strategies in the montes azules biosphere reserve, Chiapas, Mexico. Conserv Soc 12(2):175–189. https://doi.org/10.4103/0972-4923.138420

Eklund J, Cabeza M (2017) Quality of governance and effectiveness of protected areas: crucial concepts for conservation planning. Ann N Y Acad Sci 1399(1):27–41. https://doi.org/10.1111/nyas.13284

Figueroa F, Sánchez-Cordero V (2008) Effectiveness of natural protected areas to prevent land use and land cover change in Mexico. Biodivers Conserv 17(13):3223–3240. https://doi.org/10.1007/s10531-008-9423-3

Garnett ST, Burgess ND, Fa JE, Fernández-llamazares Á, Molnár Z, Robinson CJ et al (2018) Indigenous lands for conservation. Nat Sustain 1(July):369–374. https://doi.org/10.1038/s41893-018-0100-6

Geldmann J, Barnes M, Coad L, Craigie ID, Hockings M, Burgess ND (2013) Effectiveness of terrestrial protected areas in reducing habitat loss and population declines. Biol Conserv 161:230–238. https://doi.org/10.1016/j.biocon.2013.02.018

Geldmann J, Coad L, Barnes M, Craigie ID, Hockings M, Knights K et al (2015) Changes in protected area management effectiveness over time: a global analysis. Biol Conserv 191:692–699. https://doi.org/10.1016/j.biocon.2015.08.029

Hockings M (2003) Systems for Assessing the effectiveness of Management in Protected Areas. Bioscience 53(9):823–832

Hockings M, Stolton S, Leverington F, Dudley N, Courrau J (2006) Evaluating effectiveness: a framework for assessing management effectiveness of protected areas, 2nd edn. IUCN, Gland/Cambridge UK

Holmes G (2007) Protection, politics and protest: understanding resistance to conservation. Conserv Soc 5(2):184–201

IPCC (2018) Summary for policymakers. In: Masson-Delmotte V, Zhai P, Partner HO, Roberts D, Skea J, Shukla PR, Pirani A, Moufouma-Okia W, Pan C, Pidcock R, Connors S, Matthews JBR, Chen Y, Zhou X, Gomis MI, Lonnoy E, Maycock T, Tignor M, Waterfield T (eds.) Global warming of 1.5°C. an IPCC special report on the impacts of global warming of 1.5°C above pre-industrial levels and related global greenhouse gas emission pathways, in the context of strengthening the global response to the threat of climate change, sustainable development, and efforts to eradicate poverty. World Meteorological Organization, Ginebra

Jiménez-Sierra C, Torres-Orozco D, Martínez JC, Toledo-Guzmán AD (2017) A socioecosystemic view to the natural reserves: Case Study of Metztitlan's Biosphere Reserve. In Calderón-Contreras (Coordinator) In: Los sistemas socioecológicos y su resiliencia: casos de estudio. Mexico City Mexico

Jones NA, Shaw S, Ross H, Witt K, Pinner B (2016) The study of human values in understanding and managing social-ecological systems. Ecol Soc 21(1)

Jones KR, Venter O, Fuller RA, Allan JR, Maxwell SL, Negret PJ, Watson JEM (2018) One-third of global protected land is under intense human pressure. Science 791(May):788–791

Kellert SR (1996) The value of life: biological diversity and human society. Island Press, Washington, DC, 263 p

Leverington F, Costa KL, Courrau J, Pavese H, Nolte C, Marr M et al (2010) Management effectiveness evaluation in protected areas – a global study second edition – 2010 (Second). University of Queensland, Brisbane

LGEEPA (2018) Diario Oficial de la Federación. Ultima reforma publicada DOF 05–06-18

Mace GM (2014) Changes in the perception and goals of nature conservation require a solid scientific basis. Science 345:1558. https://doi.org/10.1126/science.1254704

Martínez-Esponda FX, García-Maning G, Vázquez-Quesada B, Colmenero S, Bracamontes-Nájera L (2019) In: Tornel C (Coordinator) Alternativas para limitar el calentamiento global en 1.5°C. Henrich Böll-Stiftung, Mexico City, México

Merçon J, Vetter S, Tengö M, Cocks M, Balvanera P, Rosell JA, Ayala-Orozco B (2019) From local landscapes to international policy: contributions of the biocultural paradigm to global sustainability. Global Sustain 2. https://doi.org/10.1017/sus.2019.4

Mitchell R (2006) Environmental governance in Mexico: two case studies of Oaxaca's community forest sector. J Lat Am Stud 38:519–548

Monterrubio-Solís C (2019) Formalización de Áreas Destinadas Voluntariamente a la Conservación en territorios comunitarios e indígenas, avances y reveses. EntreDiversidades Revista de Ciencias Sociales y Humanidades 6(12):79–110. https://doi.org/10.31644/ed.12.2019.a03

Naidoo R, Balmford A, Ferraro PJ, Polasky S, Ricketts TH, Rouget M (2006) Integrating economic costs into conservation planning. Trends Ecol Evol 21(12):681–687. https://doi.org/10.1016/j.tree.2006.10.003

Nolte C, Agrawal A, Silvius KM, Britaldo SSF (2013) Governance regime and location influence avoided deforestation success of protected areas in the Brazilian Amazon. Proc Natl Acad Sci U S A 110(13):4956–4961. https://doi.org/10.1073/pnas.1214786110

Oldekop JA, Holmes G, Harris WE, Evans KL (2016) A global assessment of the social and conservation outcomes of protected areas. Conserv Biol 30(1):133–141. https://doi.org/10.1111/cobi.12568

Olsson P, Folke, Berkes (2004) Adaptive comanagement for building resilience in social-ecological systems. Environ Manag 34:75–90. https://doi.org/10.1007/s00267-003-0101-7

Paz-Salinas MF (2005) La participación en el manejo de áreas naturales protegidas. Actores e intereses en conflicto en el Corredor Biológico Chichinautzin, Morelos. Centro Regional de Investigaciones Multidisciplinarias, Cuernavaca

Reser JP, Bentrupperba JM (2005) What and where are environmental values? Assessing the impacts of current diversity of use of 'environmental' and 'world heritage' values. J Environ Psychol 25:125–146. https://doi.org/10.1016/j.jenvp.2005.03.002

Robson JP (2007) Local approaches to biodiversity conservation: lessons from Oaxaca, southern Mexico. Int J Sustain Dev 10(3):267–286. https://doi.org/10.1504/IJSD.2007.017647

Rodrigues ASL (2006) Are global conservation efforts successful? Science 313(5790):1051–1052. https://doi.org/10.1126/science.1131302

Ruiz-Mallén I, Corbera E (2013) Community-based conservation and traditional ecological knowledge. Ecol Soc 18(4):12. https://doi.org/10.5751/ES-05867-180412

Saunders CD (2003) The emerging field of conservation psychology. Hum Ecol Rev 10(2):137–149

Shackleton S, Campbell B, Wollenberg E, Edmunds D (2002) Devolution and community-based natural resource management: creating space for local people to participate and benefit? Nat Resour Persp 76(76):1–6. https://doi.org/10.1016/j.ecolecon.2012.09.001

Stern PC, Dietz T, Abel TD, Guagnano GA, Kalof L (1999) A value-belief-norm theory of support for social movements: the case of environmentalism. Res Human Ecol 6(2):81–97. https://doi.org/10.2307/2083693

Stolton S, Hockings M, Dudley N, MacKinnon K, Whitten T, Leverington F (2007) Management effectiveness tracking tool (second). WWF International, Gland

Swiderska K, Palmer M (2015) Climate change and biocultural adaptation in mountain communities. Second international learning exchange of the international network of mountain indigenous people. International Institute for Environment and Development, London

Toledo VM (2005) Repensar la conservación : ¿áreas naturales protegidas o estrategia bioregional ? Gaceta Ecológica 77(77):67–83

Toledo VM (2015) ¿De qué hablamos cuando hablamos de sustentabilidad? Una propuesta ecológico política. Interdisciplina 3(7):35–56

Toledo VM, Barrera-Bassols N (2008) La memoria biocultural. Junta de Andalucía, Icaria editorial, Perspectivas Ecológicas, Barcelona

UNEP-WCMC (2020) Protected area country profile for Mexico from the world database of protected areas. Available at: www.protectedplanet.net

UNEP-WCMC IUCN & NGS (2018) Protected planet report 2018. UNEP-WCMC IUCN NGS, Cambridge UK/Gland/Washington, DC

UNEP-WCMC IUCN & NGS (2020) Protected planet live report 2020. UNEP-WCMC, IUCN and NGS, Cambridge UK/Gland/Washington, DC

Vitousek PM, Mooney HA, Lubchenco J, Melillo JM (1999) Human domination of earth' s ecosystems. Science 277(5325):494–499. https://doi.org/10.1126/science.277.5325.494

Walkid E (2011) Revolutionary Parks. Conservation, social justice, and Mexico's National parks 1910–1940. The University of Arizona press, Tucson

Watson JEM, Dudley N, Segan DB, Hockings M (2014) The performance and potential of protected areas. Nature 515:67–73. https://doi.org/10.1038/nature13947

West P, Igoe J, Brockington D (2006) Parks and peoples: the social impact of protected areas. Annu Rev Anthropol 35(1):251–277. https://doi.org/10.1146/annurev.anthro.35.081705.123308

Chapter 5
Changing the Paradigm for Better Conservation: Conceptual Proposals from the Environmental Humanities

Micheline Cariño-Olvera, René Moreno-Terrazas-Troyo, and Ananda Monteforte-Cariño

Abstract The current conservation paradigm emerged at the 1992 Rio Summit at the same time as the institutionalization of global environmental policy. Consequently, national institutions appeared that promoted conservation through various instruments, especially in the signatory countries of the Convention on Biological Diversity. Generally, these processes have been top-down with little or no participation by the population that lives in or uses the ecosystems that the institutions claim to conserve. This occurs in a context of contradictions between governmental institutions and policies, which on the one hand promote extractivism and on the other wish to conserve the environment and end poverty. The results have protected neither biodiversity nor social well-being. It is time for a transformative change that begins with a different paradigm to overcome the obstacles of the current paradigm and that has as its basis the restoration of sustainable relationships between societies and the ecosystems on which all living beings depend. Based on the environmental humanities, this chapter hopes to contribute to this urgent transformation by proposing several concepts, values, and practices for a new paradigm.

Keywords Transformative change · Socioecological sustainability · Biocultural diversity · Environmental ethics

---

M. Cariño-Olvera (✉) · R. Moreno-Terrazas-Troyo
Universidad Autónoma de Baja California Sur, La Paz, Mexico
e-mail: rtroyo@uabcs.mx

A. Monteforte-Cariño
Universidad Nacional Autónoma de México, Mexico City, Mexico

A. Ortega-Rubio (ed.), *Socio-ecological Studies in Natural Protected Areas*,
https://doi.org/10.1007/978-3-030-47264-1_5

5.1 Introduction

On May 6, 2019, the Intergovernmental Science-Policy Platform on Biodiversity and Ecosystem Services (IPBES) published a Summary for policymakers of the global assessment report on biodiversity and ecosystem services[1]. The four points on which the report is structured are:

A. Nature and its vital contributions to people, which together embody biodiversity and ecosystem functions and services, are deteriorating worldwide.
B. Direct and indirect drivers of change have accelerated during the past 50 years.
C. Goals for conserving and sustainably using nature and achieving sustainability cannot be met by current trajectories, and goals for 2030 and beyond may only be achieved through transformative changes across economic, social, political, and technological factors.
D. Nature can be conserved, restored, and used sustainably while simultaneously meeting other global societal goals through urgent and concerted efforts fostering transformative change[2] (IPBES 2019:2–7).

The report of the IPBES cited here demonstrates that the most authoritative institutions have recognized and proclaimed the urgent need for changing the paradigms that have sustained international conservation policy since the Rio de Janeiro Summit in 1992. This essay seeks to contribute several concepts to enrich the formation of a new paradigm to reorient the processes of conservation, based on a fundamental criticism of the neoliberal context in which this policy has developed:

> The commodification of nature and the idea of its exclusive use by some actors only are among the principal characteristics of neoliberal or hegemonic conservation that are now clearly inoperative. Nature is vast, complex, dynamic in multiple timeframes, subject to diverse representations and infinite uses, and above all sustaining all forms and expressions of life on the Earth. To think that it can be reduced to a profit-oriented value is not only futile, but absurd and unjust. Even worse is to consider certain spaces and ecosystem components as available for extractivism and as sinks for all kinds of waste, under the pretext of economic growth and development; that is to make them zones of sacrifice. (Machado Aráoz 2015:21)

In the first section of this essay, we present a brief historical and critical review of the neoliberal context in which the current conservation paradigm arose, in order to explain its limited scope and consequently its necessary transformation. But as the authors of the 2019 IPBES report maintain in the fourth point mentioned above, *transformative change*:

> by its very nature, could expect opposition from those with interests vested in the status quo, but such opposition can be overcome for the broader public good. If obstacles are overcome, commitment to mutually supportive international goals and targets, supporting actions by indigenous peoples and local communities at the local level, new frameworks for

[1] https://www.ipbes.net/sites/default/files/downloads/spm_unedited_advance_for_posting_htn.pdf

[2] "A fundamental, system-wide reorganization across technological, economic and social factors, including paradigms, goals and values."

> private sector investment and innovation, inclusive and adaptive governance approaches and arrangements, multi-sectoral planning and strategic policy mixes can help to transform the public and private sectors to achieve sustainability at the local, national and global levels. (IPBES 2019:7)

In the second section of the essay, we propose some concepts from the environmental humanities that can contribute to this transformative change from a critical perspective. They begin with the understanding that conservation should be an intrinsic part of our perceptions of and relations with nature and not a separate end in itself. This would change the ways in which we use the elements afforded by nature for the support of human and beyond-human life. This transformation would be guided by the three following premises:

1. The perceptions as well as the forms of ecosystem exploitation should guarantee the possibility of its long-term use, that is, be based on socioecological sustainability.
2. Any use, transformation, and protection of ecosystems and their services should first and foremost benefit and involve the populations that live in and from them, that is, to prioritize life in its place.
3. Conservation policies should not only be focused only on spaces and species, but should take many different forms (productive, recreational, educational, research, and spiritual). That is to say, to include conservation in our minds, words, practices, both individual and collective.

Our aim is to discuss how ecological criteria are necessary but in no way sufficient to conserve the ecosystem base that sustains human and beyond-human life. The protection of spaces and species is extremely limited in both geographical and ecological terms, not to mention social and political terms. A new conservation paradigm requires recognizing that conservation is also political and philosophical. It should be centered in the defense and (re)valuing of territory and its biocultural diversity, while sustaining natural and social reproduction, including cultural identities. The construction of such a paradigm implies rethinking our form of living in the world, our social relations, and our ideas of nature in order construct sustainable societies, possible futures.

5.2 Brief Historical Review of the Context in Which the Neoliberal Conservation Paradigm Arose

> The current status and projected trajectories for all levels of biodiversity are dire. Some 11,000–36,000 species may go extinct each year, and across much of the globe, local species richness has been reduced to below the threshold needed to ensure the long-term maintenance of ecosystem functions and services. Many conservation efforts have attempted to stem the loss of biodiversity, and without such actions, the situation of biodiversity would undoubtedly be worse. (Gavin et al. 2018:1)

The loss of biodiversity was one of the main themes discussed in the 1992 second Earth Summit in Rio. The Convention on Biological Diversity (CBD), a multilateral agreement signed on the spot by 168 countries, was mandated to contend with biodiversity loss. The Convention came into force in 1993 and by 2018 had organized 14 ordinary and 2 extraordinary Conferences of the Parties (COPs), in which representatives of the signatory countries come together, along with international organizations involved with biodiversity conservation. The annual financial support for this global effort comes to tens of billions of dollars, in addition to the financing of research projects in the global North and South. With this support, a great diversity of strategies, actions, and policies have been implemented over almost 30 years. And, in spite of everything, biodiversity is diminishing at an unprecedented rhythm on a world level. The rate of species extinction has accelerated, as has the probability of serious impacts on human populations throughout the world (IPBES 2019)[3].

The evidence in undeniable: the way in which conservation efforts have handled biodiversity loss has not functioned as hoped for. It is time to change the strategy. It is essential to construct a new paradigm. But we also need to understand why conservation has been undertaken the way it has, in order to overcome mistakes and design new pathways. History allows us to understand the present and orient ourselves toward the future. "Concern about the future paradoxically leads to a renewed attraction to the past. It seems as if the eyes turn desperately towards history, searching for the keys that allow for an understanding of a future that is seen as uncertain" (González de Molina and Toledo 2011:13–14).

Reviewing social processes allows us to understand not only the sudden shifts that mark their points of inflection but also the cultural, political, and ideological contexts in which they arise. A decontextualized criticism of the failure of international conservation policies based only on their failure to achieve desired results is not only partial, but can lead us in the wrong direction in making proposals for the future that claim to overcome the failures. It is essential to know the context in which these policies arose in order for a critique to be well substantiated and to allow us to understand the complexity in which their successes as well as their errors and limitations have arisen.

Nevertheless, we should not lose sight of the fact that knowing the context which produced the world environmentalist concerns does not justify abuses committed against various human and non-human actors, but it explains them.

It is thought by those with a short historical horizon that the concern for nature conservation began with one of the two first Earth Summits, Stockholm 1972 and Rio 1992. This concern is much older, however. A reaction to the overexploitation of natural resources dates to the eighteenth century: Hans Carl von Carlowitz in 1713 used for the term "sustainability" (*nachhaltigkeit* in Saxon dialect) for the first time, as a practice for the continuous extraction of wood from a single forest (Caradonna 2014:35). Toward the end of the nineteenth century, the motivations for

---

[3] https://www.unenvironment.org/es/news-and-stories/comunicado-de-prensa/la-naturaleza-esta-en-un-declive-peligroso-y-sin-precedentes

conservation were less materialistic, but the cause of degradation was the same: the devastating effects that progress had on nature.

The first national parks were created in the USA based on a romantic idea of nature and a concrete concern for its protection, Yellowstone in 1872 and Yosemite and Sequoia in 1890. Civil society associations joined forces rapidly to help protect these natural spaces and the flora and fauna they fostered. Thus the Sierra Club was founded in 1892 and the Audubon Society in 1905. In 1916, the US federal government created the National Park Service to administer these reserves.

From that time the environmental movement has not stopped growing and diversifying. We can summarize the formation of environmental consciousness in three phases. Between 1850 and 1914, we see a serious consideration of the risks that the accelerated transformation of nature posed for the future of human societies. The first half of the twentieth century was a period in which environmental themes were silenced, given the technological maelstrom brought on by petroleum and the unprecedented violence of the two world wars. The consequences of these excesses rebounded with the force of a boomerang, activating a critique of the processes of devastation, pollution, and massive death of both humans and ecosystems, warning about the seriousness of the risk to the environments on which life is based. Thus around 1960 we see the beginning of what Worster (2006) calls the Age of Ecology, whose turning point was the publication of *Silent Spring* (Carson 1962).

International conservation policy began in this third phase, based on the experience of the first period and confronting the obstacles of the second. From the first phase came the idea of protecting spaces and species, considering human societies as a whole a latent threat to nature. This is why international accords concerning the environment began at the end of the nineteenth century and until the decade of the 1990s targeted the protection of certain ecologically distinguished regions, leading to the first natural protected areas (NPAs) and also to the preservation of certain charismatic species. Some examples of this type of accord are the prohibition of seal hunting in the Bering Sea (1885), the Convention for the Regulation of Whaling (1931) for Africa, the Convention Relative to the Preservation of Fauna and Flora in the Natural State (1933), and the International Plant Protection Convention (1951) implementing control measures to avoid the introduction of plant and plant product pests.

In 1948, under the auspices of the United Nations, governments and civil society organizations created the International Union for the Conservation of Nature (IUCN) as a forum to reach consensus for decision-making. The IUCN promotes joint actions to conserve the integrity and diversity of nature, as well as to assure the equitable and sustainable use of natural resources. Currently it has more than 1300 members including states, governmental agencies, civil society organizations, economic development agencies, and academic and scientific institutions, as well as business associations. The greater part of global environmental policy has been developed under the aegis of the IUCN[4].

[4] https://www.iucn.org/about/iucn-a-brief-history

Pressure from the international environmental movement and its confrontation with the development policies of the 1950s, 1960s, and 1970s culminated in the celebration of the first Earth Summit in Stockholm in 1972 and in the gradual construction of institutions and authorities dedicated to the conservation of the human environment. UNESCO created the *Man and the Biosphere* (MAB) program in 1971, which was the first global instrument for the conservation of biological diversity in relation to human societies that was allocated support for technical and financial assistance:

> As of January 2018, it had 669 biosphere reserves in 120 countries, counting 20 transboundary sites including: 75 in 28 countries in Africa 31 in 11 countries in the Arab States 147 in 24 countries in Asia and the Pacific 287 in 36 countries in Europe and North America 129 in 21 countries in Latin America and the Caribbean The total terrestrial and marine area covered by biosphere reserves around the world amounts to over 735,000,000 hectares. More than 250 million people live in a biosphere reserve. (UNESCO 2018)

The greatest singular benefit of this program was to demonstrate that one of the best strategies for conservation is for it to be carried out by the native peoples who ancestrally or traditionally have sustained the biological diversity of the areas they inhabit, since they depend on it for their life and culture (Toledo and Barrera-Brassols 2008). This conservation strategy by and with the people is the opposite of the US idea of natural parks and of most natural protected areas (NPAs) based on the principle of protecting nature in its *pristine* state, that is, reserves without people.

The same agency of the United Nations approved the Convention concerning the Protection of World Cultural and Natural Heritage in 1972, which, even if it is not a conservation strategy in itself, contributed to the recognition that the listed sites should be particularly cared for by the countries in which they are located. The Convention on International Trade in Endangered Species of Wild Fauna and Flora (CITES) was signed in 1973 and came into effect in 1975 for the purpose of regulating the international wildlife trade to prevent trafficking of protected species. In this same period, conservationist forces continued their orientation toward particular spaces and problem areas, as in the case of the sea[5], wetlands (Ramsar Convention 1971), the atmosphere (Montreal Protocol 1987), and the fight against desertification (since 1974).

The deep causes of ecological degradation and its consequences were analyzed more systematically in the same decade through scientific research conducted by groups of expert coming together for that purpose. One of the first and most famous studies was the silenced Meadows Report called *The limits to growth* and commissioned by the Club of Rome[6] in 1972. This study showed that sustaining historical economic growth trends would lead to exceeding the limits of planetary resources, inevitably provoking a collapse within an estimated hundred-year timeframe. This critical focus was the tone kept

---

[5]The first Convention on the Continental Shelf was adopted in 1958 and ratified as a United Nations instrument in 1971. In 1961, the Convention on the Conservation of Antarctic Marine Living Resources was signed (taking effect in 1982), and the International Convention for the Conservation of Atlantic Tunas was signed in 1966 (coming into effect in 1969).

[6]https://www.clubofrome.org/

up through the decade of the 1970s, as in the case of the Cocoyoc Symposium (Mexico), convened by the United Nations Environment Programme (UNEP) in 1974, which declared the *Charter on Economic Rights and Duties of States*. This charter emphasized the importance of correlations among population, natural resources, environment, and economic growth. It adopts the concept of ecodevelopment (Sachs 1993) as an alternative to conventional development, which it considers the cause of our main environmental and social problems, that is to say, of the global crisis. It stressed that development policies should be oriented toward satisfying basic needs in nutrition, health, education, labor, and housing without overexploiting ecosystems. Environmental education was another element that made progress in this decade. In 1977, the First Intergovernmental Conference on Environmental Education in Tbilisi defined the function, objectives, guiding principles, target groups, contents, and methods of environmental education, seeing this as a strategy for solving environmental problems.

In 1975 a new international strategy for conservation and sustainable resource use emerged when the president of Zaire (now D. R. Congo), Mobutu Sese Seko, proposed to the United Nations General Assembly a Charter for Nature under the responsibility of the IUCN. The project was prepared by an international group of experts and presented in 1977. In October, 1980, the plan was presented to the United Nations General Assembly as the *World Conservation Strategy* (UICN, 1980).

In brief, during the decades of the 1960s and 1970s, the results of scientific research and reports of diverse governmental and nongovernmental agencies aroused, widened, and deepened environmental awareness in various arenas for diverse social actors. The necessity of caring for the ecosystem base on which human existence and all living beings depend became understood conclusively. This perspective suggests that ecosystem conservation and the sustainable use of natural resources should take precedence over economic growth and profit. Various authors agree that it was an extraordinary opportunity that, had it not been overshadowed and overcome by developmentalism, would really have been able to mitigate the ecological and civilizational crisis that was already out of control in the twenty-first century (Riechmann 2018:250). But the economic interests of transnational companies and the unlimited greed of politicians and business people were stronger than environmentalist warnings.

The politically correct way was to continue development while taking environmental issues into account, though in a superficial way, since they were no longer possible to conceal. Thus in both discourse and practice, it was necessary to make economic growth compatible with environmental conservation and social justice. The wand of this magic act was the concept of *sustainable development*, which appeared for the first time in the 1980 *World Conservation Strategy: Living Resource Conservation for Sustainable Development* proposed by the IUCN with the support of the United Nations Environment Programme (UNEP) and the World Wildlife Foundation (WWF). The *Strategy* recognizes the obligation to recognize and respect the intrinsic value of all forms of life and that humanity needs to base its conduct on a moral code. Likewise, conservation of nature and its resources should contribute to justice and peace, in accordance with the United Nations Charter. States should incorporate into their legislation the principle that biological resources should not be used beyond their capacity for regeneration. The strategy tried to give conservation a practical and concrete dimension since the purpose of conservation was for the Earth to maintain its capacity to achieve economic development to support life.

These efforts would continue in the proclamation of the United Nations World Charter for Nature (1982) and the creation in 1983 of the World Commission for Environment and Development, better known as the Brundtland Commission. Its mission was to study the interrelations between development and conservation, as well as to provide solutions to achieve their compatibility. The work of the Commission's experts resulted in the report entitled *Our Common Future*. It is a political instrument that recognizes the threats we face as a species and proposes steps to assure the continuity of human progress. Its key concept is *sustainable development*, which it defines as "development that meets the needs of the present without compromising the ability of future generations to meet their own needs"[7]. Starting with the Rio de Janeiro Conference in 1992, the idea of "sustainable development" was disseminated to the point where it was accepted – and generated controversy – more than its predecessor: simply "development" with no adjective.

In this chapter we do not address the debates about this highly polysemic concept, lest we distract ourselves from our object of study and because many articles and books have been written on the subject[8]. What interests us is the impact that the concept of sustainable development and the 1992 Rio Conference had on world conservation policy and its realization in diverse strategies. In the first place, it is useful to emphasize that international environmental policy was consolidated; concern for the environment became institutionalized as a binding mandate for the states that make up the United Nations.

This policy deals principally with two problems that pose a severe risk for the continuity of life on the planet: the struggles against the loss of biodiversity and climate change. The first was institutionalized through the Convention on Biological Diversity (CBD), which was signed on June 5, 1992, and came into effect in December of 1993. Its goal is to conserve biological diversity, promote the sustainable use of its components, and distribute fairly and equitably the benefits derived from the use of genetic resources. The second gave rise to the United Nations Framework Convention on Climate Change (UNFCCC), signed on May 9, 1992, and entering into effect on March 21, 1994. It is the principal international juridical instrument for facing the challenge of climate change and seeks to stabilize the concentrations of greenhouse gases in the atmosphere.

From the beginning of these policies in the first decade of this century, climate change has gotten as much attention as the loss of biological diversity. This is reflected in the number of Conferences of the Parties (COPs). The COP is the supreme body of these two Conventions and represents of all the signatory countries, called "parties." In the case of the Convention on Climate Change, it meets

[7] https://www.un.org/es/ga/president/65/issues/sustdev.shtml

[8] Gudynas, E. (2011a, b), Escobar, A (2007), Pierri, N (2005), and Reichmann, J. (1995), among others

annually to adopt decisions that enable it to achieve agreed-upon objectives. Decisions can be taken by unanimity or consensus of the parties. The first COP of the UNFCCC took place in Berlin, Germany, in 1995 and the most recent, the 24th, in Katowice, Poland, in 2018. This year, 2019, is the year of COP number 25, important not only because it marks a quarter of a century of struggle against climate change but also because the coming into force of the Paris Accords (taken in COP 21 in 2015) will be discussed. On the contrary, the CBD has only had 13 ordinary COPs, plus 1 extraordinary meeting which resulted in the Cartagena Protocol on Biosafety in January of 2000, ratified in 2003. The first three COPs of the CBD took place annually between 1994 and 1996, but then have occurred every 2 years since 1998. Protocols ratified in the COPs bind the parties to meet set goals, which influences the orientation of public policies in the different countries.

Actions to mitigate climate change and prevent the loss of biodiversity are strictly linked. As Steffen and his research team (2015) have shown, the core planetary boundaries of climate change and biosphere integrity are profoundly interrelated and highly linked with other planetary boundaries, given that, as subsystems of the geosystem, they have coevolved for almost four billion years:

> Furthermore, large changes in the climate or in biosphere integrity would likely, on their own, push the Earth system out of the Holocene state. In fact, transitions between time periods in Earth history have often been delineated by substantial shifts in climate, the biosphere, or both. These observations suggest a two-level hierarchy of boundaries, in which climate change and biosphere integrity should be recognized as core planetary boundaries through which the other boundaries operate. The crossing of one or more of the other boundaries may seriously affect human well-being and may predispose the transgression of a core boundary(ies) but does not by itself lead to a new state of the Earth system. (Steffen et al. 2015:10)

Given that the existing feedback between climate change and biodiversity loss is so close, mitigating one implies mitigating the other. Nevertheless, the fact that the former has attracted greater attention that the latter can be explained by the capitalist logic of the World System (Wallerstein 2005) in which the 1992 Rio Summit took place, giving rise to both Conventions.

Mitigating climate change offers a good opportunity for technological innovation and financial engineering, both fields generating substantial profits. On the other hand, preventing biodiversity loss, with the exception of the use of genetic resources[9], is accomplished by limiting the growth of economic activities in which powerful transnational interests are involved. We mention only a few examples of the economic political cost involved in conserving two of the biomes with the greatest biodiversity: rainforests and coral reefs. To protect the former, it is necessary to combat the expansion of monoculture in the tropics, which would slow deforestation caused by the cultivation of soy beans (Amazonia), oil palms (Malaysia and

[9] Currently regulated by the Nagoya Protocol on Access to Genetic Resources and the Fair and Equitable Sharing of Benefits Arising from their Utilization to the Convention on Biological Diversity which is an international agreement adopted at the 10th COP held in 2010 and entered into force on October 12, 2014 https://www.cbd.int/abs/about/

Indonesia), pineapples (Costa Rica and Mexico), and bananas (Central America), among others. Avoiding the destruction of rainforests also involves curbing open pit mining, which has taken over in Honduras, Mexico, Colombia, Ecuador, Brazil, etc., as well as limiting the development of intensive livestock raising. To protect coral reefs – whose days are numbered as we know because of ocean acidification caused by climate change – it is necessary to combat the industry without smokestacks that is the mass tourism that destroys beaches and dunes and sinks large amount of pollutants into the sea. It would also be necessary to curb the proliferation of large cruise ships which in addition to polluting the ocean with their waste require the construction of ports and docks, as well as the dredging of ocean depths; all these actions damage the most highly vulnerable reef zones.

No. To combat biodiversity loss does not offer opportunities to create business. But undertaking it is indispensable not only for the continuity of human and beyond-human life on the planet but also to mitigate climate change. The neoliberal capitalist orientation (in its new version of Rio +20 as the Green and Blue Economy) that underlies national and international environmental policy has not been able to keep the voices of environmentalists and the results of an enormous effort by the committed scientific community from influencing all levels of public policy, discourse on world environmental governance, and education. Without doubt, environmentalism has formed a part of the culture of our times, and the United Nations has promoted a sophisticated scaffolding to incorporate environmentalism into its priorities.

Nevertheless, the results have been insufficient, and we are now at the edge of a civilizational as well as ecological collapse (Riechman 2018; Taibo 2016; Tainter 1988, among others), since damage to the functioning of the geosystem puts the continuation of human and beyond-human life at risk in the coming decades. The future generations that are to be considered in the definition of sustainable development have already been born, including young adults who will face the water crisis (2030) and the petroleum crisis (2050), to which we can add the devastating effects of climate change and the Sixth Extinction (Kolbert 2014; Leakey and Lewin 1996).

As we mentioned at the beginning of this chapter, a transformative change is urgently required. This does not mean abandoning national and international environmental governance, but rather deepening and reinforcing its actions. At the same time, it is imperative to design many more and new forms of work in the struggle for life on this our mistreated, vulnerable, and generous spaceship (Boulding 1966). In the following two sections, we present some concepts and ideas that can enhance this work. We in no way claim to be exhaustive, but rather demonstrative and explanatory. Our goal is to contribute to the efforts of millions of conscious and sentient beings who we hope will survive the Century of the Great Test (Riechmann 2013) and to construct much better worlds to overcome – but also learn from – the terrific experience of the past three centuries.

5.3 Proposals for a Conceptual Base of a New Paradigm to Conserve Life

The origin of the modern Western rationalist paradigm is usually attributed to Descartes. Cartesian dualism has been one of the main influences in epistemology and in the construction of hegemonic knowledge. In producing a dualism between mind and body and between mind and nature, Descartes managed to claim a knowledge that is unsituated, universal, and omnipresent (Grosfoguel 2006). This paradigm, in which science was converted into the universal criterion for dominating nature, produced an objectification of the world as a hegemonic form off knowledge in which the subject is separate from its setting. The domain of power, the exploitation of nature, and the thirst for accumulation are seen in the Eurocentric colonial expansion imposed as a worldview on the colonized and oppressed peoples. Colonialism is an authoritarian exercise not only of political power but also of epistemological power based on absolute truths and narratives alien to the conquered areas. It caused the loss of local pluriverses which are characterized by a distinct culture of nature that determines an ontology (worldview) and an epistemology (traditional knowledge) linked with to the environment-habitat.

For Leff (2011), the socioenvironmental problem is the result of a profound crisis of knowledge which is fundamentally a global civilizational crisis. This crisis is seen everywhere the World System (Wallerstein 2011) has been imposed since the fifteenth century. Under a logic based on the usefulness and commodification of nature to benefit humans, the ethics constructed privileged a pyramidal hierarchy of man above other living beings and other elements in the environment. Control and dominance, two devastating forces, have grown exponentially with the extractivist-developmentalist model imposed on subaltern and "underdeveloped" countries since the middle of the twentieth century.

The current and declining World System is maintained in the first place by the assumptions of neoclassical economics: 1) the free market will always maximize profits (the basis of capital accumulation) in the interest of social well-being, and 2) there is an infinite supply not only of natural resources but also of "sinks" where the waste resulting from the exploitation of natural resources can be deposited. This model is based on the (false) premise that the free market is always operative, that total support and freedom are given to technology, and that prices are allowed to fluctuate as needed to stimulate the search for substitutes, "so absolute scarcity can be postponed to the indefinite future" (Porter and Welsh 1991:27). From a critical perspective, David Harvey (referring to Marx's description of primitive accumulation) explains that the World System also has as a (realistic) basis accumulation through dispossession. Some of the phenomena that he considers part of this process are:

> the commodification and privatization of land and the forceful expulsion of peasant populations; conversion of various forms of property rights – common, collective, state, etc. – into exclusive private property rights; suppression of rights to the commons; commodification of labor power and the suppression of alternative, indigenous, forms of production and

> consumption; colonial, neo-colonial and imperial processes of appropriation of assets, including natural resources; monetization of exchange and taxation, particularly of land; slave trade; and usury, the national debt and ultimately the credit system. The state, with its monopoly of violence and definitions of legality, plays a crucial role in both backing and promoting these processes. (Harvey 2004:74)

The paradigm that sustains this system can also be characterized as anthropocentric. In this perspective, the norm is to locate man – white, occidental, and Christian – as the point of reference for all moral considerations. He will always be first in importance; the image of the world is affected by the self-projections he uses to explain it. In contrast, narratives proper to other cultures of nature – in particular those of the disposed and colonized peoples – place more weight on biocentrism understood as the extension of moral consideration to all forms of life and/or ecocentrism extended as the maximum consideration for respect and care of the *oikos*[10] (Yang 2010).

A (re)appreciation of indigenous ethics has begun, ethics which reflect biocentric worldviews in which human beings do not have a superior status and are not ontologically separate from the other forms of life with whom they share the oikos. The crisis of the modern Western rational paradigm has led scientists and social activists to consider these previously denigrated worldviews as alternatives to the predatory culture of nature of the collapsing World System, since they often have wisdom that does not destroy the environment and that respects inhabited places.

The focus of these revaluated ethics for environmental movements is based on recuperating and (re)constructing values such as caring, empathy, relationality, and the cultivation of solidarity. Making these *other* ethics visible promotes the knowledge of *other* existences that support the critique of habits imposed by capitalist Western civilization and judged for centuries to be quite convenient, superior, and inoffensive. The homogenizing normality of the World System (models of consumption, use of time, the search for gratification, discourses that naturalize and reinforce colonialism, etc.) makes that which is different invisible. To live according to hegemonic values of triumphant financial capitalism since 1970 (Harvey 2004) hinders criticism of habits that have become normal. Under the tyranny of custom, it is more difficult to realize how toxic the culture in which globalization has immersed almost all of humanity really is.

The transition from an ecocidal to a biocentric and ecocentric paradigm that really defends biodiversity involves a forceful critique of established values. If we continue in this seemingly only "true" form of existence, for production, for economic growth, for the satisfaction of unlimited desires created by an individualist

[10] In a thorough study, María Dolores Mirón (2005) explains that the concept has been normally translated as "house," "estate," or "family." Very limited terms, which barely define parts of the Greek concept, without reaching the general meaning of the word. The difficulty increases if we consider that the term oikos was already quite broad and ambiguous in Greek. In this way, it could be used to refer to this basic cell of Greek society as a whole, but also separately, depending on the context, in the different meanings that, as we have indicated above, are translatable into English. That is, that oikos can also designate only the physical space of the house on one occasion; in another, to the properties; and in others, to the family.

and hedonistic culture, it will be very difficult to overcome the destructive practices that stem from the cult of capital accumulation.

What needs to be done then is to promote a revolution of counterhegemonic values about our relationship with the oikos. With the ecological crisis, new and unusual ethical questions are coming to the fore, especially questions of solidarity, responsibility, and justice (Riechmann 2005:43). Every transformative change begins with a revolution in thought. When people finally realize how we *inhabit* the *oikos* is when we can begin to change our behavior. This is one of the roles of the deep environmental ethics that we champion: to know the structures that lead us to reproduce the ecocidal system in order to modify the way we act in the world.

There are distinct hybrid approaches that promote bio- or ecocentrism as alternatives to the anthropocentrism that prevails in the hegemonic world vision, which in turn governs neoliberal conservation. In this chapter, we have opted to work with three of these approaches that we believe allow us to reorient both our understanding and the unfolding of the work of conservation. We do this under the assumption that a paradigm change (transformative change of the IPEBES) starts with criticism of the values of the World System and is based on values that allow us to live with and not against nature.

First, we discuss the concept of *biocultural* directed at environmental conservation and education that allows us to understand that (cultural and biological) interrelationships do not exist separately, but are indissoluble, since each human group expresses its sense of the world through environmental features. Then we will review some ideas based in political ecology that propose a new form of appropriation, (re)valuation, and defense of territories seen as the *place* where we live. Finally, we will look at the *produce to live* approach, which breaks with the productivist paradigm of the capitalist model to emphasize that productive processes should be oriented principally to meeting basic needs and to have the least environmental impact possible so these needs can be met in the long run.

5.3.1 The Ethics of Biocultural Memory

Ethics or, better, biocultural ethics are found within discourses that express ancestral world visions of the places (habitats) that maintain biological diversity and are part of the resistance against ecocidal and epistemicidal instrumental rationalities that threaten to eliminate ecophilic[11] knowledge and practices. The base of this concern is sustaining life: one of the purposes of biocultural ethic is to assure the environmental base so the living can reproduce. Its proposals from Latin America are nourished by decolonial, postdevelopmentalist, and liberation traditions. Advocating

[11] Biophilia understood as love of life was taken up by Erich Fromm to define the essence of his humanist ethics in contrast to necrophilia, or love of death. Edward O. Wilson uses the concept to describe the innate affinity for all the living, the connection with nature, with other living beings, with the habitat, and with the environment (Hernández Rosas 2016:8).

for a loving pedagogy, this philosophy seeks biocultural restoration. Against all forms of violence, it constructs a sociocultural fabric that promotes a culturally and biologically diverse blossoming. The defense of life is the fertile valley where biocultural perspectives like those of Ricardo Rozzi grow. One of the assumptions is that theoretical knowledge, habits, and life values reverberate in the form of impacts on the environment in which they unfold, with its biological and cultural diversity (Rozzi 2001:294).

The biocultural ethics developed by the ecologist and philosopher Ricardo Rozzi is a proposal of environmental philosophy that combines the protection of natural areas with a concern for keeping alive the indigenous groups who maintain an intimate relation with other beyond-human beings. It is a field of research that deals with problems that are ecological, social, and cultural at the same time. That is to say, the interaction between cultural practices of living in a place not only among humans but with other beings that also depend on an ecosystem that determines the form of life of all the cohabitants. This focus allows for concepts and approaches that open up a holistic perspective for describing the dynamics among biodiversity, culture, and the inhabited space.

The roots that nourish his proposal come from the traditions of Amerindian peoples (indigenous narratives), the intercultural philosophy of Fornet-Betancourt, the philosophical anthropology of Rodolfo Kusch, ethnoecology, the world views of the ecological sciences, the ideas of kinship coming from evolutionary theory, the earth ethics of Aldo Leopold, the bioethics of Potter, the systemic perspective of ecology, ecofeminism, and the Latin American decolonial perspective, among others.

What led to the Biocultural Conservation Program in which Rozzi[12] participated was the UNESCO Man and the Biosphere Program (MAB) which, as we explained in the first section, has as one of its objectives to not exclude humans from the processes of conservation of NPAs. A fundamental element in the method of the MAB Program was the establishment of an international network of biosphere reserves. This network includes a mosaic of unique sites that represent the principal ecosystems of the planet, protected through research programs monitoring education, conservation, and sustainable development. Today the MAB Program has three central objectives: 1) to minimize biological diversity loss, 2) to generate awareness about the interdependence of biological and cultural diversity, and 3) to promote environmental sustainability through the World Network of Biosphere Reserves (Rozzi et al. 2010).

Biocultural ethics affirm the links that are formed among diverse co**h**abitants (humans and beyond-humans), their life **h**abits, and the **h**abitats in which these take place. "The conceptual model of the '3Hs' reveals that in human cultures there are tendencies that lead towards biocultural homogenization and others that lead towards biocultural conservation" (Rozzi 2018:42). Under this formal idea of the "3Hs" of biocultural ethics, habitat conservation and the possibility of continuing cohabiting in them becomes an ethical imperative associated with respect for the

[12] See Sub-Antarctic Biocultural Conservation Program: http://www.chile.unt.edu/.

joint life of living beings and the diversity of human beings (Rozzi 2013). The 3Hs form an ethical as well as ecosystemic unit (Rozzi et al. 2008a). This makes it possible to approach case studies from both philosophy and ecology, creating an epistemic hybrid.

To introduce the term "cohabitants," human beings are only one part of a complex interrelationship that affects other elements. This is an indispensable idea that enriches the forces of thousands to assure the sustainability and continuity of life in the *oikos*. Characterizing beings as cohabitants makes it possible to propose biocultural conservation methods and activities that bring us back to the archaic meaning of the word ethos as a "den, the dwelling of an animal" (Rozzi et al. 2008b). This is most relevant in places where communities are deeply rooted to their "den," that is to say, to the inhabited place. These communities, with their knowledge of their surroundings, better understand the rhythm of nature and how to allow the regeneration of biodiversity, using only what they need to live.

The most important, for all forms of life, is to safeguard what they vitally need, such as food and protection. In order to protect communities that maintain their lifestyle in their native culture and living language, a priority is to respect their habitat and to give a status, or category, to their land (whether it be in a biosphere reserve or a NPA).

Through this method, ecologists, philosophers, and other professionals can participate in biocultural conservation projects over the long term, having direct "face-to-face" encounters with biocultural diversity. Through these in situ experiences, biocultural conservation can stop being just a concept and start to become an experience of cohabitation with diverse living beings, life histories, and institutions that are often outside the usual scope of formal education and decision-making (Rozzi et al. 2010).

In this sense, certain actions stand out that have permitted different human groups to survive for centuries by maintaining a greater equilibrium with their ecosystems, careful relationships that have been highly beneficial for sustainable habitats. These direct experiences have allowed the development of ancestral knowledge that is unique to the territory in which they live. To address the ethics derived from biocultural interactions is indispensable for uncovering the issues they deal with, to know them, and to prevent their disappearance. We need to be inspired by them in order to develop biocultural conservation plans in places that are threatened by the voracious and unlimited appetite of extractivism, real estate development, or the homogenization of lifestyles. In the words of the Chilean ecophilosopher, to deconstruct the unidimensional and monocultural neoliberal discourse that has taken over formal education and decision-making, promoting an unsustainable culture (Rozzi 2015:32).

Victor Toledo and Narciso Barrera-Bassols also propose the biocultural paradigm as a basis for conservation. In their *Memoria biocultural. La importancia ecológica de las sabidurías tradicionales* (2008), they defend the cultural diversity of indigenous peoples, their wisdom, and collective practices. For them, it is a form of resisting modern developmentalist economic rationality. The possibility of overcoming the adversity of the modernizing assault on their lands is in great measure the result of knowing the environment, being able to connect with nature and to learn the cyclical time of the seasons. The legacy of this knowledge is maintained, thanks to the memory of the community. As they explain:

> If Homo sapiens has managed to remain, colonize, and expand their presence on the earth, it is due to their ability to recognize and use the elements and processes of the natural world, a universe characterized by an essential characteristic; diversity. This ability has been made possible by the presence of a memory, individual and collective, that has been able to extend itself through the different societal configurations that form the human species. (Toledo and Barrera-Brassols 2008:15)

Rozzi, Toledo, and Barrera-Bassols, as well as many other humanists and social and natural scientists[13], have shown that the peoples who have not lost their awareness of ecodependence (Riemann et al. 2018:10) know that their lives depend on their care of their environment[14]. Consequently, conservation is not the result of a series of actions that seek to protect biodiversity. Rather, care of ecosystems is an intrinsic part of life and is imbued in daily practices of production, consumption, and exchange (material life), as well as in the conception of the world (worldview) and of ourselves (ontology). For these peoples judged as backward, savage, and ignorant (and much more), conservation is not a result, a goal; it is a means (in the sense of Agustin Berque[15]) for maintaining ecosystem balance, and it is the relationship that links them to nature, their culture, and themselves.

5.3.2 *Political-Ecological Paradigms Centered in the (Re) valuing and Defense of Territories and the Theory of "Place"*

If the civilizational crisis is a crisis of knowledge with ontological roots that determine our perception of reality, a first step in reeducating ourselves subjectively is to understand that we have constructed ourselves under the rule of the egocentric "I." Hegemonic values have captured the attention of the individualist subject in order to offer it happiness (uncertain and fictitious since it is impermanent and based on constant dissatisfaction) acquired through success, effort, and personal achievement. This generates consumer aspirations to show a certain social status through our purchasing power. Societies with problems like stress, anxiety, and depression are common under the economic pressure of the World System. Trapped into roles and social expectations, we usually conceal and sacrifice our emotions in false

---

[13] Among those who we must cite are A. Ararwal, M. Altieri, E. Berlin, F. Berkes, J. Colding, C. Folke, P. Descola, M. Gadgil, and V. Shiva.

[14] Understood as the set of relationships of societies with their environment and between their human and beyond-human members

[15] For Berque, the concept of (human) means is the relationship of a human group with a certain terrestrial extension, this is with a space and with nature (Berque 2000:13).

motivational placebos, such as discourses of personal growth and the maelstroms of consumerism (e.g., Black Friday, Christmas, Mother's Day).

As a countercurrent to this materialist and banal world, critical pedagogy is a proposal that sees education as an exercise in freedom. Along this line, Orlando Fals Borda suggested the term to "feel-think" (*sentipensar*) as an ethical-political act because it expands compassion toward the habitat and the feeling of our fellow cohabitants. Without these emotions, it is difficult to form environmental awareness. *Feel-think* is helpful for making explicit the reasons, motivations, passions, and valuing that we feel and think. *Feel-think* emphasizes the role of the emotions as energies that generate change and knowledge; it invites us to think more in terms of training than instruction (De La Torre and Moraes 2005).

One of our goals in writing this chapter is to promote a compassionate coexistence that has an awareness of caring, of responsibility for other beings, of the self-limitations of our consumer impulses: a *dwelling in the heart*. We find in *feel*-think a concept that carries an ethical intentionality. To cultivate emotions such as compassion toward our fellow travelers on the journey (life), toward those with whom we live, is a necessary step to change our habits and begin to build sustainable societies. As Riechmann (2004) says, we should base ourselves on three principles: efficiency (dematerialize production), sufficiency (reduce production), and insolence (build a world without artificial and useless needs).

Escobar (2017) explains that the One World or Eurocentric Universe has excluded multiple worlds, ontologies, and realities. On the other hand, the proliferation of struggles in defense of territories and cultural diversity suggests that from such struggles emerge more complete and complex worlds, much richer in ontologies compared with the modern Western mono-discourse. From the idea of postdevelopment, he advocates for deconstructing development, reevaluating vernacular cultures, and constructing other worlds out of social movements and resistance (re-existence). It is possible to change our practices of know-how (Cariño and Castorena 2015): there is a vast difference between meeting material needs through a capitalist market economy and doing it through non-capitalist practices and institutions (as most human communities have down throughout history) (Escobar 2005:24).

In his work *Sentipensar con la tierra* [*Feel-think with the earth*], Escobar (2017) defends what he defines as *relational ontologies* linked to forms of living in the world and in a deep-rooted place. Awareness of place makes it possible to defend activities that strengthen community links and in general are connected with the environment: knowledge acquired through their interrelations with the ecoregions that they transit with other living beings, including spiritual worlds above and below the earth. Activities such as the search for food, the harvest, singing, and weaving have a sense of belonging to the inhabited space. These relational ontologies with shared stories are different from work activities that are disconnected from place, which, instead of uniting, separate human beings into individuals. *Feel-think* with the land implies thinking from the heart and from the mind (Escobar 2017).

In recent years, migratory exoduses have unleashed social problems such as refugee camps, migration in search of work, family separation caused by deportation, and deterritorialization. One of the most recent causes of the expulsion of people from their land is the increase of ecological and distributive conflicts. The need to migrate has accelerated because of conflicts derived from resource scarcity or because of extractive projects that rely on armed groups to enforce their power. Political ecology touches on these situations: one of the lines of research is to find ways to organize and defend people's land. So political ecology, through the theory of place and the study of environmental conflicts, is linked with the objectives and processes of biosphere conservation, provided that they are not the neoliberal (or hegemonic) type, that is to say, those conservation processes which, like extractivism, also exclude people from their land and objectify nature (Klier and Folguera 2017).

The concern for space comes from its invisibility within the processes of economic globalization, which instead privileges the Eurocentric description of the world, abstract space, communication flows, rapid urban transformation made into the breeding ground for the markets, and financial speculation (Harvey 2004). To regain awareness of space is key to developing re-existence and resistance to monocultural homogenization and to not losing our rootedness in our "community den."

Escobar discusses his theory of place in various of his works. In his essay, *El lugar de la naturaleza y la naturaleza del lugar ¿globalización o postdesarrollo?* [*The place of nature and the nature of place: Globalization or postdevelopment?*], he begins with a key argument that reveals the importance it has for political ecology: a discussion of place should offer an important perspective for rethinking globalization and the question of alternatives to capitalism and modernity [...] Considering that place can be resistance, an alternative to the homogenizing practices of global capitalist culture (Escobar 2000:127). To develop his argument, he uses the definition of place given by Lefebvre: a form of lived and grounded space, whose reappropriation should be part of any radical political agenda against capitalism and spaceless and timeless globalization (Escobar 2000:128).

In another work, *Una minga para el postdesarrollo...* [*A 'minga' for postdevelopment*], he goes deeper into the concept, including a genealogy of its use in the West. In the chapter, "La culture se asienta en lugares [Culture sits in places]," based on a critique of development, he touches on the asymmetries between the global and the local; he considers that these imposed models have induced a separation of local life from place (Escobar 2012:131). This position is similar to Rozzi's, though the Chilean uses the concept of habitat instead of place; both criticize homogenization and both defend the plurality of traditional knowledges.

For the emancipatory projects of social movements, any course of action should consider place-based models of nature, culture, and politics (Escobar 2000). Based on the phenomenology of Edward Casey, Escobar underscores the following premises of the US philosopher: (1) To live is to live locally, and to know is first of all to know the places one is in. (2) Place, it is clear, is constituted by embedded social structures and cultural practices. (3) Feeling and moving are not presocial; the body in which one lives is the result of habitual cultural and social processes. Thus, it is imperative that we "get back into to place" (Escobar 2012:135). Traditional knowledge, know-how, and practices take place with the commitments made with the world lived in place. With no contradictions, paradoxes, or complexity, what Escobar stresses are ecopolitical strategies for defending places, culture, and nature in threatened territories, in spite of the fact that place is hybrid, possibly violent, porous, capable of using modern technologies such NTICs, and incapable of being understood in deterministic terms.

The languages of biodiversity, sustainability, traditional systems of production, cultural rights, and ethnic identities are interwoven by environmental movement activists into a discourse to defend space and into a political ecology framework that allows them to articulate a strategy of resistance and re-existence (Escobar 2012:166). Alternatives to devastating modernity are needed more than ever, especially those that have managed to become grounded (rooted through biocultural links) in place. As Escobar says, a politics based in place would allow academics and activists to bring back a concept that is contextualized and situated in human practice, in contrast with the de-sited and detached gaze of people promoted by Cartesianism, modern science, and the civilization that they have generated (Escobar 2012:173), and of course by capitalism.

5.3.3 *Alternative Forms of Production That Respect Ecosystemic Bases*

To reflect on the features that should be included in a new paradigm for conservation, it is necessary, even briefly, to touch on the theme of alternatives to massive capitalist production. We argue that this is crucially important for two fundamental reasons. Firstly, production is essential for meeting the basic needs of social reproduction, and this is always carried out in ecosystems (primary activities) or based on ecosystems (secondary and tertiary activities). Secondly, one of the main reasons for biodiversity loss is the overexploitation and destruction of ecosystems. As productive systems have become more complex and have dramatically increased their goods as a result, both demand for natural resources and pollution and waste have increased exponentially; both processes (input and output) cause major impacts, some irreversible, on the entire geosystem. As Guillermo Castro indicates, it has to do with an economy that by all accounts works *against* nature and whose ravages accumulated over three centuries have put the integrity of the biosphere at risk (including obviously the human species). Consequently, to be able to conserve the little that remains, it is necessary to "encounter new forms of social life that make it possible, in its turn, to begin the development of a work relationship *with* nature" (Castro 2000:54–55).

In referring to alternatives that would allow production for living and not for capital accumulation, we propose a change toward a form of production that would (De Sousa Santos and Rodriguez 2002:49–58):

1. Put emphasis on being not just an economic alternative, but a potential emancipator derived from the integration of economic processes with cultural, social, and political processes
2. Depend on its insertion in networks of collaboration and mutual support, for example, in economic solidarity experiences
3. Be promoted inside and outside the state: inside through institutions that support autonomous productive processes and outside through diverse experiences and forms of the collective organization of production
4. Concentrate not just on the local scale, but rather use a combination of scales
5. Widen the field of democratic action from the political to the economic, erasing the artificial division between political and economic that capitalism and the liberal economy have created
6. Maintain a direct connection between struggles for alternative production and struggles against patriarchal society
7. Prioritize alternative forms of knowledge
8. Use criteria for evaluation of success or failure that are gradual and inclusive
9. Enter into a relationship of synergy with alternatives in other spheres of the economy and society

With regard to alternative production, in Latin America there are two streams: the solidary economy and the popular economy. The solidary economy works through cooperatives that organize, or can organize, many people who do not necessarily have primary relations with each other; that generally cover a particular branch or sector of economic activity; that are articulated in a systematic manner with the market; and that, consequently, require a relatively clear division of labor and an effective administration to reproduce themselves and grow. Their difference with capitalist enterprises is that their agents explicitly identify the instruments of production, the resources or objects of production, and the products themselves and that it is a system of self-management by the workers. Consequently, the distribution of products, goods, services, and benefits is done or should be done in agreement with the workers and for purposes decided by them, as well as, of course, to their benefit (Quijano 2002:379).

The popular economy consists of heterogeneous institutions for the organization of production, distribution, and relationship to the market, linked, perhaps at the same time, to heterogeneous economic activities of production and distribution. These institutions are made up of people who have "primary" relationships among themselves, and therefore they are preferably small groups. They tend to be organized socially according to what some authors have called a "community logic" (Razeto et al. 1990 in Quijano 2002:380). Unlike the solidary economy, the popular economy doesn't always necessarily or explicitly involve the ideological and political self-identification of its agents nor their revolutionary vision of the world. What characterizes the popular economy is that relationships regarding work and the distribution of resources and products are basically organized around reciprocity, social life, daily social practices, and the community (Quijano 2002:380).

It is important to remember that, in spite of the fact that capitalism is the dominant system, currently there are anticapitalist productive experiences in action, they are growing, and some present themselves consciously as alternatives to the system. But it is relevant to note that there is not an "alternative economy" or "alternative production systems" that are not subject to the structures of global capitalism (Quijano 2002:397). As with all historical systems, a long time is needed for structural changes to occur; and, yes, they are always driven by actors who are conscious of their role in constructing an alternative system (Wallerstein 2003). In the case of these forms of production, the actors in question are the communities and enterprises that practice horizontal forms of production; the principles that govern these new productive structures are reciprocity and the meeting of needs.

5.4 Conclusion

It is urgent to put our all our efforts, intellectual and activist, in making a transformative change in the processes of conservation, as in all the inherent aspects of human thought, speech, and action, if we wish to overcome the challenges of this century of the Great Test. The grievances committed against nature and the majority of the world population under the pretext of progress, civilization, and development should serve as counterexamples to undertaking a relationship that restores "unity among human beings and nature. Historical ecology is dedicated to this work" (González de Molina 1993:cover page), as are the other environmental humanities.

The diversity of themes and problems it is necessary to confront in order to construct a new paradigm that orients the relationships of societies with their nature has barely been outlined in this brief essay. Next year, 2020, is of crucial importance for the future of the biosphere in part because the Paris Accords will begin to be applied and in part because the goals of the 2011–2020 strategic plan for biological diversity will be revised, with final decisions made about the framework in which the CBD will function in the future. In both agreements (UNFCCC and CBD), it is hoped that an international, multi-actor, and multiscale synergy will be found for the achievement of life on the planet. Despite adverse conditions and many opposed interests, one can only hope that both commissions will more than fulfill their duties. Systemic forces are absolutely necessary to slow down the devastating consequences of developmentalism, neoliberalism, and excessive greed.

Lessons Learned Nevertheless, as we have explained in this text, these commendable efforts are not enough. In parallel, it is necessary to construct new strategies and diverse processes to be able to conserve life, through a collective effort, in general terms favoring bottom-up processes, and in the widest spectrum possible. This must be carried out in our daily action, through a desire for a profound change in the prevailing culture of nature, which implies broadening considerably hegemonic ideas about conservation. In a new paradigm, these concern at least the following spheres that intersect relations of societies with nature: ontology (concept of the world and worldviews), ethics, epistemology (science and traditional knowledge), economics (forms of producing, consuming, and exchanging), social relations and their organization, political action and thought, education (in all its forms), and spirituality.

Although this could seem illusory, in fact it is a reality applied in various conservation processes. In order to encourage hope, we refer to just three examples in Latin America. In Magallanes Province in the south of Chile, the Omora Ethnobotanical Park was created in 2000 based on biocultural ethics and with the double purpose of rescuing the native culture of the Yagán people (in serious danger of extinction) and the exceptional biodiversity of the unfragmented Sub-Antarctic forest. In addition to being a center for scientific research associated with the Cape Horn Biosphere Reserve, the park promoted various cultural, educational, and ecotourism activities supported by the Sub-Antarctic Biocultural Conservation Program of the University of Magallanes, the University of North Texas, and the Institute of Ecology and Biodiversity[16]. Supported by a civil association, but also with the support of scientific research and higher education organizations (in this case US and Mexican), the Kaxil Kiuic Biocultural Reserve in the state of Yucatan, Mexico, is also governed by the principles of conserving simultaneously the biodiversity of the lowland forest and the culture of the Puuc Maya Yucateca people. In this case, it is a private reserve of 1800 ha, one of the largest private reserves in Mexico, and indeed of Latin America as a whole. In addition to its conservation objectives, this reserve includes in its goals research, education, and the maintenance of Maya cultural identity, as well as being open to tourism[17]. Lastly, we mention the successful program of the Panamanglar Network, whose objective is to conserve one of the largest mangrove ecosystems in the world while at the same time building an active population that demands and supports a policy of well-being for the common good through a solidary economy in a territory shared by human and beyond-human beings. It also promotes access to information and both traditional and scientific knowledge so they can be used in collective action. The network includes coastal communities (especially the most vulnerable), base community organizations (BCOs) that use the mangrove, fishers, charcoal makers, shell finders, leather tanners, and small- and medium-sized business people who use the mangrove sustainably (with both consumption and non-consumption, as in tourism). It was established in 2013 with 19 nongovernmental organizations and 20 BCOs. By 2017, the network had brought together four universities and three international organizations; the number of BCOs had grown to 40, and it had been enriched with a scientific advisory committee[18].

We deeply appreciate the support of Dr. Frederick Conway in the translation of this text from Spanish into English.

---

[16] http://www.umag.cl/facultades/williams/?page_id=855

[17] www.kaxilkiuic.org

[18] www.panamanglar.org

References

Berque A (2000) Écoumène. Introduction à l'étude des milieux humains. Berlin, Paris, p 271

Boulding K (1966) The economics of the coming spaceship earth. In: Jarrett H (ed) Environmental quality in a growing economy, resources for the future. Johns Hopkins University Press, Baltimore, pp 1–14

Caradonna JL (2014) Sustainability. A history. Oxford University Press, New York, p 331

Cariño M, Castorena L (2015) Saberes para la sustentabilidad. Icaria Editorial, Barcelona, p 214

Carson R (1962) Silent spring. Houghton Mifflin Company, Boston, p 368

Castro G (2000) La crisis ambiental y las tareas de la historia en América Latina. Papeles de población 6(24):37–60

Crutzen P, Stoermer E (2000) The 'Anthropocene'. Global Change Newsletter 41:17–18

Torre S. de la, Moraes MC (2005) Sentipensar. Fundamentos y estrategias para reencantar la educación. Archidona (Málaga): Editorial Aljibe.

De Sousa SB, Rodríguez C (2002) Introducción. Para ampliar el canon de la producción. In: De Sousa Santos B (Coord) Producir para vivir. FCE, México, pp 49–58

Escobar A (2000) El lugar de la naturaleza y la naturaleza del lugar: ¿globalización o postdesarrollo? In: Biblioteca virtual CLACSO, La colonialidad del saber: eurocentrismo y ciencias sociales. Perspectivas latinoamericanas. http://biblioteca.clacso.edu.ar/clacso/sur-sur/20100708045100/7_escobar.pdf

Escobar A (2005) El "postdesarrollo" como concepto y práctica social. In: Mato D (coord) Políticas de economía, ambiente y sociedad en tiempos de globalización). Facultad de Ciencias Económicas and Sociales, Universidad Central de Venezuela, Caracas, pp 17–31

Escobar A (2007) La invención del Tercer Mundo. Construcción y deconstrucción del desarrollo. Fundación Editorial el perro y la rana, Venezuela, p 424

Escobar A (2012) Una minga para el postdesarrollo: Lugar, medio ambiente y movimientos sociales en las transformaciones globales. Ediciones desde abajo, Bogotá, p 177

Escobar A (2017) Sentipensar con la Tierra. Nuevas lecturas sobre desarrollo, territorio y diferencia. UNAULA, Medellín, p 184

Gavin et al (2018) Effective biodiversity conservation requires dynamic, pluralistic, partnership-based approaches. Sustainability 10:1846. https://doi.org/10.3390/su10061846

González de Molina M (1993) Historia y ecología. In: González de Molina M and Martínez Alier J (eds) Historia y ecología. Ayer 11, Marcial Pons, Madrid, p 251

González de Molina M, Toledo VM (2011) Metabolismos, naturaleza e historia. Hacia una teoría de las transformaciones socioecológicas. Icaria Editorial, Barcelona, p 375

Grosfoguel R (2006) La descolonización de la economía política y los estudios postcoloniales: Transmodernidad, pensamiento fronterizo y colonialidad global. Tabula Rasa 4:17–48

Gudynas E (2011a) Debates sobre el desarrollo y sus alternativas en América Latina: una breve guía heterodoxa. In: Fundación Rosa Luxemburgo (ed) Más allá del desarrollo. Abya Yala, Quito, pp 21–54

Gudynas E (2011b) Desarrollo y sustentabilidad ambiental: diversidad de posturas, tensiones permanentes. In: Matarán Ruiz, A. y López Castellano, F. (eds) La tierra no es muda: diálogos entre el desarrollo sostenible y el posdesarrollo). Editorial Universidad de Granada, Granada, pp 69–96

Harvey D (2004) The 'new' imperialism: accumulation by dispossession. Socialist Register. https://socialistregister.com/index.php/srv/article/view/5811/2707

Hernández Rosas H (2016) Biofilia. El clima como experiencia artística. Universidad Complutense de Madrid. Colección Palabras e Imágenes, Madrid, p 76

IPBES (Intergovernmental Science-Policy Platform on Biodiversity and Ecosystem Services) (2019) Summary for policymakers of the global assessment report on biodiversity and ecosystem services of the Intergovernmental Science-Policy Platform on Biodiversity and Ecosystem Services https://www.ipbes.net/sites/default/files/downloads/spm_unedited_advance_for_posting_htn.pdf

Klier G, Folguera G (2017) ¿Caras de una misma moneda? Conservación de la biodiversidad y extractivismo en América Latina. Letras Verdes. Revista Latinoamericana de Estudios Socioambientales 22:182–204. https://doi.org/10.17141/letrasverdes.22.2017.2704
Kolbert E (2014) The sixth extinction: an unnatural history. Macmillan Publishers, London, p 336
Leakey R, Lewin R (1996) The sixth extinction: biodiversity and its survival. Weindenfeld and Nicolson, London, p 288
Leff E (2011) Sustentabilidad y racionalidad ambiental: hacia "otro" programa de sociología ambiental. Rev Mex Sociol 73(1):5–46
Machado Aráoz H (2015) Ecología política de los regímenes extractivistas. De reconfiguraciones imperiales y re-exsistencias decoloniales en nuestra América. Bajo el Volcán 15(23):11–51
Mirón MD (2005) Oikos y oikonomia: El análisis de las unidades domésticas de producción y reproducción en el estudio de la Economía antigua. Gerión Revista De Historia Antigua 22(1):61–79. https://revistas.ucm.es/index.php/GERI/article/view/GERI0404120061A
Pierri N (2005) Historia del concepto de desarrollo sustentable. In: Foladori, G. y Perri, N. Sustentabilidad ¿desacuerdos sobre el desarrollo sustentable? Porrua, México, pp 27–81
Porter G, Welsh Brown J (1991) Global environmental politics. Westview Press, London
Quijano A (2002) ¿Sistemas alternativos de producción? In: De Sousa Santos B (coord) Producir para vivir. FCE, México, pp 374–397
Riechmann J (1995) Desarrollo sostenible. La lucha por la interpretación. In: Riechmann J et al (eds) De la economía a la ecología. Trotta, Madrid, pp 11–36
Riechmann J (2004) Gente que no quiere viajar a Marte. Ensayos sobre ecología, ética y autolimitación. Catarata, Madrid, p 247
Riechmann J (2005) Un mundo vulnerable. Ensayos sobre ecología, ética y tecnociencia. Catarata, Madrid, p 421
Riechmann J (2013) El siglo de la gran prueba. Textos del Desorden, Madrid, p 162
Riechmann J (2018) El colapso no es el fin del mundo: pistas para una reflexión estratégica. In: Riechann J et al (eds) Para evitar la barbarie. Trayectorias de transición ecosocial y de colapso. Editorial Universidad de Granada, Granada, pp 246–312
Riechmann J et al (2018) Para evitar la barbarie. Trayectorias de transición ecosocial y de colapso. Editorial Universidad de Granada, Granada, p 323
Rozzi R (2001) Éticas Ambientales Latinoamericanas: Raíces Y Ramas. In: Primack R et al (eds) Fundamentos De Conservación Biológica: Perspectivas Latinoamericanas. Fondo De Cultura Económica, México, pp 311–362
Rozzi R (2013) Biocultural ethics: from biocultural homogenization toward biocultural conservation. In: Rozzi R, Pickett S, Palmer C, Armesto J, Callicott J (eds) Linking ecology and ethics for a changing world. Ecology and Ethics, vol 1. Springer, Dordrecht
Rozzi R (2015) Ética biocultural: una ampliación del ámbito socioecológico para transitar desde la homogeneización biocultural hacia la conservación biocultural. In: Bustos B et al (eds) Ecología Política en Chile: Naturaleza, Propiedad, Conocimiento y Poder. Ediciones Universidad de Chile, Santiago de Chile, pp 89–117
Rozzi R (2018) Biocultural homogenization: a wicked problem in the anthropocene. In: Rozzi R. et al. (eds) From biocultural homogenization to biocultural conservation. Springer, Ecology and Ethics, Vol 3., Switzerland, pp 21–48
Rozzi R et al. (2008a) Integrando las Ciencias Ecológicas y la Ética Ambiental en la Conservación Biocultural de los Ecosistemas Templados Subantárticos de Sudamérica, article, 2008; [Denton, Texas] (https://digital.library.unt.edu/ark:/67531/metadc102294/: accessed November 12, 2019), University of North Texas Libraries, Digital Library, https://digital.library.unt.edu; crediting UNT College of Arts and Sciences
Rozzi R et al (2008b) Field environmental philosophy and biocultural conservation: the Omora Ethnobotanical Park educational program. Environmental Ethics 30(3):325–336
Rozzi R et al. (2010) Filosofía ambiental de campo y conservación biocultural en el Parque Etnobotánico Omora: Aproximaciones metodológicas para ampliar los modos de integrar el

componente social ("S"). En Sitios de Estudios Socio-Ecológicos a Largo Plazo (SESELP). Revista chilena de historia natural [online] 83(1):27–68

Sachs I (1993) L'Ecodéveloppement. Stratégies de transition vers le XXIe siècle. Syros, Paris, p 119

Steffen et al (2015) Planetary boundaries: guiding human development on a changing planet. Science 347(6223):1–13. https://doi.org/10.1126/science.1259855

Taibo C (2016) Colapso. Capitalismo terminal, transición ecosocial, ecofascismo. Catarata, Madrid, p 217

Tainter J (1988) The collapse of complex societies. Cambridge University Press, Cambridge, p 262

Toledo VM, Barrera-Bassols N (2008) La memoria biocultural. *La importancia ecológica de las sabidurías tradicionales.* Icaria Editorial, Barcelona, p 230

UICN PNUMA WWF (1980) Estrategia Mundial para la Conservación. La conservación de los recursos vivos para el logro de un desarrollo sostenido. https://portals.iucn.org/library/sites/library/files/documents/WCS-004-Es.pdf

UNESCO (2018) Man and the biosphere programme. English Brochure. http://www.unesco.org/new/fileadmin/MULTIMEDIA/HQ/SC/images/Eglish_MAB_leaflet_2018.pdf

Wallerstein I (2003) Utopística o las opciones históricas del siglo XXI. Siglo XXI, México, p 91

Wallerstein I (2005) Análisis de sistemas mundo. *Una introducción.* Siglo XXI, México, p 153

Wallerstein I (2011) El moderno sistema mundial I. Siglo XXI, México, p 640

Worster D (2006) Transformaciones de la Tierra. Ensayos de historia ambiental. EUNED, Costa Rica, p 246

Yang T (2010) Hacia una ética global igualitaria. In: Henk AM ten Have (ed) Ética ambiental y políticas internacionales. UNESCO, París, pp 25–50

Chapter 6
Inclusion of the Human Factor in Protected Natural Areas

Manuel Jesús Pinkus-Rendón and Cielo María Ávila-López

Abstract This paper examines the links of the social dimension of sustainable development in the protected natural areas. Usually, the natural Reserves have imposed a series of limitations and restrictions over the use of natural resources by the communities located in the areas of influence of the Reserves. Public policies must guarantee sustainable development, including economic, environmental, and social dimensions and must not exclude social factor in order to avoid the configuration of conflicts arising from the use of natural resources. Public policies must integrate the theories of ecological economy and political ecology in order to guarantee economic benefits for the populations in order to reconcile the conservation of biodiversity with the use of natural resources, in order to economically benefit the population.

Keywords Ecological economy · Political ecology conflicts derived from the use of natural resources · Public environmental policy · Sustainable development

6.1 Introduction

Tankuché community in the state of Campeche, Mexico, is in the area of influence of Protected Natural Area (PNA) of Los Petenes, since 1996 (Períodico Oficial del estado de Campeche 1996). Subsequently, this protected area changes its status from state PNA to Biosphere Reserve in 1999 (DOF 1999). In this sense, from the creation of the PNA, activities around the exploitation of natural resources that made the inhabitants of the communities immersed in the PNA are tight to environmental public policies implemented by the State.

Due to more than three decades have passed since the establishment of the PNA, the interest to know the situation of sustainable development from the perspective

M. J. Pinkus-Rendón (✉) · C. M. Ávila-López
Universidad Autónoma de Yucatán, Mérida, Yucatán, Mexico
e-mail: mpinkus@correo.uady.mx

A. Ortega-Rubio (ed.), *Socio-ecological Studies in Natural Protected Areas*,
https://doi.org/10.1007/978-3-030-47264-1_6

of the population of Tankuché arises. In this manner, it could be seen if the environmental public policy in the Biosphere Reserve of Petenes, considers or not the community involvement of its population (e.g. Tankuché); that is to say if factor human is taken into account, which is an essential part of sustainable development.

Therefore, the objective of this work is to present the results from the analysis of environmental public policy. Specifically, those associated with the social dimension of sustainable development, environmental public policies implemented since the establishment of the Biosphere Reserve Petenes and to which the population of Tankuché, located in the zone of influence of the Reserve, it has had to adapt to the limitations or restrictions on the use of natural resources, that it caused and on which have generated perceptions.

In this regard, it should be noted that this work was rooted in conflicts arising from the use of natural resources, which become the intersection between the theories of ecological economics and political ecology, along with the concepts of environmental public policy, sustainable development, and community participation.

Regarding the above-mentioned conflicts, based on the theory of ecological economics, Martínez-Alier and Roca postulate that environmental problems and public policies cannot be analyzed separately from the social conflicts that they contain (2013); therefore, instead of focusing on the market, it focuses on ecological social movements. For its part, the theory of political ecology considers that those affected by such conflicts are compensated by issues such as exclusion in the use and exploitation of natural resources (Martinez 2004). Thus, political ecology stresses the importance of the social aspect of environmental issues.

Social conflicts with environmental content emerge from the intersection between the theory of ecological economy and political ecology. The conflicts derived from the use of natural resources, which for the methodological purposes of this work, are understood as those associated with the restrictions that the State implements, through environmental public policies, to the use and exploitation of natural resources and indigenous populations, without taking into account their perceptions about nature, its uses, interests, needs and economic benefits.

Since, such conflicts are issues that require attention by the State, and they are becoming important instruments through which they are addressed, which are nothing more than environmental public policies focused on PNA.

As a first point, it should contextualize that, in our country, environmental matters is enshrined in Article 27 of the Constitution of the United Mexican States. Environmental history dates from March 23, 1971, when it published the Federal Law to Prevent and Control Environmental Pollution (DOF 1971). After that time, Mexico ventured to environmental life. Subsequently, this rule is replaced by the law on environmental matters currently in force, which is the General Law of Ecological Balance and Environmental Protection (LGEEPA), published in January 1988 to promote the rational and sustainable exploitation of natural resources.

In this regard, Gil-Corrales says that environmental public policies are instruments seeking sustainable use of natural resources and protection of biodiversity and areas covered by the LGEEPA, such as PNA (2007). In addition, the environmental public policy associates with the PNA and the LGEEPA (Cámara de

Diputados del H. Congreso de la Unión 2018), along with the Brundtland Report, found that sustainable development meets the present needs without compromising those of future generations (ONU 1987); it follows that environmental public policy focuses on PNA seeks for sustainability in environmental, economic, and social dimensions (Foladori 2002) to improve the quality of life of individuals while seeking preservation of ecological balance and protection of the environment and use of natural resources.

It is for this reason that the environmental public policies are focused on PNA, addressing the three dimensions of sustainability. Thus, for environmental dimension, the goal is the conservation of biodiversity; for economic dimension, equitable sharing of benefits arising from the use of natural resources; and for social dimension, community participation and management of PNA, taking into account their perceptions, uses, and activities with natural resources and economic benefits.

Consequently, environmental public policies for PNA must prohibit or limit the use of certain natural resources, such as considering that this will impact the use that they normally give to natural resources, the populations that make up the PNA and with the economic activities that perform. To do so, these policies should also contemplate, as compensation, economic benefits for populations; because, otherwise, the conservation of biodiversity, and needs and interests of the populations are dissociated.

To ensure this, LGEEPA states that the Secretariat of Environment and Natural Resources (SEMARNAT) is obliged to create cooperation agreements with indigenous peoples to establish, administer, and manage PNA and in turn give them advice regarding the sustainable use of natural resources (Chamber of Deputies of the Mexican Congress, 2018).

Pinkus et al., (2014), point out that community participation is necessary for institutions in charge of conserving biodiversity and activities related to the use of natural resources to form links with PNA populations. For that reason, community participation is what should be promoted with PNA populations, to contemplate their perceptions and practices regarding the use of natural resources; therefore, the approach that the authorities promote with them is vital, so as not to conflict their needs and interests, with the limitations on the use of natural resources.

6.2 Materials and Methods

As for the study site, the Tankuché community (in the municipality of Calkiní, corresponding to the State of Campeche) is located in the area of influence of the Petenes Biosphere Reserve (see Fig. 6.1).

In the Tankuché community, the former hacienda can be found, which has the main house from English architectural style and in the past worked the henequen (Padilla 2002). Also according to the latest population and housing census, Tankuché has a bilingual Mayan population, which includes a total of 1006 people, of which 514 are men and 492 are women (INEGI 2010).

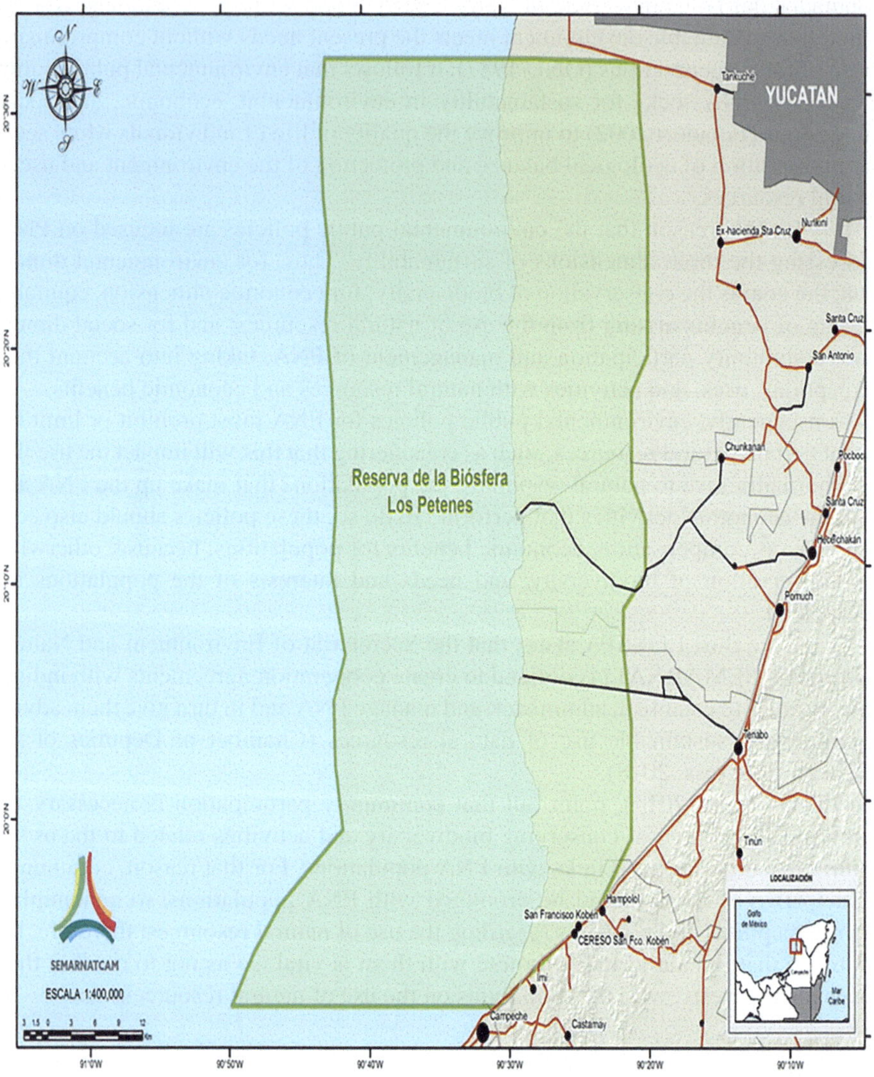

Fig. 6.1 Location of the study area. (Source: Secretaría de Medio Ambiente y Recursos Naturales de Campeche 2018)

On the other hand, the present work corresponds to the social sciences, since it analyzes the transformations to the use of natural resources, that promote public policies, in the towns on which they are applied, particularly for the groups that inhabit PNA populations and the perceptions that generate in them. Thus, the theoretical framework mentioned in the introduction is supported by various disciplines such as anthropology, economics, sociology, and ecology.

Similarly, being a qualitative investigation, ethnographic work was carried out in Tankuché, in 40 days distributed during November 2017, March, April and May 2018. In this regard, a total of 19 informants were interviewed, selected from of the snowball method, with semi-structured interviews, about the changes brought by the creation of the referred Reserve, in the use of natural resources and their perceptions generated.

6.3 Results and Discussion

6.3.1 Knowledge of the Biosphere Reserve of Petenes and the Link Between the Authorities and the Community of Tankuché

The analysis of data obtained, through semi-structured interviews applied to the population of Tankuché, showed that 100% of the informants associated the knowledge held on the Biosphere Reserve of Petenes, to issues around care from nature, flora and fauna, to plants that provide oxygen, to ban burnings and restrictions, e.g. mangrove cutting, mahogany logging and hunting deer. Among the wildlife mentioned found in the PNA population, they pointed out flamingos, parrots, ducks, tigers and pumas, among others.

In this regard, the respondents stated that the creation of the Reserve in question led to restrictions in the use of natural resources in order to care for nature. As noted by one respondent: "We know of some prevention measures … on some plant species, such as the mangrove is damaged … the mangrove area is untouchable"; which in turn information as that provided by other respondents, who claimed to know that it is forbidden hunting and cutting down mangroves, mahogany and cedar, as well as the spread of fire, because it is a PNA linked.

Now, contrasting this with the link and contact that the authorities responsible for PNA have tried to establish directly with individuals from the Tankuché community, 63% indicated that they have not had such an approach. Therefore, the communication is between the authorities and the common land commissioner. Similarly, 16% of the informants stated that they have had direct contact with the authorities, specifically with the Director of the Petenes Biosphere Reserve and that they have even asked him for support; another 16% of the interviewees, indicated that they are not even invited to participate in the meetings, in which the authorities and the remaining 5%, said that they not attending the meetings because they do not understand the issues being addressed, or not knowing what to say (see Fig. 6.2).

From the above, it was possible to observe that most of the respondents (63%) reported that the communication link between the Reserve authorities and them is between the authorities and the ejido commissioner, as described by an interviewee by specifying the following: "they give the Ejido commissioner information ... and they let us know, a meeting is held to see what we are going to do, if we go into that or not go into the programs" (59-year-old interviewee).

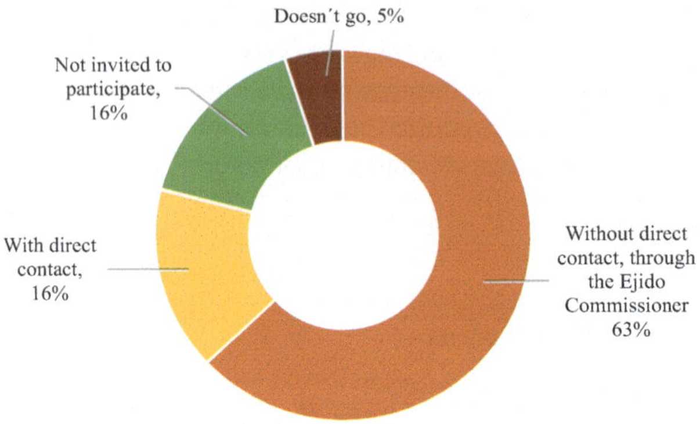

Fig. 6.2 Percentage of responses from informants on the link between the authorities responsible for the Biosphere Reserve of Petenes and individuals in the Tankuché community. Source: Prepared from the semi-structured interviews conducted

However, there were individuals who acknowledge having direct contact with the authorities responsible for the Reserve. One in particular person told to get along with the Director of the Reserve and had just done a process to ask for support for training for the conservation of PNA through Programs Sustainable Regional Development (PRODERS), Programs Conservation for Sustainable Development (PROCODES), and the Temporary Employment Program (PET), due to the time elapsed, its request had not been approved.

Also, two interviewees had direct contact with the PNS authorities, and it was presumed that they were benefited from the granting of various supports, derived from PRODERS-PROCODES; due to the signs that their land and notorious infrastructures showed, such as honey packaging areas and a shredder, it was inferred that they have been favored by these programs. Thus, the two interviewees revealed a privileged treatment by the authorities, since, unlike other individuals in the Tankuché community, they are presumed to have been one of the few to whom such support has been granted.

In terms of programs and subsidies, the National Commission for Natural Protected Areas (CONANP), once established, applied for the first time, in 2001, program resources that were designed in particular for the benefit of PNA populations: PRODERS, PROCODES, and PET. The first two (PRODERS and PROCODES) were directed to financially favor these populations, with projects for the conservation and use of natural resources, which would derive economic income for them (CONANP 2003), and the last (PET), to grant of wages for performing environmental sanitation activities; merit programs, which are still in force.

6.3.2 Perceptions About the Changes in the Use of Natural Resources and the Benefits Derived from It

On this point, 37% of respondents said that in terms of the changes brought about by the creation of the Biosphere Reserve of Petenes in the exploitation of natural resources, they have noticed it did not take care of the nature; the other 37% indicated that they are affected by the restrictions on hunting and logging, along with the decline of fisheries; and finally, the remaining 26% said that the aforementioned restrictions have benefited from the care of nature (see Fig. 6.3).

As a first point, it should be noted that even when it was reported that the reserve has turned green and thereby caring the environment, which benefits health, the truth is, that the informants tended to link the idea of conservation, with economic benefits. As a result, one of the informants said: "... perhaps the authorities would reached an agreement with the hunters ... to sit down and talk ... stop hunting, we will pay them to take care of the animals ..." (46-year-old interviewee).

This allowed warning that due to limitations imposed on hunting and logging, population demands, in exchange for doing conservation work, payment or compensation. This feeling, made infer that, in exchange for the aforementioned restrictions, compensation for the Tankuché community was not contemplated; even the supports of PRODERS-PROCODES and PET, which are programs created expressly to addressing to the aforementioned situation, according to the information provided by the interviewees, did revile that they have not sought sufficient resources in their favour and that even the number of people who have been favoured by them has been limited.

In short, it was demonstrated that the creation of the Biosphere Reserve of Petenes brought rigged restrictions on the use of natural resources; that is the case in the community of Tankuché, in which some activities are limited, such as logging and hunting. In those activities, the respondents recognized that there are

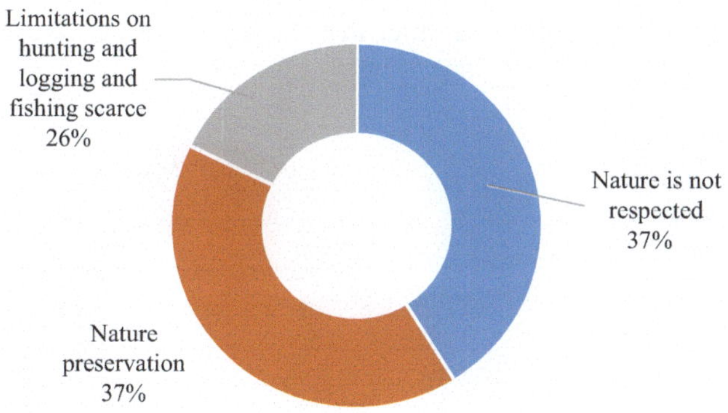

Fig. 6.3 Percentage of responses and changes resulting from the creation of the Biosphere Reserve of Petenes in the use of natural resources. Source: Prepared from the semi-structured interviews conducted

various prohibitions, such as logging of mahogany, cedar and mangrove in the PNA, as well as deer hunting; however, at the same time, the respondents acknowledged that they have violated these prohibitions, under the excuse of necessity and self-consumption. Situation in which currently Isla Arena is involved (a place where they have used to fish); since due to depredation, the increase in fuel costs and in the number of fishermen, fishing has been reduced.

Therefore, the existence of the situations described denoted absence of surveillance in the area, since such behaviours continue to be carried out, like others, that cause environmental deterioration, as the frequent garbage strip in the blanquizales area and lack of control and fire prevention, which invade the region.

Thus, it is identified that in the individuals of the Tankuché community, there is the logic of the vicious circle between depletion of natural resources, environmental deterioration and poverty (DOF 2007). Given the absence of productive alternatives for the communities and the lack of sufficient support from the PROCODERS-PROCODES and PET programs, reported by the informants and intended for PNA populations, as in the Tankuché community, (CONANP 2003), they have not been able to benefit the populations with productive projects that allow them to generate income to subsist and simulteneously ensure the conservation of biodiversity; which has led to depredation by the Tankuché community.

Regarding the perception of the respondents about the creation of the Petenes Biosphere Reserve benefited them or not and why, it came off that 53% indicated that it has not benefited them, since fishing is scarce and few have been favored; besides that they can no longer hunt and cut mangrove and mahogany as in the past. On the other hand, 47% of respondents indicated that they do consider themselves to be beneficiaries, for the conservation of flora and fauna, which protects them from pollution and provides them with oxygen, and for the support they received in previous years for the work on the short fire gaps (see Fig. 6.4).

For the 53% of the respondents who did not consider themselves to be benefited with the establishment of the PNA, they highlighted the shortage of fishing on Isla Arena, since the number of fishermen is now increased, so this activity is saturated and affected for the overexploitation and depredation derived from the non-observance of the veto. This made infer that the fishing in Isla Arena, became an

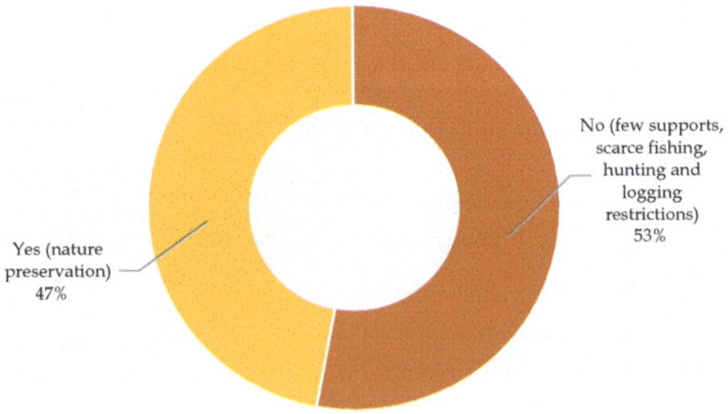

Fig. 6.4 Percentage of responses on whether or not the informants living near the Biosphere Reserve of Petenes are benefited. Source: Prepared from the semi-structured interviews conducted

alternative of income for the individuals of diverse communities, and not only of Tankuché, causing the saturation of this activity.

Similarly, it was highlighted that the respondents who reported having benefited from the work carried out in previous years of short fire gaps, have now ended and the requests have not been attended to, for their continuation; this situation was reinforced by the assumption that the population of Tankuché is willing to carry out activities related to conservation, with the condition that they receive financial support to do so. It should also be noted that individuals who reported feeling benefited from the delivery of the aforementioned supports, not about those created specifically for communities within PNA, as Tankuché; viz.: PRODERS-PROCODES and PET; but various other as PROCAMPO and supports low catch by, among others, which, questions financial failure and scope.

6.4 Conclusions

From the inferred perceptions of the population of Tankuché, it was argued that the reality they live and experience, from the creation of the Petenes Biosphere Reserve, is diametrically opposed to what they say to promise environmental public policies. This was true, since, according to information obtained through semi-structured interviews, it was demonstrated that the establishment of the PNA and knowledge around it is linked with ideas about conservation and care of nature, flora, and fauna.

Similarly, it was noted that there was a tendency to consider that its creation did not benefit the Tankuché community in any way; this only imposed a series of restrictions on hunting and logging, without the accompaniment of sufficient financial compensation, through specific programs, such as PROCODES-PRODERS and PET, and productive alternatives, around the use of natural resources, in favor of the aforementioned community. These circumstances have pushed the population of Tankuché to lean towards breaking the prohibitions imposed on logging and hunting, the overexploitation of natural resources, and in addition to the depredation, that all this causes.

Also, the fact that the approach that the Reserve authorities should seek towards the individuals of the Tankuché community occurs with the Ejidal Commissioner, prevents information about the ANP from being clearly shared and disseminated with the entire population of Tankuché. Therefore, this circumstance did not ensure the promotion of community participation, providing environmental public policy as part of sustainable development that seeks to integrate the social dimension of it which is vital, in order to determine that, indeed, the said development is achieved, without excluding the social factor. Similarly, the separation between authorities and population prevents knowing and considering the needs and interests of the population.

Consequently, in the non-inclusion of the human factor in Tankuché, it is presumed that environmental public policy, which according to the LGEEPA, constrains it to achieve sustainable development, along with its environmental, economic and social dimensions, did not achieve such development. This was so because social dimension was not properly addressed, by proper promotion of community participation in Tankuché, which resulted in ignorance and non-inclusion by the environmental public policy, interests, needs, and economic benefits of the population.

Therefore, as postulated by the intersection between the theory of ecological economy and political ecology, it is argued that they had verifying the conflicts derived from the use of natural resources. This is because the imposition of restrictions on the use of natural resources, from the establishment of the Petenes Biosphere Reserve, did not entail the consideration of the perceptions, interests, economic benefits and management of the environment of the Tankuché community.

This led to the creation of the Petenes Biosphere Reserve not representing any benefit for the Tankuché community, and continuing the lack of jobs and higher incomes in the area; therefore, it is presumed that the referred PNA did not mean a space of opportunities to the population of Tankuché.

Given the circumstances addressed, it is necessary to promote new consultation processes to learn and consider new updates of the Conservation and Management Program of the Petenes Biosphere Reserve, and the interests, needs and activities for the use of natural resources of the populations living in that area. It is also necessary to carry out revisions of programs that provide resources for the promotion of sustainable productive activities, such as the case of PRODERS-PROCODES, in order to determine if their coverage with the beneficiaries (as in the case of the Tankuché population) is sufficient and if, in effect, they generate enough financial solvency to subsist, based on the development of productive activities for which they are supported.

References

Cámara de Diputados del H. Congreso de la Unión Diario Oficial de la Federación (2018) Ley General del Equilibrio Ecológico y la Protección al Ambiente. Consultado: 3-08-18. En: http://www.diputados.gob.mx/LeyesBiblio/pdf/148_050618.pdf

Comisión Nacional de Áreas Naturales Protegidas (CONANP) (2003) Estrategia de Conservación para el Desarrollo. CONANP. D.F., México. 53 pp.

Diario Oficial de la Federación (DOF) (1971) Ley Federal para Prevenir y Controlar la Contaminación Ambiental. Consultado: 03-09-18. En: http://dof.gob.mx/copias.php?acc=ajaxPaginas&paginas=8&seccion=UNICA&edicion=204687&ed=MATUTINO&fecha=23/03/1971

Diario Oficial de la Federación (DOF) (1999) Decreto por el que se declara área natural protegida, con el carácter de reserva de la biosfera, la región del Estado de Campeche conocida como los Petenes. Consultado: 05-09-18. En: http://dof.gob.mx/nota_detalle.php?codigo=4948767&fecha=24/05/1999

Diario Oficial de la Federación (DOF) (2007) Plan Nacional de Desarrollo 2007-2012. Consultado: 07-09-18. En: http://dof.gob.mx/copias.php?acc=ajaxPaginas&paginas=todas&seccion=CUARTA&edicion=211163&ed=MATUTINO&fecha=31/05/2007

Foladori G (2002) Avances y límites de la sustentabilidad social. Economía, Sociedad y Territorio. III (12): 621–637

Gil-Corrales MÁ (2007) Crónica ambiental. Gestión pública de las políticas ambientales en México. Fondo de Cultura Económica (FCE). D.F., México. 559 pp.

Instituto Nacional de Estadística y Geografía (INEGI) (2010) Censos y conteos de población. Censo 2010 resultados por población (ITER). Consultado: 03-09-18. En: http://www.inegi.org.mx/sistemas/consulta_resultados/iter2010.aspx?c=27329&s=est

Martínez-Alier J (2004) Los conflictos ecológico-distributivos y los indicadores de sustentabilidad. Revista Iberoamericana de Economía Ecológica (REVIBEC) 1:21–30

Martínez-Alier J, Roca J (2013) Economía ecológica y política ambiental. Fondo de Cultura Económica de España. Ciudad de México, México. 641 pp.

Organización de las Naciones Unidas (ONU) (1987) Informe de la Comisión Mundial sobre el Medio Ambiente y el Desarrollo. Consultado: 05-09-18. En: http://www.un.org/es/comun/docs/?symbol=A/42/427

Padilla-Ramos R (2002) Deportación, castigo y sufrimiento. Los yaquis en Tankuché, Campeche. 464–471 pp. En: Memorias del XI Encuentro de Investigadores de la Cultura Maya, noviembre, Campeche (Vol. 2). Campeche, México: Universidad Autónoma de Campeche (UAC)

Periódico Oficial del Estado de Campeche (1996) Decreto del Ejecutivo del Estado por el que se Declara Zona Especial de Protección de Flora y Fauna Silvestre y Acuática del Área conocida como "Los Petenes". Campeche, México: Periódico Oficial del Estado de Campeche, 4 de julio de 1996

Pinkus-Rendón MJ, Pinkus-Rendón MÁ, Ortega-Rubio A (2014) Recomendaciones para el manejo sustentable en las áreas naturales protegidas de México. Investigación y Ciencia 22(60):102–110

Secretaría de Medio Ambiente y Recursos Naturales de Campeche (SEMARNACAMP) (2018) Áreas Naturales Protegidas. Consultado: 05-07-19. En: http://www.semarnatcam.campeche.gob.mx/anp/reserva-de-la-biosfera-los-petenes/

Chapter 7
Uncontrolled Urban Growth: The Crisis of Protected Natural Areas Near Cities in Mexico

Julio César Errejón-Gómez and Alfredo Ortega-Rubio

Abstract Urban areas in Mexico are growing at an accelerated rate and in a disorderly manner, fueling deforestation in the ecosystems that surround them. At different points in Mexico's history and throughout the twentieth century, attempts have been made to generate legal and political strategies to guarantee the protection of ecosystems that are affected by the growth of cities. The legal instrument most frequently used to mitigate these processes is the decree of protected natural areas. In their application, however, these decrees, laws, and regulations are subject to the private economic interests and the decisions of different levels of the Mexican government (municipal, state, and federal), which generates processes of degradation of the natural resources of legally protected areas. The main objective of this work was to develop a study of Mexican urban growth and its relationship to the degradation of terrestrial ecosystems in protected areas near cities. It was found that the decrees of protected areas ended up being obstacles to overcome in order to satisfy the economic interests of private construction companies, which manipulate each territory's legal, political, and planning regulations to repeal or modify the protection decrees.

Keywords Mexico · Protected natural areas · Urban growth

J. C. Errejón-Gómez (✉)
Environmental Planning and Conservation Program, Centro Investigaciones Biológicas Noroeste (CIBNOR), La Paz, BCS, Mexico

A. Ortega-Rubio
Red Nacional Áreas Naturales Protegidas RENANP-CONACYT, Centro de Investigaciones Biológicas del Noroeste CIBNOR-CONACYT, La Paz, Baja California Sur, Mexico
e-mail: aortega@cibnor.mx

A. Ortega-Rubio (ed.), *Socio-ecological Studies in Natural Protected Areas*,
https://doi.org/10.1007/978-3-030-47264-1_7

7.1 Introduction

Areas occupied by human activities are growing at alarming rates, reducing the distribution of terrestrial ecosystems, generating a loss of biodiversity, and causing irreversible damage to ecosystem services and the evolutionary processes that develop there (Joppa et al. 2008; Nelson and Chomitz 2009). One of the global strategies to counteract these processes of deterioration is the creation of protected natural areas (PNAs), which currently cover 14.7% of land area. However, according to the results of Jones et al. (2018), more than six million km^2, that is, 32.8%, of the total surface of PNAs is under intensive human pressure that generates its degradation. These disturbances can create a scarcity of natural resources for meeting the needs of human life and can affect socioeconomic processes as we know them today.

The degradation processes of the PNAs are caused by the current development model, which is built on anthropocentric bases, placing humans as a species on an evolutionary pedestal while viewing nature as something that must be subjugated and exploited. Human activity thus becomes a determining factor behind several environmental problems, such as the exponential increase in urbanized areas and their relationship with deforestation and environmental deterioration (Seto et al. 2012).

The economic liberalism paradigm is leading to an unprecedented environmental crisis due to the imposition of an unbridled productivity and urban growth model, which omits, in a premeditated way, the finitude of natural areas and worldwide resources. According to Victor Toledo (2015), big corporations have caused this civilizational crisis by using economic and political pressure to subordinate the government institutions in charge of designing and managing policies and laws.

In Mexico, urban areas have increased significantly over the last 30 years. In these three decades, the surface area occupied by cities has increased at a rate of 50 ha per day—nearly tenfold (Lara et al. 2017). During this period, urban growth in the country has been mainly determined by irregular construction and, in some cases, by the decisions of the construction companies who plan and design the cities in question. This strategic power was delegated during the second period of the twentieth century due to the government's inability to manage the territory, allowing private companies and the free market to impose and control urban growth patterns (Cortés 2015; Lara et al. 2017).

This makes the present work especially relevant: it is crucial to address the problem of unlimited urban growth on a planet of finite land areas and resources. For Mexico in particular, it is essential to analyze the country's current urban development model, the precepts of which are based on unlimited exponential growth a situation that increasingly affects PNAs and ecosystems surrounding cities. This study also investigates private construction companies' manipulation of the legal, political, and planning instruments within these territories to eliminate conservation strategies and, in some cases, to repeal or modify protection decrees.

To achieve the above objectives, data collection and information techniques were used on written sources, mainly books and specialized articles. This process was important because the analysis sought to have reliable and valid information (Martínez 2007) as well as a complete an overview as possible of the subject to be studied.

7.2 Urban Growth and Resource Deterioration

During the last decade, it has been estimated that 45% of the world's population live in cities. Zhou et al. (2019) have shown that the area occupied by cities across the world grew from 722,225 km^2 in 2000 to 1,169,718 in 2012, with an annual average growth of 4.1%. According to Lahoz (2010), there will be a substantial increase in urban populations in the next few years. By 2025, the number of people living in cities worldwide is expected to double; that is, from 2.4 billion in 1995, it will increase to 5 billion, representing 86% of the world's population.

The rapid expansion of cities has had significant environmental impacts. Different forms of pollution, exaggerated energy consumption, massive worldwide biodiversity loss, and the degradation of soils and terrestrial ecological systems stand out among these detrimental effects. Deforestation and the reduction in vegetation cover are particular matters of concern.

With the goal of countering these human impacts, more than 202,000 protected areas were created that now cover 14.7% of the world's land area. However, the objective of conservation is far from being achieved because 32.8% of the total area of PNAs is subject to intense human pressure (Jones et al. 2018), where agricultural activities, livestock, urban growth, and infrastructure construction stand out.

In Mexico, the problem is bigger because the environmental regulations and land planning laws are poorly applied, which encourages the application of a model of urban expansion based on the transformation and elimination of agricultural spaces, natural ecosystems, and PNAs, turning them into new industrial, service, and housing areas.

According to data from the National Forestry Commission of Mexico (CONAFOR 2013), the increase in urban stain, among other factors, contributes to the loss of 155,000 ha of forested areas each year. In rural areas, this can be explained by the low prices assigned to "rustic" spaces, mainly peripheral to cities, that acquire a substantial increase in value when they are transformed into urban land. This contempt for ecosystems generates economic incentives for construction companies, who see low ground costs as a reason to change the use of these areas as well as to destroy their vegetation cover.

7.3 When the Law Unprotects

At different stages in the history of Mexico, society has tried to generate strategies that limit the growth of cities toward certain portions of the territory. These proposals have arisen mainly from sectors of the urban population who appreciated these spaces for their landscapes and natural resources. In addition, many citizens were concerned about the loss of ecosystems and the environmental services that they provided to urban areas.

At the beginning of the twentieth century, a group of activists led by Miguel Ángel de Quevedo managed to apply the national park conservation model that was used in the United States in Mexico. Based on various political negotiations, de Quevedo's faction and similar groups influenced the decision-making of some presidents, such as Porfirio Diaz, Francisco I. Madero, and Lázaro Cárdenas, with regard to creating forest management laws and presidential decrees for the conservation of various natural areas.

Despite the pressures and achievements of this reduced environmental sector over a period of 40 years, it was not possible to obtain the necessary economic resources to lay the minimum institutional bases that would guarantee the application of the legal conservation framework. This background cemented the conservation model that the country would follow, which was replicated in stages from around 1950 to 2000. This 50-year stage is plagued by governmental initiatives of legal dye, which, in fact, acquired little force due to the lack of interest of the governments in turn.

The above can be understood in the context of the predominant economic policies of the time, the objectives of which focused on the rapid expansion of agricultural land and livestock areas and a developing industrial sector. The Mexican political elite were not willing to enforce laws that restricted the use of natural resources and thus could have limited the growth of these sectors.

During this period, the Mexican government used conservation instruments only as acts of simulation to reduce the social unrest generated by the deterioration of ecosystems and, in particular, to appease environmental activists in cities, who had noticed the decrease in forest spaces and the natural environment around them. This is most evident from policies enacted after the 1960s, which show little or no economic investment or interest in natural resource conservation (Simonian 1999).

By the end of the twentieth century, the low state investment registered for more than 50 years had rendered the entire legal framework that regulated "protected" spaces inoperative. It was not until 2000 that state investment significantly increased for the procurement of the country's PNAs. These resources arrived too late, however, for some spaces that had been decreed decades earlier and had experienced great degradation of their ecosystems.

The twentieth century brought new developments in Mexico's conservation policies with the presence of a legal framework still in place today, the creation of government institutions specialized in PNAs, and greater economic resources invested in conservation, as well as a significant increase in social interest. Despite these factors, however, the degradation of protected spaces has not been substantially reduced. Even today, the current economic model is superimposed, and the conservation of natural resources is subordinated (Sánchez and Figueroa 2007).

Much of the pressure on PNAs is caused by Mexico's real estate sector. There are also certain geographical areas that stand out for their greater investment attractiveness, which causes exponential pressure on the construction companies in these spaces. These sectors offer the most representative cases of intentional de-protection in the "protected" natural areas. Pressures on these areas can also be observed in the actions of the country's government when trying to meet the demand for urban expansion space; for instance, politicians see PNAs as borders that must be torn down. They achieve this by repealing protective decrees or manipulating management programs to allow certain economic activities or the development of human settlements inside these areas. This is greatly encouraged by construction companies and developers, who likewise see in the conservation and protection measures legal elements that must be drawn or eliminated, regardless of environmental ethical codes, degradation of PNA ecosystems, or risks of losing the environmental services that PNAs provide to the cities.

The current problem of urban growth into PNAs extends further because some stakeholders or actors linked to urbanization companies become public officials, legislating or making decisions regarding the maintenance and permanence of the PNAs. They create laws and regulations that conflict with protective decrees and original conservation interests. It is common for Mexican legislators to overlap economic and urban interests, destroying or subordinating any legal framework that can reduce the generation of profits or restrict the progress of construction projects.

We can find reference to the above in the maneuvers that these political actors perform and their interests, which are in favor of the excessive and disorganized progress of the cities. They are responsible for executing legal and political expertise to superimpose permissible laws, prepared at their convenience, in order to subjugate Mexican environmental legislation.

From our analysis of urban development and land planning policy, it can be seen that political practice in Mexico is far removed from environmental and sustainability interests. This is clearly indicated by the horizontal growth of the country's urban areas, which has shown, in recent years, a disorderly dynamic devoid of planning and makes it one of the main deforestation factors in the country (Fig. 7.1).

7.4 Pressures of Construction Companies in Mexican PNAs

There are many representative cases of construction companies putting pressure on Mexico's PNAs (Table 7.1). The coastal regions of Mexico, for example, undergo real estate harassment due to policies promoting massive beaches and sun tourism that the government has been implementing since the 1970s. This situation places protected coastal spaces at risk, since urban environments, infrastructures, and services are constantly growing.

An emblematic example is the coast of the State of Quintana Roo, which has experienced exponential growth in recent decades. According to Rubio et al. (2010), the influx of tourists to Quintana Roo increased from 2,600,000 in 1990 to 6,720,164 in 2005. This accelerated population growth and spurred the deforestation of mangroves and forests as well as increasing the strain on PNAs due to the accelerated advance of urban spaces and the construction of infrastructure in regions of touristic importance. Urban growth is the third leading cause of the loss of forest cover in Quintana Roo, coming closely behind agricultural activities and forest fires (Ellis et al. 2017).

Fig. 7.1 Construction of housing areas in the Bosque de la Primavera. (Author: J.C. Errejón Gómez)

Tulum National Park, also located in Quintana Roo, has historically suffered different forms of real estate harassment for the development of tourism infrastructure. This siege sometimes had drastic effects on the flora and fauna of the protected area. Urban development projects in the park have threatened the permanence of its natural ecosystems. In the last two decades, human intervention in this space has

Table 7.1 Some examples of PNAs close to cities that suffer from the siege of the developers

Protected natural area	State of Mexico
Cerro de las Culebras	Veracruz
Cerro de la Galaxia	Veracruz
El Bosque de la Primavera	Jalisco
Sierra Fría	Aguascalientes
Cumbres de Monterrey	Nuevo León
Tulum	Quintana Roo
San Juan de Guadalupe	San Luis Potosí
Paseo de la Presa	San Luis Potosí
Cerro de la Estrella	Mexico City

intensified mainly due to the construction of hotels and other tourism infrastructures, ignoring the entire legal conservation framework in force.

This situation has continued in recent years. On more than one occasion, there has been considerable deforestation. The last case of damage to forested areas in Tulum National Park occurred in 2018 when 2000 square miles of tropical forest were cleared for the construction of hotels. Deforestation risks remain latent due to the interests of individuals seeking to benefit from the demand for tourist services within the park.

Other cases worth analyzing are found in the peripheral areas of Mexico's cities. In multiple cases, citizens, together with local and federal governments, have proposed protection decrees and land management plans to guarantee the permanence of certain emblematic ecosystems and landscapes in their regions. However, these spaces still experience, to a greater or lesser extent, the same pressures of urban growth and real estate speculation.

One particular case worth analyzing is the San Juan de Guadalupe Urban Park in San Luis Potosí, since it gives us a clearer picture of what is happening in other areas of the country. In the southwest area of the city, there is a mountain massif called the Sierra de San Miguelito. Over time, the populations of San Luis Potosí have given the mountain a socioenvironmental value due to the flora and fauna it hosts, the environmental services it provides, and the landscapes it contains. This high valuation has led to various attempts to protect the area in different periods of Mexican history.

During the first attempt, Miguel Ángel de Quevedo, who headed the Autonomous Department of Forestry, Fishing, and Hunting, succeeded in decreeing the Protective Zone of the City of San Luis Potosí in 1937, during the government of Lázaro Cárdenas, due to the environmental services that this mountain range provided to the city. However, the decree did not endure, and over the years it was forgotten by authorities and the general public. This may have been due to the limited topographic references given to define the area—some of which no longer exist today, which makes it difficult to identify them precisely.

It was not until 1996 that another attempt was made to conserve the Sierra de San Miguelito, this time by state authorities who decreed a PNA in the massif's northeastern portion. This area of 1208 ha was named San Juan de Guadalupe Urban Park. This decree arose, as before, to prevent the urbanization of the natural area and to preserve the environmental services that it provided to the city. As with the previous decree, however, the state authorities did very little to follow up on or to apply the regulations mentioned in the document.

Over the last decade, real estate developers have increased their economic interest in the Sierra de San Miguelito. Initially, they began with the construction of residential areas, shopping centers, and other infrastructure (Fig. 7.2). However, construction companies have projected future growth toward other areas in the region—for instance, a state and municipal government project to develop urban roads, which would literally split the mountain in two. This puts the area's environmental processes at serious risk.

The constant deterioration that the Sierra de San Miguelito experiences has generated widespread social discontent as well as activism in defense of the area's ecosystems and landscapes (Fig. 7.2). Local authorities have ignored the demands of the population, however, and have exhibited a clear stance in favor of the economic interests of construction companies and the urbanization of these spaces.

In 2018, the real estate developers took the first step to urbanize the protected area of San Juan de Guadalupe Urban Park. The construction companies, accompanied by some *ejidatarios*, challenged the state conservation decree generated in

Fig. 7.2 Urban growth in the Sierra de San Miguelito, San Luis Potosí, Mexico. (Author: J.C. Errejón Gómez)

1996. The judge in charge of the case ruled in favor of the construction companies and then ordered the repeal of the PNA designation. This laid the groundwork for construction companies to buy properties in the former PNA, displacing the local farmers who owned the land. In addition, with the elimination of this legal barrier, the companies could obtain permits for land-use changes and thus begin with real estate and urban development in the area.

Today, the activities of the construction companies continue, creating a socioenvironmental struggle that is growing in terms of the protection of the Sierra de San Miguelito. As of 2019, the federal government has taken a position in favor of the Sierra's conservation and, together with local stakeholders, will seek to procure another decree that guarantees the protection of this area. Despite this turnaround, however, the construction companies show new aggression toward this natural space, looking for other ways to urbanize via the formation of building belts that will gradually "suffocate" the Sierra de San Miguelito. As Domínguez (2009) observes, this cancels the existence of buffer areas, whose function is to lessen the pressures of human activities on ecosystems.

It is clear that there is a crisis around PNAs throughout Mexico, within which those located near tourist and urban development spaces stand out. It is necessary to apply legal and political measures that really prevent the deterioration of the natural and landscape resources that these areas contain. Currently, the outlook is not encouraging due to the exponential and haphazard increase in Mexico's urban development. This raises a question: What is the limit of horizontal urban growth in a country of finite spaces?

7.5 Lesson Learned

Throughout different stages of Mexico's history, attempts have been made to generate legal and political strategies to protect nature. To date, the most frequently used instrument by the state is the natural protected areas, whose objective is to ensure the conservation of ecosystems with low anthropogenic alterations. However, the implementation of protection decrees, laws, and regulations is subjugated to economic interests and government institutions.

As a result of the analysis, it was observed that for decades the authorities showed little interest in issues of protection and in the application of regulations in NPAs. This situation was aggravated when, on some occasions, the decrees of the NPAs ended up being obstacles for the governments in power to overcome in order to satisfy economic and private interests. Seeing NPAs as borders that must be broken in order to expand urban areas, urban development companies generate political and legal strategies to realize their objectives, preventing conservation strategies from being implemented in areas close to cities.

Mexico currently lacks legal instruments that are effectively applied for the comprehensive management of the territory. There is therefore an urgent need for an objective assessment of the impacts of real estate and construction companies on habitats inside and outside protected areas in Mexico. These efforts must be transversal, where citizens in general, academia, and the state are jointly responsible for improving management practices in PNAs, as well as for monitoring compliance with environmental laws.

It is necessary that as a country we turn our gaze to the integral management of the territory, with the objective of guaranteeing the long-term conservation of nature. This highlights the requirement of slowing down the current model of exponential and disorderly growth of the cities in Mexico.

The various forms of legal transgression against PNA protection, as well as the collaboration of construction companies with the authorities responsible for monitoring compliance with the laws, do not leave an encouraging panorama for ecosystem conservation. There is growing environmental awareness, however, sparking more and more frequent and numerous citizen mobilizations and laying the foundation for maintaining the hope for change in the future.

References

CONAFOR (Comisión Nacional Forestal de México) (2013) 82 percent of deforestation in Mexico is due to land use change. Boletin 113. http://www.conafor.gob.mx:8080/documentos/docs/7/4724El%2082%20por%20ciento%20de%20la%20deforestaci%C3%B3n%20en%20M%C3%A9xico%20es%20por%20cambio%20de%20uso%20del%20suelo.pdf

Cortés J (2015) The urban growth of cities: developmental, authoritarian, neoliberal and sustainable approaches. Paradig Econ 7(1):127–149

Domínguez E (2009) Biological and social connectivity: areas of influence of protected natural areas, Serie Conocimientos 5. Comisión Nacional para el Conocimiento y Uso de la Biodiversidad

Ellis E, Romero J, Hernández I (2017) Deforestation processes in the state of Quintana Roo, Mexico: the role of land use and community forestry. Trop Conserv Sci 10:1–12. https://doi.org/10.1177/1940082917697259

Jones K et al (2018) One-third of global protected land is under intense human pressure. Science 360(6390):788–791. https://doi.org/10.1126/science.aap9565

Joppa L, Loarie S, Pimm S (2008) On the protection of "protected areas". Proc Natl Acad Sci 105(18):6673–6678. https://doi.org/10.1073/pnas.0802471105

Lahoz E (2010) Environmental reflections of urban expansion. Geogr Noteb 46:293–313

Lara J et al (2017) The costs of urban expansion: an approach based on a hedonic price model in the Metropolitan Area of the Valley of Mexico. Estudios demográficos y urbanos 32(1):7–63. http://www.scielo.org.mx/scielo.php?script=sci_arttext&pid=S0186-72102017000100037

Martínez J (2007) Instrumental developments of a monitoring and evaluation system for results management. Universidad Autónoma del Estado de México, p 167

Nelson A, Chomitz K (2009) Protected area effectiveness in reducing tropical deforestation a global analysis of the impact of protection status. The World Bank, Washington, DC, p 29. http://documents.worldbank.org/curated/en/369821468154781292/pdf/533700BRI0prot10box345607B01PUBLIC1.pdf

Rubio E, Murad M, Rovira J (2010) Environmental crisis on the coast of Quintana Roo as a result of a limited vision of what sustainable development represents. Arguments 23(63):161–185

Sánchez V, Figueroa F (2007) Effectiveness of biosphere reserves to prevent land use and land cover change in Mexico. In: Halffter G, Guevara S, Melic A (eds) Hacia una cultura de conservación

de la diversidad biológica. Sociedad Entomologica Aragonesa, Zaragoza, pp 161–171. http://seaentomologia.org/PDF/PDFSM3MVOL6/Pdf18161172018SanchezCordero.pdf
Seto K, Güneralp B, Hutyra L (2012) Global forecasts of urban expansion to 2030 and direct impacts on biodiversity and carbon pools. Proc Natl Acad Sci 109(40):16083–16088
Simonian L (1999) Defending the land of jaguar. A history of conservation in Mexico. SEMARNAP, México
Toledo V (2015) Ecocide in Mexico: the final battle is for life. Grijalbo, México
Zhou Y, Varquez A, Kanda M (2019) High-resolution global urban growth projection based on multiple applications of the SLEUTH urban growth model. Sci Data 6(1):34. https://doi.org/10.1038/s41597-019-0048-z

Chapter 8
Dynamic Simulation Models and Participatory Approaches to Support the Sustainable Management of Social-Ecological Systems in Natural Protected Areas

Alicia Tenza-Peral, Vianney Beraud-Macías, Julia Martínez-Fernández, Irene Pérez-Ibarra, Aurora Breceda, and Andrés Giménez-Casalduero

Abstract The management of natural protected areas should embrace the complexity of social-ecological systems that are often embedded in these areas, and they depend on their ecosystem services. Recently, the study of social-ecological systems has emerged with a growing theoretical framework. However, the complexity of social-ecological systems and abstraction of theoretical approaches challenge the application of social-ecological frameworks to support the management of protected areas. In this chapter, we introduce a methodological framework, based on the system dynamics approach, that includes participatory workshops, expert

A. Tenza-Peral (✉)
Área de Ecología, Departamento de Biología Aplicada, Universidad Miguel Hernández, Elche, Spain

Center for Behavior, Institutions and the Environment, Arizona State University, Tempe, AZ, USA
e-mail: atenza@umh.es

V. Beraud-Macías
Universidad Autónoma de Aguascalientes, Aguascalientes, Mexico

J. Martínez-Fernández
Fundación Nueva Cultura del Agua, Zaragoza, Spain
e-mail: julia@fnca.eu

I. Pérez-Ibarra
Agencia Aragonesa de Investigación y Desarrollo (ARAID), Zaragoza, Spain
e-mail: perezibarra@unizar.es

A. Breceda
Northwestern Center for Biological Research (CIBNOR), La Paz, Mexico
e-mail: abreceda@cibnor.mx

A. Giménez-Casalduero
Área de Ecología, Departamento de Biología Aplicada, Universidad Miguel Hernández, Elche, Spain
e-mail: agimenez@umh.es

A. Ortega-Rubio (ed.), *Socio-ecological Studies in Natural Protected Areas*,
https://doi.org/10.1007/978-3-030-47264-1_8

panels, and simulation tools to help improve our understanding of the structural causes of problematic behaviors and to evaluate alternative management options, supporting the adaptive management of protected areas. This methodological framework could help us to test in advance the effectiveness and robustness of alternative policies and management options, avoiding the waste of economic resources, time, and efforts based on unrealistic expectations or the lack of knowledge. The participation of local stakeholders is needed to encourage the appropriation of the results and the execution and monitoring of the management measures or action proposed.

Keywords Dynamic models · Resilience · Sustainability · Participatory workshops · Scenario planning

8.1 Introduction

Natural protected areas (NPAs) are essential for biodiversity conservation. In 2018, 14.9% of the world's land and 7.3% of the oceans were covered by NPAs (UNEP-WCMC, IUCN and NG 2018). Our perspective about environmental conservation and how to manage natural protected areas has changed over time, from segregative approaches that excluded human beings from the natural areas, considering them as ecological islands or pristine areas, to more integrative approaches that embrace the idea of "human-in-nature" (Phillips 2004). Social rejection, conflicts, environmental degradation, and poverty have been associated with the surroundings of strictly protected areas (West et al. 2006; Brockington and Wilkie 2015). The new paradigm seeks to minimize these issues, reconciling conservation, and sustainable development. It aims to preserve biodiversity and natural habitats but at the same time to promote the sustainable use of natural resources by local populations, improve socioeconomic wellbeing, maintain cultural values, and respond to local needs. The Biosphere Reserves, from the Man and the Biosphere (MAB) program, are good examples of this approach (UNESCO 2016).

Societies are embedded in ecosystems, which provide essential goods and services for them, and at the same time, both components are drivers of change for each other (i.e., the environment limits human activities, and anthropic activities modify landscapes and biophysical dynamics). This indissoluble relation is depicted in the concept of the social-ecological system (SES, Berkes and Folke 1998), which includes the interaction between social, economic, institutional, and ecological factors at different scales from global to local.

To effectively manage NPAs, it is necessary to attend specific problems, avoiding universal solutions, using place-based approaches (van der Leeuw et al. 2012; Carpenter et al. 2012; Balvanera et al. 2017), and considering the NPAs as complex and adaptive SESs, integrating the effects of socioeconomic and ecological changes at different scales and the inherent uncertainty coming from natural variability and lack of knowledge about complex dynamics (Palomo et al. 2014; Therville et al. 2018).

Despite the advances in the study of SESs in the last decades, most scientific literature lay on theoretical and conceptual studies, which hinder the analysis of the resilience and sustainability of SESs in an operative way. Making policy recommendations from qualitative or conceptual models is risky because of the existence of feedbacks, delays, trade-offs, and synergies that challenges our understanding of the system's behavior and can lead to misleading conclusions (Vennix 1996; Challies et al. 2014; Tenza et al. 2019). The need to quantitatively modeling SESs is clear, but frequently involves serious challenges like bridging epistemologies across disciplines, combining qualitative and quantitative methods and data sources, and integrating the human dimension (Elsawah et al. 2020).

Here we introduce a methodological framework that uses the system dynamics approach combined with social research techniques and participatory approaches to study the sustainability of SESs in NPAs, surpassing some of the challenges of modeling SESs, and providing a decision-supporting tool to enhance the management of NPAs effectively.

8.2 System Dynamics and Computational Models as a Methodological Approach

System dynamics is a methodological approach that, from a holistic point of view, aims to analyze and respond to real-world problems. It was developed by Jay W. Forrester in the 1950s. Initially applied to industrial problems (Forrester 1958, 1961), it has greatly extended its application in recent decades to various fields such as water management (Sanga and Mungatana 2016), the analysis of land-use changes (Baur and Binder 2015; Vidal-Legaz et al. 2013), and sustainability of SESs (Banos-González et al. 2016, 2018; Tenza et al. 2017, 2019).

System dynamics approach focuses on the structure of complex systems and their behavior. The complexity emerges from nonlinear relationships between system components, feedbacks, and material or information delays (Roberts et al. 1983; Vennix 1996). Dynamic simulation models are "causal-descriptive" or "white box," which means that they explicitly describe the relationships between system components. Among their advantages are their transparency (all the interactions are described by differential equations allowing the revision of the structural hypothesis); the wide range of sources of information that can be used as inputs to build, calibrate, and validate them (e.g., empirical data, statistical data, qualitative data); and their flexibility that allows to address complex problems unfeasible for analytical models, especially when data are scarce (Roberts et al. 1983; Vennix 1996; Tenza et al. 2019).

Modeling SESs allows simulating long-term dynamics and interactions (i.e., synergies and trade-offs) between social, economic, institutional, and ecological factors at different scales (Tenza et al. 2019). These models could be used as decision-support tools to evaluate the results and effectiveness of alternative management options, directed to solve specific problems under different socioeconomic or climatic scenarios (Banos-González et al. 2016, 2018; Pérez-Ibarra et al. 2016).

According to various authors, the modeling process can be divided between three and seven stages (Roberts et al. 1983; Sterman 2000; Luna-Reyes and Andersen 2003). The difference between the proposed stages is reduced to the level of aggregation or segregation of the methodological steps included in each of them. To improve the understanding of this process, we choose to divide it into three main stages, specifying the main steps within each stage and the possible methods to fulfill them (Fig. 8.1).

8.2.1 *Conceptualization: Building a Qualitative Model of SESs*

Qualitative models are the result of conceptualizing the problem under study (Tenza et al. 2017). We specified "problem" rather than system because it helps delimit the study in time and space and helps determine the elements that form part of the model. Delimitation of real systems is difficult given their complexity. Identifying a problematic behavior is a key issue for determining the variables and relationships involved (Vennix 1996).

Qualitative models are often represented by causal loop diagrams (see Tenza et al. 2017 for more details), which include the main system variables linked to the problematic behavior, the connections between these variables by means of arrows, and the sign of these interactions (i.e., positive or negative). They represent the hypothesis about the system structure and behavior (i.e., the response to "What structure is responsible for the observed behavior?").

The conceptualization phase is an iterative process with a wide range of sources of information, like scientific literature review, expert panels, participatory workshops with stakeholders, in-depth interviews, surveys and participant observation, statistical official data, and GIS and satellite images.

8.2.2 *Quantitative Modeling*

To build a quantitative model, we need to define the relationships between system variables using differential equations, algorithms, or logical functions. Software like *Vensim* or *Stella* facilitates the creation of dynamic simulation models drawing flowcharts and defining the relationships between variables with a friendly interface.

Before using the model for analytic purposes, we need to validate it properly. In order to determine the confidence level in the model, usually, we must check for robustness, reliability, and validity of the model by means of structural tests, sensitivity analyses, and behavioral tests (Barlas 1989, 1996; Solecki and Oliveri 2004).

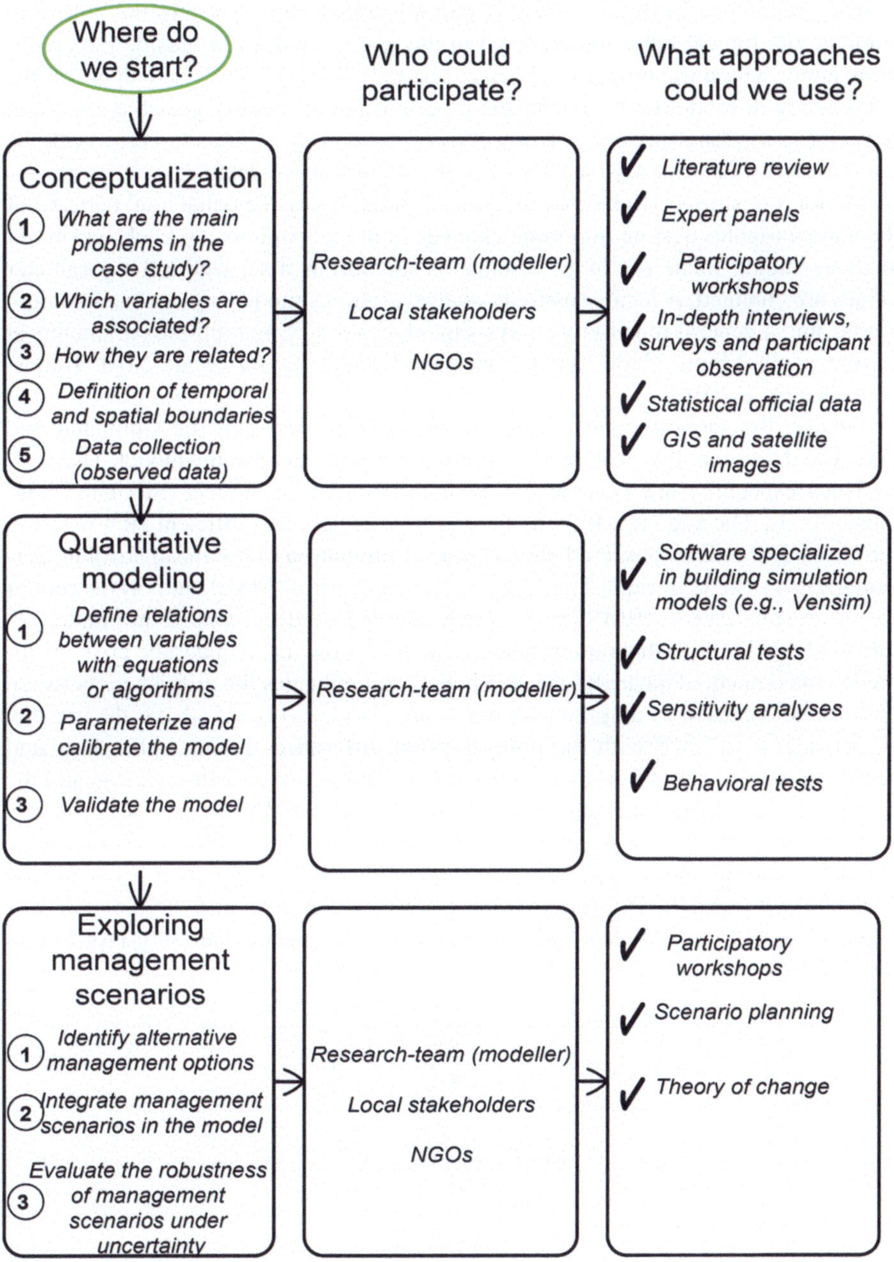

Fig. 8.1 Methodological framework

The confidence in these models is partially measured regarding the ability to simulate the logical behavior expected in the real system, even outside the conditions under which the model has been calibrated. To verify this, we can check the consistency in model units, extend the time horizon to identify possible abnormal behavior in the long run, and perform a set of extreme conditions tests to check the internal model consistency (see Tenza et al. 2019 for more details).

Model robustness is measured in terms of maintaining the behavioral patterns of the main variables despite important changes in model parameters. Each sensitivity analysis should include a large number of simulation runs, where the parameter values are changed. A local sensitivity analysis implies modifying one parameter at a time, and a general sensitivity analysis involves modifying multiple parameters at a time using Monte Carlo simulations (Ford 1990; Taylor et al. 2010; Banos-González et al. 2016).

Finally, the measurement of the goodness of fit between the simulated and observed data is another feature to determine the confidence in the model. Theil's U statistics especially are suitable for dynamic simulation models (Sterman 1984; Oliva 2003). The statistics tests include (1) a correlation coefficient (R^2) between the simulated and the observed data, (2) the computation of the mean absolute percentage error (MAPE, Eq. 8.1), (3) the mean square error (MSE), and (4) the root of the mean square error (RMSE). The MSE (Eq. 8.2) is decomposed into three factors: (i) UM (Eq. 8.3) that represents the fraction of the mean quadratic error attributed to bias (unequal means); (ii) US (Eq. 8.4) that indicates the differences between the variances caused by amplitude or noise; and (iii) UC (Eq. 8.5) that is the unequal covariance, associated with the point-to-point differences between the simulated and the observed data (Baur and Binder 2015). To Theil's statistics, we can add the computation of the normalized root mean square error (NRMSE, Eq. 8.6) proposed by Jamieson et al. (1991), which serves as an indicator of the difference between the simulated and the observed data (Andarzian et al. 2011). Simulation is considered excellent with an NRMSE under 10%; good between 10% and 20%; fair if it is higher than 20, but less than 30%; and poor if it is higher than 30% (Andarzian et al. 2011).

$$\mathrm{MAPE} = \frac{1}{n}\sum_{n}^{t=1}\left|\frac{S_t - A_t}{A_t}\right| \tag{8.1}$$

$$\mathrm{MSE} = \frac{1}{n}\sum_{n}^{t=1}\left(S_t - A_t\right)^2 \tag{8.2}$$

$$U^M = \frac{\left(\bar{S} - \bar{A}\right)^2}{\mathrm{MSE}} \tag{8.3}$$

$$U^S = \frac{\left(S_S - A_A\right)^2}{\mathrm{MSE}} \tag{8.4}$$

$$U^{C} = \frac{2 \times (1-r) \times S_S \times A_A}{\text{MSE}} \tag{8.5}$$

$$\text{NRMSE} = \frac{1}{\bar{A}} \sqrt{\frac{1}{n} \sum_{n}^{t=1} \left(S_t - A_t\right)^2} \tag{8.6}$$

where S_t and A_t are the simulated and the observed (actual) data at time t, respectively; $\bar{S}$ and $\bar{A}$ are the mean values of the simulated and the observed data, respectively; S_S and A_A are the standard deviation of the simulated and the observed data, respectively; r is the correlation coefficient between both series (simulated and observed data); and n is the number of observations.

8.2.3 Exploring Management Scenarios

With the quantitative model correctly validated, it is time to explore alternative management scenarios. Local needs and the desired future state of the social-ecological system should be defined by the stakeholders. From the scenarios or management options proposed by the stakeholders, which could be based even on narratives, we need to translate and integrate these management scenarios into the dynamic simulation model. These scenarios may involve structural changes (e.g., new variables, changes in feedbacks), changes in external conditions (e.g., external variables like climate or market prices), or quantitative changes in the internal parameters of the system.

We highlight scenario planning as an attractive, useful, and extended method to lead participatory workshops in the context of modeling SESs (Peterson et al. 2003; Palomo et al. 2011; Butler et al. 2016). This methodology has been proven useful to reflect on the possible futures of SESs and, in the face of uncertainty, decide on policies and management measures for getting closer to desirable objectives and also minimizing the most unfavorable aspects. This is an interesting exercise to perform in participatory processes since it invites reflection on the role of local actors in the future of their communities and they can help resolve conflicts or at least place them on the table, seek consensus, and also give a perspective on the different scales that affect us, from the local to the global.

Integrating and analyzing alternative management scenarios under different socioeconomic and climate scenarios allow us to test in advance the effectiveness and robustness of these policies, avoiding the waste of economic resources, time, and efforts based on unrealistic expectations.

8.3 Social Research Techniques and Participatory Processes in the Study of SESs

Social research techniques and participatory processes are critical for the study of SESs. In-depth interviews, surveys, and participant observation are essential to reach and understand local ecological knowledge (Gadgil et al. 1993; Becker and Ghimire 2003; Lunas-Reyes and Andersen 2003; Tengö and Belfrage 2004). This knowledge can help to understand the general dynamics of SESs qualitatively and to identify local adaptive strategies to cope with natural disturbances (Forrester 1992; Lunas-Reyes and Andersen 2003; Tengö and Belfrage 2004). Also, in many cases, it is a valuable source of high-quality information (quantitative and semi-quantitative) that complements the more conventional techniques in scientific research (Huntington 2000; Anadón et al. 2009; Parry and Peres 2015; Tenza et al. 2019).

Communication between researchers and stakeholders, transparency, and the establishment of a good relationship of trust have been identified as decisive factors for the success of participatory processes aimed to support decision-making processes in SESs (Palomo et al. 2011; Hanspach et al. 2014; Balvanera et al. 2017). The active participation of local communities in this process is an indispensable condition for the appropriation of the results and for the proposed strategies to be executed and monitored effectively (Reed 2008).

Getting close to local communities and developing participatory processes take time. That is why the study of SESs needs long-term research approaches (Carpenter et al. 2012; Balvanera et al. 2017). This is a challenge within conventional scientific schemes since much of the funding and valuation of scientific work depends almost exclusively on scientific publication in high-impact journals, which may enter into a conflict with the dedication of time and effort to the immersion process in the case study and organization of participatory dynamics and ensure feedbacks between local communities and scientists (Balvanera et al. 2017).

8.4 Final Remarks and Lessons Learned

Dynamic simulation models allow us to embrace the complexity of social-ecological systems (SESs), where social, economic, institutional, and ecological factors interact across different scales from global to local. Dynamic simulation models are an optimal methodological approach to study the sustainability and resilience of natural protected areas (NPAs) under different scenarios (e.g., management scenarios, socioeconomic scenarios, climate change scenarios) because of their transparency, the possibility of including different sources of information (i.e., qualitative, quantitative, and semi-quantitative) from different disciplines, and their flexibility even when data is scarce. This methodological framework could help us to test in advance the effectiveness and robustness of alternative policies and management options, avoiding the waste of economic resources, time, and efforts based on unrealistic expectations or the lack of knowledge. The participation of local stakeholders is needed to encourage the appropriation of the results and the execution and monitoring of the management measures or action proposed. The participation of stakeholders can take place throughout the entire modeling process. However, it is usually carried out in the conceptualization stage, at the beginning, and in the exploration of future management scenarios, at the end. However, the participation of stakeholders could be also useful to validate the quantitative model.

References

Anadón JD, Giménez A, Ballestar R, Pérez-Ibarra I (2009) Evaluation of local ecological knowledge as a method for collecting extensive data on animal abundance. Conserv Biol 23:617–625. https://doi.org/10.1111/j.1523-1739.2008.01145.x

Andarzian B, Bannayan M, Steduto P, Mazraeh H, Barati ME, Barati MA, Rahnama A (2011) Validation and testing of the AquaCrop model under full and deficit irrigated wheat production in Iran. Agric Water Manag 100:1–8. https://doi.org/10.1016/j.agwat.2011.08.023

Balvanera P, Daw TM, Gardner TA, Martín-López B, Norström AV, Chinwe Ifejika Speranza CI, Spierenburg M, Bennett EM, Farfan M, Hamann M, Kittinger JN, Luthe T, Maass M, Peterson GD, Perez-Verdin G (2017) Key features for more successful place-based sustainability research on social-ecological systems: a Programme on Ecosystem Change and Society (PECS) perspective. Ecol Soc 22:#14. https://doi.org/10.5751/ES-08826-220114

Banos-González I, Martínez-Fernández J, Esteve MA (2016) Using dynamic sustainability indicators to assess environmental policy measures in Biosphere Reserves. Ecol Indic 67:565–576. https://doi.org/10.1016/j.ecolind.2016.03.021

Banos-Gonzalez I, Martínez-Fernández J, Esteve-Selma MA, Esteve-Guirao P (2018) Sensitivity analysis in socio-ecological models as a tool in environmental policy for sustainability. Sustainability 10:#2928. https://doi.org/10.3390/su10082928

Barlas Y (1989) Multiple tests for validation of system dynamics type of simulation models. European Journal of Operational Research 42 (1):59–87

Barlas Y (1996) Formal aspects of model validity and validation in system dynamics. Syst Dyn Rev 12:183–210. https://doi.org/10.1002/(SICI)1099-1727(199623)12:3<183::AID-SDR103>3.0.CO;2-4

Baur I, Binder CR (2015) Modeling and assessing scenarios of common property pastures management in Switzerland. Ecol Econ 119:292–305. https://doi.org/10.1016/j.ecolecon.2015.09.019

Becker CD, Ghimire K (2003) Synergy between traditional ecological knowledge and conservation science supports forest preservation in Ecuador. Conserv Ecol 8:#1. [online] URL: http://www.consecol.org/vol8/iss1/art1/

Berkes F, Folke C (1998) Linking social and ecological systems: management practices and social mechanisms for building resilience. Cambridge University Press, New York

Brockington D, Wilkie D (2015) Protected areas and poverty. Philos Trans R Soc B Biol Sci 370:20140271. https://doi.org/10.1098/rstb.2014.0271

Butler JRA, Bohensky EL, Suadnya W, Yanuartati Y, Handayani T, Habibi P, Puspadi K, Skewes TD, Wise RM, Suharto I, Park SE, Sutaryono Y (2016) Scenario planning to leap-frog the Sustainable Development Goals: an adaptation pathways approach. Clim Risk Manag 12:83–99. https://doi.org/10.1016/j.crm.2015.11.003

Carpenter SR, Folke C, Norström A, Olsson O, Schultz L, Agarwal B, Balvanera P, Campbell B, Castilla JC, Cramer W, DeFries R, Eyzaguirre P, Hughes TP, Polasky S, Sanusi Z, Scholes R, Spierenburg M (2012) Program on ecosystem change and society: an international research

strategy for integrated social–ecological systems. Curr Opin Environ Sustain 4:134–138. https://doi.org/10.1016/j.cosust.2012.01.001
Challies E, Newig J, Lenschow A (2014) What role for social–ecological systems research in governing global teleconnections? Glob Environ Chang 27:32–40. https://doi.org/10.1016/j.gloenvcha.2014.04.015
Elsawah S, Filatova T, Jakeman AJ, Kettner AJ, Zellner ML, Athanasiadis IN, Hamilton SH, Axtell RL, Brown DG, Gilligan JM, Janssen MA, Robinson DT, Rozenberg J, Ullah IIT, Lade SJ (2020) Eight grand challenges in socio-environmental systems modeling. Soc Environ Syst Model 2:#16226. https://doi.org/10.18174/sesmo.2020a16226
Ford A (1990) Estimating the impact of efficiency standards on the uncertainty of the Northwest electric system. Oper Res 38:580–597. https://doi.org/10.1287/opre.38.4.580
Forrester JW (1958) Industrial dynamics: a major breakthrough for decision makers. Harv Bus Rev 36:37–66
Forrester JW (1961) Industrial dynamics. Pegasus Communications, Waltham
Forrester JW (1992) Policies, decisions and information sources for modeling. European Journal of Operational Research 59 (1):42–63
Gadgil M, Berkes F, Folke C (1993) Indigenous knowledge for biodiversity conservation. Ambio 22:151–156
Hanspach J, Hartel T, Milcu AI, Mikulcak F, Dorresteijn I, Loos J, von Wehrden H, Kuemmerle T, Abson D, Kovács-Hostyánszki A, Báldi A, Fischer J (2014) A holistic approach to studying social-ecological systems and its application to southern Transylvania. Ecol Soc 19:#32. https://doi.org/10.5751/ES-06915-190432
Huntington HP (2000) Using traditional ecological knowledge in science: methods and applications. Ecol Appl 10:1270–1274. https://doi.org/10.1890/1051-0761(2000)010[1270:UTEKIS]2.0.CO;2
Jamieson PD, Porter JR, Wilson DR (1991) A test of computer simulation model ARC-WHEAT1 on wheat crops grown in New Zealand. Field Crop Res 27:337–350. https://doi.org/10.1016/0378-4290(91)90040-3
Luna-Reyes LF, Andersen DL (2003) Collecting and analyzing qualitative data for system dynamics: methods and models. Syst Dyn Rev 19:271–296. https://doi.org/10.1002/sdr.280
Oliva R (2003) Model calibration as a testing strategy for system dynamics models. Eur J Oper Res 151:552–568. https://doi.org/10.1016/S0377-2217(02)00622-7
Palomo I, Martín-López B, López-Santiago C, Montes C (2011) Participatory scenario planning for protected areas management under the ecosystem services framework: the doñana social ecological system in southwestern Spain. Ecol Soc 16:#23
Palomo I, Montes C, Martín-López B, González JA, García-Llorente M, Alcorlo P, García-Mora MR (2014) Incorporating the social-ecological approach to protected areas in the Anthropocene. Bioscience 64:181–191. https://doi.org/10.1093/biosci/bit033
Parry L, Peres CA (2015) Evaluating the use of local ecological knowledge to monitor hunted tropical-forest wildlife over large spatial scales. Ecol Soc 20:#15. https://doi.org/10.5751/ES-07601-200315
Pérez-Ibarra I, Janssen MA, Anderies JM (2016) Food security in the face of climate change: adaptive capacity of small-scale social-ecological systems to environmental variability. Glob Environ Chang 40:82–91. https://doi.org/10.1016/j.gloenvcha.2016.07.005
Peterson GD, Cumming GS, Carpenter SR (2003) Scenario planning: a tool for conservation in an uncertain world. Conserv Biol 17:358–366
Phillips A (2004) Turning ideas on their head: the new paradigm for protected areas. Environ Hist 9:173–198
Reed MS (2008) Stakeholder participation for environmental management: a literature review. Biol Conserv 141:2417–2431. https://doi.org/10.1016/j.biocon.2008.07.014
Roberts N, Andersen DF, Deal RM, Garet MS, Shaffer WA (1983) Introduction to computer simulation: the systems dynamics approach. Addison-Wesley, Reading

Sanga GJ, Mungatana ED (2016) Integrating ecology and economics in understanding land-use changes externalities internalization in water catchments. Ecol Econ 121:28–39. https://doi.org/10.1016/j.ecolecon.2015.11.011
Solecki W, Oliveri C (2004) Downscaling climate change scenarios in an urban land use change model. J Environ Manag 72:105–115. https://doi.org/10.1016/j.jenvman.2004.03.014
Sterman JD (1984) Appropriate summary statistics for evaluating the historical fit of system dynamics models. Dynamica 10:51–66
Sterman JD (2000) Business dynamics: systems thinking and modeling for a complex world. Irwin/McGraw-Hill, Boston
Taylor TR, Ford DN, Ford A (2010) Improving model understanding using statistical screening. Syst Dyn Rev 26:73–87. https://doi.org/10.1002/sdr.428
Tengö M, Belfrage K (2004) Local management practices for dealing with change and uncertainty: a cross-scale comparison of cases in Sweden and Tanzania. Ecol Soc 9:#4
Tenza A, Pérez I, Martínez-Fernández J, Giménez A (2017) Understanding the decline and resilience loss of a long-lived social-ecological system: insights from system dynamics. Ecol Soc 22(2):#15. https://doi.org/10.5751/ES-09176-220215
Tenza A, Martínez-Fernández J, Pérez-Ibarra I, Giménez A (2019) Sustainability of small-scale social-ecological systems in arid environments: trade-off and synergies of global and regional changes. Sustain Sci 14:791–807. https://doi.org/10.1007/s11625-018-0646-2
Therville C, Mathevet R, Bioret F, Antona M (2018) The challenges and opportunities of transboundary cooperation through the lens of the East Carpathians Biosphere Reserve. Reg Environ Chang 18:607–618. https://doi.org/10.1007/s10113-017-1231-4
UNEP-WCMC, IUCN and NGS (2018) Protected Planet Report 2018. UNEP-WCMC, IUCN and NGS, Cambridge/Gland/Washington, DC
United Nations Educational, Scientific and Cultural Organization (UNESCO) (2016) Ecological sciences for sustainable development: World Network of Biosphere Reserves (WNBR). UNESCO, Paris. [online] URL: http://www.unesco.org/new/en/natural-sciences/environment/ecological-sciences/biosphere-reserves/world-network-wnbr/
van der Leeuw S, Wiek A, Harlow J, Buizer J (2012) How much time do we have? Urgency and rhetoric in sustainability science. Sustaina Sci 7:115–120. https://doi.org/10.1007/s11625-011-0153-1
Vennix JAM (1996) Group model building. Facilitating team learning using system dynamics. Wiley, Chichester
Vidal B, Martínez-Fernández J, Sánchez-Picón A, Pugnaire F (2013) Trade-offs between maintenance of ecosystem services and socio-economic development in rural mountainous communities in southern Spain: a dynamic simulation approach. J Environ Manag 131:280–297. https://doi.org/10.1016/j.jenvman.2013.09.036
West P, Igoe J, Brockington D (2006) Parks and peoples: the social impact of protected areas. Annu Rev Anthropol 35:251–277. https://doi.org/10.1146/annurev.anthro.35.081705.123308

Part II
Methodological Aspects

Chapter 9
The Payment of Environmental Services as an Economic and Governance Mechanism for the Conservation and Management of Natural Protected Areas

Martín Enrique Romero-Sánchez and Victor Javier Arriola-Padilla

Abstract According to the 2006 Millennium Ecosystem Assessment (MA), ecosystem services (ES) are defined as "the benefits people obtain from ecosystems" which are divided into four categories of ecosystem services: supporting, provisioning, regulating and cultural. As a strategy to protect and promote conservation of ecosystems and the services they provide, the payments for environmental services (PES) were conceive as part of a new and more direct conservation paradigm, explicitly recognising the need to bridge the interests of landowners and outsiders. The financing and payment mechanism are circumscribed according to the legal conditions of each country. The PES concept in natural protected areas (NPAs) arises as a tool to adequately internalise in individual and social decision-making the value they possess, and the well-being provided to individuals and society by the environmental services safeguarded in NPA. The resources invested in the natural protected areas have favoured their inhabitants and conservation. Likewise, they have managed to protect them through compensation to the locals, who are empowered to ensure the governance of these lands. The Payment for Environmental Services (PES) programme in Mexico has been an effort carried out by the National Forestry Commission (CONAFOR) and several important partners such as the National Commission of Natural Protected Areas (CONANP). This programme aims to promote the recognition of the value of environmental services provided by forest ecosystems, agroforestry and natural resources, in addition to supporting the creation of markets for these services inside the NPA. The implementation of PES has been an important economic and governance mechanism for the conservation and management of natural protected areas.

M. E. Romero-Sánchez · V. J. Arriola-Padilla (✉)
National Institute of Forestry, Agricultural and Livestock Research, National Centre for Disciplinary Research on Conservation and Improvement of Forest Ecosystems (INIFAP), Mexico City, Coyoacan, Mexico
e-mail: romero.martin@inifap.gob.mx; arriola.victor@inifap.gob.mx

A. Ortega-Rubio (ed.), *Socio-ecological Studies in Natural Protected Areas*,
https://doi.org/10.1007/978-3-030-47264-1_9

Keywords C6 · Deforestation · Forest conservation · REDD+

9.1 Introduction

9.1.1 Conceptualising Ecosystem Services and Environmental Services

The concept of ecosystem services (ES) was born from the environmental movement in the United States in the 1970s (Daily 1997), as well as from the growing conceptualisation of nature as a set of integrated systems (Odum 1989).

ES are defined as the components and processes of ecosystems that are consumed and enjoyed or lead to increased human well-being taking into account the demand of the beneficiaries as well as the dynamics of the ecosystems (Daily 1997; Boyd and Banzhaf 2007; Quétier et al. 2007; Luck et al. 2009; Quijas et al. 2010).

The transformation of the planet has meant a considerable contribution in terms of economic development; however, it has generated degradation in natural resources, diminishing the SE, and with these losses, it can significantly modify the current and future well-being of humanity (Daily 1997; Costanza and Farber 2002; EEM 2005; Figueroa et al. 2009).

SE and environmental services (ES) are partially equivalent. The first is used in academic contexts and some international programmes to emphasise that services are a product of the interaction between different components of ecosystems. Environmental services refers to "environment" or "environment" to harmonise with the lexicon of secretariats or ministries in the field (SEMARNAT 2003; Balvanera and Cotler 2007; Balvanera et al. 2015).

There are numerous definitions of SE or SA; Boyd and Banzhaf (2007) define them as the components of nature directly used, consumed or enjoyed bringing about human well-being. However, this definition forgets the services provided by nature that provide well-being to people indirectly, for example, carbon sequestration by forests, which provide well-being through the quality of air or the composition of the atmosphere that humans have at their disposal and use directly (Balvanera et al. 2015).

In very general terms, any good or service provided by nature, and which provides welfare to some or many people, can be considered as ES or SE (Figueroa et al. 2009). They receive a value according to the benefits they provide both to societies and individuals; thus, the concept of payment for environmental services (PES) arises, which can be described by five basic criteria:

(1) They are a voluntary transaction, where (2) a well-defined ES is purchased by (3) at least one user from (4) at least one supplier, if and only if (5) the supplier ensures the continuous provision of the ES. This last condition implies legal certainties regarding the framework of ownership and dominion of natural resources,

as well as a regulated market regime and political stability and consensus among the social actors involved (Wunder et al. 2007).

Although there are some differences between the different authors on the definition of PES, the basic principle that environmental service providers should be compensated for the cost of providing such services is maintained (Figueroa et al. 2009).

A definition proposed by Wunder (2005) states that PES constitute a voluntary transaction, where a well-defined environmental service is purchased by at least one buyer from an environmental service provider, and only if the latter ensures the provision of the traded service. Besides, it requires environmental service monitoring to determine acceptable levels of compliance and success. Currently, four types of SAs stand out (Landell-Mills and Porras 2002; Wunder 2006):

1 **Carbon sequestration and storage:** For example, a northern hemisphere power company pays farmers in the tropics to plant and maintain trees.
2 **Protection of biodiversity:** e.g. donors paying local people to protect and restore areas to create a biological corridor.
3 **Watershed protection: for** example, downstream users pay upstream landowners to adopt land uses that limit deforestation, soil erosion and flood risks, among others.
4 **Scenic beauty and recreation: for** example, a tourism company pays a local community not to hunt in a forest used for wildlife viewing tourism.

The flow and type of services provided by ecosystems depend on multiple factors, including the ES provider unit (forest, wetland, sea, grassland, agricultural land, among others) and the state of its conservation. Thus, while services can occur in any ecosystem, not all ecosystems necessarily provide services of the same quality or quantity (Kremen 2005).

PES is one of the economic incentives for conservation that promises the highest potential for replication in the future, especially in Latin America (Figueroa et al. 2009). The PES concept in protected areas (PA) arises as a tool to adequately internalise in individual and social decision-making the value they possess and the well-being provided to individuals and society by the environmental services safeguarded in PA. The main objective is to force and make explicit an adequate valuation of the ES provided through conservation carried out in PA, thus producing an efficient allocation of these services, like what a properly functioning market would do. Also, payments for the services used generate income that constitutes an incentive for those who manage the ecosystems and decide on their uses and conservation and represent a way to finance the control, surveillance and management activities of PA and their ecosystems, necessary to restrict free access to them and make possible the conditions for their conservation over time (Figueroa et al. 2009).

PA is characterised by the conservation or preservation of ecosystems in each geographical area and thus also maintain the capacity of ecosystems to continue producing the ES they generate. The International Union for the Conservation of Nature (IUCN) defines a PA as an area of land and sea primarily dedicated to the protection and maintenance of biological diversity, as well as natural and cultural resources, and managed through legal or other effective means (UICN 2008; Figueroa et al. 2009).

The protection of natural ecosystems in the PA extends for almost 140 years. Since the creation of Yellowstone Park in the United States in 1872, there has been worldwide recognition of the importance of protected natural areas, as a result of which most countries in the world have established natural areas for protection purposes (Figueroa et al. 2009).

In many countries, the establishment of PA constitutes one of the leading environmental management tools for the protection of ecosystems. Thus, PA represents an essential source of ES supply, which, in many cases, is used as an input by the tourism and hydroelectricity and water production sectors, among others, without paying the administration of protected areas for the flow of such services (Figueroa et al. 2009).

In Latin America and the Caribbean, as well as in the world, the area protected by PAs has increased. South America, for example, has more than 4000 PAs with an area of more than four million km2, representing 20% of the world's total units (UICN 2007). Likewise, Central America and the Caribbean increased the number of PAs from 261 to 709 units between 1992 and 2003 (UICN 2003; Figueroa et al. 2009).

9.2 Methodology

To obtain information about the PES, we consulted electronic sources and government agency websites and reviewed technical reports and scientific articles. A summary of PES programmes with an emphasis on Mexico's natural protected areas is given, and some examples are presented.

9.3 Results

9.3.1 The International Context of PES

In Latin America, there are several PES projects, including water services (watershed protection), carbon capture and sequestration, biodiversity conservation, recreation and scenic beauty and bioprospecting (Fig. 9.1). The financing and payment mechanism are circumscribed according to the legal conditions of each country, so it is not standardised since it is mediated between the parties that buy and sell environmental services (Figueroa et al. 2009).

The downstream market for hydrological services has been based more on local and regional schemes. This market recognises relationships by taking the watershed as a unit. This type of service is recognised in countries such as the United States,

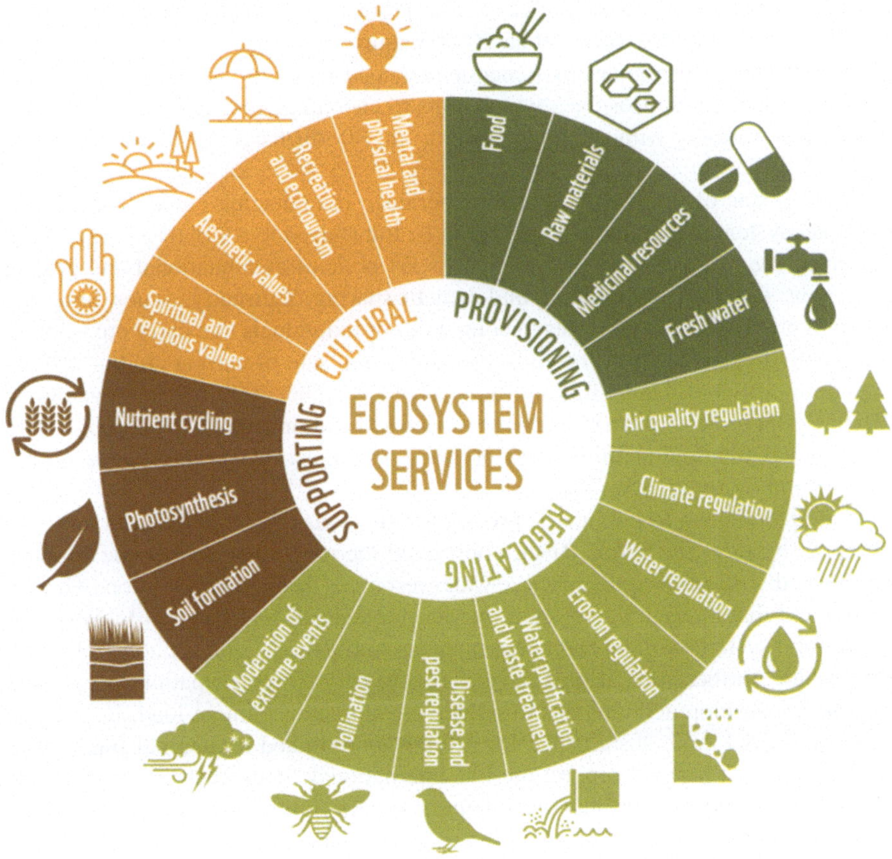

Fig. 9.1 Ecosystem services. (Source: WWF 2016)

Ecuador, Costa Rica, Mexico, Brazil and Honduras; this system recognises the availability and quality of water for human consumption, but also hydroelectric generation (Vargas-Guillén et al. 2009).

The ES market for biodiversity has been based on indirect actions, such as the recognition of conservation areas, since the aim is that, through their permanence, flora and fauna of the ecosystems are maintained or biological corridors are promoted as connectors that favour the mobility of species and their distribution, which is expected to reduce the risk of their loss. However, the experience of this market is not extensive, since in some cases the pharmaceutical industry and non-governmental conservation organizations (NGOs) are the primary beneficiaries of this service (Vargas-Guillén et al. 2009; Esquivel 2013).

Costa Rica was the first country to develop the PPSA. In 1997, through FONFIFO (Fondo Nacional Financiamiento Forestal), it promoted its PPSA programme, a fund created with taxes collected through fuel. Forest law 7575 recognises the following environmental services: climate change mitigation actions; protection of water for urban, rural and hydroelectric use; protection of biodiversity; and protection of scenic beauty (Malavasi et al. 2003; Esquivel 2013).

In 2008, the Government of Ecuador launched its environmental service programme called "Socio Bosque", which aims to encourage resource owners to conserve their forests, with a compensation of 30 dls/ha/year (Manzano 2010; Esquivel 2013).

Brazil is one of the countries with experience in managing environmental services (ecosystem) programmes, which can be transferred to a REDD+ programme. Due to its size, this country has defined its ES policies at subnational and project level, thus ensuring compliance and benefit-sharing. Currently, states have the power to grant commercial rights to companies or individuals that carry out reforestation activities and with it the right to manage the sale of carbon credits derived from these activities. Under this authority, as of December 2009, at least seven PES/REDD+ pilot projects were moving forward to sell carbon credits, this through state and national funds, donations from private companies, NGOs and philanthropic organisations (Costenbader 2011).

In the Dominican Republic, the project for the management and conservation of the upper basin of the Yaque River, considered the most important in the country, was located. The aim was to conserve water resources through the creation of a PES system and that through this system the development of forestry, agroforestry and agricultural systems will be promoted in an environmentally and economically viable manner; furthermore, it will have a system of local participation to promote community development, all this to reduce erosion and sediment load. In order to make the model more sustainable, local funds were linked to national funds from the Dominican Energy Corporation and international funds such as the German Cooperation Fund. The project considered a system of analysis and monitoring to create transparency and efficiency of the measures adopted in the project (González and Riascos 2007; Ezquivel 2013).

El Salvador has had some PES experiences despite not having consensual government support. One of these is El Imposible National Park, where families who benefit from the drinking water system pay a monthly fee. With the resources collected, it is possible to finance the administration, operation and maintenance of the system and the work of two park rangers, who are environmental promoters (Ezquivel 2013).

Latin America has considerable geographic variation in the estimated value ecosystem services. According to Siikamäki et al. (2015), there is a potential for PES in Latin America where the value per ES ranges from near zero to several thousand per hectare (Fig. 9.2).

China's "Sloping Land" programme considers a spatial selection of land and the application of differentiated payments. It considers farmers who transform their arable land, which is located in river basins, into forests. For the selection of the lands, the spatial component and the level of payment they will receive are combined; the combination of indicators is justified under the need to optimise resources

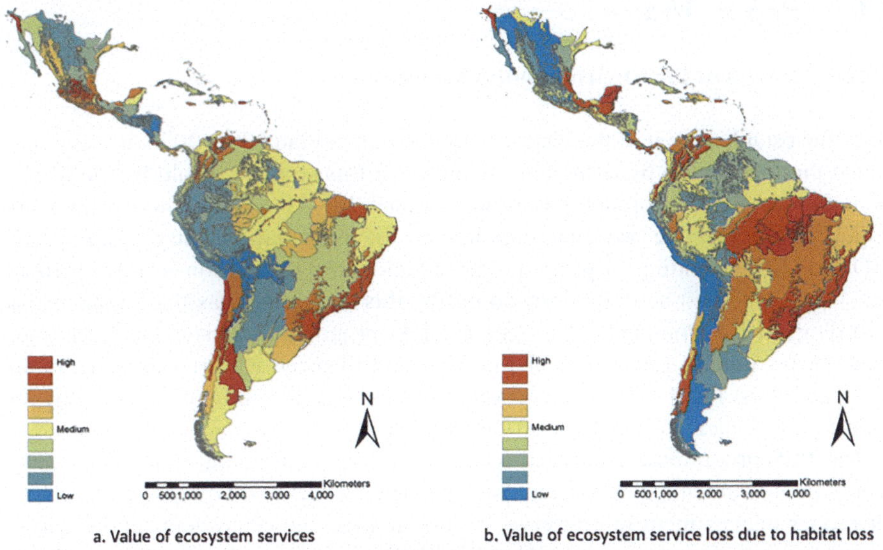

Fig. 9.2 Predicted value (per hectare) of ecosystem services and loss in ecosystem services due to habitat loss, 2000–2010. (Source: Siikamäki et al. 2015)

and apply benefits with the measurement of opportunity costs, to provide more significant environmental benefits at a low cost. Programme implementers contemplate that, by applying this cost differentiation, they increase their transaction costs; however, they also recognise that the increased profitability of differentiated payments could cover the increase in transaction costs (Bennet 2008; Herbert and Tepper 2012; Esquivel 2013).

Concerning funders, the international public sector is an essential avenue for injecting funds to cover PES programmes. In this regard, the World Bank has provided sustainable sources to finance several Latin American experiences, such as the BioCarbon Fund, which is an international source of payments to offset carbon emissions from land-use change, including payments for activities permitted by the Kyoto Protocol (FAO 2007). Similarly, the Global Environment Facility (GEF), which acts on behalf of the United Nations Framework Convention on Climate Change (UNFCCC) to conserve global public goods, provides funding to entities that support PES programmes (Figueroa et al. 2009).

In terms of the proportion of the total environmental budget devoted to conservation issues in PAs, Costa Rica and Ecuador allocate between 10 and 20% and Colombia and Uruguay more than 20% (UICN 2003; Figueroa et al. 2009).

9.3.2 PES in Mexico

9.3.2.1 Payment for Environmental Services

Since the establishment of the National Development Plan 2001–2006, it was determined that one of the fundamental principles of this strategy would be the implementation of a new approach based on sustainability. One of the modifications to achieve this objective was the creation of the National Forestry Commission (CONAFOR), intending to promote conservation and restoration in forestry matters. One of the first activities carried out by this institution was the Payments for Hydrological Environmental Services (PHES) programme, designed by the then National Institute of Ecology (INE) in 2003 and financed with resources from the Mexican Forest Fund for the protection of basins from the amount collected by the collection of Federal Water Rights (CIDAC 2014).

The PES programme emerges as a public policy instrument aimed at creating economic instruments to reward and compensate owners and holders of forest land for proper management to conserve, restore and use the ecosystems and environmental services they provide (CONAFOR 2018).

In the beginning, it focused on highlighting the importance of preserving the forest soil for the supply of the aquifers, since, otherwise, the capture and filtration of water is hindered. To that end, payments were directed to the preservation of temperate and tropical forests and, particularly, cloud forests or mesophilic mountain forests, which are associated with the maintenance of hydrological and nutrient cycles. However, at present, the HSP maintains a broader range of objectives (CIDAC 2014):

1. To reduce poverty rates in forest areas through a subsidy that also promotes adequate natural resource management
2. To promote the conservation and sustainable use of forest resources
3. To contribute to the productivity of forest goods without damaging the conservation of resources, as well as to improve the quality of life of their owners

With the implementation of the Kyoto Protocol, a body responsible for reviewing and taking decisions on the implementation of agreements issued by the United Nations Framework Convention on Climate Change (UNFCCC), called the Conference of the Parties (COP), which has met annually since 1995, was integrated. This body issued the REDD concept through "a group of scientists who developed the mechanism as a national approach to reducing deforestation, called compensated reductions" (UN-REDD Programme 2009a, b). The basic concept of the REDD (reducing emissions from deforestation and forest degradation) proposal is that countries would be compensated for the measurable reduction in their deforestation rate (compared to a historical national reference level of deforestation), where the country would generate credits that it could sell on the carbon markets if its deforestation rate were below the reference rate. This mechanism succeeded in arousing the interest of the carbon markets, which were mainly directed towards commercial plantations since at the same time as it added to the forest area, it captured CO2 in one of the stages of highest tree growth. However, it soon became apparent that a broader mechanism needed to be designed to prevent leakage, including the possibility of stimulating natural forest restoration processes. In this context, REDD was promoted, whose basic approach is that "those countries willing and able to reduce emissions by avoiding deforestation and degradation of their forests should be financially compensated for such actions" (UN-REDD Programme 2009a, b; Gerez-Fernández and Pineda-López 2011).

Every year, around 13 million hectares of forests are destroyed worldwide, making deforestation one of the most important sources of emissions, as it releases more carbon dioxide into the atmosphere than the entire transport sector (land, air and sea) worldwide. That is why CONAFOR designed a strategy to reduce emissions from deforestation and forest degradation in Mexico (known internationally as REDD+); through which the causes and agents that cause deforestation and degradation of our forests will be addressed while promoting carbon conservation in forest ecosystems and its increase through sustainable forest management (CONAFOR 2011).

With the implementation of the UN-REDD Programme, the governments of Norway and Mexico signed a collaboration agreement in 2010 on environmental, forest and climate change issues, which provides for cooperation in specific areas involving the implementation of strategies and policies to reduce emissions from degradation and deforestation and focuses on conservation, sustainable forest management and the enhancement of forest carbon stocks. The agreement is based on three main actions: (a) the development and implementation of a reliable methodology and standards for monitoring, reporting and verification as part of REDD+ and the post-Kyoto (post-2012) climate regime; (b) enabling Mexico to play a critical role in the Latin American region, specifically in Central America as a centre for South-South cooperation to promote the exchange of successful experiences and capacities in forest monitoring systems (MRV) and "lessons learned" on the implementation of the REDD+ mechanism; and (c) identifying local incentives and investigate financing options for REDD+ based on case studies and experiences available in Mexico. This cooperation project with Norway focuses on the development of an accurate and reliable methodology and standards for VMR, based on Mexican experiences, which can be later transferred to other countries in the region, including an intense training and capacity building component (UN-REDD Programme 2009a, b; CONAFOR 2011; Gerez-Fernández and Pineda-López 2011).

In conclusion, REDD+ is one of the issues under negotiation within the United Nations Framework Convention on Climate Change, which seeks to promote the development of comprehensive strategies that encourage adequate management of forest ecosystems and provide sufficient incentives to local communities, which own these ecosystems, to slow down the processes of deforestation and degradation (CONAFOR 2011; Gálmez 2013).

This strategy, based on the history of deforestation and degradation over the past 10 years at least, will establish the expected levels of reduction, through the implementation of activities either at the national or regional level, as deemed appropriate. The results in the reduction of emissions will be monitored through a national system based on the use of satellite images in combination with the National Forest and Soil Inventory, which is updated annually by CONAFOR; likewise, the accounting of reduced emissions will be done at the national level, thus seeking to ensure that there are no leaks. Besides, options for obtaining the necessary funding for the implementation of activities may come from markets, funds or a combination of both, with distribution mechanisms that, on a national scale, would replicate the experiences of the Payment for Environmental Services Program and the use of the Mexican Forest Fund (CONAFOR 2011).

There are two distinct elements of PES in Mexico that contribute as learning to strengthen the development of benefit-sharing mechanisms for REDD+ in the country: On the one hand, the differentiation of payments by ecosystem type allows for compensation of the opportunity cost incurred by forest landowners in avoiding deforestation or forest degradation caused by the pressure of land-use change due to various economic activities. These payments are made in different amounts depending on the risk of loss of the hedge (CONAFOR 2011).

On the other hand, the strategy for the development of local mechanisms of payment for environmental services allows the financing of conservation and management activities of forest ecosystems from the contributions of the direct users of environmental services. Although the PES in Mexico has been successful for forest conservation and represents an instrument for the development of REDD+ in the country, it should be considered that the latter has a broader scope, which involves cross-cutting strategies to create active institutional and governance structures in a sustainable rural development logic that allows for the reduction of deforestation and forest degradation (CONAFOR 2011).

Given the above, there is an urgent need to develop and test cost-effective and reliable methodologies to implement MRV systems at different scales: national, subnational and local. Mexico provides an ideal pilot area for the development and testing of methodological efforts that can lead to guidance documents for MRV systems and REDD+ readiness activities; it also has an operational national forest inventory, but no data on land-use change. One of the objectives of Norway's project, in the medium term, is to bring Mexico into the second phase of REDD+ implementation by enabling the design and implementation of a forest monitoring system for MRV based on remote sensing. This system should make it possible to report annual changes in forests and land use and changes.

During the 16th Conference of the Parties (COP 16) in Cancun, Mexico, the host country presented the document REDD+ Vision ("Mexico's Vision on REDD+: Towards a National Strategy"). It includes targets for zero net land-use change emissions and significant reductions in forest degradation rates by 2020.

This last strategy offers a new window of opportunity to influence the conditions of Mexican forests since forest owners and associated professionals could access funds for projects that link the conservation of environmental services with an orderly and long-term use of forests, generating local jobs and income. The relevance of this mechanism is that it offers the possibility of articulating several priority aspects for the country, such as social and community development; diversification of alternatives for biological conservation, soil, water, landscape and forest; as well as the production of goods such as wood and others that generate regional production chains (CEIBA 2010). However, for this to be operational, it will be necessary for intervention actions, as well as objectives and goals, to be established following regional and even local conditions, as there are calls for attention to the negative social implications of a lack of active participation and commitment by forest landowners to control land-use change and other causes of deforestation and degradation (Gerez-Fernández and Pineda-López 2011).

Recently, national legislation, in the Official Gazette of the Federation (DOF 2018), issued the General Law of Sustainable Forestry Development and amended the first paragraph of Article 105 and added a second paragraph to the same article of the General Law of Ecological Balance and Environmental Protection. In Articles 13 numeral XXII, 27 and 139, it refers to the payment of environmental goods and services. According to the last one, the Mexican Forest Fund will be the instrument to promote "...promote the conservation, increase, sustainable use and restoration of forest resources and their associated resources, facilitating access to financial services in the market, promoting projects that contribute to the integration and competitiveness of the productive chain and developing the mechanisms of collection and payment of environmental goods and services. To ensure more efficient management of the Fund's resources, the services of private banks may be used".

Historically, Mexico has established a diversification of its operating models, which include public and private entities, the Biodiversity Heritage Fund with international donations and local PES mechanisms through concurrent funds. In 2003, the PHES started activities with US$20 million derived from water taxes under Article 233 of the Federal Law transferred by the National Water Commission (Conagua) to CONAFOR. The PHES considers payments to ejido landowners and agrarian communities, as well as individual landowners, to maintain forests in hydrologically important areas (Esquivel 2013; CONAFOR 2018).

In the context of the PES, brigades are integrated to prevent and fight fires and monitor pests and diseases; also, part of the resources can be invested in sustainable, productive projects, with a focus on zero deforestation according to the conditions of the property and other activities that pay for their development needs (CONAFOR 2018).

After operating the National PES for several years, CONAFOR detected that a single scheme could not cover all the needs and conditions caused by the country's cultural, ecosystem, economic and political diversity. With this perspective, in 2008 the scheme of Local PES Mechanisms through Concurrent Funds (MLPSA-FC) was designed, which allowed putting into operation schemes with innovative guidelines that provided flexibility in the agreements between environmental service providers with the private sector and local governments. The local PES mechanisms that are established or strengthened should have a vision of watersheds and biological corridors, and it is recommended that the proposed mechanism be based on The Economics of Ecosystems and Biodiversity (TEEB) initiative (CONAFOR 2018).

This Commission operates its strategy with different policies and schemes: (I) the national programme of payment for environmental services, (II) the biodiversity

Table 9.1 Payments for environmental services (PES) programmes

Year of operation	Collaboration and financing schemes
2003	PHES: Payment for environmental hydrological services
2004	PSA-CABSA: Programme for the development of markets for environmental services for carbon capture and biodiversity derivatives and to promote the establishment and improvement of agroforestry systems
2006	PSAB. Payment for environmental services biodiversity, aimed at promoting markets or similar mechanisms to channel payment from users to providers of forest environmental services using a World Bank loan and a grant from the global environment facility (GEF)
2007	ProArbol: Programme to restore, manage and conserve forest resources, as well as the orderly and sustainable use of them. Collects and improves the experiences of PHES, CABSA and PSAB to include them as four of its five support concepts: hydrological environmental services, conservation of biodiversity, agroforestry systems with shade crops and development of the idea of the carbon sequestration project
2008	MLPSA-FC: Local payment mechanisms for environmental services through concurrent funding
2011	BPF: Biodiversity trust fund (area 1)
2012	WB: World Bank
2015	PRONAFOR: National Forestry Programme
2018	BIOCOMUNI: Guide to community monitoring of biodiversity

Source: Torres (2010), Iglesias et al. (2010), Esquivel (2013), CONAFOR (2018) and CONANP (2019a)

patrimonial fund and (III) the creation of local mechanisms of payment for environmental services through concurrent funds, all within the framework of ProArbol (CONAFOR 2011). Historically, in Mexico, various collaboration and financing schemes have been implemented to strengthen PES (Table 9.1).

In Mexico, unlike other Latin American countries, the PES programme has diversification in its operational models, being these the Payment for Environmental Services Program; the Local PES Mechanisms through Concurrent Funds (MLPSA), where public and private entities collaborate; and the Biodiversity Patrimonial Fund with international donations. In 2004, the Program to Develop the Market for Environmental Services through Carbon Capture and Biodiversity Derivatives and to Promote the Establishment and Improvement of Agroforestry Systems (PSA-CABSA) was adhered to; however, the latter was only in operation for 3 years, with only the PASH remaining in effect (Esquivel 2013; CONAFOR 2018; CONANP 2019a).

Starting in 2006, with partial financing from the World Bank and the GEF (Global Environment Facility), Mexico signed the Forest Environmental Services Project (FESP), in order to strengthen existing programmes and promote new schemes with local, national and international financing. With ProArbol, 2007, attention to the forestry sector has been increasingly strengthened to date. One indicator of this is the unprecedented increase in the budget allocated to the sector, the most significant proportion of which corresponds to direct incentives to support the incorporation of areas into conservation schemes through payment for environmental services; the incorporation of areas into sustainable forest management and the diversified use of ecosystem goods and services; support for the establishment of commercial forest plantation projects; protection against fires, pests and forest diseases; and the restoration of degraded forest areas and the recovery of their plant cover through reforestation and soil conservation works (Torres 2010). With these elements, the aim is to value the goods and services provided by forest ecosystems and to offer development alternatives and improvement in the quality of life of the owners and holders of forest resources, with an adequate forest management scheme, in order to discourage changes in land use.

Both programmes (ProÁrbol and Cabsa) were designed to recognise the environmental services provided by forest ecosystems, which usually are not rewarded, such as water quality, climate regulation, landslide prevention, soil formation, maintenance of biodiversity, carbon sequestration and scenic beauty, among others. In this way, economic incentives are provided to forest landowners (ejidos, communities and smallholders) to promote conservation practices and avoid land-use change (deforestation and degradation) in forest areas, through the design of the best management practices programme in each of the participating forest communities (Esquivel 2013; CONAFOR 2018; CONANP 2019a).

The Biodiversity Patrimonial Fund (area 1) was constituted to generate a long-term financing scheme that allows the conservation of forest ecosystems that shelter biodiversity of global importance. This fund began with a seed capital of ten million dollars, coming in equal parts from the Global Environment Facility (GEF) and the Government of Mexico through CONAFOR. In 2011, it will increase by $10 million more, in the same proportion. This innovative scheme will only use the interest generated by the capital to carry out PES in the eligible areas identified with a regional focus and an emphasis on biological corridors. The fund will act through investment packages for each regional eligible area, seeking to trigger more significant investment from other private or public sources in the same area for conservation purposes. Resources from the Biodiversity Heritage Fund will be allocated through a technical committee, which has been formed by both public and private sector institutions with extensive experience in biodiversity conservation (CONAFOR 2011).

By 2018, 218 collaboration agreements have been signed that have contributed to the conservation of more than 726,000 hectares in the country (CONAFOR 2018).

The National Forestry Programme (PRONAFOR) 2014–2018 is aligned with the Sectoral Program for the Environment and Natural Resources 2013–2018 (Promarnat). Under its scheme, annual calls are issued through the Rules of Operation, inviting the participation of environmental service providers (owners or possessors of land) interested in becoming beneficiaries of the PES. Participants with positive applications (subject to the programme's annual authorised budget) sign a 5-year agreement with CONAFOR, with which they commit to implementing a work programme based on the technical document for participatory planning, called the Best Management Practices Guide (GMPM). During the time of support,

Table 9.2 PES implemented in different entities of the country

State, entity or municipality	PSA
Chiapas jungle	Carbon capture payment
Oaxaca	The beauty of the coastal landscape
Oaxaca	Carbon capture payment
Quintana Roo	Hydrological performance
Sierra de Coahuila	Hydrological
State of Mexico and Michoacán	Monarch fund
State of Queretaro and San Luis Potosí	Carbon capture payment
Veracruz	Hydrological

Source: Hesselbach (2009), Galindo (2010), CONAFOR (2011) and Charchalac (2012)

it is considered that the supported participants can count on technical assistance, allowing the development of capacities of the beneficiaries and their families (CONAFOR 2018).

In Mexico, different PES initiatives have been created in different entities (Table 9.2). Likewise, different projects have been developed such as in the Monarch Butterfly Biosphere Reserve, in indigenous communities in Oaxaca; in the Alto Nazas Basin, Durango, the Yucatan aquifer; and in the Pixquiac River Basin, Veracruz, among others (CONAFOR 2011).

As an example, the State of Mexico initiated a Payment for Water Services Program in 2007 (FIPASAHEM), to ensure water supply for one of the most populated entities in the country, protecting and conserving forests and maintaining the recharge capacity of aquifers (GEM, Gaceta del Gobierno 2007; Zarza 2010). This programme was structured with the background of CONAFOR's PES programmes and with references from programmes operating in Costa Rica. The objective of payment for environmental services in the State of Mexico is to maintain forest conditions and provide the necessary hydrological service for areas with high population density and industrial and tertiary activities, including the livelihood of the inhabitants. Although due to the deteriorating conditions suffered by the forests of the State of Mexico and the volumes of drinking water required to meet the demand of its inhabitants, it was essential to establish short-term actions to guarantee the provision of hydrological services; it is also true that this programme, like others in the Latin American region, did not have a robust environmental diagnostic base that would make it possible in the medium and long term to calibrate the operation of the programmr and indicate, quantitatively and qualitatively, the maintenance or increase in the production of the hydrological service subject to payment (Alberto-Villavicencio 2009; Alberto-Villavicencio 2011; Zarza 2010).

A methodology for a PES system has been proposed in a novel natural resource management policy given the environmental challenges of the entity, the constant increase in users of environmental services and the levels of marginality of the populations established in the producing ecosystems. The principles of the system are the guidelines to follow for its development, monitor its results, evaluate its efficiency and reorient its processes (Alberto-Villavicencio 2009; Zarza 2010).

It is conceived as a set of concatenated elements that interrelate generating cycles and feedback processes. It is made up of elements that form integrated phases with each other because of the relationships they have and their attributes. Each phase provides information and analysis inputs for others. The PES system is characterised by the relationships established between components and phases, levels of activity processes that develop between them and their degree of complexity. The proposed system can be understood as the arrangement of interrelated components that form a complex whole, which alludes to an evolutionary and iterative process (Alberto-Villavicencio 2009, 2011).

It also meets multiple criteria: environmental, social, economic, cultural, political, administrative (land tenure), legal, scientific and democratic. The system identifies the direct and indirect relationships that exist between its components and phases. The processes should also been identified within and between the phases; the components, which intervene in each phase; the cycles that make up the system; its inputs, outputs and feedback processes (Alberto-Villavicencio 2009).

The ultimate goals of the system are to maintain or increase the production of environmental services, to promote local development and to define a new and sustainable long-term policy for the management and conservation of ecosystems (Alberto-Villavicencio 2009).

It is conceived as an open system, where information inputs, analysis or incorporation of other variables and adjustments to the phases or processes of feedback are necessary, according to the evolutionary process and the transformations of the ecosystems, as well as the development processes of the population that interacts in them (Alberto-Villavicencio 2009).

The proposed system is composed of four phases: geographic information system; operation of payment for environmental services (which includes evaluation and monitoring); a market for environmental services; and efficiency of payment for environmental services. The phases do not follow a strict order, each requiring inputs or outputs from others, thus establishing input, output or feedback relationships between them and their components (Alberto-Villavicencio 2009).

Technological instruments and tools are another component of the system, necessary for environmental studies, economic valuation and administration of financial resources. Political management and governance are fundamental because decisions on resource administration and development of actions of the various stages, among others, depend on the political decisions taken by the administrators or managers of a PES system. We point out some critical elements for the development of the PES system proposed here: a solid information base, environmental, population, economic, social, legal and political-administrative. Technological instruments, scientific methodologies, equipment and highly qualified personnel (Alberto-Villavicencio 2009).

In the period 2013–2018, 2.6 million hectares have been incorporated into conservation, distributed in the following ecosystems: forests (49%), jungles (35%), drylands (13%) and Mangroves (3). CONAFOR allocated more than 5.4 billion pesos that benefited 3653 properties. Due to the active conservation, management and implementation of good forestry practices, the beneficiaries invested more than 600 million pesos to generate approximately one million wages that generate economic and social welfare (CONAFOR 2018). Based on the Land Use and Vegetation Charts prepared by INEGI, CONAFOR's analysis shows that the rate of net forest loss has decreased by half between 2005 and 2010 in Mexico (CONAFOR 2011; Ruíz-Jimenez and Valtierra-Pacheco 2017).

However, one of the main problems faced by the PHES is the lack of alternatives for the forest owners, once the programme has come to an end. There are very few cases where local payment for environmental services schemes have been generated or where access to voluntary payment for environmental services schemes has been facilitated. Therefore, after the 5 years of programme support, there is no guarantee that producers will continue to maintain the supported land and not return to the conditions they had before participating in the programme or that they will begin to harvest wood (Ruíz-Jimenez and Valtierra-Pacheco 2017).

Ruíz-Jiménez and Valtierra-Pacheco (2017) analysed the situation of the ejido forest lands of Texcoco, State of Mexico, after having participated in the PHES from 2005 to 2010. In the beginning, the three forest lands were classified as "In Rest"; that is, they were not harvested for timber. During this period, the forest conditions of the three properties participating in the PHES improved due to soil conservation and reforestation works and because during the 5 years no timber was harvested. In 2011, the ejidos decided to restart timber harvesting on two of the benefited properties when the payment for environmental services ended. However, the third property was preserved as a conservation area. Then, the payment made by CONAFOR through the PHES does not guarantee the conservation of forests or the provision of environmental services in the long term.

The ejidos decide to use their land again because the money they receive from the PHES is very little and they only manage to pay for some conservation activities and works that are not enough to give the necessary care and maintenance to the forest. None of the resources granted by the PHES was distributed directly to the ejidatarios' pockets. They were only paid wages to carry out conservation work on the benefited properties. Here, the ejidatarios' dilemma is to conserve without receiving significant economic benefits or to use these forest-oriented lands to harvest the wood and receive income. The national forestry policy does not support the idea that "a forest or a piece of land that is being harvested is capable of providing the same environmental benefits as one that is not touched. That is, 'Active Conservation' or under proper management is not acceptable for CONAFOR's PHES" (Ruíz-Jiménez and Valtierra-Pacheco 2017).

Arid regions in Mexico, on the other hand, are excluded from the benefits of the CONAFOR PES scheme because of their low forest cover. Priority areas for payment for environmental services equivalent to 911 ha have been identified, with characteristics of a high and very high risk of losing the ecosystem services provided by biodiversity conservation (Aguilar-Sánchez and González-Vizcarra 2019).

9.3.2.2 Natural Protected Areas and PES

In Mexico, natural protected areas (NPAs) are under the administration of the National Commission of Natural Protected Areas (CONANP). It currently manages 182 areas of federal character (Table 9.3), representing 90,839,521.55 hectares, and supports 339 areas destined voluntarily to Conservation (CONANP 2019b).

The establishment of NPAs is the central pillar of Mexico's strategy for the conservation of ecosystems and their biodiversity. Recently, complementary options have been opened for different sectors of society, owners of land with natural ecosystems, to voluntarily participate in conservation and sustainable management, with the recognition of the federal authority. Initially this was done extensively through the implementation of the Wildlife Conservation Management Unit (WCMU) programme and recently with the certification of land for conservation, empowering landowners to establish, manage and administer their own natural protected areas. This second category of NPAs has a different legal nature, being established by a certificate issued by Semarnat. The main reasons why proponents apply for certification are the conservation of local environmental services, the achievement of a premium in agroforestry products, to facilitate support from the CONAFOR PES Program, to achieve a "green" image in ecotourism projects or for the conviction of conserving ecosystems outside the legal framework of federal decrees (de la Maza 2010) (Fig. 9.3).

Today, PNAs have been resignified from being pure sources of raw material and scenarios where only the protection and conservation of biodiversity matters, to the offer of intangible benefits such as cultural ecosystem services, reflecting new links between society and nature. For example, there is currently a strong tendency to create infrastructure and conditions to attract and promote tourism in the NPAs, so that "passive conservation" can generate local income and encourage conservation. However, given the cultural and social conditions prevailing in most PNAs, it is imperative to carefully analyse the approaches, scope and impacts of these ecotourism projects, because of the implications they have on local relationships, especially in areas that can be very sensitive due to their socio-ecological characteristics. This is the case of the Monarch Butterfly Reserve, or the Copper Canyon Tarahumara,

Table 9.3 Natural protected areas in Mexico

Natural protected area	Quantity
Biosphere reserve	45
Flora and fauna protection areas	40
National Parks	66
Sanctuaries	18
Natural resource protection areas	8
Natural monuments	5
Total	182

Source: CONANP (2019b)

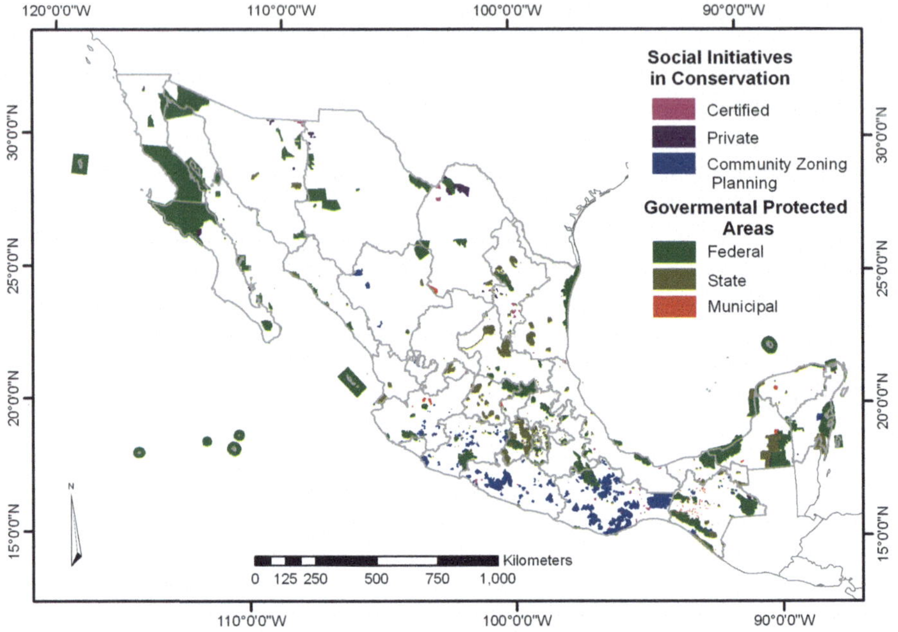

Fig. 9.3 Protected areas of Mexico. (Source: Ochoa-Ochoa et al. 2009)

among others, where the socio-economic development objectives exceed the biological conservation objectives (Galicia et al. 2018).

In order to integrate this new socio-ecological perspective in the PNAs, it is necessary to consider the causes of the environmental problems in order to influence management for protection and conservation, that is, to understand how humans (the different actors) and our institutions are the drivers of change. Therefore, the inclusion of a socio-ecological analysis framework in protection and conservation plans is crucial for at least three factors: (1) it would provide more information about the trajectory of change of the landscape and its drivers of change; (2) it would facilitate the identification of productive and traditional activities so that management plans would have to be adjusted to community exploitation interests; and (3) management plans should recognise the importance of conserving specific areas and species, but also activities and areas of cultural relevance for cultural heritage conservation purposes, where social practices and perceptions of resource use are included (Galicia et al. 2018).

CONANP has supported the holders of certified areas so that they can have access to financing from government programs, funds from foundations and advice from various institutions for productive, tourism and research projects. Among them are the incentives provided by CONAFOR, with the Proárbol Programme, particularly the PSA, which a large number of certificate holders have signed up. The challenge for these certified areas to be maintained in the long term is to ensure that the incentives received by the certifiers are long enough for the productive projects they are starting to mature, based on ecosystem management until they generate the income needed to make them sustainable (de la Maza 2010; Torres 2010).

CONANP is an essential partner for CONAFOR in the execution and development of the ProArbol PES. It has also supported the dissemination of the programme, support to the ejidos and communities for the creation of files and the follow-up of specific cases in the Comité Tecnico Nacional. It has managed to attract resources from this programme to generate sustainable regional development, directly benefiting the inhabitants of the areas and consolidating in these communities the value of conservation for the sustainability and continuity of the mentioned services (CONANP 2019a).

Since the beginning of the PHES programme in 2003 and throughout its development and transformation, the participation of NPAs has been increasing. In 2003 it started supporting projects in 15 NPAs, 42 in 2004, 31 in 2005 and 27 in 2006, reaching a maximum of 55 NPAs in 2007. As of 2008, there were a total of 50 NPAs that have benefited from at least one of the concepts included in the programme (CONANP 2009; CONANP 2016).

The regions with the most significant presence in the PES programme are the Central and Neovolcanic Axis as well as the Western and Central Pacific, due to the coincidence between the eligibility areas designated by CONAFOR each year and the location of the NPAs (CONANP 2009).

The total number of projects executed has been increasing from 2003 with 103 projects executed to 2008 with 463, being the Regional Coastal Plain and Gulf of Mexico the one that developed more projects in 2008 (129 projects) and where 70.5% of the projects of that year were executed within the Biosphere Reserve Pantanos de Centla (CONANP 2009).

The PNA that has developed the most projects over the 6 years of the PES programme is the Sierra Gorda Biosphere Reserve with 127 approved projects (CONANP 2009).

The total implementation amounts approved for the nine regionals in the period 2003 to 2008 exceed 795 million pesos. Of these, the regional ones that received the most support in this period were the Central and Neovolcanic Axis with 27.2%, followed by the Western and Central Pacific with 26.58% of the resources. The year 2008 has been the year that more support has been received with a total of 339′142,130.49 pesos. Moreover, the PNA that has executed the most resources in 6 years is the Natural Resources Protection Area of the Feeding Basin of the National Irrigation District 043, State of Nayarit (CONANP 2009).

From 2003 to 2008, a total of 996 projects have been approved and executed in a total area of 455,909 hectares and with a total investment amount of $795,350,419.47. The PNA in which most projects have been developed during the 6 years of the PES programme is the Sierra Gorda Biosphere Reserve (127 projects). The PNA that benefited most in terms of surface area from this programme was the Natural Resources Protection Area of the Feeding Basin of the National Irrigation District 043, State of Nayarit (32,162.87 ha), as well as the area that has received the most resources ($61′275,993.08). The total number of projects implemented has been increasing throughout the project from 2003 (103 projects) to 2008 (463 projects) (CONANP 2016).

Below we present PES experiences in two natural protected areas:

Monarch Butterfly Biosphere Reserve (MBR) The reserve is one of the most important protected areas in Mexico and, at the same time, one of the most complex due to its social conflicts. In this area of 56,259 hectares, the population of monarch butterflies migrating from the United States and Canada spends 5 months a year. Since 2000, most of the owners of the core zone have been given financial incentives by the Monarch Fund to conserve their forests. This strategy was built with the participation of all stakeholders, the federal government, the ejidatarios who own the land, the World Wildlife Fund (WWF) and the Mexican Fund for the Conservation of Nature (FMCN), and began its operation with a donation from the Packard Foundation of the United States, the federal government and the governments of Michoacán and the State of Mexico (Galindo 2010). In response to the increase in illegal logging documented in 2003, WWF's Monarch Butterfly Program has promoted several initiatives. In addition to monitoring activities of the forest and butterfly colonies, the Regional Monarch Butterfly Forum was organised as an annual space where all institutional and social actors meet to analyse, discuss and propose coordinated solutions to regional problems. The Monarch Forum has increased institutional collaboration, the joint proposal of priorities in aspects of social participation, productive projects, research and environmental education. It has allowed direct interaction of communities in the region with the authorities and the continuous request to intervene against organised illegal logging. Also, it allowed for more significant state investment and finally for financial competition between various organisations, including payment for environmental services from the National Forestry Commission (CONAFOR 2011).

For every peso that CONAFOR puts per hectare of well-conserved forest, the Monarch Fund puts 1.21 pesos. The Monarch Fund is a mechanism to channel economic incentives to owners who conserve their forests in the core zone of the Monarch Butterfly Biosphere Reserve. Of the 93 properties, 40 are located in the core zone of the reserve, of which 32 participate in the Monarch Fund. Of these, 11 belong to the State of Mexico and 21 to the State of Michoacan. In the 2010 forest monitoring conducted by the Monarch Fund, a significant reduction in deforestation was observed in the core zone, which allows for the survival of various species and the maintenance of processes such as water recharge in the upper parts of the Cutzamala System, in addition to being the site of the monarch butterfly's hibernation (CONAFOR 2011).

The experience generated and lessons learned in Mexico's Monarch Fund are highly relevant to REDD+, as discussions and experimentation with nested approaches to managing national strategies increase. Nested approaches in REDD+ and national carbon accounting are linked to subnational activities and accounting. Although there is no national biodiversity accounting in Mexico, the Monarch Fund demonstrates that subnational co-financing activities can contribute to the achievement of national objectives, a central idea in the nested approaches (Fonafifo 2012; Ezquivel 2013).

Pico de Orizaba, Veracruz In Veracruz is located the Hydrological Region number 28 Papaloapan where is located the basin of the Rio Jamapa-Cotaxtla, which is in the central part of the state. There is currently a payment mechanism for environmental services in the region, in which the drinking water operating system of the municipalities of Veracruz-Boca del Río-Medellín called the Water and Sanitation System (SAS) charges citizens a maximum fee of 10 pesos on their water bill. However, such a charge is not mandatory as the citizen can choose not to pay it. The annual resource collected is used to make payments to the landowners in the upper part of the basin to implement actions that allow the recovery of the forests in the Pico de Orizaba area and to contribute to the recharge processes of the Jamapa and Cotaxtla river basins (CONAFOR 2011).

Likewise, an inter-institutional and multidisciplinary working group was formed in which the various entities of the different orders of government, as well as representatives of private initiative, civil society and universities, analyse and determine the projects and lines of action to be carried out in the Pico de Orizaba (CONAFOR 2011).

A communication system or network has been established that allows the dissemination of the various activities undertaken, as well as the provision of environmental education workshops and promotion of alternative productive activities (CONAFOR 2011).

Of the total area of PES, 42% is located in natural protected areas, and an estimated 43% of the area corresponds to native peoples' territories (CONAFOR 2019).

In 2013 the project "Coastal Basin Conservation in the Context of Climate Change" (C6) was developed, which is an instrument to address the impacts associated with C6. It consists of a series of criteria to be followed for the analysis of activities, which will be determined in detail at the beginning of the implementation of C6. The MGAS includes a series of principles and procedures for the evaluation of environmental and social impacts of these activities. These guidelines focus on measures to develop plans to reduce, mitigate or compensate for adverse impacts and to enhance positive impacts. The MGAS for this C6 includes an Indigenous Peoples Planning Framework (IPPF) and a Procedural Framework (PF) as additional documents. The MPPI is used for the activities that C6 will develop in populations with indigenous peoples. The MP seeks to support populations living in natural protected areas (NPAs), whose access to natural resources may be limited by restrictions on the use of these resources by Management Programs focused on the conservation of biodiversity and environmental services of these spaces. The CONANP will strengthen existing NPAs and create two new ones in the C6 focus basins (Severino and González 2018).

The National Forestry Commission will support climate change mitigation actions through forest conservation through PES in the basins selected for C6, as well as sub-projects for sustainable forest management and against soil degradation. The National Institute of Ecology and Climate Change (INECC), together with

CONANP and CONAFOR, will develop action plans for integrated watershed management (PAMIC) based on the results of watershed monitoring, focusing on monitoring changes in land use, ecosystem degradation, reduction in carbon emissions and socio-economic factors. For its part, the Mexican Fund for the Conservation of Nature, A.C. (FMCN), will contribute its experience in the management and channelling of financial resources, allocation to sub-projects by call, fundraising and its relations with civil society (Severino and González 2018).

The C6 consists of five components. Component 1, which consists of the creation and consolidation of natural protected areas (NPAs), will be executed by CONANP and FMCN following the model developed in previous SINAP projects supported by the World Bank. Component 2 supports the creation of new NPAs and strengthening the management effectiveness of new and existing NPAs through the funding of biodiversity conservation activities included in Annual Operational Plans (AOPs). Components 3, 4 and 5 aim for the implementation of fundraising activities aimed at obtaining resources additional to those of the GEF, which are deposited with the FCC in order to finance biodiversity conservation activities in the NPAs and selected basins (Severino and González 2018).

9.4 Discussion and Conclusions

The progress made in PES schemes is significant; however, evaluations show limitations ranging from lack of reliable and precise information on the services provided by ecosystems, non-existence of studies on flows and hydrological balances, measurement of flows that reflect the impact of payment on the environmental service or knowing the origin of the services received by the user (Alberto-Villavicencio 2009). According to Quintero and Estrada (2006) Camacho (2008) and Alberto-Villavicencio 2009) PES systems have often been established without prior diagnostic studies, starting with the definition of regulatory frameworks or negotiations between service providers and buyers, without knowing the adequate provision of services.

According to CONAFOR (2018), the PES has proved to be a useful tool for maintaining forest ecosystems in good condition, resilient to climate change and with a lower rate of deforestation (2005–2010) and land-use change, in addition to maintaining and increasing social capital in the participating agrarian nuclei; however, it has been indicated that it does not guarantee the self-sufficiency of the communities and even, in the end, the inhabitants may incur in forest exploitation.

Likewise, it has been pointed out that the PHES in Mexico has helped to reduce poverty gradually, since it gives priority to those sites with certain conditions of marginalisation, as well as to areas owned by indigenous populations, and recognises the participation of women, all this in combination with the fact that poor and indigenous communities own most of the forest lands. According to Muñoz-Piña, et al. 2008 (Fonafifo 2012), 78% of the payments made so far have been in forests owned by people in a high or very high marginalisation situation. However, according to the National Council for the Evaluation of Social Development Policy (CONEVAL), from 2008 to 2018, the poverty rate increased by 0.5%, although it decreased by 3.6% (CONEVAL 2019).

Payments for environmental services programmes motivate the conservation of natural resources, but they are perfectible. The economic instrument that underlies this environmental policy must be accompanied by other elements of evaluation and design to achieve the objective of ensuring the care of the forest resource and the consequent provision of environmental services that promote the development of productive activities for multiple communities.

CONAFOR points out that the logic of the PES programme is an incentive to boost local and national ES markets, since it is expected that, at the end of the first 5-year period, the owners of the resources will be able to seek investors (private or public) to become the users of this ES (Ezquivel 2013).

Some of the current challenges to implement PES schemes in protected areas are the identification and quantification of the different services produced by the ecosystem or ecosystems present in the PA, the creation of sustainable financing mechanisms, the design and implementation of payment mechanisms that guarantee adequate incentives, the development and adaptation of adequate institutional frameworks and, finally, the equitable distribution of costs and benefits among stakeholders (Pagiola 2002; Figueroa et al. 2009).

The adequate conservation of the extensive areas and environmental services provided by protected areas requires the mobilisation of financial resources. These resources must be invested in ecosystem maintenance works to guarantee their operation and the continuous flow of environmental services, for which purpose such works must moderate the impact on PA ecosystems of the various changes occurring in the planet as a result of the growing human population and the increasing activities derived from them (Figueroa et al. 2009). Besides, PES in PA areas should be of international interest and benefit from international funds and, if necessary, be supported for more than 5 years to avoid the occurrence of lousy conservation and management practices.

The maintenance of the PA is done through various programs, including environmental education, tourism, monitoring and control, recovery and restoration of ecosystems. In the case of the National Systems of Protected Areas, the primary source of funding for the development of these activities corresponds to the financial contribution from the annual budget of the states, whereas the PA of private and public-private origin also receive contributions from donations and direct charges (Barzev 2005).

Accordingly, a PES will only achieve its objectives if it succeeds in influencing the way land users use the land. To this end, Pagiola (2002) suggest the following principles: continuity, direction and avoiding creating perverse incentives. Concerning the principle of continuity, benefits must be maintained over a long period (where land users must be paid once a year to maintain the incentive). As regards the principle of steering, a system of undifferentiated payments should be sought, which pays all land users equally, although it is usually less efficient (requiring higher payments to achieve the same level of benefits) than a system of targeted payments. However, a targeted payment system can be more costly to implement than a non-targeted payment system. Finally, about the principle of avoiding perverse incentives, it is noted that practices that threaten the ecosystem are not encouraged; for example, making payments for reforestation may encourage land users to cut trees in the first place, so that they can receive payment when reforestation takes place (Figueroa et al. 2009).

Despite the principles described, there are variations of PES in practice. The differences include the type of environmental service, the management model, the collection and payment mechanisms, among other elements. However, all PES schemes share the objective of ensuring the provision of environmental services, revealing the value that they have for those who demand them and estimating the willingness of the latter to pay for them, establishing institutional systems that allow for the setting of prices that adequately reflect the value of environmental services and guarantee their cancellation, in order to achieve changes that ensure the future use of the land and the use and conservation practices of the ecosystems that allow for maintaining the provision of environmental services indefinitely in time. In the case of PAs, PES then allows financing alternatives to paying for the ecosystem conservation activities implemented in them (Figueroa et al. 2009).

Discrimination about which services should be covered by payment proves to be the cornerstone in formulating an effective PES plan (Figueroa et al. 2009). This requires an assessment of the possible environmental services that would contribute to the achievement of environmental, social and economic objectives. This assessment, in turn, should be based on an understanding of the biophysical sciences and the underlying economic interests of suppliers, as well as an assessment of demand.

An ecosystem can provide a large number of services, but its economic value will ultimately depend on the availability and ability to pay by the demanders of that environmental service. Without a doubt, the first step before elaborating a PES scheme and with the objective of not generating unrealistic expectations is the analysis of the possible demanders of the environmental service(s) generated (Porras 2003; Figueroa et al. 2009).

Therefore, one of the biggest challenges for a PES system is to have a mechanism that accurately determines the environmental service to be involved in the payment, the value of this service, the beneficiaries and the willingness of buyers to pay (Figueroa et al. 2009).

In the case of public schemes, there is no direct link between buyers and sellers, as the state acts on behalf of the individual beneficiaries. However, there are other intermediaries that coordinate purchases of environmental services, including NGOs and product certifiers (Figueroa et al. 2009).

For its part, the private sector is playing an increasingly active role in payment programmes, as its incentive includes both interests in maximising sales to environmentally aware consumers and pressure from shareholders and consumers for greater corporate social responsibility (FAO 2007).

In the case of payment for biodiversity conservation, recreation or scenic beauty, compliance with the limits on visitors established by the carrying capacity of the ecosystems should be monitored, so that excess visitors and activities carried out by them do not compromise the health of the ecosystems and the provision of services (Figueroa et al. 2009).

The implementation of public policy instruments, such as NPAs and PES, requires the establishment of interactions between the different biophysical and social elements and the economic and political instruments that make up a socio-ecological system. Ecosystem services must have the capacity to demonstrate the causal relationship between change in an ecosystem attribute and a resulting measure of human or social well-being. For example, it is vital to recognise how management decisions, NPAs or PES can contribute to the provision of ecosystem services, such as carbon sequestration in soils for climate mitigation, among others (Galicia et al. 2018).

Finally, the NPAs have been resignified, going from being single sources of raw material and scenarios where only the protection and conservation of biodiversity matters, to the offer of intangible benefits such as cultural ecosystem services, reflecting new links between society and nature. However, few studies concerning protected areas have focused on ecological and socio-cultural variables in these ecologically fragile territories, to guide plans and programmes aimed at achieving both the conservation of natural resources and the use and enjoyment of resources (Galicia et al. 2018).

REDD+ can be understood as an international financial transfer mechanism to reduce net GHG emissions from forest sector activities in developing countries. The mechanism directs payments to forest owners and users, either directly or through governments at the national level in order to reduce deforestation and improve forest management (Angelsen 2009). The international debate on the most appropriate way to implement REDD+ and how global mechanisms could fit with local, national and private sector initiatives is still ongoing (Mahanty et al. 2013; Gálmez 2013).

REDD+ has followed a rapid evolutionary process and in only 8 years (2004–2011) has managed to double stakeholder involvement leading to the development of new ideas and concepts on REDD+. This growing involvement of actors and ideas responds to specific institutional processes that occurred before this period. In turn, the expansion of actors and ideas was followed by a new moment of progress towards processes of institutionalisation (Gálmez 2013; den Besten et al. 2014).

In 2005, the PHES had a "passive conservation" approach, and therefore only admitted properties considered "at rest" in the Management Plan of the three ejidos studied, although they were not the ones most at risk of being deforested. It has been demonstrated in other studies that a well-managed forest estate can provide the same or a greater quantity and quality of environmental services than a resting estate (Ruiz-Jimenez and Valtierra-Pacheco 2017), so it is essential not to discriminate between areas.

The participation of ejido lands in the PHES does not guarantee that they will be conserved in the long term because the PHES did not contribute to creating a permanent local mechanism for payments for environmental services. Despite this, the ejidos continue to carry out conservation activities on the land (Ruíz-Jimenez and Valtierra-Pacheco 2017).

In theory, if all the components of the PES scheme work, poverty eradication, development and biodiversity conservation could be incorporated into one model, tested, and benefit all sectors involved. The implementation of PES schemes is not a local issue, but an international one. Some ES such as air purification, carbon sequestration, water capture and purification are of global competition, as they affect entire regions of the planet. Also, the use and degradation of resources such as air and water are a consequence of global production processes. Therefore, efforts and funding of PES schemes should be international (Macip and Macip, 2013).

CONANP is the leading institution coordinating the Mexican Government's effort before the GEF for C6. Its role is also relevant due to the incorporation of nine NPAs in the Gulf of Mexico to C6 and others to be defined in the Gulf of California, in which CONANP will carry out the most strategic conservation activities ensuring effective and efficient use of resources. The advisory councils and sub-councils of the NPAs will be critical forums for the mobilisation of the stakeholders required for the work in the basins (CONANP 2013).

References

Aguilar-Sánchez G, González-Vizcarra R (2019) Zonas con Potencial para el Pago de Servicios Ambientales en el Valle del Mezquital, Hidalgo. Revista de Estudios Andaluces 38:162–184

Alberto-Villavicencio Á (2009) Propuesta metodológica para un sistema de pago por servicios ambientales en el Estado de México. Cuadernos Geográficos 44:29–49

Alberto-Villavicencio Á (2011) Actores y procesos del programa Pago por Servicios Ambientales Hidrológicos del Estado de México. Spanish J Rural Dev 37:46

Angelsen A, (2009) Realising REDD+. National Strategy and Policy Options. CIFOR, Bogor, Indonesia

Balvanera P, Cotler H (2007) Acercamientos al estudio de los servicios ecosistémicos. Gaceta ecológica 84–85:8–15

Balvanera P, Castillo A, Avila P, Caballero K, Flores A, Galicia C, Galindo LM, Lazos-Chavero E, Martínez Y, Maass M, Martínez L, Quijas S, Saldaña A, Sánchez M, Sarukhán J (2015) Marcos conceptuales interdisciplinarios para el estudio de los servicios ecosistémicos en América Latina. El valor ecológico, social y económico de los servicios ecosistémicos. Conceptos, herramientas y estudio de casos, pp 39–67

Barzev R (2005) Guía Metodológica: Estudios de Valoración Económica de Bienes y Servicios Ambientales y el Diseño de Propuestas de Mecanismos Financieros para la Conservación. Proyecto GEF/PNUD

Bennett MT (2008) China's sloping land conversion program: institutional innovation or business as usual? Ecol Econ 65(4):699–711

Boyd J, Banzhaf S (2007) What are ecosystem services? The need for standardized environmental accounting units. Ecol Econ 63:616–626

Camacho DC (2008) Esquemas de pagos por servicios ambientales para la conservación de cuencas hidrográficas en el Ecuador. Investigación Agraria: Sistemas y recursos forestales 17:54–66

CEIBA (Centro Interdisciplinario de Biodiversidad y Ambiente) (2010) Aplicación de mecanismos de reducción de emisiones por deforestación y degradación (REDD) en ejidos y comunidades. Programa de Competitividad de la USAID/México, México, D.F., 49 pp

Charchalac S (2012) Experiencias en compensación por servicios ambientales en América Latina (PSA o REDD+). Descripción de casos relevantes. Forest Trends, NORAD, AVINA, RRI, Washington, DC

CIDAC (Centro de Investigación para el Desarrollo, A.C.) (2014) Pagar para conservar nuestros bosques: ¿Funcionan los pagos por servicios ambientales en México? México, D.F., 62 pp

CONAFOR (Comisión Nacional Forestal) (2011) Servicios ambientales y Cambio climático. Comisión Nacional Forestal (CONAFOR). Guadalajara, Jal., México, p 76. Disponible en: http://www.conafor.gob.mx:8080/documentos/docs/24/2727DOSSIER.pdf, fecha de consulta: 30 de octubre de 2018

CONAFOR (Comisión Nacional Forestal) (2018) Capítulo 5. Pago por servicios ambientales. In: CONAFOR (ed) La Comisión Nacional Forestal y El desarrollo Sustentable de México – 2012–2018, pp 110–126

CONAFOR (Comisión Nacional forestal) (2019) Celebra la CONAFOR 16 años de la creación del programa Pago por Servicios Ambientales. Disponible en: https://www.gob.mx/CONAFOR/prensa/celebra-la-CONAFOR-16-anos-de-la-creacion-del-programa-pago-por-servicios-ambientales, fecha de consulta: 19 de diciembre de 2019

CONANP (Comisión Nacional de Áreas Naturales Protegidas) (2009) Informes de Logros [Report of achievements] Comisión Nacional de Áreas Naturales Protegidos, México

CONANP (Comisión Nacional de Áreas Naturales Protegidas) (2013) Marco de gestión ambiental y social (MGAS). Proyecto "Conservación de Cuencas Costeras en el contexto del cambio Climático (C6). CONANP, SEMARNAT, FMCN, INECC, 67 pp

CONANP (Comisión Nacional de Áreas Naturales Protegidas) (2016) Programa de pago por servicios ambientales en Áreas Naturales Protegidas. Disponible en: https://www.gob.mx/CONANP/acciones-y-programas/programa-de-pago-por-servicios-ambientales-en-areas-naturales-protegidas?idiom=es, fecha de consulta: 2 de diciembre de 2019

CONANP (Comisión Nacional de Áreas Naturales Protegidas) (2019a) Pago ́por servicios ambientales en Áreas Naturales Protegidas. Disponible en: https://www.gob.mx/conanp/documentos/pago-por-servicios-ambientales-en-areas-naturales-protegidas, fecha de consulta: 2 de enero de 2019

CONANP (Comisión Nacional de Áreas Naturales Protegidas) (2019b) Áreas Naturales Protegidas decretadas. Disponible en: http://sig.CONANP.gob.mx/website/pagsig/datos_anp.htm, fecha de consulta: 2 de enero de 2019

CONEVAL (Consejo Nacional de la Política de Desarrollo Social) (2019) Medición de la pobreza. Disponible en: https://www.coneval.org.mx/Medicion/Paginas/PobrezaInicio.aspx, fecha de consulta: 3 de enero de 2019

Costanza R, Farber S (2002) Introduction to the special issue on the dynamic and value of ecosystem services: integrating economic and ecological perspectives. Ecol Econ 41(3):367–373

Costenbader J (2011) REDD+ benefit sharing: a comparative assessment of three national policy approaches

Daily GC (1997) Introduction: what are ecosystem services. Nature's services: societal dependence on natural ecosystems. 1(1):1–10

De la Maza R (2010) Áreas naturales certificadas. In: Carabias J et al (coords) (2010) Patrimonio natural de México. Cien casos de éxito. Comisión Nacional para el Conocimiento y Uso de la Biodiversidad, México, pp 18–19

Den Besten JW, Arts B, Verkooijen P (2014) The evolution of REDD+: an analysis of discursive-institutional dynamics. Environ Sci Policy 35: 40–48

DOF (Diario Oficial de la Federación) (2018) DECRETO por el que se abroga la Ley General de Desarrollo Forestal Sustentable, publicada en el Diario Oficial de la Federación, el 25 de febrero de 2003, se expide la Ley General de Desarrollo Forestal Sustentable; y se reforma el primer párrafo al artículo 105 y se adiciona un segundo párrafo al mismo artículo de la Ley General del Equilibrio Ecológico y la Protección al Ambiente es uno de los principales. Disponible en: http://www.diputados.gob.mx/LeyesBiblio/pdf/LGDFS_050618.pdf, fecha de consulta: 15 de noviembre de 2018

EEM (Evaluación de los Ecosistemas del Milenio) (2005) Ecosistemas y bienestar humano: Marco para la evaluación. Informe del Grupo de Trabajo sobre Marco Conceptual de la Evaluación de Ecosistemas del Milenio Islans Press, Washington, DC, 235 pp

Esquivel E (2013) Mecanismos nacionales e internacionales de pago por servicios ambientales (PSA) existentes. Disponible en: http://www.Monitoreoforestal.gob.mx/repositoriodigital/files/original/dd542543089a388798fe0cbcdd3692c9.pdf, fecha de consulta: 20 de octubre de 2018
FAO (Food and Agriculture Organization of the United Nations) (2007) The state of food and agriculture: paying farmers for environmental services. Rome, 222 pp
Figueroa B, Reyes V, Rojas V (2009) Pago por servicios ambientales en áreas protegidas en América Latina. Programa FAO/OAPN. Fortalecimiento del Manejo Sostenible de los Recursos Naturales en las Áreas protegidas de América Latina. FAO. Disponible en: http://www.fao.org/3/a-i0822s.pdf, fecha de consulta: 20 de octubre de 2018
FONAFIFO (Fondo Nacional de Financiamiento Forestal) (2012) Lecciones aprendidas para REDD+ desde los programas de pago por servicios ambientales e incentivos para la conservación. Ejemplos de Costa Rica, México y Ecuador. Fonafifo, CONAFOR y Ministerio de Medio Ambiente del Ecuador, 176 pp
Galicia C-VB, Kolb M, Jasso-Flores RI, Rodríguez-Bustos LA, Solís EL, Guerra de la Cruz V, Pérez-Campuzano E, Villanueva A (2018) Perspectivas del enfoque socioecológico en la conservación, el aprovechamiento y pago de servicios ambientales de los bosques templados de México. Madera y Bosques 24:2. https://doi.org/10.21829/myb.2018.2421443
Galindo C (2010) Participación social y gobernanza para la conservación: el Foro Mariposa Monarca. In: Carabias J et al (coords) Patrimonio natural de México: cien casos de éxito. CONABIO, México, D.F, pp 232–233. Disponible en: https://www.biodiversidad.gob.mx/pais/cien_casos/pdf/cap98.pdf, fecha de consulta: 3 de noviembre de 2018
Gálmez V (2013) Retribución por servicios ambientales. Consultoría de Helvetas Swiss Intercooperation/ASOCAM para el Programa Manejo Forestal Sostenible (MFS). Disponible en: http://repiica.iica.int/docs/B3395e/B3395e.pdf, fecha de consulta 18 de noviembre de 2019
GEM (Gobierno del Estado de México) (2007) Gaceta del Gobierno: Periódico Oficial del Gobierno del Estado Libre y Soberano de México. N.° 31. Tomo CLXXXIV. Toluca de Lerdo, Méx., 13 de agosto del 2007, Estado de México, México, pp 1–3
Gerez-Fernández P, Pineda-López MDR (2011) Los bosques de Veracruz en el contexto de una estrategia estatal REDD+. Madera y Bosques 17(3):7–27
González Á, Riascos E (2007) Panorama latinoamericano del pago por servicios ambientales. Gestión y Ambiente 10(2):129–144
Herbert T, Tepper D (2012) Lecciones aprendidas para REDD+ desde los programas de pago por servicios ambientales e incentivos para la conservación. Un financiamiento sostenible para PSA/REDD+. FONAFIFO, CONAFOR y Ministerio de Medio Ambiente del Ecuador, 176 pp
Hesselbach H, Galindo LM, Caballero K, de la Maza R, Sánchez LF, Guevara Y (2009) Estrategia de pago de servicios ambientales por desempeño hidrológico en Quintana Roo. Comisión Nacional para el Conocimiento y Uso de la Biodiversidad p 74. Disponible en: http://centro.paot.mx/documentos/conabio/estrategia_pago.pdf, fecha de consulta: 16 de octubre de 2018
Iglesias L, Martínez ER, Graf S, Muñoz C, Gutiérrez J, Flores F, Bauche P (2010) Pago de servicios ambientales para conservar la biodiversidad. In: Carabias J, Sarukhán J, de la Maza y Galindo C (coords) Patrimonio natural de México: Cien casos de éxito. Conabio. México, D.F, pp 176–177
Kremen C (2005) Managing ecosystem services: what do we need to know about their ecology? Ecol Lett 8(5):468–479
Landell-Mills N, Porras IT (2002) Silver bullet or fools' gold?: a global review of markets for forest environmental services and their impact on the poor. Instruments for sustainable private sector forestry series. International Institute for Environment and Development, London, 272 pp. Available at: https://core.ac.uk/download/pdf/48032188.pdf
Luck GW, Harrington R, Harrison PA, Kremen C, Berry PR, Bugter R, Terence P, Dawson TP, de Bello F, Díaz S, Feld CK, Haslett JR, Hering D, Kontogianni A, Lavorel S, Rounsevell M, Samways MJ, Sandin L, Settele J, Sykes MT, van den Hove S, Vandewalle M, Zobel M (2009) Quantifying the contribution of organisms to the provision of ecosystem services. Bioscience 59:223–235. https://doi.org/10.1525/bio.2009.59.3.7

Macip R, Macip RF (2013) Pago por servicios ambientales (ecosistémicos) en México ¿Una alternativa para la conservación de la biodiversidad y el desarrollo? Biocyt biología, ciencia y tecnología 6:375–387

Mahanty S, Suich H, Tacconi L (2013) Access and benefits in payments for environmental services and implications for REDD+: lessons from seven PES schemes. Land Use Policy 31:38–47

Malavasi EO, Mora LS, Carvajal CB (2003) Impacto del programa de Pago de Servicios Ambientales en Costa Rica como medio de reducción de la pobreza en los medios rurales

Manzano DI (2010) Pagos por Servicios Ambientales. Caso Ecuador REDD 74. Libro II Régimen Forestal. Glosario de Términos. Dctyo. Ejecutivo no 3516, R. O. S. especia 2:255–254. Disponible en: https://www.revistajuridicaonline.com/wp-content/uploads/2010/10/255a264_pagos_por.pdf, fecha de consulta. 16 de octubre de 2018

Muñoz-Piña C, Guevara A, Torres JM, Braña J, (2008) Paying for the hydrological services of Mexico's forests: analysis, negotiations and results. Ecol Econ 65(4):725–736

Ochoa-Ochoa L, Urbina-Cardona JN, Vázquez L-B, Flores-Villela O, Bezaury-Creel J (2009) The effects of governmental protected areas and social initiatives for land protection on the conservation of Mexican amphibians. PLoS One 4(9):e6878. https://doi.org/10.1371/journal.pone.0006878

Odum EP (1989) Ecology and our endangered life support system. Sinauer Associates, Inc. Publishers, Sunderland. EE.UU

Pagiola S (2002) Paying for water services in Central America: learning from Costa Rica. In: Selling forest environmental services: market-based mechanisms for conservation and development, London: Earthscan pp 37–62

Porras I (2003) Valorando los servicios ambientales de protección de cuencas: consideraciones-metodológicas. En: Foro Regional Sistemas de Pago por Servicios Ambientales en Cuencas Hidro-gráficas. 9 al 12 de Junio 2003. INRENA – REDLACH – FAO. Arequipa, Perú. http://www.rlc.fao.org/prior/recnat/foro/porras.pdf

Quétier F, Tapella E, Conti G, Cáceres D, Díaz S (2007) Servicios ecosistémicos y actores sociales. Aspectos conceptuales y metodológicos para un estudio interdisciplinario. Gaceta Ecológica 84–85:17–26

Quijas S, Schmid B, Balvanera P (2010) Plant diversity enhances provision of ecosystem services: a new synthesis. Basic Appl Ecol 11:582–593

Quintero M, Estrada RD (2006) Pago por servicios ambientales en Latinoamérica y sus perspectivas en los Andes: una visión desde la práctica. CIP, Perú, 47 pp. Disponible en: https://cgspace.cgiar.org/bitstream/handle/10568/21888/21888.pdf?sequence=1, fecha de consulta: 25 de octubre de 2018

Ruiz-Jiménez M, Valtierra-Pacheco E (2017) Impacto del pago por servicios ambientales hidrólogoicos en los bosques de tres ejidos de Texcoco, México. Agricultura, sociedad y desarrollo 14(4):511–531

SEMARNAT (Secretaría de Medio Ambiente y Recursos Naturales) (2003) Introducción a los servicios ambientales. Secretaria de Medio Ambiente y Recursos Naturales (SEMARNAT) y Hombre naturaleza, México, D.F, 71 pp

Severino RS, González F del PR (2018) Proyecto: territorios Productivos Sostenibles México. SEMARNAT, SAGARPA, CONABIO, INECC, FIRA, GEF, AFE, GIZ. Disponible en: https://www.gob.mx/cms/uploads/attachment/file/291845/Marco_de_procedimientos_para_la_restriccion_en_ANPs.pdf, fecha de consulta: 25 de octubre de 2018

Siikamäki JV, Peter V, Rebecca E-N, and Francisco S-A (2015) Mapping the Value of Ecosystem Services in Latin America and the Caribbean. Resources 188(20). https://www.resourcesmag.org/archives/mapping-the-value-of-ecosystem-services-in-latin-america-and-the-caribbean/

Torres JM (2010) ProÁrbol, un programa para restaurar, manejar y conservar bosques. In: Carabias J, Sarukhán J, de la Maza J, Galindo C (coords) Patrimonio natural de México. Cien casos de éxito. México, Comisión Nacional para el Conocimiento y Uso de la Biodiversidad

UICN (Unión Internacional para la Conservación de la Naturaleza) (2003) Áreas Protegidas en Latinoamérica – De Caracas a Durban. Un vistazo sobre su estado 1992–2003 y tendencias

futuras. Disponible en: https://portals.iucn.org/library/node/8337, fecha de consulta: 12 de diciembre de 2019
UICN (Unión Internacional para la Conservación de la Naturaleza) (2007) Declaración de Bariloche. II Congreso Latinoamericano de Parques Nacionales y otras Áreas Protegidas. Disponible en: https://www.iucn.org/sites/dev/files/import/downloads/declaracionbariloche.pdf, fecha de consulta: 15 de noviembre de 2019
UICN (Unión Internacional para la Conservación de la Naturaleza) (2008) Directrices para las categorías de manejo de áreas protegidas. Gland, Suiza, 96 pp
UN-REDD (2009a) UN-REDD. https://www.un-redd.org/. Accessed on 13 Nov del 2009
UN-REDD (2009b) Framework for action 2009–2014 on measurement. Reporting and verification (MRV). FAO, UNDP, UNEP, 13 pp
Vargas-Guillén A, Aguilar-Martínez S, Castillo-Santiago MA, Esquivel-Bazán E, Hernández-Vázquez MA, López-Gómez AM, Quechulpa-Montalvo S (2009) Programa estatal para la compensación por servicios ecosistémicos. Una propuesta para Chiapas. Comisión Nacional para el Conocimiento y Uso de la Biodiversidad. Corredor Biológico Mesoamericano México. Serie acciones (5). México, D.F., México, 58 pp
Wunder S (2005) Payments for environmental services: some nuts and bolts
Wunder S (2006) Pagos por servicios ambientales: principios básicos esenciales. CIFOR, Lima
Wunder S, Wertz-Kanounnikoff S, Sánchez RM (2007) Pago por servicios ambientales: una nueva forma de conservar la biodiversidad. Gaceta ecológica 84:39–52
WWF (2016) Living planet report 2016. Risk and resilience in a new era. WWF International, Gland
Zarza EB (2010) El costo de oportunidad como instrumento de apoyo para el pago por servicios ambientales. Caso de estudio: programa de pago por servicios ambientales hidrológicos en el Estado de México. Tijuana, México, 124 pp

Chapter 10
An Integrated Dynamic Model for Beach Zoning in Natural Protected Areas

Isaac Azuz-Adeath and Pamela Castro-Figueroa

Abstract A new methodological proposal for an integrated (socioeconomic, governance, and environmental) model for beach zoning is defined. The model takes into account the dynamic character of the beach geomorphology, the legal aspects related to natural protected areas in Mexico, the biodiversity status, and the connection with surrounding communities. The model looks at analytical relationships among variables to generate plausible zoning scenarios automatically. A case study will be presented.

Keywords Dynamic model · Coastal zoning · Natural protected areas · Mexico

10.1 Introduction

This chapter presents the characteristics of a dynamic model for beach zoning in natural protected areas applied in one Mexican study case. The model has an integral character, that means, it considers several management dimensions in its composition, mainly environmental, social, economic, institutional, legal, and regulatory elements. For its operation, the contributions, decisions, and knowledge of the local populations are required. From a spatial point of view, it focuses on the zoning of beaches located in natural protected areas, which in Mexico present a clear regulation on the activities that are allowed. The dynamic nature of the model allows adjustments to be made by the user or decision-maker in accordance with the regular monitoring or survey programs carried out in the study area.

Mexico is considered a mega-diverse country (UNEP/CBD 2002, 2016), part of the group of countries with 70% of global biodiversity. This rich biological diversity

I. Azuz-Adeath (✉)
Centro de Excelencia en Innovación y Diseño, Colegio de Ingeniería, Centro de Enseñanza Técnica y Superior (CETYS Universidad), Ensenada, Baja California, Mexico
e-mail: Isaac.azuz@cetys.mx

P. Castro-Figueroa
Maestría en Innovación en Ingeniería, Colegio de Ingeniería, Centro de Enseñanza Técnica y Superior (CETYS Universidad), Ensenada, Baja California, Mexico

A. Ortega-Rubio (ed.), *Socio-ecological Studies in Natural Protected Areas*,
https://doi.org/10.1007/978-3-030-47264-1_10

is mainly due to its location with coast in both the Atlantic and the Pacific oceans; latitudinal extension (14°–32° North); continental, insular, and maritime size (approximately 5 million square kilometers); physiographic variety (15 physiographic regions from sierras to coastal plains); geomorphology (including two peninsulas and two semi-enclosed seas, the Sea of Cortez and the Gulf of México); and climatology (from humid to desert). The Mexico's coastal zone has a coastline extension of 11,200 km, shared among 17 coastal states, and includes most of the coastal environments such as sandy, muddy, and rocky coast, wetlands, estuaries, coastal lagoons, tidal inlets, deltas, barrier coasts, and several coral reef areas with the second largest in the world (Azuz-Adeath and Cuevas 2018).

The natural terrestrial capital and biodiversity of the 17 Mexican coastal states could be considered one of the biggest in Mexico. The highest numbers of several species in the country, such as pteridophytes, gymnosperms, angiosperms, amphibious, reptiles, birds, and mammals, are located in states of the coastal region (SEMARNAT 2014). The largest densities of endemic species occurred in this area, most of them in the Pacific coast (Koleff and Soberón 2008; SEMARNAT-DGEIA 2016), and six of seven terrestrial ecoregions in Mexico could be observed (INEGI-CONABIO-INE 2008). Specifically, the coastal states Veracruz (VER), Oaxaca (OAX), and Chiapas (CHIS) have the highest number of arthropods, vascular plants, and vertebrate species in the country (Llorente-Bousquets and Ocegueda 2008; SEMARNAT 2013).

According to Azuz-Adeath and Cuevas (2018), one of the most important element for biodiversity conservation considered by the Mexican environmental law is the possibility to create and manage natural protected areas (NPAs). The establishment of protected areas around the world has been recognized as a cornerstone process to preserve global biodiversity, ecological integrity, environmental services, and livelihoods which bring direct benefits to the human population (Ervin et al. 2010; Pullin et al. 2013). In Mexico, the economic and social benefits related to the creation of protected natural areas have been evaluated and documented among others by Bezaury (2009), Bezaury-Creel and Gutiérrez (2009), and Méndez-Barrera (2012); the effectiveness of natural protected areas in preventing land use and land cover changes has been documented extensively by Figueroa and Sánchez-Cordero (2008) and Blackman et al. (2015); the territorial extension, geographic distribution, and managerial status of the Mexican NPAs could be found in CONANP (2016a, b); and several site-specific, threat-oriented, management-focused, or species-/ecosystem-oriented studies could be found among many others in Carr (2007), Méndez-Contreras et al. (2008), Durand and Lazos (2008), Lara-Lara et al. (2008), Coates et al. (2015), and Ávila-Canto et al. (2017).

In 2016, a very important increase in the extension of NPAs occurred in Mexico, adding 43.6 million hectares (Ha) with the decree of two marine NPAs, the "Deep Mexican Pacific" and the "Caribbean," covering the Aichi targets. Actually, the country has 182 federal NPAs that protect 90.8 million hectares (CONANP 2017). Considering only the NPAs located in the 17 coastal states, the territorial extension of the federal NPAS covers 13.4 million Ha, from which 9 million Ha correspond to coastal municipalities (Azuz-Adeath and Cuevas 2018).

In this context, marine spatial planning and beach zoning are two powerful tools to sustainably preserve and manage some of the most valuable regions of protected natural areas such as beaches.

This chapter is structured as follows. The following section presents, on the one hand, the framework of management based on scientific knowledge and, on the other, the conceptual principles that underpin the dynamic model. The methodological section shows the procedures followed to integrate the different elements and variables of the dimensions considered as well as the characteristics of the study area used to apply the model. In the results section, several runs of the model will be presented, and finally, the document shows the conclusions obtained and the lessons learned.

10.2 General Framework and Model Design Principles

The following two sections describe the theoretical principles that guide the creation of the dynamic model and the conceptual and operational bases on which the model was built.

10.2.1 General Framework

During the last three decades, different theoretical frameworks and action guides have been proposed and adopted internationally to sustainably manage marine and coastal areas. Among the most important ones are the Ecosystem-Based Approach (EBA), the Large Marine Ecosystems (LME) strategy, and the Integrated Coastal Zone Management (ICZM).

The EBA was proposed in 1995 by the Convention on Biological Diversity (CBD) in the framework of the United Nation Environment Programme (UNEP) and considers 12 principles and implementation guidelines. The conceptual basis could be defined as follows: "The ecosystem approach is a strategy for the integrated management of land, water and living resources that promotes conservation and sustainable use in an equitable way. An ecosystem approach is based on the application of appropriate scientific methodologies focused on levels of biological organization, which encompass the essential structure, processes, functions and interactions among organisms and their environment. It recognizes that humans, with their cultural diversity, are an integral component of many ecosystems. The ecosystem approach requires adaptive management to deal with the complex and dynamic nature of ecosystems and the absence of complete knowledge or understanding of their functioning" (CBD 2004).

The effectiveness of the use of EBA has been proven around the world in marine spatial planning (Clarke and Jupiter 2010; Foley et al. 2010; Sardá et al. 2014; Thomas and Bhola 2017; Schmidtbauer et al. 2017), in fisheries and aquaculture management (Garcia et al. 2003; Jennings 2005; FAO 2009; Espinoza-Tenorio et al. 2012; Gullestad et al. 2017; O'Farrell et al. 2017; Link and Marshak 2019), as well as in natural protected areas preservation, considering among other issues the environmental services, stakeholders visions, activities and climate change impacts, and adaptation strategies (Keller et al. 2009; Espinoza-Tenorio et al. 2010; Portman 2016; Wei et al. 2019; Maestro et al. 2019).

According to the LME portal (http://lme.edc.uri.edu/index.php/lme-introduction), Large Marine Ecosystems are relatively large areas of ocean space of approximately 200,000 km^2 or greater, adjacent to the continents in coastal waters where primary productivity is generally higher than in open ocean areas. The LMEs produce about 80% of the annual world's marine fisheries catch. Globally, they are the centers of coastal ocean pollution and nutrient overenrichment, habitat degradation, overfishing, biodiversity loss, and climate change effects.

There are two important features in the LME approach to the improved management of LME goods and services. First, the physical extent of the LME and its boundaries are based on ecological, rather than political or economic, criteria. Based on ecological criteria, 64 distinct LMEs have been delineated around the coastal margins of the Atlantic, Pacific, and Indian Oceans. The second important feature of the LME approach is the application of five strategies for measuring the changing states of LMEs and for taking remedial actions toward the recovery and sustainability of degraded goods and services. From a management perspective, it is essential to establish a baseline condition against which to measure the success or failure of actions to recover depleted fish stocks, restore degraded habitats, and reduce and control coastal pollution and nutrient overenrichment. The strategies are focused on the application of suites of indicators for measuring LME (i) productivity and oceanography, (ii) fish and fisheries, (iii) pollution and ecosystem health, (iv) socioeconomics, and (v) governance.

Triggered by the deterioration of world fisheries and global ocean pollution, the LME vision incorporates in early stages the EBA (Duda and Sherman 2002). With the international commitment and support, the LME strategy has produced an important body of scientific knowledge and managerial options for the oceans' sustainability (Sherman et al. 2009, 2010; Sherman and McGovern 2011, 2012; Sherman and Adams 2013; Sherman 2015) (for a comprehensive bibliographic review, see Kelley (2016)).

In the confluence region between ocean and continents, Integrated Coastal Zone Management is the principal methodological option for the management of this dynamic regions and for the improvement of its governance. The basis of ICZM are supported by the EBA, systems theory, adaptive management, sectors and government harmonization, public participation in decision-making, legal and institutional strengths, and economic support (Cicin-Sain and Knecht 1998; Salomons et al. 1999; Kay and Alder 2005).

ICZM is a dynamic process in which a coordinated strategy is developed and implemented for the allocation of environmental, sociocultural, and institutional resources to achieve the conservation and sustainable multiple use of the coastal zone (Sorensen 1993). To attain this goal, ICZM proposes a series of techniques and methodologies in the administrative, socioeconomic, and technical spheres and seeks to comply with a series of basic principles (Ahlhorn 2018): (i) the protection of natural resources, (ii) the management of coastal development, (iii) prioritization of coastal activities, (iv) public access to common coastal goods, (v) development and preservation of urban and historic places, (vi) governmental coordination and consultation, (vii) public participation, and (viii) comprehensive planning, conservation, and management strategies and programs.

10.2.2 Model Design Principles

The integrated dynamic model for beach zoning in natural protected areas (BeZo), presented in this chapter, is based on the EBA, LME, and ICZM principles, concepts, and techniques and uses elements from two wide-accepted spatial planning methodologies: marine spatial planning (MSP) and coastal zoning (CZ), according to Fig. 10.1.

Five dimensions of analysis were included in BeZo's model: (a) environmental, (b) social, (c) economic, (d) governance, and (e) climatic change. Every dimension has several variables which are modeled by analytical, statistical, probabilistic, or empirical methods, depending on the data availability. The chosen variables used or the incorporation of new ones could be done by the modeler or decision-maker. The spatial extension of the modeled area and morphological characteristics also need to be defined as an input for the model, the beach geometry (onshore, offshore, and lateral limits), the presence of coastal landforms, wave climate, sediment size, and artificial structures. The time extent of the zoning scenario also can be defined by the model's operators. Figure 10.2 shows the dimensions considered in the model as well some examples of variables.

The model shows a list of variables (elements of the system) for each dimension (see Table 10.1), and users can define new ones that would not have been considered

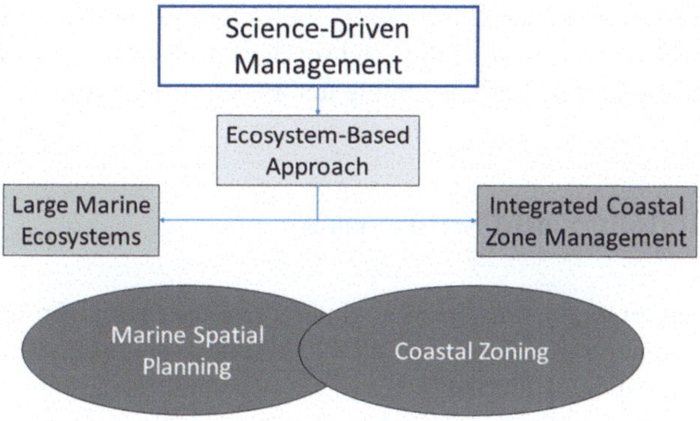

Fig. 10.1 Conceptual elements and spatial planning methodologies considered in the BeZo model

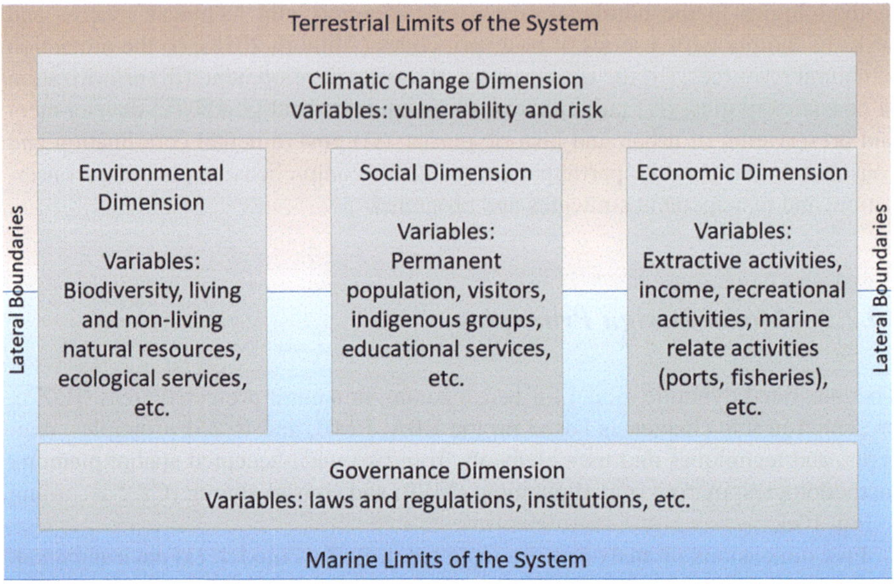

Fig. 10.2 General characteristics of the dimensions considered in BeZo's model and some examples of variables incorporated

or that respond to specific characteristics of the study area. In a participatory manner, users, actors, and zoning managers can assign weights to the different variables and dimensions. As a restriction, no dimension can have a greater weight than governance because of this dimension conditioning the zoning structure in natural protected areas.

After the definition of variables and ponderation of dimensions, the most important and complex step is the incorporation of zoning criteria. Some elements are considered from the governance dimension, particularly laws and regulations; also, the constructive or destructive interactions among variables were used as a key element. The second element that is taken into account is the precautionary principle; prioritizing the environmental conservation, the well-being of the local populations, and the sustainable economic activities and services is another criterion; and finally, the scenarios of minimum conflicts among actors are evaluated.

10.3 Methodology and Study Area

BeZo's model architecture consists of six interactive user-defined modules and two noninteractive elements (see Fig. 10.3). In the module 1, the users define the general physical characteristics of the study area: geometry and geographic limits, beach characteristics, wave conditions, and relevant coastal elements (natural or artificial). In the module 2, the users define the variables to be used in the model, incorporate

Table 10.1 Variables considered in the model BeZo

Governance dimension	Environmental dimension	Social dimension	Economic dimension	Climate change dimension
Laws	Biodiversity distribution	Permanent population	Extractive activities	Social vulnerability
Regulations	Protected species	Visitors	Income	Economic vulnerability
Institutions	Living resources	Indigenous groups	Recreational activities (tourism)	Environmental vulnerability
Management programs	Nonliving resources	Educational services	Port activities and facilities	Risk
Scientific research	Ecological services	Health services	Fishery activities and facilities	
Environmental monitoring and surveillance	Turtle nest areas	NGOs	Power plants or renewable energy installations	
		Clean water availability	Agriculture	
		Sanitation services	Aquaculture/ mariculture	
		Violence	Cattle raising or feeding	
Other	Other	Other	Other	Other

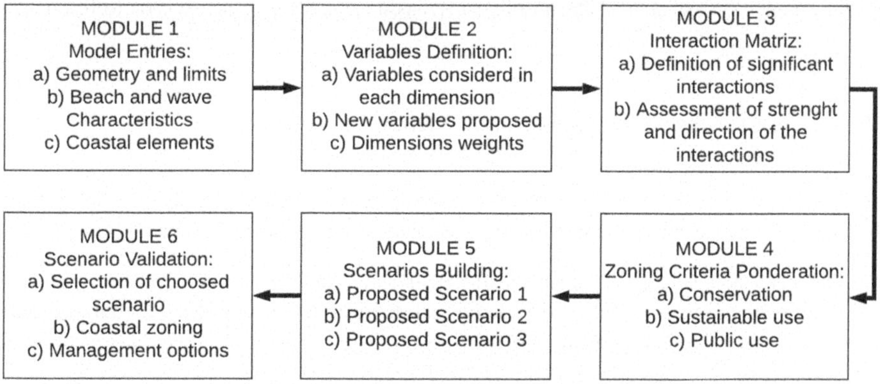

Fig. 10.3 Model BeZo's computational architecture diagram (interactive elements)

new ones, and give the ponderations for each dimension considered. In the module 3, BeZo's model proposes an interaction matrix according to the selected variables and the users (decision-makers, actors, local population, etc.) and given values for the strength and direction (constructive, destructive, or neutral) of the interactions. In module 4, the weights of the zoning criteria were defined by the users (i.e., conservation, sustainable use, and public use). In module 5, some zoning scenarios were proposed by the model in plant (2DH) and profile view (2DV), and finally in module 6, the users validate the chosen zoning structure.

10.3.1 Module 1: Beach Form and Characteristics

Module 1 defines the plan view and profile shape of the beach. Two options are proposed for the plan view: (a) linear beach and (b) crenulate-shaped bay (Silvester and Hsu 1993). The beach profile form considered the Dean's square berm profile with backshore (Kriebel and Dean 1993; Dean and Dalrymple 2002). As input information for the configuration of the beach, the model requires the length and width of the beach (linear form), dimensions of the control line and its orientation, angle of approach of the waves, and width of the beach (crenulate form), and for the beach profile shape, medium sediment size, berm height, backshore width, beach face slope, and some significant parameter for wave height (e.g., $H_{1/3}$, H_{12s}). The last subsystem of module 1 allows users to define the artificial and natural structures present on the beach. The predefined natural elements are upper and lower headline, permanent rivers, seasonal rivers, deltas, coastal bodies, tidal inlets, coastal dunes, and beach cusp. The artificial elements considered are groins, detached breakwaters, seawalls, piers, energy-related infrastructure, "palapas" (simple handmade elements to provide shadow, built with local materials usually wood or palm), showers, recreational facilities, beachfront houses, and scientific facilities. Information about the location, extension, and frequency of use (days/year) is requested from the users.

10.3.2 Module 2: Definition of Study Variables

Module 2 allows the user to define the variables that will be taken into account in the zoning scheme. The general categories can be grouped as follows:

(a) Governance: Laws, regulations, institutions, management programs, scientific research, environmental monitoring, environmental surveillance, management categories, NGOs, community organizations, educational centers, climate change-related plans, and others (locally defined)
(b) Environmental: Significant ecological areas, significant ecological species, significant ecological communities, conservation objectives, ecosystem priorities, ecosystem vulnerabilities, protected species, coastal commercial species, coastal construction natural materials, and others (locally defined)
(c) Social: Permanent population, groups of age, temporal visitors, indigenous groups, educational services, health services, clear water availability, sanitation services, violence, and others (locally defined)
(d) Economic: Extractive activities, income, recreational activities (tourism), port activities and facilities, fishery activities and facilities, power plants, renewable

energy facilities, agriculture, aquaculture/mariculture, cattle raising or feeding, and others (locally defined)

(e) Climate change: Tropical systems impacts, flooding, droughts, coastal erosion, sea level rise evidence, coral bleaching evidence, abnormal species mortality, abnormal species presence, lack of pollinators, changes in flowering periods, and others (locally defined)

For each variable, the program asks about the existence, availability, nonexistence, or nonavailability of the information, the phenomenon, or the activity. For positive responses, the program requests specific information of a quantitative nature and, in some cases, its location and spatial distribution. A key issue in this process is the participation of the local community, stakeholders, authorities, and researchers.

10.3.3 Module 3: Interaction Matrix

Module 3 of BeZo's model presents a proposal of an interaction matrix according to the variables defined in previous step (module 2). The users express the interaction sense by means of three colors: a) green represents a positive interaction among variables, b) orange means a negative interaction between variables, and c) blue, no interaction at all. The yellow cells are designed for the weight of each dimension. The sum of the five cells needs to be one with the additional restriction that the governance dimension needs to have the largest ponderation. Table 10.2 shows the general structure of the interaction matrix.

10.3.4 Module 4: Zoning Criteria

As a first approximation, the BeZo model considers only three zoning categories:

1. Conservation: This category looks at the preservation of natural resources and landscapes. Only contemplation and observation activities are allowed for tourists and non-extractive activities for the local population.
2. Sustainable use: This category permits the sustainable use of the natural resources and low-impact activities. Regulated low-density tourism activities and facilities are allowed in land and sea, as well as extractive activities by local communities under agreed management programs.
3. Public use: This category allows the construction of public or private infrastructure under criteria of low visual impact, energy and water efficiency, zero pollution and waste, sewage treatment, low carbon footprint, and similar sustainable standards. Regardless of the applicable national laws, the community must participate in the evaluation of potential projects. Some types of extractive activities are allowed to improve the living conditions of local people, like wood, seashells, pebbles, or sand extraction, use of medicinal plants, and artisanal fishing, among others.

Table 10.2 Interaction matrix between variables defined by the users in BeZo's model

	Gov	G1	Gn	Env	E1	En	Soc	S1	Sn	Eco	Ec1	Ecn	CC	CC1	CCn
Gov															
G1															
Gn															
Env															
E1															
En															
Soc															
S1															
Sn															
Eco															
Ec1															
Ecn															
CC															
CC1															
CCn															

The yellow cell is for the ponderation of the corresponding dimension. Green color can be used for positive interactions (constructive), orange color for negative or destructive interactions, and blue color for neutral interactions. *Gov* **governance;** *Env* **environment;** *Soc* **social;** *Eco* **economic;** *CC* **climate change. The subindices correspond to different variables (1 to n)**

To define the spatial distribution of these zoning categories, the first elements of analysis that the model considers are the legal and regulatory aspects applicable to the study area. If there are protected species or coastal environments under some protection status (e.g., wetlands of global importance) inside the zoning area, the users need to define the general distribution or spatial limits of these elements. After this step, the model analyzes the interaction matrix to define compatible (positive interactions) or incompatible relationships among the variables considered. If the existing information of the variables allows it, the proposed methodology makes use of cluster analysis and its visual representation of the dendrograms (Everitt 1974; Romesburg 2004) to define the location of each category. The longshore and cross-shore extension of each category are determinate by the data (mostly by incompatible variables according to the interactions matrix).

10.3.5 Modules 5 and 6: Proposed Scenarios and Selected Scenario

Modules presents as output the bidimensional longshore (2DH) zoning and beach profile configuration (2DV) from the backshore to the limit depth of sediment movement. In some special cases (if the information for the underwater beach profile is available), the zoning criteria could be applied in the 2DV configuration, but the zoning cannot be extended until open sea because of the limited dynamic ocean process considered in the model. In its initial stage, the BeZo model can give only a regular distribution of zoning areas (square or rectangular sections). The location and extension of zoning areas need to be validated by the users, stakeholders, authorities, community, and researchers, as well as their temporal validity. The dynamic model BeZo allows the seasonal actualization of the information and thus changes in the zoning structure. The seasonal term was considered optimum for the participatory process – costs and time investment of the participants – and also because of the dynamic of tourism flow.

After the presentation of different zoning scenarios (three options), the people involved in the process chose the best one under their own considerations. This zoning scheme was adopted for the time frame agreed, and some management elements are displayed, taking into consideration the governance elements and environmental, social, and economic restrictions for each zoning categories and spatial areas.

10.3.6 Study Area

The model was validated with information from Cabo Pulmo, a natural protected area located in the Gulf of California in Baja California Sur State. The extension of this area is around 7,000 hectares, most of them located in the marine portion (99%) and only a small portion in the coastal zone (see Fig. 10.4). In 1995, Cabo Pulmo was decree a National Marine Park because the rich biodiversity and quality of its coral reef (the only coral reef in the Gulf of California). The management program follows an exemplar technical and participatory process with the involvement of the local community, academy, and government, and it can be seen as a world example

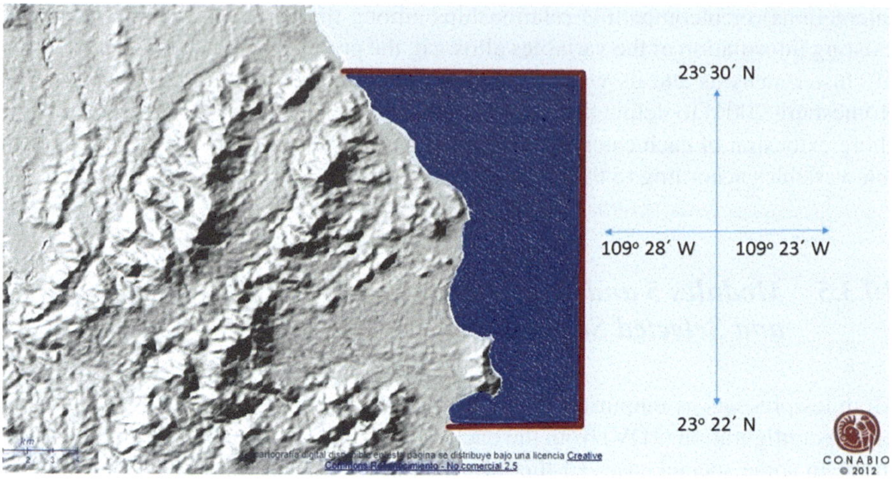

Fig. 10.4 Outline of the natural park Cabo Pulmo area. (Source of the base map CONABIO 2012)

of conservation (Arizpe et al. 2013). Local people benefited by changing their lifestyle from exploitation of marine resources to their protection and the offer of sustainable tourism services.

This area was chosen because of the low density of population and infrastructure in the coastal zone and the clear regulatory elements that norm the activities in the area and because of the detailed technical information about the environmental elements obtained in the diagnosis of the official program (CONANP 2006).

The most important achievements that have been attained with the park management program, as well as some threats, will be presented in the following list (CONABIO 2012; Arizpe-Covarrubias and Arizpe-Vicencio 2015):

- Achievements: Monitoring shark activities and reef ecosystems, assessment and valuation of ecosystem services, development of management strategies for the coral reef, and the participatory program for the public use of the national park
- Threats: The development of large tourism resorts in the vicinity of the protected area, the destruction of coastal dunes around the national park, and nonregulated commercial and sport fishing outside bot in the vicinity of the marine protected area of the park

The official management program of Cabo Pulmo establishes five management categories: (a) two preservation zones, (b) two sustainable use zones, and (c) one public use zone. For each zone or subzone, the permitted and prohibited activities have been clearly established in joint processes between local people, authorities, and academics. And the most important fact is that for several years, the community has respected, enforced, and encouraged them.

To validate BeZo's model, a coastal portion of Cabo Pulmo area was selected. This portion goes from the northern part of Cabo Pulmo town to the headland limit in the south of Playa Arbolitos. According to the official management program, this section comprises two categories of management: public use (the main part) and a small portion of preservation type 2 zone, in the southern part.

10.4 Results

In this section, the results obtained as outputs from BeZo's model will be presented, starting with the 2DV and 2DH beach configuration, followed by the interaction matrix and the proposed zoning schemes. Public participation processes need to be involved through the process after the definition of the beach morphology.

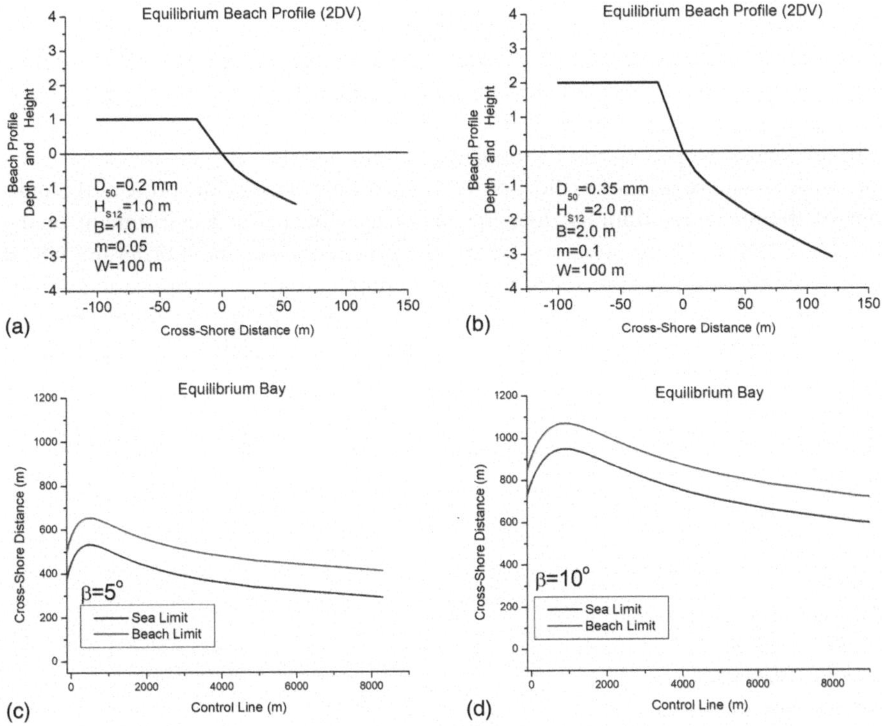

Fig. 10.5 Examples of equilibrium beach profiles (upper) and equilibrium bay shapes (lower) under different sedimentary, wave, and morphologic conditions

10.4.1 Beach Morphology 2DV and 2DH

Two examples of equilibrium beach morphology (i.e., beach profile and crenulate-shaped bay) for different sediment (D_{50}), wave (H_{S12}, β), and beach conditions (B, m and W) can be seen in Fig. 10.5 in which D_{50} is the mean sediment size (mm), H_{S12} the yearly significant wave height (m), β the wave approach direction (degrees), B the berm height (m), m the beach face slope, and W the backshore wide (m).

For the upper part of Fig. 10.5, two conditions were chosen: a) low energy (right) and b) high energy (left). The lower part shows two different incident wave approach angles (β). Under seasonal changes in wave energy, beach morphology, or sediment characteristics, BeZo's model can generate easily the corresponding beach structure (2DV and 2DH) to adjust the zoning configuration.

10.4.2 Construction of the Variables and Interaction Matrix

A theoretical public participation exercise based on the conditions of Cabo Pulmo could have generated the following characterization of variables (only the governance dimension is shown in Table 10.3).

In this case, the perception of the Cabo Pulmo population recognizes the existence of laws and regulations (most of them federal because of the protected conditions of the area), institutions (like the Mexican Ministry of Environment or the State University), management programs (the official management program, which has been constructed and presented to the local population), environmental monitoring campaigns in which the villagers have actively participated, management categories according to the management program, NGOs which have a permanent

Table 10.3 Characterization of variables from the governance dimension (example)

Variables	Existing	Available	Nonexistent	Not available	Unknown
Laws	1				
Regulations	1				
Institutions	1				
Management programs	1				
Scientific research	1				
Environmental monitoring	1				
Environmental surveillance			1		
Management categories	1				
NGOs	1				
Community organizations	1				
Educational centers	1				
Climate change-related plans			1		
Other					

presence in the community, community organizations of which they are part, and educational centers in which they participate regularly.

For the "environmental surveillance" variable, even when federal authorities have to perform this function by law, the local community perceives that it is not done. The same could be thought about the "climate change plans" until specific information is not shared with the community. The same exercise could be done for the variables in the other dimensions, giving information like the one shown in Fig. 10.6.

With all this information available, an interaction matrix like the one shown in Table 10.2 could be constructed. After this step, the community needs to define, by consensus, the weights of each dimension and the magnitude and direction (neutral, constructive, or destructive) of each of the interactions.

In parallel, the spatial characterization of the elements present in the zoning area must be carried out according to the matrix presented in Table 10.4. This information can be taken by the administrators or technical personnel.

For the case of Cabo Pulmo area, the upper and lower headlines are rocky points (see the number 1 in red), and there are no permanent or seasonal rivers, either deltas, coastal bodies, tidal inlets, or coastal dunes. During the survey process, there were, possibly, no beach cusps, which, if presented later, can be characterized in the updating of the information. The zoning area presents artificial elements (see number 1 in red), and these elements are related to the town facilities (e.g., beachfront houses, scientific facilities) or beach recreational installation like "palapas," a temporal palm or wood roof with umbrella shape to provide shadow. The location and extension of all these elements need to be done by means of GPS in the field.

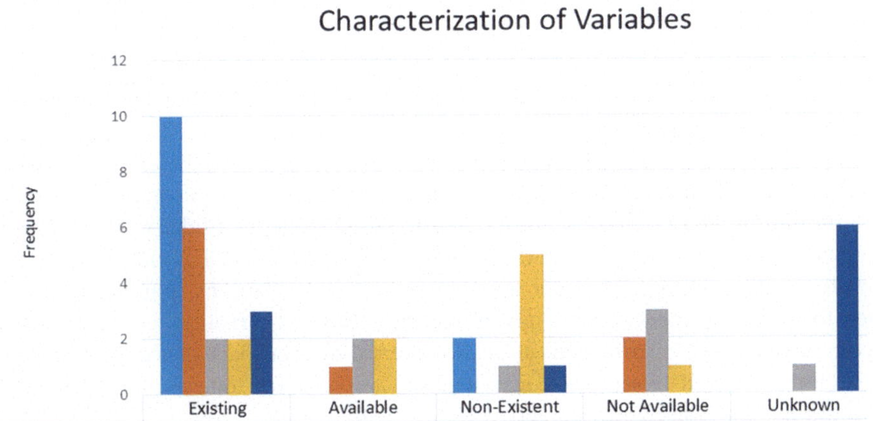

	Existing	Available	Non-Existent	Not Available	Unknown
Governance	10	0	2	0	0
Environment	6	1	0	2	0
Social	2	2	1	3	1
Economic	2	2	5	1	0
Climate Change	3	0	1	0	6

Fig. 10.6 Results from the characterization of variables from Cabo Pulmo theoretical case

Table 10.4 General format for the definition of the elements present in the zoning area

Natural Elements:	Presence	Location	Extension	Frequency of Use (days/year)
Upper Headline	1	LS and CS	(m)	
Lower Headline	1	LS and CS	(m)	
Permanent Rivers	0			
Seasonal Rivers	0			
Deltas	0			
Coastal Bodies	0			
Tidal Inlets	0			
Coastal Dunes	0			
Artificial Elements:				
Groins	0			
Detached Breakwaters	0			
Sea Walls	0			
Piers	0			
Energy Related Infrastructure	0			
"Palapas"	1	LS and CS	(m)	
Showers	0			
Recreational Facilities	0			
Beach Front Houses	1	LS and CS	(m)	
Scientific Facilities	1	LS and CS	(m)	

The part for data capture in the table (e.g., coordinates) was omitted for clarity purposes

10.4.3 The Proposed Zoning Schemes

As example of the zoning proposal generated by BeZo's model, Fig. 10.7 shows one potential 2DV configuration.

The proposed zoning scheme permits the sustainable use of the emerged part of the beach profile until the backshore and the conservation of the submerged part of the profile. In the exposed part of the beach, swash zone, and tide pools, regulated low-density tourism activities and facilities are allowed, as well as extractive activities by local communities under agreed management programs (e.g., shell collection for sailing purposes, turtle nest monitoring, birds observation, consumption of rocky mollusk and crustacean, etc.). For the submerged part of the profile, only contemplation and observation activities are allowed for tourists and recreation, swimming, and non-extractive activities for the local population.

This vertical configuration needs to be superimposed and complemented with the horizontal zoning proposal (2DH). Figure 10.8 shows the equilibrium bay shape and the proposed zoning scenario.

In Fig. 10.8, the proposed zoning schemes need to be considered between the two lines that define the bay shape, conservation for the lower part of the bay and public use for the upper. This zoning scheme also gives the local population,

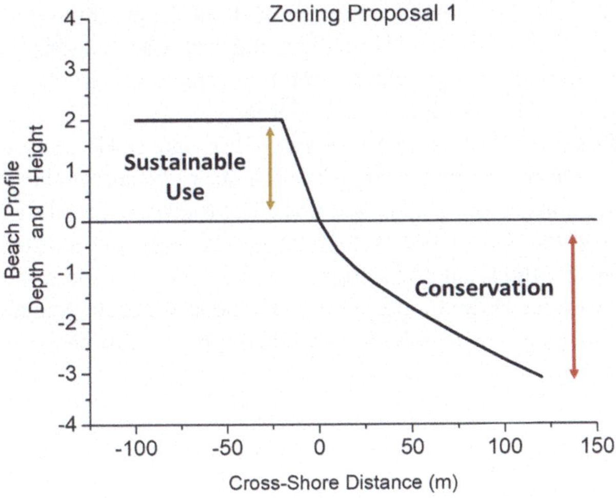

Fig. 10.7 Zoning proposal for the vertical configuration of the beach

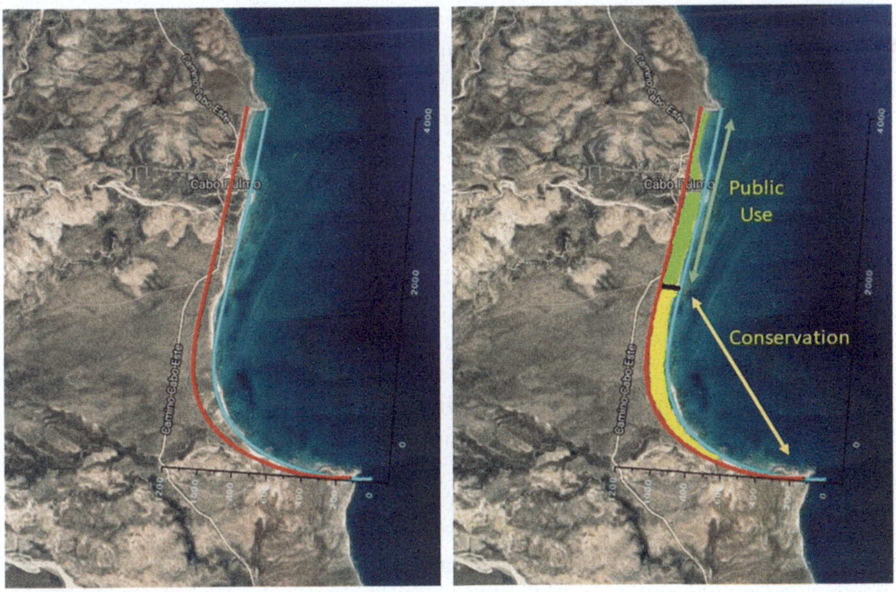

Fig. 10.8 Equilibrium bay shape (left) and zoning proposal (right). (Source of the base map: Google maps, 2019)

managers, and technical personnel information about the potential points (places) of erosion and accretion. In the middle part of the bay, one section of the line goes beyond the blue line, this is a potential zone for erosion, and in the future, this area needs to be monitored to take preventive actions.

The combination of these zoning schemes (2DV and 2DH) can give elements to the local population and manager to select the best option for the protection and sustainable development of the analyzed area. The transparency of the methodology also contributes to the acceptance of the chosen schemes, the reinforcement of the zoning categories by the local population, and the respect of the same by the users. The dynamic character of the model permits the adaptive management and the easy construction of new zoning scenarios under changing conditions.

10.5 Conclusions

10.5.1 General Conclusions

This chapter presented the results of the application of a dynamic model, based on broad public participation, which allows local communities located in natural protected areas to build and select zoning schemes for the management of their beaches and coastal resources.

The flexibility of the model allows the adaptation of management schemes quickly and simply, so it is recommended to update them seasonally or when local conditions so require (e.g., tourism season, species migration, storm season). The management categories may be more specific (i.e., subcategories) according to the requirements of the study area.

The most complex elements to define within the model are the boundaries between management categories in the 2DH dimension. As a future line of study, the definition of buffer or transition areas between zones will be analyzed.

The study area considered in this theoretical exercise has very particular conditions: a) difficult access, b) extensive scientific knowledge of the place, c) awareness and involvement of local people, d) well-defined and respected management program, and e) low population density, limited infrastructure, and very localized economic activities. The application of this methodology to more complex scenarios is required to fully validate the model.

BeZo's model seeks to be a technological tool that contributes to better management of beaches in protected areas, based on the principle of using the largest and best available scientific information and traditional knowledge, in the framework of effective and regular public participation processes, with the commitment of local people, specialists, and managers.

10.5.2 Lessons Learned

Natural protected areas can be considered as one of the main international tools of environmental policy that have effectively contributed to the preservation of global natural capital. The proper and effective management of these areas requires broad social participation and the use of the best available scientific information and the appropriate technological tools. The construction of a dynamic model for zoning the beaches of these sensitive regions has represented a challenge in relation to the incorporation of elements related to decision-making. However, the results obtained in this first approach are promising and allow the incorporation of the human dimension (e.g., decision-making, public participation, governance) as a key element in the operation of models based on analytical functions for zoning.

In the development of the model, the incorporation into the zoning criteria of the boundaries between categories was particularly difficult. A broad validation of the model is still required in areas with other characteristics than those studied in the selected study case.

References

Ahlhorn F (2018) Integrated coastal zone management. Status, challenges and prospects. Springer Vieweg, Wiesbaden

Arizpe O, Urciaga J, Arizpe OA (2013) Tourism in natural protected areas of Mexico. In: Transactions on ecology and environment, WIT Press. https://doi.org/10.2495/SC130952

Arizpe-Covarrubias O, Arizpe-Vicencio M (2015) Prospective of the natural protected areas of the south of the Baja California Peninsula. In: Transactions on ecology and environment, vol 199, WIT Press. https://doi.org/10.2495/RAV150041

Ávila-Canto JE, Velázquez-Mendóza C, Castelblanco-Martínez N, Niño-Torres C, Córdova-Tapia F (2017) Is the Antillean manatee (Trichechus manatus manatus) back in town? Presence of the species at the "Área de Protección de Flora y Fauna YumBalam", Quintana Roo, Mexico. Revista Mexicana de Biodiversidad 88:999–1002. https://doi.org/10.1016/j.rmb.2017.10.036

Azuz-Adeath I, Cuevas A (2018) Potential impacts of decadal climate variability on coastal biodiversity and societal important productive activities: a case study in Mexican coastal states, 319-345. In: A Ortega-Rubio (ed), Mexican natural resources management and biodiversity conservation, Springer. https://doi.org/10.1007/978-3-319-90584-6_15

Bezaury JE (2009) El valor de los bienes y servicios que las áreas naturales protegidas proveen a los mexicanos. The Nature Conservancy-CBD-Comisión Nacional de Áreas Naturales Protegidas, México, D.F

Bezaury-Creel JE, Gutiérrez D (2009) Áreas naturales protegidas y desarrollo social en México, en Capital natural de México, vol. II: Estado de conservación y tendencias de cambio. Conabio, México, D.F

Blackman A, Pfaff A, Robalino J (2015) Paper park performance: Mexico's natural protected areas in the 1990s. Glob Environ Chang 31:50–61

Carr DL (2007) Resource management and fertility in Mexico's Sian Ka'an Biosphere Reserve: Campos, cash, and contraception in the lobster-fishing village of Punta Allen. Popul Environ 29:83–101

CBD (2004) The ecosystem aproach. Convention on Biological Diversity (CBD), Montreal, Canada.

Cicin-Sain B, Knecht RW (1998) Integrated coastal and ocean management. Concepts and practices. Island Press, Washington, DC/Covelo

Clarke P, Jupiter S (2010) Principles and practice of ecosystem-based management a guide for conservation practitioners in the tropical western pacific. Wildlife Conservation Society, Bronx

Coates R, Aguirre A, Campos-Villanueva A (2015) Plant use by artisans within the Los Tuxtlas biosphere reserve, southern Mexico. Hum Ecol 43:621–631

CONANP (2006) Programa de conservación y manejo Cabo Pulmo México. Comisión Nacional de Áreas Naturales Protegidas de México. Secretaría de Medio Ambiente y Recursos Naturales (SEMARNAT), México, D.F

CONANP (2016a) Prontuario estadístico y geográfico de las Áreas Naturales Protegidas de México. Secretaría de Medio Ambiente y Recursos Naturales (SEMARNAT)-Comisión Nacional de Áreas Naturales Protegidas (CONANP), México, D.F

CONANP (2016b) Fichas de evaluación ecológica de áreas naturales protegidas del noroeste de México. https://simec.conanp.gob.mx/evaluacion.php?menu=4

CONANP (2017) Áreas Naturales Protegidas decretadas. Comisión Nacional de Áreas Naturales Protegidas (CONANP) http://sig.conanp.gob.mx/website/pagsig/datos_anp.htm

CONABIO (2012) Cartografía digital. Comisión Nacional para el Conocimiento y Uso de la Biodiversidad (CONABIO). http://www.conabio.gob.mx/informacion/gis/

Dean RG, Dalrymple RA (2002) Coastal processes with engineering applications. Cambridge University Press, Cambridge

Duda AM, Sherman K (2002) A new imperative for improving management of large marine ecosystems. Ocean Coast Manag 45:797–833

Durand L, Lazos E (2008) The local perception of tropical deforestation and its relation to conservation policies in Los Tuxtlas biosphere reserve, Mexico. Hum Ecol 36:383–394

Ervin J, Sekhran N, Dinu A, Gidda S, Vergeichik M, Mee J (2010) Protected areas for the 21st century: lessons from UNDP/GEF's portfolio. New York: United Nations development Programme and Montreal: convention on biological diversity. New York

Espinoza-Tenorio A, Montaño-Moctezuma G, Espejel I (2010) Ecosystem-based analysis of a marine protected area where fisheries and protected species coexist. Environ Manag 45:739–750

Espinoza-Tenorio A, Wolf M, Taylor MH, Espejel I (2012) What model suits ecosystem-based fisheries management? A plea for a structured modeling process. Rev Fish Biol Fish 22:81–94

Everitt B (1974) Cluster analysis. Heinemann educational books (HEB) and the social research council, London

FAO (2009) Practical implementation of the ecosystem approach to fisheries and aquaculture. Food and agriculture organization of the united nations regional office for Asia and the Pacific, Bangkok

Figueroa F, Sánchez-Cordero V (2008) Effectiveness of natural protected areas to prevent land use and land cover change in Mexico. Biodivers Conserv 17:3223–3240

Foley MM, Halpern BS, Micheli F, Armsby MH, Caldwell MR, Crain CM, Prahler E, Rohr N, Sivas D, Beck MW, Carr MH, Crowder LB, Duffy JE, Hacker SD, McLeod KL, Palumbi SR, Peterson CH, Regan HM, Rukelshaus MH, Sandifer PA, Steneck RS (2010) Guiding ecological principles for marine spatial planning. Mar Policy 34(5):955–966

Garcia SM, Zerbi A, Aliaume C, Do Chi T, Lasserre G, (2003) The ecosystem approach to fisheries. Issues, terminology, principles, institutional foundations, implementation and outlook. FAO fisheries technical paper no. 443, Rome

Gullestad P, Abotnes AM, Bakke G, Skern-Mauritzen M, Nedreaas K, Søvik G (2017) Towards ecosystem-based fisheries management in Norway – practical tools for keeping track of relevant issues and prioritizing management efforts. Mar Policy 77:104–110

INEGI-CONABIO-INE (2008) Ecoregiones terrestres de México. Instituto Nacional de Estadística y Geografía (INEGI), Comisión Nacional para el Conocimiento y Uso de la Biodiversidad (CONABIO), Instituto Nacional de Ecología (INE). Map available at: http://www.conabio.gob.mx/informacion/gis/layouts/ecort08gw.png

Jennings S (2005) Indicators to support an ecosystem approach to fisheries. Fish Fish 6:212–232

Kay R, Alder J (2005) Coastal planning and management. Taylor & Francis, London

Keller BD, Gleason DF, McLeod E, Woodley CM, Airamé S, Causey BD, Friedlander AM, Grober-Dunsmore R, Johnson JE, Miller SL, Steneck RS (2009) Climate change, coral reef ecosystems, and management options for marine protected areas. Environ Manag 44:1069–1088

Kelley E (2016) Large marine ecosystems of the world: an annotated bibliography. NOAA technical memorandum NMFS-F/SPO-167, Silver Spring

Koleff P, Soberón J (2008) Patrones de diversidad espacial en grupos selectos de especies, 323–364 pp. En Capital Natural de México, vol. 1: Conocimiento actual de la biodiversidad. CONABIO, México, D.F. http://www.biodiversidad.gob.mx/pais/conocimientoActual.html

Kriebel DL, Dean RG (1993) Convolution method for time-Dependent Beach-profile response. J Waterw Port Coast Ocean Eng 119(2):204–226

Lara-Lara JR, Arreola JA, Calderón LE, Camacho VF, de la Lanza G, Escofet A, Espejel MI, Guzmán M, Ladah L, López M, Alf E, Moreno P, Reyes H, Ríos E, Zertuche JA (2008) Los ecosistemas costeros, insulares y epicontinentales, en Capital natural de México, vol. I: Conocimiento actual de la biodiversidad. Conabio, México, D.F

Link JS, Marshak AR (2019) Characterizing and comparing marine fisheries ecosystems in the United States: determinants of success in moving toward ecosystem-based fisheries management. Rev Fish Biol Fish 29:23–70

Llorente-Bousquets J, Ocegueda S (2008) Estado del conocimiento de la biota, en Capital natural de México, vol. I: Conocimiento actual de la biodiversidad. Conabio, México, D.F

Maestro M, Pérez-Cayeiro ML, Chica-Ruiz JA, Reyes H (2019) Marine protected areas in the 21st century: current situation and trends. Ocean Coast Manag 171:28–36

Méndez-Barrera A (2012) Natural protected areas in Mexico. Int J Wilderness 18(3):25–30

Méndez-Contreras J, Dickinsion F, Castillo-Burguete T (2008) Community member viewpoints on the Ría Celestún biosphere reserve, Yucatan, Mexico: suggestions for improving the community/natural protected area relationship. Hum Ecol 36:111–123

O'Farrell H, Gruss A, Sagarese SR, Babcock EA, Rose KA (2017) Ecosystem modeling in the Gulf of Mexico: current status and future needs to address ecosystem-based fisheries management and restoration activities. Rev Fish Biol Fish 27:587–614

Portman ME (2016) Environmental planning for oceans and coasts, Methods, tools and technologies. Springer. https://doi.org/10.1007/978-3-319-26971-9

Pullin AS, Bangpan M, Dalrymple S, Dickson K, Haddaway NR, Healy JR, Hauari H, Hocley N, Jones JPG, Knight T, Vigurs C, Oliver S (2013) Human well-being impacts of terrestrial protected areas. Environ Evid 2(19):1–41

Romesburg HC (2004) Cluster analysis for researchers. Lulu Press, Morrisville

Salomons W, Turner RK, de Lacerda LD, Ramachandran S (1999) Perspectives on integrated coastal zone management. Springer, Berlin/Heidelberg

Sardá R, O'Higgins T, Cormier R, Diedrich A, Tintoré J (2014) A proposed ecosystem-based management system for marine waters: linking the theory of environmental policy to the practice of environmental management. Ecol Soc 19(4):51

Schmidtbauer J, Ruskule A, Kopti M, Kappeler B, Dael S, Wesolowska M (2017) The ecosystem approach in maritime spatial planning –a check list toolbox. European Union, Baltic SCOPE

SEMARNAT (2013) Informe de la situación del medio ambiente en México. Compendio de Estadísticas Ambientales. Indicadores Clave y de Desempeño Ambiental. Edición 2012. México, D.F

SEMARNAT (2014) El Ambiente en Números. Secretaría de Medio Ambiente y Recursos Naturales (SEMARNAT), México, D.F. http://www.semarnat.gob.mx/sites/default/files/documentos/estadisticas/numeralia_2014.pdf

SEMARNAT-DGEIA (2016) Informe de la situación del medio ambiente en México 2015. Compendio de estadísticas clave, de desempeño ambiental y de crecimiento verde. Dirección General de Estadística e Información Ambiental (DGEIA), Secretaría de Medio Ambiente y Recursos Naturales (SEMARNAT), México, D.F

Sherman K (2015) Food for thought. Sustaining the world's large marine ecosystems. ICES J Mar Sci 72(9):2521–2531. https://doi.org/10.1093/icesjms/fsv136

Sherman K, Adams S (2013) Stress, sustainability, and development of large marine ecosystems during climate change: policy and implementation. United Nations development Programme (UNDP) and Global Environment Facility (GEF), New York

Sherman K, McGovern G (2011) Toward recovery and sustainability of the world's large marine ecosystems during climate change. IUCN, Gland

Sherman K, McGovern G (2012) Frontline observations on climate change and sustainability of large marine ecosystems. United Nations Development Programme (UNDP) and Global Environment Facility (GEF), New York

Sherman K, Aquarone MC, Adams S (2009) Sustaining the World's large marine ecosystems. International Union for the Conservation of Nature (IUCN), Gland

Sherman K, Adams S, Aquarone MC (2010) Scope and objectives of global environment facility supported large marine ecosystems projects, NOAA large marine ecosystem program report, Narragansett

Silvester R, Hsu JRC (1993) Coastal stabilization, Advanced series on ocean engineering 14. World Scientific, Singapore

Sorensen J (1993) The international proliferation of integrated coastal zone management efforts. Ocean Coast Manag 21(1–3):45–80

Thomas H, Bhola N (2017) Delivering ecosystem-based marine spatial planning in practice: an assessment of the integration of the ecosystem approach into UK and Ireland marine spatial plans. World wide Fund for Nature (WWF) and UN environment world conservation monitoring Centre (UNEP-WCMC)

UNEP/CBD (2002) Cancun declaration of like-minded megadiversity countries. United Nations Environmental Program-Convention on Biological Diversity (UNEP-CBD). UNEP/CBD/COP/6/INF/33, The Hague

UNEP/CBD (2016) Like-minded mega-diverse countries carta to achieve Aichi biodiversity Target 11. United Nations Environmental Program-Convention on Biological Diversity (UNEP-CBD). UNEP/CBD/COP/13/INF/45, Cancún

Wei F, Wang S, Fu B, Liu X (2019) Representation of biodiversity and ecosystem services in East Africa's protected area network. Ambio 49:245. https://doi.org/10.1007/s13280-019-01155-4

Chapter 11
Managing the Galapagos National Park: A Systemic Approach Based on Socio-ecological Modeling and Sustainability Indicators

Julia Martínez-Fernández, Miguel Ángel Esteve-Selma, Isabel Banos-Gonzalez, Carolina Sampedro, Carlos Mena, and Jorge Carrión-Tacuri

Abstract The assessment of the sustainability should be tackled with a systemic perspective that enables an integrated analysis of the environmental, social, economic, and institutional factors and their interactions characterizing protected areas, as well as other complex socio-ecological systems. An integrated framework for such assessment is presented with the following key components: (i) the hierarchical definition of sustainability goals and indicators; (ii) a dynamic system model taking into account the key socioeconomic and environmental factors and their interactions, in which the most representative indicators and their sustainability thresholds are integrated; (iii) the analysis of vulnerabilities to exogenous drivers (scenario analysis) and the exploration of available management and planning options (policy assessment); (iv) a multicriteria procedure, in which indicators, sustainability thresholds, model outputs, and scenario and policy analysis are integrated to guide decisions for an improved sustainability in complex socio-ecological

J. Martínez-Fernández (✉)
Fundación Nueva Cultura del Agua, Zaragoza, Spain
e-mail: julia@fnca.eu

M. Á. Esteve-Selma
Departamento de Ecología e Hidrología, Universidad de Murcia, Murcia, Spain
e-mail: maesteve@um.es

I. Banos-Gonzalez
Departamento de Didáctica de las Ciencias Experimentales. Facultad de Educación, Universidad de Murcia, Murcia, Spain
e-mail: ibbg1@um.es

C. Sampedro · C. Mena
Instituto de Geografía, Diego de Robles y Pampite (Cumbaya), Quito, and Universidad San Francisco de Quito, Quito, Ecuador
e-mail: csampedro@usfq.edu.ec; cmena@usfq.edu.ec

J. Carrión-Tacuri
Dirección del Parque Nacional Galápagos, Santa Cruz, Islas Galápagos, Ecuador

A. Ortega-Rubio (ed.), *Socio-ecological Studies in Natural Protected Areas*,
https://doi.org/10.1007/978-3-030-47264-1_11

system. The whole framework integrates a participative approach, mainly for the initial and final steps. In this chapter, some of these components, specifically indicators, dynamic model, and scenario analysis, are applied to one of the most valuable and challenging protected sites in the world: the Galapagos Islands (Ecuador).

Keywords Galapagos National Park · Water indicators · Sustainability · Dynamic model · Scenario analysis

11.1 Introduction

11.1.1 Integrated Tools to Socio-ecological Systems Sustainability

Sustainability is considered a concept that offers a solution to socio-ecological problems, which requires integrated strategies to guide the development of the system so as to allow to meet their needs within the limits of the earth's carrying capacity (Benitez-Capistros et al. 2014). Nevertheless, despite the increasing awareness about the need to move faster forward the implementation of more sustainable policies, their real application is quite far from required (Liu et al. 2015), even within protected areas.

Several barriers and difficulties explain this gap between knowledge and action. One major obstacle is the lack of adequate tools to understand, assess, and communicate the best options for more sustainable systems and to share visions among policy-makers, stakeholders, and other agents regarding key sustainability issues, based on sound scientific knowledge.

Many approaches for sustainability have been based on indicators (Bell and Morse 2005; van Zeijl-Rozema et al. 2011; Benitez-Capistros et al. 2014), since indicators provide a reasonably simple tool that allows the analysis and communication of complex ideas by condensing their multifaceted nature into a manageable amount of meaningful information (Singh et al. 2012). However, in complex socio-ecological systems (SES), the analysis of the interactions between indicators cannot be addressed using traditional, static catalogues of indicators, since they have limited utility to assess policies and plans. To properly assess sustainability, an integrated dynamic systems approach is required (Constanza et al. 2016). Moreover, there is a remarkable need of dealing with long-term dynamics, since sustainable policies can only be successful if they consider longtime horizons. Long-term planning is especially important when short-term decisions have long-term consequences, and it makes it necessary to visualize key issues that may otherwise be missed. Scenario development and policy assessment is one of the major tools to compare the potential outcomes of a variety of alternatives and to anticipate the long-term consequences of scenarios, policy decisions, and actions (Banos-González et al. 2016a, 2018).

In addition, a good system understanding is essential to build shared diagnosis among all actors and to facilitate the involvement of nontechnical stakeholders in the decision-making process and in the management (Perez-Cayeiro and Chica-Ruiz 2015; Tsoukala et al. 2018).

Therefore, the application of an integrated approach in the management of complex protected areas is essential, and it necessarily includes tools that facilitate (i) the conceptualization and comprehension of the complex and dynamic interrelations of the system, as well as the participation of stakeholders and agents to identify the sustainability objectives of the area under study; (ii) the monitorization and assessment of policy and scenarios to generate useful information for decision-making (Kelly et al. 2013; Voinov and Shugart 2013); and (iii) procedures to assist decisions for improving the sustainability and to increase the co-responsibility of stakeholders in their application.

In the following sections of this chapter, we present an integrated methodological approach which answers the gaps launched along the introduction. This framework is general for wide applicability, although flexible enough to include distinct characteristics of the physical, socioeconomic, or cultural environment of any protected site and other socio-ecological systems, as shown here for the case of Galapagos.

11.1.2 The Socio-ecological System of the Galapagos Islands

The Galapagos Islands, made up of 13 main islands and other small islands and islets of volcanic origin, are located in the Pacific Ocean on the 0° parallel, about 1000 km off the coast of Ecuador where they are part of. The isolation, their special climate, and other characteristics have contributed to generate a biodiversity characterized by a relatively small number of species, but with a high proportion of endemisms, extremely important from the scientific point of view (it was one of the main inspirations for Darwin's theory of evolution) and for the preservation of the world genetic heritage. This has been recognized at the national and international levels by different protection status: In 1959, Galapagos was designated by the Ecuador government as a national park; in 1976, as a World Heritage Site by UNESCO; and in 1984, as a biosphere reserve also by UNESCO. In 1986, its inland waters were designated by the Ecuador government as marine resources reserve; in 1990, these waters also became a whales' sanctuary, as designated by UNESCO; and in 2001, Galapagos was also designated by UNESCO as Ramsar site for the protection of wetlands.

The spectacular and accessible terrestrial and marine fauna – with species such as the giant tortoise, the Galapagos penguin, the flightless cormorant, and the marine iguana – have generated a nature tourism that constitutes one of the main economic bases. The human colonization of Galapagos began in the middle of the last century, but it is from 1970, coinciding with the beginning of ecotourism, when a high immigration rate begins. The national park occupies 97% of the archipelago. The remaining 3% constitutes the human settlements area, which encompasses the agricultural and urban areas of the four inhabited islands (Santa Cruz, San Cristobal, Isabela, and Floreana).

However, the human settlements in the Galapagos Islands, even if outside the limits of the national park and occupying only a 3% of the total area in the Galapagos Islands, are source of key socio-ecological impacts that are threatening the natural resources of the islands, unique in the world, as well as the well-being of local communities' development (Martínez et al. 1996; Martínez-Fernández and Esteve Selma 2002; Benitez-Capistros et al. 2014; Pizzitutti et al. 2016). In the last decades, the explosive tourism development posed serious threats to the integrity of the irreplaceable insular ecosystems and to the well-being of local Galapagos populations (Ministerio de Relaciones Exteriores 2007; UN 2007). Pressures associated with tourism, introduction of invasive species, demographic growth, illegal fishing, and governance issues threaten to destroy the endemism of the Islands and its unique biodiversity (Mena et al. 2013). In fact, in 2007, UNESCO included Galapagos within the List of World Heritage in Danger, where it remained until 2010.

On the other hand, Galapagos, as other volcanic insular systems, in addition to a valuable biodiversity and a very high sensibility to alien species, is characterized by the scarcity of water resources. Main annual rainfall in the lower parts of the inhabited islands, where population concentrates, is usually below 100 mm. Main water problems are urban water supply in terms of water quantity and the poor water quality (Mateus et al. 2019), as well as the treatment of wastewaters. Water scarcity also limits the agricultural and livestock production, which reduces the potential food production of the island. This makes it necessary to import a considerable part of food needs, which is one of the main sources of alien species entrance into the islands and, hence, one of the main threats to the conservation of Galapagos biodiversity. Water resources are, therefore, key for the overall sustainability of Galapagos.

How can the 3% of area outside the protected site threaten the biodiversity of the whole Galapagos Islands? To understand this, it is necessary to apply a systemic approach, considering Galapagos as a particular type of socio-ecological systems (SES). This concept refers to systems characterized by social, economic, and environmental factors interacting in a nonlinear fashion, presenting reinforcing mechanisms, which tie the social and ecological system together in patterns of coevolution (Berkes and Folke 1998; Ostrom 2009; Pazmiño et al. 2018).

As other socio-ecological systems, Galapagos might face systemic changes, arising either from an external hazard event or from gradual endogenous change (Filatova and Polhill 2012; Pazmiño et al. 2018). Systems thinking and a holistic approach are essential to integrate the understanding of different causes and impacts (Pizzitutti et al. 2016). The analysis of this dynamics requires a holistic approach, which considers wide time and spatial scales to address the connections between the different factors of the entire SES (Martínez et al. 1996; Hodbod and Adger 2014; Pizzitutti et al. 2016; Elsawah et al. 2017; Sampedro et al. 2018).

Nevertheless, despite the increasing acknowledgment about the need for such holistic approaches, the application of integrated perspectives in protected areas is less frequent than desirable. Among the difficulties behind this, (i) the need for a new conceptual perspective concerning the relationships between science and the management of real systems and (ii) the lack of tools to manage the inherent complexity of such systems should be emphasized. In the following sections, these two difficulties are further discussed with the case study of Galapagos National Park, with a focus on two key issues: (i) the water resources, by means of the development of a Galapagos-specific water indicators system, and (ii) the overall dynamics of this socio-ecological system, by means of a dynamic system model.

11.2 Methodological Approach

11.2.1 General Framework

In this chapter, we propose an overall integrated methodological framework with the following components and tools, in order to understand and improve the overall sustainability of socio-ecological systems, included in the management of protected areas:

1. The hierarchical definition of sustainability objectives, along with their indicators and thresholds
2. A dynamic simulation model, where the above indicators are integrated
3. The assessment of policy options and the analysis of vulnerabilities to external scenarios
4. A procedure to assist decisions for improving the sustainability

Figure 11.1 presents the relationships among the basic components.

This overall approach requires the participation of involved actors, particularly in the first and last stages. Involved actors (local and regional public administrations, economic sectors, technical managers, civil society entities, and citizens) have a key role: (i) at the beginning, in the identification of key problems and sustainability goals to be achieved, (ii) in the initial conceptualization of the dynamic model, and (iii) at the end, in the establishment of valuation preferences, within a decision-making process where results arising from the dynamic model, indicators, and policy and scenario analysis are considered and valued by the involved actors, in order to build consensus on the priority options, on the basis of the best available knowledge.

Usually stakeholders and citizens only participate, if so, in the final stages of any policy implementation. However, it is crucial that such participation takes also place in the definition of problems and objectives and in the assessment of available options to address such problems. This is particularly important to accelerate the adoption of changes (Derkzen et al. 2017). The early and wide engagement of involved actors is essential to achieve some key objectives:

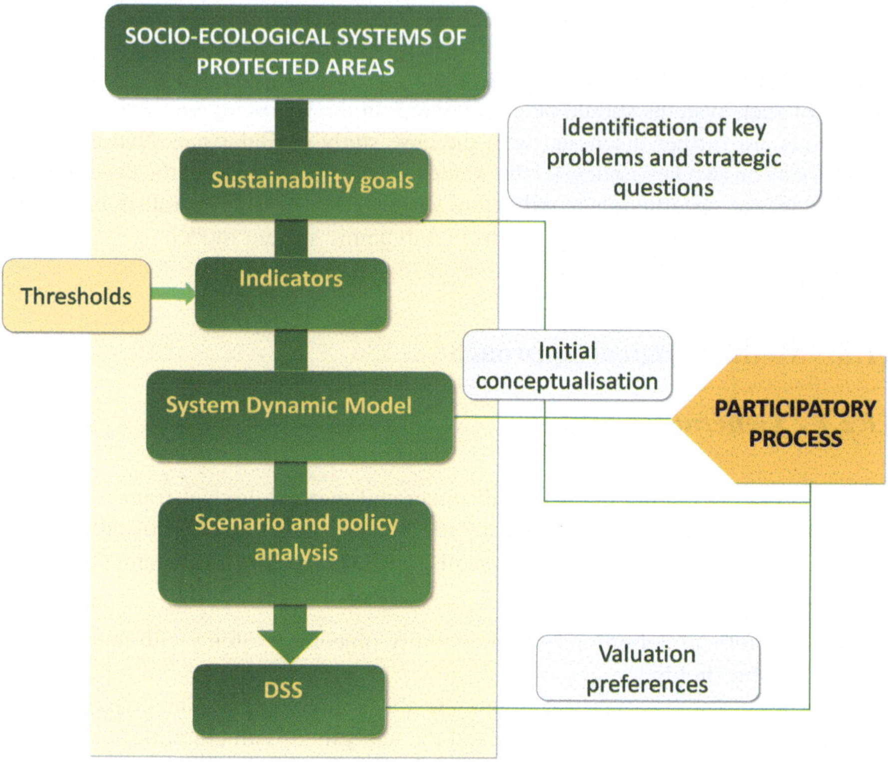

Fig. 11.1 Overall methodological framework to assess socio-ecological systems sustainability, as those of protected sites

- First, a participatory process is required to deliberate and agree on the goals to be achieved in a concerned protected site.
- Second, local actors can contribute to the analysis of problems, causes, consequences, and potential measures to be implemented, by means of the so-called coproduction of knowledge (Brugnach and Ingram 2012; Lepenies et al. 2018). In the integrated methodological framework proposed here, the main stage of knowledge coproduction would be the contribution of different actors to the initial model conceptualization, by means of collaborative modeling and other techniques using participatory processes. The participation of local actors in the initial model conceptualization (by means of group model building and other techniques) not only improves the system understanding by including the available local knowledge but also contributes to build a shared diagnosis on problems, a basic step to later facilitate the agreement on potential solutions.
- Third, an advanced, deliberative, and participative participatory process based on the best available knowledge is required to adopt decisions with the necessary legitimacy.

- Fourth, the wide involvement of stakeholders and other concerned actors along the process is essential to acquire a high level of co-responsibility around the goals to be achieved and the measures to be implemented in order to reach such goals. As Pazmiño et al. (2018) has pointed out, to achieve sustainability goals in socio-ecological systems, including protected sites, a broad collaboration among many actors is needed, from public administrations to stakeholders and different social groups. Such collaboration will be strongly facilitated when all actors feel co-responsible on goals and required actions.

In the following sections, the components and tools showed in Fig. 11.1 are described in detail, and some of them are applied to the case of the Galapagos.

11.2.2 Indicators as a Tool for a Systemic Perspective: The IDIS Approach

One of the difficulties for an adequate understanding of sustainability is the overwhelming amount of information referred to a wide number of aspects, frequently very heterogeneous in terms of level of detail, spatial and temporal scales, and other basic properties. Tools as indicators, helping to reduce and organize the relevant information, are required, so it can be transformed into useful knowledge (Benitez-Capistros et al. 2014).

Indicators allow to monitor and assess key aspects of sustainability in a quantitative way, to carry out comparative analysis, and to provide relevant knowledge in a nontechnical language to policy-makers, managers, stakeholders, and the general public (Kristjánsdóttir et al. 2018). For these purposes, Spangenberg (2002) underlined that indicators should be (i) general (not dependent on a specific situation, culture, or society), (ii) relevant and indicative (representative of the phenomenon they are intended to characterize), (iii) sensitive (reactive to changes in what they are monitoring), and iv) easily computed and understood, useful for communication, and with a minimum of overlap with other indicators (Adriaanse 1993; Bell and Morse 2005).

In recent years, an increasing effort is devoted to the development of sustainability indicators. However, it is claimed that indicators just have a moderate weight on the adoption and assessment of sustainable policies and practices (Reed et al. 2006; Moldan et al. 2012). Among the limitations of conventional catalogues of indicators, we highlight:

- Its static dimension, which do not allow to consider synergies and trade-offs between indicators and the assessment of their future pathways under different management options
- Its reduced significance for each specific case of study
- The "top-down" approach, which reduces the involvement and responsibility of different agents in monitoring objectives through such indicators

- The frequent absence of thresholds to determine whether the changes in the indicators are acceptable or not in terms of sustainability (Moldan et al. 2012)

The IDIS approach, presented here, addresses such limitations.

11.2.2.1 The IDIS Hierarchical Approach

In order to define the indicators, we propose the hierarchical approach graphically described in Fig. 11.2 and outlined as follows:

- Establishment of the *overall sustainability goals* to be achieved.
- Identification of the *dimensions or components of the sustainability* of the concerned protected site or socio-ecological system.
- For each dimension, some *specific objectives* are defined.
- For each specific objective, it is necessary to formulate some *strategic questions*, relevant for the policy-making process, to be answered. The answers to these strategic questions determine whether the system is moving toward achieving the specific objective. These strategic questions must be relevant at the level of policy formulation, communicable in nontechnical language, and of interest to politicians, entities involved in management, stakeholders, and other local actors and citizens in general.
- Finally, to answer each strategic question, one or few relevant indicators are selected.

The final system of indicators derives from the objectives to be achieved; it is specific for each system, applies an integrative perspective, and includes only the required indicators, avoiding redundancy. This approach also aims at carefully

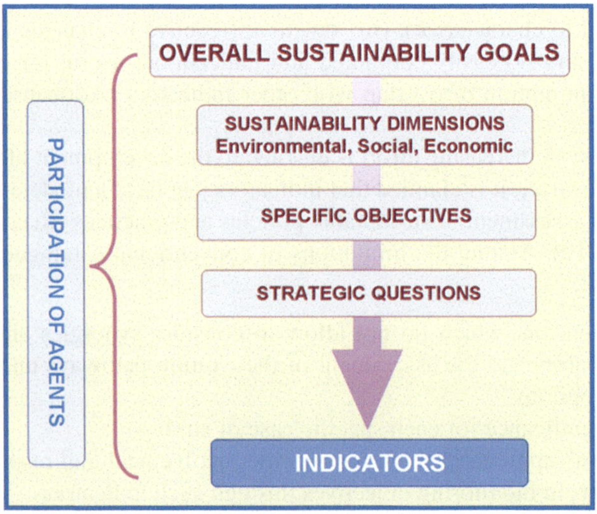

Fig. 11.2 Hierarchical approach for the definition of sustainability indicators

selecting a restricted number of indicators, since a limited and manageable number of indicators create a more useful tool than a large number of unselected ones or indicators not connected with specific objectives (Lancker and Nijkamp 2000).

The participation of the concerned actors contributes to build confidence on the final indicators system and to increase the co-responsibility of policy-makers and stakeholders in its application. To achieve this, the IDIS hierarchical procedure is carried out by means of several participative workshops, using specific techniques to guide the participants along the process.

The hierarchical connections between objectives, strategic questions, and indicators help to define indicator systems that are truly consistent with the overall sustainability goals to be achieved. For instance, efficiency indicators are very important, but an indicators system mainly based on these type of relative indicators (such as per capita or other per unit indicators) might improve while, at the same time, the system is moving away from sustainability (Banos-González et al. 2016a). For example, total water consumption might increase despite the per capita water use showing a reduction. Therefore, these efficiency indicators and their changes over time should be taken with caution (Hanley et al. 2009) to avoid misunderstandings and errors in the diagnosis (Figge and Hahn 2004; Mori and Christodoulou 2012).

11.2.2.2 Sustainability Thresholds

In terms of interpretation of the indicators and to be a useful tool for the decision processes, a quantitative notion of what is acceptable for sustainability – a threshold – is needed (Rodríguez-Rodríguez and Martínez-Vega 2012; Banos-González et al. 2016a). Without thresholds, indicators can describe but are less useful to help decisions. Thresholds allow not only to track the direction and magnitude of changes but also to determine whether such changes are acceptable or not in terms of sustainability (Lancker and Nijkamp 2000; Moldan et al. 2012; Proelss and Houghton 2012).

The identification or definition of sustainability thresholds may be a difficult task but is essential. For some authors, a threshold may be a background value or it can be a meaningful reference value related to the irreversibility of the system (Dahl 2012). Depending on the nature of the indicator, threshold values can be provided by mandatory legal standards, guidelines from different institutions, benchmarking (best practices and experiences from other sites), and reference values taken from historical values of the system.

11.2.3 System Dynamic Model

System dynamic models (SDM) allow to understand the structure and behavior of complex systems, by means of the causal relationships, feedback loops, delays, and other processes of the system (Martínez-Moyano and Richardson 2013). Negative feedback loops, which tend to absorb disturbances and maintain the overall behavior within certain ranges, are also essential features for the sustainability of socio-ecological systems (Suárez et al. 2016).

The application of system dynamic modeling tools allows to facilitate the comprehension of complex systems (Kelly et al. 2013; Martínez-Moyano and Richardson 2013; Elsawah et al. 2017) aimed at generating useful information for decision-making (Jakeman and Letcher 2003; Voinov and Shugart 2013). Another important feature of SDM is its context-specific approach. Context-adapted models are needed to be able to address the concrete problems, challenges, and needs of real protected sites and, therefore, to provide proper solutions (Banos-González et al. 2016b). SDM are particularly appropriate to visualize the overall system, to consider a long-run perspective, to present factors and relationships in a transparent way, and to integrate sustainability indicators. All this makes dynamic models valuable tools for a participatory management, helping in the communication among the scientific-technical, management, and social agents' sides. Furthermore, system dynamic models have been applied to a wide range of problems and issues, including the understanding of entire socio-ecological systems (Vidal-Legaz et al. 2013; Banos-González et al. 2015, 2016a; Pizzitutti et al. 2016; Sampedro et al. 2018; Tenza et al. 2019).

The modeling process involves several iterative stages (Fig. 11.3): development of a conceptual model, formulation of model equations, calibration against the

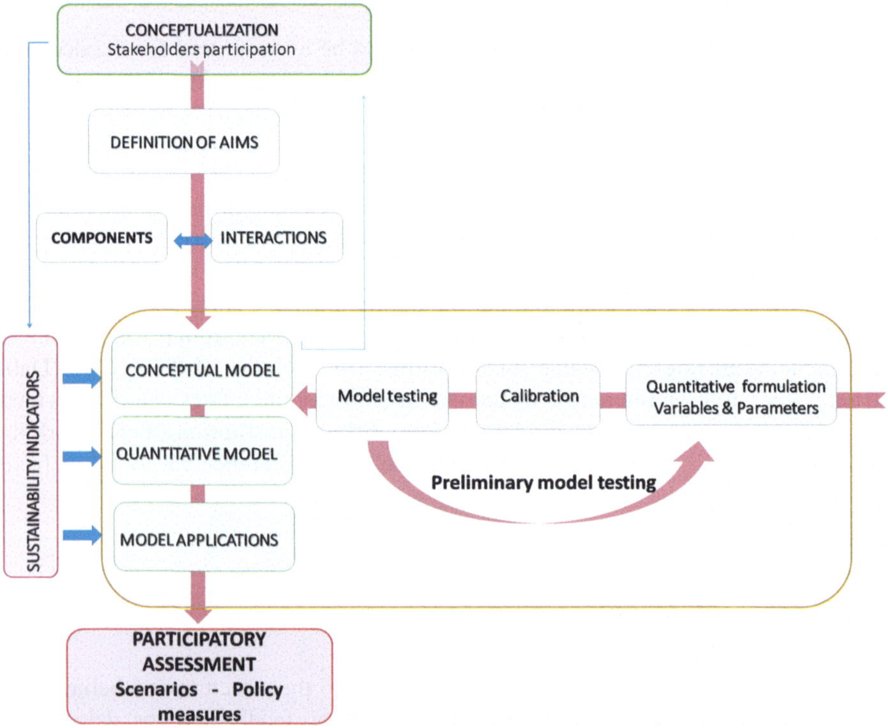

Fig. 11.3 Methodological steps to develop dynamic system models

observed data of main variables, and model testing, by means of structural tests (Barlas 1996), including dimensional consistence tests, sensitive analysis, and extreme condition tests. After successful testing, the model can be used to assess the expected effects of different policies and scenarios.

11.2.4 Assessment of Policies, Scenarios, and Vulnerability

System dynamic models are particularly useful as prospective tools. Scenario development, including policy options, is one of the major tools to visualize and compare the potential outcomes of a variety of policies to meet sustainability and conservation objectives, as well as to anticipate the long-term consequences of policy decisions and actions (Zhang et al. 2015).

In coherence with this, dynamic models are not developed to tell what will happen but to answer "what-if" questions (Han et al. 2009). In the case of SES, including protected areas, these "what-if" questions can refer to actions that can be taken within the modeled system or to changes in the boundary conditions, that is, the factors that condition the system but whose behavior cannot be determined or decided within the system. These two different situations are usually distinguished with the terms *policies* (actions which can be implemented within the system to achieve certain objectives) and *scenarios*. Dynamic models can be applied to assess the expected effects of different policy measures and to explore the vulnerability of the system to certain scenarios, such as climate change or an economic recession.

In relation to policy measures, SDM can be applied to (i) analyze the measures proposed by different agents or test the effectiveness of actions contained, for example, in a conservation plan, (ii) quantify their effects in terms of sustainability indicators and thresholds, (iii) identify side effects and trade-offs among objectives, (iv) determine the degree of uncertainty of the simulation results, and (v) prioritize among measures, as shown in Banos-González et al. (2016a, b).

11.2.5 Procedure to Assist Decisions for Sustainability

Decision Support Systems (DSS) are tools and procedures to assist, facilitate, and support decision-making (Farhan and Lim 2010), which need to cope with multiple aims. These tools and procedures must take into account local carrying capacities, meet the flexibility and complexity of dynamic systems, conduct multicriteria analysis, and integrate multidisciplinary approaches (Touksala et al. 2018). DSS allow to assess policies in evolving environments, as protected areas (de Medeiros et al. 2019).

As opposed to conventional DSS, frequently based on assigning weights to each criteria, which are then combined into mixed indexes to find an optimal solution, in the integrated framework presented in this chapter, a different approach is proposed.

This approach allows to explicitly deal with the complex nature of real problems and the interactions and trade-offs among variables, by means of the indicators embedded in the model. The establishment of threshold values for each indicator provides a way to identify those policies which would exceed the concerned sustainability thresholds. Following the rule "*Threshold out, measure out*," any policy exceeding a threshold should be rejected or assigned the lowest priority (Banos-González et al. 2016a). This approach avoids the use of indexes mixing non-reductible dimensions and, instead, keeps track of the positive and negative effects of each policy or scenario on the different environmental, economic, and social factors involved in the concerned socio-ecological system.

11.3 Results

In this section, we present results for the case of Galapagos National Park regarding two of the steps outlined in Sect. 11.2.1: the development of an indicators system and the elaboration of a dynamic system model. The indicators system addresses the water issues, one of the main sustainability problems in the inhabited islands of Galapagos. The Galapagos Water Indicators System (SIAG) is applied and tested in the case of Santa Cruz, the island with highest population and where water problems are more severe. The dynamic model integrates a number of different submodels to take into account the dynamics of the entire socio-ecological systems of the Galapagos National Park.

11.3.1 Application of the IDIS Approach to Develop the SIAG (Galapagos Water Indicators System)

11.3.1.1 Synthetic Diagnosis of Water Problems

In order to establish the sustainability goals regarding water resources in Galapagos, it is important to carry out a diagnosis on the key water problems, along with their main causes and consequences. A conceptual model, which is also a first step in the development of a dynamic simulation model, constitutes a good basis for such diagnosis. Figure 11.4 shows a preliminary conceptual model of key factors of water resources in Santa Cruz island, whose dynamics is described as follows:

The scarcity of water resources and their low availability have resulted in the public supply pipe network not working in continuous time, leading to intermediate storage of water, which favors pathogen recontamination and prevents the effectiveness of purification. This leads to a very poor quality of public water supply and to the need of buying purified water, relegating the use of public water to secondary uses such as cleaning. In short, what was initially a quantity problem ends up generating or aggravating a quality problem. On the other hand, the low quality of the

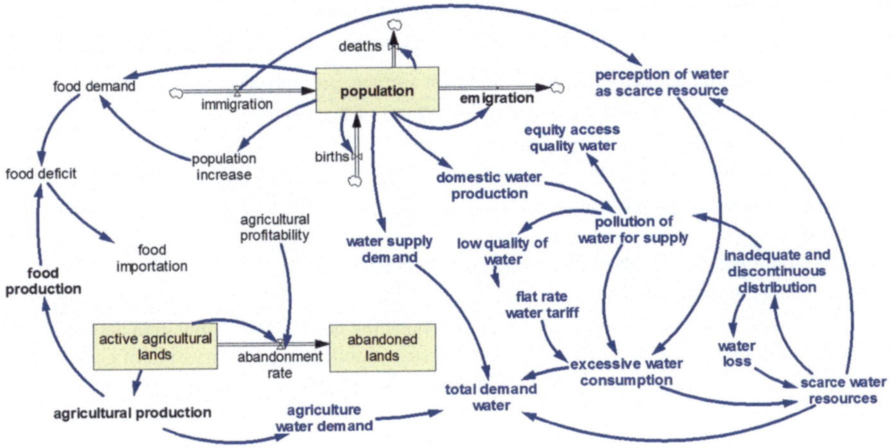

Fig. 11.4 Relationships between quality, quantity, and efficiency in the use of water supply in Santa Cruz, Galapagos (Source: Elaborated by the authors. The interactions between quality, quantity and efficiency in the use of water supply are shown in blue color)

distributed water contributes to maintaining flat rates of water tariff. This flat rate, independent of consumption, along with the low quality of distributed water favors wasteful behaviors, which closes the vicious cycle between scarcity, pollution, and low efficiency in the use of water resources in Santa Cruz.

In Santa Cruz, the synergies between quantity, quality, and efficiency in the use of water resources emerge with a special clarity. These interactions make it difficult for partial solutions to achieve the desired success. All these key factors have been considered in the application of the IDIS methodology to develop the water indicators system regarding the water sustainability goals to be defined (Table 11.1), the strategic questions to be answered, and the specific indicators to be included in the SIAG. Table 11.1 shows these components for the environmental sustainability dimension.

Following the IDIS approach, once the indicators were selected, a target value was identified for each indicator. Various strategies have been used for the establishment of the target values or thresholds, including the adoption of regulatory limit values, recommendations of different international organizations, and benchmarking techniques, that is, comparison with good practices in other territories where a value that can be considered desirable has been reached or established.

Using these threshold values as a reference, a valuation scheme for each indicator has been applied similar to that of the EU Water Framework Directive, which has five status or categories: Very Good, Good, Bad, Very Bad, and Without Enough Information. Each of these categories is displayed with a color code that goes from blue (Very Good Condition) to red (Very Bad Condition), as shown in Table 11.2.

The SIAG (Galapagos Water Indicators System) includes 34 indicators which address four dimensions of sustainability: environmental, social, economic, and

Table 11.1 Galapagos Water Indicators System. Specific goals

Environmental sustainability of water
Goal A1. To maintain the good ecological status of surface, groundwater, and coastal waters
Goal A2. To minimize the energy and emissions impact of the water cycle
Goal A3. To purify all wastewater in a properly way
Social sustainability of water
Goal S1. To guarantee the human right to water for the basic needs of the entire population of Galapagos
Goal S2. To protect the health of the population of Galapagos
Economic sustainability of water
Goal E1. To use water as efficiently as possible to meet the needs of the population and promote agricultural production, a key factor in the overall sustainability and conservation of Galapagos
Goal E2. To move toward the economic and financial sustainability of the integral water management system
Institutional sustainability of water
Goal I1. To implement a public control of the entire integral water cycle
Goal I2. To perform adaptive water management
Goal I3. To establish networks of autonomous and continuous monitoring of the key indicators and to implement systems that centralize the results for consultation by managers, entities, and citizens
Goal I4. To improve the training of the inhabitants of Galapagos regarding the integral water cycle and good citizen practices for its sustainable use

institutional. Table 11.1 shows the specific objectives of each of these dimensions, whereas Table 11.2 displays the strategic questions, indicators, and target or threshold values of the environmental sustainability dimension.

A series of participatory workshops were held to provide inputs for the identification of goals, strategic questions, and potential indicators to be included and, at the end of the process, to validate the Galapagos Water Indicators System. Participants in these workshops included, among others, representatives of different institutions of Galapagos Islands (Galapagos National Park, Council of Galapagos Local Government, Municipality Galapagos Council, Galapagos Water Office), research institutions (Charles Darwin Foundation, universities), and NGOs (World Wildlife Fund for Nature [WWF], Conservation International, and Galapagos Conservancy).

11.3.1.2 Application of the SIAG to Santa Cruz Island

A preliminary application of SIAG was carried out on Santa Cruz island, which is the most populated one and where the water-related problems appear more acutely (Mateus et al. 2019). Due to the important deficiencies regarding the available information, this preliminary application of SIAG on Santa Cruz island only allowed the determination of approximately half of the indicators, as collected in Table 11.3. The application of SIAG on Santa Cruz island has revealed serious information

Table 11.2 Galápagos Water Indicators System (SIAG). Environmental sustainability dimension

Dimension	Sustainability goal	Strategic questions	Indicators	Threshold values
Environmental sustainability	Goal A1. To maintain the good ecological status of surface, underground, and coastal waters	Do we maintain or move forward to achieve a good quality of surface, underground, and coastal waters?	I1. Polycyclic aromatic hydrocarbons	<0.0003 mg/l (TULAS normative)
			I2. Heavy metals (worst value of three parameters): Pb, Cd, and Hg	Pb< 7.2 μg/l
				Cd < 0.15 μg/l
				Hg< 0.05 μg/l
				(directive 2008/105/EU)
			I3. Phenols	<0.001 mg/l (TULAS normative)
			I4. Total nitrogen	<0.2 mg/l (Queensland)
			I5. Total phosphorus	<0.2 mg/l (Queensland)
		Do we manage the basal aquifer in a way that avoids or reduces marine intrusion?	I6. Percentage of control points with saline intrusion	0%
	Goal A2. To minimize the energy and emissions impact of the water cycle	Do we use renewable resources, which require low or no energy contributions?	I7. Proportion of homes with rainwater collection systems	100% (Bermudas' regulations, mandatory system for all homes)
			I8. Proportion of total water demand covered with renewable resources	Potential value to be established by a specific study
		Do we use renewable resources, which reduce the energy dependence of the integral water cycle?	I9. Proportion of energy used in desalination from renewable sources	100% (100% renewable El Hierro project)
	Goal A3. To purify all wastewater in a properly way	Do we avoid untreated wastewater flows?	I10. Proportion of the total population with adequate treatment of wastewater by conventional or unconventional systems	100%
		Do we perform an environmentally adequate management of purines?	I11. Proportion of purines with adequate environmental management	100%

Source: Elaborated by the authors. TULAS: Unified Text of Secondary Environmental Legislation, Ecuador; Queensland: Queensland Government Legislation for coastal waters gaps, since only 15 out of 34 indicators belonging to SIAG were able to be properly determined. Information is lacking particularly regarding the institutional sustainability dimension.

Moreover, preliminary results reveal the magnitude of the pending challenges, since those indicators for which information is available are pretty far from their sustainability thresholds, as shown in Table 11.3. In most of cases, urgent and convincing or forceful measures are required.

11.3.1.3 Sustainability Indicators as a Tool to Manage Protected Sites

The application of the IDIS approach to develop the water indicators system in the Galapagos Islands (SIAG) allows to discuss some of the limitations of conventional indicator catalogues, as explained below.

Table 11.3 Application of SIAG (Galapagos Water Indicators System) to Santa Cruz island. Results of water indicators for which information is available

Dimension	Indicator	Value	Sustainability thresholds	Assessment/Valuation
Environmental sustainability	I2. Heavy metals	In general, inconclusive with the used methods	Pb< 7.2 μg/l Cd < 0.15 μg/l Hg< 0.05 μg/l	Insufficient information. Occasional peaks of Hg
	I3. Phenoles	0.02 mg/l. Mean value 200–2007 y 20011 in 5 Las Ninfas lagoon)	0.001 mg/l	Very bad. Far above the threshold. It requires forceful measures
	I4. Total nitrogen in coastal water	Pick values below 0.1 mg/l in 2007–2008 in Muelle Artesanal	0.2 mg/l	Very good. Far below the threshold. Without immediate risks
	I5. Total phosphorus	0.215 mg/l. Mean value 2005–2012 in Laguna de Las Ninfas	0.02 mg/l	Very bad. Far above the threshold. It requires forceful measures
	I6. Percentage of control points with saline intrusion	Inconclusive trends in the recent period 2005–2013	0%	Insufficient information. Evidence of saline intrusion in previous periods
	I7. Proportion of homes with rainwater collection systems	32% mean value in the island 6% mean value in Puerto Ayora (the urban center)	100%	Very bad. Far below the threshold. It requires forceful measures
	I9. Proportion of energy used in desalination from renewable sources	0% in 2014	100%	Very bad. In 2014 away from the value – objective. Advanced status for commissioning of photovoltaic plants
	I10. Proportion of the total population with adequate treatment of wastewater by conventional or unconventional systems	5% in 2014	100%	Very bad. Far below the threshold. It requires forceful measures
	I11. Proportion of purines with adequate environmental management	0% in 2014	100%	Very bad. Far below the threshold. It requires forceful measures
Social sustainability	I12. Population without access to potable water, at affordable price	66%	0%	Very bad. Far above the threshold. It requires forceful measures
	I13. Fecal coliforms in final consumption	Between 12 y 39 UFC/100 ml in final consumption	Below 1.1 NMP/100 ml in final consumption	Very bad. Far above the threshold. It requires forceful measures
Economic sustainability	I17. Proportion of losses in the urban supply network	70%	10%	Very bad. Far above the threshold. It requires forceful measures
	I21. Proportion of population with double gray/black water system	0%	100%	Very bad. Far below the threshold. It requires forceful measures
	I22. Proportion of purified wastewater reused in agricultural irrigation	0%	100%	Very bad. Far below the threshold. It requires forceful measures
Institutional sustainability	I32. Proportion of water indicators and parameters accessible online in the Water Information System	0%	100%	Very bad. Far below the threshold. It requires forceful measures

Static Versus Dynamic Indicators The static catalogues of indicators cannot take into account the interactions and trade-offs among indicators, by which the improvement in some indicators may cause a worsening in others (Banos-González et al. 2016a). Therefore, it is important to integrate the indicators within a dynamic model (Vidal-Legaz et al. 2013; Banos-González et al. 2015; Liu et al. 2015). Therefore, the next step would be to integrate the SIAG indicators within the Galapagos dynamic model covering the key sustainability dimensions.

Need of Context Specificity The SIAG includes several indicators regarding the resilience to climate change in the urban areas of Galapagos, characterized by a very arid climate. One of these indicators is the proportion of households and other buildings, which have devices for rainwater collection, a traditional water system that is being lost during the last decades. This system of water supply has other advantages, as the provision of high-quality water for basic human needs with no or very low input of energy. However, some studies have included as indicator of poverty in Galapagos the existence of water supply systems different to the public network (Granda Leon et al. 2013), since this is one of the indicators of poverty being applied in the continental Ecuador. The consideration of rainwater collection as an indication of poverty contributes to a negative perception of this device (Guyot-Tephany et al. 2013) and, therefore, is counterproductive to increase the overall resilience of the urban areas in Galapagos. This exemplifies how the direct translation of indicators from general catalogues can be inadequate to specific contexts, such as the Galapagos Islands, where rainwater collection is not related to poverty and, in fact, it should be promoted to increase Galapagos resilience to climate change.

Participatory Versus Top-Down Approach The involvement of stakeholders is crucial in the development of sustainability indicators (Verma and Raghubanshi 2018). A top-down, nonparticipatory approach does not facilitate the co-responsibility of all actors in the effective application and follow-up of the indicators. In the case of SIAG, water managers and stakeholders have participated in its development, which has contributed to improve the initial proposal and to increase the interest of involved agents on an effective application of the indicators system.

Need of Thresholds A frequent weakness of many catalogues of indicators is the lack of reference values. Without thresholds, the indicators can describe but are less useful as valuation tools to guide decisions. In addition to monitor the direction and magnitude of change that indicators can carry out, thresholds allow to determine whether such change is acceptable or not regarding the specific sustainability goals to be achieved (Lancker and Nijkamp 2000; Moldan et al. 2012; Proelss and Houghton 2012). In the case of SIAG, a threshold was identified for each of the 34 indicators what allowed to get some measure of the distance to goal and, therefore, to better prioritize the actions to be taken.

As explained earlier, the next step would be to integrate the water indicators within the Galapagos socio-environmental model, described below, as part of a water sector, in order to capture the links between water and the other social and environmental components of the Galapagos dynamics. The next section presents an overview of the Galapagos dynamic model, where the water sector will be implemented in future works.

11.3.2 Socio-environment Systems Model for the Galapagos: A Linkage Between Tourism, Biodiversity, and Infrastructure

11.3.2.1 Overview

The Galapagos Archipelago and its ecosystems evolved through a natural process in isolation, becoming a unique place with high endemism (Jackson 2007). Additionally, its terrestrial and marine ecosystems are in relatively good conservation status (Le Saout et al. 2013). External pressures, including tourism, invasive species, climate change, and endogenous drivers, including local population growth, have increased the degree of vulnerability of the Galapagos National Park and Marine Reserve (González et al. 2008). Galapagos is a fundamental place for conservation and science.

The Galapagos Islands is one of the 24 provinces of Ecuador, with a total population of 25,244 inhabitants in 2015 (INEC 2015), accounting for 0.18% of Ecuador's total population (INEC 2012). Approximately 95% corresponds to local residents and 5% can be considered short-term migration (Salvador Ayala 2015). The population is divided into four inhabited islands: 62% in Santa Cruz, 28.5% in San Cristobal, 9% in Isabela, and 0.5% in Floreana (INEC 2015).

Local population growth has been 4.54% per year during the period of 1974–2015 (INEC 1974, 2015). Several factors have influenced the rapid population rate, the special national and local development policies, and, mainly, the spontaneous economic growth of the tourism industry, which has stimulated immigration toward the islands and has changed the social, economic, and ecological dynamics.

Tourism industry has followed an exponential growth from 3056 tourists in 1962 to 182,037 tourists in 2018, becoming the main economic driver of the islands. Thus, in the early 2000s, the tourism industry employed, through direct or indirect employment, 40% of Galapagos resident's population and represented 65.4% of the local economy (Kerr et al. 2004; Taylor et al. 2009). In addition, tourism explained the 74% of the 78% increase in the GDP from a base of USD$41 million between 1999 and 2005 (Espín 2016).

Despite of the importance of tourism industry as driver of the dynamics in the Galapagos Islands, it has been poorly studied. The magnitude of the direct and indirect impacts on the natural and protected ecosystem, as well as in the inhabited areas and local residents and on the tourism industry itself, remains mostly unknown. Some of the factors to be considered are the population growth, introduction of invasive species, increased local consumption, pollution, pressure on basic services infrastructure, and illegal fishing, among others. These factors threaten to destroy the endemisms of the islands, risking the survival of its unique biodiversity (Walsh and Mena 2013).

In 2010, the local and national government asked several institutions for the generation of future scenarios that would support decisions about the socio-environmental system. Consequently, the project "Determination of the Social, Environmental, and Economic Scenarios for Sustainability of the Socio-ecological System of the Galápagos Islands" was created. The goal of the project is to contribute with a scenario-planning tool to be used by policy-makers to examine tourism management in the Galapagos Islands through the analysis of the impacts of tourism and local population growth on different subsystems of the islands.

System dynamic modeling was applied to represent components and relationships of this coupled human and environmental system as it allows the representation of the variables that form the system and quantify its relationships and establishes present and future behavioral patterns. Several studies present a model-based approach to the management of tourist destinations, with emphasis on natural environment conservation (Walker et al. 1998; Georgantzas2003; McGrath 2010; Law et al. 2012), and others are specifically applied to study tourism in islands (Wiranatha and Smith 2000; Hernández and León 2007; Xing and Dangerfield 2011), proving to be an efficient approach in policy analysis and decision-making in development planning and resources management.

The SDM generated very interesting insights about the Galapagos system. It generated i) a conceptual model of the relationships between the main subcomponents of the system and ii) a set of impacts, resulting from the social and environmental interactions from a broad perspective, to assess possible results of different public policy approaches in the social, environmental, and economic subsystems. The sub-models included in this SDM model are tourism and population growth, the introduction of alien species, protected area tourism management, land occupation, land tourism infrastructure, goods imports, electrical production, carrying capacity of the Galapagos National Park, population and employment, and agricultural production and imports (Espín 2016; Pizzitutti et al. 2016; Sampedro 2018).

11.3.2.2 Methods

The modeling process was divided into three stages: (1) conceptualization of the dynamic relationships, (2) model building in Vensim, and (3) model testing and validation. The first phase addresses the definition of variables, limits, and scale of the study system, as well as the description of the problem to be represented. We identified the main variables to be included in the model and designed its structure based on a theoretical framework. The set of dynamic relationships are represented by means of diagrams which guided the modeling efforts and data collection.

The second phase is devoted to build the simulation model based on a conceptual framework and available data. It includes a model that represents components and dynamic relationships using mathematical equations to represent the behavior of each variable along time. Vensim software was used for modeling, using a 5 years' time period of historical data and accounting for 20 years in the future. The third phase is intended to generate checks in the dimensional and mathematical consistency of the model and to validate the model by comparing the simulated behavior to the actual data about the system.

A key challenge of the system implementation was to find a way to include the views and concerns of different stakeholders. This tool was intended for participation of decision-makers in the process during the three phases – conceptualization and design, construction, and verification of the model. This time, the institutions that were part of the project meetings were the Ministry of Environment, Ministry of Tourism, Galapagos National Park, Council of Galapagos Local Government, and the Municipality Galapagos Council, nongovernmental organizations such as WWF and Conservation International, as well as representatives of several resident groups and community institutions.

11.3.2.3 Model Implementation

The approach presented in this chapter, through a participative methodology, generates three possible scenarios of tourism growth for the next 20 years (2013–2033): (1) zero growth, (2) moderate growth, and (3) accelerated growth (Pizzitutti et al. 2016). In the zero growth scenario, the tourist arrivals remain frozen in time to the level reached in 2012. This scenario hypothesizes that the flow of tourist arrivals froze as a result of improved management of the Galapagos Islands. The second scenario, moderate growth, assumes that the number of tourists will change by the same number of persons each year in the future as the average annual absolute change observed over the base period, which is 7066 tourists per year. The third scenario, the accelerated growth, considers that the number of tourists will grow following the same average annual growth rate during each year in the future as it did during the base period – which is 8% for foreign tourists and 7.6% for national tourists and is based in an exponential function. The last two scenarios were calibrated with respect to the historical database of tourist arrivals recorded in the last 20 years (1992–2012). These three scenarios were the base of the model to build the resident population projections assuming a direct relationship between the number of tourist arrivals and the growth in the local population, through an ordinary least squares regression of census counts of residents from the Ecuadorian National Institute of Statistics and Census (INEC 1982, 1990, 2001, 2010). For the scenario of zero growth, the population growth is calculated as the natural demographic growth.

As for the number of tourists, the model considers two categories: (1) land-based tourism, which depends on the local products, services, and labor and has shown a rapid tendency to increase, and (2) boat-based tourism, which responds to an all-inclusive package dynamic that keeps the tourist on board most of the time, consuming onboard products, which are usually collected on the continent.

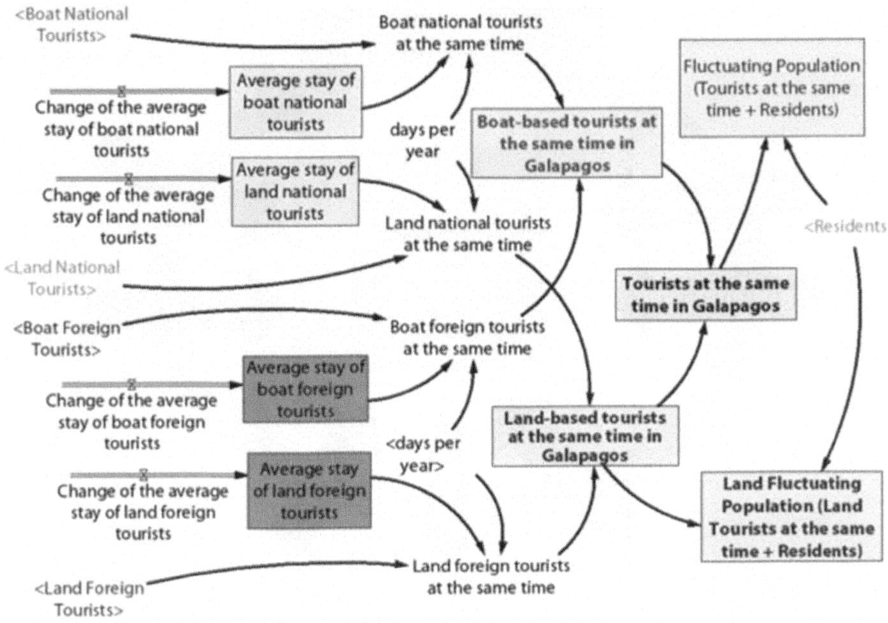

Fig. 11.5 Stock flow diagram of calculations of the number of tourists at the same time in Galapagos (Pizzitutti et al. 2016)

To account for the impacts and demands of goods and services due to the presence of tourists, the variable "tourists at the same time" was created, which calculates the average number of tourists that are in any day of the year in Galapagos (Pizzitutti et al. 2016). Thus, the average number of people in Galapagos (total fluctuant population) corresponds to the number of residents plus the number of tourists at the same time. Refer to Fig. 11.5 to address the sub-model of total arrivals and resident population and tourists at the same time in Galapagos.

There are nine sub-models that are derived and related to the calculation of the number of tourists and residents and allow to identify the future tendency of the Galapagos in important issues as environmental conservation, consumption of food and goods and basic services, among others (Fig. 11.6).

Here, the nine sub-models are briefly described:

1. The ***Introduced Species*** module is based on the Charles Darwin Foundation database of registered species in Galápagos since 1960 (Charles Darwin Foundation 2014). The system generates an alteration index – expressed as the fraction or percentage over time of invasive species with respect to the total species, found each year.
2. The ***Protected Area Tourism Management*** sub-model uses the values of maximum daily total capacity for boats and land visit sites established by the Galapagos National Park. We established the numerical relationship between the

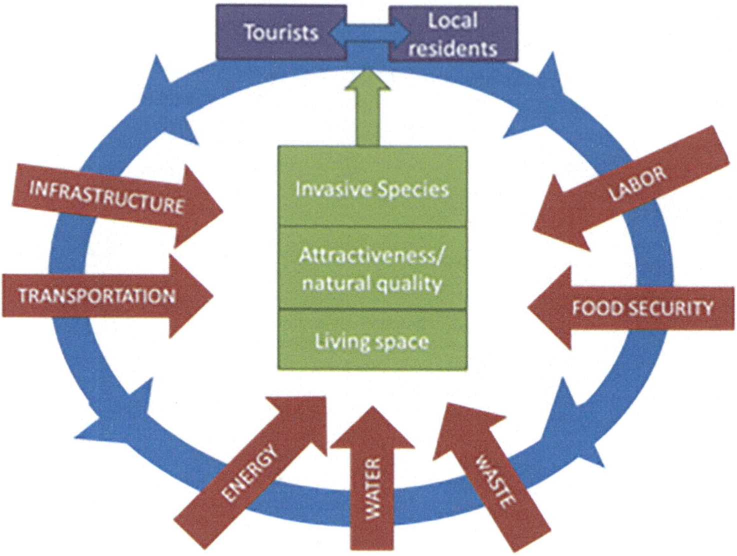

Fig. 11.6 Model conceptualization

level of tourists' arrivals and the capacity of the Galapagos National Park. Then the "load" can be directly compared with the projections of tourist arrivals to determine the ability of the national park to receive tourists under each projection.

3. ***Land Occupation*** represents the dynamics by which land covered by natural vegetation will be transformed into land managed for human activities, such as urbanization, roads, and agriculture. Satellite image processing was key to gather multi-temporal information.
4. ***Land Tourism Infrastructure*** simulates the dynamics of hotel development in Galapagos. The most important variables shown in this subsection are number of hotels and hotel occupancy rate.
5. ***Electrical Production*** explains the amount of energy generated by fossil fuels and by clean energy and also capture the dynamics of consumption tendencies due to the evolution of tourism and population growth.
6. The ***Transportation Module*** models the dynamics of the different types of transportation in the Galapagos, including the dynamics of ships for importation of goods and materials from the mainland, interisland maritime cabotage, land transport, and air transport.
7. ***Solid Waste*** captures the dynamics of the solid waste generation per person and per segment of populations linked to the final destination and represented by the amount of solid waste recycled and the waste accumulated in landfills.
8. ***Immigration and Labor*** characterizes population, labor, and its links to environmental and social variables. This module computes two aggregated indexes that

represent the overall progress in terms of income, social capital, and perceptions about conservation. An integrated socio-ecological evaluation of the sustainability of tourism strategies is carried out through the cross comparison of these indexes.

9. The ***Agricultural Production and Food Imports*** module focuses on the agricultural and livestock production, based on food consumption data and the influence of tourism in agricultural land abandonment. Also, this module looks at nutrition and food transition of population and projects the consequences on food imports.

11.3.2.4 Results of the Galapagos SDM

The simulations and projections of the Galapagos SDM indicate that the levels of alteration will continue to grow according to the level of local and tourist population increase. In addition, under current conditions, Galapagos, being a nature tourist destination that attracts thousands of tourists a year, is close to its saturation level, in different processes. If tourism continues to grow at the pace of the last two decades, depending on the type of growth, in a few years, there will be no physical space to be able to ensure every tourist who visits Galapagos a nature tourism experience with high-quality standards.

On 2016, an actualization of the model with historical data available up to 2015 showed that the accelerated scenario is the one that is representing best the behavior of tourism and population growth in the islands. Thus, following the tendency for 2035 is expected a total fluctuant population of 86,861 people, from which 15,927 are tourists, explained by a tourism increase rate of 8.1% per year. Therefore, the reception capacity of the Galapagos National Park may probably saturate in the early 2020s, and the hotel occupancy rate will grow to 213% on 2035, from 27% in 2015.

In addition, the tendency in species introduction in the islands will increase from 24% in 2015 to 49% in 2035, largely due to the increase by almost 95% in the import of agricultural food to the islands. In this regard, the Galapagos food security will be at risk, becoming fully dependent on the imports from the continent. Therefore, the number of ships for imports will raise from 4 ships to 16 on 2035, to satisfy the food and goods requirement. The fraction of electricity produced by clean energy will decrease from 12% in 2015 to 3.3% in 2035, while the electricity consumption will increase. In addition, the production of solid waste also will increase from 8454 tons/year in 2015 to 38,496 tons/year in 2035.

The results of the different scenarios for the sub-models are more than a prediction: They show us the path that the Galapagos is following. They allow us to understand how the social and environmental systems will react and interact to current decisions and policies, identifying the most efficient interventions to redirect this path to the future objective we are aiming for this special place.

11.4 Lessons Learned

The case of the Galapagos National Park and the work presented in this chapter have provided three important learned lessons that we summarize as follows:

(i) ***The conservation of protected sites strongly depends on the dynamics outside such protected sites***. As it has been shown, the area devoted to human settlements, all of them outside the Galapagos National Park, only represents a 3% of total territory of Galapagos Islands, whereas 97% corresponds to the Galapagos National Park. Despite this, the socio-environmental dynamics put in motion in this 3% outside the national park is seriously threatening the conservation of Galapagos. This is explained by the interactions among the different factors, particularly by the growth of the tourist industry, leading to an increased immigration rate, therefore to higher demands of food and finally to higher imports from the continent, which increases the rate of introduction of foreign species into Galapagos, the most serious threat to the Galapagos biodiversity.

(ii) ***Integrated methodological frameworks are needed for a deep understanding of problems, their causes, and the expected benefits, limitations, and unintended effects of different policies and management options***. We have shown how indicators, sustainability thresholds, simulation modeling, and scenario analysis can contribute to guide decisions based on the best available knowledge. Indicators are required to capture and reduce the complexity of socio-ecological systems, as shown with the water indicators system of Galapagos (SIAG). Indicators should be specifically developed to address the goals to be achieved in each protected site. Moreover, sustainability thresholds help to decide on the acceptability of indicator values regarding the desired situation. Finally, dynamic simulation models allow a comprehensive understanding of the whole socio-ecological system and constitute a helpful tool to explore the system response under different scenarios and policy options.

(iii) ***Participation is crucial for an advanced governance of protected sites.*** Stakeholders, public administrations, NGOs, and other local actors have provided valuable inputs to develop the water indicators system and the Galapagos dynamic model, showing how local knowledge can be integrated to improve the diagnosis to be built and the tools to be applied. Besides improving diagnosis by integrating local knowledge, participatory approaches also contribute to build a shared vision on problems and their causes, which constitutes a first step toward a shared vision on potential solutions. Finally, participatory approaches are crucial to support decisions and to engage all actors on the goals to be reached and on the measures to be adopted in order to achieve such goals.

References

Adriaanse A (1993) Environmental policy performance indicators. Nature indicators survey. Ministry of Environment and Energy, Denmark

Banos-González I, Martínez-Fernández J, Esteve-Selma MA (2015) Dynamic integration of sustainability indicators in insular socio-ecological systems. Ecol Model 306:130

Banos-González I, Martínez-Fernández J, Esteve-Selma MA (2016a) Using dynamic sustainability indicators to assess environmental policy measures in Biosphere Reserves. Ecol Indic 67:65–576

Banos-González I, Martínez-Fernández J, Esteve-Selma MÁ (2016b) Tools for sustainability assessment in island socio-ecological systems: an application to the Canary Islands. Isl Stud J 11(1):9–34

Banos-González I, Martínez-Fernández J, Esteve-Selma MÁ, Esteve-Guirao P (2018) Sensitivity analysis in socio-ecological models as a tool in environmental policy for sustainability. Sustainability 10(8):2928

Barlas Y (1996) Formal aspects of model validity and validation in system dynamics. Syst Dyn Rev 12(3):183–210

Bell S, Morse S (2005) Holism and understanding sustainability. Syst Pract Action Res 18(4):409–426

Benitez-Capistros F, Hugé J, Koedam N (2014) Environmental impacts on the Galápagos Islands: identification of interactions, perceptions and steps ahead. Ecol Indic 38:113–123

Berkes F, Folke C (1998) Linking sociological and ecological systems for resilience and sustainability. In: Berkes F, Folke C (eds) Linking sociological and ecological systems: management practices and social mechanisms for building resilience. Cambridge University Press, New York, pp 1–25

Bosch OJ, Nguyen NC, Maeno T, Yasui T (2013) Managing complex issues through evolutionary learning laboratories. Syst Res Behav Sci 30(2):116–135

Brugnach M, Ingram H (2012) Ambiguity: the challenges of knowing and deciding together. Environ Sci Policy 15:60–71

Charles Darwin Foundation (2014) Charles Darwin Foundation. Retrieved from Galapagos Species List website: https://www.darwinfoundation.org/en/datazone. Accessed 20 Aug 2019

Chen H, Chang YC, Chen KC (2014) Integrated wetland management: an analysis with group model building based on system dynamics model. J Environ Manag 146:309–319

Constanza R, Daly L, Fioramonti L, Giovannini E, Kubiszewski I, Mortensen LF, ... & Wilkinson R (2016) Modelling and measuring sustainable wellbeing in connection with the UN Sustainable Development Goals. Ecological Economics, 130, 350–355. https://doi.org/10.1016/j.ecolecon.2016.07.009

Dahl AL (2012) Achievements and gaps in indicators for sustainability. Ecol Indic 17:14–19

De Medeiros M, de Carvalho CRS, Montalván RAV et al (2019) Land use of the environmental protected area of the coastal environment of Serra do Tabuleiro State Park-Palhoça/SC, Brazil: zoning and environmental restrictions. Environ Dev Sustain 21:1225

Derkzen ML, van Teeffelen AJA, Verburg PH (2017) Green infrastructure for urban climate adaptation: how do residents' views on climate impacts and green infrastructure shape adaptation preferences? Landsc Urban Plan 157:106–130

Elsawah S, Pierce SA, Hamilton S et al (2017) An overview of the system dynamics process for integrated modelling of socio-ecological systems: lessons on good modelling practice from five case studies. Environ Model Softw 93:127–145

Espín P (2016) Model-based assessment of tourism sustainability in island ecosystems: understanding the dynamics of the Galapagos Islands. Dissertation, Universidade Nova de Lisboa

Espin PA, Mena CF, Pizzitutti F (2019) A model-based approach to study the tourism sustainability in an island environment: the case of Galapagos Islands. In: Kvan T, Karakiewicz J (eds) Urban Galapagos, social and ecological interactions. Springer, Cham, pp 97–113

Farhan AR, Lim S (2010) Integrated coastal zone management towards Indonesia global ocean observing system (INA-GOOS): review and recommendation. Ocean Coast Manag 53(8):421–427
Figge F, Hahn T (2004) Sustainable value added—measuring corporate contributions to sustainability beyond eco-efficiency. Ecol Econ 48(2):173–187
Filatova T, Polhill G (2012) Shocks in coupled socio-ecological systems: what are they and how can we model them?. International Environmental Modelling and Software Society (iEMSs). In: International Congress on Environmental Modelling and Software Managing Resources of a Limited Planet, Sixth Biennial Meeting, Leipzig
Georgantzas NC (2003) Tourism dynamics: Cyprus' hotel value chain and profitability. Syst Dyn Rev 19(3):175–212
González JA, Montes C, Rodríguez J, Tapia W (2008) Rethinking the Galapagos Islands as a complex social-ecological system: implications for conservation and management. Ecol Soc 13(2)
Granda Leon M, González Camba S, Calvopiña Carvajal V (2013) Medición de pobreza en Galápagos. In: Informe Galápagos 2011–2012. DPNG, CGREG, FCD y GC, Puerto Ayora, Galáagos
Guyot-Tephany J, Grenier C, Orellana D (2013) Usos, percepciones y manejo del agua en Galápagos. In: Informe Galápagos 2011–2012. DPNG, CGREG, FCD y GC. Puerto Ayora, Galápagos, pp 67–75
Han J, Hayashi Y, Cao X, Imura H (2009) Application of an integrated system dynamics and cellular automata model for urban growth assessment: a case study of Shanghai, China. Landsc Urban Plan 91(3):133–141
Hanley N, McGregor PG, Swales JK, Turner K (2009) Do increases in energy efficiency improve environmental quality and sustainability? Ecol Econ 68(3):692–709
Hernández JM, León CJ (2007) The interactions between natural and physical capitals in the tourist lifecycle model. Ecol Econ 62(1):184–193
Hjorth P, Bagheri A (2006) Navigating towards sustainable development: a system dynamics approach. Futures 38(1):74–92
Hodbod J, Adger WN (2014) Integrating social-ecological dynamics and resilience into energy systems research. Energy Res Soc Sci 1:226–231
INEC (1974) Censo de Población y Vivienda. Quito
INEC (1982) Censo de Población y Vivienda 1982. Quito
INEC (1990) Censo de Población y Vivienda 1990. Quito
INEC (2001) Censo de Población y Vivienda 2001. Quito
INEC (2010) Censo de Población y Vivienda 2010. Quito
INEC (2012) Ecuador: Proyección de Población de Población por Provincias. Quito
INEC (2015) Censo de Población y Vivienda. Ecuador
Jackson M (2007) Galapagos: a natural history. University of Calgary, Calgary
Jakeman AJ, Letcher RA (2003) Integrated assessment and modelling: features, principles and examples for catchment management. Environ Model Softw 18(6):491–501
Kelly RA, Jakeman AJ, Barreteau O et al (2013) Selecting among five common modelling approaches for integrated environmental assessment and management. Environ Model Softw 47:159–181
Kerr S, Cardenas S, Hendy J (2004) Migration and the environment in the Galapagos: an analysis of economic and policy incentives driving migration, potential impacts from migration control, and potential policies to reduce migration pressure. In: Motu Working Paper Series No. 03-17. Motu Economic and Public Policy Research, Wellington
Kristjánsdóttir KR, Ólafsdóttir R, Ragnarsdóttir KV (2018) Reviewing integrated sustainability indicators for tourism. J Sustain Tour 26(4):583–599
Lancker E, Nijkamp P (2000) A policy scenario analysis of sustainable agricultural development options: a case study for Nepal. Impact Assess Proj Apprais 18(2):111–124
Law A, De Lacy T, McGrath GM et al (2012) Towards a green economy decision support system for tourism destinations. J Sustain Tour 20(6):823–843

Lepenies R, Hüesker F, Beck S, Brugnach M (2018) Discovering the political implications of coproduction in water governance. Water 10(10):1475

Le Saout S, Hoffmann M, Shi Y et al (2013) Conservation. Protected areas and effective biodiversity conservation. Science 342(6160):803–805

Liu J, Mooney H, Hull V, Davis SJ et al (2015) Systems integration for global sustainability. Science 347(6225):1258832

Lotze-Campen H (2008) The role of modelling tools in Integrated Sustainability Assessment (ISA). Int J Innovat Sustain Dev 3(1–2):70–92

Martínez J, Esteve MA, Cruz E (1996) Islas Galápagos: problemas bajo un caparazón paradisíaco. Ecosistemas 16:38–45

Martínez-Fernández J, Esteve Selma MA (2002) Las dificultades prácticas del desarrollo sostenible en las Islas Galápagos (Ecuador). El papel de la participación pública y la cooperación. In: Montes A (ed) Debates en Cooperación al Desarrollo. Servicio de Publicaciones Universidad de Murcia, Murcia, pp 195–215

Martínez-Moyano IJ, Richardson GP (2013) Best practices in system dynamics modeling. Syst Dyn Rev 29(2):102–123

Mateus C, Guerrero CA, Quezada G et al (2019) An integrated approach for evaluating water quality between 2007–2015 in Santa Cruz Island in the Galapagos Archipelago. Water 11(5):937

McGrath GM (2010) Towards improved event evaluation and decision support: a systems-based tool. Proc Annu Hawaii Int Conf Syst Sci 7(1):1–10

Mena CF, Walsh S, Pizzitutti F et al (2013) Determinación de las relaciones sociales, ambientales y económicas que permitan desarrollar, en base a procesos de modelación, potenciales escenarios de sostenibilidad del sistema socio-ecológico de las Islas Galápagos con énfasis en la dinámica del flujo de visitantes al archipiélago. Ministerio del Ambiente Ecuador, Parque Nacional Galápagos. Galapagos Science Center

Ministerio de Relaciones Exteriores (2007) UNESCO inscribe a las islas Galápagos en la listade patrimonios en peligro. http://www.cancilleria.gob.ec/unesco-inscribe-a-las-islas-galapagos-en-la-lista-de-patrimonios-enpeligro. Accessed 20 Aug 2019

Moldan B, Janoušková S, Hák T (2012) How to understand and measure environmental sustainability: indicators and targets. Ecol Indic 17:4–13

Mori K, Christodoulou A (2012) Review of sustainability indices and indicators: towards a new City Sustainability Index (CSI). Environ Impact Asses 32(1):94–106

Ostrom E (2009) Institutional rational choice: an assessment of the institutional analysis and development framework. In: Sabatier P (ed) Theories of the policy process, 2nd edn. Westview Press, Boulder, pp 21–64

Pazmiño A, Serrao-Neumann S, Low Choy D. (2018) Towards Comprehensive Policy Integration for the Sustainability of Small Islands: A Landscape-Scale Planning Approach for the Galápagos Islands. Sustainability, 10(4), 1228. https://doi.org/10.3390/su10041228

Perez-Cayeiro ML, Chica-Ruiz JA (2015) Evaluation of a programme of integrated coastal zone management: the Ecoplata Programme (Uruguay). Mar Policy 51:527–535

Perz SG, Muñoz-Carpena R, Kiker G, Holt RD (2013) Evaluating ecological resilience with global sensitivity and uncertainty analysis. Ecol Model 263:174–186

Pizzitutti F, Walsh SJ, Rindfuss RR et al (2016) Scenario planning for tourism management: a participatory and system dynamics model applied to the Galapagos Islands of Ecuador. J Sustain Tour 9582(November):1–21

Poveda CA, Lipsett MG (2014) An integrated approach for sustainability assessment: the Wa-Pa-Su project sustainability rating system. Int J Sustain Dev World 21(1):85–98

Proelss A, Houghton K (2012) The EU common fisheries policy in light of the precautionary principle. Ocean Coast Manag 70:22–30

Quiroga D (2009) Crafting nature: the Galapagos and the making and unmaking of a "natural laboratory". J Polit Ecol 16(1):123–140

Reed MS, Fraser ED, Dougill AJ (2006) An adaptive learning process for developing and applying sustainability indicators with local communities. Ecol Econ 59(4):406–418

Rodríguez-Rodríguez D, Martínez-Vega J (2012) Proposal of a system for the integrated and comparative assessment of protected areas. Ecol Indic 23:566–572
Safavi HR, Golmohammadi MH, Sandoval-Solis S (2015) Expert knowledge based modeling for integrated water resources planning and management in the Zayandehrud River Basin. J Hydrol 528:773–789
Sahin O, Mohamed S (2014) Coastal vulnerability to sea-level rise: a spatial temporal assessment framework. Nat Hazards 70(1):395–414
Salvador Ayala G (2015) Análisis del sistema de producción y abastecimiento de alimentos en Galápagos. Dissertation, Universidad San Francisco de Quito
Sampedro C, Pizzitutti F, Quiroga D et al (2018) Food supply system dynamics in the Galapagos Islands: agriculture, livestock and imports. Renew Agric Food Syst:1–15
Schwartz MW, Cook CN, Pressey RL et al (2018) Decision support frameworks and tools for conservation. Conserv Lett 11(2):1–12
Singh RK, Murty HR, Gupta SK, Dikshit AK (2012) An overview of sustainability assessment methodologies. Ecol Indic 15(1):281–299
Spangenberg JH (2002) Institutional sustainability indicators: an analysis of the institutions in Agenda 21 and a draft set of indicators for monitoring their effectivity. Sustain Dev 10(2):103–115
Suárez M, Gómez-Baggethun E, Benayas J, Tilbury D (2016) Towards an Urban Resilience Index: A Case Study in 50 Spanish Cities. Sustainability, 8, 774. https://doi.org/10.3390/su8080774
Taylor JE, Hardner J, Stewart M (2009) Ecotourism and economic growth in the Galapagos: an island economy-wide analysis. Environ Dev Econ 14(02):139–162
Tenza A, Martínez-Fernández J, Pérez-Ibarra I, Giménez A (2019) Sustain Sci 14:791–807
Tsoukala A, Spilanis I, Banos-González I et al (2018) An exercise in decision support modelling for islands: a case study for a'typical'Mediterranean island. Isl Stud J 13(2)
UN (2007) Galápagos Islands added to UNESCO list of endangered heritage sites. UN News Centre: http://www.un.org/apps/news/story.asp?NewsID=23036&Cr=heritage&Cr1=#.V1VuEJPhCRt. Accessed 20 Aug 2019
van Zeijl-Rozema A, Ferraguto L, Caratti P (2011) Comparing region-specific sustainability assessments through indicator systems: feasible or not? Ecol Econ 70(3):475–486
Verma P, Raghubanshi AS (2018) Urban sustainability indicators: challenges and opportunities. Ecol Indic 93:282–291
Vidal-Legaz B, Martínez-Fernández J, Sánchez Picón A, Pugnaire F (2013) Trade-offs between maintenance of ecosystem services and socio-economic development in rural mountainous communities in southern Spain: a dynamic simulation approach. J Environ Manag 131:280–297
Voinov A, Shugart HH (2013) 'Integronsters', integral and integrated modeling. Environ Model Softw 39:149–158
Walker P, Greiner R, McDonald D, Lyne V (1998) The tourism futures simulator: a systems thinking approach. Environ Model Softw 14(1):59–67
Walsh SJ, Mena CF (2013) Perspective for the Study of the Galapagos Islands: complex systems and human – environment interactions. In: Walsh SJ, Mena CF (eds) Science and conservation in the Galapagos Islands. Springer, New York, pp 49–67
Wiranatha A, Smith P (2000) A conceptual framework for a dynamic model for regional planning: towards sustainable development for Bali, Indonesia. In: 1st international conference on object-based image analysis international conference on systems thinking in management. Deaking University, Geelong
Xing Y, Dangerfield B (2011) Modelling the sustainability of mass tourism in island tourist economies. J Oper Res Soc 62(9):1742–1752
Zhang J, Ji M, Zhang Y (2015) Tourism sustainability in Tibet–Forward planning using a systems approach. Ecol Indic 56:218–228

Chapter 12
Local Stakeholders' Perception as a Contribution to the Identification of Negative Impacts on Protected Areas: A Case Study of Torres del Paine National Park

Macarena Fernández Génova, Germaynee Vela-Ruiz Figueroa, Fiorella Repetto-Giavelli, Juliana Torres Mendoza, Nicolás Recabarren Traub, Inti González Ruiz, and Romina López Márquez

Abstract Tourism in protected areas has experienced an important growth in the last decade generating positive impacts but has also increased pressures on biodiversity conservation. Managing impacts generated by tourism is a permanent challenge for protected areas' managers and involved stakeholders. Therefore, it is important to implement participatory approaches that recognize local knowledge as an input to define negative impacts caused by tourism. This chapter studies the local stakeholders' perception about tourism impacts generated over mountain hiking circuits of Torres del Paine National Park, Chilean Patagonia. The problems identified are related to environmental and ecological issues, public use infrastructure, and management aspects which coincided with scientific, technical, and governmental views. In conclusion, the participatory approach is highlighted as a tool that provides relevant information and views from stakeholders involved in the management and decision-making process of a protected area.

Keywords Tourism · Participation · Perception · Torres del Paine National Park Impacts

---

M. Fernández Génova · G. Vela-Ruiz Figueroa (✉) · F. Repetto-Giavelli · J. Torres Mendoza
I. González Ruiz · R. López Márquez
Centro de Estudios del Cuaternario de Fuego - Patagonia y Antártica (Fundación CEQUA), Punta Arenas, Región de Magallanes y de La Antártica Chilena, Chile

N. Recabarren Traub
Ridepatagonia, Punta Arenas, Región de Magallanes y de La Antártica Chilena, Chile

A. Ortega-Rubio (ed.), *Socio-ecological Studies in Natural Protected Areas*,
https://doi.org/10.1007/978-3-030-47264-1_12

12.1 Introduction

Protected areas (PAs) represent today the most important strategy to achieve effective and real conservation of biological diversity and cultural heritage (UNEP-WCMC and IUCN 2016; Worboys 2015). Ecosystem goods and services provided by PAs include recreation and tourism (Stolton et al. 2015; Pabon-Zamora et al. 2008). In a context of worldwide tourism growing, several PAs have experienced a strong increase in visitation, which according to trends will continue to increase (Hvenegaard et al. 2018).

Within this growth scenario, undeniable benefits are offered by PAs' tourism options, which include biodiversity conservation and socioeconomic development of these territories (Eagles et al. 2002, Secretariat of the Convention on Biological Diversity 2004; Buckley 2010; Spenceley et al. 2018). However, to achieve positive outputs of tourism, the activity needs to be properly managed (Leung et al. 2018a) and not affecting the main objectives of biodiversity and cultural heritage conservation of each protected area (Stolton et al. 2015). This is a complex situation for managers since there is usually a tension between two approaches: on one hand, where the natural conditions of protected areas should be maintained for future generations, and on the other, where it is aimed to enhance the development of a tourism destination (Spenceley et al. 2015). To reduce this conflict, managers and stakeholders should promote a sustainable tourism within and around PAs (Leung et al. 2018a). Sustainable tourism is defined as "the tourism that takes fully into account current and future economic, social and environmental impacts, addressing the needs of visitors, the environment and host communities" (UNWTO and UNEP 2005).

Balancing the conservation objectives with the socioeconomic needs will allow to achieve an effective and sustainable tourism development in PAs and to promote greater participation by local communities (Andrade and Rhodes 2012). In this context, it is essential to develop participatory planning processes that recognize the perception that local stakeholders have about costs and benefits of tourism development in PAs (Alkan et al. 2009; Salerno et al. 2010; Nunkoo and Ramkissoon 2011; Sharpley 2014; Robinson et al. 2019). That is why the perception of local stakeholders may be a key to propose measures and solutions to improve current models, particularly in PAs where there is a rapid growth in visitation, generating changes and negative impacts on ecosystems (Xu et al. 2006; Pelegrina-López et al. 2018). Based on its psychological definition, perception is understood as the cognitive process of consciousness that consists of the recognition, interpretation, and significance of judgments obtained from the physical and social environment, in which other psychic processes influence such as learning, memory, and symbolization (Melgarejo 1994). Following this, the concept of social perception arises, which is designated as the one in which social and cultural factors influence, including the physical and social environment (Melgarejo 1994). In relation to PAs, the local population perception depends on the perceived costs and benefits of the PA, the local resources' dependence, and their knowledge about the PA management (Xu et al. 2006).

12.1.1 Tourism's Impacts on Protected Areas

Some tourism direct benefits on behalf of local development are job generation, local businesses creation, education, and investment opportunities, among others. Likewise, sustainable tourism generates significant opportunities to achieve effective conservation of protected areas, such as additional revenue to implement conservation strategies and increasing the people connection with and caring for nature through recreation, education, and environmental interpretation (Eagles et al. 2002; Secretariat of the Convention on Biological Diversity 2004, 2015; Spenceley et al. 2018).

The increase visitation rates in many PAs worldwide are having repercussions on its management and are turning into important challenges for biodiversity conservation, planning, quality recreational opportunities, and especially defining strategies which minimize and mitigate the negative impacts produced by tourism and visitation (Eagles et al. 2002; Leung et al. 2018a).

The environmental effects of inadequate tourism management are widespread and internationally recognized and include different impact's scales. Some of the environmental problems are alteration of habitats and ecosystems due to infrastructure, increase of fire, soil erosion and compaction, creation of unplanned trails, multiple treads, trampling on vegetation, fauna disturbance, and water contamination, among others (Marion and Leung 2001; Secretariat of the Convention on Biological Diversity 2004; Cole 2004; Marion et al. 2006; Manning 2007). Reduction on the quality of the experience generated by the amount of visitors has also been documented. In this sense, agglomerations and social conflicts, as well as aesthetic deterioration, are generated (National Park Service 1997; Leung and Marion 2000; Manning and Lime 2000). In terms of tourism's sociocultural impacts, they are diverse and can affect the access to resources (such as water, energy, and food), traditional livelihoods, crime spread, inequitable benefits, and increase in living costs for local inhabitants, among others (Eagles et al. 2002; Nunkoo and Ramkissoon 2011; Spenceley et al. 2015; Spenceley et al. 2018).

Therefore, it is essential to focus efforts on planning tourism activities and allocating financial and human resources properly, in order to avoid, mitigate, or minimize problems generated to the PAs' natural and cultural resources (Secretariat of the Convention on Biological Diversity 2015). In order to overcome the challenge of tourism development, there are several tools and methodologies available which focus on sustainability's criteria and visitors experiences' quality in PAs (Cole et al. 1997; Leung and Marion 2000; Cole 2004; Laven and Krymkowski 2005; Manning et al. 2011; Salerno et al. 2013; Interagency Visitor Use Management Council 2016; Vela-Ruiz 2017; Halpenny et al. 2018). These methodologies allow the development of planning processes, which are mainly based on relationship between three variables: the environmental resources that might be maintained and conserved, the quality of the recreational experience that might be delivered, and the extent and type of management measures that can be implemented (Manning 2007). Nonetheless, the main difficulty lies in the ways to determine how much change should be allowed within each of the three variables of tourism management on PAs and how to add value and incorporate the perception of local stakeholders in these processes.

12.1.2 Participatory Tourism's Planning Processes in Protected Areas

Achieving an effective conservation and sustainable tourism's development in PAs requires planning processes that incorporate the stakeholders participation and how they are linked and/or affected by the tourism (Pirot et al. 2000; Hung et al. 2011; Andrade and Rhodes 2012; Interagency Visitor Use Management Council 2016; Leung et al. 2018b). The participation of different stakeholders allows active, efficient, and transparent processes, reducing the possibilities of conflicts, increasing trust, and generating opportunities for different interested institutions and organizations (Pomeroy 1995; Pirot et al. 2000; Involve 2005). Likewise, stakeholders' participation is fundamental for governance's development processes that support effective protected areas' conservation (Borrini-Feyerabend et al. 2013).

Usually, local inhabitants and stakeholders are linked closely to PA, using ecosystem services, and in many cases benefiting themselves from tourism as an economic activity (Xu et al. 2006; Hung et al. 2011). This has an impact on the stakeholders' knowledge and visions of the protected area, being essential for managers to know and understand the synergies and conflicts that exist between the different users so it can be managed in the best possible way (Stolton et al. 2015; Salerno et al. 2010). In relation to tourism's sustainability, it is essential to recognize local's perception on the positive and negative impacts generated by the activity (Nunkoo and Ramkisoon 2005; Robinson et al. 2019), since they have potential implications in the decision-making process (Turner et al. 2014), in the tourism development's funding (Ramseook-Munhurrun and Naidoo 2011), and in local tourism policies (Brida et al. 2014). Recognition and knowledge of local stakeholders' visions in the planning processes allow to give greater sustainability to the tourist activity, making possible an integrated management approach, taking in consideration the objectives of managers and the interests of those who are benefitted from the area (Nunkoo and Ramkisoon 2011; Salerno et al. 2010; Salerno et al. 2013; Robinson et al. 2019).

Recognizing local stakeholders' perception on the current and future development of tourism allows the better management processes' development and a more collaborative and sustainable PAs' planning over the time (Pirot et al. 2000; Villasante 2006; Xu et al. 2006; Salerno et al. 2010; Halpenny et al. 2018). Involving stakeholders in planning processes from the beginning enables better and long-lasting decisions (Reed 2008). This requires continuous and planned processes that ensure active participation and contribute to conflict reduction, taking advantage of local stakeholders' contributions to conservation, and which ensure livelihoods through local natural resource management strategies (Borrini-Feyerabend et al. 2013).

The objective of this chapter is to expose the perception of local stakeholders on negative impacts of tourism in Torres del Paine National Park, which were raised through a participatory methodological approach. These results are considered part of a process that sought to provide relevant input for the decision-making process of the tourism planning of one of Chile's most emblematic PAs.

12.2 Area of Study: Mountain Circuits in TPNP

Torres del Paine National Park (TPNP) has an area of 181,414 ha, and it is located in the southern part of Chile, in the Region of Magallanes and Chilean Antarctica, on the eastern slope of the Andes Mountains, between latitudes 50°45′ and 51°20' S and longitudes 72°31′ and 73°22' W. The TPNP is part of Chilean National System of Protected Wildlife Areas (SNASPE for its Spanish acronym) and is managed by the National Forestry Corporation (CONAF) under the Ministry of Agriculture. Cerro Paine Reserve (CPR) is a private protected area (IUCN category VI) of approximately 4, 400 ha located within the national park. Despite not sharing administrations, both protected areas are closely linked through the protection of the Paine Mountain Range and its ecosystems, as well as allowing visitation and trekking on the mountain circuits W and Macizo Paine (Vela-Ruiz and Repetto-Giavelli 2017).

The TPNP is the country's most important protected area in tourism's development terms, having 304, 947 visitors in 2019, contributing significantly to the region's socioeconomic development. This park has maintained a sustained growth of visitors, with an annual average increase of 12% between 2013 and 2019. The growing tourism activity on TPNP and CPR has caused people saturation in trails and tourist infrastructure, forest fires, solid and liquid pollution, and fauna's disturbances, among other impacts, threatening the PA conservation and the visitors' quality experience (Farrell and Marion 2001; Vidal 2012; Repetto-Giavelli and Cabello-Cabalín 2015; Vela-Ruiz and Repetto-Giavelli 2017).

The area of study (Fig. 12.1) involves circuits known as "Macizo Paine" and "W" which are part of both the TPNP and CPR. These circuits contain the best known and most used trekking trails in the park, reaching a total of 119 km, which are divided into 13 sections. In order for visitors to be able to travel these trails, there are 12 overnight areas with shelters and camping facilities, operated by CONAF, by concessionaires, and by private companies (in the private sector, CPR).

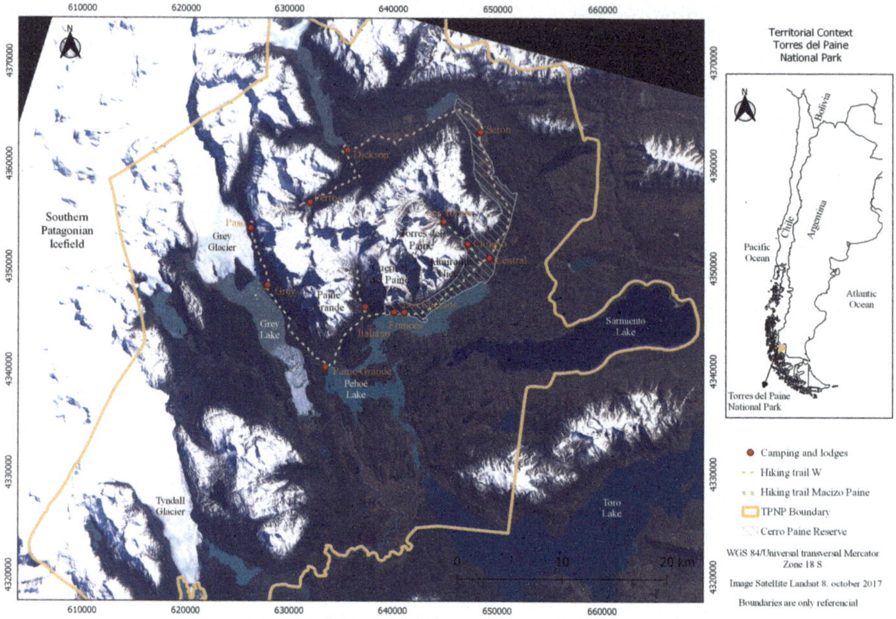

Fig. 12.1 Area of study: W (green) and Macizo Paine (orange) mountain's circuits of Torres del Paine National Park and Cerro Paine Reserve. Red dots indicate accommodation places, which can be campsites, shelters, or both

12.3 Methodology to Identify Perception of Negative Impacts Generated by the Tourism Activity in Torres del Paine National Park

In order to define a tourism planning process in the TPNP, a methodology was developed to identify the stakeholders' perception, which lasted 2.5 years and established a Tourism Management System for mountain circuits in TPNP and CPR. It was needed and defined due to a high demand of visitation, diverse stakeholders with rights and interests, and a poorly planned tourism activity.

The proposed methodology was a dynamic and adaptive process, which allowed including the locals' perceptions and definition of problems or critical points (CP) on those trekking circuits with a greater tourist demand. During the years 2014–2015, the participatory work allowed to identify impacts.

The methodology developed consisted of four chronological stages, which allowed the gathering of information of the TPNP stakeholders. They are shown in Fig. 12.2 and explained below.

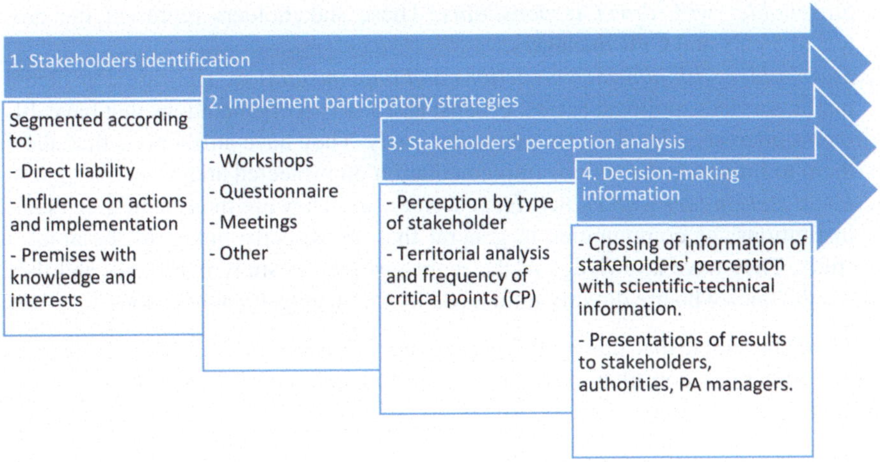

Fig. 12.2 The four chronological stages implemented during the participatory work in the TPNP

12.3.1 Stakeholders' Identification

As a first stage, the identification of stakeholders was considered. Stakeholders are defined as all those who have a direct link with the area of study. They can be individuals, local and indigenous communities, public and private institutions, as well as formal and informal organizations, which are important for the planning, design, implementation, or evaluation of a specific project, which can be positively or negatively affected according to each one's interests (Freeman 1984; Borrini-Feyerabend et al. 2013). Likewise, they are people settled inside the area of study who have different roles, interests, and knowledge and are essential on the economic development and social and cultural growth (Pirot et al. 2000; Salerno et al. 2010; INGEP 2012; ICF – IHT – USAID ProParque 2016). Stakeholders' interaction in tourism planning processes of a protected area provides support for the manager's decisions, since the process should involve and consider the diverse stakeholder's opinion as well as interested formal and informal institutions (Graham et al. 2003; Whittingham 2010; Borrini-Feyerabend et al. 2013). The stakeholders' identification as the first methodological step in a participatory process is essential. From the beginning, it allows to identify who should and could be part of the process (Reed 2008). Stakeholders are associated to different levels and types, those with specific knowledge, and those with expectations according to their own interests (Choy 2005; Gawler 2005; Tapella 2007; USAID-Ecuador 2010; Ministerio de Obras Públicas 2018; INGEP 2012). For this case study, the identification of stakeholders was categorized under three levels according to their role within the protected area:

1. *Stakeholder with direct responsibility.* These stakeholders represent the protected area's and CPR managers.
2. *Stakeholder with influence on actions and implementation.* They are considered public services, political actors, groups, and tourism organizations that have different interests related to the area under study. They have attributions that allow them to influence the actions implementation on protected areas.
3. *Local stakeholder with knowledge and interests.* They are those natural persons, institutions, or communities in general that are directly linked to the area of study. They have knowledge of the reality where the study is located, and they are the ones who are directly influenced by the management decisions.

It should be noted that during the participatory process, it was not necessary to apply indigenous consultation because ancestral indigenous communities do not live in the area of study.

12.3.2 Participatory Strategies' Implementation

During the study and considering the local reality, different methods were applied. Participatory strategies aimed at a common objective, which was exposed to participants and which later allowed a comparative and unifying analysis to complement and strengthen the data collected (Reed 2008). The participatory strategies were:

12.3.2.1 Participatory Workshops

As a group technique, a workshop is a practice or a set of steps tested, systematized, and oriented aimed at achieving a particular objective (Zarzar Charur 2000). The workshops are tools that allow the collection and/or validation of information with representative groups of the community on certain topics through working together in which all participate (Ministerio de Obras Públicas 2007). The workshops in this research were designed in a way that they allowed stakeholders to contribute with their visions and experiences on tourism impacts (positive and/or negative) on the TPNP.

Because of the diversity and number of stakeholders, the workshops were divided by type, in order to allow a better participation and obtain more information. The workshops were dynamically structured in working groups, guided by the question: *Which are the critical points (CPs) that you identify in the mountain circuits of the TPNP?* Maps of the area of study were used to graph the perception of the stakeholders in a territorial way, organizing the perceived impacts by section or sector of the W circuits and Macizo Paine of the protected area. The group results were introduced and discussed during the same workshop.

12.3.2.2 Stakeholder Perception Questionnaire

This tool was used to collect the stakeholders' perception of those who could not participate in workshops or other initiatives, especially park rangers and tourism workers of TPNP and CPR. The objective of the questionnaire was to know the stakeholders' perception and to determine different aspects that would contribute to the tourist planning of the protected area. The information requested in the questionnaire covered the same objectives as the workshop activities so they can be compared and unified under the analysis.

The stakeholders' perception questionnaire was designed by the work team considering the following specific topics: (1) characterization of the tourist operation in the TPNP mountain circuits, (2) identification of CPs in the TPNP mountain circuits, (3) vision of the TPNP in the next 25 years, and (4) observations, comments, annotations, and suggestions.

These questionnaire were applied between December 2014 and March 2015 through the following strategies: (i) printed distribution to hotels and tour operators of Punta Arenas and Puerto Natales cities; (ii) direct submission in the protected area to park rangers, guides, porters, and workers of tourism companies; and (iii) sent by e-mail to different hotels.

12.3.2.3 Meetings and Presentations

Meetings were complementary strategies. This tool is useful to inform population about aspects related to the decision-making process and/or to generate an instance of exchange and raise opinions, which generated commitments among different stakeholders involved (Ministerio de Obras Públicas 2018). Meetings are based on an open conversation oriented to a reflective guided discussion. The importance of this method is that it involves stakeholders with influence in actions and implementation who may not always be able to attend participatory activities.

In our case, the meetings were held with public and political entities, as well as local organizations, and were guided by the semi-structured interview. The interview is a data collection technique used to collect qualitative data through a face-to-face communication between two or more individuals, establishing an asymmetry between the roles (interviewee-interviewer) (Aguirre Baztán 1995).

12.3.3 Stakeholders' Perception Analysis

12.3.3.1 Perception Analysis by Type of Stakeholder

A word clouds tool was used to analyze the perception by stakeholder type. It enables to represent in a visual way the most important words that appear. Therefore, the most significant topics are visualized through an image according to the number of times the words were repeated, which are shown with different densities and typographic number. The larger the word size appears, the more times it was mentioned.

The analysis of the information from the participatory workshops was carried out using the QSR Nvivo software, which made it possible to compare the perceptions of the CP by group of local stakeholder. It was worked on the gathered information from the workshops focused on identifying the CPs (workshops with park rangers, guides and porters, concessionaires and tour operators, and workers from the tourism services company associated with CPR). The visualized words respond, on one hand, to the vision and the link that the stakeholders have with the protected area and, on the other hand, to the priority and interest aspects. In this way, the word cloud graphs show the priority aspects from the stakeholders' perception.

12.3.3.2 Identification of Critical Points (CPs)

The perception's analysis of the tourism's impacts on the TPNP mountain circuits through participatory strategies was carried out according to the following steps:

1. *Identification of the main issues of the stakeholders' perception*. Researchers identify before, during, and after the data collection the topics that are abstract constructs (Fernández Núñez 2006). In this case, the topics are linked to the stakeholders' perception of the CPs.
2. *Topics' classification into categories of analysis.* The core content of the information collected was analyzed to determine what was significant, then patterns were identified in those qualitative data and were transformed into meaningful categories and themes (Patton 2002). This work was done considering that classifying and coding qualitative data produces a framework to organize and describe what was collected during the fieldwork (Fernández Núñez 2006). Thus, from the information systematization, the following five categories were determined: (1) Dimension, (2) Variable, (3) Critical point, (4) Section, and (5) Observation. The organization of the categories is shown in Table 12.1.

The main topics' identification allowed similar concepts to be grouped as they were mentioned during the participatory strategies. For example, the words *treads, multiple treads, different paths on a trail,* and *bypass* were categorized as a single

Table 12.1 Systematization table of the information collected during the participatory process

Dimension	Variable	Critical point	Section	Observation	Participatory strategy		
					Workshop	Questionnaire	Meeting

theme called "multiple treads." Then the identified issues were classified in a central category of analysis called Critical Point (CP). The construction of these categories was based on the problems identified in TPNP's mountain circuits. Following the example, the topic identified as multiple treads was categorized as CP since it gives an account of a bad maintenance, bad state, or another problem that is worried by the stakeholders. The number of times indicated gives an account of the prioritization.

From the category CP, other categories were incorporated: each CP was organized in a common Variable understanding like the concept or criterion that unifies different CPs. For example, the multiple treads CP corresponds to the variable classified as *trail.* It is followed by the category Section which indicates the section of trail or accommodation zone where the CP was identified. Finally, the different variables were organized according to the Dimension category considering the variables defined by Manning (2007) for the management of impacts generated by tourism in protected areas: *1. Environmental-Ecological, 2. Environmental-Infrastructure,* and *3. Management Capability.* These dimensions made it easier to cross-check, compare, and validate the information gathered by professionals and scientists who also identified environmental impacts during the project. Likewise, it made possible to orient problems to generate management solutions always understanding that a CP could be linked or have consequences in more than one dimension.

3. *Territorial analysis and frequency of CPs*: Once previous steps were completed, the CPs were plotted on maps, considering the presence of one CP per stretch of trail or public use zone of the protected area. This facilitated the visualization of the CPs in a territorial way, helping to understand the magnitude of the problem, limit them geographically, and focus the generation of management measures.

   Cartography was developed for each of the CPs identified in the participatory process using the programs Quantum Gis 3.1 and Landsat 8 (October 2017), which was elaborated to recognize the sections and the areas where visitors stay overnight in the mountain circuits that have the greatest number of CPs, according to the stakeholders' perception.

12.3.4 Decision-Making Information

The last methodological stage included analyzing the obtained and systematized results through the word clouds and CPs maps. They were socialized and compared with the scientific-technical results generated in the tourism planning process of the TPNP and the CPR. These results included impacts on trails, flora, fauna, landscape, enabling infrastructure, and visitor management. These results were presented and used by managers and the TPNP, CPR, and the CEQUA Foundation Regional Center technical teams as an important input for the tourism planning process. Results were socialized in instances of political discussion, meetings, and workshops with local actors on the TPNP's tourism planning.

12.4 Results

12.4.1 *Stakeholders' Identification*

Twenty-seven stakeholders in Torres del Paine National Park were identified at different levels.

1. *Stakeholder with direct responsibility*: CONAF as administrator of the Torres del Paine National Park and the board of directors of the Cerro Paine Reserve (CPR) in the private area.
2. *Stakeholder with influence in actions and implementation:* Stakeholders were represented by the public entities which included the principal of the Region of Magallanes and Chilean Antarctica; the mayor of Puerto Natales and the mayor/ess of Torres del Paine; the governor of the Province of Última Esperanza; Regional Ministerial Secretaries of Agriculture, National Assets, Economy, Health, and Environment; Public Roads Department, National Tourism Service (SERNATUR), ProChile, Economic Development Agency (CORFO); as well as tourism and trade organizations of the Provinces of Magallanes, Tierra del Fuego, and Última Esperanza.
3. *Local stakeholder with knowledge and interests*: At this level was found the TPNP park rangers, guides, porters, the Torres del Paine tourism association, tourism concessionaires within the TPNP, workers from the concessionaires, as well as inhabitants of the neighboring cities of Puerto Natales and Cerro Castillo.

12.4.2 *Implementation of Participatory Strategies*

Six participatory workshops were held between November 2014 and April 2015, in the closest localities to the protected area (Puerto Natales and Villa Cerro Castillo). A total of 123 people participated. The workshops were aimed at (i) public and private actors, (ii) TPNP guides and porters, (iii) the community of Villa Cerro Castillo, (iv) park rangers and workers of TPNP, (v) TPNP concessionaires and tour operators of Torres del Paine commune, and (vi) workers from the tourism company associated with CPR.

The opinions of 116 people were compiled in the questionnaire of stakeholders' perception. A total of 34 questionnaires were answered by park rangers, 38 by TPNP guides, 11 by TPNP porters, 16 by workers from the tourist services company associated with the CPR, 13 by workers from the concessionaire of lodging services in the TPNP mountain circuits, 1 by workers from the kayak concession, and 1 by TPNP volunteers.

The opinion of the local stakeholders was considered through five meetings and presentations, carried out in the cities of Punta Arenas, Puerto Natales, and Villa Cerro Castillo. Meetings were held with the Governorate of Última Esperanza, the

Municipality of Torres del Paine, various Regional Ministerial Secretariats, SERNATUR, ProChile, Chambers of Tourism and Associations of Tour Guides of Puerto Natales and Punta Arenas, and Chamber of Tourism of Última Esperanza.

12.4.3 Stakeholder' Perception Analysis

12.4.3.1 Perception by Type of Stakeholder

Four word clouds were developed, summarizing the stakeholders' perception. The ten aspects most mentioned in the TPNP guides' and porters' workshop were trails, camping, staff, lack (mentioning different deficiencies), bathroom, CONAF, Paine, information, infrastructure, and Torres (camping) (Fig. 12.3). The word cloud from the TPNP park rangers' workshop shows that the most frequently mentioned topics were trail (highlighting significantly more than the next word), camping, Torres (camping), multiple treads, people, horses, bathrooms, and enabling infrastructure such as lookouts, passageways, and boardwalk (Fig. 12.4). Regarding the workshop

Fig. 12.3 Cloud of words generated from the workshop developed with tour guides and porters

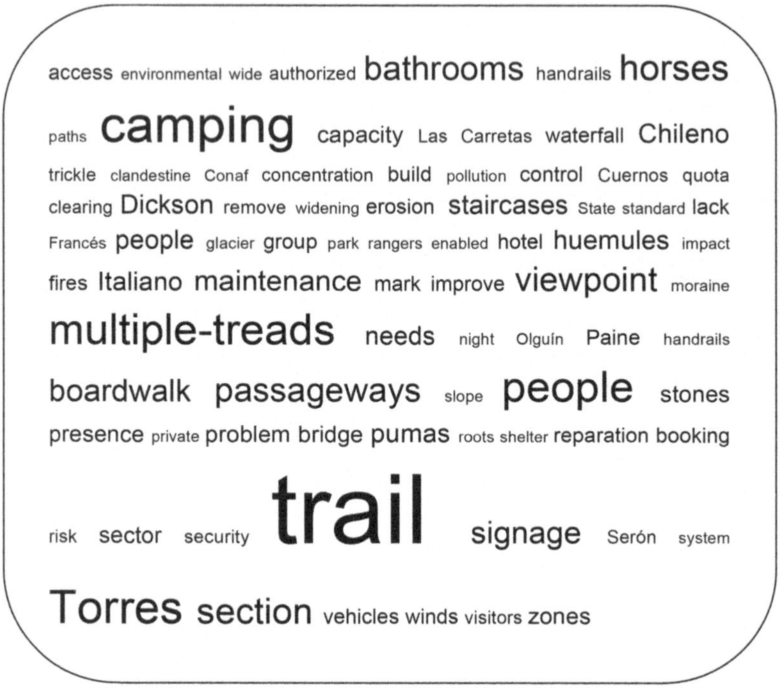

Fig. 12.4 Cloud of words generated from the workshop developed with TPNP's rangers

with TPNP concessionaires and tour operators of Torres del Paine commune, the following words stand out in order of importance: staff, infrastructure, information, CONAF, bathrooms, camping, lack (mentioning different deficiencies), people, circuits, and hotel (Fig. 12.5). Finally, the ten aspects most mentioned in the workers' workshop from the tourism services company associated with CPR were camping, lack, people, trash, bathrooms, cooking shelter, capacity, staff, people, and group (of people) (Fig. 12.6).

12.4.3.2 Critical Points' Perception (CPs)

The breakdown of the perception of local stakeholders' analysis gave a total of 46 CPs classified in three dimensions and distributed territorially in the 13 sections of the TPNP mountain circuits, which are presented in Table 12.2 and described below. In total for the Environmental–Ecological dimension, the stakeholders perceive 11 CPs, in the Environmental–Infrastructure 19 CPs, and in Management Capacity of TPNP 16 CPs.

In order of frequency, the ten most frequently mentioned CPs (independently of the category) were (i) state of bathrooms (toilets), showers, and washrooms, (ii) signage, (iii) existence of erosive processes, (iv) inadequate design of trails, (v) state

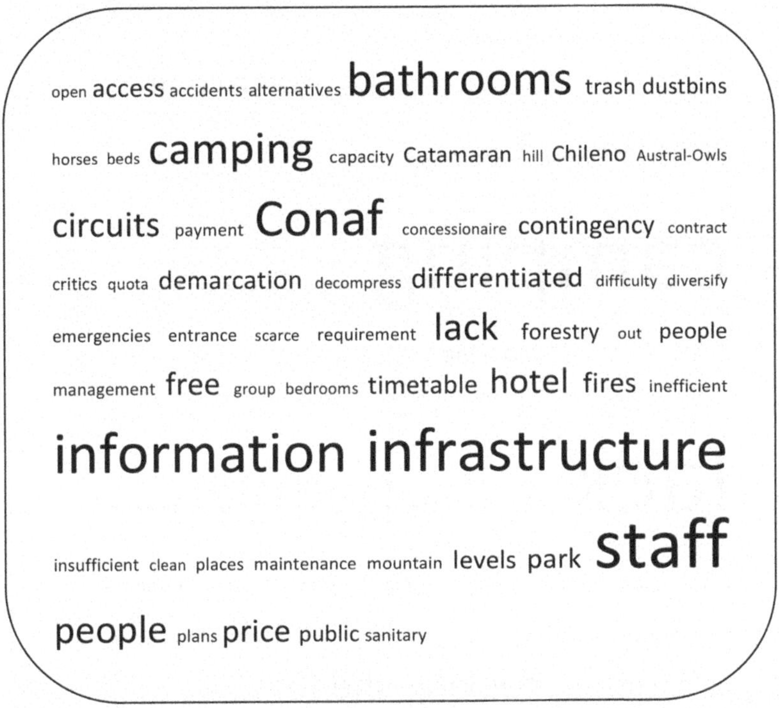

Fig. 12.5 Cloud of words generated from the workshop developed with concessionaires and tour operators

of passageways and boardwalks, (vi) concentration of people, (vii) inadequate waste management, (viii) state of bridges, (ix) state of steps and stairways, and (x) multiple treads presence.

The results were broken down by the dimension as follows.

Environmental–Ecological Dimension
Three variables were identified in this dimension: the *trail* variable that counts four CPs, the *vegetation* variable that considers three CPs, and the *fauna* variable four CPs. In general, from the stakeholders´ perception, the trails' poor condition and the presence of erosive processes and multiple treads are relevant, generating a direct impact on elements related to natural resources such as flora, fauna, and scenic beauty. The high concentration of exotic flora and fauna species is also perceived as CP, as it is the presence of domestic livestock, in the case of horses in CPR, which move on the same trails as visitors.

Environmental–Infrastructure Dimension
The infrastructure aspects emphasized mainly on two areas: (i) paths and their enabling infrastructure (*improvement of the path* considers five CPs, the second variable *signage* includes two CPs as well as the *viewpoint* variable) and (ii) camping (*sanitary facilities* has two CPs and *equipment* has eight CPs). In relation to

Fig. 12.6 Cloud of words generated from the tourism company associated with the Cerro Paine Reserve

trails, the lack of maintenance and the poor state of the stairs, bridges, boardwalks, markings, signage, etc. is of concern, which is why it is proposed to improve and incorporate new infrastructure. Concerns is also felt in the poor state of the enabling infrastructure, increasing visitor risk of accidents, which could lead to a negative image of TPNP. Finally, it is highlighted the lack of trails design and maintenance, with no standards, layout, directionalities, attractions, and/or viewpoints. The signage is considered as CP because of the poor quality of information provided, the lack of uniformity in terms of design and materiality, as well as the lack of geographic and historical information for visitors.

In relation to campsites and shelters, it was emphasized that toilets, showers, cooking shelters, tables, and platforms have become insufficient for the number of people who are arriving. The price–quality ratio of the services is not according to the visitor expectations, forcing managers to think how to manage the visitors' demand. At the same time, the stakeholders consider that the existing infrastructure should be limited by a defined capacity, improving design material in order to reduce environmental and visual impact. The stakeholders agree that the TPNP priority should be to maintain its biodiversity, scenic beauty, and natural landscape.

Table 12.2 Critical points raised in the TPNP on ecological, infrastructure, and management aspects

Dimension	Variable	Critical points (CPs)
Environmental ecology	Trails	- Multiple treads - Erosion - Inadequate trail design - Waterlogged areas
	Vegetation	- Damage to flora - Exposed roots - Burned forest
	Fauna	- Huemul (*Hippocamelus bisulcus*) presence - Pumas (*Puma concolor*) presence - Domestic livestock - Wildlife impacts
Infrastructure-environmental	Trail improvement	Deterioration or bad state of: - Handrails, railings - Passageways, boardwalks - Steps, stairways - Bridges - Rivers, lakes, creeks
	Signage	Inadequate: - Brands, beacons, arrows - Signage
	Viewpoints	- Bad state of formal viewpoints - Wide presence of viewpoints
	Sanitary installations	Bad state of: - Showers, toilets, laundries - Saturation of sanitary pits
	Equipment	State, nonexistence, and noncompliance with regulations, among others: - Platforms - Cooking shelters - Tables, benches - Reception - Machine room, cellar - Lighting - Campsite areas - Dock

(continued)

Table 12.2 (continued)

Dimension	Variable	Critical points (CPs)
Management capacity	Visitor planning	- Concentration of people - Illegal campsites - Improper conduct - Off-circuit path - Inadequate TPNP access control - Quality and price of services offered
	TPNP planning	- Inadequate waste management - Insufficient administrative control - Overdemand camping, shelters - Need for medical assistance for visitors - Lack of contingency plans - Inefficient vessel - Motor vehicles
	Human resources planning	- Inadequate conduct of workers - Lack of training - Lack of personnel within the TPNP

Handling Capacity Dimension

In terms of management, the stakeholders described (i) TPNP lack of management (*TPNP planning considers* seven CPs and the *human resources planning* variable three CPs) and (ii) increased visitation (*visitor planning* with six CPs). It was highlighted institutional planning needs for both TPNP and CPR. In this area, funding support is perceived as an important problem, since the entrance fees do not return in a full way to the protected area, but it goes to a national system of income distribution for all the SNASPE. The stakeholders' perception is that this does not allow the park to fund important aspects such as human resources and the parks planning needed.

In general, insufficient administrative control was observed in the tourism activity. In the mountain trekking circuits, the high tourist demand in summer generated an overdemand for accommodation services, generating management and health problems. It was emphasized that there should be greater institutional presence and control on the mountain trails by increasing the frequency of patrolling. It was also expressed that staff should training continuously in different topics such as fire control, environmental education, and first aids. It was proposed to make a greater orientation for visitors in different aspects concerning the trekking circuit, environmental education, and rules to follow within the park.

12.4.3.3 Cartography Generation with Critical Points

The 31 CPs with territorial representation were mapped. As an example, the participatory maps of the three most mentioned CPs in the TPNP and CPR mountain circuits are shown. Red circles and red stripes indicate the sector in which the perceived CP is observed. The image A in Fig. 12.7 shows the most frequently mentioned CP

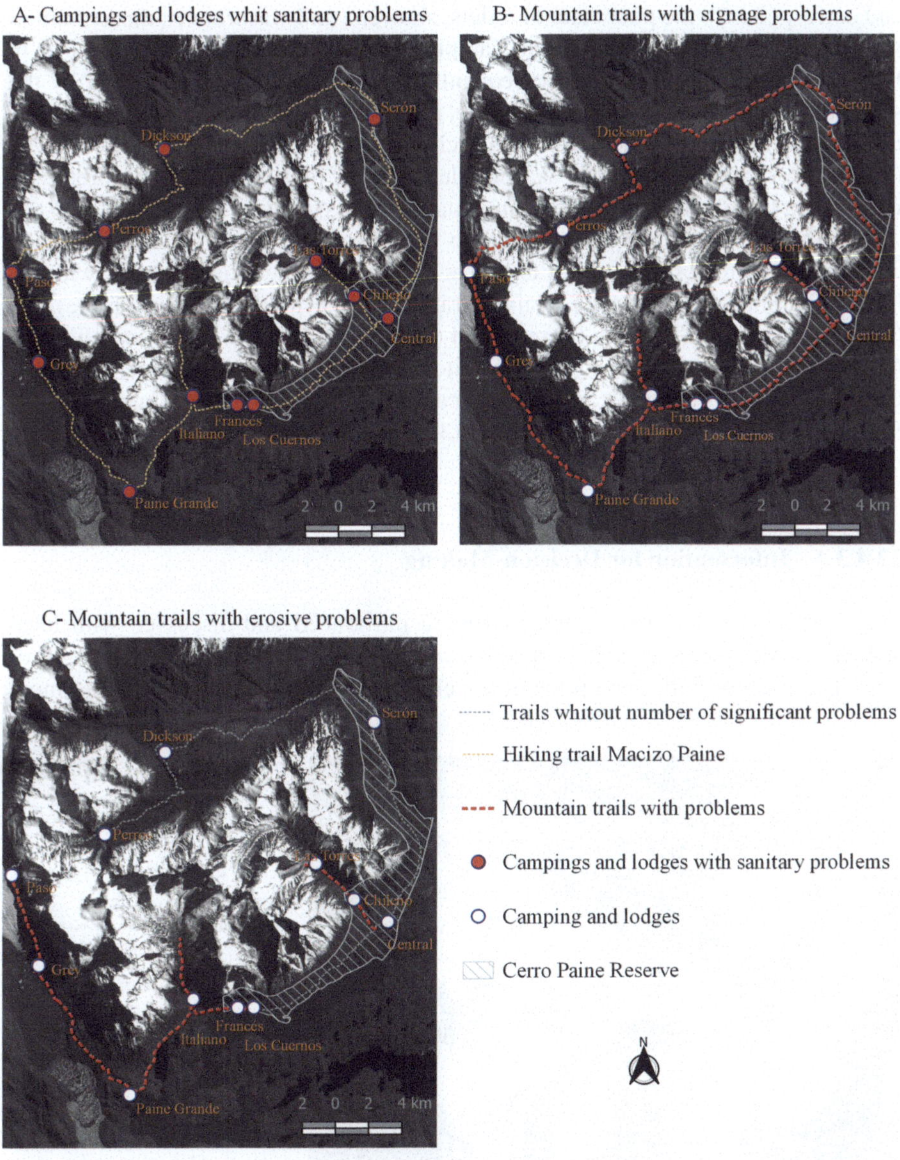

Fig. 12.7 Participatory maps of the three most mentioned CPs in the TPNP and CPR mountain circuits. Image A shows the most frequently mentioned CP: poor state of toilets, showers, and washrooms. The red color indicates sanitary problems in all camping areas. Image B shows the second most frequent problem: signage in the mountain trail circuits. The red-colored trails indicate that the problem is widespread in all circuits W and Macizo Paine. Image C shows the third most mentioned CP: the presence of erosive processes in the trails, which are indicated in red color

and corresponds to the poor state of toilets, showers, and washrooms; this condition was perceived by local actors in all camping and shelter sectors. The image B in Fig. 12.7 shows the second most frequently mentioned problem identified by the stakeholders. It is the state, presence, type, materiality, and information provided by the signage throughout the W mountain circuit and the Macizo Paine circuit. The third CP most mentioned by the stakeholders was the presence of erosive processes in the most travelled trails of the W mountain circuit (image C in Fig. 12.7).

Finally, Fig. 12.8 identifies the sections of trails of the W and Macizo Paine circuits with the greatest number of CPs. The results identified that the trails with the highest number of CPs are (1) Paine Grande–Grey with 14 CPs (in purple), (2) Cuernos–Central with 9 CPs (in red), and (3) Central–Chileno–Las Torres with 10 CPs (in orange). In relation to camping sites, Fig. 12.8 shows that those with more CPs are (1) Perros camping and Francés camping, each with 9 CPs (in purple), (2) Italiano camping with 8 CPs (in red), and (3) Paso camping and Paine Grande camping, each with 7 CPs (in orange).

12.4.3.4 Information for Decision-Making

The word clouds and maps produced were introduced and discussed with local stakeholders in meetings and workshops during the TPNP's tourism planning process. These allowed the perception of stakeholders to be valued as an effective input

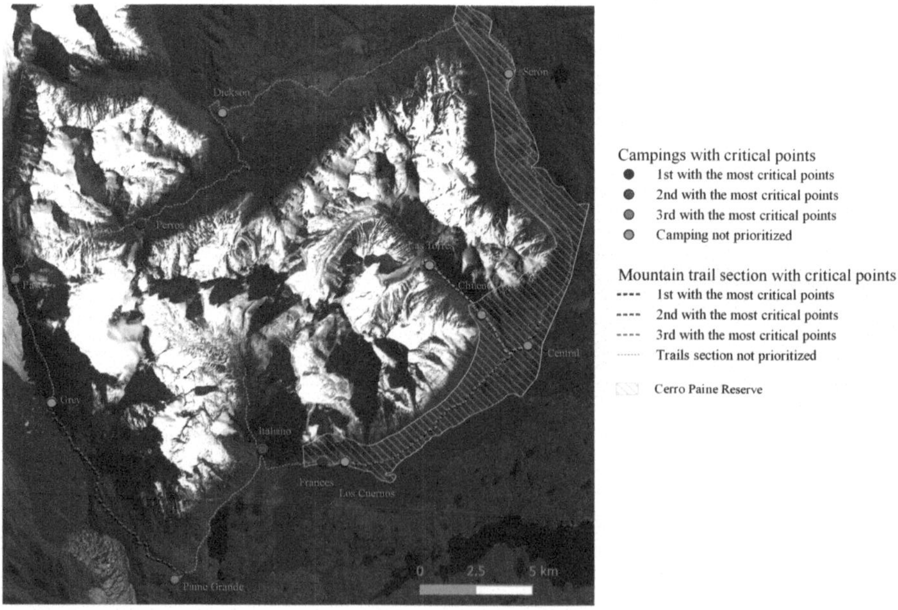

Fig. 12.8 Mountain circuit sections and camping sites of TPNP and CPR with the highest number of CPs

for the decision-makers. The results were considered for tourism management decision-making in TPNP and CPR together with the gathering of information considered by the specialists of the different dimensions addressed. Collected information was used to prioritize management actions in the most frequent CPs and in the sections and overnight areas with the greatest presence of CPs.

12.5 Discussion

The results of the stakeholders' perception analysis gave a critical view of tourism in the Macizo Paine and W mountain trekking circuits of the TPNP and the CPR. Despite the fact that the stakeholders indicate that among the benefits perceived by tourism in TPNP, the Region of Magallanes and Chilean Antarctica stand out as a world-renowned tourist destination, receiving more and more tourists who leave economic benefits that have repercussions for the commune, province, and region. In addition, it allows encouraging new tourist options taking advantage of the high demand of visitors who want to stay longer periods in the territory. This perception agrees with what has been presented by different authors who have emphasized the close link between increase in tourism in the TPNP and the socio-economic development of Puerto Natales city and the Province of Última Esperanza (Villarroel 1996; Ferrer 2003; Araya 2007; Vela-Ruiz 2009; Vela-Ruiz and Delgado 2010). This positive perception of the benefits generated by tourism has led local stakeholders to agree with this productive activity, a reality that is in line with what was researched by Nunkoo and Ramkissoon (2011).

Participatory work developed showed an uncontrolled demand of visitors and inadequate management, which need measures to reduce environmental impacts that are affecting the conservation of the biodiversity of the TPNP and the quality of the visitors' experience. In order to improve and maintain the tourism activity in this known park, a collaborative and responsible plan is needed, according to the needs and objectives of the national park.

The most frequent CPs relate to issues of infrastructure, inadequate trails and camping design and maintenance, management capacity, and planning. In general, the stakeholders' perception agree on needs to improve the negative effects of the sustained and significant growth of visitation, which support the implementation of tourism management measures in this protected area.

In relation to the workshops, there are common aspects that suggest the reality faced by the TPNP managers and that concern those who work there. In contrast, it can be seen that in all the workshops, "camping" and "bathrooms" were mentioned as one of the outstanding words, being the sanitary services of the mountain circuits the most mentioned CP. In addition, the words related to infrastructure (in camping sites and trails), garbage and people, are considered. It should be noted that for the park rangers, guides, and porters, the most mentioned aspects were very similar. The aspects that stakeholders mention are related to the reality that they live daily, concerning the trails design and maintenance, the visitors' management, and the trails and camping infrastructure quality. The vision of the concessionaires and tour operators is different, since the keywords were linked to their conditions as service providers, and they focus on words such as prices, camping, infrastructure, CONAF, and staff; similar to what happens to the tourism services, workers associated with CPR, in which the aspects fall on the concern for their working environment.

The most mentioned CPs were those related to the enabling infrastructure such as bathrooms, showers, laundry rooms, and signage of mountain circuits. Likewise, ecological problems such as the erosive processes presented in the trails and lack of waste management were frequently mentioned. What it is expected of the TPNP is to maintain its quality of "wild beauty," since people who visit are looking for a natural and wild experience.

The perception of local stakeholders shows that although the TPNP is considered a successful case of tourism development at the national level, the fact that it has grown without major restrictions, regulations, and management resources has generated a negative impacts for the conservation of the protected area and for sustainability as a tourist destination.

The stakeholders have similar visions: if the sustained increase in visitation is not managed, there will be negative repercussions for the conservation of the protected area such as erosion and destruction of trails. They also agree that more private options such as hotels, shelters, and camping sites will harm the TPNP's flora and fauna. Therefore, the national park and CRP should be based on shared and collaborative responsibilities, which should consider economic, environmental, and social aspects. Efforts should aim at a collaboration between tourism and conservation based on a strategic plan.

It is perceived that the budget allocated to the TPNP is not sufficient to cover all the problems found, and stakeholders emphasize that the total income generated or a percentage of the entrance to the park should remain in this protected area. This perceived situation in the TPNP coincides with the vision of the managers of other SNASPE units in Chile (Repetto-Giavelli et al. 2018).

In relation to the methodology developed, it allowed the perception of stakeholders to be gathered at different levels, and it also made it possible to demonstrate that the perception is coherent and not very distant from what scientific studies describe about the impacts generated by the tourism use in the TPNP and CPR (Vela-Ruiz and Repetto 2017; Torres et al. 2018, 2019). This means that the perception of the stakeholders linked to the TPNP is in line with reality and favors knowledge of the dynamics and impacts in the mountain circuits of this protected area. This is supported by Reed (2008), who points out that local and scientific knowledge can be integrated to understand in a better way natural systems and processes, allowing to evaluate solutions to better face environmental problems.

In general, the stakeholders are critical to define and guide tourism especially when the activity has been developed without planning and is generating environmental impacts, not just affecting the national park biodiversity but also having important issues on the socioeconomic system of the whole area. All the CPs mentioned above should aim to improve the environmental quality of the trekking circuits and to improve the quality of the visitors' experience.

Local stakeholders question how tourism has developed in the TPNP because problems are evident in different ways: environmental, infrastructure, and management capacity especially during the peak of the tourism season (December to February). This has led to shortcomings in management aspects and in the quality of the visitor experience (Ferrer 2003 and AMBAR 2004).This local stakeholders' negative view in relation to the impacts of tourism in a protected area corresponds with other international research, such as the one presented by Alkan et al. (2009) and Robinson et al. (2019). This demonstrates the need to improve the sustainability of the tourism operation in the TPNP and on other national parks in Chile (Repetto-Giavelli et al. 2018).

In the decision making process, for the first time the managers of TPNP considered the perception and knowledge of the stakeholders. According to Borrini-Feyerabend et al. (2013), the governance of a protected area is shaped by history, culture, and interaction between local, subnational, national, and international stakeholders and institutions. It is important that the technical teams of the TPNP and the CPR can maintain the instances of effective participation due to the great economic and cultural interest and dependence that the population of the Province of Última Esperanza has on this territory (Ferrer 2003; Araya 2007; Vela-Ruiz 2009; Vela-Ruiz and Delgado 2010). Considering also that local-scale transformations if are not accompanied by changes at a higher scale can be suffocated by economic, political, or cultural structures that from above prevent them from consolidating (Villasante 2006).

Stakeholders' perception gathered is relevant data from the experience and interest of actors working for and in the TPNP. It is important to include and achieve proposals with participatory approaches that have been slowly developed and implemented in Chile. This research has demonstrated that considering the local stakeholders' perception in the planning processes of protected areas allows obtaining important information to prioritize management actions and to improve the management on protected areas (Pelegrina-López et al. 2018), hence the importance of institutionalizing the participation (Reed 2008) and effectively linking the perception of stakeholders in the decision-making processes.

12.6 Lessons Learned

The used methodology allowed us to include different local stakeholders for its flexibility to gather information including the perception of those who work in the mountain circuit and who were not necessarily linked to each other. This methodology enables us to collect the opinion of a large number of local stakeholders which enriches the process. However, it requires time, willingness, and budget. In this sense, the possibility of carrying out a complete participatory process in part is defined for the period, the work team, and the project budget. If funds had not been available to address the different participatory strategies, the gathered information could have had greater difficulties. Participatory processes are challenging to face but can be combined. So the invitation is to innovate and create a specific strategy to be able to collect information according to the reality of each PA. It should be kept in mind that regardless of the participatory strategy(s), what it is important is to aim for a common objective during the process and then use and complement the information in the analysis process.

Although participatory processes are long and complex, this proposal sought to link different local stakeholders throughout the process through the instruments described above. These instances facilitated spaces for discussion, joint work, and transparency where the perception of the local community was considered in terms of knowledge and voice in decision-making in the mountain circuits of the TPNP. It is considered that the methodology used can be replicated in other protected areas, so it is recommended to implement it from the beginning of the planning process.

It is important to manage the expectations of the local stakeholders since in cases where it is an external agent who carries out the participatory processes once this stage is over, it is the function and duty of the managers to consider and/or continue with the processes and implementation which are often lost in the planning process itself.

Finally, it should be noted that the participation degree and the stakeholders' empowerment toward the protected area also have an influence since if they are not willing to participate or work collaboratively, the process becomes more complex and can become completed. In our case study, local stakeholders were actively involved and willing to give their vision of the impacts that tourism is generating in the protected area and to seek together possible solutions. Furthermore, participation makes it possible to validate issues or visualize others that have not been previously considered.

Acknowledgments This research was carried out through the project Innova Corfo of Bienes Públicos para la Competitividad Nacional Cód. 14BPC4-28654 "Sistema de Manejo Turístico en Áreas Protegidas de Chile: caso piloto Parque Nacional Torres del Paine." The financing support and collaboration of CONAF Magallanes y Antártica Chilena and the Cerro Paine Reserve are gratefully acknowledged. Special thanks to the different public and private actors that were part of the different participative instances that sustain this work. Finally, we would like to thank Sanober Mirza for her contribution to the article.

References

Aguirre Baztán A (1995) Etnografía. Metodología cualitativa en la investigación sociocultural. Boixareu Universitaria-Marcombo Editores. Barcelona

Alkan H, Korkmaz M, Tolunay A et al (2009) Assessment of primary factors causing positive or negative local perceptions on protected areas. J Environ Eng Landsc Manag 17(1):20–27

AMBAR S.A (2004) Desarrollo e implementación de un sistema de gestión del uso público en los Parques Torres del Paine y Bernardo O'Higgins XII Región de Magallanes y Antártica Chilena. Estudio contratado por EUROCHILE y financiado por CORFO

Andrade GSM, Rhodes JR (2012) Protected areas and local communities: an inevitable partnership toward successful conservation strategies? Ecol Soc 17(4):14. https://doi.org/10.5751/ES-05216-170414

Araya P (2007) El impacto del turismo en la conservación de una reserva de la biósfera y en el desarrollo de su zona de influencia. El caso de la reserva Torres del Paine. En Halffeter G, Guevara S, Melic A (eds) Hacia una cultura de conservación de la diversidad biológica. Capítulo, vol 13, pp 115–124

Borrini-Feyerabend G, Dudley N, Jaeger T et al (2013) Governance of protected areas: from understanding to action. Best practice protected area guidelines series N° 20. IUCN, Gland, p xvi+124

Brida JG, Disegna M, Osti L (2014) Residents' perceptions of tourism impacts and attitudes toward tourism policies in a small mountain community. Toursimos 9:37–71

Buckley RC (2010) Conservation tourism. CABI Publications, Wallingford, p 214

Choy M (2005) Como incidir en políticas públicas. Fundación Centro de Información y Recursos para el Desarrollo (CIRD). Asunción, Paraguay

Cole DN (2004) Monitoring and management of recreation in protected areas: the contributions and limitations of science. Working Pap Finnish Forest Res Inst 2:10–17

Cole D, Watson A, Hall T et al. (1997) High use destinations in wilderness: social y biophysical impacts, visitor responses y management options. Research paper. INT – RP – 496. Forest Service. Intermountain Research Station

Eagles PFJ, McCool SF, Haynes CDA (2002) Sustainable tourism in protected areas: guidelines for planning and management. IUCN, Gland/Cambridge, p xv+183

Farrell T, Marion JL (2001) Trail impact and trail impact management related to visitation at Torres del Paine National Park, Chile. Leisure/Loisir 1–2:31–59

Fernández Núñez L (2006) ¿Cómo analizar datos cualitativos?. Butlletí LaRecerca Universidad de Barcelona, Institut de Ciències de l'Educació, Secció de Recerca, p 13. https://www.ub.edu/idp/web/sites/default/files/fitxes/ficha7-cast.pdf. Accessed on 9 Ene 2020

Ferrer D (2003) Conservación de la naturaleza y el territorio en Chile. El Parque Nacional Torres del Paine y su área de influencia socioeconómica. Tesis Doctoral, Universidad Autónoma de Madrid, España

Freeman RE (1984) Strategic management: a stakeholder perspective. Pitman, Boston, p 13

Gawler M (2005) Project design in the context of project cycle management: sourcebook. World Wildlife Fund Introductory course. Artemis Services for Nature Conservation and Human Development, Prevessin-Moens

Graham J, Amos B, Plumptre T (2003) Governance principles for protected areas in the 21st century, a discussion paper. Institute on Governance in collaboration with Parks Canada and Canadian International Development Agency, Ottawa

Halpenny E, Salenieks T, Manning R et al (2018) Adaptive management for sustainable tourism. In: Leung Y, Spenceley A, Hvenegaard G et al (eds) Tourism and visitor management in protected areas: guidelines for sustainability, Best practice protected area guidelines series N°27. IUCN, Gland, pp 41–62

Hung K, Sirakaya-Turk E, Ingram L (2011) Testing the efficacy of an integrative model for community participation. J Travel Res 50(3):276–288

Hvenegaard G, Buckley R, Spenceley A et al (2018) The future of protected area tourism. In: Leung Y, Spenceley A, Hvenegaard G et al (eds) Tourism and visitor management in protected areas: guidelines for sustainability, Best Practice protected area guidelines series N° 27. IUCN, Gland, pp 89–95

ICF – IHT – USAID ProParque (2016) Guía metodológica para la elaboración de Planes de Uso Público en Áreas Naturales Protegidas del SINAPH. Instituto Nacional de Conservación y Desarrollo Forestal, Áreas Protegidas y Vida Silvestre, Instituto Hondureño de Turismo y Proyecto USAID ProParque. Tegucigalpa, MDC

INGEP (2012) Mapeo de actores para la política pública migratoria. Guatemala

Interagency Visitor Use Management Council (2016) Visitor use management framework, a guide to providing sustainable outdoor recreation. Edition One. U.S. Department of the Interior, p 130

Involve (2005) People & participation: how to put citizens at the heart of decision-making. Beacon Press, London, p 116

Laven DN, Krymkowski DH (2005) The relationship between visitor-based standards of quality y existing conditions in Parks and Outdoor recreation. Leis Sci 27(2):157–173. https://doi.org/10.1080/01490400590912060

Leung Y, Marion T (2000) Recreation impacts and management in wilderness: a state of knowledge review. USDA Forest Service Proceedings RMRS-P-15, vol 5, pp 23–45

Leung Y, Spenceley A, McCool S et al (2018a) Tourism and visitation in protected areas: the sustainability challenge. In Leung Y-F, Spenceley A, Hvenegaard G, Buckley R (eds) Tourism and visitor management in protected areas: guidelines for sustainability. Best practice protected area guidelines series No. 27. IUCN, Gland, pp 1–8

Leung Y, Halpenny E, Salenieks T et al (2018b) Adaptive management for sustainable tourism. In: Leung Y, Spenceley A, Hvenegaard G et al (eds) Tourism and visitor management in protected areas: guidelines for sustainability, Best practice protected area guidelines series N° 27. IUCN, Gland, pp 41–62

Manning R (2007) Parks and carrying capacity – commons without tragedy. Island Press, p 328

Manning R, Lime D (2000) Defining and managing the quality of wilderness recreating experiences. In: Cole DN, McCool SF, Borrie WT et al (eds) Wilderness science in a time of change conference. Vol 4: Wilderness visitors, experiences, and visitor management; 1999 May 23–27; Missoula, MT. Proceedings RMRS-P-15, Vol 4. U.S. Department of Agriculture, Forest Service, Rocky Mountain Research Station, Ogden, pp 13–52

Manning R, Valliere W, Anderson L et al (2011) Defining, measuring, monitoring, and managing the sustainability of parks for outdoor recreation. J Park Recreat Adm 29(3):24–37

Marion J, Leung Y (2001) Trail resource impacts and an examination of alternative techniques. J Park Recreat Adm 19(3):17–37

Marion J, Leung Y, Nepal SK (2006) Monitoring trail conditions: new methodological considerations. Visitor impact monitoring. George Wright Forum 23(2):36–49

Melgarejo L (1994) Sobre el concepto de percepción. Alteridades 4(8):47–53

Ministerio de Obras Públicas (2007) Manual Participación Ciudadana para iniciativas del Ministerio de Obras Públicas. Chile

Ministerio de Obras Públicas (2018) Guía para la Gestión de Participación Ciudadana. Dirección General de Obras Públicas. Santiago – Chile, p 136

National Park Service (1997) The Visitor Experience and Resource Protection (VERP) framework: a handbook for planners and managers. U.S. Department of the Interior, p 108

Nunkoo R, Ramkinssoon H (2011) Developing a community support model for tourism. Ann Tour Res 38(3):964–988

Pabon-Zamora L, Bezaury J, Leon F et al (2008) Valorando la Naturaleza: Beneficios de las áreas protegidas. Serie Guía Rápida. In: Arlington JE (ed) The Nature Conservancy, p 34

Patton MQ (2002) Qualitative research and evaluation methods, 3rd edn. Sage Publications, Thousand Oaks

Pelegrina-López A, Ocaña-Peinado FM, Henares-Civantos I et al (2018) Analyzing social perception as a key factor in the management of protected areas: the case of the Sierra Nevada Protected Area (S Spain). J Environ Plan Manag 61(1):124–142. https://doi.org/10.1080/09640568.2017.1291413

Pirot JY, Meynell PJ, Elder D (2000) Ecosystem management: lessons from around the world. A guide for development and conservation practitioners. IUCN, Gland/Cambridge, p 132

Pomeroy R (1995) Community-based and co-management institutions for sustainable coastal fisheries management in Southeast Asia. Ocean Coast Manag 27(3):143–162

Ramseook-Munhurrun P, Naidoo P (2011) Residents' attitudes toward perceived tourism benefits. Int J Manag Mark Res 4(3):45–56

Reed M (2008) Stakeholder participation for environmental management: a literature review. Biol Conserv 141:2417–2431

Repetto-Giavelli F, Cabello-Cabalín JL (2015) Ecological restoration potential in areas of public use in Torres del Paine National Park. Anales Instituto Patagonia (Chile) 43(1):115–121

Repetto-Giavelli F, Vela-Ruiz G, Díaz Beros M, Fernández Génova, M., López, R. y González, B. 2018. Diagnóstico sobre la planificación e implementación del uso público en el Sistema Nacional de Áreas Silvestres Protegidas del Estado de Chile. Biodiversidata 7: 116–124.

Robinson D, Newman SP, Stead SM (2019) Community perceptions link environmental decline to reduced support for tourism development in small island states: a case study in the Turks and Caicos Islands. Mar Policy 108:103671

Salerno F, Cuccillato E, Caroli P et al (2010) Experience with a hard and soft participatory modeling framework for social-ecological system management in Mt. Everest (Nepal) and K2 (Pakistan) protected areas. Mt Res Dev 30(2):80–93

Salerno F, Viviano G, Manfredi E et al (2013) Multiple carrying capacities from a management-oriented perspective to operationalize sustainable tourism in protected areas. J Environ Manag 128:116–125

Secretariat of the Convention on Biological Diversity (2004) Guidelines on biodiversity and tourism development: international guidelines for activities related to sustainable tourism development in vulnerable terrestrial, marine and coastal ecosystems and habitats of major importance for biological diversity and protected areas, including fragile riparian and mountain ecosystems. (CBD Guidelines) Montreal, p 29

Secretariat of the Convention on Biological Diversity (2015) Tourism supporting biodiversity – a manual on applying the CBD guidelines on biodiversity and tourism development. Montreal, p 56

Sharpley R (2014) Host perceptions of tourism: a review of the research. Tour Manag 42:37–49

Spenceley A, Kohl J, McArthur S et al (2015) Visitor management. In: Worboys GL, Lockwood M, Kothari A et al (eds) Protected area governance and management. ANU Press, Canberra, pp 715–750

Spenceley A, Hvenegaard G, Bushell R et al (2018) The impacts of protected area tourism. In: Leung Y, Spenceley A, Hvenegaard G et al (eds) Tourism and visitor management in protected areas: guidelines for sustainability, Best practice protected area guidelines series N° 27. IUCN, Gland, pp 9–26

Stolton S, Dudley N, Avcıoğlu Çokçalışkan B et al (2015) Values and benefits of protected areas. In: Worboys GL, Lockwood M, Kothari A et al (eds) Protected area governance and management. ANU Press, Canberra, pp 145–168

Tapella E (2007) El mapeo de actores claves, documento de trabajo del proyecto "Efectos de la biodiversidad funcional sobre procesos ecosistémicos, servicios ecosistémicos y sustentabilidad en las Américas: un abordaje interdisciplinario". Universidad Nacional de Córdoba, Inter-American Institute for Global Change Research (IAI)

Torres J, Repetto-Giavelli F, Quezada F et al (2018) Evaluación de senderos de montaña y medidas de manejo para mitigar impactos en el Parque Nacional Torres del Paine, Chile. Áreas Naturales Protegidas Scripta 4(2):7–28

Torres J, Vela-Ruiz G, Olave C et al (2019) Análisis de la dinámica de visitantes en los senderos de montaña del Parque Nacional Torres del Paine. Biodiversidata 8:87–94

Turner RA, Fitzsimmons C, Forster J et al (2014) Measuring good governance for complex ecosystems: perceptions of coral-reef dependent communities in the Caribbean. Glob Environ Chang 27:22–30

UNEP-WCMC and IUCN (2016) Protected planet report 2016. Cambridge/Gland, p 84

UNWTO and UNEP (2005) Making tourism more sustainable: a guide for policy-makers. Madrid/Paris, UNWTO and UNEP, p 209

USAID-Ecuador (2010) Mapeo de actores para el ordenamiento y certificación de playas en la zona costera de la cuenca del río Ayampe, Ecuador. USAID Costas y Bosques Sostenibles. Agencia de los Estados Unidos para el Desarrollo Internacional (USAID)

Vela-Ruiz G (2009) Contribución desde el enfoque de capitales a la comprensión de la inclusión de las comunidades en los procesos generados por áreas protegidas, Región de Magallanes. In Tesis País 2009 Piensa un País sin Pobreza. Fundación para la Superación de la Pobreza. Chile, pp 138–159

Vela-Ruiz G (2017) Desafíos para manejar el turismo en el Parque Nacional Torres del Paine. In: Vela-Ruiz Figueroa G, Repetto-Giavelli F (eds) Guía de conocimiento y buenas prácticas para el turismo en el Parque Nacional Torres del Paine. Ediciones CEQUA, Punta Arenas, pp 113–135

Vela-Ruiz G, Delgado M (2010) Contribución del enfoque de desarrollo territorial rural a la comprensión de los procesos generados en torno a áreas protegidas en la Patagonia Chilena. Revista Chilena de Estudios Regionales 2(1):83–94

Vela-Ruiz Figueroa G, Repetto-Giavelli F (eds) (2017) Guía de conocimiento y buenas prácticas para el turismo en el Parque Nacional Torres del Paine. Ediciones CEQUA, Punta Arenas, p 140

Vidal O (2012) Torres del Paine, ecoturismo e incendios forestales: perspectivas de investigación y manejo para la biodiversidad erosionada. Revista Bosque Nativo 50:33–39

Villarroel P (1996) Efecto del Turismo en el Desarrollo Local. Ambiente y Desarrollo 12(4):58–64

Villasante TR (2006) Procesos de planificación participativa para la sustentabilidad. Cuadernos CIMAS-Observatorio Internacional de Ciudadanía y Medio Ambiente Sostenible, España

Whittingham M (2010) ¿Qué es la gobernanza y para qué sirve?. Revista Análisis Internacional (RAI), 2:219–235

Worboys GL (2015) Concept, purpose and challenges. In: Worboys GL, Lockwood M, Kothari A et al (eds) Protected area governance and management. ANU Press, Canberra, pp 9–42

Xu J, Chen L, Lu Y et al (2006) Local people's perceptions as decision support for protected area management in Wolong Biosphere Reserve, China. J Environ Manag 78:362–372

Zarzar Charur C (2000) La didáctica grupal. Col. Santa María de la Rivera. Ed. Progreso, p 256

Chapter 13
ICZM Strategy for the Socioecological System of the Mar Menor (Spain): Methodological Aspects and Public Participation

Juan Manuel Barragán Muñoz, Javier García Sanabria, and María de Andrés García

Abstract Lessons learned from the process of the ICZM Strategy as part of the socioecological system of the Mar Menor (Murcia, Spain) may well be of interest to coastal managers. ICZM Strategy was one of the instruments chosen by the regional government of Murcia to deal with the critical situation of the largest lake in the western Mediterranean. Human activities that exert most pressure are those related to agriculture, urban growth and tourism-leisure. The Mar Menor is identified as a socioecological system defined through its river basin. The conceptual frameworks followed for the model of public management of the lagoon and its surroundings were DPSIWR and the Decalogue of ICZM. The scheme of work provides some lessons learned from previously analysed parts: (a) background and institutional arrangements, (b) operational diagnostic model, (c) strategic diagnosis, (d) proposals and (e) the participatory process. Additionally, the process of institutionalisation allows for other lessons to be learned. The document was finalised in 2016, and it has been further developed in order to become institutionalised. During this period of time, the regional government has already implemented actions that coincide with the proposals of the strategy, informing us that some instruments can start working before being officially approved.

Keywords ICZM Strategy · Mar Menor · DPSIWR · Participative process · Public policy

J. M. Barragán Muñoz (✉) · J. García Sanabria · M. de Andrés García
Research Goup on Integrated Coastal Zone Management, University of Cadiz, Cádiz, Spain
e-mail: juan.barragan@uca.es; javier.sanabria@uca.es; maria.deandres@uca.es

A. Ortega-Rubio (ed.), *Socio-ecological Studies in Natural Protected Areas*,
https://doi.org/10.1007/978-3-030-47264-1_13

13.1 Introduction

The Mar Menor is the largest coastal lagoon in the western Mediterranean and forms part of the most valuable natural capital of the autonomous region of Murcia and the Spanish east coast. It is a large coastal lagoon, separated from the Mediterranean Sea by a fine sandy strip 21 km in length, denominated "La Manga," which translates as "sleeve" in English (Fig. 13.1). The maximum width of the water body of the lagoon reaches 11 km, which gives an approximate surface area of 135 km^2. It is shallow with its average depth being 3.6 meters, with bathymetric maximums of 6.5–7 meters (LAGOONS 2012; Consejería de Fomento Obras Públicas y Ordenación del Territorio. Dirección General de Transportes Costas y Puertos 2014). The relationship between the two bodies of water – that of the Mar Menor and the Mediterranean Sea – is evident through three gaps in the shoal, or *golas* according to the local denomination. In the interior of the Mar Menor, there are small islands of low altitude and volcanic origin (Isla del Barón, El Ciervo, Perdiguera-Esparteña, Redonda, Sujeto, Islote de la Galera). The municipalities that share the same boundary as the lagoon are San Pedro del Pinatar, San Javier, Los Alcázares and Cartagena. Others that run through its river basin are Murcia, Torre Pacheco, Fuente Álamo and La Unión.

The "environment" of the Mar Menor is almost as important as the lagoon itself. The shoal and associated wetlands contain a high marine coastal biodiversity characteristic of these environments (Mintzberg et al. 2002). So deep and dependent is their relationship. The watershed of the Mar Menor is about 1300 km^2, and there is a network of streams and minor channels that feed the wadi running through the

Fig. 13.1 The Mar Menor and its surroundings

Campo de Cartagena, which then flows into the main body of water of the lagoon. The limits of this basin are marked to the Northwest by the Sierras de Columbares, Villares and especially Carrascoy and to the Southwest by Fausilla, Gorda and Algarrobo (DG de CH del Segura 2007). Undoubtedly, in the territorial area of the Mar Menor, water is one of its fundamental elements: comparable to blood circulating through a very vulnerable living system (Clark 1996; Mintzberg et al. 2002).

For several decades, the serious deterioration and ecological disaster of the lagoon have caused alarm. Additionally, this has been putting pressure on water resources in the basin in several ways. On the one hand, in times of drought (when the transfer of the Tagus River to the Segura River is not sufficient to cover the demand for water), there has been excessive exploitation of aquifers. On the other hand, aquifer water is brackish and must be treated before being used for irrigation, which has led to the appearance of brine discharges into the lagoon. And last, but not least, chemical and phytosanitary fertilizers from agriculture come into the lagoon.

The relationship between loss of water quality and increasing input of agricultural nutrients in the Mar Menor is clear (Perez-Ruzafa et al. 2005). Indeed, recent research aimed at the creation of predictive models indicates that the contribution of nitrogen to the lagoon from the basin could be around 1000–1300 tonnes per year (Martínez Fernández et al. 2013). In addition, there are other sources of water pollution: several decades ago, researchers (De León et al. 1982) found that heavy metals from former mining operations were accumulating in large concentrations in the sediments in the centre and south of the lagoon.

To all these pressures, the construction of large infrastructures and facilities that have altered the natural processes must be added. In the Mar Menor, in addition to the creation of several artificial beaches, one can find 90 dikes or breakwaters and up to 12 marinas. The largest of the marinas (more than 1600 moorings) was due to the expansion of the "Estacio" canal that connects the lagoon with the Mediterranean. This caused a major change in both the salinity and temperature conditions of the lagoon, both in its hydrodynamics and erosion/sediment deposition processes.

Given the explosive population growth, demand for infrastructure located on the edge of the Mar Menor has risen, replacing its banks and associated humid zones with residential areas, infrastructures and facilities. This situation of population concentration around coastal ecosystems of special value is repeated throughout the Spanish territory (de Andrés et al. 2017), as well as on the coasts of the planet as a whole, the consequences of which are already being studied by researchers (Barragán and de Andrés 2015).

The pressures described have emerged in a series of changes, for example, the substitution of seagrass beds for the proliferation of algae as a consequence of sediment changes and increased turbidity (Pérez-Ruzafa et al. 2012). Changes in benthic fish have also occurred due to dredging and the installation of rigid structures (Pérez-Ruzafa et al. 2006). Some researchers (Carreño et al. 2008; Moreno González et al. 2013; Salas et al. 2006) point to the intensive agriculture developed in the Cartagena Field as the origin of many of the problems found in the lagoon – those being the same agricultural activities that employ up to 80% of the population in some municipalities of the area.

Naturally, the disclosed situation translates into a loss of human wellbeing (Díaz et al. 2006), and coastal lagoons constitute one of the ecosystems that provide the most ecosystem services (UNEP 2006). In particular, regulation services are most affected, for example, morpho-sedimentary regulation (causing the erosion processes), "buffer" zones originating from disturbances from the marine environment (of special importance in the Manga del Mar Menor) or those of biological regulation (Marcos et al. 2015). Additionally, cultural services are greatly affected, meaning that certain touristic and recreational activities, such as sunbathing, etc., are incompatible with the appearance of millions of jellyfish (Cotylorhiza tuberculata and Rhizostoma pulmo) during the summer months.

In this context, the institutional concern for the evolution of the lagoon is reflected in the development of multiple instruments with limited results or, sometimes, with insufficient or non-development of them. This situation is enhanced by a scenario of conflict and disagreement between the institutions themselves and the users and between them and the numerous, very active, social agents. Additionally, it cannot be forgotten that from a legal and administrative point of view, the lagoon in its entirety belongs within the Public Maritime-Terrestrial Domain.

The technical work of the strategy took place between December 2015 and the first five months of 2016. They were carried out through institutional agreements between CARM, the University of Cadiz and the consulting firm Atlántida S.L. (Barragán and Sanabria 2016). During 2017, the corresponding Environmental Impact Study was carried out. In March 2018, the Initial Approval was received and the final result can be consulted at http://sitmurcia.carm.es.

13.2 Objective, Conceptual Framework, Method and Sources of Information

The main objective of this article is to inform coastal managers on an actual ICZM case study. The perspective offered due to the time that has elapsed since the formulation of the strategy allows us to carry out analysis and obtain some lessons learned. The practical and applied nature of ICZM recommends studying cases from which conclusions can be extracted.

The ICZM Strategy of the Mar Menor was the instrument employed to confront the serious ecological crisis of this lagoon. The aim was to develop understanding of aspects of management of the Mar Menor and, to a much lesser extent, to increase management of ecological knowledge (we will see later that there is very valuable information to be gathered from a technical-scientific point of view). The responsible for the formulation of the strategy was the Autonomous Community of the Region of Murcia (CARM in Spanish). Within this institution, the strategy came under the responsibility of the Ministry of Development and Infrastructure (MDI), to which the Directorate General of Transport, Coasts and Ports pertains. The strategy was developed and directed towards all departments at regional level as well as local and national administration.

This management tool was conceived to establish guidelines and also for long-term planning. The philosophy was to include the following premises: corporate (the strategy is for the CARM and not just a specific department), co-operative (based on the institutional commitments of the three levels of the administration) and participatory (where social agents express their will democratically). With these terms, the viability of the instrument is also sought (Mintzberg et al. 2002).

The conceptual framework followed was twofold (Fig. 13.2) and had already been used in other previous work on a national scale (Barragán and Lazo 2018). On the one hand, it was decided that the Mar Menor and its surroundings constituted a socioecological system (SSE-MM), due to the intense relationship between human activities and the natural ecosystems (de Jonge et al. 2012). For the analysis of this relationship, the DPSIR conceptual framework (Driver, Pressures, State, Impact, Responses) was employed (Cooper 2013). Added to this model were the changes in human wellbeing derived from the loss of ecosystem services (DPSIWR) (de Groot 1992; de Groot et al. 2002; MA 2003, 2005; Wang et al. 2013).

The conceptual framework followed to analyse the management model was the Decalogue for ICZM, which had already been used in previous case studies (Barragán et al. 2003, 2005; Barragán 2010). Basically, it consists of studying the various elements of a coastal management system: policy, regulations, institutions, coordination mechanisms, instruments, managers, participation, education, information and economic resources. It is worth clarifying that the proposals for action

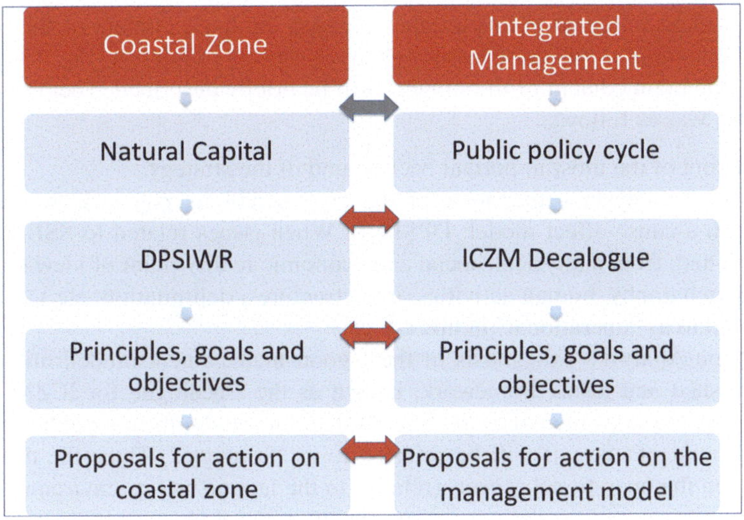

Fig. 13.2 Conceptual framework for the ICZM Strategy of the Mar Menor

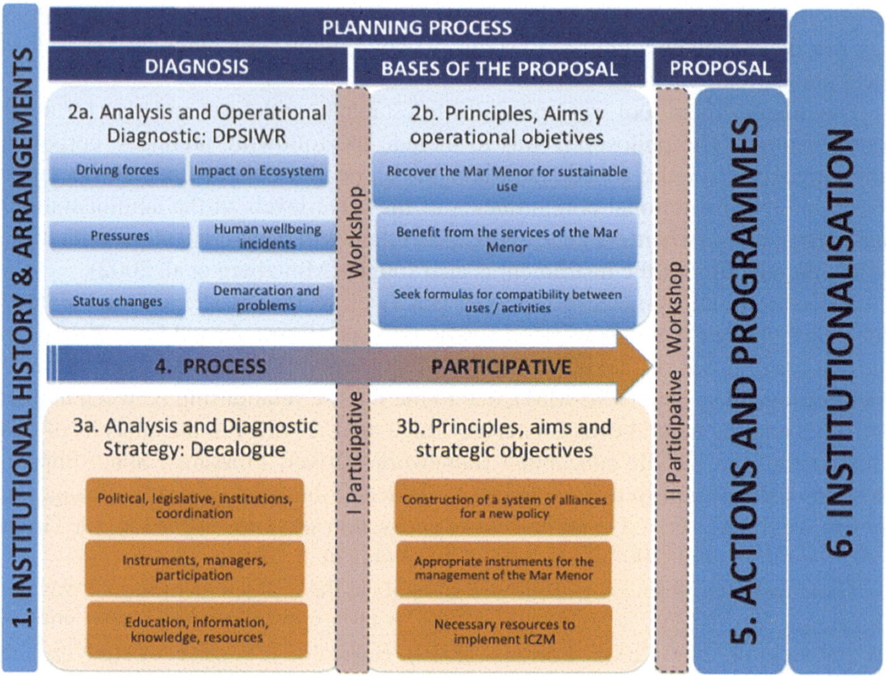

Fig. 13.3 Methodological plan

on the coastal zone and on the management model are equivalent to the Responses (R) sought in the DPSIWR model.

The method followed in this article combines the basic content of the strategy with the intention of obtaining lessons on some aspects of interest (Fig. 13.3):

First, the main content of the strategy will be briefly described. The programme developed was as follows:

1. Gathering of the most important background of the strategy.
2. Operative analysis – diagnosis of the lagoon and its physical environment, through a cause-effect model: DPSIWR. When issues related to SSE-MM are mentioned, from a physical, social and economic reality point of view (environment, geography, human activities, infrastructures, delimitation, etc.), it will be referred to as "operational" in this article.
3. Strategic Analysis – diagnosis of the lagoon management model, utilising an established and tested framework, known as the Decalogue for ICZM. When referring to issues related to the SSE-MM in this study, from a public management point of view, it will be referred to as "strategic". Thus, the difference between the operational aspects (relative to the lagoon and its environment) and the strategic aspects (relative to management in the strict sense) is established.

4. The proposals that facilitated the formulation of the strategy.
5. The participatory process, which has been identified as one of the most influential issues for practical purposes.

Secondly, we will provide a series of reflections of a synoptic nature for education purposes. A reasonable number of lessons learned from the case study will be sought. It is likely that coastal managers will benefit from a number of these lessons. The order of work will be the same as before (background, analysis–operative diagnosis, analysis–strategic diagnosis, strategic proposals and participatory process).

In addition to the above, a sixth section is added: the process of institutionalisation of the strategy (June 2016 to March 2018). Here, two different but complementary points of view will be taken into account: (1) legal/administrative and (2) executive. The first includes understanding the claims made in the Development Strategy, the Environmental Impact Study and the Initial Approval of the Strategy (February 2018). The second requires assessment of measures carried out in accordance with recommendations from the strategy albeit it was not definitively approved.

The sources of information have been of three types: (1) bibliographic, (2) documentary and (3) personal interviews. On the first, it can be said that the Mar Menor has been the object of numerous studies of great scientific value (Espejo Marín 2011; García García 2004; Pérez-Ruzafa et al. 2005, 2012; García-Pintado et al. 2009). The large number of publications in books and specialised magazines can be accounted for the fact that in the region of Murcia, there are two major public universities (the University of Murcia and the Polytechnic University of Cartagena) and research centres (the Spanish National Research Council, the Spanish Institute of Oceanography, the Geological and Mining Institute of Spain, the Murcian Institute of Agricultural and Food Research and Development). Among the publications of interest are those of ecology, hydrology and geography. In addition, in 2009, the Euro-Mediterranean Water Institute published a compilation of various interesting studies entitled "The Mar Menor. The Current State of Scientific Knowledge" (Cabezas Calvo-Rubio 2009). Also, a large number of both regional and state administration departments have written documents and statistics of great technical interest.

The third source of information consisted of three rounds of open and face-to-face interviews with 42 different social and institutional agents. Each interview was structured in two blocks with five questions each: (a) one referred to the opinion of the interviewee regarding those matters in which he was most involved or interested (as an entrepreneur, as an ecologist, as a neighbour, as a technician of the administration, etc.) and (b) one referred to their interpretation of what had happened in the lagoon in the past, what is currently happening and what might happen in the future. The result of these interviews served as material for diagnostics as well as preparation of the participatory workshops.

13.3 Results

13.3.1 Background and Institutional Arrangements of the Mar Menor Strategy

In Spain, several ICZM strategies can be cited: those referring to the coasts of the whole country and those of coastal Autonomous Communities such as Valencia, Catalonia, Andalusia or the Canary Islands. The public administration did not regard any of these strategies as useful instruments, and in most cases, they were not officially approved (Barragán 2010). Consequently, only technical documents commissioned by the public administration without legal administrative value should be considered.

The concern of the General State Administration (AGE in Spanish) for the Mar Menor goes back at least three decades. In 1985, the Ministry of Public Works and Transport through the CEOTMA (Centre for Territorial Planning and Environment Studies) conducted a study entitled "Territorial Planning of the Mar Menor Zone and Its Surroundings". It was concluded that there was no correspondence between the functioning of the ecosystem and the fragmented model of institutional management (CEOTMA 1985). Since that date, and for decades to come, the AGE has executed or authorised a considerable number of actions that affect the lagoon, accounting for a substantial percentage of its annual investments. Table 13.1 summarises the major initiatives and studies carried out since 2003 until the start of the formulation of the ICZM Strategy of the Mar Menor.

In 2013, in order to create a strategy for the integrated management of the Mar Menor and its surroundings, regional and state institutions agreed to create the following instruments:

- The signing of a protocol of collaboration between the Ministry of Agriculture, Food and Environment (AGE) and the Autonomous Community of the Region of Murcia
- Creation of the Joint Monitoring Commission composed of administrative political authorities
- Creation of an Inter-administrative Technical Commission composed of officials and technicians from the different administrations of the State and CARM

In the Collaboration Protocol, the outline of the strategy process was finalised:

- Demarcation of the geographic scope and the creation of several inventories (legislative, agents and administrations involved, problems, etc.)
- Participation of social and institutional stakeholders to discuss problems identified and proposed measures for their solution
- Formulation of the strategy
- Implementation and monitoring of the strategy

Finally, a particular reflection of the political interest of the regional government has been the initiative to use the European figure of "Integrated Territorial

Table 13.1 Institutional studies and initiatives prior to the Mar Menor Strategy

Year	Action	Institutions involved	Content
2003	Feasibility study for an Integrated Management Programme of the Mar Menor and its area of influence (CAMP Mar Menor)	Mediterranean Action Plan, Priority Actions Programme, AGE and Ministry of Agriculture, Water and Environment of the Council of the CARM	Framework, budgets and organisation to include the Mar Menor as a CAMP project
2007	First Approximation Report (FAR) for the agreement of an action programme in the Mar Menor area	General Directorate of Coasts and Segura River Hydrographic Confederation (AGE)	Diagnosis of the situation of the Maritime-Terrestrial Public Domain and Hydraulic Public Domain of the Mar Menor basin
2008	First Approximation Report (FAR) for the agreement of an action programme in the Mar Menor area	General Directorate of Coasts, General Secretary for Territory and Biodiversity and Ministry of the Environment (AGE)	Analysis of the proposals made by social and institutional agents in relation to the 2007 First Approximation Report (FAR)
2013	General protocol between the Ministry of Agriculture, Food and Environment (AGE) and the CARM on integrated management of the Mar Menor	Ministry of Agriculture, Food and Environment and CARM	Agreement to set up the Joint Protocol Monitoring Commission and the Inter-administrative Technical Commission
March 2014	Agreement for Integrated Territorial Investment (ITI). Measures for an integrated approach in the use of European Structural and Investment Funds (EIE) for territorial development of the Mar Menor	General Directorate of Community Funds, General Sub-Directorate for Territorial Programming and Evaluation of Community Programmes (AGE) and CARM	Implementation of the part of the strategy that corresponds to CARM, through the Integrated Territorial Investment (ITI)
April 2014	Conference on Integrated Coastal Zones Management of the Mar Menor and its environment	Social and institutional stakeholders involved	Consultative process for an action plan
May 2014	Report on the Integrated Coastal Zone Management Strategy in the Mar Menor and its surroundings	General Directorate of Transport, Coasts and Ports of the Ministry of Public Works, Public Works and Territorial Planning (CARM)	The process and the status of the work done is summarised.

Investment" (hereinafter ITI) for the development of the Integrated Management Strategy of the SSMSM (2015). This means that the government will be able to invest in the implementation of the strategy by making investments co-financed by various Structural and Investment Funds, specifically using the ERDF (European

Regional Development Fund), EAFRD and FEMP European Maritime and Fisheries Fund. This situation has two possible readings:

- The strategy is of interest in the political field to capture European investments that will be derived from the figure of ITI. This situation provides for continued political support in both the formulation and development of the strategy.
- The starting point of the strategy is very promising, since there are few instruments of this type that have economic resources, at least for the development of the first stage of implementation, before being formulated.

13.4 Analysis, Diagnosis and Operational Bases (of the Mar Menor and Its Surroundings)

The results obtained for the SSE-MM using the model (DPSIWR) have been summarised in Fig. 13.4. Significant data has been summarised in Table 13.2. This analysis begins by studying the drivers observed in the SSE-MM. The most important are population growth, tourism-leisure development and innovation of agricultural techniques. The first two originate from the pressure from intense urbanisation and the construction of numerous infrastructures (highways, railways, ports, coastal defence installations) as well as massive construction of houses for second residence and tourism. Also, agriculture, one of the most intensive and competitive in Europe, creates pressures on the lagoon. The tourist and leisure activity in the lagoon cause pressures on the ecosystem due to the anchoring and navigation of thousands of motor boats, diving, sport fishing, etc.

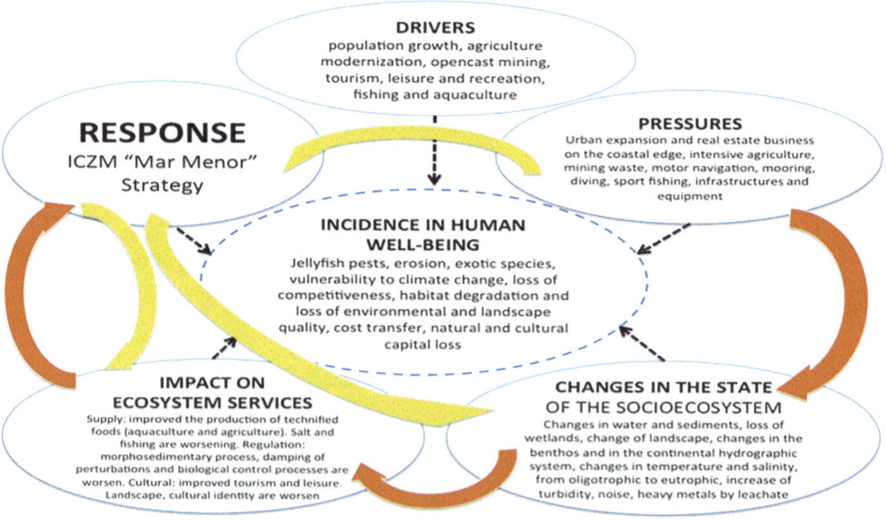

Fig. 13.4 DPSIWR model adapted to SSE-MM (Source: Barragán and Sanabria (2016))

Table 13.2 Some indicative figures on human activities in the Mar Menor

	Indicators	Sources
Population	The SSE-MM population goes from 158,000 to 358,000 inhabitants between 1950 and 2014.	National Institute of Statistics
	65% of the contracts in the SSE-MM pertain to agriculture.	Regional Statistics Centre of Murcia (2015)
	The artificial surface of the edge of the lagoon is 63%.	Prepared by the authors
Agriculture	The agricultural space of the Campo de Cartagena occupies around 900 km^2 with 82,000 cultivated hectares (rainfed and irrigated).	Soto et al. (2014)
	The area irrigated in the Campo de Cartagena is 34,000 ha according to official figures; however, according to studies based on satellite images of land use coverage, this figure is between 55,000 and 60,000 ha.	Martínez et al. (2013)
	In Campo de Cartagena, there is one of the largest irrigation communities in Europe with highly developed technology.	Soto et al. (2014)
	There are 4092 small reservoirs and more than 4000 greenhouses in the Campo de Cartagena.	COAG (Coordinator of Farmers and Livestock Organizations)
	The average concentration of nitrates in the lagoon amounts to 28 mg/l in 2003–2004. The average concentration of nitrates between July and October 2013 shows average values of 153.6 mg/l and maximum values of 249 mg/l.	García-Pintado et al. (2009) Measurement of the Segura River Hydrographic Confederation (2013)
	It is estimated that the contribution of nitrogen from the basin could be around 1000–1300 tons per year.	
Fishing	In the interior of the lagoon, there are 18 areas of artisanal fisheries.	
	In 2012, 593,000 kg of species were landed from the lagoon. The average production of salt during the previous decade was 80,000 metric tonnes.	Regional Statistics Centre of Murcia (2013a)
	Changes to the bed of the lagoon due to seasonal anchoring of more than 2000 sports boats.	

(continued)

Table 13.2 (continued)

	Indicators	Sources
Tourism and leisure	The total number of homes is 234,000, but almost 79,000 are secondary residences and 31,000 are empty.	Regional Statistics Centre of Murcia (2013a)
	Hotel accommodation is scarce: only 8053 hotel beds in 2013.	Regional Statistics Centre of Murcia (2011)
	There are 12 marinas and yacht clubs in the Mar Menor. And the largest of them, Puerto Tomas Maestre, has 1600 moorings.	
	The extension of the Estacio channel in 1973, 30 meters wide and 5 meters deep, has caused the invasion of Mediterranean species in the lagoon.	Gimenez Casalduero et al. (2012)

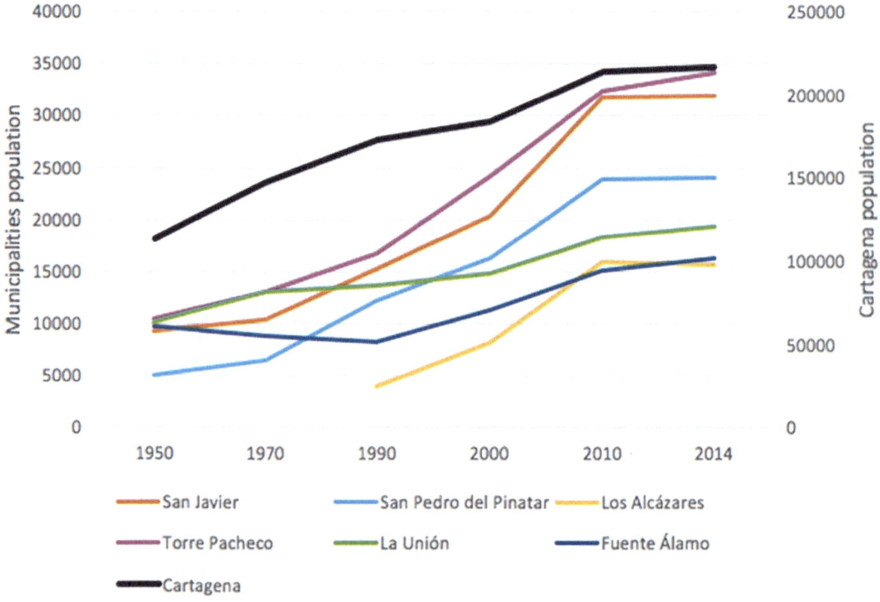

Fig. 13.5 Evolution of the actual population of the Mar Menor environment (Source: National Institute of Statistics)

The population statistics reveal that the SSE-MM municipalities have gone from 158,000 inhabitants in 1950 to 358,000 in 2014 (Fig. 13.5). Some cases are striking, such as those of San Pedro del Pinatar, which multiplied its population almost five times in that period of time, or those of San Javier and Torre Pacheco where inhabitants increased 3.4 times.

Changes in the status of the SSE-MM have manifested in various forms: by altering the continental hydrographic system (due to drastic changes in the circulation of water and sediments), replacing natural vegetation with crops, degrading the landscape by large engineering works, loss of biodiversity due to the disappearance of

Table 13.3 The importance of the socioecological system of the Mar Menor to provide a service and service flow trend.

Type of Service			Importance and evolution of service
Provision	Traditional food production	Traditional fishing	↓→ (Low Importance)
		Traditional agriculture	↓ (Low Importance)
		Salt	↓→ (Low Importance)
	Technical food production	Intensive Agriculture	↑ (Very High Importance)
		Aquaculture	↑ (Medium Importance)
	Water Supply for human use		↓ (Very High Importance)
Regulation	Hydro regulation		↓ (Very High Importance)
	Regulation morpho -sedimentary		↓ (High Importance)
	Buffering against disturbances		↓ (Very High Importance)
	Biological regulations		↓ (High Importance)
Cultural	Scientific knowledge		↑ (Medium Importance)
	Cultural identity, belonging		↓ (Very High Importance)
	Landscape – aesthetic enjoyment		↓ (High Importance)
	Recreational activities – tourism		↑ (Very High Importance)
	Environmental education		↑ (Medium Importance)

Low Importance	Medium Importance	High Importance	Very High Importance

Deterioration in service	Maintenance of service	Improvement to service
↓	→	↑

Source: Barragán and Sanabria (2016)

coastal wetlands, modification of the lagoon bed, changes in salinity and temperature, the evolution of an oligotrophic lagoon into a eutrophic lagoon, increase in turbidity, increase in noise due to the engines of vessels and the presence of heavy metals due to leachate from the slopes of old open pit mines.

The impacts on ecosystem services were also analysed. The summary is as follows (Table 13.3):

- Supply Services

In the long term, the supply service of food such as fishing (traditional) is deteriorating (Marcos et al. 2015). The production of sea salt is stable. Technical food production service is increasing. Also the production of food from aquaculture is increasing. The water supply from underground aquifers is deteriorating due to progressive salinisation, as a consequence of irrigation pressure.

- Regulation Services

In general terms, it is possible to view the regulation services of the lagoon and La Manga as deteriorating over the last decades. In the first place, morpho-sedimentary regulation is being affected by the barrier effect that certain port infrastructures cause. The damping service of disturbances from the marine environment has also been reduced due to the disappearance of a large part of the sand and the dune field of La Manga, mainly due to the changes in the hydrodynamics of the lagoon and inadequate urban planning. The result is that this coastal stretch is in the process of regression, which will undoubtedly lead to significant socio-economic and environmental impacts. The lagoon also offers nursery services for fish species of fishing interest that find shelter or food in the first stages of their development. It is possible that the opening of the Estacio, by facilitating a tendency of equal conditions in salinity and temperature, has caused a decrease in this valuable service. On the other hand, there is a marked decline of the most emblematic species of the lagoon: the seahorse (Hippocampus guttulatus). This situation has led to their inclusion in the Red Book of Vertebrates as "Critically Endangered Species". The seahorse is an iconic species of the Mar Menor, which has given rise to special attention for its conservation and recovery by diverse public and private institutions, as well as civil society groups.

- Cultural Services

Substantial improvement is observed in some and a worsening in others. For example, indicators of recreational fishing licenses issued, valid per year, should be expected to increase (Santos-Martín et al. 2015). There is also no doubt that there have been increases in the number of tourists and visitors from the process of urbanisation of La Manga.

The positive evolution of some of the above indicators is likely to explain the worsening of other ecosystem services. Thus, those linked to local ecological knowledge, cultural identity and sense of belonging, landscape and aesthetic enjoyment could be suffering for the success of previous ones. More updated detailed studies on ecosystem services of the Mar Menor and its public perception can be found in Velasco (2017).

The scope of the strategy was also considered. In Fig. 13.6, the geoterritorial scheme of the proposed SSE-MM can be observed. The following units were

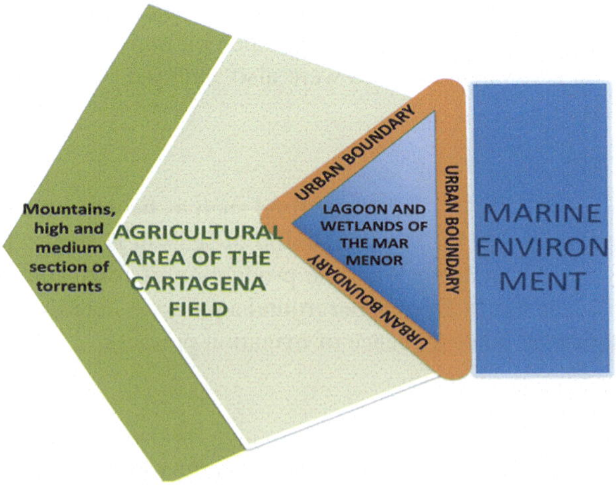

Fig. 13.6 Geoterritorial overview of the socioecological system of the Mar Menor

Table 13.4 Problems of the SSE-MM and their possible causes

Environment	Manifestation	Possible causes
Social/ environmental	Alterations to the biophysical process of the SSE-MM	Construction of coastal infrastructures (dams, ports), landfills, dredging and widening of water inlets/outlets. Radical changes to the water system
	Pollution of lagoon, wetland and aquifer waters	Agrarian residues (Albujón and other wadis) and urban waste (partially solved), tourism and navigation, brine from desalination, leachates and particulate material from residual mining
	Deterioration of natural habitats and loss of biodiversity	Intense urbanisation of the lagoon perimeter, changes in land use in the Campo de Cartagena, the entry of exotic species, solid waste, pollution, soils and lagoon sediments with heavy metals from mining waste
	Increased vulnerability to climate change	Occupation of the coastal edge that eliminates natural barriers of La Manga (dune field)
Social/cultural	Deterioration of the quality of the environment and the landscape	Development of certain uses and economic activities: urbanisation and intensive agriculture
	Loss of competitiveness due to decrease in attractiveness and increase in maintenance cost	Quality of the waters of the lagoon. Banalisation of the landscape. Plague of jellyfish. Costs that increase due to the maintenance of facilities and infrastructures
	Loss of natural capital and cultural heritage	Degradation of public, natural and cultural heritage, which can support economic, cultural and leisure activities. Loss of cultural identity and affection for the Mar Menor as a "family" space to be defended and protected
	Transfer of costs between activities and users	Predominance of some economic activities over others that generate externalities

Based on CARM (2016)

included: Laguna del Mar Menor with associated wetlands and salt marshes, urban contour of the lagoon and agricultural region of the Campo de Cartagena (where human activity is more intense), hills and slopes of the high and middle sections of streams and torrents and marine environment.

Finally, it is worth summarising the problem of SSE-MM. Table 13.4 illustrates the relationship between how problems arise and what appear to be the associated causes.

13.5 Analysis, Diagnosis and Strategic Bases (of the Management Model): The Decalogue

Table 13.5 displays the main ideas related to the strategic diagnosis or model of the management system. Each element of the Decalogue was extensively discussed. One of the main findings was in fact contradictory: the Mar Menor has registered

Table 13.5 Summary of the Decalogue for the ICZM in the Mar Menor and its surroundings

Policy	The most effective public policies have been those aimed at promoting and facilitating infrastructure and productive activities in an intensive manner (tourism and leisure, real estate, agriculture). Additionally, these have focused on developing standards but not on ensuring their application. To date, an explicit public policy of ICZM has not been identified.
Legislation	There is a considerable amount of regulation of a sectoral nature, but there is currently no specific regulatory instrument aimed at managing the Mar Menor and its surroundings in order to move towards a new management model adapted to its uniqueness.
Institutions	In spite of the complex institutional structure that the three scales of the administration have implemented, there is no institution that has been specifically designed to manage the SSE-MM in an integrated manner.
Coordination	The protocol signed between the AGE and the CARM for the development of the Mar Menor Strategy is an important step towards improving institutional coordination and cooperation. There are also general coordination instruments between the different regional institutions, but these are not specific enough for the management of the lagoon.
Instruments	Considerable quantities of instruments exist that apply to the Mar Menor, but there is great difficulty in putting them into practice. The aim is therefore to evaluate the different initiatives to be able to correct the problems that explain the lack of efficiency of the instruments used.
Administrators and managers	It is necessary to pay more attention to the training offered to public and private managers, related to the Mar Menor Administration and its surroundings, to foster better responses to the needs of ICZM. This must also be continuous, objective and practical.
Public participation	There are no specific mechanisms of public participation associated with the management of the lagoon, and this is despite increased citizen awareness and mobilisation in relation to the problems of the Mar Menor.
Education and awareness of sustainability	The existing environmental education programmes are oriented towards knowledge and assessment of protected natural spaces. There is a lack of a periodic environmental education programme that deals with issues related to the problems of the lagoon.
Information and knowledge	There is sufficient knowledge of the lagoon and its problems, and there is also a series of scientific institutions that can continue to provide information and knowledge.
Economic resources	Economic resources are assigned by the different administrations with sectoral criteria, in insufficient quantities and without coordination criteria to address the critical situation of the lagoon. The considerable quantity of extraordinary resources is European originating from Integrated Territorial Investment with the aim of creating an integrated management model.

critical deterioration in recent years despite the fact that there are a large number of environmental protection instruments. In fact, there are six international, European, national and regional protection bodies: Wetland of International Importance (RAMSAR), Specially Protected Areas of Importance for the Mediterranean (ZEPIM), Protected Landscape of Open Spaces and Mar Menor Islands, Regional Park of Salinas and Arenales of San Pedro del Pinatar, Site of Community Importance (SCI) and Special Protection Area for Birds (ZEPA).

13.6 The Participative Process

The design of the public participation processes of the SSE-MM ICZM Strategy was especially cautious, including two distinct stages: diagnosis and proposals (Fig. 13.7). At the diagnostic stage, the participation of the agents was incorporated from the beginning. More than 40 face-to-face interviews were carried out with institutional agents from different scales of public management, entrepreneurs, scientists from universities and research centres, non-governmental organisations and civil society through neighbourhood associations. In so doing, different versions of the draft of the diagnosis were created with the improvements that the different agents contributed. The result was again sent and submitted to telematic consultation by all the agents interviewed and others with whom it was not possible to have a face-to-face meeting. The different versions of the draft document were sent to more than 60 people. Thus, a third version of the diagnostic document of the strategy was obtained before the first participatory workshop.

The advantages of this method were multiple: on the one hand, the different agents (who were consulted) felt their views were taken into account, thus generating support for the process of formulating the strategy. On the other hand, the diagnostic document was enriched by the different visions and information provided by the agents. In this manner, a favourable climate was created for the first participatory workshop.

Participatory workshops were carefully designed. First, the appropriate level or degree of public participation was determined for the decision being considered, thus setting realistic expectations of the participants. For the Mar Menor Strategy, the levels of participation were established between section three (involve the public to ensure that their concerns are considered through the decision-making process; Workshop 1) and section four (collaborate with the public to develop decision-making and alternative criteria, as well as identifying preferences for solutions; Workshop 2), according to the classification used as a reference model, which was proposed by the Environmental Protection Agency (2012).

As a general criterion, it was proposed that the workshops be attended by a total of 100 people, with a distribution that sought the balance of interests among the participants: 25% from public administration, 25% from science, 25% from social organisations and 25% from economic activity.

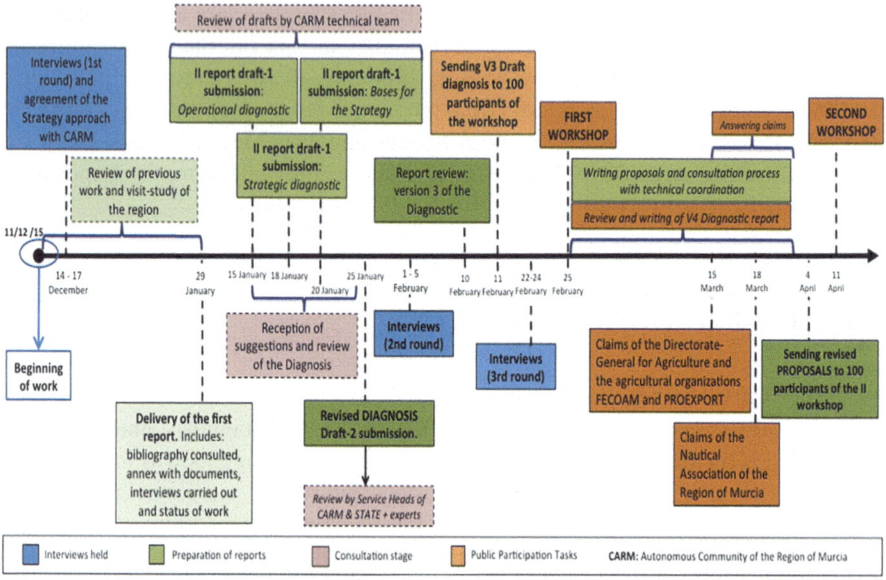

Fig. 13.7 Evolution of the participatory process

The objective of the first workshop was to understand the degree of agreement that existed regarding the material and the ideas sent to the participants on the diagnosis of the Mar Menor and its management model. The result was conceived as a first step to discuss options and proposals in a second workshop.

Thereof, the participants were divided into four working groups. The first worked on the operational diagnosis, the second and third worked on the strategic diagnosis and the fourth on the bases for the development of the future strategy. To facilitate the work, each group had a tab with the main ideas of the diagnosis already done, and questions were put to the participants by a facilitator who organised the work in each of the groups. The groups were then brought together to discuss the results so that they could influence the work of others in an open debate.

Workshop I had high participation, reflecting the interest that the Mar Menor provokes in the different social, institutional, economic or scientific areas of the society of Murcia. In fact, although 100 representatives of the different sectors were contacted for their participation, the total attendance was 110 people (Fig. 13.8). This was due to the fact that some of the social or business organisations decided to send several representatives.

The balance between the different interests in the first participatory workshop, and therefore the representativeness and legitimacy of the process, was very satisfactory. The public administration was represented by 25% of participants; 29% came from the corporate sphere, 28% from social groups and 18% from the scientific and academic field.

One of the most important issues of the process was the fact that the diverse interests of those attending the workshop were well represented in the four

Fig. 13.8 Participatory workshop of the ICZM Strategy of Mar Menor and its surroundings

workshops. There were some differences due to the necessary flexibility of the process, which forced the assignment of new assistants in groups not initially foreseen.

In general, there was great interest on the part of the different social and institutional representatives in addressing the problems and proposing solutions to improve the state of the Mar Menor. It was also possible to reach a high level of agreement of the participants regarding the ideas reflected in the diagnosis of the strategy.

The second participatory workshop sought to collect ideas and thoughts regarding the proposals of the strategy, which were prepared by the drafting team according to the objectives agreed upon at the first workshop and were sent to the participants in advance. The second workshop attracted high levels of participation, reflecting the interest that the Mar Menor raises in the different social, institutional, economic or scientific areas of the society of Murcia. One hundred representatives of the different sectors were invited to participate, and total attendance was 77 people.

The balance between the different interests in the participatory workshop, the representativeness and legitimacy of the process, was again very satisfactory. Twenty-two percent of the participants came from the public administration, 31% from the business group, 27% from social groups and 20% from scientific and academic fields. It is clear there is heightened interest in the Mar Menor between the social and economic agents of the area, who attended the workshops in a greater percentage (58%) than the collective representing administrations and science (42%). In the second workshop, all the proposals of the participants were collected, and the majority of modifications were accepted.

13.7 Proposals for the Formulation of the Strategy

The proposals of the strategy were agreed among the participants and organised in two different sections, following the same scheme as the Operational and Strategic Analysis-Diagnosis of the SSE-MM: The first refers to the object, that is, the Mar Menor as a physical and social reality, while the latter focused exclusively on the management system, looking for a new model of governance.

With this double scheme agreed upon, principles of action were then addressed. Goals and general objectives were established and described in the strategy. Then 21 operational plans and programmes, linked to subsystems, were agreed upon. These are laid out in Fig. 13.7. The lagoon and its associated wetlands were designated as the Critical Zone, and the participants proposed a Management Plan for the Protected Areas (coinciding with those the CARM already had in progress) and a Plan for the Regulation, Promotion and Control of Facilities and Activities of the Mar Menor. The urban contour of the lagoon and the agricultural area of the Campo de Cartagena were designated as Coastal Lands of Intensive Activities. Here, the proposed plans were linked to the reduction of polluting contributions from crops and cities and to substantial improvements of territorial and urban planning. Finally, the forest areas of the mountains, or Littoral Influence Areas, were linked with river basin plans. Finally, for the marine space or coastal waters, plans related with the control of tourist and fishing activities and adaptations to climate change were proposed.

Ten specific strategic objectives were agreed, one for each element of the ICZM Decalogue. These resulted in 27 tangible actions. For example, one of these specific strategic objectives was: "No.3 To ensure that the management of the SSE-MM stands out as a model of social participation. Actions: (a) Creation of the Forum of the Mar Menor, (b) Preparation of a Directory of all managers of the Mar Menor, (c) Creation of a Mar Menor Communication and Information System". Each of the actions was summarised so that the interested parties and participants would be aware of the details of each action. Finally, the strategy raised other interesting issues: institutional leadership and responsibility for the execution of each action; assessment of the importance, urgency and achievability of each action; plans and action programmes; system of indicators; and detailed exposition of the participatory process.

13.8 Lessons Learned

13.8.1 Background Information Such as Previous Agreements and the Formulation Process of the Strategy Were Developed in a Political-Technical Framework. The ICZM Is Not an Exclusively Technical-Scientific Discipline

Due to the experience of the ICZM Strategy of Andalusia (Barragán et al. 2008), the Mar Menor participants paid close attention to previous institutional agreements and studies. A first lesson was learned: in the Murcia region, any ICZM project would be absolutely influenced by the existing conflict between political parties. Proof of this can be found in the failed attempt in 2003 to include the Mar Menor as a pilot case in the Coastal Area Management Programme (CAMP) of the

Mediterranean Action Plan, part of the Priority Action Programme of the United Nations. There was even an agreement between the State Administration and the CARM to carry out a feasibility study, and although the result was positive, a change of government in Spain in 2004 meant that the pilot case for the CAMP went to the Coast of Almeria (Andalusia). The reason for this change is simple: Until 2003, Spain and Murcia were both governed by the conservative party. In 2004, the elections gave the government of Spain to the socialist party, who in turn chose another coastal region, also governed by the socialist party, for the CAMP Programme: the Coast of Almeria in Andalusia.

From the time in which the project took place and up until 2018, the situation in Murcia has not been ideal to execute a strategy requiring consensus among citizens and political parties. In this case, the executive power belongs to the conservative party, but the legislative power is in the hands of a coalition of centre-left parties, the consequences of which are evident. For example, in February 2018, the Regional Parliament of Murcia approved a law on "Urgent measures to guarantee the environmental sustainability of the Mar Menor environment". This replaced a decree of the executive power of 2017 with the same title and purpose.

On a positive note, it is worth mentioning the Collaboration Protocol described in Sect. 16.3.1 between the government of Spain and that of the autonomous region of Murcia. In this case, it was an agreement reached between political-administrative authorities of the same political party. The agreement established the creation of political and institutional commissions. Above all, the latter facilitated the initiation and real progress of the strategy. One of the results was to streamline the relations between technicians and officials of both administrations. This involved facilities for meetings, information exchanges, participation in workshops, etc. Thanks to this agreement, the strategy was formulated in 2016 and obtained Initial Approval in 2018. Consequently, the lesson learned is that although the ICZM may have a technical-scientific profile, in reality, it is the political and institutional environment that marks the pace of progress and even its success or failure.

It is also worth mentioning the European instrument of Integrated Territorial Investment (ITI) for the Mar Menor. From our experience and observation, it is clear that the significant contribution of financial resources was a constant incentive for the political-administrative authorities.

In addition to the general background described, there was another more specific aspect that turned out to be fundamental leading up to and during the whole process. It is that of the technical support that an initiative of this nature requires. On the one hand, the CARM officials acting as supervisors of the strategy (General Directorate of Transport, Coasts and Ports) wrote the focus for the terms of reference of the subsequent public contract in a very sound manner. They incorporated the new approaches (DPSIWR, Decalogue) and worked decisively to make the strategy participatory and transparent. Another key aspect is that these officials continued working on the strategy when the politico-administrative authorities changed. This idea of continuity in a project as unique as this within the Spanish public administration is fundamental and must be taken into account. Additionally, the trust placed in consultants specialised in coastal management from a university strongly oriented towards coastal marine affairs gave the work greater consistency and credibility. This aspect is also important. It cannot be forgotten that the deterioration of the lagoon due to political and institutional inaction created a very tense social climate.

13.8.2 New Conceptual Approaches to the Field of Study Could Be Employed Due to the Thorough Explanation to Officials from Other Departments, Political-Administrative Authorities and Participants

In the SSE-MM Strategy, the conceptual framework used was a synthesis of several approaches. On the one hand, the DPSIR (Driver, Pressures, State, Impact, Responses) and, on the other hand, the wellbeing associated with the ecosystem services proposed by the MA were taken into account (MA 2003, 2005). The Mar Menor and its surroundings were interpreted as a socioecological system from the beginning. In this case, intense human activity makes it very difficult to separate natural and social realities.

It is necessary to recognise that in Spain, these conceptual frameworks are used mainly in academic and scientific fields. To a lesser extent, they have managed to become relevant in coastal plans and programmes of the public administration. For this reason, the authors of the strategy made an extraordinary effort to explain to the different stakeholders the benefits of this conceptual framework: in the initial meetings, this new approach was briefly explained, and in the participative workshops, didactic presentations were made using examples from the Mar Menor to clarify this new approach. The conceptual framework was proposed as a socio-environmental interpretation of what had happened in the Mar Menor.

The result was very successful. There was no rejection on the part of any stakeholder, at least publicly. On the contrary, in the meetings, the DPSIWR model was understood as an explanatory cause-effect discourse.

The approach utilised resulted in criticism of the limitations established in the 2014 Report prepared by the CARM. In this report, the land limits were marked, to a large extent, by the so-called "Vulnerable to Nitrate Pollution Zone" (Fig. 13.10). In this case, it is considered vulnerable due to the surface of land whose run-off flows into the lagoon. In reality, the waters are indeed susceptible if no action is taken, due to the contamination of nitrates from agricultural sources (art 3.2 Directive 91/676/CE, relative to the Protection of Water against Pollution produced by Nitrates from Agrarian Sources).

As shown in Fig. 13.9, the 2014 boundary did not account for the upper and middle part of watercourses that flowed into the Mar Menor. For this reason, a proposal was made in 2016 to extend the scope of the strategy to include the entire Mar Menor watershed. The justification was simple: under the 2014 boundary, the upper and middle part of the watercourses was occupied by forest uses and traditional agriculture. However, one could also imagine an unwanted future scenario for those

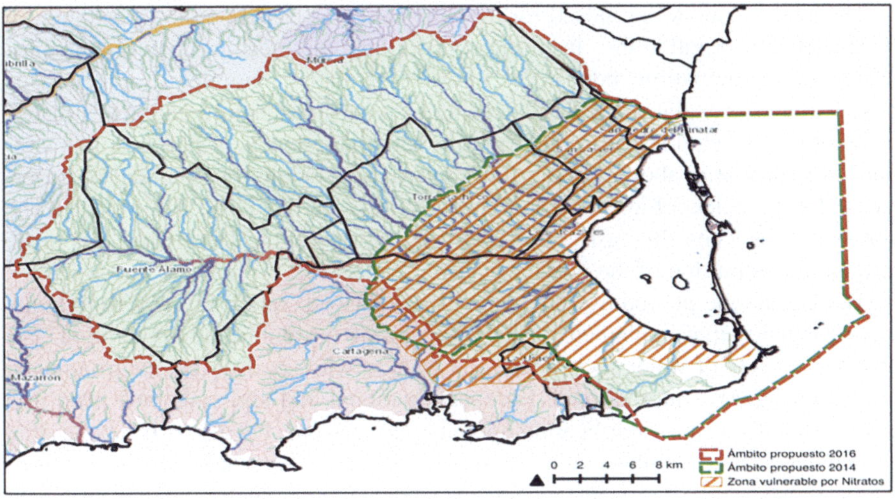

Fig. 13.9 Boundaries of the socioecological system of the Mar Menor

areas: one that would expand irrigated agriculture or urban areas to those areas. At first, there was some resistance to change; however, after the first participatory workshop, where the citizens were clearly in favour of this expansion, the new SSE-MM boundary was accepted and is now in place.

Additionally, the presentation of the concerns mentioned in Table 13.4 established a new approach for the Mar Menor. Its division into socio-environmental (4) and sociocultural (4) made it clear that the connection between natural and social ecosystems is as intense as it is permanent. The manifestation and possible causes of the eight problem groups were agreed among the participants of the participatory process. Although it is necessary to stress that on the whole, human activity was the cause of the problems described.

13.8.3 It Is Necessary to Clearly Focus on the Institutional Management Model When Working with ICZM. A Concise Outline Is the Best Way to Address It

The analysis of a public management model is not normally an easy task. Often, the issues are not familiar to many citizens. It is not uncommon to encounter plans or programmes that are too focused on the operational aspect (dealing fundamentally with what needs to be done regarding ecosystems) and, to a lesser extent, those that are skewed towards strategic aspects (how and what administrations or departments should do). It is clear that a reasonable balance must be sought in this regard.

The analysis model proposed for the SSE-MM was based on the Decalogue for ICZM: policy, regulations, institutions, coordination mechanisms, instruments, managers, participation, education, information and economic resources. This model was also explained carefully owing to its abstract nature and also because many of the stakeholders were not familiar with these details of public administration. In reality, it is about looking for an alternative or improved model of governance for a socioecological system (the answers of the DPSIWR model). The connection between this last conceptual framework and the Decalogue was made through the definition of the problems.

The Decalogue proved useful for the task described. And it brought to light the inconsistencies that explained the inefficiency of the public administration. For instance, a body of water such as the Mar Menor, with six different environmental protection figures (international, European, national and regional), had been in a critical state for years – implicating not only public agents but also private ones that carried out certain activities. Other examples appeared when analysing the results of public policies that had been applied over the previous decades, the insufficient existing coordination mechanisms, the weakness of the participatory system within the framework of a mobilised local society or the dispersion of the large amount of information and existing knowledge on the lagoon.

13.8.4 An Adequate and Flexible Participatory Process Strengthens the Instrument That Is Being Formulated

The participatory process was key from the beginning of the study. Even the diagnosis of the SSE-MM (operational and strategic) was called "Participatory Diagnosis". The same was applied at the proposals stage, which was debated and selected in a public and open procedure. This was possible, thanks to the officials of the CARM and the drafting team, "the technicians", who also acted as facilitators. The procedure was simple: the main documents were always sent to the participants so they were informed but, above all, so that they could contribute suggestions.

In this sense, a very transparent process could be developed. The feedback obtained from citizens or administration, in the form of suggestions or disagreements, was raised in the participatory workshops. To channel other opinions, some cards were produced where each participant could write their support, rejection or nuances in response to what was written by the technical team. All of the above was incorporated into the working documents for the conduct of the general discussions in which majority decisions were taken. For this reason, the authorship of the result is not only the technical consultant team. In fact, the consultants produced drafts: well structured, with the best information available, trying to synthesise existing knowledge, explained in the most didactic way possible; however, in essence, this is what they were – drafts. This is how trust and support of the participants were obtained.

When the work to develop the strategy began, a considerable number of the stakeholders of the Murcia region, and, above all, of the Mar Menor environment, were irritated and mobilised. The main reason is being related to the accountability of public administrations that were not considered very effective with respect to the critical situation of the lagoon. A citizen association called "Pact for the Mar Menor" (https://pactoporelmarmenor.blogspot.com.es) had been created. This association had been very active in promoting demonstrations, conferences, studies, etc. They were also present in the participatory process. Special attention was also given to the media. For this reason, representatives of newspapers of the most widely read press in the region were invited (as well as other participants). This brought more transparency and strengthened the participatory process.

This strength precisely did not always prevail for some political-administrative authorities at certain moments of the strategy, especially when it came to supplying all the information available to the participants or when some consensual proposals did not coincide with those of these political authorities.

The protagonism of farmers' associations in the participatory process should be highlighted. This cohort was one of the most active and best-organised groups. Their participation had a great influence on the final text of the strategy. They had two specific goals: to place more emphasis on the impact caused by other human activities in the lagoon and to downplay the impacts of intensive agriculture in the Mar Menor.

For the above reasons, in some debates, a very simple technique was used to make the influence of each stakeholder group more transparent in the strategy. This involved highlighting changes accepted in the drafts of the strategy, both in the diagnoses and in the proposals with different colours (a different colour for each group). Subsequently, all pages of the strategy were reduced in size until a single image was obtained. In this final image, the colours that represented the changes made to the draft by group could be easily identified. The result was very clear: the farmers achieved the most changes in their favour. Despite farmers' efforts to continually modify the text of the strategy, the participants continued to designate clear responsibility for the dumping of nitrates into the lagoon. With the use of this technique, the writing team demonstrated transparency to the suggestions of the participants.

Finally, two more considerations should be made. The first is very specific: in a tense political and social environment, the representation of 25% of the group by scientists and academics contributed towards objectiveness and a calming influence to the discussions. Therefore, from the point of view of knowledge and social-political debate, their contribution was very positive. The second consideration is more general: the participatory and transparent process developed contributed towards public opinion being in favour of the strategy. It is difficult to quantify by how much, but it is worth bearing in mind that none of the other ICZM Strategies in Spain were approved from an administrative point of view.

13.8.5 For the Formulation of the Proposals, it Was Beneficial to Insist on a Method That Took into Account: The Philosophy of the Project, a Sequential or Chained Character of Reasoning, and the Socioecological Context

The result of the Operational (Mar Menor and environment) and Strategic Diagnostics (management model) presented a considerable number of different proposals. The participation of numerous social and institutional agents guaranteed an intense debate. To ensure this part of the process was more objective, three points were insisted upon. The first pertained to the philosophy of the project. That is, as we had worked with certain conceptual frameworks (DPSIWR and ecosystem services, Decalogue for the ICZM and public policy), we had to continue with that same logic. To this end, the dual but complementary scheme with which the work was initiated continued: the operative versus the strategic, functioning connection between ecosystems versus ten elements of its management system in the search for answers (R).

Next, steps were established for a chain of reasoning: the Problems were agreed, then the Principles to sustain the proposals were established, the Goals were set and the Objectives were proposed. All of the above was the premise for the answers (R) to be specified with proposals (Actions or Measures).

Finally, the best way to contextualise the proposals was to associate them with one of the four main units of the socioecological system: Critical Zone (or Water Mirror of the lagoon and associated wetlands), Coastal Lands of Intensive Activities (agricultural area of the Campo de Cartagena and the urban belt of the Mar Menor), Areas of Coastal Influence (mountain area of great slope, high and middle sections of watercourses) and Coastal Waters (those affected by the Water Framework Directive, which included small islets, etc.).

13.8.6 The Process of Institutionalisation Teaches Us That a Strategy Can Advance Even Without Being Officially Approved by an Institution

In Spain, the approval of some public management instruments is a long administrative process. The ICZM Strategy of the Mar Menor employed the same procedure as any urban or territorial instrument. It took 21 months to resolve the allegations made by the interested parties (the farmers were the ones who presented the most allegations), to carry out the Environmental Impact Study and to gain Initial Approval. The strategy is expected to have its Final Approval in the second half of 2018.

However, the above is actually the administrative process. The most interesting question for this article is: What happened in the CARM management model during those 21 months? During that time (June 2016 and March 2018), there were significant changes: A General Directorate of the Mar Menor was created, a Law on Urgent Measures was approved to guarantee the environmental sustainability of the Mar Menor environment and a Scientific Advisory Committee and a Committee on Social Participation were created. In addition, work is being done to create a Mar Menor Observatory and a Research Plan.

It is true that the strategy is not yet definitely approved (it has Initial Approval). But it should be noted that all the actions referred to in the previous paragraph are described in the text of the 2016 Strategy (not exactly with the same names but in terms of tasks). What this real case study teaches is that, sometimes, the influence of some instruments can begin before they are formally approved.

13.9 Conclusions

Real-life case studies constitute an important source for the knowledge of ICZM as a technical-scientific discipline. These experiences that contribute towards plans, programmes or strategies are necessary for progress in its advancement. In this sense, deductive knowledge methods (ICZM models applied to a case) and inductive methods (experiences gained from specific cases to the ICZM model) are extremely important. The second feeds back into the first.

The ICZM Strategy of the socioecological system of the Mar Menor has technical-scientific bases, but an important part of its development has taken place in the political and social arena. In fact, it is the political and social sciences that have provided arguments that help explain the success or failure of ICZM instruments. In the present case, the great political and social tension provided the conditions in which the technical work was carried out.

Furthermore, innovations in conceptual frameworks are possible and desirable, but they need consistent didactic explanation. In the case of the Mar Menor, it was of great importance that the technical team (officials and consultants) knew how to interpret the new conceptual frameworks, as well as knowing how to develop them within political-administrative scenarios. This leads us to believe that it is desirable that ICZM practitioners should have skills of institutional and social relationships, in addition to knowledge related to ecosystems. This conclusion would be very positive if it were put into practice in ICZM manuals and university teaching.

Finally, the influence of a sound participatory process was another key element of the strategy. Without doubt, this contributed greatly in the process of Initial Approval and in the anticipated execution of certain actions. It is quite probable that a well-organised, flexible and transparent participatory process will assist in channelling the institutional and social energy of the Mar Menor environment towards creative and constructive ends.

Acknowledgements We would like to express our gratitude to the officials of the Autonomous Community of the Region of Murcia (in particular to Amparo Suárez and Francisco Marín de la Dirección General de Transportes, Costas y Puertos). We are also immensely grateful to the General State Administration and local corporations, as well as all organisations and social agents involved, for their contributions and comments that greatly improved the strategy and overall for all the effort and enthusiasm they dedicated to our work. This publication and research have been partially granted by INDESS (Research University Institute for Sustainable Social Development), University of Cadiz, Spain.

References

Barragán JM (2010) Coastal management and public policy in Spain. Ocean Coast Manag 53:209–217

Barragán JM, de Andrés M (2015) Analysis and trends of the world's coastal cities and agglomerations. Ocean Coast Manag 114:11–20

Barragán JM, Lazo O (2018) Policy progress on ICZM in Perú. Ocean Coast Manag 157:203–216

Barragán JM, Sanabria J (2016) Estrategia de gestión integrada de zonas costeras para el sistema socio-ecológico del Mar Menor (SSEMM), Comunidad Autónoma de la Región de Murcia, Consejería de Fomento e Infraestructuras, Dirección General de Transportes, Costas y Puertos, 170 pp

Barragán JM, Dadon JR, Matteucci SD, Morello JH, Baxendale C, Rodríguez A (2003) Preliminary basis for an integrated management program for the coastal zone of Argentina. Coast Manag 31:55–77

Barragán JM, Castro C, Alvarado C (2005) Towards integrated coastal zone management in Chile. Coast Manag 33:1–25

Barragán JM, Chica Ruíz JA, Pérez Cayeiro ML (2008) Estrategia Andaluza de Gestión Integrada de Zonas Costeras. Consejería de Medio Ambiente de la Junta de Andalucía, 255 pp

Cabezas Calvo-Rubio F (2009) El Mar Menor. Estado actual del conocimiento científico, Murcia. Instituto MEditerráneo del Agua, 540 pp

Carreño MF, Esteve MA, Martínez J, Palazón JA, Pardo MT (2008) Dynamics of coastal wetlands associated to hydrological changes in the watershed. Estuarie, Coast. Shelf Sci 77:475–483

Centro Regional de Estadística de Murcia (2011) Turismo. http://econet.carm.es/

Centro Regional de Estadística de Murcia (2013a) Agricultura, Ganadería, Silvicultura y Pesca. http://econet.carm.es/

Centro Regional de Estadística de Murcia (2013b) Construcción y vivienda. http://econet.carm.es/

Centro Regional de Estadística de Murcia (2015) Economía. http://econet.carm.es/

CEOTMA (Centro de Estudios de Ordenación del Territorio y Medio Ambiente) (1985) Ordenación Territorial de la Zona del Mar Menor y su Entorno, Serie Divulgación y Síntesis, vol 4. Ministerio de Obras Públicas, Madrid

Clark JR (1996) Coastal Zone Management Handbook. CRC Press, Boca Raton

Consejería de Fomento, Obras Públicas y Ordenación del Territorio. Dirección General de Transportes, Costas y Puertos (2014) Estrategia de Gestión Integrada de Zonas Costeras en el Mar Menor y su entorno. CARM

Cooper P (2013) Socio-ecological accounting: DPSWR, a modified DPSIR framework, and its application to marine ecosystems. Ecol Econ 94:106–115

de Andrés M, Barragán JM, García Sanabria J (2017) Relationships between coastal urbanization and ecosystems in Spain. Cities 68:8–17

de CH del Segura DG (2007) Primera aproximación para la concertación de un Programa de Acción en el Área del Mar Menor, Diagnóstico previo de afecciones al Dominio Público y el Patrimonio Natural

de Groot R (1992) Functions of nature: evaluation of nature in environmental planning, management and decision making. Wolters-Noordhoff BV, Groningen
de Groot R, Wilson M, Boumans R (2002) A typology for the description, classification and valuation of ecosystem functions, goods and services. Ecol Econ 41:393–408
de Jonge V, Pinto R, Turner RK (2012) Integrating ecological, economic and social aspects to generate useful management information under the EU Directives 'ecosystem approach'. Ocean Coast Manag 68:169–188
De León AR, Guerrero J, Farazo F (1982) Evolution of the pollution of the coastal lagoon of Mar Menor. In: VI Journées Étud. Pollutions. C.I.E.S.M., Cannes, pp 355–358
Díaz S et al (2006) Biodiversity regulation of ecosystem services. In: Ecosystem Human well-being. Current state and trends. Island Press, Washington, DC, pp 297–329
Environmental Protection Agency (2012) Caja de Herramientas para la Participación Pública, EPA, USA, 71 pp
Espejo Marín C (2011) Innovación para la competitividad turística en la Manga del Mar Menor (Murcia). Cuadernos de Turismo 27:321–339
García García C (2004) Impacto y riesgo ambiental de los residuos minero-metalúrgicos de la Sierra de Cartagena-La Unión (Murcia-España). Universidad Politécnica de Cartagena
García-Pintado J et al. (2009) Caracterización y evaluación de fuentes antrópicas de nutrientes transportados por la Rambla del Albujón hacia el Mar Menor, El Mar Menor. Estado actual del conocimiento científico, Fundación Instituto Euromediterráneo del Agua, pp 131–166
Gimenez Casalduero F et al (2012) Aplicación de modelos ecológicos conceptuales para identificar indicadores de seguimiento en la laguna costera del Mar Menor (Murcia, España), Congreso Iberoamericano de Gestión Integrada de Áreas Litorales, pp 1022-1032
LAGOONS (2012) The Mar Menor Lagoon – current knowledge base and knowledge gaps
MA (Millennium Ecosystem Assessment) (2003) Ecosystems and human well-being: a framework for assessment. Island Press, Washington, DC
MA (Millennium Ecosystem Assessment) (2005) Ecosystems and human well-being: synthesis. Island Press, Washington, DC
Marcos C, Torres I, López-Capel A, Pérez-Ruzafa A (2015) Long term evolution of fisheries in a coastal lagoon related to changes in lagoon ecology and human pressures. Rev Fish Biol Fish 25:689–713
Martínez Fernández J, Fitz C, Esteve Selma MA, Guaita N, Martínez López J (2013) Modelización del efecto de los cambios de uso del suelo sobre los flujos de nutrientes en cuencas agrícolas costeras: el caso del Mar Menor (Sudeste de España). Ecosistemas 22:84–94
Mintzberg H, Quinn James B, Ghoshal S (2002) The strategy process: concepts, contexts, cases. Prentice Hall, Englewood Cliffs
Moreno González R, Campillo JA, León VM (2013) Influence of an intensive agricultural drainage basin on the seasonal distribution of organic pollutants in seawater from a Mediterranean coastal lagoon (Mar Menor, SE Spain). Mar Pollut Bull 77:400–411
Perez-Ruzafa A, Fernández AI, Marcos C, Gilabert J, Quispe JI, García-Charton JA (2005) Spatial and temporal variations of hydrological conditions, nutrients and chlorophyll a in a Mediterranean coastal lagoon (Mar Menor, Spain). Hydrobiologia 550:11–27
Pérez-Ruzafa A, García-Charton JA, Barcala E, Marcos C (2006) Changes in benthic fish assemblages as a consequence of coastal works in a coastal lagoon: the Mar Menor (Spain, Western Mediterranean). Mar Pollut Bull 53:107–120
Pérez-Ruzafa A, Marcos C, Bernal CM, Quintino V, Freitas R, Rodrigues AM, García-Sánchez M, Pérez-Ruzafa IM (2012) Cymodocea nodosa vs. Caulerpa prolifera: causes and consequences of a long term history of interaction in macrophyte meadows in the Mar Menor coastal lagoon (Spain, southwestern Mediterranean). Estuarine, Coast. Shelf Sci 110:101–115
Salas F, Marcos C, Neto JM, Patrício J, Pérez-Ruzafa A, Marques JC (2006) User friendly guide for using benthic ecological indicators in coastal and marine quality assessment. Ocean Coast Manag 49:308–331

Santos-Martín F et al (2015) La evaluación de los servicios de los ecosistemas aplicada a la gestión pesquera
Soto M, Martínez V, Martín B (2014) El regadío en la Región de Murcia. Caracterización y análisis mediante indicadores de gestión, Universidad Politécnica de Cartagena para la Comunidad de Regantes del Campo de Cartagena y Sindicato Central de Regantes del Acueducto Tajo-Segura, 267 pp.
UNEP (2006) Marine and coastal ecosystems and human well-being: a synthesis report based on the findings of the millennium ecosystem assessment
Velasco AM, Pérez-Ruzafa A, Martínez-Paz JM, Marcos C (2017) Ecosystem services and main environmental risks in a coastal lagoon (Mar Menor, Murcia, SE Spain): the public perception. J Nat Conserv 43:180–189
Wang S, Fu B, Wei Y, Lyle C (2013) Ecosystem services management: an integrated approach current opinion. Environ Sustain 5(1):11–15

Chapter 14
Training for Aquaculture and Fishery Activities for the Conservation and Sustainable Use of Biodiversity

José Manuel Mazón-Suástegui, Abel Betanzos-Vega, Mercedes Isla-Molleda, Romina Alzugaray-Martínez, Milagro García-Bernal, and Araceli Avilés-Quevedo

Abstract In Cuba, 211 protected natural areas (PNAs) that constitute 20% of the national territory are under eight management categories. This chapter describes "Las Picúas-Cayo Cristo" (LPCC) categorized as a wildlife refuge (55,970 ha) located in north-central Cuba in the Sabana-Camagüey Archipelago. The management plan of this PNA includes the conservation of mangrove ecosystems, seagrasses and coral reefs, terrestrial species, migratory birds, and marine species of ecological interest (turtle, manatee), among others of commercial interest (fish and lobsters, queen conch, sponges, and oysters). Authorities must be attentive to avoid conflicts of interest between conservationists and fishers, so experimental oyster and sponge aquaculture farms have been established. Native population recognizes that protecting the existing natural resources guarantees a sustainable source of employment, food, and family welfare.

In contrast, approximately 700 fishers who live nearby could have a negative impact on the PNA because their families depend economically on extracting marine resources. The short-, medium-, and long-term objectives are linked to aquaculture, conservation, and public use programs. This chapter describes actions of the fisheries sector of the Ministry of Food Industry (MINAL) and the Center for Fisheries Research (CIP) that derived from national and international scientific projects and consultancies, which have contributed to the evaluation and protection of fishery resources of commercial interest, their aquaculture, and environment.

J. M. Mazón-Suástegui · M. García-Bernal
Centro de Investigaciones Biológicas del Noroeste, La Paz, Baja California Sur, Mexico
e-mail: jmazon04@cibnor.mx

A. Betanzos-Vega (✉) · M. Isla-Molleda · R. Alzugaray-Martínez
Centro de Investigaciones Pesqueras (CIP), La Habana, Cuba
e-mail: abetanzos@cip.alinet.cu; merisla@cip.alinet.cu; ralzugaray@cip.alinet.cu

A. Avilés-Quevedo
Freelance Services for R & D in Marine Aquaculture, La Paz, Baja California Sur, Mexico

A. Ortega-Rubio (ed.), *Socio-ecological Studies in Natural Protected Areas*,
https://doi.org/10.1007/978-3-030-47264-1_14

Keywords Protected areas · Cuba experiences · Sustainable fishing · Aquaculture

14.1 Introduction: Natural Protected Areas in Cuba and Aquaculture Development

At global level and with the purpose of conservation, societies have declared terrestrial or marine sites as a protected natural area (PNA) where exceptional flora, fauna, and environmental values have historically concentrated and developed. FAO (2018) has declared that aquaculture has unchained an impressive growth worldwide in providing fish and seafood for human consumption, and conservation of oceans, seas, and marine resources guarantees food assurance and sustainable development in the future.

The techniques and traditional fishing gear used in aquaculture also function as artificial reefs or fishing ground for fish, crustaceans, and other invertebrates searching for refuge, substrate, and food, which increases richness and diversity in the aquaculture area, positively impacting the environment. In Cuba, the National System of Protected Areas (SNAP) has a national strategy on biological diversity, in which all conservation and sustainable use of biodiversity actions are included. A participative management of the local communities, traditional users of natural resources and other actors, has been promoted, so they can equitably benefit from the conservation of such natural spaces (Ruiz-Plasencia 2017). The final objective of conservation is to achieve a greater understanding of man-nature interaction as a way of developing knowledge and capabilities that allow integrating an environmental, economic, and social link as cornerstone of sustainable development (FAO 2018).

The Ministry of Science, Technology and Environment (CITMA) was created in 1994, and it is the official entity responsible for environmental management in Cuba. The National Center for Protected Areas (CNAP) was created a year later to carry out the integrated management of the National System of Protected Areas (SNAP) and promote its optimal performance. These three entities relate and coordinate specific issues with the Cuban Ministry of Agriculture (MINAGRI).

Cuba has a surface extension of 179764.59 km^2 where 20.20% of its territory is found protected under different categories of protected natural areas (PNAs), including marine water, in such a way that the National System of Protected Areas (SNAP) maintains 17.16% of the terrestrial part and 24.96% of the marine shelf under protection coverage (CNAP 2013). This system is integrated by 211 areas with special characteristics and relevant natural values that distinguish them as potential PNAs for the 2014–2020 period. Only 120 of them have a formal administration and are grouped in eight management categories: three Natural Reserves (NR), 14 National Parks (NP), 23 Ecological Reserves (ER), 13 Featured Natural Elements (FNE), 15 Managed Floristic Reserves (MFR), 26 Wildlife Refuges (WR), 12 Protected Natural Landscapes (PNL), and 14 Protected Areas of Managed Resources (PAMR).

One of the coastal marine regions of Cuba with the greatest anthropic-environmental affectation and biodiversity vulnerability is the north-central shelf in the country (Alcolado et al. 1999; Puga et al. 2018). This region is known as the Sabana-Camagüey Archipelago (SCA), which extends from west to east from Punta Hicacos in Varadero to Bahía de Nuevitas, along 465 km approximately with a maximum width of 50 Km, covering a marine area of 8311 km^2 (Alcolado et al. 2007). The SCA region has 2515 keys and islets distributed basically in the shelf border (Fig. 14.1), which limits free communication of the shallow waters with the external part of the shelf and open ocean; at the same time this interior water body is subdivided into shallow "bays" (1.5 m in average). The available resources show a great variety of options for anthropic activities, including those related with fishing, tourism, coastal traffic, industrial development, and population, among others, in addition to several natural-type factors that influence their respective ecosystems in quality (Alcolado et al. 1999; Perigó et al. 2004; Betanzos-Vega et al. 2013; Montalvo-Estévez et al. 2014; Cobas et al. 2015).

The coastal communities whose main source of employment depends on fishing activity are La Panchita, Carahatas, and Isabela de Sagua in the province of Villa Clara and Punta Alegre and Turiguanó in the province of Ciego de Ávila (Fig. 14.1). Moreover, three cities, Cárdenas, Caibarién, and Nuevitas, offer other working alternatives because of their social importance and economic development; although they do not depend much on fishing, this activity contributes with significant income, especially in Caibarién (Betanzos-Vega and Valle 2015).

Marine fisheries in the SCA region respond to different sectors of society, such as state and private commercial fishing, self-consumption fishing by state enterprises, and sport-recreational fishing, without ruling out illegal fishing or poaching.

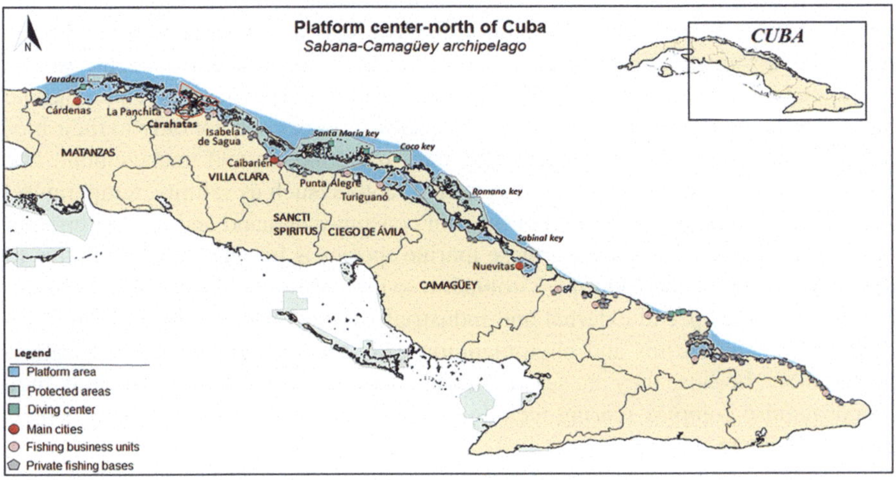

Fig. 14.1 Sabana-Camagüey Archipelago in north-central Cuba from the 200-m isobath northward from Matanzas, Villa Clara, Sancti Spíritus, Ciego de Ávila, and Camagüey provinces

The responsible governing authorities of some of these coastal communities have tried to diversify the working activities to generate other work sources that are not linked to the fishing sector. These jobs are in the majority of services to the community, except for some local agriculture productions; nevertheless, fishing is the best option to generate sustenance and economic income, both at personal and community levels (Betanzos-Vega and Valle 2015).

In the north-central shelf or the SCA region, 22 PNAs are located: two National Parks (NP), five Ecological Reserves (ER), nine Wildlife Refuges (WR), one Featured Natural Element (FNE), three Protected Areas of Managed Resources (PAMR), and two Managed Floristic Reserves (MFR). Depending on the conservation objectives, these PNAs are located on coastal sections, keys, islets, or marine waters, but the marine component is found in all of them, to a point where at least in 20 of them the marine ration is greater than the terrestrial one. The marine ecosystems present in the SCA are linked not only by hydrodynamics but also by population exchange among the different habitats (Alcolado et al. 1999; Claro et al. 2004; Puga et al. 2009). This form of connectivity makes identifying, assessing, and protecting critical habitats necessary because of the fishery resources in this region with the purpose of assuring their conservation and stability or recovery of their natural biological populations.

This chapter documents the national and local effort directed to contribute to the protection of biodiversity and marine fishery resources emphasizing on a sustainable development raising awareness in fishers and other users of environmental services. It also documents training and creation of alternatives for developing eco-friendly fisheries, such as aquaculture of native filtering organisms of Las Picúas-Cayo Cristo (LPCC-PNA), Fauna Refuge of great importance for the marine biodiversity of Cuba.

The species with greater potential for developing artisanal rearing in this region are sessile bivalve mollusks, such as oysters of the family Ostreidae, mussels of the family Mytilidae, and clams of the families Veneridae and Arcidae that live buried on sandy substrates. Other native species with high potential for aquaculture and good commercial value exist, such as the sea cucumber, sponges, and some marine algae. The aquaculture potential of these species is associated not only to their biological parameters, such as growth rate, easy reproduction and management, and tolerance to limiting values in environmental factors, such as salinity, temperature, and dissolved oxygen, but also relevance in consumer demand and market product price. The advantage of farming these marine species is because they are native to the region, found in the natural environment, so they do not require a habitat adaptation process and special external and industrialized feed because they obtain it by filtering organic detritus and phytoplankton from the surrounding marine waters. Furthermore, because they are sessile benthic organisms with little movement, they do not require complex rearing techniques, but they need protection against their natural predators.

14.2 The Governance in Natural Protected Areas and Contribution of the Fisheries Sector to Biodiversity Protection

The National System of Protected Areas (SNAP) is integrated by a combined group of entities and institutions, which in a collegiate manner contribute to the conservation in situ of the natural Cuban patrimony through three management coordination levels: national, provincial, and local (Resolution No. 146/2009 of CITMA). The coordination at national level is performed by a National Coordination Board that allows member entities a better performance in rectory, management, and control functions related to the protected areas. This board is chaired by the National Center of Protected Areas (CNAP) guided by the SNAP Plan. This plan has been designed as a strategic, normative, and methodological instrument, in which by means of objectives, norms, and programs, the actions to be performed in short, medium, and long term are established (CNAP 2013). According to this plan, at least 17% of the terrestrial areas and continental waters and 10% of marine and coastal zones, important for the biological diversity and ecosystem services, are expected to be conserved, protected, and managed efficiently and equitably, establishing temporal or permanent closed seasons, according to the species and respecting their reproductive periods.

The mechanisms of managing fisheries in Cuba show the following order: (1) deriving from scientific research projects, Centro de Investigaciones Pesqueras (CIP) contributing technical-scientific elements to the Dirección de Regulaciones Pesqueras y Ciencia, of the Ministerio de la Industria Alimentaria (MINAL): the CIP reports allow producing and proposing regulatory measurements that must be discussed and approved by the evaluating and consultancy body (Comisión Consultiva de Pesca), which involves other institutions not related with the fishing sector (scientific, conservationist, state inspection body, coastguard, etc.). Once approved, the resolutions of the maximum MINAL direction must be attended obligatorily by the fishing sector of the Grupo Empresarial de la Industria Alimentaria (GEIA) and the control entity, which is the Oficina Nacional de Inspección Estatal.

The mission and vision of the national fishing sector are projected in agreement with the objectives at medium and long term of the PNA, which implies periodical and sequential monitoring of abiotic and biotic quality of the critical habitats for the fishing resources in the different stages of their life cycle. The previous information allows preserving the natural conditions in the biological corridors to protect reproductive migrations of free-life species, massive spawning processes of sessile species and other biological movements, as well as the adequate (sustainable) management of the fishing extraction activity. The creation and control of protected marine areas, as well as promotion, development, and protection of eco-friendly marine rearing, are crucially important. These actions are included in the new Estrategia Nacional para el Desarrollo del Maricultivo (National Strategy for the Development of Marine Culture) in Cuba, which should contribute key elements to conserve and increase abundance and diversity of fishing resources.

14.2.1 Contribution of the Fishing Sector to Conservation and Protection of Fishing Resources and Marine Biodiversity: Emphasis on the North-Central Shelf of Cuba

***Caribbean Spiny Lobster** (Panulirus argus)* The lobster fishery is the most important in Cuba, ant it is developed on the basis of creating artificial refuges that provide punctual aggregations and sustainable increase of the populations with fishery target; they do not affect in any way the natural banks of resources found in the reef zones, which are reserved as genetic reproductive banks and generally located at a great depth, so they are safe from traditional fishing.

The main actions of the fishing sector related to management of lobster fishery respond to the investigations and recommendations of CIP. The regulations applied include prohibiting commercial capture for all out-of-state fishers and total closing season for the breeding areas. There are regulations of minimum and maximum legal capture size, a six-month reproductive closed season, and capture quotes previously defined in function of the maximum sustainable capture data in function with annual biomass production, control, and follow-up measurements, among others (Puga et al. 2013, 2018), which jointly conform a strategy that allows assuring stability of lobster fishery in time. For this purpose, fishing gear that allows 95% of the lobsters with size inferior to the legal minimum to escape from the lobster traps has been implemented in the north-central region of Cuba. Additionally, biodegradable cords for tying the harvesting trap doors, which allow eliminating, "ghost fishing" in case of losing the trap (Palma et al. 2010), or alternatively a large 5-cm mesh should be used to make the traps or fishing gear.

Commercial Fish Fisheries of commercial fish are mainly multi-specific because of the diversity of fish species that can be captured with the same fishing gear and the high volume of bycatch species. This situation has obligated promoting strategies to assess, control, and recover fish populations by applying models directed to the representative species with different points of reference. To achieve this objective, annual population studies have been performed starting from the use of different vulnerability and stock evaluation models, considering the fish species with greater presence in capture and commercial value (Puga et al. 2018). In the analyses of the SCA region, ecosystem health indicators have been taken into account, such as species diversity, richness, mortality by fishing, effects of natural disturbances (frequency and intensity), and those of anthropic origin. These indicators have allowed knowing the state of fisheries and habitats, as well as applying measures for an ecosystem management and sustainable use of the fishery resources in Cuba.

In the last years, newer and stricter regulations have been established, such as eliminating site nets for fish (enclosures, corrals) and prohibiting trawl fishing, highly destructive on marine bottoms and benthic species. Furthermore, hydroacoustic studies have been performed in the north-central region to identify fish aggregation and spawning areas, which allow determining spawning biomass

(Betanzos-Vega et al. 2015). These studies are based on experiences of Hernández-Corujo et al. (1998) and Linares et al. (2009) with an outstanding result in locating and evaluating biomass of a spawning population of "biajaiba" (lane snapper) (*Lutjanus synagris*, Linnaeus 1758), at exterior side of Boca de Sagua channel, northwest of Cayo Esquivel, 20–40 m in depth. Likewise, an average size was determined in furcal length (FL) of 19.7 ± 3.4 cm and average density of 5 ind/m^2 and 0.6 kg/m^2 with biomass of 1.3 t in this fish aggregation for the assessed site. Based on the total area (0.935 km^2), known as spawning aggregation zone for this species, total biomass estimations of 600 t were performed (Betanzos-Vega et al. 2015). This information could be useful to define and establish technical and administrative criteria to promote and sustain biodiversity and removable fish biomass in the area of study.

Queen Conch The queen conch (*Lobatus gigas*), known in Cuba as "cobo," is among the mollusks susceptible to capture and commercialization in SCA and a resource whose exploitation and commercialization are regulated by the Convention on International Trade in Endangered Species of Wild Fauna and Flora (CITES). The management measurements applied by the scientific fishery sector of this species are: (1) fish under well-defined permissible quotas for each zone according to the monitoring and evaluation results; (2) observe closed reproductive season from May to September and a commercial size of ≥10 mm in width of the exterior lip; (3) capture only by free diving in "apnea"; (4) observe fishing limit in a depth range from 3 to 10 m to avoid the presence of juveniles and brood stock; (5) prohibit throwing empty shells in fishing areas and species habitat; (6) control by state fishery inspectors of all these measurements, and detect and control poaching and use of this mollusk as "bait"; and (7) commercialize only with export destination in agreement with CITES authorities (Formoso 2008).

***Antillean Mangrove Oyster** (Crassostrea rhizophorae)* The regulations for fishing this resource are the following: (1) its extraction with commercial destination is prohibited to all fishers not linked with the state oyster sector; (2) closed reproductive season from May 1 to July 31 (Resolution No. 336/95) for the north-central region of Cuba; (3) minimum legal size of 40 mm in anteroposterior length; and (4) commercialization in pickled mass or animal without its conch, establishing 4.9% of clean mass weight with respect to total oyster weight in conch, as yield in minimum meat to guarantee an adequate industrial efficiency.

14.3 Aquaculture Training as an Alternative to Sustainable Fishing and Biodiversity Protection

In the north-central region of Cuba or Sabana-Camagüey Archipelago, the introduction of cultured marine species is essential due to the significant reduction of captures caused by a high anthropic pressure on fishery resources and the environment. The development of economically viable and sustainable fishery production alternatives constitutes an additional contribution to environmental sustainability, a working alternative for fishers, the creation of new work sources of communal employment, and generation of economic income. A progressive conviction exists on the need to evolve from the current collector level to the level of producer, as a measure of advance toward social, economic, and cultural development.

14.3.1 Sponge Culture in the Protected Natural Area "Las Picúas-Cayo Cristo"

An experimental farm for sponge culture of the family Spongiidae was built at LPCC-PNA using the methodology of "free sponges" (Quirós et al. 2009). This project was sponsored by the Programa de Pequeñas Donaciones (PPD) of the United Nations World Environment Fund (UN), in support of the fishing sector. Similar projects have been performed in the north-central region with different culture methodologies, which have allowed training fishers in aquaculture techniques of the operational procedure for this type of marine culture (Fig. 14.2), laying the foundations for the development of sponge culture at medium term (Betanzos-Vega et al. 2019).

As a result of these activities, sponge culture in Cuba offers a safe and predictable production of a higher quality product than that offered by natural capture. The technique has a low investment cost because sponges are filtering organisms, which,

Fig. 14.2 Cultured sponge harvest according to the free-sponge method in Las Picúas-Cayo Cristo Protected Natural Area in Cuba. Photographs provided by Abel Betanzos-Vega

just like oysters, do not need additional food from the natural environment. Marine sponge culture in LPCC-PNA constitutes a sustainable alternative to reduce fishing pressures on natural wild sponge populations and promotes increase in natural surrounding banks. Scaling-up sponge culture would guarantee greater availability of employment and foreign currency income through export because of the high marketing price (15–50 USD/kg) of the commercial sponges (Betanzos-Vega et al. 2019).

14.3.2 Oyster Culture in the Protected Natural Area "Las Picúas-Cayo Cristo"

The oyster of Bahia de Sagua has been traditionally considered of good flavor and quality for human consumption since their natural banks in exploitation are not located in the coastal zones as in the rest of the country. The natural oyster banks are located at keys or islets located from 2 to 5 nautical miles from the coast far from the potential urban contamination sources and terrestrial contribution. In the region of study, the fishing enterprise unit dedicated to commercial oyster exploitation is located in Isabela de Sagua, off the territorial frame of the PNA. However, an important part of oyster collection is performed from aquaculture at different zones of the PNA.

The cultural importance and tradition in the production and local consumption in this region, involving fishers and women in the process lines for commercialization, has favored the fishing sector and conservationists to focus on searching for financial support for local oyster rearing development. In consequence, different activities have been performed in the PNA region and in other adjacent areas always linked to different international projects with the support and co-financing of the national fishing sector and consultancy of Centro de Investigaciones Biológicas del Noroeste, S.C. (CIBNOR), Baja California Sur, Mexico (www.cibnor.gob.mx). Among the main results obtained in oyster rearing, the following projects are included:

"Sustainable Development of the Fishing Sector in Villa Clara" (2006)
This project was sponsored by WWF, with advice from the CIP, and the most relevant results were (a) improvement of fishing infrastructure in the communities La Panchita, Carahatas, and Isabela de Sagua and the electricity communal net of Carahatas; (b) introduction to Environmental Education in elementary school environment; (c) support in material and equipment resources for the Biological Station of Las Picúas-Cayo Cristo PNA; (d) delivery of a boat to the corps of inspectors for vigilance of the LPCC-PNA; (e) analysis of the potential oyster rearing development and environmental quality study of Santa Clara, Carahatas, and Sagua la Grande bays; and (f) assessment of the state of mangrove oyster production in the north-central region of Cuba, variability, and natural and anthropic impacts. A component related to oyster farming was documented by Betanzos et al. (2009, 2010) and Betanzos and Arencibia (2010).

"Sustainable Culture and Management of the Mangrove Oyster (*Crassostrea rhizophorae*) as Alternative of Ecosystem Protection in the Bahía de Sagua la Grande, Provincia de Villa Clara" (2010–2012)
This project was performed with the consultancy of CIBNOR-México and resources of the fund for the World Environment Program of Small Donations (PPD) of the United Nations for Development (PNUD), whose most relevant results were the following:

(a) Direct participation of the local government and fishing sector (Grupo Empresarial de la Industria Alimentaria, GEIA) in a communal development for sustainability in Isabela de Sagua, integrated by fishers with experience in oyster culture and youth of the community with the support and participation of environmental organizations and local government
(b) Development of an experimental oyster culture with community participation, application of eco-friendly artisanal techniques, and mangrove protection
(c) Experimental farm with a production capacity at medium term of 10 t of shelled oyster eastward from Bahía de Sagua la Grande
(d) Community training and academic development by means of a workshop that included eight oyster fishers and other residents of the community Isabela de Sagua, offering training in new techniques in management and artisanal mangrove oyster culture
(e) Training at CIBNOR-México and developing a master's thesis (Rivero-Suárez 2012) of the responsible technician of the project by the Cuban fishing sector
(f) Environmental assessment and biological monitoring in oyster farming areas and grow-out in the region of Sagua La Grande (Rivero-Suárez, 2012) and economic-environmental culture feasibility (Betanzos-Vega et al. 2014)

"Protection of the Sabana-Camagüey Ecosystem: Environmental Sustainability in Oyster Production in Isabela de Sagua" (2011–2015)
The "Oyster Cultivation" subproject was carried out with the consultancy of CIBNOR-México. Overall project activities were financed by the United Nations Development Program (GEF-PNUD) and the Cuban Environment Agency (AMA). Some of its activities were documented by Mazón-Suástegui et al. (2011), Betanzos et al. (2013, 2014), and Betanzos-Vega and Valle (2015) whose most relevant results were the following:

(a) Sustainable oyster culture with ecosystem focus, emphasis on development and marine production expansion by means of oyster culture
(b) Delivery of material resources for maintenance and expansion of an oyster farm eastward of Bahía de Sagua, including a boat motor for an oyster fishing vessel
(c) Support for introducing a new oyster farm westward of Bahía de Sagua la Grande, at the inner part of the PNA in the locality of Cayo Esquivel, with an initial production capacity of more than 10 t of shelled oyster

(d) Financial training mechanisms for the conservation of natural resources, incorporating an economic value (eco-value) to the mangrove according to the environmental goods and services offered
(e) Financial support of the Secretary of State (Secretaría de Relaciones Exteriores de México) for consultancy and training to fishers in situ
(f) Training technical staff at CIBNOR-México by Dr. José Manuel Mazón-Suástegui, focused on adopting new culture and management zoo-techniques for artisanal oyster farms within the Project GEF/PNUD
(g) Introducing new culture technologies in Isabela de Sagua, including measures made with empty oyster shells and oyster baskets for grow-out and individual oyster production
(h) Analysis of environmental economic feasibility of mangrove oyster rearing according to different variabilities of production, extractive fishing, artisanal rearing, and technical artisanal rearing with introduction to cost-benefit analysis according to culture variants and environmental cost (eco-value)

In general terms, the results of these projects have been satisfactory although still discrete. It is worth mentioning that in the last years (2014–2017), the annual average commercial oyster production was 103.6 t, 27% higher than that obtained in the previous period (2010–2013) with an annual average of 75.6 t. This result is related with the reintroduction and expansion of oyster farms in Enfermería Keys (westward from Bahía de Sagua) and in Punta Pargo (Cayo Esquivel del Sur) in the LPCC-PNA (Betanzos-Vega and Valle 2015).

14.4 Las Picúas-Cayo Cristo Protected Natural Area. Case Study

14.4.1 Geographic and Environmental Characteristics

Las Picúas-Cayo Cristo (LPCC-PNA) is a Wildlife Refuge with a total surface of 55,970 ha (40,250 marine, 15,720 terrestrial hectares); it is located northwest of the Villa Clara Province (Fig. 14.3) in the special sustainable development region of the Sabana-Camagüey Archipelago (SCA). It includes the northern area from Cayo Blanquizal to Cayo Cristo and the adjacent marine zone to the 200-m isobath. It extends to the south from Cayo Atravesado in Bahía de Santa Clara (to the west of Cayo Las Picúas), bordering the coasts from Bahía de Carahatas to Cayo Bamba in Bahía de Sagua la Grande.

The coast of Bahía de Carahatas has been affected notably by damming the Sagua la Grande River, the one with the greatest volume of water and extension in the region, which flows into the Bahía de Sagua. This measure has generated a reduction of the fluvial contribution, increasing salinity in coastal waters and affecting notably the Las Picúas-Cayo Cristo PNA located 7 nautical miles to the east. Other affectations are associated to shallow waters (2 m in average) in the bays of

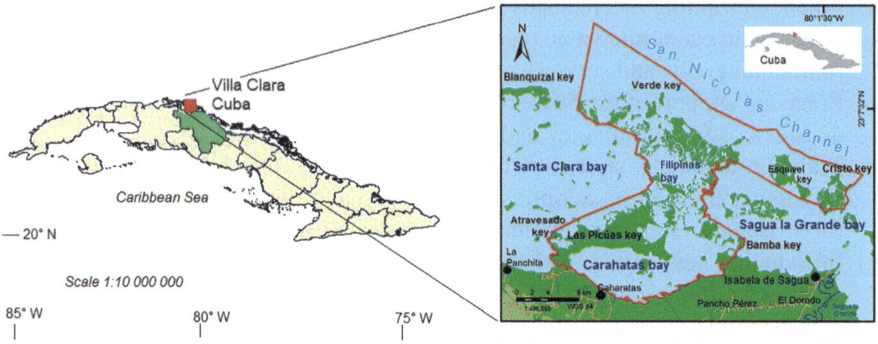

Fig. 14.3 Localization of Las Picúas-Cayo Cristo Protected Natural Area (LPCC-PNA), a Wildlife Refuge area in the Sabana-Camagüey Archipelago, Cuba

Santa Clara, Carahatas, and Isabela de Sagua; moreover, the frequent meteorological events generate abrupt changes in oceanographic variables, while the marine currents affect the coastline with sea-level rise or fall, taking different courses depending on the direction of the canals between keys or islets. In general, the PNA has been affected by natural processes, such as hurricanes and cold fronts and other anthropic processes, such as damming the Sagua la Grande River, generating changes in quality of this ecosystem. In this respect, it is important to point out an increase in salinity and variability in its spatial distribution from 36.0 UPS in the line of contact with oceanic water to higher than 50.0 UPS in Bahía de Carahatas (Siam et al. 2007).

14.4.2 Biodiversity

The bird community of the LPCC-PNA includes 114 species that belong to 18 orders, 41 families, and 65 genera. Of the total species, 64 nest in the area; the most represented orders are Passeriformes, Ciconiiformes, Anseriformes, Pelecaniformes, and Charadriiformes; 70% of the species are permanent residents. Because of its biogeographic location, Cuba receives an important influx of migratory individuals, which makes us suppose it is also an important center for the reproduction of aquatic birds of North America (Garrido and Kirkconnell 2012). Thus, the ornithological values of Cuba not only have local importance but also regional because of the significant exchange of individuals among resident and migratory populations.

The LPCC-PNA has important reproduction, feeding, refuge, and resting sites for numerous species of aquatic birds. According to the characterization report of the Fauna Refuge and Management Plan (MINAGRI-CNAP 2014), Cayo Las Picúas, located in Punta Arena, stands out as a nesting site for the Caribbean pink flamingo (*Phoenicopterus ruber*) (Fig. 14.4a), which maintains an adult nesting

Fig. 14.4 (**a**) Pink flamingo *Phoenicopterus ruber*; (**b**) pelican *Pelecanus occidentalis*. Photographs provided by Abel Betanzos-Vega

population of approximately 8000 individuals from March to July and the pelican or gannet (*Pelecanus occidentalis*) with several nesting and habitat areas (Fig. 14.4b).

The LPCC-PNA shows a great number of lowlands and keys covered with mangroves that provide an important protection from wind and surge, which guarantees refuge and food to a dissimilar marine fauna and flora associated to this ecosystem. The mangrove forest is distributed in all the keys and along the PNA coastline, where the four species in Cuba are found: red mangrove (*Rhizophora mangle*), black mangrove (*Avicennia germinans*), white mangrove (*Laguncularia racemosa*), and the evergreen shrub (*Conocarpus erectus*). In general, red mangrove summons up great importance for fishery resources since this species borders the coastline and many keys. The aerial roots of the mangrove constitute a natural intertidal habitat for mangrove oyster (*C. rhizophorae*) that uses it as fixation substrate (Fig. 14.5a), and mangrove roots that are always flooded serve as refuge for lobster (*Panulirus argus*) and several commercial fish (Fig. 14.5b).

Due to the previous characteristics, LPCC-PNA is a natural breeding spot of the Caribbean spiny lobster with important grow-out areas in the northern portion of the Bahía de Filipinas and in the group of keys east and southeast from Cayo Verde. These areas have an additional breeding area (170 Km^2) to the east of Bahía de Santa Clara, which are adjacent to the PNA, considered as the main lobster breeding spot in all the northern coast of Cuba (Cruz et al. 1990).

With respect to marine fish, they are all subjected to an intense exploitation in all the north-central shelf region of Cuba due to the use of massive fishing gear and the inexistence of closed seasons. LPCC-PNA functions as control area because it has three spawning sites, snappers (family Lutjanidae), mojarras (family Gerreidae), and sea basses (Serranidae). Additionally, other fishery resources have protection,

Fig. 14.5 (**a**) Juveniles (seeds) of mangrove oyster *Crassostrea rhizophorae* in mangrove roots; locality: Cayo Esquivel, Cuba; (**b**) lobsters (*Panulirus argus*) sheltered in mangrove roots; locality: eastern and southeastern keys from Cayo Verde, Cuba. Photographs provided by Abel Betanzos-Vega

such as queen conch (*Lobatus gigas*), crabs *Menippe mercenaria* and *Callinectes sapidus*, and sponges (*Spongia obscura*, *Spongia barbara*, *Hippospongia lachne*). Commercial sponge abundance in Bahía de Santa Clara has promoted its commercial capture, which has caused low densities of the species *H. lachne* and *S. barbara* because they reach greater sale price than *S. obscura* (Blanco and Formoso 2009; Betanzos-Vega et al. 2019). In this respect, the development of sponge culture projects has a special compensatory effect.

Preliminary studies of marine invertebrates in the ecoregion indicate the presence of 43 species of echinoderms, which belong to 29 genera, 23 families, and 13 orders with a greater number of species in the northeastern part of the PNA and the area east of Bahía de Santa Clara, adjacent to LPCC-PNA (MINAGRI-CNAP 2014).

Several keys have sandy beaches; Cayo Esquivel has the best and the most extensive beaches in good conservation because of no human intervention where nesting sites are found for marine turtle, such as *Caretta caretta* and *Chelonia mydas*.

The prairies of phanerogams or seagrasses are well preserved in the PNA, above all those of *Thalassia testudinum* with greater density and health in the Bahía de Filipinas, north of the PNA, which support the greatest fishing volume. In samplings performed in the region, 80 phytobenthic species were identified, of which 38 were chlorophytes, three phaeophytes, 35 rhodophytes, and four phanerogams (MINAGRI-CNAP 2014).

The reef areas with the greatest number of species are located to the north between the exterior border of the keys and the insular bank since corals are structural species of such habitat. The most represented species are (in ascending order) *Colpophyllia natans*, *Siderastrea radians*, *Diploria labyrinthiformis*. *Agaricia agaricites*, *Montastraea cavernosa*, *Montastraea faveolata*, *Millepora alcicornis*, *Siderastrea siderea*, *Porites astreoides*, and *Porites porites* (MINAGRI-CNAP

2014). In some reef areas, live coral coverage has been threatened by algal growth that proliferates because of their scarce natural biologic controls, such as sea urchin and herbivore fish. Nonetheless, the staghorn (*Acropora cervicornis*) coral and also the elkhorn (*Acropora palmata*) coral recover and grow behind the reef.

Biodiversity is not only of vital importance for the future of important biological, ecological, fishing, and aquaculture communities but also essential for the sustainable use of human settlements in the region that the LPCC-PNA hosts. The authorities are conscious and make important efforts to transmit clear messages to the population; in these sense, they raise awareness that in mangrove ecosystems, seagrass prairies, and coral reefs, a high marine fauna diversity reproduces, feeds, and then disperses, supporting the coastal fisheries of all the SCA region.

14.4.3 Pressures and Threats

The main pressures and threats are related with natural and anthropic factors. In the last years, the region has been the object of a high rate of socioeconomic development, so it has been subjected to anthropic impacts (cultural contamination, damming rivers, tourism development), which add up with the traditional fishing activity (Claro et al. 2007; Alcolado et al. 2015; Cobas et al. 2015). These human impacts act in synergy with natural factors that have grown in frequency and intensity such as hurricanes, temperature increase and solar radiation, as well as extreme drought events. The sum of these factors affects coral abundance and in consequence the availability of refuge for diverse pelagic and benthic organisms that constitute the food of larger organisms, which will be expressed, for example, in density and removable fish biomass by the fishing industry (Alcolado et al. 2007).

According to MINAGRI-CNAP (2014), the main active causes of pressure on marine habitats in the area of study are: (a) harmful impacts and diseases in coral banks (whitening, breakage by the effect of fishing and tourism activity, increase of macroalgae); (b) river damming and increase in salinity (decrease of trophic levels); (c) contamination and dumping residual waters (industrial and domestic effluents); (d) unsustainable fishing (purse-seine net, massive fishing gear, biodiversity affectation); (e) extreme meteorological events (hurricanes, excessive rainfall and extreme drought); (f) presence of exotic species (*Pterois antennata*); (g) illegal fishing and poaching species of high commercial value (mainly lobster); and (h) bycatch of protected species (marine turtles and others).

14.4.4 Legal Framework and Conflicts

According to the 2014–2018 Management Plan, the LPCC-PNA was created in 1991 by the Ministry of Agriculture (MINAGRI) although without legal approval, and by resolution 520/1995 of MINAGRI, it was reverted to the National Enterprise for the Protection of Flora and Fauna (NEPFF) for its administration. The species object of monitoring had the status of hunting or capture prohibition, such as some aquatic birds, with special emphasis on the pink Caribbean flamingo (*Phoenicopterus ruber*); some terrestrial mammals, such as the Cuban "jutía" (*Capromys pilorides*); and reptiles such as the Cuban iguana (*Cyclura nubila*), marine turtles such as the caouana "caguama" (*Caretta caretta*), green turtle (*Chelonia mydas*), and Carey (*Eretmochelys imbricata*), same as with the manatee population (*Trichechus manatus*), a regional endemic species.

The PNA was approved by the Agreement 4262 of the Executive Committee of the Council of Ministers (Comité Ejecutivo del Consejo de Ministros), adopted on December 14, 2001, with the management category of Fauna Refuge, which corresponds to a Category IV of the International Union for Conservation of Nature (IUCN). Furthermore, sites under special use and protection regime have been approved legally to protect the reproduction colonies of flamingoes and other birds; any anthropic activity is prohibited there, and strict regulations have been established for fishing in the spawning sites of fish of commercial interest:

> Classified as Fauna Refuge: land, marine, or a combination of both where the protection and management of habitats or species result essential for subsistence of significant wild, migratory, or resident fauna populations. Fauna refuges do not necessarily require to be natural territories in totality, which is why human activity linked to their resource management can exist, as long as it does not contravene the established regulations and is in function of the specific objectives of the area.

The administration (Biological Station) of the LPCC-PNA, which belongs to NCPFF, is located in the coastal community of Carahatas, municipality of Quemado de Güines, province of Villa Clara. Carahatas was founded in 1800, and it currently has 700 inhabitants that have a bearing directly on the LPCC-PNA. Marine fishing constitutes the main source of employment of this community, and it has a Management Plan whose focal objectives of ecological conservation are mangrove ecosystems, prairies of marine grass (phanerogams), and coral reefs. The previously mentioned faunistic resources are protected equally, such as the pink flamingo, Cuban iguana, marine turtles, and manatees, among other marine species of commercial interest, such as the common lobster (*P. argus*), fishes, queen conch (*L. gigas*), sponges (of the family Spongiidae) and oysters (*C. rhizophorae*), migratory birds, and other terrestrial species of flora and fauna.

The 2014–2018 Management Plan considered commercial fishing of crustaceans, fishes, and mollusks. This activity is in conflict with conservation of critical habitats (spawning, reproduction, breeding) at protected areas, since the management category of "wildlife refuge," associated to the LPCC-PNA, allows regulated fishing activities. Nonetheless, these regulations are not always fulfilled in lobster capture when smaller sizes than the legal minimum were extracted, as well as when using massive fishing gear for other species. Failure to comply with fishing regulations could compromise fishing stocks, since they could capture considerable volumes of species, both commercial and non-commercial without complying with the minimum commercial size. A collateral problem is that the minimum legal capture size in several commercial fish species is inferior to size at first sexual maturity, which complicates the protection scenario of those resources if, additionally, their reproduction periods are not respected (MINAGRI-CNAP 2014).

14.4.5 Socioeconomic Characterization of the Area and Its Environment

The LPCC-PNA is divided into zones and has in its management objectives at short, medium, and long term reflected in the program of public use (recreation and eco-tourism, dissemination and environmental education, scientific investigation and monitoring). Part of these actions involve not only the PNA administration but also scientific institutions, enterprises of the fishing, agriculture and tourism sectors, schools, and general population The marine part is used in totality by the fishing units of the communities La Panchita, Carahatas, and Isabela de Sagua where a good coverage and quality of the technical and social infrastructure (Table 14.1) exists, including telephone, postal, and transport services, among others. The state fishing establishments of these three communities have ice factories and docking bays in good conditions (supervised by the Cuban Boat Registry [Registro Cubano de Buques]), cold storage, repair shops for boats and land transport units.

The marine area of LPCC-PNA is located in a very important region of fishing resources, whose economic basis is given by its capture and commercialization. State professional fishing, private commercial fishing, and sport fishing are performed (Table 14.2). The most frequently used natural resources by the three main coastal communities are related with fishing with a negative bearing on coastal communities related to fishing and thus a negative bearing on conservation due to the need of applying a maximum fishing exploitation. This exploitation regime could reach the limit of sustainability; therefore, staff and authorities of the PNA should always be attentive to the compliance of the official norm and also contribute to raising collective awareness of conservation and sustainable use of the PNA resources. Private commercial fishing has time, distance, and fishing gear regulations, and only commercial fish extraction is authorized. The Federación Cubana de Pesca Deportiva (FCPD) rules sport and recreational fishing licenses, only authorizing the use of hand-fishing lines and hook, and the capture is limited to a total from 15 to 30 kg, according to the fishing area.

The fishing sector, according to the state fishing units registered in the LPCC-PNA, has 136 workers (14 women and 122 men), of which 32 fishers are dedicated to extracting lobster, 48 to fish and other resources, and 56 are indirect workers in the town Carahatas. In the town La Panchita, there are 150 workers and 87 are fishers. In Isabela de Sagua, there are 139 workers, of which 117 are fishers.

Both the administration of the protected area and the head of the fishing enterprise, Empresa Pesquera Industrial de Caibarién (EPICAI), are projected toward a sustainable development region with the establishment at the short term of new regulations and the progressive elimination of the fishing effort, both in boats as in

Table 14.1 Characterization of the communities with direct bearing on Las Picúas-Cayo Cristo Protected Natural Area (MINAGRI-CNAP 2014)

Communities		Carahatas	La Panchita	Isabela de Sagua
Essential economic activities		Fishing	Fishing	Fishing
Population	Total	676	890	3007
	Men	342	448	1432
	Women	334	442	1575
Number of homes		283	500[a]	657
Schools	Kindergarten			1
	Elementary	1	1	1
	High school			1
Nursing home			1	
Medical consultants		1	1	2
Pharmacies		1	1	1
Recreation	Social club	1	1	1
	Library	1	1	1
	Video club	1	1	1
	Culture house	1		1
	Restaurant	1	2	2
Aqueduct		1	1	1
Electricity		1	1	1
Dump		1	1	1
Accessibility by land roads		Regular	Good	Good

[a]203 of these homes are intended for vacational use

Table 14.2 Distribution of fishing vessels according to the type of fishery resource and by communities in Las Picúas-Cayo Cristo Protected Natural Area (MINAGRI-CNAP 2014)

Fishing fleet according to species		Carahatas	La Panchita	Isabela de Sagua
	Lobster	6	7	6
State fishing	Oyster			6
Fish and other resources		13	8	7
Private commercial	Fishes	31	28	45
Sport fishing	Fishes	16	9	50
Total, fishing boats		66	52	114

aggressive fishing gear that depress marine fauna populations. It is important to highlight that fishing in this region contributes to food supply for local sale and export, as well as for attending tourism. It is only necessary to point out EPICAI has unavoidable commitments with society, such as fish delivery for medical diets of the population and public hospitals.

Through environmental education and the participation in environmental festivals, inhabitants have accepted progressively the necessity of protecting the resources that exist there and recognize they are a source of food, pleasure, and welfare (MINAGRI-CNAP 2014). On the other hand, a large number of fishing workers depend economically on marine fauna extraction, mainly fish, mollusks, and lobster, an activity that occasionally has generated conflicts of interest among fishers and conservationists. Conceptually, fishing is a collecting activity, and at world level, these types of activities have been progressively substituted by real productive activities, such as agriculture and livestock. Thus, training in aquaculture activities offers new production alternatives to the community of LPCC-PNA, not only for collection but also generating new sources of working opportunities for the population from adjacent and nearby communities. It should be noted that aquaculture is also a source that generates foreign currency for the country, which finally comes back in infrastructure and services to the community as part of the commitments of the state.

14.4.6 Socioeconomic Problems in the Protected Natural Area and Its Environment

The level of instruction and environmental education starting from the work of the technicians of the Biological Station that administers LPCC-PNA and other sectors of society (schools, communication media) has increased, mainly in the locality of Carahatas and, in lesser measure, in La Panchita and Isabela de Sagua. This is the last locality that has less direct actions of environmental education, which affects the behavior of the population in relation to domestic waste, care, and protection of the environment.

In accordance to the ecological and environmental problems, the main topics that the administration of the PNA deals with in relation to the marine species are marine ecosystem health and its fragility and importance for the sustainability of local and also regional fisheries. The main groups to work with are fishers and children, the first ones because of the action directed to exploitation of fishing natural resources and the second ones for their capacity and potential in having a bearing on their parents and other adults in their environment.

According to surveys performed by LPCC-PNA specialist, the great majority of the fishers are aware of the environmental problems of the region; they acknowledge that massive fishing gear is a threat for future fisheries and that closed seasons must be established during the massive reproduction periods of the species of commercial interest. On the other hand, conflicts of interest are expected among fishermen and governmental staff in decision taking according to the official norm in force, related to the equilibrium that should be achieved between the need of conservation of natural resources and the economic need for fishermen subsistence.

14.4.7 Current Management Problems and Concerns

Illegal or bycatch fisheries of species in danger of extinction (CITES), such as manatees, turtles, and queen conch, among other species with temporal presence, is a permanent concern of the LPCC-PNA Management Plan, as well as the capture of prohibited commercial species carried out by nongovernmental actors (MINAGRI-CNAP 2014).

The historical evolution of fish capture in the north-central shelf of Cuba has shown a significant reduction after 1980, with levels considered as overfishing. During the economic crisis of 1990, a decline occurred due to the increase in the use of trawling nets without the compliance of regulations in length and selectivity. This situation increased species bycatch of the families Rajidae and Dasyatidae and those with sizes smaller than the commercial size (Claro et al. 2004; Obregón et al. 2007).

This trend was maintained until the beginning of 2012 when the use of trawling nets was prohibited for fishes and a legislation was applied to all fishing gear in the country. The reduction of fishing effort and adoption of new regulations facilitated a partial recovery and certain stability of the traditional fishing population. Nonetheless, other collateral causes continued and still continue having a negative bearing on the fishing activity in the region, such as illegal fishing and the lack of regulated fishing gear. This situation has recently gotten worse due to the loss of legal and authorized fishing gear and the effects of Hurricane Irma (September 2017). Finally, it is worth mentioning that fishers have also been affected by the reduction of fishing areas because the authorities decided to increase protected marine areas applying a sustainable future vision.

Contamination is a problem to be considered, including management of solid or liquid waste in the adjacent communities to LPCC-PNA. Currently, part of that waste reaches the sea due to a deficient design of septic tanks, making it necessary to create a domestic and industrial waste dump according to the official norms already established. An important part of the housing units located along the coastline continues using toilets installed directly on the sea. Unfortunately, in some occasions, oily water discharge from bilge pumps of the boats anchored or in movement occurs in the bays Carahatas and Isabela de Sagua.

For the conservation of nature, the development of fisheries with more eco-friendly methods is greatly important; furthermore, not only the strict application of protection strategies of massive biological resources is greatly important but also biodiversity because there are noncommercial species that have a great ecological value. It is necessary to reduce fishing efforts and increase action for developing projects of sustainable aquaculture and with an ecosystem focus. Different from fishing, which is a collection activity, aquaculture is a productive activity, and in this case, it should focus totally on native species of LPCC-PNA, preferably filter feeders and of high commercial value, such as oysters and sponges that have a rapid growth. With an adequate management, aquaculture could guarantee positive environmental impacts: (1) contributing with larvae for restocking at local level and adjacent areas; (2) "treating" the water body receptors by biological filtration of the particulate organic matter; (3) offering new production alternatives; and (4) contributing to new sources of employment (Mazón-Suástegui et al. 2017; Betanzos-Vega et al. 2018). Aquaculture activity "generates employment" because with one direct employment, it creates three indirect ones (FAO 2010). However, with the increase in aquaculture production, these values have increased at world level in 15 or 22 indirect jobs per direct employment (FAO 2018).

14.5 Lessons Learned

The pressure exerted by the coastal communities with direct influence on the Las Picúas-Cayo Cristo Protected Natural Area (LPCC-PNA) and on other adjacent communities (not analyzed in this study) has caused a negative impact on natural resources and ecosystems. Besides the internal impact, the adjacent communities exert some pressure on the PNA, both in recreational activities, such as the use of beaches, and socioeconomic activities of the productive type, mainly fishing. Eliminating or decreasing these impacts (contamination, non-sustainable fishing, illegal fishing, illegal hunting, logging, and poaching, among others) depends in great measure on environmental education, understanding, commitment, and participation of the different local and regional actors that have direct or indirect bearing on the PNA. On the other hand, it is absolutely necessary to place the needs of the local and nearby population in a balance or those aspects related with their feeding habits and the improvement of their housing fund and social welfare. Many of the villagers have land to grow and raise animals, but their habits and customs, as well as the market price and demand for fish and shellfish, tip their activities toward the sea.

The management plan established by the Biological Station that administers the LPCC-PNA with consultancy and support from the Empresa Nacional para la Conservación de la Flora y Fauna (ENCFF) overlaps with the needs of the state and private commercial and recreational sport fishing sectors. The common and desired result is no other than guaranteeing protection to marine biodiversity and its habitats to reach a recovery of fish, crustacean, and mollusk populations of commercial value with the fundamental objective of satisfying economic, social, and individual demand of the biological resources in general and fishery and aquaculture marine resources in particular.

Undoubtedly, this PNA may offer ecotourism destinations with a great potential of economic income by the efficient and regulated use of all the natural goods and services the PNA has, which would allow contributing to its self-financing and a substantial betterment in life quality of local inhabitants. The accessibility and relative closeness of the LPCC-PNA to important tourism poles such as Varadero and Cayos del Norte de Caibarién of the SCA region, the presence of spectacular and visually attractive marine and terrestrial fauna, the existence of pristine and practically virgin beaches in keys and islets, and the striking artisanal fishery practice in that place could turn out to be the motor of sustainable ecotourism development where the same biological diversity conservation, both terrestrial and marine, might make an economic and social sustainability possible.

For this purpose, it is essential to apply all the knowledge acquired surrounding the efficient management of fishery resources and the technological development reached in matter of aquaculture. Involving all actors and users of the PNA of the terrestrial and coastal areas is necessary besides providing a holistic and efficient governance, introducing solid and interactive mechanisms of environmental education and training in responsible aquaculture techniques with good management practices and follow-up. The establishment of experimental farms, mainly marine filter-feeder species, such as sponges, oysters, and other bivalve mollusks, has shown to be a good alternative for the productive and community sectors. The fishing gear and facilities of suspended or floating cultures function as fish and invertebrate aggregates and also as reproductive banks given that organisms in culture mature and spawn in the culture site, but their larvae disperse in the surroundings by the effect of marine currents. The appropriation of the self-sustainability principles and self-governance, added to knowledge and technologies already developed in situ and thus available, shall set the basis for an ecosystem management of the region, guaranteeing protection and enrichment of LPCC-PNA biodiversity.

References

Alcolado PM, García EE, Espinosa N (1999) Protección de la biodiversidad y desarrollo sostenible en el Ecosistema Sabana-Camagüey. Proyecto PNUD/GEF Sabana-Camagüey CUB/92/G31, CESYTA S.L. Madrid. 145p

Alcolado PM, García EE, Arellano-Acosta M (Eds.). (2007) Ecosistema Sabana-Camagüey. Estado actual, avances y desafíos en la protección y uso sostenible de la biodiversidad. Proyecto PNUD/GEF Sabana-Camagüey CUB/98/G32; CUB/99/G81. Editorial Academia. 183 pp

Alcolado PM, Claro R, Valle S (2015) Arrecifes coralinos: por la pesca, el turismo y la protección de las costas. p 151-159. En: Menéndez-Carrera L, Arellano-Acosta M, Alcolado PM (eds) *¿Tendremos desarrollo socioeconómico sin conservación de la biodiversidad? Experiencias del proyecto Sabana-Camagüey en paisajes productivos*. La Habana, Editorial AMA. ISBN: 978-959-300-105-2. Impresos dominicanos, S.R.L. 223 p

Betanzos A, Arencibia G, Delgado G, Nodar R (2009) Caracterización de la calidad del agua al norte de Villa Clara, Cuba, para definir zonas de cultivo del ostión de mangle (Crassostrea rhizophorae, Güilding, 1828). Revista Argentina de Ecotoxicología y Contaminación Ambiental 1(4):1–9

Betanzos A, Arencibia G (2010) Tensores naturales y antrópicos al norte de Villa Clara, Cuba; efectos en la producción de ostión (Crassostrea rhizophorae, Guilding, 1828). Memorias del VI Taller Internacional CONyMA 2010: CD-ROM ISBN 978-959-300-008-6. 9pp. http://www.oceandocs.org/handle/1834/3628

Betanzos-Vega A, Siam C, Arencibia G (2010) Variación de la salinidad y su relación con la distribución por talla del ostión de mangle, Villa Clara, Cuba. Rev Cuba Investig Pesq 27(1):41–46

Betanzos-Vega A, Capetillo N, Lopeztegui A, Martínez-Daranas B (2013) Variación espaciotemporal de la turbidez y calidad en cuerpos de agua marina de uso pesquero, región norcentral de Cuba, 2008–2010. Serie Oceanol 12:24–35

Betanzos A, Rivero-Suárez S, Mazón-Suástegui JM (2013) Factibilidad de sostenibilidad ambiental en el cultivo de ostión de mangle *Crassostrea rhizophorae* en la Isabela de Sagua, Cuba., p 57-67. En: Potenciando la conservación de la biodiversidad mediante la evaluación económica y ambientalmente sostenible de actividades productivas en el ecosistema Sabana Camagüey, Cuba. Proyecto PNUD/GEF/Sabana Camagüey, 131pp. ISBN 978-959—300-037-6. Editorial AMA

Betanzos Vega A, Rivero Suárez S, Mazón-Suástegui JM (2014) Factibilidad económico-ambiental para el cultivo sostenible de la ostra de mangle *Crassostrea rhizophorae* (Güilding, 1828), en Cuba. Lat Am J Aquat Res 42(5):1148–1158

Betanzos-Vega A, Valle S (2015) La subsistencia de los pescadores: alternativas y prácticas sostenibles. p 174-185. En: Menéndez-Carrera, L., M. Arellano-Acosta, P. M. Alcolado (Eds) *¿Tendremos desarrollo socioeconómico sin conservación de la biodiversidad? Experiencias del proyecto Sabana-Camagüey en paisajes productivos*. La Habana, Editorial AMA. ISBN: 978-959-300-105-2. Impresos dominicanos, S.R.L. 223 pp

Betanzos Vega A, Linares Pérez E, O., Martí HA, Valle S, Tizol Correa R Hernández López, JL (2015). Prospección hidroacústica en zonas de desove de peces de la plataforma nororiental de Cuba. Caso: biajaiba (Lutjanus synagris, Linnaeus, 1758). Serie Oceanológica, 14: 46–57

Betanzos-Vega A, Mazón-Suástegui JM, Arencibia-Carballo G (2018) La ostricultura: una alternativa de desarrollo pesquero para comunidades costeras en Cuba. Universidad Autónoma de Campeche. 104 p. ISBN: 978-607-8444-43-4. https://www.redicomar.com/wp-content/uploads/2018/11/OSTR-Cuba-1.pdf

Betanzos-Vega A, Mazón-Suástegui JM, Formoso-García M, Avilés-Quevedo MA (2019) Sponge fishery and aquaculture in Cuba: impacts and challenges. Intech Open, Chapter aquatic invertebrates. 16 p. https://www.oceandocs.org/handle/1834/15239

Blanco JC, Formoso M (2009) Estimado de la abundancia de esponjas comerciales en áreas correspondientes a las plataformas suroccidental (Golfo de Batabanó) y norcentral (Archipiélago Sabana-Camagüey) de Cuba. Rev Invest Mar 30(2):99–106

Claro R, García-Arteaga JP, Gobert B, Cantelar K, Valle S, Pina F (2004) Situación actual de los recursos pesqueros del archipiélago Sabana-Camagüey, Cuba. Bol Invest Mar Cost 33:49–67

Claro R, Cantelar K, Pina F, García-Arteaga JP (2007) Cambios en las comunidades de peces de los arrecifes coralinos del Archipiélago Sabana-Camagüey, Cuba. Rev Biol Trop 55(2):537–547

CNAP (2013) Plan del Sistema Nacional de Áreas Protegidas 2014–2020, Centro Nacional de Áreas Protegidas, Ministerio de Ciencias Tecnología y Medio Ambiente, La Habana, Cuba. 366 p. ISBN:978-959-287-049-9

Cobas S, Morales O, Puga R, Piñeiro R, de León ME (2015) Factores estresantes al hábitat de la langosta espinosa en la zona norcentral de Cuba. Rev Cub Inv Pesq 32(1):48–54

Cruz R, Baisre JA, Díaz-Iglesia E, Brito R, Blanco W, García C, Carrodeguas C (1990) Atlas biológico – pesquero de la langosta en el archipiélago cubano. Centro de Investigaciones Pesqueras: La Habana, Cuba. 125 p

FAO (2010) El estado mundial de la pesca y la acuacultura 2010. Roma. ISBN 978-92-5-306675-9, 242 p

FAO (2018) El estado mundial de la pesca y la acuacultura 2018. Cumplir los objetivos de desarrollo sostenible. Roma. Licencia: CC BY-NC-SA 3.0 IGO. 234 p

Formoso M (2008) Biología, ecología, pesquería y proceso industrial del Cobo *Lobatus gigas*. Aspectos generales y situación actual en Cuba. Memorias del Encuentro de Pesca Sostenible en Cuba., del 3 al 5 de septiembre, 2018. Playa Santa Lucía, Cuba. Proyecto SOS Pesca "Sostenibilidad de las Pesquerías"

Garrido OH, Kirkconnell A (2012) Aves de Cuba. Comstock Publishing Associaties, Estados Unidos, 287p

Hernández-Corujo C, Linares EO, Barros OZ (1998) Medición de la fuerza del blanco acústico de dos especies de la familia Lutjanidae: *Lutjanus synagris* y *Lutjanus chrysurus*. Fondo de Manuscritos ACC, 8 p

Linares EO, Caballero PI, Guillard J, Sierra L, Hernández JL (2009) Medición de la fuerza del blanco acústico de dos especies de peces marinos de las familias Lutjanidae y Haemulidae (pargo criollo, *Lutjanus analis* y ronco amarillo, *Haemulon sciurus*). Serie Oceanol 5:99–110

Mazón-Suástegui JM, Avilés-Quevedo MA, Rivero-Suárez S (2011) Bases tecnológicas para el cultivo sostenible del ostión nativo *Crassostrea rhizophorae*, en el Ecosistema Sabana Camagüey, República de Cuba. Manual Técnico, Proyecto Piloto para el Cultivo de Ostión, GEF/PNUD "Protección al Ecosistema Sabana-Camagüey", 69 pp. http://www.oceandocs.org

Mazón-Suástegui JM, Rivero-Suárez SE, Betanzos-Vega A, Saucedo PE, Rodríguez-Jaramillo C, Acosta-Salmón H (2017) Potential of sites in northern Cuba for developing an industry of the native mangrove oyster (*Crassostrea rhizophorae*). Lat Am J Aquat Res 45(1):218–222

MINAGRI (2014) *Plan de Manejo del Refugio de Fauna Las Picúas-Cayo Cristo (2014–2018).* Estación Biológica de Carahatas, Empresa Nacional para la Protección de la Flora y la Fauna, MINAGRI, Villa Clara, Archivo del CNAP, 151 p y Anexos

Montalvo-Estévez JF, García-Ramil I, de los A, Almeida-Rodríguez M, Betanzos-Vega A, García-García N (2014) Modelación de la eutroficación e índice de calidad del agua en algunas bahías del archipiélago Sabana Camagüey. Tecnología Química 34(3):184–196

Obregón MH, Pérez I, Valle S, Alcolado PM (2007) El sector de la pesca. En: Ecosistema Sabana-Camaguey. Estado actual, avances y desafíos en la protección y uso sostenible de la biodiversidad. 124-129. Proyecto PNUD/GEF Sabana-Camagüey, CUB/98/G32; CUB/99/G81

Palma R, Betanzos-Vega A, Puga R, Alzugaray R (2010) Dispositivos de escape en nasas langosteras, una alternativa ecológicamente responsable en la protección de recursos marinos. Memorias del VI Taller Internacional CONyMA 2010: CD-ROM, 9 p. http://www.oceandocs.org/handle/1834/3628

Perigó E, Montalvo JF, Cano M, Martínez C, Niévares A, Pérez DM (2004) Principales fuentes contaminantes en la eco-región norcentral de Cuba (Archipiélago Sabana–Camagüey). Impactos y respuestas. Contribución a la Educación y la Protección Ambiental 5:14–26

Puga R, Piñeiro R, Capetillo N, de León ME, Cobas S (2009) Caso de estudio 2: Estado de la pesquería de la langosta espinosa (*Panulirus argus*) y su relación con factores ambientales y antrópicos en Cuba. En: Hernández A et al (2009) *Evaluación de las posibles afectaciones del cambio climático a la biodiversidad marina y costera de Cuba.* 25 p. ISBN: 978-959-298-017-4. http://www.redciencia.cu/

Puga R, Piñeiro R, Alzugaray R, Cobas LS, de León ME, Morales O (2013) Integrating anthropogenic and climatic factors in the assessment of the Caribbean spiny lobster (*Panulirus argus*) in Cuba: implications for fishery management. Int J Mar Sci 3(6):36–45

Puga R, Valle S, Kritzer JP, Delgado G, de León ME, Giménez E, Ramos I, Moreno O, Karr KA (2018) Vulnerability of nearshore tropical finfish in Cuba: implications for scientific and management planning. Bull Mar Sci 94(2):377–392. https://doi.org/10.5343/bms.2016.112

Quirós A, Perdomo ME, Rodríguez E, Silverio LC (2009) Cultivo de esponjas marinas. Manual de buenas prácticas. Resultado del Proyecto CUB/OP/2/07/09 "Uso Sostenible de los Recursos Naturales en la Comunidad Costera de Carahatas", Villa Clara, Cuba: CESAM-CITMA

Rivero-Suárez S (2012) Potencial de cultivo del ostión en Isabela de Sagua, a partir de la fijación natural en colectores de mangle. Tesis de Maestría, Universidad de La Habana, La Habana, 136 pp. Director de Tesis: Dr. José Manuel Mazón-Suástegui, CIBNOR, S.C., La Paz, México

Ruiz-Plasencia I (2017) Las Áreas Protegidas de Cuba. Centro Nacional de Áreas Protegidas, La Habana, Cuba, 394 p. ISBN: 978-959-287-079-6

Siam C, Betanzos-Vega A, Giménez E (2007) Características Oceanográficas del Sector Costero Sierra Morena – Isabela de Sagua, Norte de Villa Clara, Cuba. Memorias del III Taller Internacional Pesca 2007. Ciudad de la Habana, Cuba; del 5 al 9 de marzo del 2007

Chapter 15
Applying Epistemic Approach to Analyze Bio-intercultural Relationships Among Local Indigenous People and Nature

Magdalena Lagunas-Vázques

Abstract This chapter presents a review of the socio-historic and cultural knowledge that conditions the ways of understanding, interpreting, and relating in and with the natural world from an intercultural perspective on the part of the local indigenous inhabitants of the natural protected areas (NPA) of Latin America. The Good Living theory will be the philosophical approach to develop the present analysis, from the ancestral indigenous American perspective, which is considered an integrating theoretical framework with a cosmic approach and not anthropocentric.

Keywords Interculturality · Intercultural dialogue · Biocultural conservation · Latin America

15.1 Introduction

The knowledge that conditions the ways of thinking, interpreting, and connecting in and with the natural world, from an intercultural perspective on the part of the local indigenous inhabitants of natural protected areas (NPA) of Latin America, is found in at least two large regions where two civilizations developed. The Mesoamerican and the Andean civilizations are two of the few thousand-year-old civilizations which modern society has inherited from the ancient world. The former encompasses center and southeast Mexico and practically all Central America (Díaz and Escobar 2006), while the latter, also known as Inca or Incan, is located in South America in the current nations of Peru, Chile, Bolivia, Colombia, Ecuador, and Argentina (Milla 1983; Mann 2006). Below, certain ancient historic generalizations about both will be described. This will concisely allow envisaging relationships, interactions, and knowledge said civilizations had about the nature that surrounded

M. Lagunas-Vázques (✉)
CONACYT-CCGS Centro de Cambio Global y La Sustentabilidad,
Villahermosa, Tabasco, Mexico

A. Ortega-Rubio (ed.), *Socio-ecological Studies in Natural Protected Areas*,
https://doi.org/10.1007/978-3-030-47264-1_15

them. These correlations lead to worldviews and cosmogonies about nature and culture that still thrive today in various socioculturally diverse native peoples currently living in different regions of Latin America. In some cases, the lands of these peoples are now within protected areas.

The development of pre-Columbian cultures stems from a metaphysical, mythological world, where myths, rituals, and symbols are as essential for the creation of these civilizations as they are for the construction of the worlds of their ideas and the several cultural representation of said peoples (León-Portilla 1956; González 2003). From this perspective, the structure, creation, and recreation of pre-Columbian life were imbued with sacred aspects; every social expression was ritualized and held sacred (León-Portilla 1956; González 2003; Rivera 2018). From this idea, one can put together what was sacred of natural elements in their respective civilizations and cultures, since their lifestyles were closely related to nature (Fig. 15.1). For instance, there were more than 2000 gods in Mesoamerican civilization (León-Portilla 1956), among which goddesses and gods inherently related with natural elements dominated. A few of these are rain (*Tlaloc*), wind (*Ehecatl*), the sun (*Tonatiuh*), the moon (*Tecciztecatl*), corn (*Centeotl*), jaguar (*Ocelotl*), feathered snake (*Quetzalcoatl*) (Fig. 15.1), agave (*Mayahuel*), hummingbird (*Huitzilopochtli*), and coyote (*Huehuecoyotl*) (León-Portilla 1956). Other examples are the idea of a major, comprehensive, and absolute entity in the Andean civilization, *Pacha*, the earth that connects to everything existing on it and the cosmos (Rivera 2018); and the deity *Viracocha*, its duality and its feminine symbols of moon and winter, and those masculine ones of sun and lightning (Kusch 1962).

This chapter presents a review of the knowledge that conditions the ways of thinking, interpreting, and relating in and with the natural world from an

Fig. 15.1 Snakes from the replica of the door to the Temple of Quetzalcoatl in Teotihuacan, Mexico. (Author: Bertram Bilek)

intercultural perspective on the part of the local indigenous inhabitants of protected natural areas (NPA) of Latin America. The epistemic approach of *Buen vivir* ("Good Living") is presented as an integrating theoretical framework, with a cosmic, non-anthropocentric property, which enables guidance of interactions between cultural and natural diversity in NPA of Latin America in order to favor better biocultural conservation practices. Additionally, it is also intended that the information provided in the chapter will provide a better understanding about this ancient indigenous philosophical perspective from the Americas, as well as the interactions between nature and current rural native societies in Latin America.

15.1.1 Mesoamerican Civilization and Its Relation with the Natural World

Great cosmological myths[1] were the base of Nahua culture and thought. Specialist scholars consider this profound and rich thought, which was called *tlamatiliztli* by its creators, a philosophy (León-Portilla 1956; Mann 2006). Nahua culture developed literature and philosophical thought in *Calmecac* and *Telpochcalli* schools (This was possibly the first compulsory education program in history, since every male citizen of the Triple Alliance had to attend, regardless of social class, a kind of school until aged 16.) (Mann 2006). Ancient Mexican people cultivated history, poetry, rhetoric arts, arithmetic, astronomy, and every science of which they left proof (León-Portilla 1956). The *tlamatini*, who was in and of itself writing and wisdom, had to write and preserve codices and live as a role model to the community. Several *tlamatinime* (plural of the word) taught priests, teachers, and administrators of the next generation in elite academies (Mann 2006).

Besides physical, medicinal, and theological knowledge (León-Portilla 1956), Clavijero (1945) observes abstract concepts and metaphysical language in this ancient Mexican tongue. Philosophy, religion, and astronomical and physical ideas were painstakingly linked to one another (Fig. 15.2) (León-Portilla 1956). The aesthetic-artistic analysis of the *Coatlicue* (Fig. 15.3) greatly indicates the worldview and cosmological ideas of the Aztecs (Fernández 1972). Justino Fernández (1959) reads all the Aztec (Nahua) cosmogony on the sculpture, on which he finds pyramid, cross, and human shapes. This reflects the Aztec understanding, with its cosmic space and all its dimensions. It even mentions a Nahua aesthetic nature, for instance, as these ancient poets considered poetry the highest of languages, the truest, most beautiful way of saying something (León-Portilla 1956).

There is a theological conception of the *tlamatinime*, supported by the supreme dual principle (*Ometeotl*, god of duality). They conclude that there is a dual principle, their god, giver of everything that exists, female-male *Ometeotl* of twin

[1] *Myth* in the sense of providing models for human behavior and conferring, for the same reason, meaning and value upon existence, from the perspective of Mircea Eliade (1991).

Fig. 15.2 The Sunstone of the Aztec Calendar, Mexico City Museum of Anthropology. (Author: Bertram Bilek)

Fig. 15.3 La Coatlicue god-goddess of earth, death, and fertility (idea that death is the generator of life) in Mexica mythology. Exhibited at the National Museum of Anthropology, Mexico City. (Author: Bertram Bilek)

divine nature (León-Portilla 1956). *Ometeotl* (god-goddess) and its relation with all natural things: rain, fire, water, earth, light, darkness, sky, cosmic forces. There are some existential characteristics of *Ometeotl*: lord of water and jade skirt, father-mother, and he who brings forth the light. This cosmic generation-conception (*Moyocoyani*) gives rise to the universe and constitutes being itself. He is the Lord who thinks or invents himself, he who no one shaped, and who exists beyond all time and place. These interpretations form the supreme climax of Nahua philosophical thought, which exalts thoughts and philosophical contemplation about divinity (León-Portilla 1956). With the philosophical conception about: Action and cosmic presence of pantheism *Ometeotl* (God creator of the whole universe); this civilization has an intellectual thought about the origin of the universe, the origin of nature, and origin of the world (León-Portilla 1956). They also possessed a vast polytheism in and with a symbolic myth language and symbolic references (Fig. 15.4) of natural phenomena of the origin of the afterlife and the divinity. Holders of lofty philosophical and cosmological ideas, metaphors, and myths, as those every thousand-year-old person in the world has (Clavijero 1945; León-Portilla 1956).

15.1.2 The Andean Civilization and Its Relation with the Natural World

The Andean civilization has developed a way of thinking of the world—Andean philosophy—which is precisely the *relationality* of everything, the network of nexuses and connections that is the life force of everything in existence (Milla 1983; Estermann 1998). There is nothing (in a very vital way) without this transcendental condition. It is a dialectical Andean rationality, both due to its underlying structure of conceiving reality and the abundance of phenomenological indications (Estermann 1998). For the *runa/jaqi*, sky and land, sun and moon, light and dark, truth and

Fig. 15.4 Art Wixárika woven textile (Nierikas), Nayarit, Mexico. (Author: Bertram Bilek)

falsehood, day and night, good and evil, and male and female are not mutually exclusive, but complements necessary to declare a superior and comprehensive entity (Estermann 1998, Rivera 2018). This complementarity principle appears on every level and in every aspect of life, both in cosmic and anthropologic dimensions, as well as in ethical and social ones (Estermann 1998). The Andean ideal is not one of two extreme opposites, but the harmonious integration of both (Estermann 1998).

Maybe the highest thought of major, comprehensive correlation of this culture is the *Pacha* (Quechua/Aymara). This cosmogony and worldview is the symbolic representation of interconnected cosmos (*Pacha*/earth) through several cardinal axes; it is not about the western sense of "theory" or "view" (Milla 1983). The Andean philosophy of *Pacha* is a comprehensive conception-reflection of the cosmic relationality as a manifestation of the collective Andean experience of reality (everything that exists) (Kusch 1962; Estermann 1998; Rivera 2018). Andean time is not unidirectional (from past to future), but bio-multidirectional (Rivera 2018). For cyclical rationality, future is actually behind, and past is forward; but it is also all the way around. For the *runa/jaqi*, history is not the space where new things, progress, or development toward something better is made; rather, it is a cyclical repetition of an organic process, corresponding to the cosmic order and its relationality (Rivera 2018). Basic tasks and skills of the *runa/jaqi* consist of the power of symbolic presentation of the cosmos through ritual and celebratory ways. For the Andean human being, celebrations and rituals are gnosiologic and ethical acts of the first order. To celebrate the cosmic order (in a symbolic way) means to know it and preserve it (Estermann 1998; Rivera 2018).

In his book *Génesis de la cultura andina*, Carlos Milla (1983) mentions, among other things, the existence of Amauta wise men, who could be considered men of science or knowledge. The *Camasca Amauta Runa* were architects, who also appeared as great mathematicians and astronomers in this grand civilization. The *Capaccuna* was a sort of Andean communal confederate state. The *Tawantinsuyo* were communities or communal systems located in four regions related to the four stars of the Southern Cross constellation, a complex indigenous astronomical system of orientation directed southward (Milla 1983).

The Aymara nation had a helio-lunar 13-month calendar (Milla 1983). There are myriad astronomical and ritual observatories (astronomical ceremonial centers, circular observatories, wells, and other structural ways of measurement and observation of stellar phenomena) all over the vast Andean lands (thousands of kilometers) (Milla 1983). Their mathematical knowledge was captured in the extensive number of figurines and structures where geometrical and astronomical relations are plentiful (Milla 1983). In the southern hemisphere, the Southern Cross, which comprises four points that merge in one, determines the collectivist thought and the synthesis ability of Andean civilization. Archaeological sites suggest a cultural continuum and the existence of the communal state organized by the *Capaccuna* of the *Amautas*, which lasted for at least 16 centuries (from preceramic period). It had elements of the same scientific and cultural core (Andean Unitarian culture) (Milla 1983).

The Andean human being is then defined by relationships. It is, by itself, a *chakana*, a bridge or node of multiple connections and relations (Rivera 2018;

Estermann 1998). Firstly, the center of gravity is not the individual or a completely isolated person (*kat'authon*), but a meta-individual entity, which is not necessarily universal (Estermann 1998). The western individual is "nothingness," a "total void," a non-being for Andean philosophy. Individuality, if it were a value, is just conferred in a derivative and secondary sense, but in no way as a primordial, foundational fact (Estermann 1998).

In the Andes, the fundamental collective entity (one could even say it is transcendental) and the essential principle for identity is the *ayllu*, which is the ethnic unit of farmer communities and economic basis of survival and domestic trade (Rivera 2018). There is no exact translation for the Quechua/Aymara word *ayllu*. It is the ethnic unit of an Andean farmer community, but it can also refer to the extended family. It can even be the town or village in a geographical sense (*mark'a*) (Milla 1983). Andean ethics has the cosmic order, the universal relationality of everything in existence, as axiological foundation (cosmic ethics) (Estermann 1998). This order complies with the principles of correspondence, complementarity, and reciprocity (Estermann 1998; Rivera 2018).

Incan civilization domesticated potatoes. They made the frost turn from curse to blessing by inventing the *Ch'uñu*. They built a vast territory of agreed cohabitation, where cultural and material production, as well as its physical, geographical space, allows archaeological and sociocultural observation of local cultural expressions by identifying connections with the sun and moon cycles (Rivera 2018). Everyday nature showed creative acts, and acts of desire and imagination rooted in the landscape and the living memory of people. There is even an untranslated aspect due to other epistemological horizons (Estermann 1998): Aymara language, for instance, has four grammatical persons (I, you, he, she), and *jiwasa*, which is singular and collective (not the plural "we," "you," or "they"), is everyone and me. The collective singular *jiwasanaka* is a "we-human" that includes even the one that is not present (Rivera 2018).

The synthetic Mesoamerican and Andean cosmogonic, theogonic, and symbolic interpretation mentioned above can be considered Latin American indigenous thought. This allows identifying the sacredness of nature on these native human groups, illustrated in their extensive agrarian rites, knowledge of climatic and rain seasons, and a wide range of traditional ecological knowledge they implemented in their daily lives (Figs. 15.5 and 15.6). All this continues nowadays among the socioculturally diverse human groups who currently inhabit these regions (González 2003; Toledo and Barrera-Bassols 2008; Calle et al. 2016; Naranjo 2018).

15.1.3 Andean Philosophical Approach: Is Buen vivir *the Same as* Sumak Kawsay*?*

Academic interest in researching the term *Sumak Kawsay* comes out of its inclusion in the Constitution of Ecuador in 2008. Its debatable translation as *Buen vivir* ("Good Living") has roused several epistemological and ontological discussions.

Fig. 15.5 Milpa corn plant, main agro-food of the Mesoamerican culture. (Author: Bertram Bilek)

Fig. 15.6 *Mazorca* (cob, tender corn), main agro-food of the Mesoamerican culture. (Author: Bertram Bilek)

Currently, three schools of thought study the emerging paradigm of *Buen vivir*, as stated by Cuestas-Caza (2019): indigenous-culturalism, post-developmentalism-ecologism, and socialism-statism. According to Cuestas-Cazas' analysis (2019), each of them has interpreted the indigenous knowledge related to *Sumak Kawsay* in different ways, which has led to several questionings. Among them, there is a possible epistemic extractivism in the usage and content of the term *Buen vivir*. Moreover, in practice, the inclusion of *Buen vivir* in the Constitution of Ecuador in 2008 and the acknowledgment of the rights of nature have given rise to a series of critiques and political-environmental dilemmas due to the neo-extractivist public policies implemented by the government of Rafael Correa (Cuestas-Caza 2019).

No one knows precisely when and how the words *Sumak Kawsay* appeared to refer to Andean system of life (Oviedo 2017a). Truth is, this thousand-year-old system of life of Andean people did and does exist, with its principles, models, institutions, and a kind of Andean consciousness (*Tawantin*) that envelops the system of life of the native peoples who have inhabited the Andes for more than 10,000 years (Oviedo 2017a). An expression in Kichwa language that might be similar to Buen vivir is *Alli Kawsay*, not *Sumak Kawsay*. *Alli Kawsay* expresses the integrality and the wish to improve the quality of life interdependently with beings in the environment, at personal (*runa*), family (*ayllu*), community (*llakta*), and intercommunity level within a territory. Involved social individuals come and go searching for *Alli Kawsay* through everyday actions (Cuestas-Caza 2017).

As a result of a wide documentary review, Cuestas-Caza (2019) describes, from a phenomenological perspective, that *Sumak Kawsay* represents an indigenous way of life, different from the Euro-American "way of life," and that would have been present for hundreds of years, and that would still prevail nowadays. Unlike the western model, *Sumak Kawsay* centers on what is communal and on reproduction and care of life (*Kawsay*), which would relate to the proposals of care economy and feminist economy. The concept of *Sumak Kawsay*, or *Buen vivir*, refers to an ancient social phenomenon that continues today: the way of life of the Ecuadorian Amazonian native (Kichwa, Achuar, and Shuar). It is based on the search and the keeping of the harmony among oneself, the community, and all other living beings in nature, in the protection of the Amazon rainforest, through the combination of spiritual and material elements (Cuestas-Caza 2019).

According to Oviedo (2017a), the Andean consciousness does not conceive the dichotomy between "good" and "evil" (justice-injustice, peace-war, developed-underdeveloped). Firstly, because it does not accept the existence of evil. Secondly, because it does not value elements as good or evil, but just as they exist, living and coexisting. This sort of thought is not concerned by an assessment of good-evil, positive-negative, for it considers there are no good, bad, positive, nor negative experiences. There are only experiences, without any description or classification whatsoever (Oviedo 2017a; Rivera 2018). Consequently, *Alli Kawsay* is considered superfluous and minute to the Andean world. Hence, mixing *Buen vivir* with *Alli Kawsay* is "dangerous," and it is even more with *Sumak Kawsay* (Oviedo 2017a).

Although *Buen vivir* and *Sumak Kawsay* are not synonyms, the debate about them has certainly opened up doors for the West and the Andes to start an intercultural dialogue on equal conditions, where two different civilizing nuances can establish an epistemic exchange (Quijano 1992; Grosfoguel 2013a, b), even when their principles may be considered incompatible (Tavares 2013). This dialogue is essential to compensate the exhaustion *Bien vivir* and Andean thought suffer after years of neo-extractivist governmental policies (Cuestas-Caza 2019).

In a case study in Imbabura, Cuestas-Caza (2019) found that native peoples understand and live *Sumak Kawsay* in a profound relation with nature. They have an institutionalized community sense, which they express mainly in rituals, parties, and diverse celebrations. The researched communities understand *Sumak Kawsay* as a "recreated tradition" that represents their social life project, a utopia if you will. Although this is a recent term in their vocabulary, they have started to appropriate it (Cuestas-Caza 2019).

Among the contributions of literature specialized in definitions of *Buen vivir* from the perspective of ancient native peoples of the Americas, the following conditions stand out: every definition is about integrative, cosmic non-anthropocentric aspects, in some way (Estermann 1998; Oviedo 2017a; Rivera 2018). Below, there is a translated quote about the term *Buen vivir*:

> *Buen vivir* advocates a world "made of many worlds", where different cultural rationalities coexist, confront and dialogue. It is based on the acknowledgment and respect of difference and diversity. Difference as wealth of life and diversity as the most beautiful expression of existence (Oviedo 2017b).

If *Buen vivir* is an ethical approach, in an essential sense of the term, it is important to advise from the beginning that it is not a thoroughly complete, undebatable one (Jiménez 2011; Dussel 2013; Cuestas-Caza 2019). On the contrary, it is in a process of pluralistic, complex, and controversial construction. Thus, to understand it in a useful way, it is necessary to have previous knowledge of the profound, vital contexts of ancient Andean culture, from which *Buen vivir* is product and expression (Jiménez 2011).

Following Cuestas-Caza's contributions (2019), *Sumak Kawsay* is still a very abstract concept that needs to be fueled by the empirical study of practices and knowledge of native peoples. These studies must be based on real coexistence and mutual learning, putting aside the characteristics of classic ethnography to introduce elements of critical, intercultural, decolonializing ethnography (Fornet-Betancourt 2009; Cuestas-Caza 2019; Grosfoguel 2019). Therefore, the Academy has a great responsibility that goes beyond westernized ethnography and the intellectual comfort of cross references.

Oviedo's proposal (2017a) are indispensable here:

> The least, but most appropriate thing to do, for a responsible, serious researcher, is to try to enter the consciousness of the people, and then, dare to make hypotheses, even though internalizing a culture to speak with property and profoundness would be of uttermost importance. If a philosophy is not lived personally, it becomes susceptible to manipulation and deformation. Thus, even when the best of intentions encourage some individuals, and despite being descendant of native peoples, their positions may end up being part of the indoctrination and perennial civilizational catechesis. (Oviedo 2017a).

It is in the interest of this work to contribute guiding methodological elements to establish ways to interrelate the philosophical approach of *Buen vivir* with the native peoples of the NPA of Latin America, in favor of biocultural conservation.

15.2 Biocultural Conservation: Epistemic Diversity and Ecosystemic Biodiversity

15.2.1 Relationship Between the Local Native Inhabitants of NPA with Nature and Environmental Conservation

The lands of indigenous peoples all around the world intertwine with approximately a 40% of all protected areas and over 65% of the most distant and least inhabited territories on Earth (Garnett et al. 2018). There is a clear overlap between the areas with the most biodiversity in the planet and the regions densely inhabited by the indigenous population of the world, since a high biological and linguistic diversity coincides there (Toledo 2001, 2005; Maffi 2005). According to the World Wildlife Fund, 4635 ethnolinguistic groups live in the 238 land eco-regions that are most important due to their biological diversity. This figure represents 67% of all ethnolinguistic groups identified on Earth (WWF 2000).

Berkes et al. (2001) consider there is a coevolution between culture and nature. According to Pretty et al. (2009), this interaction is a process that has been taking place for thousands of generations. Therefore, the conservation of biodiversity is impossible without taking into account the sociocultural factors that, as a whole, condition it (Mascia et al. 2003). Many traditional rural societies that coexist with adjacent nature (Fig. 15.7) and depend on it for their survival live together in such way that allows the conservation of biodiversity (MMBT 2004, Luque and Doode 2007, Pretty et al. 2009).

Fig. 15.7 *Nopales* (cactus) and local inhabitants of San Nicolás Tetelco, Tláhuac Delegation, in Mexico City. (Author: Bertram Bilek)

During the last decades, there has been considerably more academic information endorsing Traditional Ecological Knowledge (TEK), as well as local knowledge about the use of natural resources by the inhabitants of rural and indigenous environments. These include conceptualizations of transdisciplinary terms, such as socioecological system (Burger et al. 2001; Berkes et al. 2001; Toledo 2001, 2005; Pretty et al. 2009; Merino and Martínez 2014), socionatural systems, socioecosystems (Robertson and McGee 2003; Folke et al. 2005; Johnson et al. 2016), biocultural heritage and Mesoamerican trend or school of thought of Ethnoecology (Toledo and Barrera-Bassols 2008; Toledo et al. 2018; Alarcón-Cháires and Toledo 2018).

15.2.1.1 Indigenous Peoples, Traditional Ecological Knowledge, and Biocultural Conservation

Indigenous peoples comprise approximately 5% of world population and manage 11% of world forest lands. They usually own, occupy, or use between 22% and 65% of land surface of the Earth (UNDP 2011; RRI 2015). It has been estimated that native lands and other areas protected to safeguard rights on land, indigenous means of substance, biodiversity, and other values contain over 312 trillion tons of carbon (Joly, 2009). Recent analysis reveals that the indigenous territories of the Amazon basin, the Mesoamerican region, the Democratic Republic of the Congo, and Indonesia contain over 20% of all the carbon stored in the surface of the planet, all on their own (AMPB et al. 2015).

Indigenous peoples and local communities depend directly on their immediate environment to meet their basic needs of survival; thus, they are affected by the global environmental change (Reyes-García et al. 2019). Indigenous peoples play a fundamental role in the conservation of biological diversity and protection of forests and other natural resources. Due to their close relationship to their environment, they are keepers of the learning and knowledge about how to successfully deal with weather variability at local level and how to respond effectively to great environmental changes, such as natural disasters (Ramos-Castillo et al. 2017).

A study to assess conservation of the forest in the Bolivian Amazonia has identified a prominent overlapping between TEK (spatially overlaying 624 homes in 59 villages to estimate TEK) and conservation of the forest (using remote sensing data) in the Bolivian Amazonia. Therefore, the use of TEK in development, conservation, and climate-related policies is highly recommended. Besides taking urgent action to protect indigenous cultural systems, it is crucial to create policies that are more effective in these fields (Paneque-Gálvez et al. 2018).

A multi-criteria participatory analysis with native communities from northern Rupununi, Guyana, where local participants analyzed their own situation by defining indicators of meaningful successful strategies for them, defined six good practices that are environmentally sustainable. These strategies are closely bound to the themes of native knowledge, governance and local values, and associations and networks. Said work identified that reinforcing and debating sustainable practices, and displaying them in communities, makes people take pride on what is local (Mistry et al. 2016).

Many studies all around the world demonstrate that indigenous and local peoples play a fundamental role in successful relief efforts. It should even be considered that current global responses to climate change will fail unless they are based on the acknowledgment of the rights of indigenous peoples and local communities, their territory, and their means of subsistence (Ramos-Castillo et al. 2017; Blackman and Veit 2018; Rights and Resources Initiative 2015). Indigenous knowledge can be used in the practice of climate governance (Chanza and De Wit 2016). Case studies in diverse places, such as China, Kenya, and the Bolivian Andes, show that the use and knowledge of the varieties of traditional farming are essential for adapting to an unpredictable weather (Nakashima et al. 2012).

It is recognized that diversity of life includes both biological and cultural-linguistic diversity (Berkes et al. 2001; Boege 2008, 2009; Toledo 2001, 2005; Pretty et al. 2009; Maffi 2005; Maffi and Woodley 2010). It was more common to divide nature and culture some years ago (Tylor 1871; Lévi-Strauss 1987). This partly comes from a dualist conception and a desire to control nature, both of which are typical of western culture (Descola and Pálsson 2001; Hviding 2001; Desmet 2014; Galceran-Huguet 2016). The great variety of academic subdisciplines that have appeared in the last decades confirms this recent way of thinking of biological and cultural diversity as interconnected and interacting (Pretty et al. 2009; Lagunas-Vázques et al. 2017). The conception, interaction, and complementation of interdiscipline, as well as the acknowledgment and validation of other knowledge, are helping with understanding the degree at which biological diversity is bound to cultural (Leff 1994, 2006; Shiva 1995; Dussel 2014; Lander 2000; Grosfoguel 2013a).

Conservation of nature has been conceived within western epistemic discourse (Hviding 2001; MacDonald 2004); thus, biological diversity has been written in anthropocentric (Haraway 1995; Ollantay 2014), Eurocentric terms (Gudynas 2011; Desmet 2014) within capitalist discourse (Leff 2006; Dussel 2014). Some authors currently use the term of western conservation within the theme of conservation (Desmet 2014; MMBT 2016). The western conception of science is presently the only dominant way of creating, consolidating, and legitimizing knowledge (Kuhn 1975; Feyerabend 1986; Quijano 1992, 1995, 1999, 2000a, 2000b, 2007; Castro-Gómez and Grosfoguel 2007; Walsh 2007; De Sousa Santos 2010; Dussel 2014; Galceran-Huguet 2016). This is only one way of seeing and understanding or interpreting the world (De Sousa Santos 2010; Grosfoguel 2013b; Dussel 2014).

From the perspective of acceptance of other knowledge, the ancient philosophical thought of *Buen vivir* can be considered an epistemic, theoretical, and methodological approach that can contribute to conservation of nature in NPA of Latin America. This is especially true concerning cultural and natural diversity, multicultural and intercultural interaction, and acceptance of other knowledge and its integration in activities that favor biocultural conservation.

15.2.2 Indigenous Worldview of Interrelation with Nature

Before most NPAs were declared as such in Mexico, there were communities already inhabiting them and using their natural resources (Bezaury-Creel and Gutiérrez-Carbonell 2009). Alcorn (1994) claims that a big part of the biodiversity of the planet is within indigenous territories, which is also the case in Mexico according to Boege (2008, 2009). Regarding human population in NPA in Mexico, 46% of them are home to indigenous population as of April 2016 (Lagunas-Vázques et al. 2017). Out of the total land surface that comprises NPA in this country, which is 25,628,238 ha (CONANP 2016), over 60% is social property (Bezaury-Creel and Gutiérrez-Carbonell 2009), including more than 30% of indigenous territory (Boege 2009).

Latin America and the Caribbean are estimated to have around 800 cultural groups and a population of 43 million (Delgado 2004). Calculations for the same region consider that natives inhabit at least 80% of NPA (Alcorn 1994). Countries such as Bolivia, Guatemala, and Ecuador are prominent with indigenous populations of 70%, 47%, and 38%, respectively (Delgado 2004). The number of native languages in some countries of South and Central America is noticeable in Brazil (185), Peru (75), and Colombia (47). There are at least 216 indigenous groups in Brazil, of which there were almost a thousand in the sixteenth century (GEO 2002). Most of these indigenous peoples gather around tropical rainforest zones (Toledo 2001).

The nature in the present Americas is, in most of its areas, a social construction, since man has altered it throughout time (Dachary and Burne 2009). Recent studies mention that the population of the Americas before 1492 was much larger than that of Europe during the age of Spanish Conquest (Mann 2006). The jungles and forests of the Americas have lasted because there are peoples that coexist with this reality to which they are so bound. Therefore, not only is nature without these peoples a false idea, but it is also anti-ethical, because those who live in and maintain it are not taken into account (Dachary and Burne 2009).

The Pristine Myth, as coined by geographer William Denevan (Mann 2006), is the false belief that the Americas were practically intact before the arrival of the Spaniards. According to Mann (2006), in spite of the fact that the aforementioned statement is false, it laid down the foundations for the Wilderness Act of 1964 in the United States. This document is foundational for the global environmental movement.

The pre-Columbian Americas included, in their two largest civilizing regions, vast inhabited territories, and their populations were larger than dozens of millions of people (Mann 2006). In Mesoamerica, the Triple Alliance, led by the Mexicas or Aztecs in the Central Mexican Plateau, had a population of 5 (Semo 2019) to 25.2 million people; as of the age of the Conquest of Mexico, it was the most densely populated region in the whole planet (Mann 2006). Mesoamerican cultural peoples invented a dozen different writing systems, established very extensive trading routes, registered the orbits of the planets, created a calendar with 365 days a year (which was much more exact than those in Europe, at that time), and registered their own history in books made of bark. One of their greatest intellectual feats was the invention of zero (Mann 2006). The Olmecs, the Mayas, and other Mesoamerican societies were world pioneers in mathematics and astronomy. The Maya territory, a collection of around 60 cities and kingdoms that formed a complex network of alliances, was home to one of the most sophisticated intellectual cultures in the ancient world of the Western Hemisphere before 1492. According to present information, it was a thriving place of amazing diversity, with outstanding social, political, and economic systems, hundreds of languages, and dozens of millions of inhabitants (Mann 2006; Díaz and Escobar 2006).

As of 1491, the Incan had the vastest empire on Earth. Inca territory spread throughout 32° of latitude. This empire covered all imaginable types of terrain, ranging from rainforest in the Ecuadorian Amazonia to the deserts of the Peruvian coast or the Andean peaks at 6000 meters above sea level. The potential of this empire, in terms of environmental adaptability, shows how the Incas were the most impressive builders of empires of their time (Mann 2006).

15.2.3 Biocultural Perspective to Conservancy and Inhabit Nature from the Approach of Buen vivir

This new biocultural paradigm is propelling a new idea: not to separate the study and conservation of biodiversity from the study and conservation of cultures (Toledo and Barrera-Bassols 2008; Toledo 2013; Gavin et al. 2015). Among other premises, it establishes that there were around 12,000 languages in the world 5000 years ago, but the European expansion to Africa, Asia, and the Americas—for instance, the Conquest of what is now known as Latin America—wiped out almost half of them. Nowadays, there are 18-thousand-year-old cultures in Africa, and 40-thousand-year-old ones in Australia. The Pigmies, considered the oldest living culture in the world, are 60,000 years old (Toledo and Barrera-Bassols 2008).

In actuality, there are diverse cultures in different countries of Latin America (Fig. 15.8). Among them are the thousand-year-old hunter-gatherer Seri from Sonora, who existed without agriculture, based solely on fishing, hunting, and gathering. The Huastecs, the Mayas from Yucatan, and the Waraos from Venezuela are all 3000 years old. In Mesoamerica, the Wixarikas (Huichols) (Fig. 15.9) have made their pilgrimage to the desert of San Luis Potosi for 2500 years (Toledo and Barrera-Bassols 2008). According to some authors (Toledo and Barrera-Bassols 2008), the greatest intercultural questions that would reveal relevant information for sustainability and conservation of nature are: How have these communities been able to survive for so long? What mechanisms do they have for survival? What is their view on the knowledge and appropriation of local resources? How do they see the world?

Some scientists (Toledo and Barrera-Bassols 2008) currently consider these ancient cultures, with a plethora of millenary, long ignored, unappreciated, or

Fig. 15.8 People, Tzotzil; locality, Marcos E, Becerra, Chiapas. (Author: Bertram Bilek)

Fig. 15.9 Art Wixárika woven textile (Nierikas), Nayarit, Mexico. (Author: Bertram Bilek)

misunderstood wisdom, as keepers of the key to overcome the present ecological and social crisis (Toledo and Barrera-Bassols 2008; Alarcón-Cháires and Toledo 2018). The conservation of nature and environmental sustainability have been dealing with the problem of biological richness or biodiversity. Therefore, given the ecological and social crisis in the contemporary world, identifying and acknowledging the biocultural memory of human kind is essential, since it allows having a lasting historical perspective, revealing the epistemological, technical, and economic limits and bias of modern life, and visualizing solutions to current problems at a civilizing scale (Toledo and Barrera-Bassols 2008).

15.2.4 Guiding Methodological Elements with the Philosophical Approach of Buen vivir *for Biocultural Conservation*

In the NPA of Latin America, it is important to consider the significance of biocultural diversity and to establish management and governance from intercultural interactions (Toledo 2015; Zalles 2017) and within the conceptual framework of *Buen vivir*. Below, some recommendations are given as guiding elements to establish ways to correlate the philosophical approach of *Buen vivir* with the indigenous peoples of the NPA in Latin America, in favor of biocultural conservation.

This ancient indigenous perspective from the Americas is considered a theoretical integrating framework with a cosmic, non-anthropocentric approach. As Oviedo (2017a) mentions, there is no way to *Sumak Kawsay*, but *Sumak Kawsay* is the way (just as there is no way to love, but love is the way). The journey of *Sumak Kawsay* is at least 10,000 years old, so starting to speculate with new hypothesis is not the idea (Oviedo 2017a). Syncretism is not valid, nor is minimizing, inclusion or mixture. Hybridization can only happen between elements of the same paradigm. Unifying the different views of *Buen vivir* is not the intention, either. To dogmatize through ideological theories and rhetoric that do not come from living and experiential feeling-thought, or to make individualists views when this is a communal millenary process, would be to undervalue the historical and experiential process of those who shaped it throughout a very long lifetime. Reductionist, civilizing paradigms should not be applied to a multiversal, vitalist archetype either. Likewise, it should not be romanticized by believing this system can become paradise. *Sumak Kawsay* is a philosophical theory with ideals and principles that regulate and constitute it. Nonetheless, this is the challenge or motivation to apply it in its entirety and profoundness (Oviedo 2017a).

It is important that all societies, cultures, and human groups that opt for a radical change (paradigm shift) in human relations and their relationship with nature allow indigenous philosophical roots to guide them on the way of *Sumak Kawsay*. As Oviedo (2017a) mentions, it is not valid that some groups or peoples want to appropriate it and claim it as theirs. *Sumak Kawsay* has nothing to do with peoples, ethnic groups, geographies, or communities that want its exclusivity in any sense or way. Throughout the history of humanity, every culture has known and practiced vitalism in different periods of their existence, and in different regions of Mother Earth, with different names within the same essence of coexisting with nature in a harmonious way (Oviedo 2017a).

(a) *Content Ethics: It Is Experiential, from a Community*

Buen vivir delights (ecstasy) in an immense cosmos, and its being and doing in the world come from there (Estermann 1998). *Buen vivir* reflects a vast expression of experiences, a collective interpretation of "reality," and, at the same time, a hermeneutical interpretation of phenomena (Estermann 1998). *Buen vivir* can be considered an everyday activity (full development of a community), which is also harmonious, anthropocosmic, solidary, bioethical, and biopolitical (Dussel 2013). From these philosophical life proposals, only certain projects, based on respecting the *Pacha*/Earth (Dussel 2013), can be developed within the territories of the NPA: those that are anti-extractivist (respecting the nature), multi-intercultural (respecting among cultures), and non-bourgeois (political system of complete participation and self-management, participative democracy).

(b) *Communal and Cosmic Life*

Life is communal and cosmic in the mythical world, that is, in the symbolic narrative of a determined human group, where nature is Mother Earth or *Pacha* (Dussel 2013). This includes the entirety of cosmos with all due respect (León-Portilla 1956; Estermann 1998; Dussel 2013). Therefore, those who live there respect the ecosystems, take care of water, and protect the forest in real life, in their everyday life. Some peoples native to Mexico show examples of this kind of communal life: Cherán (Lemus-Jiménez 2013; Campos and Partida 2015), other native groups in the Sierra Norte de Puebla (Gledhill 2013), and some towns in Chiapas (Marcos 2010; Mujeres and La Sexta 2017a, b).

(c) *Reconfiguration of Vital Principles in the Natural Protected Areas*

To the common westerner, nature is natural resources; while to the thought of native groups from the Americas, the *Pacha* is nature. It is a living entity, a living being interconnected with everything; between humans, everything live and the cosmos (Estermann 1998; Dussel 2013). It also has a bio-ecological interrelation between human being and the cosmos (Dussel 2013; Rivera 2018), which is symbolic and sacred, but not religious, since it is about complete dignity and unquestionable values—loyalty, integrity, justice, freedom, peace, equality, and so on (Dussel 2013). *Buen vivir* coincides with some conventional principles or basic life values (universal human rights), but it adds others, which complements and reconfigures the entire unique, universalist conception of human rights. Thus, it adds reciprocity, correspondence, harmony, balance, stability, polarity, cyclicity, dynamic stability, and complementarity to equality, freedom, peace, solidarity, social equity, and social justice (Oviedo 2017b).

(d) *Other Biocentric Valuation*

There is a fracture between the civilizing mechanicist paradigm and the archetype of vital consciousness, between the western thought and that of the indigenous people of the Americas, which contrasts two very different life systems (Milla 1983; Estermann 1998; Lajo 2003; Dussel 2013). Indigenous cultures do not live by quantifiable, discriminatory, penalizing parameters, but by pragmatic, sensitive, relative archetypes that respect complementary diversity (Oviedo 2017a). To the Andean world, politics and economy are not core issues, but consciousness (or totality) is (Oviedo 2017a; Rivera 2018). Regarding economy, it is "not anthropocentric, nor market centric, but *biocentric*, that is, with life and the conservation of the live at its core." Since there is no sector or level that is completely motionless (or lifeless) in the Andean worldview, indigenous economy is also cosmocentric (or "*Pacha*-centric") (Oviedo 2017a; Rivera 2018).). Vitalism does not work with democracy either, but in consensual synergy or *biocracy* (Oviedo 2017a).

Consequently, valuations that foster harmonious reconciliation between two conflicting parties should be included. Both parties have to seek consensus through mediation until they reach an agreement that is relatively satisfactory for them (Oviedo 2017a). This is engaging in a dialogue (De Sousa Santos 2010; Dussel 2013). That means no authority from the outside imposes a truth, nor is there a punishing system. The same parties decide under the observance of all the community, which serves as guarantor of the reestablishment of the comprehensive balance of everybody. Hence, in the Andean world, the *Qhapaq Ñan* is not the "way of the just, the righteous, the noble," but the motivational way that takes to complete balance and harmony (Oviedo 2017a).

(e) *The Meaning of Life: Kapak Ñan and Yachak Ñan*

In the Andean philosophy, *Kapak Ñan* and *Yachak Ñan* can help solve problems with agreeing on civilizing matters, such as peace, autonomy, independence, justice, self-determination, freedom, and politics. *Kapak Ñan* and *Yachak Ñan* are ancient teachings for new generations to have the possibility of following the sacred path toward a new consciousness—of the meaning of life (Lajo 2003; Oviedo 2017a). One could even think of it as a kind of high consciousness, such as the one practiced in Buddhism through meditation. It is the ability a human being and a society have to rediscover the secrets and mysteries of the laws and models of life, that is, of nature. It is then possible to move through all its powers, talents, and skills in complete harmony and balance: a conscious way.

This understanding is known in the Andean case as *Kapak Ñan* or *Qhapaq Ñan* (way of spiritual beings), whose opposite complement is *Yachak Ñan* (way of the wise beings) (Oviedo 2017a). They are the consciousness of being lucid to awaken or reactivate. They are complementary to each of the elements and manifestations of life altogether (Oviedo 2017a). They are a learning.

(f) *Conscience of Life: Sacred and Holistic Dimension*

Kawsay ("coexist/get along") is not a trivial or common living, but one that implies a sacred and holistic dimension (Estermann 1998; Oviedo 2017a; Rivera 2018). *Kawsay* is not a mundane and disjointed act, but an integral-integrative sacred coexistence of life. It also has a subtle state of the living energy (vital *elan*) that other cultures around the world observe in philosophical and mystical traditions. Some examples of this essential force of life are *prana* for the Hindus, *aether* for ancient European cultures, *chi* for the Chinese, *ki* in Japanese culture, *ushai* in Ecuadorian Kichwa, *sama* for the Aymaras, and *kawsay* for the Quechuas. For Andean priests, the world is made up of a variety of living energies, which receive the collective name of *kawsay* (Oviedo 2017a).

The word *sumak/suma* is very special in the Andean world, since it comprises several meanings: harmony, balance, whole, wholeness, aware, culture, quality, beautiful, healthy, pure, art, equity, wisdom, totality, sublime, sacred, mutual, complementary, corresponding, comprehensive, holistic, symbiotic, synergic, and homeostatic (Oviedo 2017a). To coexist lovingly (*Kuyay Kawsay*) and to coexist wisely (*Yachay Kawsay*) are necessary to coexist in a complementary way, recreating harmony and balance in every manifestation of life. Therefore, *Sumak Kawsay* is also "to know how to coexist and support each other." This view is similar to the definition ancient Greek philosophers had of the word philosophy: wisdom of love, or love for wisdom (philos: love; sophos: wisdom) (Oviedo 2017a).

Sumak Kawsay is the knowledge of the thought-feeling of life, or more precisely, it is conscious life, conscious coexistence, life philosophy (to know how to love life), or the art of living in complement to one another (Oviedo 2017a; Rivera 2018). Hence, this concept should be only one word, *Sumakawsay*, so as to also express that nothing is separate and because words are agglutinative in Kichwa, especially when expressing models of life. This is the case of *Sumagamaña* in Aymara (Oviedo 2017a).

15.3 Conclusion

The ancient history of (pre-Columbian) Latin America, concerning nature and natural resources that exist in the region nowadays, guides current contexts of socioculturally diverse human groups that inhabit it now. This applies to the knowledge of nature they have, as well as the traditional uses and interactions they have with their surrounding nature. Consequently, it is in dispensable to consider the significance of biocultural diversity, the integration of intercultural correlations, and the approach of *Buen vivir* in any human activity that relates with nature. It is an urgent necessity for humankind to opt for biocultural memory (Toledo and Barrera-Bassols 2008) or ancient memory (Oviedo 2017a). Fortunately, it is there, in our collective unconscious, which we must recover, as indicated by Oviedo (2017a), after having strayed from the natural and conscious pathway.

Acknowledgments To the *Centro del Cambio Global y la Sustentabilidad* (CCGS) for the facilities they provided to conduct this work. To the program Cátedras CONACyT and the project Cátedras CCGS no. 945 "Socio-environmental vulnerabilities and measures for the adaptation to climate change in Southeast Mexico." The author would also like to thank Emanuel Alejandro Cruz for his translation service and Dr. Alfredo Ortega-Rubio for the invitation to participate in his editorial project.

References

Alarcón-Cháires P, Toledo V (2018). Tópicos bioculturales: reflexiones sobre el concepto de bioculturalidad y la defensa del patrimonio biocultural de México. Universidad Nacional Autónoma de México, 120 pp

Alcorn JB (1994) Noble savage or noble state? Northern myths and southern realities in biodiversity conservation. Etnoecológica 3:7–19

AMPB (Mesoamerican Alliance of Indigenous Peoples and Forests), COICA, AMAN, REPALEAC, Woods Hole, EDF (2015) Tropical forest carbon in indigenous territories: a global analysis. https://www.edf.org/sites/default/files/tropical-forest-carbon-in-indigenous-territories-a-global-analysis.pdf

Berkes F, Colding J, Folke C (2001) Rediscovery of traditional ecological knowledge as adaptive management. Ecol Appl 10:1251–1262

Bezaury-Creel JD, Gutiérrez-Carbonell (2009) Áreas naturales protegidas y desarrollo social en México, Capital natural de México, vol. II: Estado de conservación y tendencias de cambio. Conabio, México. 819 pp

Blackman A, Veit P (2018) Amazon indigenous communities cut forest carbon emissions. Ecol Econ 153:56–67

Boege E (2008) El patrimonio biocultural de los pueblos indígenas de México. INAH, CDI, México. http://www.cdi.gob.mx/biodiversidad/biodiversidad_0_preliminares_1-31_eckart_boege.pdf

Boege E (2009) El reto de la conservación de la biodiversidad en los territorios de los pueblos indígenas, Capital natural de México, vol. II: Estado de conservación y tendencias de cambio. Conabio, México. 819 pp.

Burger J, Ostrom E, Norgaard RB, Policansky D, Goldstein B (2001) Protecting the commons: a framework for resource management in the Americas. Island Press, Washington, DC, 328 pp

Calle BE, Maurice ME, Miranda RV (2016) Del meta-paisaje en el paisaje sagrado y la conservación de los lugares naturales sagrados. Sociedad y Ambiente 1(9):1–25

Campos A, Partida G (2015) Cumple Cherán 4 años de autogobierno, sin partidos. Periódico La Jornada. http://www.jornada.unam.mx/2015/04/20/estados/029n1est

Castro-Gómez S, Grosfoguel R (2007) El giro decolonial: reflexiones para una diversidad epistémica más allá del capitalismo global. Siglo del Hombre editores

Chanza N, De Wit A (2016) Enhancing climate governance through indigenous knowledge: case in sustainability science. S Afr J Sci 112(3–4):1–7

Clavijero FJ (1945) Historia antigua de México. Colección de escritores mexicanos. Volumen 4. Editorial Porrua Mexico

CONANP (Comisión Nacional de Áreas Naturales Protegidas) (2016) Áreas Naturales Decretadas. http://www.conanp.gob.mx/que_hacemos/

Cuestas-Caza J (2017) Sumak kawsay: el buen vivir antes de ser buen vivir. In: Congreso El Extractivismo en América Latina: Dimensiones Económicas, Sociales, Políticas y Culturales. Universidad de Sevilla, Seville, pp 356–368

Cuestas-Caza J (2019) Sumak Kawsay entre el postdesarrollismo occidental y la Filosofía Andina. Ánfora: Revista Científica de la Universidad Autónoma de Manizales 26(47):109–140

Dachary AC, Burne SMA (2009) Pueblos originarios y turismo en América Latina: La conquista continúa. Estudios y perspectivas en turismo 18(1):69–91
De Sousa Santos B (2010) Descolonizar el saber, reinventar el poder. Ediciones Trilce. Uruguay. http://www.boaventuradesousasantos.pt/media/Descolonizar%20el%20saber_final%20-%20C%C3%B3pia.pdf
Delgado GC (2004) Biodiversidad, desarrollo sustentable y militarización: esquemas de saqueo en Mesoamérica. UNAM. México. 196 pp
Descola P, Pálsson G (2001) Naturaleza y sociedad: perspectivas antropológicas. Siglo XXI. México
Desmet E (2014) Conservación y pueblos indígenas: un análisis socio-jurídico, vol 75. Universidad de Deusto, Bizkaia
Díaz N, Escobar S (2006) Articulación de actividades didácticas con algunos aspectos históricos de la cultura y matemática maya en el desarrollo del pensamiento espacial y sistemas geométricos del grado séptimo (Doctoral dissertation. Universidad de Nariño). Colombia
Dussel E (2013) Primer Encuentro del Buen Vivir https://www.youtube.com/watch?v=ieRwuIurppo
Dussel E (2014) 16 Tesis de economía política. Interpretación filosófica. Siglo XXI Editores México, 424 pp
Eliade M (1991) Mito y realidad Editorial Labor España, 100 pp
Estermann J (1998) Filosofía Andina estudio intercultural de la sabiduría autóctona andina La Paz, Bolivia, 433 pp
Fernández J (1959) Coatlicue Estética del arte indígena antiguo Editorial: Centro de Estudios Filosóficos Instituto de Investigaciones Estéticas Edición 1ª México, 285 pp
Fernández J (1972) Estética del arte mexicano: Coatlicue. El retablo de los reyes. El hombre Universidad Nacional Autónoma de México, Instituto de Investigaciones Estéticas, 599 pp
Feyerabend PK (1986) Tratado contra el método. Editorial Tecnos S. A. Madrid España, 337 pp
Folke C, Hahn T, Olsson P, Norberg J (2005) Adaptive governance of social-ecological systems. Annu Rev Environ Resour 30:441–473
Fornet-Betancourt R (2009) Interculturalidad en procesos de subjetivización Consorcio Intercultural Coordinación General de Educación Intercultural y Bilingüe México, 77 pp
Galceran-Huguet M (2016) La bárbara Europa: Una mirad desde el postcolonialismo y la decolonialidad. Traficantes de Sueños España, 380 pp
Garnett ST, Burgess ND, Fa JE, Fernández-Llamazares Á, Molnár Z, Robinson CJ, Collier NF (2018) A spatial overview of the global importance of indigenous lands for conservation. Nature Sustain 1(7):369
Gavin MC, McCarter J, Mead A, Berkes F, Stepp JR, Peterson D, Tang R (2015) Defining biocultural approaches to conservation. Trends Ecol Evol 30(3):140–145
Geo (2002) Gentes de la Amazonia Documento 180 España
Gledhill J (2013) Límites de la autonomía y autodefensa indígena: experiencias mexicanas. OXÍMORA Revista Internacional de Ética y Política 2:1–21
González F (2003) El simbolismo precolombino: cosmovisión de las culturas arcaicas. Editorial Kier Argentina, 320 pp
Grosfoguel R (2013a) Transmodernidad y Pluriversalismo https://www.youtube.com/watch?v=pJfUhI7qCw8
Grosfoguel R (2013b) Racismo/sexismo epistémico universidades occidentalizadas y los cuatro genocidios/epistemicidios del largo siglo XVI. Tabula Rasa 19:31–58
Grosfoguel R (2019) Curso Conocimiento transdisciplinar en América Latina https://youtu.be/BsvjOJ6_PI0
Gudynas E (2011) Desarrollo, Derechos de la Naturaleza y Buen Vivir después de Montecristi. En: Debates sobre cooperación y modelos de desarrollo. Perspectivas desde la sociedad civil en el Ecuador. Gabriela Weber, editora. Centro de Investigaciones CIUDAD y Observatorio de la Cooperación al Desarrollo Quito Ecuador, pp 83–102
Haraway DJ (1995) Ciencia, cyborgs y mujeres: la reinvención de la naturaleza Ediciones Catedra Madrid España, 442 pp

Hviding E (2001) Naturaleza, cultura, magia, ciencia. Sobre los metalenguajes de comparación en la ecología cultural. En Naturaleza y sociedad. Perspectivas antropológicas. Philippe Descola y Gísli Pálsson (Eds.). México: Siglo XXI Editores S. A., 192–213 pp

Jiménez R (2011) Rescatar y valorar otros pilares éticos. El Buen Vivir http://www2.world-governance.org/article689.html?lang=en

Johnson JT, Howitt R, Cajete G, Berkes F, Louis RP, Kliskey A (2016) Weaving indigenous and sustainability sciences to diversify our methods. Sustain Sci 11(1):1–11

Joly CA (2009) Editorial: synergies between the Convention on Biological Diversity (CBD) and the United Nations Framework Convention on Climate Change (UNFCCC). Biota Neotropica, 9(1). https://doi.org/10.1590/S1676-06032009000100001

Kuhn T (1975) La estructura de las revoluciones científicas. Fondo de cultura Económica México

Kusch R (1962) América profunda. Editorial Biblos, 180 pp

Lagunas-Vázques M, Bobadilla-Jiménez M, Beltrán-Morales LF, Ortega-Rubio A (2017) Bases Antropológicas y Sociológicas para la Conservación en Áreas Naturales Protegidas Latinoamericanas con un enfoque Pluricultural e Intercultural. Capítulo III. pp. 51–76 En: Espitia-Moreno I.C., ArriolaPadilla V.J. y Ortega-Rubio A (Editores) Gestión Manejo y Conservación en Áreas Naturales Protegidas. Universidad Michoacana de San Nicolás de Hidalgo Morelia Michoacán México, 178 pp

Lajo J (2003) Qhapaq ñan: La ruta Inka de sabiduría Editorial Amaro Runa Lima, 196 pp

Lander E (2000) La Colonialidad del saber Eurocentrismo y ciencias sociales CLACSO Argentina, 249 pp

Leff E (1994) Ecología y capital racionalidad ambiental democracia participativa y desarrollo sustentable México Editorial Siglo XXI, 437 pp

Leff E (2006) Complejidad racionalidad ambiental y diálogo de saberes Ponencia presentada en el I Congreso internacional interdisciplinar de participación, animación e intervención socioeducativa, celebrado en Barcelona en noviembre de 2005. https://www.miteco.gob.es/es/ceneam/articulos-de-opinion/2006_01eleff_tcm30-163650.pdf

Lemus-Jiménez A (2013) La transmisión de la historia a través de la oralidad por mujeres p'urhepecha de Cherán. En: Senti-pensar el género. Perspectivas desde los pueblos originarios. Georgina Méndez Torres, Juan López Intzín, Sylvia Marcos y Carmen Osorio Hernández (Editores). Guadalajara, Jalisco, México Red Interdisciplinaria de Investigación de los Pueblos Indios de México AC/Taller edit. La Casa del Mago/Red de Feminismos Descolonia. 2013. Estudios Latinoamericanos, 185–208 pp

León-Portilla M (1956) La filosofía náhuatl. Instituto Indigenista Interamericano México, 344 pp

Lévi-Strauss C (1987) Antropología estructural: mito, sociedad, humanidades Siglo XXI México, 352 pp

Luque D, Doode S (2007) Sacralidad, territorialidad y biodiversidad comcáac (seri) Los sitios sagrados indígenas como categorías de conservación ambiental. Relaciones Estudios de historia y sociedad 28(112):157–184

MacDonald KI (2004) Conservation as cultural and political practice. Policy Matters 13:6–17

Maffi L (2005) Linguistic, cultural, and biological diversity. Annu Rev Anthropol 34:599–617

Maffi L, Woodley E (2010) Biocultural diversity conservation: a global sourcebook. Earthscan, London, 304 pp

Mann C (2006) 1491 Una nueva historia de las Américas antes de Colón Santillana Ediciones Generales, S. L. (Taurus) España, 632 pp

Marcos S (2010) Cruzando Fronteras. Mujeres indígenas y feminismos, abajo y a la izquierda. Universidad de la Tierra. Chiapas México, 326 pp

Mascia MB, Brosius JP, Dobson TA, Forbes BC, Horowitz L, McKean MA, Turner NJ (2003) Conservation and the social sciences. Conserv Biol 17(3):649–650

Merino L, Martínez AE (2014) A vuelo de pájaro: las condiciones de las comunidades con bosques templados en México. Comisión Nacional para el Conocimiento y Uso de la Biodiversidad (CONABIO) México, 196 pp

Milla VC (1983) Génesis de la cultura andina. Fondo Editorial del Colegio de Arquitectos del Perú. Colección Bienal Lima, Perú, 266 pp

Mistry J, Berardi A, Tschirhart C, Bignante E, Haynes L, Benjamin R, Jafferally D (2016) Community owned solutions: identifying local best practices for social-ecological sustainability. Ecol Soc 21(2):42

Movimiento Mundial por los Bosques Tropicales (MMBT) (2004) Áreas Protegidas: ¿Protegidas contra quién? Uruguay. MMBT. 432 pp

Movimiento Mundial por los Bosques Tropicales MMBT (2016) Las raíces coloniales racistas de la conservación occidental de los bosques: una mirada a un proyecto REDD en Kenia. http://wrm.org.uy/es/articulos-del-boletin-wrm/seccion1/las-raices-coloniales-racistas-de-la-conservacion-occidental-de-los-bosques-unamirada-a-un-proyecto-redd-en-kenia/

Mujeres y La Sexta (2017a) Encuentro de las Mujeres Zapatista con las mujeres del mundo (diciembre 2007). https://mujeresylasextaorg.wordpress.com/audiosfotos-videos-y-textos-del-encuentro-de-mujeres/

Mujeres y La Sexta (2017b) La Revolución de las Mujeres Zapatistas. https://mujeresylasextaorg.wordpress.com/ley-revolucionaria-de-mujeres-zapatistas/

Nakashima D, Galloway, McLean K, Thulstrup H, Ramos A, Rubis J (2012) Weathering uncertainty: traditional knowledge for climate change assessment and adaptation. Paris UNESCO and Darwin UNU, 120 pp

Naranjo AMA (2018) Sitios sagrados: protección y defensa del territorio ancestral Wiwa de la Sierra Nevada de Santa Marta. Facultad de Ciencias Sociales y Humanas Universidad Externado de Colombia, 128 pp

Ollantay I (2014) El Buen Vivir no es desarrollo, ni el desarrollo es sostenible. Rebelión. http://www.rebelion.org/noticia.php?id=191761

Oviedo FA (2017a) Sumak Kawsay Arte de Vivir en Armonía De la Revolución Verde a una Revolución Arcoíris 5ta edición Global Sur Editores Quito-Ecuador, 182 pp

Oviedo FA (2017b) Qué es el buen vivir (Sumak Kawsay). Rebelión. https://www.rebelion.org/noticia.php?id=231005

Paneque-Gálvez J, Pérez-Llorente I, Luz AC, Guèze M, Mas JF, Macía MJ, Reyes-García V (2018) High overlap between traditional ecological knowledge and forest conservation found in the Bolivian Amazon. Ambio 47(8):908–923

Pretty J, Adams B, Berkes F, De Athayde S, Dudley N, Hunn E, Sterling E (2009) The intersections of biological diversity and cultural diversity: towards integration. Conserv Soc 7(2):100–112

Quijano A (1992) Colonialidad y modernidad/racionalidad. Perú indígena 12(29):11–20

Quijano A (1995) Raza, etnia y nación en Mariátegui: cuestiones abiertas. Estudios latinoamericanos 2(3):3–19

Quijano A (1999) Colonialidad del poder, cultura y conocimiento en América Latina. Dispositio 24(51):137–148

Quijano A (2000a) Colonialidad del poder, eurocentrismo y América Latina. En Edgardo Lander (compilador) La colonialidad del saber: eurocentrismo y ciencias sociales. Perspectivas latinoamericanas, Buenos Aires: CLACSO, 201–246 pp

Quijano A (2000b) El fantasma del desarrollo en América Latina. Revista venezolana de economía y ciencias sociales 6(2):73–90

Quijano A (2007) Colonialidad del poder y clasificación social. En S. Castro-Gómez and R. Grosfoguel, El giro decolonial. Reflexiones para una diversidad epistémica más allá del capitalismo global. Bogotá, Pontificia Universidad Javeriana/Instituto Pensar/Universidad Central-IIESCO/Siglo del Hombre Editores: Pontificia Universidad Javeriana, Colombia

Ramos-Castillo A, Castellanos EJ, McLean KG (2017) Indigenous peoples, local communities and climate change mitigation. Clim Chang 140(1):1–4

Reyes-García V, Fernández-Llamazares Á, McElwee P, Molnár Z, Öllerer K, Wilson SJ, Brondizio ES (2019) The contributions of indigenous peoples and local communities to ecological restoration. Restor Ecol 27(1):3–8

Rivera CS (2018) Un mundo ch'ixi es posible: ensayos desde un presente en crisis. Buenos Aires Tinta Limón, 175 pp

Robertson HA, McGee TK (2003) Applying local knowledge: the contribution of oral history to wetland rehabilitation at Kanyapella Basin, Australia. J Environ Manag 69(3):275–287

Rights and Resources Initiative RRI (2015) Toward a global baseline of carbon storage in collective lands: an updated analysis of indigenous peoples' and local communities' contributions to climate change mitigation. Washington, DC

Semo, Enrique (2019). La Conquista. Catastrofe De Los Pueblos Originarios/Tomo 1 Siglo XXI Editores. México. 316 pp

Shiva V (1995) Abrazar la vida. Mujer, ecología y desarrollo. Editorial Horas y Horas Madrid, España, 320 pp

Tavares M (2013) A Filosofia Andina: Uma interpelaçãoao pensamento ocidental. Colonialismo, colonialidade e descolonização para uma interdiversidade de saberes (J. Estermann). Eccos Revista Científica 32:197–252

Toledo VM (2001) Indigenous peoples and biodiversity. Encycl Biodiver 3:451–463

Toledo VM (2005) La ecología rural. Ciencia y Desarrollo 174:36–43

Toledo VM (2013) El paradigma biocultural: crisis ecológica, modernidad y culturas tradicionales. Sociedad y Ambiente 1(1):50–60

Toledo VM (2015) ¿De qué hablamos cuando hablamos de sustentabilidad? Una propuesta ecológico política. Revista Interdisciplina 3(7):35–55

Toledo VM, Barrera-Bassols N (2008) La memoria biocultural: la importancia ecológica de las sabidurías tradicionales (vol 3) Icaria editorial

Toledo VM, Alarcón-Cháires P, Barrera-Bassols N (2018) ETNOECOLOGÍA MESOAMERICANA. Antología de publicaciones 1980–2018 México. UNAM

Tylor EB (1871) Primitive culture: researches into the development of mythology, philosophy, religion, art, and custom, vol 2. J. Murray, Londres

UNDP (United Nations Development Programme) (2011) Human development report 2011: sustainability and equity –a better future for all. Palgrave Macmillan, New York

Walsh C (2007) ¿Son posibles unas ciencias sociales/culturales otras? Reflexiones en torno a las epistemologías decoloniales. Nómadas. Universidad Central de Colombia 26:102–113

WWF World Wildlife Fund (2000) Indigenous and traditional peoples of the World and Ecoregion Conservation. Suiza, 120 pp

Zalles JI (2017) Conocimiento ecológico local y conservación biológica: la ciencia postnormal como campo de interculturalidad. Iconos. Revista de Ciencias Sociales 59:205–224

Part III
Social Groups and Ecological Knowledge

Chapter 16
Sociocultural and Environmental Interactions Between People and Wild Edible Plants: The Case of *Sierra la Laguna* Biosphere Reserve

Juan Fernando Pío-León and Alfredo Ortega-Rubio

Abstract The Biosphere Reserve *Sierra de la Laguna* is a continental island with a great level of endemism, and its population is characterized by a special culture not found in other places of Mexico and named as South-Californian ranchers. This research shows the wild edible plants consumed by the inhabitants of the reserve. Although there are no Indian tribes in the reserve, the number of wild edible plants reported in this research is similar or higher than those of other Indian tribes in neighboring states. A distinctive characteristic of the ranchers in the reserve is the use of some plants to prepare teas and the consumption of oak acorns. This chapter also compares the implications of life in and out of the reserve.

Keywords Ethnobotany · Edible plants · Rancher culture

16.1 Biological and Social Contexts of *Sierra la Laguna* Biosphere Reserve

The Biosphere Reserve *Sierra de la Laguna* (hereafter, called *Sierra la Laguna*) is located on the south extreme of the Baja California Peninsula in the state of "Baja California Sur" of Mexico; it was decreed as Biosphere Reserve in 1994 with an area of 112,437.02 ha (Conanp 2003). *Sierra la Laguna* comprises the only tropical ecosystem in the peninsula and includes the driest tropical deciduous forest of Mexico and the only pine-oak forest of the state. The reserve is found within the

J. F. Pío-León (✉)
Universidad Politécnica del Mar y la Sierra, La Cruz de Elota, Mexico
e-mail: jpio@upmys.edu.mx

A. Ortega-Rubio
Red Nacional Áreas Naturales Protegidas RENANP-CONACYT, Centro de Investigaciones Biológicas del Noroeste CIBNOR-CONACYT, La Paz, Baja California Sur, Mexico
e-mail: aortega@cibnor.mx

A. Ortega-Rubio (ed.), *Socio-ecological Studies in Natural Protected Areas*,
https://doi.org/10.1007/978-3-030-47264-1_16

bioregion *forests of the Cape region* (González-Abraham et al. 2010), which is considered as a continental island because it comprises a tropical ecosystem surrounded by xerophytic scrubs, the main ecosystem in the peninsula, favoring the development of a high degree of endemism of flora and fauna. Considering the flora, the higher-plants biodiversity includes around 1000 species, and about 100 are endemic (León-de la Luz et al. 2012). On the other hand, the main reason to decree *Sierra la Laguna* as a reserve was because this is the main area of collection of the water used for human consumption in the main cities of the state (*La Paz* and *Los Cabos*): its mountains and creeks are the only places in the region where superficial water streams are found during most of the year (Conanp 2003).

The inhabitants of Sierra la Laguna are themselves identified as *South-Californian ranchers*. The culture of these ranchers was originated in the Jesuits' missions who had been established in the peninsula since 1697; they introduced animal breeding and small-scale agriculture. However, the soldiers of Spain colony and ranchers of neighboring states got the legal land tenure after the Jesuits' expulsion in 1768, forming the first formal ranches (Reygadas-Dahl and Landa-Romo 2013). Prior to the Jesuits' era, Pericú's Indians were the landowners, nomads, and mainly dedicated to hunter and recollection; this tribe was practically exterminated by the conflicts with the colonizers and their diseases. Then, the lifestyle and new culture of the South-Californian ranchers resulted from the mixture of cultures and harsh conditions in the peninsula (Reygadas-Dahl and Landa-Romo 2013).

Nowadays, the population in the reserve is about 500 inhabitants, who are distributed in about 100 little ranches, composed of one to three houses and mainly located on the low lands of the sierra at 300–700 m above sea level and near to water bodies as creeks/ravines or waterholes. Their main economic activities are extensive animal breeding (sheep, bovine, or caprine) and day laboring, which are complemented with the production of vegetables and fruits in family gardens, the construction of *palapas*, leather workshop, and as nature tourism guides (Conanp 2003, Reygadas-Dahl and Landa-Romo 2013). The population is mostly adults or seniors and decreasing because the youngers left their communities to look for jobs and better living conditions in urban centers such as *La Paz* and *Los Cabos*.

16.2 The Importance of Wild Edible Plants in the World

The edible wild plants are valuable sources of nutrients and micronutrients as well as of phytogenetic resources for the populations of rural communities around the world. In the world, there are about 30,000–50,000 edible plants: nowadays, only near to 3000 of them are being used, but only about 100 species provide 90% of the food for the worldwide population (Kermath et al. 2014).

The ethnobotanical studies are the main tools to identify and to promote the sustainable use of plants (Joshi et al. 2015; Madeiros et al. 2011). In this regard, natural protected areas and mainly the biosphere reserves are very important for the protection of the phytogenetic resources and the traditional knowledge that must be used to choose and to establish new food crops. However, these resources and knowledge are at risk by the application of conservationist policies instead of the sustainable use of plants (Hill et al. 2015; Urgenson et al. 2014).

This research analyzes the diversity of the edible plants native to Sierra la Laguna and the socio-environmental interactions of its inhabitants; moreover, the results were compared to those obtained in a similar study carried out in another community found out and in the limit of the reserve, where the populations of both sites have the same origin and cultural identity.

16.3 Methodology

The presented data were obtained from a large ethnobotanical study carried out since 2013 in the Baja California Sur State of Mexico, including the Biosphere Reserve *Sierra la Laguna*, and part of them have been previously published (Pío-León et al. 2017a, b). The employed strategies included ethnobotanical surveys, multiple stays in the homes of the inhabitants of the reserve, and field trips to monitor and to collect the studied species. This research presents for the first time the full and updated list of plants with any type of edible use, including fruits, vegetables, seasonings, starchy tubers, seeds, exudates, teas, or coffee substitutes. The list was ordered based on three levels depending on the nutritional and cultural value: the most important plants were included in level I whereas the least important in level III. The order was established by employing diverse ethnobotanical indexes during the surveys and the acquired experience in the stays with the inhabitants of the ranches. Among the factors included in the study are the number of persons who consume the plant, their preferences, processing capacities, and the efforts for the plant conservation or domestication.

16.4 Results and Discussion

16.4.1 Wild Edible Plants in the Biosphere Reserve Sierra la Laguna

In *Sierra la Laguna*, there were about 100 plant species reported with at least one edible organ or use for teas (Table 16.1); this value was higher than that reported for the Seri's Indians (Felger and Moser 1976) and similar to that reported for the Mayos' Indians (Yetman and Van Devender 2002), both in the state of Sonora, Mexico, neighbor of Baja California Sur. Thus, it is clear the mestizo culture of the *South-Californian ranchers* of the Sierra la Laguna includes traditional knowledge about the use of wild plants, which is similar to that reported for Indian communities of Mexico.

Table 16.1 Wild edible plants in the Sierra la Laguna Biosphere

Scientific name	Common name	Edible part	Consumption way
Level I (most important group)			
Ficus petiolaris Kunth	Zalate (wild fig)	Fruit	Raw fresh, dry or conserves
Stenocereus thurberi (Engelm.) Buxbaum	Pitaya dulce	Fruit	Raw fresh
Cyrtocarpa edulis (Brandegee) Standl.	Ciruela de monte (wild plum)	Fruit	Raw, dry, prickles
Cnidoscolus maculatus (Brandegee) Pax & K. Hoffm.	Caribe	Seed	Raw, roasted, as a chocolate/coffee substitute
Capsicum annuum var. *glabriusculum* (Dunal) Heiser & Pickersgill	chiltepín/chilpitín (wild pepper)	Fruit	Raw fresh, dry, prickles
Matelea cordifolia (A. Gray) Woodson	Talayote	Fruit	Raw fresh or cooked as greens
Turnera diffusa Willd.	Damiana	Leaves	Tea
Level II (intermediate importance)			
Stenocereus gummosus (Engelm.) A. Gibson & K.E. Horak	Pitaya agria	Fruit	Raw
Quercus brandegeei Goldman/*Q. tuberculata* Liebm.	Bellota de encino	Seed	Roasted or as coffee substitutes
Pachycereus pecten-aboriginum (Engelm. Ex S. Watson) Britton & Rose	Cardón barbón	Fruit and seeds	Raw
Pachycereus pringlei (S. Watson) Britton & Rose	Cardón pelón	Fruit and seeds	Raw
Yucca capensis L.W. Lenz	Datilillo	Flower	Cooked as greens
Pinus lagunae (Rob.-Pass.) Passini	Pino piñonero	Seed	Raw or roasted
Pithecellobium dulce (Roxb.) Benth.	Guamúchil	Fruit	Raw or cooked as greens
Vitis peninsularis M.E. Jones	Uva de monte (wild grape)	Fruit	Raw, wine
Ferocactus townsedianus Britton & Rose	Biznaga	Fruit, flowers	Cooked as greens
Diospyros californica (Brandegee) I.M. Johnst.	Guayparín	Fruit	Raw
Randia capitata D.C.	Papache	Fruit	Raw
Quercus devia Goldman/*Q. albocincta* Trel.	Bellota de encino	Seed	Coffee substitutes
Ipomoea bracteata Cav.	Jícama	Tuber	Raw (juicy)
Tagetes filifolia Lag./*T. micrantha* Cav.	Anís	Aerial parts	Tea
Lantana velutina M. Martens & Galeotti	Confiturilla	Leaves	Tea

(continued)

Table 16.1 (continued)

Scientific name	Common name	Edible part	Consumption way
Prunus serotina subsp. *virens* (Wooton & Standl.) McVaugh	Cerezo	Bark and leaves	Tea
Porophyllum gracile Benth.	Hierba del venado	Aerial parts	Tea
Pellaea ternifolia subsp. *brandegeei* (C.C. Hall) Mickel	Helecho peyote	Bulbs	Tea
Aloysia barbata (Brandegee) Moldenke	Margarita, santimia	Leaves	Tea
Pectis multiseta var. *ambigua* (Fernald) D.J. Keil	Parra leña	Aerial parts	Tea
Lippia palmeri S. Watson	Orégano	Leaves	Seasoning food
Brahea brandegeei (Purpus) H.E. Moore	Palma de taco	Fruit	Raw
Portulaca oleracea L.	Verdolaga	Aerial parts	Raw or cooked as greens
Celtis reticulata Torr.	Vainoro	Fruit	Raw
Malpighia diversifolia Brandegee	Manzanita	Fruit	Raw
Sideroxylon peninsulare (Brandegee) T.D. Penn.	Bebelama	Fruit	Raw
Amaranthus spp.	Quelite	Leaves	Cooked as greens
Nasturtium officinale L.	Berro	Aerial parts	Raw
Amoreuxia palmatifida DC./*A. gonzalezii* Sprague & L. Riley	Saya	Tuber	Raw or cooked
Opuntia lagunae Baxter ex Bravo/*O. bravoana* E. Baxter.	Tunas/nopales	Fruits and leaves	Raw or cooked
Solanum americanum Mill.	Chichiquelite	Leaves	Cooked as greens
Solanum stoloniferum Schltdl. & Bouché	Papa silvestre (wild potato)	Tuber	Cooked
Matelea pringlei (A. Gray) Woodson	Talayote chino	Fruit	Raw or cooked as greens
Agave spp.	Magueyes	Bloom and floral stems	Bloom cooked as greens, floral stem raw
Psidium guajava L.	Guayaba	Fruit	Raw
Level III			
Passiflora foetida L./ *P. pentaschista* (Killip) H.T. Svoboda	Granadilla	Fruit	Raw
Psidium sartorianum (O. Berg) Nied.	Arrayán	Fruit	Raw
Myrtillocactus cochal (Orc.) Britton & Rose	Frutilla	Fruit	Raw
Oxalis spp.	Agritos	Leaves	Raw
Mammillaria spp.	Viejitos	Fruit	Raw
Cylindropuntia cholla (F.A.C. Weber) F.M. Knuth	Cholla	Fruit	Raw

(continued)

Table 16.1 (continued)

Scientific name	Common name	Edible part	Consumption way
Lophocereus schottii (Engelm.) Britton & Rose	Garambullo	Fruit	Raw
Vallesia glabra (Cav.) Link	Otatave	Fruit	Raw
Physalis spp.	Tomatillos	Fruit	Cooked
Chenopodium spp.	Chuale	Leaves	Cooked like greens
Parkinsonia florida subsp. *peninsulare* (Rose) Hawkins & Felger	Palo verde	Aril	Raw
Dysphania ambrosioides (L.) Mosyakin & Clemants	Epazote	Leaves	Seasoning food
Ebenopsis confinis (Stan) Barnaby & Grimes	Uña de gato/palo fierro	Seed	Coffee substitutes
Peniocereus striatus (Brandegee) Buxbaum	Raja matraca	Fruit	Raw
Tecoma stans (L.) Juss.	Palo de arco	Flor	Raw or cooked
Arbutus peninsularis Rose & Goldman	Madroño	Fruit	Raw
Simmondsia chinensis (Link) C.K. Schneid.	Jojoba	Fruit	Roasted
Hydrocotyle umbellata L./*H. verticillata* Thunb.	Patitos	Leaves	Raw
Selaginella lepidophylla (Hook. & Grev.) Spring	Siempre viva	Tender leaves	Raw
Phaseolus spp.	Frijol	Sprouts	Cooked
Parkinsonia praecox (Ruiz & Pav. ex Hook.) Hawkins	Palo brea	Bark exudate	Raw
Rubus scolocaulon Brandegee	Zarzamora	Fruit	Raw
Bourreria sonorae S. Watson	Negrito, tejocote negro	Fruit	Raw
Lysiloma divaricatum (Jacq.) J.F. Macbr.	Mauto	Bark exudate	Raw
Cyperus spp.	Tulillos	Bulbs	Raw
Typha domingensis Pers.	Tule	Bulbs	Raw
Arracacia brandegeei J.M. Coult. & Rose	Anisón, chucupate	Leaves	Tea
Lepechinia hastata (A. Gray) Epling	Chicura de la sierra	Leaves	Tea
Vachellia farnesiana (L.) Wight & Arn.	Vinorama	Bark exudate	Raw
Fragaria mexicana Schltdl.	Fresa silvestre	Fruit	Raw
Crotalaria incana L./*C. pumila* Ortega	Garbancillo	Tender seeds	Raw
Erythrina flabelliformis Kearney	Colorín, chilicote	Flower	Cooked
Karwinskia humboldtiana (Schult.) Zucc.	Negrito	Pulp fruit; seeds are toxic	Raw
Albizia occidentalis Brandegee	Palo escopeta	Tender seeds	Raw

The most consumed edible plants by the ranchers were the Cactaceae (ten taxa) and particularly the *pitayas* fruits (*Stenocereus* spp.) that were the most important foods for the inhabitants of Sierra la Laguna, as happened with the Seri and Mayo Indians (Yetman and Van Devender 2002). Level I (most important edible plants) is composed of the juicy fruit of wild plum (*Cyrtocarpa edulis*), the seed of caribe (*Cnidoscolus maculatus*) (Fig. 16.1), the spicy fruits of wild pepper (*Capsicum annuum* var. *glabriusculum*), and the zucchini-like fruits of talayote (*Matelea cordifolia*) (Fig. 16.2). It is remarkable that this group includes at least one of each type

Fig. 16.1 Seed of caribe (*Cnidoscolus maculatus*). (Author: JF Pío-León)

Fig. 16.2 Fruits of talayote (*Matelea cordifolia*). (Author: JF Pío-León)

of food plant, a fruit, a vegetable, a seasoning, and an almond-like seed, which could be associated with empirical selection by local people to meet their basic nutritional needs. A clear difference in the *Rancheros* community was the employment of several plants for teas and the consumption of oak (*Quercus* spp.) acorns; such plants were neither mentioned by the Mayo community nor by other ethnobotanical studies carried out in the Americas. This phenomenon could be explained by the Hispanic-European origin of the South-Californian ranchers; in those regions, the consumption of the oak acorns is common (García et al. 2014).

16.4.2 Sociocultural and Environmental Implications

In general, the consumption of the wild edible plants depends on the production season or is occasional; practically, the surveys did not register domestication efforts or preferences for some species but only the conservation of valuable species within the limits of their homes where land clearance is permitted. Remarkably, most of the inhabitants in the ranches have great conservation awareness for wild plants that are used only for local consumption. These plants show a significant contribution to the population nutrition, and independent of the year season, one or more plant species are available for consumption: e.g., pitayas, June–August; caribe almonds (*Cnidoscolus maculatus*), August–September; wild plums, August–October; talayotes (*Matelea cordifolia*), October–November; chiltepin, October–December; guayparín (*Diospyros californica*), January–February; and jícama (*Ipomoea bracteata*). The behavior of the inhabitants in the reserve's ranches agrees with legal regulations and with the permanent interactions with researchers of universities and research centers in the locality, e.g., Autonomous University of Baja California Sur (UABCS) and Research Center of Biological Investigations of the Northwest (CIBNOR). On the other hand, the neighbor communities in the limits of the reserve do a commercial use of several plants: e.g., selling of fresh fruits of pitaya (*Stenocereus thurberi* and *Stenocereus gummosus*) (Fig. 16.3); leaves of oregano (*Lippia palmeri*) and damiana (*Turnera diffusa*); fresh and dried fruits of chiltepin (*Capsicum annuum* var. *glabriusculum*); or fruits of wild plums (*Cyrtocarpa edulis*). These plants provide important economic resources for the inhabitants of such communities.

The wild edible plants in *Sierra la Laguna* have their beauty as an added value that is used as a tourist attraction: one of the services offered by the inhabitants of the ranches is guided visits by trails where many of the edible species are tasted. This strategy generates economic resources for the community, and its negative environmental impact is much smaller than that associated with the direct selling of products.

In *Sierra la Laguna*, the germplasm of wild relatives of important crops of Mexico and the world are preserved (Fig. 16.4): e.g., plum, potato (*Solanum stoloniferum*), blackberry (*Rubus scolocaulon*), fig (*Ficus petiolaris*), grape (*Vitis peninsularis*), and several species of beans (*Phaseolus* spp.) and bag tomatoes (*Physalis* spp.). Unexpectedly, local inhabitants classified most of these species as of medium to low importance as foods; this could be by their low abundance and distribution in

Fig. 16.3 Sale of pitayas (*Stenocereus thurberi*) in the city of La Paz. (Author: JF Pío-León)

Fig. 16.4 Some wild crop relatives found in *Sierra la Laguna*: (**a**) wild plum (*Cyrtocarpa edulis*); (**b**) wild pepper (*Capsicum annuum* var. *glabriusculum*); (**c**) wild potato (*Solanum stoloniferum*); (**d**) wild fig (*Ficus petiolaris*); (e) wild grape (*Vitis peninsularis*). (Author: **a**, **b**, **d**, and **e**, JF Pío-León; **c**, José Luis León-de la Luz)

the highlands of the sierra (> 1200 m a.s.l.) (e.g., wild potato) and far from the ranches where only some inhabitants know about them and their consumption, a situation that contributes with greater certainty to conserve these plants as phytogenetic resources.

Most of the ethnobotanical studies focus the analysis on their most popular species. Those with the larger number of mentions are who are more frequently used/known by the community. However, it is also important to attend to those that are consumed only by a few people (those with the highest knowledge) because they are in risk to knowledge erosion. For example, in Sierra la Laguna, only few people consume the flowers of talayote and palo de arco (*Tecoma stans*) or the bark exudates of several legumes as mauto (*Lysiloma divaricatum*), vinorama (*Acacia farnesiana*) and *Prosopis* spp. (Fig. 16.5).

Fig. 16.5 Some of less popular wild edible plants of Sierra la Laguna, consumed only by the people with the highest knowledge in the use of wild plants: (**a**) flowers of talayote (*Matelea cordifolia*); (**b**) rhizomes of tulillos (*Cyperus* spp.); (**c**) patitos (*Hydrocotyle* spp.); (**d**) flowers of palo de arco (*Tecoma stans*); (**e**) palo of the fruits of kakachila (Karwinskia humboldtiana) (seeds are highly toxic); (**d**) bark exudate of mauto (*Lysiloma divaricatum*) (Author: JF Pío-León)

16.5 Conclusions and Lessons Learned

In *Sierra la Laguna*, the richness of wild edible plants and traditional knowledge about their sustainable use are great: this area could be considered as a living germplasm bank. The decree of *Sierra la Laguna* as a biosphere reserve favored the conservation of its resources but limited the freedom in its exploitation. Consequently, other activities have improved the viability of the reserve: alternative tourism and interaction of the local inhabitants with students and scientists who are working to increase the knowledge about the environment and its value, as well as in the implementation of rational conservation strategies.

References

Conanp (2003) Programa de manejo Reserva de la Biosfera Sierra la Laguna. Comisión Nacional de Áreas Naturales Protegidas, México D.F

Felger RS, Moser MB (1976) Seri indian food plants: desert subsistence without agriculture. Ecol Food Nutr 5:13–27

García E, Pereira J, Tardío J, Pardo M (2014) El pan de bellota. La fertilidad de la tierra 56:66–69

González-Abraham CE, Garcillán PP, Ezcurra E, el Grupo de trabajo de Ecorregiones (2010) Ecorregiones de la península de Baja California: una Síntesis. Sociedad Botánica de México 87:69–82

Hill R, Craig M, Newell B, Dunlop M, Gordon LJ (2015) Why biodiversity declines as protected areas increase: the effect of the power of governance regimes on sustainable landscapes. Sustain Sci 10:357–369

Joshi N, Siwakoti M, Kehlenbeck K (2015) Wild vegetable species in Makawanpur District, Central Nepal: developing a priority setting approach for domestication to improve food security. Econ Bot 69:161–170

Kermath BM, Bennett BC, Pulsipher LM (2014) Food plants in the Americas: a sourvey of the domesticated, cultivated, and wild plants used for human food in North, Central, and South America and the Caribbean. https://www.researchgate.net/publication/263888295_Food_Plants_in_the_Americas_A_Survey_of_the_Domesticated_Cultivated_and_Wild_Plants_Used_for_Human_Food_in_North_Central_and_South_America_and_the_Caribbean. Accessed 10 Jan 2020

León-de la Luz JL, Domínguez-Cadena R, Medel-Narváez A (2012) Flora de vegetales superiores. In: Ortega-Rubio A, Lagunas-Vázquez M, Beltrán-Morales LF (eds) Evaluación Biológica y Ecológica de la Reserva de la Biosfera Sierra la Laguna, Baja California Sur. Avances y Retos. Centro de Investigaciones Biológicas del Noroeste, La Paz, pp 357–393

Madeiros MFT, Silva PS, Albuquerque UP (2011) Quantification in ethnobotanical research: an overview of índices used from 1195–2009. Sitientibus série Ciências Biológicas 11:211–230

Pío-León JF, Delgado-Vargas F, León-de la Luz JL, Ortega-Rubio A (2017a) Prioritizing wild edible plants for potential new crops based on deciduous forest traditional knowledge by a rancher community. Bot Sci 95:47–59

Pío-León JF, Delgado-Vargas F, Murillo-Amador B, León-de la Luz JL, Vega-Aviña R, Nieto-Garibay A, Córdoba-Matson M, Ortega-Rubio A (2017b) Environmental traditional knowledge in a natural protected area as the basis for management and conservation policies. J Environ Manag 201:63–71

Reygadas-Dahl F, Landa-Romo C (2013) Cultura del Rancho Sudcaliforniano dentro de los límites de la Reserva de la Biosfera Sierra la Laguna. In: Lagunas-Vázques M, Beltrán-Morales LF, Ortega-Rubio A (eds) Diagnóstico y análisis de los aspectos sociales y económicos en la Reserva de la Biosfera Sierra la Laguna. Centro de Investigaciones Biológicas del Noroeste S.C, La Paz, pp 135–232

Urgenson L, Schmidt AH, Combs J, Harrell S, Hinckley T, Yang Q, Ma Z, Yongxian L, Hongliang L, MacIver A (2014) Traditional livelihoods, conservation and meadow ecology in Jiuzhaigou National Park, Sichuan, China. Hum Ecol 42:481–491

Yetman D, Van Devender TR (2002) Mayo ethnobotany. Land, history, and traditional knowledge in Northwest Mexico. University of California Press, Berkeley

Chapter 17
Hunting in the Yucatan Peninsula: Knowledge and Worldviews

Miguel Pinkus-Rendón and Enrique Rodríguez-Balam

Abstract Hunting is an activity that mankind has developed since its emergence as a species. At present, there is a conjunction between local knowledge with respect to hunters' prey, such as the habitat they occupy, breeding seasons, and even eating habits. Added to this is the social and cultural part of the hunt, represented by the types of hunting that can be practiced, the age to start hunting, the links that can be established between hunters, the hunting rituals performed pre- and post hunting, and the legends and myths around the activity.

Keywords Environment · Hunting · Local knowledge · Sustainable · Mayan culture

17.1 Introduction

Since the appearance of humans, meat has been used for self-consumption through hunting. To do this, we had to manufacture tools that had auxiliary help in this activity: knives and spears, among others. Likewise, we had to develop strategies to obtain prey difficult to obtain since they were in herds or because their size represented difficulties that could hardly be resolved without a previous hunting strategy.

These early visions of the hunt can be seen reflected in the cave paintings of the Paleolithic period, such as the caves of Altamira in Spain, where different paintings of mammoths, horses, bison, and other animals are glimpsed (data from more than 40,000 years ago. Or in South America, where pictorial samples of guanacos are found with manifestations of hunter-gatherers in the caves of the Casa de Piedra Hill in the province of Santa Cruz, Argentina, which date to approximately 2130 BC (Aschero 1985).

M. Pinkus-Rendón (✉) · E. Rodríguez-Balam
Centro Peninsular en Humanidades y Ciencias Sociales, Universidad Nacional Autónoma de México, Mérida, Yucatán, Mexico
e-mail: mapinkus@humanidades.unam.mx

A. Ortega-Rubio (ed.), *Socio-ecological Studies in Natural Protected Areas*,
https://doi.org/10.1007/978-3-030-47264-1_17

The graphic representations of the hunt in the different places give us an indication of the availability of animals that are in the environment and also of the capture or hunting arts that are used in the periods represented. In the case of the Mayan area, we can see in several pages of the Madrid Code how the Maya performed the capture of the deer (*Odocoileus virginianus*) by means of traps, in which it is noticed from the moment that the animal is taken by the paw with a rope tied to a tree until it is skinned and carried by the hunter (Fig. 17.1).

In ethnozoarchaeological studies obtained during the Classic Terminal period conducted by Götz (Götz and Emery 2013 cited in Götz 2014), it was found that the best-known vertebrate animal species found in the Mayan area stand out: the white-tailed deer (*Odocoileus virginianus*), jaguar (*Panthera onca*), manatee (*Trichechus manatus*), tapir (*Tapirus bairdii*), rattlesnake (*Crotalus durissus*), black iguana (*Ctenosaura similis*), swamp and river crocodile (*Crocodylus moreletii and C. acutus*), boa (*Boa constrictor*), sea turtles (*Caretta caretta*), green sea turtle (*Chelonia mydas*), hawksbill (*Eretmochelys imbricata*), the turkey of the monte (*Meleagris*

a b

Fig. 17.1 (**a** and **b**). Pages of the Madrid Codex. (Source: http://www.famsi.org/mayawriting/codices/pdf/madrid_rosny_bb.pdf)

ocellata), hocofaisán (*Crax rubra*), quetzal (*Pharomachrus mocinno*), and royal toucan (*Ramphastos sulfuratus*), among others.

Investigations carried out promptly in two places of the Yucatan Peninsula (Chichen Itza and Xcambó) give us an indication of the species of animals that were part of the fauna of which it was used by the pre-Hispanic Maya. Thus, for the lowlands in the north of the Mayan area (Chichen Itza), most of the bones obtained from landfills are attributed to mammals (94%), followed by reptiles (4%), birds (2%), and fish bone (*Actinopterygii*) (0.03%). The majority of mammalian bones belong to the white-tailed deer (*Odocoileus virginianus*) (60%). The reptile species identified were striped iguanas (*Ctenosaura similis*) – 95% of the total reptile sample – and freshwater and land turtles. Of the birds, only few skeletal remains of hocofaisán (*Crax rubra*) and turkey (*Meleagris ocellata*) were found (Götz 2007). In contrast, for the north coast (Xcambó), the archaeofauna is different because of the availability of resources, and remains of crustaceans (7.7%), fish (*actinopterigia and condrict*) (38.8%), reptiles (34%), birds (3.7%), and mammals (15.8%) were found. Among the fish fauna, several shark and fish genera such as catfish (*Ariopsis felis*), grouper (*Epinephelus* sp.), gurrubata (*Micropogonias* sp.), and robalo (*Centropomus* sp.) stand out. The most represented reptiles are sea turtles (65.5% of the total reptile sample), followed by land and freshwater iguanas and turtles. Regarding mammals, the most frequent – as in the other place – is the white-tailed deer (*Odocoileus virginianus*), on which we will focus in this chapter (Götz and Sierra Sosa 2011 cited by Herrera and Götz 2014).

According to Götz (2011), among the multitude of species used for consumption, there is a series of vertebrates that were hunted in the pre-Hispanic times, as in the colony, and some until today. Among the species mentioned are paca or jaleb (*Cuniculus paca*), pizote (*Nasua narica*), collared peccary (*Pecari tajacu*), white-lipped peccary (*Tayassu pecari*), white-tailed deer (*Odocoileus virginianus*), temazate deer (*Mazama* spp.), turkey (*Meleagris ocellata*), and black iguana (*Ctenosaura similis*). Likewise, there are a number of species that are mentioned in the sources of colonial times and whose bones have also been found in pre-Hispanic contexts of food consumption, but which are not mentioned in terms of exploitation in modern times. Species like opossum (*Didelphis* sp.), monte rabbit (*Sylvilagus* sp.), dog (*Canis lupus familiaris*), and tapir (*Tapirus bairdii*) possibly show discontinuities of consumption, either, as in the case of the dog, because they were replaced by domestic animals only raised for food purposes or because hunting of such animals is already prohibited.

17.2 Hunting Methods

As for the ways of hunting and obtaining food during the pre-Hispanic era, different practices were developed as mentioned before: the use of traps as indicated by the Madrid Codex, blowguns, arrows, spears, and even utensils of obsidian and stone as knives, which were presumably used not only for hunting but also for cutting (Aoyama 2006).

During the colonial era, changes were made in the forms of hunting and the instruments used for it, this due to the influence of the arrival of the Spaniards who included the metals as essential materials in the bill of cutting utensils. At the same time, the introduction of firearms as tools for obtaining prey had a great impact on various aspects of the hunt. Either by adaptation that the hunters had to stealthily approach the potential prey and thus avoid the noise caused by the detonation of the weapon, and the consequent flight of the animal in case the shot fails.

17.3 From Death to Intake: Food Preparation

According to Götz (2011), the preparation of food in the pre-Hispanic era – which still persists today – was done by cooking the meat over fire, without neglecting the historical cooking in the Mayan area, which consisted of burying the meat under the burning coals:

> A traditional way of stewing meat, which is mentioned in recipes used in cities and is followed in contemporary Mayan rural contexts, corresponds to the preparation in the form of the so-called 'pib'. Although the term 'pib' comes from the Mayan verb 'pi', which means to bury, it refers in virtually all culinary applications to cooking something in an underground oven, that is, burying it. The underground furnaces consist of previously excavated cavities, in which a base of stones is placed, which are heated by means of a fire that is ignited within the same cavities. The meat or the whole animal for cooking is placed on the hot stones, either in a metal container or on a base of leaves and the cavities are closed by placing branches of the ramón tree (Brosimum alicastrum) and earth on top. When preparing parts of animals in these underground ovens, the meat is cooked, while the bones show traces of boiling due to high temperatures. (Götz 2011)

17.4 Types of Contemporary Hunting in the Yucatan Peninsula

Currently in the Yucatan Peninsula area, firearms are used as the main tools for hunting. Among them, the most used is the 12-gauge shotgun with different cartridge sizes and in less cases the 22-gauge rifle. However, the "rubber strip" is also used (as it is known in Yucatan), which is an instrument in the form of "Y" with an elastic cord and a receptacle of stones in the middle part of the rope; this is usually made of leather. The rubber strip is frequently used for the hunting of small species such as birds or iguanas since it can be used quickly when a dam appears suddenly.

Montiel and Arias (2008) point out that in Yucatan (and we could extend it to the peninsula), four different types of hunting used by peasant-hunters are established:

(1) P'uuj (in Mayan language) or batida, (2) Ts'on or nocturnal hunting (also referred to as "lampareo"), (3) Ch'uk or stalking, and (4) Ximba ts'on or opportunistic hunting (also referred to as "walking the bush"). In this regard, information on hunting types will be expanded.

The first type of hunting is collective, in which a group of hunters (usually men) is organized, with the purpose of acquiring a larger number of specimens compared to those made individually.

(1) *P'uuj (in Mayan Language) or Batida*

In some populations of Calkiní, Campeche is also known as Clamoreada, in which they shout, whistle, and make noise to attract deer or any animal that rounds them because the activity is done during the day, and this is, to say of hunters, one of the techniques to attract prey. According to Montiel and Arias (2008), the batida "shows a complex form of social organization and a community living space that deserves a more detailed semblance." About it, he says:

> Here, a group of hunters accompanied by dogs, meet at dawn at a certain point in the community and designate by consensus "teachers" (usually the most experienced hunters in the group) to coordinate the day's activities. Upon arriving at the site selected for the hunt, by means of a raffle, the participants are assigned to one of two groups – the whisk and the waiting group. To start the hunt, the waiting group surrounds the selected hunting sector forming a semicircle, in which each hunter is placed laterally twenty to thirty meters from his immediate companion. With their shotguns ready, they are waiting for the animals, which will be directed towards them by the beating group, which they will approach from the opposite end of the sector parallel to the waiting group. At the order of the teacher, and with the help of the dogs, the beating group will begin its movement within the sector hoping to detect an animal. During the tour, the hunters advance in coordination in their position to avoid hurting or being injured by a partner. At the moment of achieving a dam, the shooter communicates it to the rest of the group and the teacher, depending on the terrain that remains to be traveled to find the waiting group, decides to stop or continue with the whisk. When the groups meet, the hunters gather the beats' dams and keep them hanging on a nearby tree. In this pause the hunters rest, no more than a quarter of an hour, socialize the details of the hunt and subsequently restart the activity with another whisk. In each hunting round the waiting position alternates two to three times for each group. The duration and number of rounds on a whipping day are decided by the teachers based on the amount of prey and the fatigue of the participants in each group. After the activity, usually at sunset, the hunters return with their prey to the community and meet at the home of one of the teachers. There, the work continues until skinning and highlighting the dams, from which similar portions of meat will be obtained to be distributed among all the participants of the batida. (Montiel and Arias 2008)

For the batida, the dam is the white-tailed deer keh (in Mayan), a favorite animal for the Mayan area since pre-Hispanic times because it is a conspicuous animal and of suitable size to extract more meat. Also in the places where it is available, the temazate or yuk (*Mazama pandora*) deer is hunted. Another animal that helps in the whisk is the collared peccary (kitam), which is sometimes found in a herd, and more than one specimen can be captured. Briceño et al. (2011) pointed out for southern Yucatan that although biomass the deer is more tasted for hunting, the peccary is for number of individuals, where there is no discrimination between males and females, which for the deer if such a distinction is made when hunting, since hunters consider that this ensures reproduction and availability for the future. In addition to this, they assume that the peccary is a harmful species for different crops.

It should be noted that the distribution of meat will depend on the relevance in the participation that each individual has had during the hunt, as well as the number and species of prey that have been obtained. If a deer was hunted, the shooter would be awarded a full back leg, and the remainder would be shared simultaneously with other people. It is noteworthy that in the distribution, dogs – a fundamental part of hunting – are also considered since when they go out to the beats, they are given the internal organs of the prey to be skinned and ingested.

The next three ways of performing the hunt are individually in pairs or small groups. For each of these types of hunting, it is necessary to have extensive local ecological knowledge from hunters.

(2) *Ts'on or Night or Light Hunt*

It refers to the night hunt, known as lampareo for using hand or head lamps to carry out the hunt inside the monte[1] or hubché (in Mayan) (Mandujano and Gray 1991). For this type of hunting, hunters have to agree because it is a nocturnal activity and using the whipping technique, it is risky that they are distanced from each other since this implies the danger of being confused with the prey.

At the moment of listening to a potential target, the lamp is lit to illuminate it; that is why the hunters say that this action has the function of leaving the animal "dazzled"; that is, putting a fixed light on the animal's eyes makes it stay static for a moment with what is used to make the shot with the shotgun. Hence, even some peninsular hunters say in a colloquial way that the night hunter does not see animals to shoot but must see the gleam of eyes at night because they will indicate the place where the prey is located: "We do not hunt animals, we catch bright eyes on the monte." That is why they also mention that the way to distinguish that it is an animal and not a person is by their eyes, "the eyes are red at night," as well as their size. The lampareo generally is realized walking in the monte; nevertheless, in the north coast of Yucatan (Dzilam de Bravo), we were pointed out that there are even people who go in vans with powerful halogen lights in the upper part of the cabin and hunt with fire weapons, a practice far from the ways of hunting in the localities, a reason why it could be deduced, and to say of the same settlers, that in these cases, they are people external to the towns. As in the whisk, the target is set for deer hunting.

(3) *Ch'uk or Stalking or Spy Spot*

One of the forms of hunting that is done individually or in pairs is the spillway. For this, it is necessary to know how to recognize the traces and excreta on the roads, as well as to be able to recognize how recent they are to know how far a potential dam is located. Therefore, it is appropriate to say that hunters have a wide

[1] Tropical low jungle.

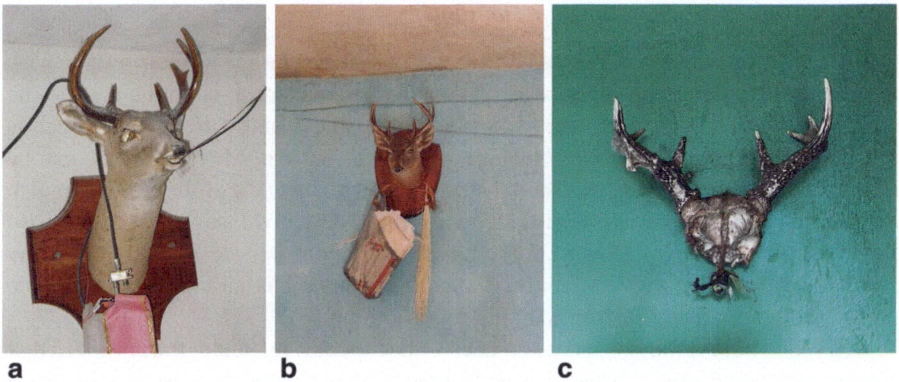

Fig. 17.2 Deer heads. (**a**) Source: Photo by authors, Dzilam González, Yucatán, Mexico. (**b**) Source: Photo by Mauricio López, Calkiní, Campeche, Mexico. (**c**) Source: Photo by authors, Yucatán, Mexico

local ecological knowledge, since they know about the main dams: from their biological cycles (breeding season, gestation time) to environments or boundaries within the monte where they can or not transit; they know how to distinguish at a distance, between males and females, in the case of deer, the former being the most valuable, especially when their antlers are three-pointed or more (Fig. 17.2).

Likewise, they recognize the different places where animals live, sleep, and feed, so that they ecologically relate the habits of the species to the place they will select to go to "throw." A hunter said: "Well, if it is in the grassland in the low monte, it says that there it eats and there it sleeps, yes, if there are deer also, yes, in the hubché that is said, there it eats the deer and there it makes a breeding ground; there it makes its nest (Hunter of Valladolid, Yucatan, personal communication)".

Another hunter also commented on the spaces with the characteristics that make it optimal for hunting:

> Well there [in the savannah] you find deer, kitam, even wild boar, tepescuincle, any kind of animal you see there because they come in at night; sometimes there are people who go, stay spying by day and stay until night, because if no deer falls during the day, tepescuincle falls at night or sometimes the deer comes at night, by the water, it has to come to drink water. (Hunter of Valladolid, Yucatan, Com. Personal)

A nearby tree is selected that can be used to settle and spy on the animal. The hunter climbs the branches, standing in a place that is free in terms of visibility, but also against the wind so that animals cannot distinguish their smell. Since they can be in the same place for several hours, they usually sit on the tree branches or put wood in such a way that it is like a bench to accommodate. There are people who even carry their hammocks to hang them between the branches and rest in them while the dam arrives having enough time to aim and shoot without being seen.

(4) *Ximba Ts'on or Opportunistic Hunting*

This type of hunting is mentioned by the peasants who go to their milpa[2] with their tirahule and their shotgun or as well as the beekeepers. It is on the road to the cornfield or the apiary where they can find animals susceptible to hunting. Such is the case of the turkey of monte *Meleagris ocellata* (Kutz in Mayan), which can be heard and hunted along the way, and the peasants who have the ability to distinguish the song or sound of several birds can determine their arrival, as would be the case with the turkeys, among others. In the case of turkey, it is necessary to take many precautions to make movements with stealth because it is an animal that can be elusive to the degree of fleeing with the minimum noise.

They also hunt if they observe keh deer. Other species related to the cornfield that eat the crops and that can be obtained in the same are the coatí (*Nasua narica*), which by walking in flocks have a great impact on the crops. Also the *tepescuincle jaleb* (C. paca), pecarí kitam (*P. tajacu*), and sereque (*D. punctata*) enter the field to eat horticultural crops. As one hunter told us: "Tepescuincles likes cassava, sweet potatoes, good sweet potatoes are in good soil, not like now. The black earth [preferably], lots, sweet potatoes, is what enters the tepescuincle, the sereke, the wild boar (Campesino de Valladolid, Yucatan Com. Pers.)".

There is even a variety of birds that eat corn cobs, which are hunted with shotguns or with tirahule. About them, a hunter tells us:

> One day I was lying down, and I heard a noise, and I tell him it is not that they are parrots, a lot, sometimes the parrots go down in a cornfield and in the middle, they get off, to see how much corn is going to bite. Right now it is possible to make the cornfield, right now parrots (*Amazona xantholora*), the Chel (*Cyanocorax yucatanicus*), and others who say pap, another that says Kau (*Quiscalus mexicanus*), all those little birds harm the milpa (Campesino de Valladolid, Yucatán Com Pers.)

Santos-Fita and collaborators (2013) report for the center of Quintana Roo, there is an activity that they called as "feeder-trap milpa," in which the peasants allocate an area where they make the traditional milpa, but on a small scale of 1 to 3 "mecates,"[3] this small cornfield is reserved for the animals of the monte; that is, the farmers do not harvest any crop products, indicating this place as a trap where they can be hunted.

On the other hand, although it is not a daily activity, the use of traps is also used in the Yucatan Peninsula, especially to catch a species that is, according to the peasants, important due to the damage it can cause in the crops. Tuza (*Orthogeomys hispidus*) (Ba in Maya), for example, is an animal that makes tunnels under the crops, leaving the soil very loose, being harmful not only for eating the crops but also for making them not secure to the earth. For its capture, wire is used, which is

---

[2] Traditional cultivation of the Mesoamerican area, in which corn is sown together with pumpkin and beans, among other species. The system of agriculture is rubbing, tomb, and burning, in which, to prepare the ground, weeds and shrubs are first cut down and trees that are not useful for peasants are demolished and finally set on fire in order to clean completely the land to cultivate.

[3] A mecate is an area used by Mayan peasants to designate a 20 × 20 m space.

linked in a circle and is introduced into a hole that has been left. It is tied to the other end with a large stone or tied to a tree, so that the animal is entangled, removed, and with a wood annihilated by hitting the head.

> Where the sweet potatoes were planted, the stones that the tuzas are taking are piled up, I set the trap in the ground, I dig, I grab a piece of stone I cover it, it is being half clear, my dad lifts me to see the traps. I get there his chan hand is caught in the trap, sometimes two, three tuzas. Only his hand is trapped, so I live everything, I get where the house is in the field, I bring two, three tuzas, on his head with a stick, he washes it well, he has it, puts it in salt, grabs salt, grabs the leaves of Banana, then the hot candle they prepare, they put it on top, and we go to work and when it is well we take it out, we grab the chile and smash it. (Campesino Valladolid, Yucatán Com. Pers.)

17.5 Rituals Around the Hunt

The perception, appropriation, and use of resources in cultures are related to the vision regarding how they understand the world and its environment. That is, they are not a cluster of isolated ecological and biological knowledge but are linked to the belief systems they have about nature (living and nonliving things), where species can have symbolic burdens. An example of this would be that of some nocturnal birds that announce a bad omen, or spaces that have a cultural connotation beyond just the territory such as the caves, sascaberas, and cavities considered by some as a kind of "entrances or connections with the underworld."

At present, there are fewer and fewer traditions that are followed due to different sociocultural changes such as migration to the capital city of Merida, the Mayan Riviera, or the USA; the gradual loss of the language, the change of productive activities as well as urbanization processes, population growth, and change of land use have modified in many ways the worldview that the inhabitants have (Baños 2004, Ramírez 2006).

Even so, in the culture of the Mayan peoples, peninsular speakers, the belief is maintained that the monte is a space that has owners, and in order to make use of this or the resources that it holds, it is necessary to request permission from the owners. In that sense, the hunting issue is no different from other activities carried out within the scope of the monte.

Before going hunting, people make an "offering," for which they use a kind of ground corn balls (nixtamalized) that are stirred with water, food and drink known as "pozol, pozole, or saká." That is put in a container, usually a jícara[4], and a prayer is made to ask for the favors of the owners of the monte as well as their permission to be able to enter to hunt some animals. These prayers are for the lords or owners of the monte or *yum dziles* or *yumdzi loob*. Elsewhere in the peninsula, the "scoop" is made after a reprimand or punishment has been received for not asking permis-

[4] Fruit of the *crescentia bush cujete* that is cut in half, the content removed and then dried, which makes it rigid and is used to contain liquids.

sion from other owners of the monte, known as *aluxes*. In this case, once you were warned by these beings, the hunter is obliged to do the *saká* and perform a prayer to be able to receive the hunting authorization.

> Now yes, then you enter this area, where the lord only makes cornfield, and you are a hunter you see that there is deer in his cornfield; then you will see the first night; the alux is going to throw you away, you will go crazy, give them his offering: kaax is a large tree that bears fruit as jicaritas; you lower the fruit as you cut it as jícara; the saka you grind it; you go to the place where the cornfield is, where you see that there is *laja*; you are cleaning, removing garbage; you put the kaax jícaras and the ground you ground is negotiating with water. You make his atole. But do not stick candle, in atole of dough, the saka, you are putting 7 jicaritas, and pebbles is your offering; you put a candle; the candle is the wax of the monte bee; you are going to make a candle and it will be in the middle. So when you put all that, he gladly did your will for him.

On the other hand, Santos-Fita (2016) collected information from the central part of Quintana Roo, specifically in the municipality of Carrillo Puerto, where a ritual practice known as *Loojil Ts'oon* or carbine ceremony is performed. In it, the shotguns used by the peasants for hunting are cleaned after they have been used several times since if they continued to do so, they could have serious consequences for the hunter. In this ceremony, the hunters, the shotguns, and the deer and peccary jaws that were prey are purified (in total, there must be 13 animals for each weapon). Once the ritual is celebrated, the latter are returned to the monte to revive the animals so that the resource does not end.

Unlike the first fruits observed above, *Loojil Ts'oon* is carried out by a *Jmen* or Mayan priest,[5] for being a ceremony in which the community participated. For this ceremony, several animals must be hunted in order to have different types of meat that are used to make in pib as part of the ritual. In addition, several edible products have to be prepared that are distributed at the end of the ceremony. An altar is made where the processed products are deposited, as well as the shotguns used and the jaws of the animals. The J men prays in Spanish and Mayan, while the zipché plant (Zip Plant or Guardian Deer) purifies and cleans both the shotguns and the hunters. Once the ceremony is over, both the hunter and the weapons are free to return to the hunt.

17.6 Myths and Legends

As noted in the previous paragraph within the beliefs that exist in the Yucatan Peninsula, there is that of a being who owns the deer; in this case, it is known as the zip, and it is a deer with large antlers between which is a hornet ek. According to Villa Rojas (1995 see pages 181–182):

[5] A person who has knowledge of therapeutic species (animals and plants) that has been designated by divine entities to heal and guess, who is responsible for performing rituals and ceremonies in communities.

[... the deer are under the protection of St. George and, especially of a class of supernatural deer designated with the term zip ... This kind of spirits usually teases the Indians by having them run after deer that, in the end, turn out to be iguanas; Men who possess a special talisman called yut are exempt from this deception, which consists of a calcareous formation stone that is occasionally found in the deer stomach. The owner of this talisman should not abuse this magical power, because, then the zip punishes him causing some disease through the evil winds that the guardian leaves in his path. Although the zip is pure wind, the Indians say they know a secret to hunt it; It consists of a magic projectile that is prepared by mixing gunpowder with some of the cottony garbage that is found in the hiding places of a certain marsupial called holi och the projectile must be, also marked with a cross. In addition, before firing it, fragments of dried leaves must be placed inside the barrel of the shotgun. The supposed efficacy of this secret does not prevent the Indians from refraining from using it due to the fear of being touched by the evil winds that the zip carries with it…].

It should be noted that in other locations, only the zip appears when it has already been hunted in excess, and the hunters are warned not to do so or when rituals have to be performed to appease it.

On the other hand, the stone of the deer or *yut* was mentioned, which appears in the stomach of a particular deer, and that gives luck at the time of hunting, even mentioning that with a single shot it demolishes the animal. There are some towns in the south of the peninsula where some hunters say that the power or gift granted to the hunter who owns the stones is such that even if the fired animal has not been a deer, when going to collect the prey, it will have become the body of a deer. In order for the talisman to work, it is necessary to keep the secret from the person who obtained it, since if it is discussed with some other hunters, the *yut* loses its effect. Likewise, if several animals have been hunted – some hunters comment on thirteen prey – it is necessary to return it to the monte so as not to receive misfortunes for overexploiting resources. Just as we talk about the stone of the deer, in Calkiní, Campeche, a white worm is mentioned that comes out of the snout of the recently depressed deer and that has the same properties as *yut*.

These myths could be part of a regulatory system from the cultural point of view, to establish a balance with the number of dams that are hunted in the monte of the Yucatan Peninsula. In that sense, care must be taken not to exceed the limits established by supernatural beings so as not to be punished. However, all these beliefs have been lost in the localities because the new generations no longer believe in them.

Finally, another legend mentioned in Tankuché, Campeche, is about a water eye known to locals as dzibichen (translated as cenote written in Maya). In this place, there are some marks that are thought to be letters or signs marked on the rocks around the mouth of the cenote. People say, a long time ago when the chewing gum was cut (approximately in the decades of 1920–1940), a group of people, after having carried out their activities and going in search of water, found the opening of the cenote; the cutters (the nombre of individuals varies according to the rapporteur) went in search of firewood to prepare the camp to stay in the place where one of them remained. The narration tells that they hunted a jaguar and that the person who stayed to take care of it began to skin the animal, removing all the skin. When the other hunters returned, this person took off his clothes and put the jaguar's skin on top with the intention of scaring his companions. However, once they were close to him, he began to shout so that they would not come closer, since the jaguar's skin was embodied in his body to begin his transformation into an animal, roaring and scraping the rocks; at the same time, he wrote with its claws on the walls letters that still remain in place.

17.7 Final Comments

Hunting today is seen as a practice plagued with negative connotations, particularly in urban societies. However, the way of seeing it among rural populations is far from political correctness, speeches, and ideologies that exclude sociocultural contexts behind an activity that for years combines all spheres of social life and cultural practices.

When approaching the study of the hunt, it is tried to investigate – not to justify under cultural relativisms, the social, ritual, and symbolic contexts that sustain the permanence of the hunt. Likewise, we try to show not only the way in which Mesoamerican groups, especially the Mayan speakers of the Yucatan Peninsula, maintain cultural elements, but also what gives them sustenance: family economy, self-consumption, commerce, and subsistence. That is why generalizing on the subject without resorting to historical and ethnographic data little helps to understand a phenomenon that, in addition to little studied, is increasingly stigmatized.

Regarding the issue of regulations or legislation around the hunt of the white-tailed deer (*Odocoileus virginianus*), we decided to put it aside because by itself it would be worthy of a separate article or investigation. However, much of what has been pointed out for this chapter also relates to the way people and the authorities interpret and, above all, how they apply the regulations in this regard. Uses and customs, conflicts between the inhabitants, past quarrels with the municipal authorities, and forms of internal organization have repercussions on the sanctions, permits, or laxity with which said regulations regarding hunting are applied. All of the above would also serve without a doubt, to understand why certain practices and beliefs related to the monte have been modified or left behind in some cases.

Therefore, we decided to favor a general approach in the first place to a minimum classification of hunting types. The batida, perhaps the best known way of hunting, night hunting or lampareo, stalking and opportunistic. All have common points. The main one has to do with the ethnobiological knowledge on the part of the hunters of both the environment and the behavior of the animals. Some respondents have such a degree of knowledge that one could well speak of an "ethology of the bush animal."

In addition to the above, another shared element is that in each case, the rituals, myths, and legends associated, either to the activity of the hunt or to the monte, do not depend on going out to hunt in groups as in the beats. Every person who enters the monte to hunt carries a deep knowledge about the biocultural environment they are in: they can read it, write it, and rewrite it according to the situation they face. This is evident if we understand that as it is necessary to ask permission from the owners of the monte, make prayers, or project beliefs under religious syncretisms, it is also true that they are crossed by social interactions that range from the configuration of spaces of male and female socialization – depending on the case or type of participation – even ideas that without proposing it could well be understood as conservationist visions, as is the case of the idea of reciprocity between man and nature.

Even if it is not expressed verbally, another element present among hunters is that everyone who extracts goods from nature is aware that they must act in accordance with a retribution with their environment. It does not mean, under a romantic look, that the misuse or excessive use of natural assets does not exist, but it is precisely beliefs, myths, and legends, which are relevant because they maintain this discourse as the foundation of an activity such as hunting.

It is always necessary to return to the environment what has been taken away from it or to look for ways in which resources can be regenerated, either from the social, from the economic, or from the symbolic. And we must remember, beliefs, rituals, and symbolic elements are relevant not for what they do or do not, but for what they communicate as the main element that builds discourses within a given culture (Fig. 17.3).

And it is that the vision of man with nature is such – also present in all activities linked to hunting – that it goes far beyond what is explained in local legends, with men who go beyond the human dimension to see from the eyes and skin of an animal to permute skin with that of a jaguar or become a deer in another temporal space

Fig. 17.3 Girl and deer. (Photo by authors)

and dimension. These are ways in which the human being builds empathy with the animal that is another and himself.

And it is that from an early age coexists with the animals of monte, among them, of course, the deer. It is not at all strange to see deer lost from his pack, cross the plots or land of people, and even get into their most domestic spaces such as the house or the space for cooking food next to the stove. However, when this type of thing happens, the prey is not hunted, since it requires being in the place that belongs to it and where the hunting practice is configured: the monte.

References

Aoyama K (2006) La guerra y las armas de los mayas clásicos: puntas de lanza y flecha de Aguateca y Copán. Estudios de cultura maya 28:27–49

Aschero C (1985) Pinturas rupestres en asentamientos cazadores-recolectores: dos casos de análisis aplicando difracción de rayos-x. Cuadernos del Instituto Nacional de Antropología. V.10. HI83–1985: 291–306. Buenos Aires

Baños Ramírez O (2004) Modernidad, imaginario e identidad rurales: el caso de Yucatán, México. Centro de Estudios Sociológicos, El Colegio de México, 299 pp

Briceño Méndez M, Montes Perez R, Aguilar Cordero W, Pool Cruz A (2011) Cacería del pecarí de collar (*Pecarí tajacu*) (Artiodactyla: Tayassuidae) en Tzucacab, Yucatán, México. Revista Mexicana de Mastozoología Nueva Época Año 1:8–18

Götz Ch (2007) El aprovechamiento de animales vertebrados en Chichén Itzá, Yucatán -uso alimenticio y ritual-. En: Chichén Itzá. Nuevas interpretaciones históricas (eds) Alexander Voss, Andreas Koechert, pp 51–74

Götz Ch (2011) Una mirada zooarqueológica a los modos alimenticios de los mayas de las tierras bajas del norte. en Héctor A. Hernández Álvarez y Marcos Pool Cab (eds) Identidades y Cultura Material en la Región Maya. Universidad Autónoma de Yucatán, pp. 89–109, ISBN 978-607-7573-76-0

Götz C (2014) La alimentación de los mayas prehispánicos vista desde la zooarqueología. An Antrop 48-I:167–199

Herrera Flores D, Götz C (2014) La alimentación de los antiguos mayas de la península de Yucatán: consideraciones sobre la identidad y la cuisine en la época prehispánica. Estudios de cultura maya 43:71–98

Mandujano S, Rico-Gray V (1991) Hunting, use, and knowledge of the biology of the white-tailed deer (*Odocoileus virginianus* Hays) by the maya of Central Yucatan, Mexico. J Ethnobiol 11(2):175–183

Montiel Ortega S, Arias Reyes L (2008) La cacería tradicional en el Mayab contemporáneo: una mirada desde la ecología humana. Avance y Perspectiva Nueva época 1(1):21–27

Ramírez Carrillo L (2006) Impacto de la globalización en los mayas yucatecos. Estudios de cultura maya 27:73–97

Santos-Fita D (2016) Documental Loojil Ts'oon. Ceremonia de la carabina. Renovando el permiso divino de la cacería. https://www.youtube.com/watch?v=KfC-7G3GAVk

Santos-Fita D, Naranjo Pinera E, Bello BE, Estrada Lugo E, Mariaca Méndez R, Macario Mendoza P (2013) La milpa comedero-trampa como una estrategia de cacería tradicional maya. Estudios de cultura maya 42:87–118

Villa Rojas A (1995) Estudios etnológicos: los mayas. Segunda Edición. Instituto de Investigaciones Antropológicas. UNAM

Chapter 18
The Nagoya Protocol, Intellectual Property, and Biodiversity Conservation in Mexico

Luis Felipe Beltrán-Morales, Eduardo Gutiérrez-Rivero, and Gerzaín Avilés-Polanco

Abstract The objective of this work was to analyze the Nagoya Protocol policy instruments in the context of the public administration in Mexico, their scope, and the instruments developed to date. In addition, we performed an analysis of the evolution of the rights of plant breeders by obtaining a breeder title granting temporary rights of proprietary exploitation and of the records entered in the National Catalog of Plant Varieties (CNVV). To this end, we analyzed policy instruments and the state of the art through the main patent search engines at an international level and reviewed the official statistics on the rights of plant breeders and the plant varieties registered in Mexico.

Keywords Nagoya Protocol · Breeder's rights · Plant species · Conservation Biodiversity

---

L. F. Beltrán-Morales
Centro de Investigaciones Biológicas del Noroeste S.C. (CIBNOR), La Paz, Baja California Sur, Mexico
e-mail: lbeltran04@cibnor.mx

E. Gutiérrez-Rivero
Ministerio de Ciencia, Tecnología y Medio Ambiente, Centro Meteorológico Provincial, Bayamo, Cuba

G. Avilés-Polanco (✉)
CONACYT-Centro de Investigaciones Biológicas del Noroeste, La Paz, Baja California Sur, Mexico
e-mail: gpolanco@cibnor.mx

A. Ortega-Rubio (ed.), *Socio-ecological Studies in Natural Protected Areas*,
https://doi.org/10.1007/978-3-030-47264-1_18

18.1 Introduction

The Nagoya Protocol to the Convention on Biological Diversity is an international agreement aiming to share the benefits arising from the utilization of genetic resources in a fair and equitable way (CBD 2011), which came into force in Mexico on October 12, 2014. Mexico was the sixth country to endorse this protocol, which has been ratified by 118 countries to date. Mexico is a megadiverse country from biological and cultural perspectives, in addition to being a user and supplier of genetic resources. Mexico and its regions can potentiate the use of genetic resources as one of the markets for broad-spectrum natural products. The objective of the Protocol is the fair and equitable sharing of the benefits arising from the use of genetic resources and acknowledges the significant contribution of technology transfer and cooperation to sustainable development, to create research and innovation capacity to add value to genetic resources in developing countries, pursuant to Articles 16 and 19 of the Convention (Evanson et al. 2010). It also considers the potential contribution of access and benefit-sharing to the conservation and sustainable use of biological diversity, as well as to the reduction of poverty and environmental sustainability, thus contributing to achieve the Millennium Development Goals. It also aims to promote equity and fairness in negotiations of mutually agreed terms between genetic resource suppliers and users. A relevant aspect is the acknowledgment of the critical role of women in access and benefit-sharing and the need for women to fully participate at all levels of the formulation and implementation of policies for the conservation of biological diversity.

18.2 Background

The diversity of genetic resources is a key component of biodiversity and is associated with potential commercial uses, as these resources have a high market value as essential raw materials for R&D in the pharmaceutical, botany, agricultural, and biotechnological industries, among others (Laird and Kerry 2002; Pisupati and Bavikatte 2014; Pushpangadan and Narayanan 2005; Williams 1993). Poor and marginalized rural communities, with limited resources and access to capital, have few opportunities to increase their income. The growing demand for genetic resources offers an opportunity for these communities to benefit from a vigorously growing market. All this is made possible by the diversity of varieties (species and breeds) that allow the use of particular populations in areas where they are uniquely adapted and are readily available (Rege and Gibson 2003).

According to the theory of property rights and the provisions set out in the Nagoya Protocol, genetic resources are sovereign properties of states, rather than resources belonging to the common heritage of humankind. Consequently, the supplier states can claim compensation to users of their national genetic resources, leading to the need for bilateral or multilateral agreements to share the commercial benefits derived from the use of genetic resources. However, these benefit-sharing agreements are frequently reached ex ante, and the potential benefits are difficult to quantify when proper instruments of intellectual protection and economic valuation of resources are not in place (Vogel et al. 2011; Nelliyet 2017). Thus, some issues have emerged from the asymmetric information on the commercial value of such resources for both suppliers and users. The Nagoya Protocol sets out the possibility of obtaining economic and social benefits to boost development and growth in regions that supply these resources. Although this treaty came into force just recently, it may be acting as negative economic and conservation externalities in both our ecosystems and society as a whole. There are no proper economic, technological, or trade valuations of genetic resources currently available before these can be marketed to obtain benefits in return.

Most of the previous studies on the assessment of genetic resources sought to estimate the potential benefits associated with the use of genetic resources for agronomic products. These studies were based on the preferences of local stakeholders to determine which genetic resources to prioritize based on the opportunity costs of maintaining certain species by using limited resources, such as the territory. However, these studies are insufficient to reflect the potential preferences of stakeholders in other countries, as is the case of an industry in demand for raw materials for production processes. On the other hand, although the expected marketing profits derived from genetic resources in various sectors are known by professionals and academics alike, the specific indicators for the assessment of genetic resources are still unclear (Lee and Sohn 2016; Pisupati and Bavikatte 2014).

18.3 Evolution of the Rights of Plant Breeders and the Catalog of Plant Varieties in Mexico

In Mexico, plant varieties can be officially registered in two ways:

(a) By obtaining a plant breeder's title granting a temporary right of exclusive exploitation
(b) Through entry in the National Catalog of Plant Varieties (CNVV[1]), which does not confer proprietary rights of use but is required in the seed-quality assessment process and the official listing of commonly used varieties

Both registrations may take place either in parallel or separately. The regulatory framework in the first case is the 1996 Federal Law on Plant Varieties (LFVV), which is consistent with the 1978 Convention of the International Union for the Protection of New Varieties of Plants (UPOV) subscribed by Mexico in July 1997. The second type of registration is based on the Mexican 2007 Federal Law on Production, Certification, and Trade in Seeds (LFPCyCS).

[1] Hereafter, unless otherwise stated, all acronyms refer to the respective name in Spanish.

The registration of varieties, in its two modalities, is an essential component to foster applied innovation and technological development, as well as for the production of strategic supplies, i.e., improved seeds. This translates into actions directly related to two of the ten pillars of change of the comprehensive strategy to increase productivity and maximize productivity in the farming sector (Programa Sectorial de Desarrollo Agropecuario, Pesquero y Alimentario 2013–2018 [Sectoral Program of Farming, Fisheries and Food Development]).

The rights of plant breeders are the legal right conferred to a natural or legal person upon being acknowledged as a breeder of a new plant variety for exclusive temporary use and exploitation, including propagation materials, seeking to promote research on and transfer of new technologies aimed at producing more and better plant varieties. The Federal Law on Plant Varieties (LFVV) regulates these matters, and its provisions set the grounds for the establishment of the Committee for the Assessment of Plant Varieties (CCVV). This committee considers the opinion of specialists in Technical Support Groups to rule on applications for a title as plant breeder and registration in the CNVV. In addition, this committee establishes the procedures for the conduct and evaluation of technical tests and contributes its opinion for the formulation of Mexican Official Standards relating to the characterization and evaluation of plant varieties.

Since the implementation of the system for the protection of the rights of plant breeders, from 1996 to 2016, applications for titles as plant breeders have increased significantly, as illustrated in Fig. 18.1.

Evidently, the trends in the number of applications and titles of plant breeder rights granted have risen in the period reviewed. The cumulative total amounts to 2314 applications for the protection of 123 crops, resulting in 1619 titles of plant

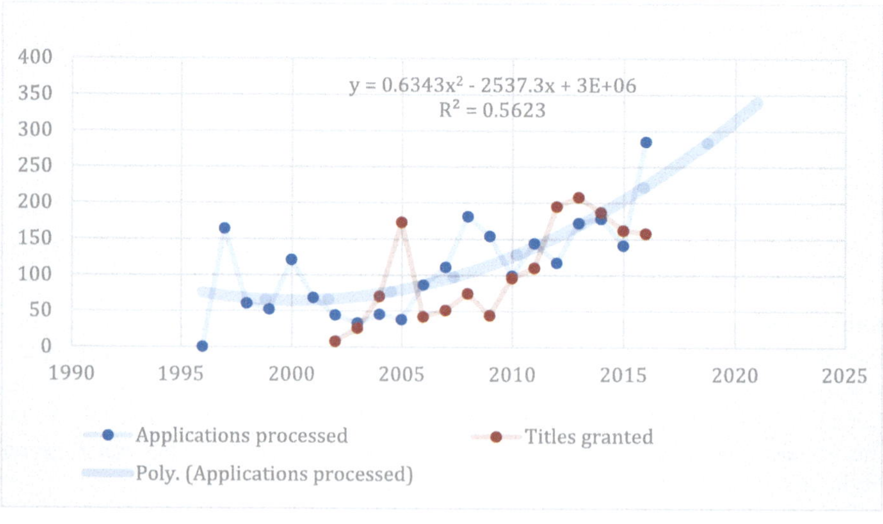

Fig. 18.1 Number of applications and titles of Plant Breeder Rights, 1996–2016. (Source: Adapted from Jefferson et al. 2018)

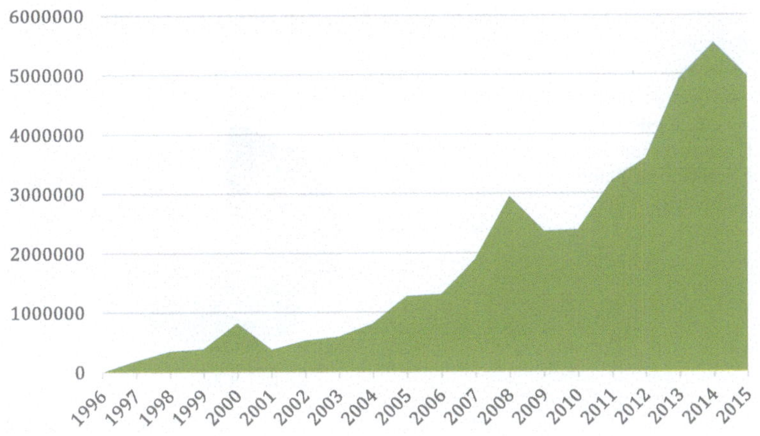

Fig. 18.2 Income received from applications and issuance of breeder titles (thousands of Mexican pesos). Own elaboration. (Source: Based on data published in the SAGARPA data on the Rights of Breeders of Plant Varieties published in the Official Gazette)

breeder rights granted and in public income from payment of fees; this increasing trend has continued through time (Jefferson et al. 2018), as shown in Fig. 18.2.

As of the third quarter (3Q) of 2015, revenues from the payment of fees related to plant breeder rights amounted to $4,974,281.00 Mexican pesos. Of this total, payments for an endorsement of breeder titles represent $2,282,574.00 pesos; the assessment and processing of applications for breeder titles rank second, with $1,843,987.00 pesos, followed by payments for issuance of breeder titles and submission certificates, with $778,444.00 pesos. These numbers illustrate the interest of breeders in keeping the rights to exclusive exploitation of the varieties registered and in submitting new applications.

To note, 84.3% of the 1619 titles of plant breeder issued as of 2006 are still in force. Figures 18.3 and 18.4 show the distribution of applications and titles of plant breeder rights in absolute numbers, as well as the percent distribution and by major countries that have requested and obtained titles.

A total of 2314 applications were registered as of 2016; of these, the number of applications and titles granted by country are 832 and 531 for Mexico, 727 and 540 for the United States, and 404 and 309 for the Netherlands (Fig. 18.4). Seventy percent of all applications have been granted a breeder title.

As of 3Q 2015, a total of 234 different breeders were recorded. Almost half (46.6%) were distributed as follows: the Instituto Nacional de Investigaciones, Forestales, Agrícolas, y Pecuarias (National Institute for Forest and Farming Research, INIFAP) of Mexico heads the list with 201 varieties with a plant breeder title, followed by Pioneer Hi-Bred International, Inc. with 111, Driscoll Strawberry Associates, Inc. with 86, Seminis Vegetable Seeds, Inc. with 78, Semillas y Agroproductos Monsanto, S.A. de C.V. with 74, Numhems BV with 39, and Rosen Tantau, Mathias Tantau Nachfolger with 29 (Fig.18.5).

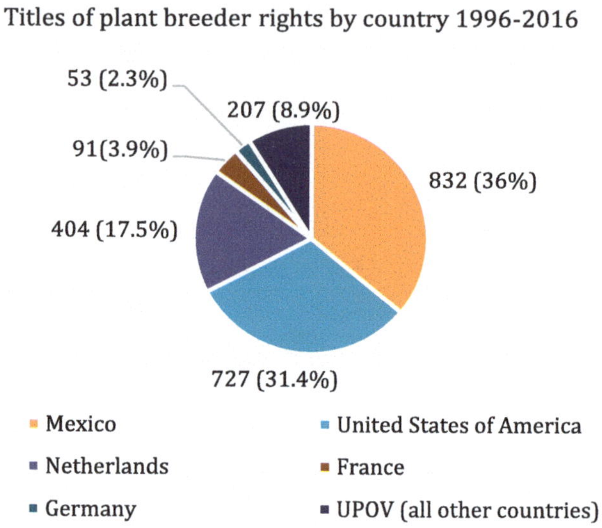

Fig. 18.3 Applications for titles of plant breeder rights by country, 1996–2016. (Source: Own elaboration based on SAGARPA data on the Rights of Breeders of Plant Varieties published in the Official Gazette)

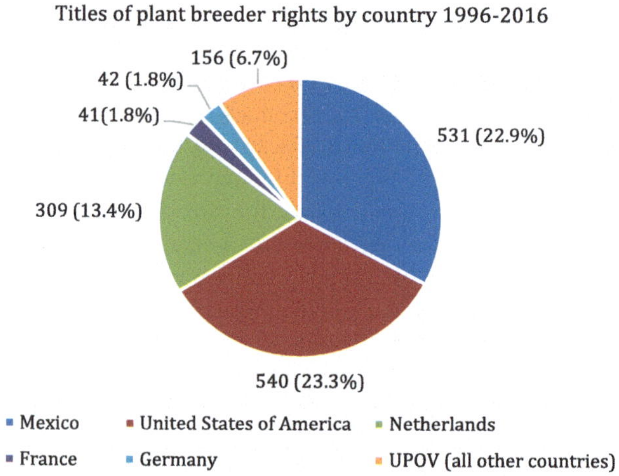

Fig. 18.4 Titles of plant breeder rights by country, 1996–2016. (Source: Own elaboration based on SAGARPA data on the Rights of Breeders of Plant Varieties published in the Official Gazette)

The number of applications for a Title of Plant Breeder Rights by crop type is shown in Fig. 18.6. Clearly, the greatest interest by companies and national and foreign institutions focuses on agricultural varieties due to the commercial activities involved. Second in importance and sharing virtually the same proportion are ornamental and fruit-producing varieties, with vegetables slightly below (Jefferson et al. 2018).

The National Catalog of Plant Varieties (CNVV) of Mexico lists 2249 varieties belonging to 59 species. These include 229 commonly used varieties that are important for being species native to Mexico, highlighting 46 varieties of prickly-pear

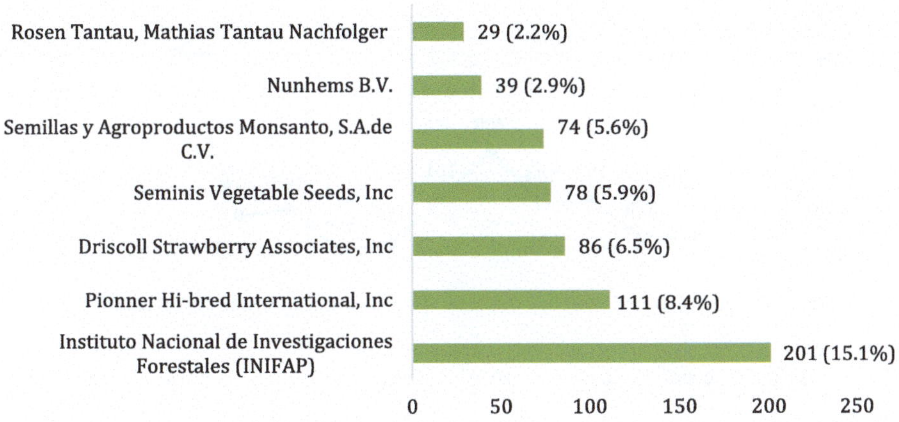

Fig. 18.5 Titles of plant breeder rights by company/institution, 1996–2015. (Source: Own elaboration based on SAGARPA data on the Rights of Breeders of Plant Varieties published in the Official Gazette)

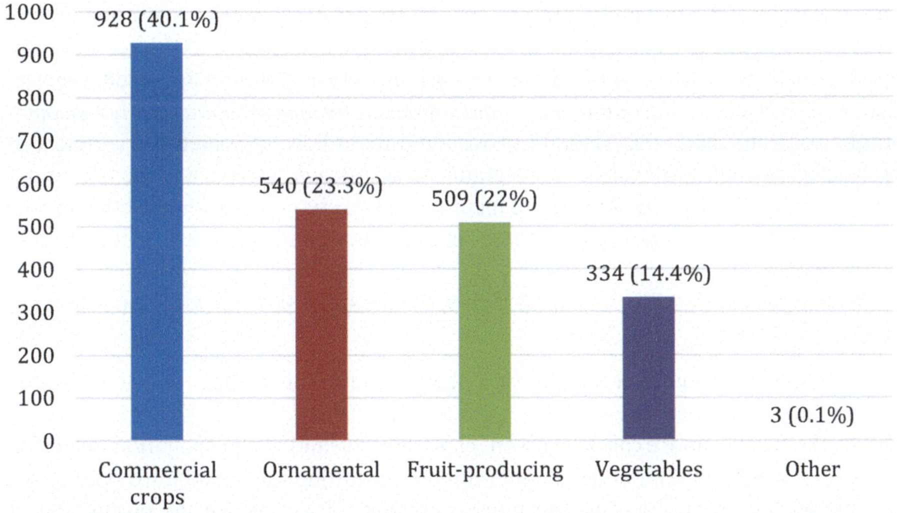

Fig. 18.6 Applications for titles of plant breeder rights by crop type, 1996–2016. (Source: Adapted from Jefferson et al., 2018)

pads, 30 of *cempoalxochitl*, 29 of *xoconostle*, and 20 of husk tomato, listed as with either provisional or final registration. Figure 18.7 illustrates annual application figures; in the period from 2006 to 2016, 1339 applications were submitted, contrasting with 910 in previous years.

According to data on plant breeder rights published in the Official Gazette, 33 new applications were submitted in 3Q 2015. For 19 varieties of four species, either

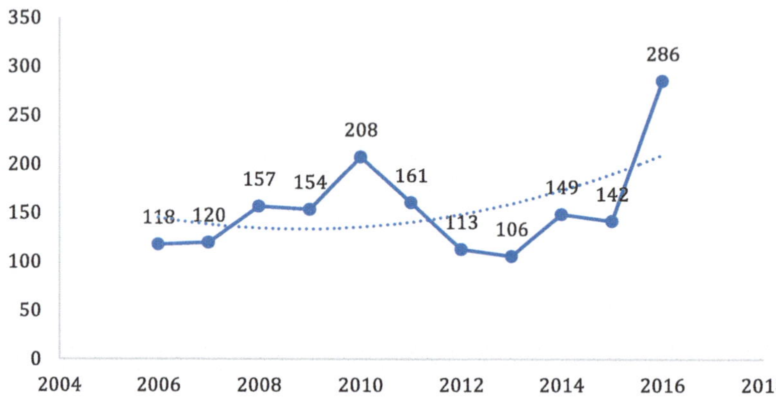

Fig. 18.7 Applications in the National Catalog of Plant Varieties for the period 2006–2016. (Source: Own elaboration based on SAGARPA data on the Rights of Breeders of Plant Varieties published in the Official Gazette)

a provisional registration was granted or additional documentation was requested to complete the registration process: maize, 12 varieties; potato, three; chickpea, two; and agave, two. Additionally, ten provisional registrations were issued: maize, six varieties; chickpea, two; and potato, two. The same period of 2014 recorded 43 applications, with titles granted for 35 varieties corresponding to seven species: maize, 27 varieties; *Jatropha*, three; and one each for apple, *habanero* hot pepper, sorghum, beans, and soy. Also, 47 provisional records were issued for varieties of eight species: maize, 30 varieties; sorghum, nine, *Jatropha*, three; and the remaining ones for five species. Eighteen permanent records were granted for varieties of five species: husk tomato, ten; apple, three; maize, three; and one each of tomato and hot pepper.

With regard to the title of breeder rights, the Mexican state protects the following rights of holders:

(a) Acknowledgment as a breeder of a plant variety. This right is non-transferable and does not prescribe.
(b) Right to use and exploit a given plant variety and its propagation material, exclusively and temporarily, either directly or through licensed third parties, for production, reproduction, distribution, or sale, as well as for the production of other varieties and hybrids for commercial purposes.

As to the period of validity of this protection, the Federal Law on Plant Varieties establishes that the breeder rights shall be valid for 18 years in the case of perennial species (trees, fruit-producing trees, grapefruit vines, ornamental varieties) and their rootstocks and 15 years for other species. These periods shall start from the date of issue of the breeder title; after the deadline, the plant variety, its use, and exploitation will pass to the public domain.

The Federal Law on Plant Varieties establishes that for a plant variety to be subject to protection, it shall comply with the following requirements:

New: The plant variety or its propagation material meets this characteristic provided it complies with the following when the application is submitted:

1. It has not been sold in the national territory, nor within 1 year prior to the date of submission of the application for a title of breeder rights.
2. It has not been sold abroad, nor was the sale completed within 6 years prior to the submission of the application, in the case of perennials (vines, trees, fruit-producing plants, and ornamental varieties), including their rootstocks, or within 4 years prior to the submission of the application, for all other species.

Distinction: The plant variety will be listed in this category if it is clearly differentiated, from a technical standpoint, by one or several characteristics that are unique to it and different from those of any other variety known when the application is submitted.

Stability: The plant variety meets this characteristic if it has maintained its relevant characteristics unchanged after successive reproduction or propagation events.

Uniformity: The plant variety meets this characteristic if it is sufficiently uniform in its relevant characteristics, considering the variation that may be expected under sexual or vegetative propagation.

Denomination: It will be considered as its generic name. To be approved, it should be different from any other denomination already in use in the country or abroad and should comply with all other requirements set forth in this law; this variety should not be identical or similar to an extent that leads to confusion relative to a previously protected variety.

18.4 Conclusions

Currently, the genetic resources and traditional knowledge of Mexico and its regions are not being properly assessed in economic and technological terms within the framework of the Nagoya Protocol, leading to negative externalities for society and ecosystems in the region. It is essential to identify the traditional knowledge associated with the use of genetic resources, identify how economic and non-economic benefits are being distributed, and assess whether these are equitable, assess the economic and technological impact on genetic resources within the framework of the Nagoya Protocol, organize prospecting and training workshops on Intellectual Property rights, and propose recommendations on protection of intellectual property protection and negotiation of benefits. One of the issues identified is that a large number of patents and plant breeder rights on genetic resources and plant varieties refer to ownership by countries, companies, and institutions that are not part of the Nagoya Protocol, as is the case of the United States. The concern is the destination of the profits from the exploitation of varieties and genetic materials from Mexico.

Acknowledgments We thank the National Council of Science and Technology of Mexico (CONACYT) for the financial support that this agency provided for this research (Grant number PDCPN-2017/6113) and Ing. Alejandro Borges Sánchez and MC Ileana Serrano Fraire for the support they provided. María Elena Sánchez-Salazar translated the manuscript into English.

References

CBD (Convention on Biological Diversity) (2011) Nagoya protocol on access to genetic resources and the fair and equitable sharing of benefits arising from their utilization to the Convention on Biological Diversity. Convention on Biological Diversity, United Nations, p 30

Evanson CK et al (2010) The Nagoya Protocol on access to genetic resources and benefit sharing: what is new and what are the implications for provider and user countries and the scientific community? Law Environ Dev J 6(3):246. Available at http://www.lead-journal.org/content/10246.pdf. Accessed 24 May 2019

Jefferson JD et al (2018) Intellectual property and the governance of plant genetic resources in Mexico: trends and implications for research and innovation. In Ortega-Rubio (ed) Mexican natural resources management and biodiversity conservation. https://doi.org/10.1007/978-3-319-90584-6_6. Accessed 23 May 2019

Laird S, Kerry TK (2002) Biodiversity prospecting: the commercial use of genetic resources and best practices in benefit sharing. In: Laird SA (ed) Biodiversity and traditional knowledge equitable partnerships in practice. Earth Scan Publication Limited, London, pp 241–286

Lee BK, Sohn SY (2016) Patent portfolio-based indicators to evaluate the commercial benefits of national plant genetic resources. Ecol Indic 70:39–48

Nelliyet P (2017) Bio-resources valuation for ensuring equity in access and benefit sharing: issues and challenges. In: Nilayangode Laladhas K, Oommen PV (eds) Biodiversity for sustainable development. Environmental challenges and solutions, vol 3. Springer, Cham

Nelliyet P, Pisupatu B (2014) Economic valuation of bio resources for access and benefit sharing. Report, National Biodiversity Authority, Chennai

Official Gazette of the Rights of Breeders of Plant Varieties of SAGARPA, 2017, taken from https://www.gob.mx/snics/acciones-y-programas/que-es-el-derecho-de-obtentor, October 30, 2018

Pisupati B, Bavikatte KS (2014) Access and benefit sharing as an innovative financing mechanism. Asian Biotechnol Dev Rev 16(2):53–70

Pushpangadan P, Narayanan NK (2005) Value addition and commercialization of biodiversity and associated traditional knowledge in the context of the intellectual property regime. J Intellect Prop Rights 10:441–453

Rege JEO, Gibson JP (2003) Animal genetics resources and economic development: issues in relation to economics valuation. Ecological Economics 45(3):319–330

Vogel JH et al (2011) The economics of information, studiously ignored in the Nagoya Protocol on access to genetic resources and benefit sharing. Law Environ Dev J 7(1):52–65

Williams JT (1993) Genetic resource conservation in natural habitats: biological and socioeconomic dimensions. In: Potter JG, Cohen JL, Janczewski D (eds) Perspectives on biodiversity: case studies of genetic resources conservation and development. AAAS Press, pp 31–42

Chapter 19
Social Participation for Implementation of Camera Traps Projects in Managed and Protected Natural Areas of Mexico

Efrén Moreno-Arzate, Juan Pablo Esparza-Carlos, María Magdalena Ramírez-Martínez, and Luis Ignacio Iñiguez-Dávalos

Abstract Wildlife is an important component among the array of natural resources that the ecosystems can provide to the human communities. In developing countries, with high levels of poverty, wild animals can be an important source of proteins or produce economic benefits through regulated hunting or other uses or receiving a compensation to maintain environmental services. However, to obtain information on diversity, population, and habitat of wildlife to generate management decisions can be a difficult process. The use of camera traps has become an important tool to produce relevant data, but certain level of training to install and operate the cameras and to analyze and interpret the information is required. It is important that communities who are the owners of the resources get involved in the process, increasing both their technical capacity to make decisions and the sense of identity and empowerment of the community. This work focuses on the importance of involvement of local or regional stakeholders and institutions to successfully conduct wildlife monitoring projects. Therefore, we will provide some ideas about the use of camera traps in different spheres of influence and levels of management and present two successful examples of empowerment of the people using this very specialized technique in Jalisco, western Mexico. Finally, we will provide some ideas, based on our experiences, about how to systematize these processes to make them functional and effective.

E. Moreno-Arzate
Doctorado en Biosistemática, Ecología y Manejo de Recursos Naturales y Agrícolas, Departamento de Ecología y Recursos Naturales, Universidad de Guadalajara, Autlán, Jalisco, Mexico

J. P. Esparza-Carlos · L. I. Iñiguez-Dávalos (✉)
Departamento de Ecología y Recursos Naturales, Universidad de Guadalajara, Autlán, Jalisco, Mexico
e-mail: liniguez@cucsur.udg.mx

M. M. Ramírez-Martínez
Departamento de Ciencias de la Salud y Ecología Humana, Universidad de Guadalajara, Autlán, Jalisco, Mexico

A. Ortega-Rubio (ed.), *Socio-ecological Studies in Natural Protected Areas*,
https://doi.org/10.1007/978-3-030-47264-1_19

Keywords Community empowerment · Monitoring · Trap cameras · Wildlife management

19.1 Introduction

Among the elements that integrate the biodiversity of a country, being considered as natural resources to preserve and use, obtaining different types of environmental services, including food and other materials, perhaps one of the most neglected is wildlife. Foresters, cattle and agricultural stakeholders, many governmental officers, and even nongovernmental organizations (NGOs) do not consider the faunistic component of the ecosystems in their plans and decisions. There are several reasons for this. Wildlife is difficult to study, manage, and use. It requires specialized techniques to evaluate populations and their effects, as well as to improve the habitat for specific species. It is easier to get benefits from crops, cattle, and trees, and in many countries, these activities are supported by public policies and economic programs. Even the hunting activity is changing, under the pressure of animal rights organizations and the response of the authorities to this pressure.

However, in many developing countries, the rural communities still depend on wildlife products for their subsistence. In some of them, the meat of wildlife is either the main or a supplementary source of animal protein. Also, some environmental services such as pollination or seed dispersal by mammals and birds could provide more resources for the people. Finally, for the cosmovision of many native cultures, some animals could be part of their spiritual interaction with nature, giving them a sense of integration and pride. Due the pressures of market economy and the globalization process, many of these tangible and intangible values are being lost, or at least diluted, because of the cultural integration and the economic inequities. One of the tools that we believe is important to regain the sense of appropriation and pride in those communities is the use of camera traps to generate information to make management decisions and regain the sense of empowerment along the community.

This work focuses on the importance of involvement of local or regional stakeholders and institutions to successfully conduct projects in protected natural areas (PNA) and "unprotected" areas under management.

Therefore, we will provide some ideas about the use of camera traps in different spheres of influence and levels of management and present two successful examples of empowerment of the people using this very specialized technique in Jalisco, western Mexico. Finally, we will provide some ideas, based on our experiences, about how to systematize these processes to make them functional and effective.

19.2 Background

19.2.1 Protected Natural Areas in Mexico

The protection of natural resources in Mexico is not new, since in pre-Hispanic cultures it was a common practice. As an example, there are archaeological evidences of it in the Yucatan Peninsula, where you can still find *Pet-koot* (live fences) that the Mayan natives of the peninsula used to designate areas with useful plants. Therefore, Mexico has a long tradition in this subject, but due the social and economic dynamic changes, from the colony to modern era, even with all these efforts, it seems that we have not made much progress (De la Maza 2000).

The National Commission of Protected Natural Areas (CONANP) has under its administration 182 natural terrestrial areas, which represent 11.14% of the national territory. Of these, 44 are categorized as Biosphere Reserve, 40 Areas of Flora and Fauna Protection, 67 National Parks, 18 Sanctuaries, eight Areas of Natural Resources Protection, and five Natural Monuments (CONANP 2019). With regard to the marine ecosystem protection, the protected area accounts for 22.05% of the oceanic exclusive economic zone. Also, there are 351 Voluntary Conservation Areas (ADVC) in 24 states, where approximately 86,000 people participate; CONANP supports 339 of these ADVC. These are the protected natural areas of federal level, whose function is of public interest; they are established by means of a decree that is issued by the President through the Ministry of Environment and Natural Resources (SEMARNAT) requested by CONANP. Protected natural areas have grown in extension in the last 79 years, from 1% in 1932 to almost 12% today.

Protected natural areas (PNA) are key in the preservation of biodiversity, and for Mexico, these are of high relevance, since we are one of the few mega-diverse countries on the planet (González-Ocampo et al. 2014). However, the impact of human activities on the environment had become a topic of relevance in the political and academic fields at all levels: local, national, and international. Monitoring is one of the cornerstones of protected natural areas, and the use of technologies has been incorporated in the past few years.

19.2.2 Managed "Unprotected" Areas, Used for Natural Resources Production

About 12% of the national territory is under federal management as protected natural areas. Along the country, another smaller percentage is under state or municipal protection and as voluntary areas under conservation, both in private and communal lands. The rest of the territory is "unprotected" and therefore under management to use the local natural resources.

However, the technical and legal regulations have been advancing, as well as the observation of their accomplishment in different parts of the country. These regulations are focused on generating a sustainable use of the natural resources, including wildlife. This has been a slow but certain change in the way of thinking of both the owners of the resources and the users, mainly due to the public pressures influencing public policies, to support and fund sustainability and enforce the observation of environmental laws. To accomplish this, ejido communities and private owners started to use several methods and techniques that include camera traps.

19.2.3 Use of Camera Traps in Mexico: Research

The study of wildlife is not a simple task; direct observation of animals in natural conditions rarely occurs. In most cases, species are difficult to observe because of their elusive behaviors related to human activity (Wilson et al. 1996; Karanth et al. 2004). The monitoring of medium and large mammals is not easy with traditional methods, including direct visual detections or methods that involve the capture, marking, and/or radio-tracking of individuals, because they can be very expensive and complicated to implement (Chavez et al. 2013). Currently, camera traps are a very efficient tool for sampling and monitoring populations of rare or elusive species because they allow to increase the number of observations or the possibility of detecting species without interfering with their behavior (Kays and Slauson 2008; Chavez et al. 2013).

Camera traps are not a new tool since they are a device that dates back to 1863, when the German explorer G. Fritsch invented the first device to be used in an expedition to South Africa (Guggisberg 1977). An important person in the story of camera traps is George Shiras, who in 1890 used camera traps to photograph wildlife; several of his photographs were published in the National Geographic Magazine. This technology evolved over time, starting from 1863 to the present, where most of the advances were developed in Europe and later in the United States. A relevant work in the middle of the twentieth century was that of Pearson (1959, 1960), who documented movements of nocturnal rodents in a meadow using a video camera and a photo-electric trigger. When photography became a popular activity, the development and use of camera traps also start to grow and improve, thanks to the efforts of many wildlife researchers and interested people to improve the record of different species, and finally, the point of developing sophisticated digital equipment with more efficient sensors, large memory capacity to store large numbers of photographs, and even wireless connectivity is reached. In the late 1980s, American hunters began to use this technology to identify and follow larger specimens that could be trophies. This encouraged an emergent market, as well as the creation of new companies that manufactured camera traps at large scale (O'Connell et al. 2011). The work of Karanth with tigers (*Panthera tigris*) in India (Karanth 1995) was very important because the relevance and importance of the use of camera traps was widely disseminated, contributing to the popularization and use of this tool in a large number of countries.

Regarding Mexico, the history of camera traps dates back to the 1930s. Tappan Gregory, a passionate naturalist and a Chicago lawyer, did several expeditions throughout the United States and Mexico. He was the one who placed the first camera trap in the north of the country, obtaining photographs of mountain lions (*Puma concolor*) in 1937 (Gregory 1939). Later, in another expedition to Mexico, Stanley P. Young in 1946 managed to photograph mountain lions, using baits for the first time to perform a more efficient detection probability (O'Connell et al. 2011).

Finally, when the camera traps began to be a common research tool because a company started to manufacture them at large scale, the first cameras in Mexico were brought in 1994 by Dr. John Laundré, who previously used them in Idaho, USA (Altendorf et al. 2001). These cameras were from the Trailtimer © camera traps brand (Trailtimer Co., St. Paul, Minnesota, USA), using a regular camera photographic roll; they were mechanically activated by an infrared motion sensor, which was placed separated from the camera and connected by a wire. With these cameras, the first paper with camera traps in Mexico was published by Carlos A. López-González (then a PhD student), Dr. Laundré, and Dr. Alberto González; the investigation was carried out in the Sierra de Chamela in the southern part of Jalisco, where they determined the home range of bobcat (*Lynx rufus*) (López-González et al. 1998). Another pioneering work was that developed by a student of Dr. Carlos López, the biology student Lorenzana-Piña. The study area was located in the geographical boundaries between Sonora and Chihuahua. The research focused on carrying out an inventory of medium and large mammals, as well as a description of their activity, back in 2004 (Lorenzana-Piña et al. 2004). Since then, the use of camera traps in Mexico has grown fast in quantitative and qualitative terms, as well as in the geographic distribution of researches (Mandujano 2019).

19.2.4 The Role of Hunters in Northern Mexico

Mexico and Latin America civilizations have a great culture around hunting since pre-Hispanic times; people have looked for wildlife as a resource for different uses, being mainly food the most sought in the country (Alvard 1995; Naranjo et al. 2010). Recently, the north of Mexico began to develop a hunting tourism industry, in the 1960 and 1970 decades, along the northern states of Sonora, Tamaulipas, Chihuahua, and Nuevo León. In the beginning, the hunt was carried out in ranches with large land extensions, of private owners who initially had contacts in the states of the US border, such as Arizona, Texas, and California (Villarreal 2008). This tourism brought with it new technologies to make hunting activity more efficient. Among the technologies were the camera traps, initially used for the selection of the best trophy (Balderrama pers. comm. 2019). Later, in 1997, SEMARNAT (the Ministry of Environment and Natural Resources) created a new entity to improve conservation and use (including hunting) of the wildlife and their habitats: the Wildlife Conservation and Management Units (UMA), in which a branch of SEMARNAT, the General Directorate of Wildlife, is the agency responsible for the implementation and regulation of it. It was a strategy of the Mexican government to provide rules to all those farms dedicated to hunting, management, and conservation throughout the country (SEMARNAT 1997). Within the UMA, the first camera traps began to be used in the northern states at the farms that already had greater hunting operations. Probably, the use of camera traps has increased (Balderrama pers. comm. 2019) because the extension of land destined to be UMA increases every year in Mexico, presently occupying approximately 28.95 million hectares (14.74%) of the country area (SEMARNAT 2008; Tessaro et al. 2009).

19.2.5 Accompaniment: Government Institutions and Technical Service Providers

In Mexico, ongoing research projects related to the use of camera traps have been created, which aim to study wildlife biology and ecology, in order to contribute to the long-term recovery and conservation of the species. Within the Ministry of Environment and Natural Resources (SEMARNAT), the National Commission of Protected Natural Areas (CONANP) administers several programs focused on the conservation of flora and fauna species, which there are three major fields of action, in which the use of camera traps is applied. Such is the case of the Recovery and Restocking Program for Species at Risk (PROCER). This program aims to provide knowledge for conservation of at-risk species and their habitat, promoting the collaboration and participation of higher education institutions, civil society organizations, ejidos, and communities, in order to reach sustainable development. The second field are the Action Programs for Species Conservation (PACE). Its objective is to consolidate, promote, and apply specific conservation strategies for populations of priority species in Mexico. Unlike other recovery programs in the past, PACEs are structured and executed with the active participation of all actors related to the specific species, in a co-participation and co-responsibility scheme (SEMARNAT 2019a). The third program where camera traps are used for monitoring is the Conservation Program for Sustainable Development (PROCODES), which aims to promote conservation of ecosystems and their biodiversity in priority regions through the improvement of living conditions of the human communities (SEMARNAT 2019b).

The Program for the Protection and Restoration of Ecosystems and Species at Risk (PROREST) is among the most recent environmental programs that use camera traps, which are considered part of the main axes to promote the conservation and restoration of the representative ecosystems into the natural protected areas, as well as the protection and preservation of their biodiversity, through the support of technical studies and derived actions (SEMARNAT 2019c). Also recently, the Biological Monitoring Program in Protected Natural Areas (PROMOBI) has the general objective to contribute to generate information on the conservation status of species, higher taxonomic groups, and specific ecosystems, through the participation of higher education institutions, as well as research and civil society organizations, in the execution of biological monitoring activities in protected natural areas and their influence areas (SEMARNAT 2019d).

As part of the strategy to promote mechanisms of payment for environmental services (PSA) in Mexico, the Federal Government, through the National Forestry Commission (CONAFOR), did create two initiatives since 2004, the Hydrological Environmental Services Program (PSAH) and the Program to Develop the Market of Environmental Services by Carbon Capture and Biodiversity Derivatives and to Promote the Establishment and Improvement of Agroforestry Systems (PSA-CABSA). The objective of both is to maintain conditions in the ecosystems that favor the generation of various environmental services, for which a contract is settled between the owner of the land and CONAFOR, where the owners agree to maintain forest cover or carry out practices to conserve the natural ecosystems inside their property, and the federal agency undertakes to pay a fixed compensation per hectare for a 5-year period. By becoming beneficiaries of these payment programs for environmental services, the owners must prevent changes in land use and are encouraged to carry out surveillance activities in the areas under payment, in order to avoid illegal logging, poaching, forest fires, as well as other activities potentially harmful to ecosystems. CONAFOR recently integrated the use of new technologies in monitoring such as sound and ultrasound recorders and camera traps in the latest version of the National Forest Inventory. With these new technologies, different wildlife species were documented more efficiently within the Mexican forest ecosystems (SEMARNAT 2019e). However, recently, the resources for Payment for Environmental Services (PSA) had diminished dramatically (CONAFOR, pers. comm. 2019). The experience shows that once this kind of supports has gone, in many of the preserved areas which were supported by PSA, forest is rapidly transformed on agricultural or cattle lands.

In the protected natural areas, many community guards have been trained, through different projects that have been able to obtain financing to pay them. Among many examples through the country, we can mention a community guard in the Sierra de Manantlán Biosphere Reserve (RBSM), who managed to take photographs and videos of a pair of jaguars mating: the female jaguar "Matlida" (with a radio tracking collar since 2013) and a male (Fig. 19.1). Also, an outstanding photo of a jaguar was taken by another community guard in the RBSM (Fig. 19.2). In some other areas of Mexico, as in the states of Durango, Oaxaca, and Yucatán, there are also community guards, financed by PSA projects, who take excellent photographs of wildlife because of their knowledge and effort. However, resources are always limited, and in the face of the sway of government resources, there is a continuous risk of losing these excellent community monitors.

19.2.6 An Example of Organization in Western Mexico

To not depend totally on political decisions, it is necessary to create funds that are not managed exclusively by governments, but through synergic alliances, including NGOs, government agencies, universities, and communities. One example of

Fig. 19.1 Photography of two jaguars after mating (the female has a radiotelemetry collar), taken by a community guard trained at the RBSM. A related video is available in the Directorship of the RBSM facebook page: https://www.facebook.com/reserva.sierrademanantlan/videos/847530725407329/

Fig. 19.2 A male jaguar. Photography taken by Elionardo Elías Bartolo, community guard at the RBSM

collaboration in this type of projects with citizen monitoring is the MiJO project (Integrated Management of the Jaguar habitat in Western Mexico, through community participation). It is developed on intervention areas of the Patrimonial Biodiversity Fund (FPB), operated by the National Forestry Commission (CONAFOR) in the states of Jalisco, Nayarit, and Durango, and aims to ensure connectivity between protected natural areas and forested lands of high ecological value. The objective of the MiJO Project is to strengthen community support, through the development of management skills and practices that preserve and improve the habitat of jaguars and the associated biodiversity. Among the beneficiaries, the aim is to raise awareness of the ecological and socioeconomic importance of felines, with emphasis on the jaguar, to reduce their hunting and that of their prey, to mitigate conflicts between cattle and felines, and to strengthen community monitoring of biodiversity. Project beneficiaries are encouraged to form a wildlife monitoring brigade. In workshops, organized by the Department of Environmental Services of Forest and Biodiversity of CONAFOR, plus academic and civil society organizations, who collaborate with the project, people from communities and ejidos are advised to keep in mind the importance of community work for conservation, in order for them to get external funds for maintaining support for these activities.

By linking the wildlife component, there is an increase on the visibility for the project of payment for environmental services. The jaguar, as a flag species, is taken to attract resources from international funders for the protection of jaguars and the biodiversity that this entails, in areas with strong infrastructure pressures, illegal activities, and marginalization, which generally do not receive support for conservation.

Within the MiJO project, several workshops are held with the support of experts from the University of Guadalajara, civil associations (FMCN, COVIDEC AC, Biocenósis AC, Nenenki AC), local municipalities (Autlán, Casimiro Castillo, Acaponeta), intermunicipal watershed boards (JIRA, JISOC), and related PNA (CONANP, Sierra de Manantlán Biosphere Reserve, Marismas Nacionales Nayarit Biosphere Reserve, Sierra de Vallejo-Río Ameca Watershed Protection Reserve). The experts explored options together with the communities, in order to change the negative perception of the jaguar and to promote the appropriation of their environmental richness and fragility, and certain activities include use of camera traps, photo trapping data generation and database management, ecology of jaguar and other felines, talks to avoid conflicts with cattle, as well as local jaguar festivals for environmental education. The project aims to ensure that community monitoring improves the ecological knowledge of both communities and scientists to allow an adaptive management at local level, strengthen governance, consolidate the identity of the community, and make management decision process more inclusive. All these goals involved specific technical knowledge for the jaguar and its prey photo trapping, but it has also been more extensive. The protocol designed and led by CONAFOR, for community biodiversity monitoring (BIOCOMUNI), has been promoted. This monitoring protocol includes vegetation description, bird recording, mammal tracks and scats, as well as photo trapping (cameras are required to be working 40 continuous days in the dry season and another 40 days in the rainy season).

In the MiJO project, it has been sought to emphasize that communities and landowners must have trained personnel for the use and installation of camera traps, to photograph medium and large mammals, especially jaguars and their prey; also they must be trained in the use of Global Positioning System (GPS) to locate monitoring stations. The ideal scenario is that the local inhabitants of the communities are those doing these tasks.The most encouraging result has been with camera traps. Camera traps encourage the formation of conservation agents: participants in general are active and passionate about capturing wildlife photographs. In the exchange of experiences, due to enthusiasm and knowledge of wildlife behavior and habitat use, people in the communities achieve excellent wildlife photographs. In the monitoring of wildlife in certain communities, there is still a strong dependence on the hired technicians: sometimes, the community people are dedicated to install the cameras, check them, and give the data to the hired technicians. Therefore, it is necessary for communities to appropriate the knowledge and learn how to use it as a management tool.

This is a complex process, within which it is important to continue training for systematization and data processing. We had a good experience organizing workshops for data management and generating results. So, these workshops have the following objectives: (a) to diminish the fear about the use of computers; (b) to create awareness that photos are information; (c) to make databases in spreadsheets, with basic monitoring data; and (d) to train on how to present and interpret results with tables and figures. The last one can be done in a simple way with dynamic tables that summarize the information: lists of species, species richness, and abundance of species by type of vegetation and so on. A very important part in community monitoring is the coexistence with other neighboring communities and even other regions: so, as the people of the communities show their photos to others, results create a positive effect and generate great pride in the people who are dedicated to monitoring in other places. In addition, they exchange experiences on how to improve the capture of wildlife photos and the effort of obtaining a single photo. In these workshop initiatives, the same participants arise, for example, that photos must be shown to the schools because the children of their communities do not even know what wildlife is in the area, a phenomenon that other academic researchers have found (Álvarez et al. 2015; Esparza-Carlos et al. 2019; Mascote et al. 2016).

There is a major challenge in community monitoring: moving forward to have the information process by themselves. In our experience with people in rural areas, they show us that with training, communities can generate their databases and obtain listings, tables, and graphs that summarize the information of wildlife monitoring. The inclusion of young people in this stage of the process makes it easier to generate more independent communities. As many of those who participate in monitoring are older people having no contact with computers, they are generally reluctant to use them. However, young people have already used computers and are attracted to new technologies, making it easier for them to understand data processing. In addition, the inclusion of young people in these groups allows the transmission of knowledge from older people to younger people. A serious problem in communities is the loss of this knowledge.

Another challenge is the labor division and empowerment. Some of the forestry technical service providers generally do all the work on information processing of the camera traps. They are probably afraid to not be indispensable anymore. However, we consider that works on the contrary: training communities in the generation of databases and processing and information does not exclude technicians, researchers, or government agents; in fact, it leaves them more time to do other activities. It is still important for them, but with less investment of time, to verify the generated information and the opportunity to perform more complex analyses or even time to seek more support for the communities. Therefore, the labor division is important to enhance capacities: many of the people in rural communities are excellent at photographing wildlife because they know the terrain, know the fauna, or have been hunters (as some who were hunters become excellent photo-trappers).

19.3 Study Cases

Rural Monitoring: The Need to Include Youth and Women in Wildlife Monitoring

In order to ensure the dialogue and exchange of knowledge in the communities, it is necessary to include young people in monitoring activities. In addition, it is necessary to include women, who contribute a different vision to that of men. A case that is a good example is Fernanda Lizeth Gabriel García; she was a 17-year-old teenager when she started monitoring wildlife and is still continuing in 2020, 5 years later. She is an inspiration to young women, who work in conservation and sustainability and in overcoming challenges. Fernanda collaborates in the project "Integrated management of jaguar habitat through community participation in western Mexico (MiJO)" implemented by CONAFOR in collaboration with the NGO Northwestern Fund (FONNOR) and financed by the U.S. Fish and Wildlife Service (USFWS). She is a young female, which is a unique case in wildlife monitoring since the majority of the participants are men over 40 years old (Fig. 19.3); her voice is now heard in the community, which is a challenge, especially in rural communities of Mexico, where macho culture is deeply rooted. She lives in the locality of Santa Monica (Ejido Ruíz Cortines, Municipality of Ayutla, Jalisco), a small town of 96 inhabitants. This ejido is located in the Sierra de Cacoma, which is part of the Sierra Madre del Sur, and together with the Sierra de Manantlán Biosphere Reserve, they are a priority area for jaguar conservation (Chavez et al. 2016) within the Central Pacific Biological Corridor of Jaguar (Ceballos et al. 2018; Mercado 2019).

Fernanda began to participate when the Biodiversity Patrimonial Fund together with Peace Corps summoned the ejidos that were beneficiaries of the CONAFOR Payment for Environmental Services program to work on photo-trapping skills. She began to participate due to the invitation of her father: "At that time my father was the ejido head and they told him to look for 2 or 3 young people to learn how to use

Fig. 19.3 Wildlife (including jaguar) monitors, technicians, and instructors of a training workshop in the MiJO project (FONNOR-PSA/CONAFOR). Fernanda L. Gabriel (pink shirt, next to the jaguar's head) is the only woman who participates directly in monitoring with camera traps. Her participation inspires other young people in the region

the computer to store the data. We were three girls, and now I'm the only one" (F. L. Gabriel com. Pers. 2019). They attended workshops, organized by the FONNOR project "Systematic registration of monitoring and participatory mapping" supported by the Patrimonial Biodiversity Fund. The participation of a second young woman, Nayelli Villegas, decreased due to lack of permission from her parents to attend workshops and fieldwork, plus her activities as a teacher in the community; the third girl only attended the workshops and did not continue.

Fernanda's motivation was to "learn new things and since I realize what was the subject, I was excited to see everything we could discover with the photo-trapping." The first time they set up camera traps, she was accompanied by her dad. After a month, while doing the first revision, she was surprised because there was a very good picture of a puma (*Puma concolor*). The surprise was because "we had heard about (the puma)," but "they had no evidence that it was there." Another surprise was because "it is not far from the community! It is 800 m straight line from the community." This photo was a trigger to continue monitoring wildlife; she comments that "I got more excited! I wanted to continue after that first photo!" Fernanda has attended all the workshops organized by FONNOR-PSA/CONAFOR project MiJO, where she has been trained, strengthening biodiversity monitoring techniques, information processing, aspects of ecology and importance of felines and wildlife in general, and strategies to reduce conflicts related to livestock, among others.

Fig. 19.4 Photographs taken by Fernanda L. Gabriel within the community biological monitoring at the MiJO project (FONNOR/CONAFOR) (ocelot) and Peace Corps/CONAFOR initiative (puma)

She comments: "I have been 5 years (with this activity) ¡Every time you get more excited because you learn new things! Looking the photos, you move the cameras where you believe they can pass through"... "with the time spent and the results motivated me to see photos of all the wildlife we had nearby ...." Along the years, in her community, they have photographed several times the puma, ocelot, bobcat, white-tailed deer, and squirrels, among others (Fig. 19.4). Fernanda tells us: "yet, with each new photo I am more excited, and I want to continue on that, working even more!" Now she continues with that enthusiasm, and her goal is to have a photograph of the jaguar, if there is still in the ejido (which is possible).

At a regional level, the story of Fernanda is known, transmitted by word of mouth, speaking about "some girls from the Sierra de Cacoma interviewed by NATGEO." This refers to an article that appeared on the blog of NATGEO in English (Blois 2016). Also, people talk about "some high school girls who won a photogame contest." With the motivation of Matt Blois, the group she leads did participate in an international photo contest (Trailcampro 2016 Photo Contest, at the International category). She says that at first, "she felt excited because we were winning," but in the end, they finished in the tenth place, although she achieved the third place in number of "likes" in the Facebook voting process. The important point is the sense of achievement, as she felt very satisfied with the results of this first tryout "since I was working barely a year with this (photo-trapping)." This has been a key impulse to ensure that Fernanda remains involved in monitoring, despite the sociocultural and administrative complications of the ejido model. Her contribution along several years strengths the community, due to continuity in the accumulated expertise of the monitoring team (usually, ejidos change monitors when the head of the ejido is replaced). Therefore, there is an increase in the possibilities that conservation work is maintained and replicated for several more years. Furthermore, the permanence promotes the participation of young people like her brother and other teens or young people of other communities; this region is where more young people participate in monitoring.

Fernanda is a conservation agent (sensu Sodhi and Ehrlich 2010) and feels satisfaction because it supports her community; she explains that she does it: "to benefit the community and the ejido; also to make them react… showing them photos: You have these varieties of wildlife in the ejido, whose, if the hunting continues, in the end the affected ones will be themselves." But it was a difficult road up to now because in Mexico, as in much of the world, the macho culture relegates women, especially in rural environments, in small and isolated locations, accentuated by the lack of work options for them: Fernanda is dedicated to the tasks related with the payment of PES to the community but also must collaborate in the farming work with her father, mostly agriculture and livestock.

She says that "The first time I stand to give a report they (men) didn't take me into account, they didn't take me seriously" … "Most of the ejido members are old people; if one told them something, they were closed in their beliefs, in their things" … "Comments due to the fact of being a woman, (like) what are you going to know about the wilderness? We have walked much more than you." However, she continued working and has managed to be a voice that people listen in her community and in the region, promoting the conservation of natural resources and the role of women and youngsters in the community: "They are already becoming aware of the care they should have (with nature), in creating awareness in all people in the community" … "When there are community meetings, I inform about what I found." In the future, she plans to continue "To keep making people aware of how monitoring is useful to us, and that we must take care of the animals' habitat" … "Raise awareness in the elderly and work with the young people, so they grow with this idea" … "Everything that is worked on means advances, and therefore any effort is worth it to do". Citizen monitoring: The work of Jesús Valle Guzmán, M.D.

Citizen monitoring: The work of Jesús Valle Guzmán, M.D.
The municipality of Tecolotlán is located in the central region of the state of Jalisco. The Sierra de Quila Flora and Fauna Protection Area (APFFSQ) is located next to the municipality urban area. The APFFSQ is an area with the typical characteristics of mountain ranges in western Mexico, in geology, soils, climate, and vegetation. In the region, there has been a constant hunting of the fauna that inhabits the Sierra de Quila, which has generated a culture and strong roots in hunting throughout the area, as in other parts of the country with forested areas adjacent to population centers.

Within the hunter community, there are people with great respect for nature; such is the case of Mr. Jesús Valle Guzmán, a 56-year-old physician. Dr. Valle (as is widely known in his home town) is recognized throughout the municipality for being a surgeon as well as a good hunter, in addition to being an excellent naturalist, with great appreciation for the biological diversity present in the region (Fig. 19.5). After practice hunting for many years, he realized that eliminating the fauna in an inappropriate way affected negatively the animal populations, so that thought became a point of inflection for him, and he decided to hunt the wildlife but through cameras, both photographic and video.

This is how he met new technologies such as camera traps; knowing and using them in the field allows him to establish a personal citizen monitoring financed with his own resources, placing the first analog camera traps in 2000, so he currently has been monitoring for about 20 years in the Sierra de Quila and surrounding areas. Presently, he continues placing analog (i.e., film roll) camera traps, as well as modern digital equipment, acquired through the years. Within his portfolio, he has a lot of information, such as photographs of his personal cameras, material of the camera

Fig. 19.5 Dr. Jesús Valle, who generated an example of long-term citizen monitoring on the wildlife of the APFF Sierra de Quila

traps, and edited videos that he has taken with special camouflage costumes within walking distance of the wild animals.

Currently, Dr. Valle continues to collaborate with different academic institutions, civil organizations, and the authorities in charge of the Sierra de Quila preservation and management. One of the most important collaborations was with us at the University of Guadalajara; together we managed to document the first jaguar records in the region and the protected area, which confirmed its presence, an issue that had been speculated since long time ago. Dr. Valle is very passionate about nature, and we consider this is why his long-term effort has generated one of the most notable examples of citizen monitoring in the western part of Mexico, contributing with his great knowledge to the preservation of the biodiversity at the local and regional scales.

19.4 Conclusion: A Methodological Proposal to Reach Appropriation and Empowerment by Communities

Based on our experiences, we present some recommendations to impulse the use of camera traps as a strategy to empower communities interested in the PSA programs or the development of wildlife management for hunting or other uses.

(a) Interest in the Use of Camera Traps

The first step should be to arouse the community interest in monitoring wild mammals in order to reach specific goals: inventories, biodiversity, abundance of particular species, behavior, habitat relationships, and species abundance, among others. In other cases, it may be for hunting purposes, based on the UMA model. This interest has generally been created for agents outside the community, such as NGOs, academic institutions, government, technical service providers, and touristic services brokers.

(b) The Community Is Allied to the Project

One key to success using this technique is the approval of community members about participation of allies in order to integrate people with wildlife knowledge, including biologists, hunters or wildlife observers interested in participating and contributing to land management and conservation.

(c) Training of the Use of Camera Traps

The personnel will require training on theoretical knowledge (wildlife biology and ecology), adequate designs to achieve the management objectives, and technical knowledge to operate the equipment. Since manuals generally come in non-native language and have a lot of technical terms, this technical training initially requires companion to explain and reinforce the theoretical and technical knowledge. However, the prospect is that in the future, people will become increasingly independent, until they are fully responsible for the use of the equipment, training new prospects and developing the technique until they become experts to discuss results and make new contributions collaborating with academic or technician collaborators.

(d) Integrate Mixed Community Groups

The participation of resource users must be open to all those who are interested, but to be highly successful, it is important to include people with different skills, experience, and perspectives:

1. Older people, to take advantage of their field knowledge, who have experience tracking and describing animal behavior; many of them are or have been hunters.
2. We recommend that young people must be integrated, whether they are college students or not, since young people are prone to be linked through technology. These youngsters can be important in the processing of information and databases and the generation of useful results to be applied by the community in order to obtain support for community promotion and related management issues.
3. The inclusion of women is important, since they have another perspective that can facilitate greater success with the camera traps technique.

(e) Integrate Community Members to Observe the Work with Cameras

A good feature about the camera traps is that they create a strong impact on people, by photographing animals that walk along the same paths that they do. This generally surprises the community members. In addition, the inhabitants are also surprised to see mammals that they had not seen in their community, since many of them are nocturnal animals difficult to observe directly. Frequently, it awakens an interest in taking the best picture of the rarest animal. People who do monitoring generally have recognition in their community and in neighboring towns.

(f) Less Participation of External Agents

We believe that conservation and management of natural resources must be in the hands of the owners of the land, and therefore, they are the direct managers, users, and beneficiaries of its own richness. The other interested allies, academics, technicians, NGOs, and government agencies, should plan that gradually the field work and its use to develop information and knowledge remain in the hands of the communities and achieve empowerment through this technique. Conservationists must focus on a regional, state, or national systematization to process and take advantage of the information, incorporating it in wider perspectives.

(g) Information Processing

To achieve empowerment, it is necessary that the community can make its own databases with the camera traps to process them and obtain results (descriptive statistics). This can be done manually, with paper and calculator or using spreadsheets in a simple way, because its logic is constant over time. Although it is possible to train community people in the use of specialized software or analysis techniques, as these require constant updating due to changes in computational versions or analytical methods, this can be promoted only with people that show a long-term compromise with the task. They can be encouraged to course higher education to develop the adequate skills to improve management.

(h) Generating Reports

The results should be expressed according to needs, such as list of species, abundance of species by habitat type, richness, diversity, community composition, trophic structure, and conservation issues. These results are presented in charts or graphs. In our experience working with communities, especially young people or young adults, they easily acquire this knowledge and elaborate useful descriptive results for the community.

(i) Use of the Information

The information should be used to support projects financing production, conservation, or tourism endeavors. Community people should take advantage if they have photos of species with risk category as this attracts the attention of funders or external community agents.

(j) Make the Information Available to the Members of Your Community

This is necessary because it is increasingly common for people, especially children, to ignore the fauna of their community. Their knowledge is indirect and it is increasingly common for people to have misperceptions of wildlife because mass media favor these distortions (Álvarez et al. 2015; Esparza-Carlos et al. 2019; Mascote et al. 2016). What the community monitors obtain must be presented (a) in schools at all levels; (b) to adults in community meetings and ejido meetings, livestock associations, and farmers, among others; (c) in environmental festivals, where the photos of the local fauna are presented, accompanied by knowledge of people who participate in the monitoring; (d) in social networks, to publicize the achievements of the communities.

(k) Meetings to Exchange Experiences, Knowledge Is Reinforced

It is necessary to organize meetings to exchange experiences. These meetings must include all the people who are involved in carrying out monitoring: landowners, ejidatarios, indigenous communities, technicians, academics, NGOs, government agencies, etc., to obtain feedback and to incorporate new knowledge plus technical improvements. In these meetings, people must present their results because it generates pride that their work is seen and valued by others and at the same time motivates them to continue with this activity.

(l) Integration of Information Must Be Available to the Public

Something that does not exist in Mexico yet, although it is a subject that has been raised in meetings with experts on jaguar and there have been initiatives of CONABIO and CONAFOR, is the creation of a platform that has all the systematized information that is generated in the different monitoring efforts (CONABIO

2012). This idea has not materialized yet. These platforms should be accessed by anyone who, through a process of registering user data and commitment to give credit, wants to study.

References

Altendorf KB, Laundré JW, López González CA, Brown JS (2001) Assessing effects of predation risk on foraging behavior of mule deer. J Mammal 82(2):430–439

Alvard M (1995) Shotguns and subsistence hunting in the neotropics. Oryx 29:58–65

Álvarez NS, Gerritsen PRW, Gómez Llamas JC (2015) Percepciones campesinas del Jaguar en diez localidades de la Reserva de la Biosfera Sierra de Manantlán en el Occidente de México: implicaciones para su conservación. Sociedad y Ambiente 1(7):35–54

Blois M (2016) Tech-savvy teenage girls are the camera-trap pros in this Mexican community. https://blog.nationalgeographic.org/2016/09/06/tech-savvy-teenage-girls-are-the-camera-trap-pros-in-this-mexican-community/

Chávez C, De la Torre A, Bárcenas H, Medellín RA, Zarza H, Ceballos G (2013) Manual de fototrampeo para estudio de fauna silvestre. El jaguar en México como estudio de caso. Alianza WWF-Telcel, Universidad Nacional Autónoma de México, Mexico. 106 pp

Chavez C, Zarza H, de la Torre JA, Medellín RA, Ceballos G (2016) Distribución y estado de conservación del jaguar en México. In: Medellín RA, de la Torre JA, Zarza H, Chavez C, Ceballos G (eds) El jaguar en el siglo XXI. La perspectiva continental. Fondo de Cultura Económica, Universidad Autónoma de México, Mexico, pp 47–92

Ceballos G, Zarza H, Cerecedo-Palacios G, Lazcano Barrero MA, Huerta M, De la Torre A, Rubio Y, Job J (2018) Corredores biológicos y áreas prioritarias para la conservación del jaguar en México. Alianza Nacional para la Conservación del Jaguar. Ciudad de México, México

CONABIO (Comisión Nacional para el Conocimiento y Uso de la Biodiversidad) (2012) Áreas protegidas decretadas de México. SEMARNAT-CONANP

CONANP (Comisión Nacional de Áreas Naturales Protegidas) (2019) Áreas Naturales Protegidas decretadas. 2019. Consulted in: http://sig.conanp.gob.mx/website/pagsig/datos_anp.htm (24 de septiembre 2019)

De la Maza R (2000) Una historia de las áreas naturales protegidas en México. Biblioteca interactiva de medio ambiente. INE-SEMARNAT 1995-2000. México

Esparza-Carlos JP, Gerritsen PRW, López-Parraguirre SA, García-Rojas MD, Peña-Mondragón JL (2019) Cómo perciben los niños el jaguar, *Panthera onca* (Carnivora: Felidae) en Jalisco, México. Revista de Biología Tropical, 67:380–395

González-Ocampo H, Cortés-Calva P, Iñiguez-Dávalos LI, Ortega-Rubio A (2014) Las áreas naturales protegidas de México Natural. Investigación y Ciencia de la Universidad Autónoma de Aguascalientes 60:7–15

Gregory T (1939) Eyes in the night. Thomas Y. Crowell Co., New York, NY

Guggisberg CAW (1977) Early wildlife photographers. Taplinger Publishing Co., New York

Karanth KU (1995) Estimating tiger (*Panthera tigris*) populations from camera-trap data using capture-recapture models. Biol Conserv 71:333–338

Karanth KU, Nichols JD, Kumar NS (2004) Photographic sampling of elusive mammals in tropical forest. In: Thomson WL (ed) Sampling rare or elusive species. Island Press, Washington, DC, pp 229–247

Kays RW, Slauson KM (2008) Remote cameras. Pp 110–140, in: Long RA, MacKay P, Zielinski WJ, Ray JC (eds). Noninvasive survey methods for carnivores. Island Press, Washington, DC

López-González CA, González-Romero A, Laundré JW (1998) Range extension of the bobcat (*Lynx rufus*) in Jalisco, Mexico. Southwest Nat 43:103–105

Lorenzana-Piña GP, Castillo-Gomez RA, Lopez-Gonzalez CA (2004) Distribution, habitat association, and activity patterns of medium and large-sized mammals of Sonora, Mexico. Nat Areas J 24:354–357
Mandujano S (2019) Analysis and trends of photo-trapping in Mexico: text mining in R. Therya 10:25–32
Mascote C, Castillo A, Peña-Mondragón JL (2016) Perceptions and knowledge of the jaguar among children in communities neighboring the Montes Azules Biosphere Reserve in Chiapas, Mexico. Tropical Conservation Science 9:1–11
Mercado A (2019) Nuestra biodiversidad: el jaguar esta en casa. Material de divulgación para beneficiarios del Fondo Patrimonial. Fondo del Noroeste A.C. 8 p.
Naranjo EJ, López-Acosta JC, Dirzo R (2010) La cacería en México. Biodiversitas 91:6–10
O'Connell AF, Nichols JD, Karanth KU (2011) Camera traps in animal ecology. Methods and analyses. Springer, Tokyo. 271 p
Pearson OP (1959) A traffic survey of *Microtus-Reithrodontomys* runways. J Mammal 40:169–180
Pearson OP (1960) Habits of *Microtus californicus* revealed by automatic photo records. Ecol Monogr 30:231–249
SEMARNAT (Secretaría del Medio Ambiente y Recursos Naturales) (1997) Programa de conservación de la vida silvestre y diversificación productiva del sector rural: 1997–2000. Ciudad de México, México
SEMARNAT (Secretaría del Medio Ambiente y Recursos Naturales) (2008) Sistema de Unidades de Manejo para la Conservación de la Vida Silvestre (SUMA). Subsecretaría de Gestión y Protección Ambiental de la Dirección General de Vida Silvestre. http://www.semarnat.gob.mx/gestionambiental/vidasilvestre/
SEMARNAT (Secretaría del Medio Ambiente y Recursos Naturales) (2019a) Programa de Recuperación de Especies en Riesgo (PROCER). https://www.gob.mx/conanp/acciones-y-programas/programa-de-conservacion-de-especies-en-riesgo
SEMARNAT (Secretaría del Medio Ambiente y Recursos Naturales) (2019b) Programa de Conservación para el Desarrollo Sostenible (PROCODES). https://www.gob.mx/conanp/acciones-y-programas/programa-de-conservacion-para-el-desarrollo-sostenible-procodes-2019
SEMARNAT (Secretaría del Medio Ambiente y Recursos Naturales) (2019c) Programa para la Protección y Restauración de Ecosistemas y Especies en Riesgo (PROREST) en. https://www.gob.mx/conanp/acciones-y-programas/programa-para-la-proteccion-y-restauracion-de-ecosistemas-y-especies-en-riesgo-prorest-2019
SEMARNAT (Secretaría del Medio Ambiente y Recursos Naturales) (2019d) Pago Por Servicios Ambientales en Áreas Naturales Protegidas. https://www.gob.mx/conanp/documentos/pago-por-servicios-ambientales-en-areas-naturales-protegidas
SEMARNAT (Secretaría del Medio Ambiente y Recursos Naturales) (2019e) Programa de Monitoreo Biológico en Áreas Naturales Protegidas (PROMOBI). https://www.gob.mx/conanp/acciones-y-programas/programa-de-monitoreo-biologico-en-areas-naturales-protegidas-promobi
Sodhi NS, Ehrlich PR (2010) Conservation biology for all. Oxford University Press, Oxford
Tessaro S, Hernández-Huerta A, Delfín-Alfonso C, González-Gallina A (2009) Unidades para la conservación, manejo y aprovechamiento sustentable de la vida silvestre en México (UMA). Retos para su correcto funcionamiento. Investigación Ambiental 1(2):143–152
Villarreal J (2008) Historia del manejo del venado cola blanca *Odocoileus virginianus* en Nuevo León, México. En: Memorias del XX Simposio sobre Venados en México. "Jorge G. Villarreal González", Facultad de Medicina Veterinaria y Zootecnia-UNAM, México, D. F
Wilson D, Cole R, Nichols J, Rudran R, Foster M (1996) Measuring and monitoring biological diversity, standard methods for mammals. Smithsonian Institution Press, Washington, DC/London

Chapter 20
Socioenvironmental Affectation of Coffee Production Activity in Tributaries of La Suiza River at El Triunfo Biosphere Reserve, Chiapas

Maria Eugenia González-Ávila, Guiyer Eulselmar Vázquez-González, and Walter López-Báez

Abstract Coffee production in the high areas of Chiapas has a great socioeconomic and environmental importance because of the quality of the coffee bean produced in the region. Nonetheless, during the washing process of coffee beans, great volumes of water are required and are generally discharged without treatment, causing effects on health and the environment. Thus, this study aimed to assess the potential socioenvironmental affectations caused by the discharge of mead or *honey* and the pulp derived from coffee production in the *humid benefit stage* (*HBS*) in La Suiza Microbasin. La Suiza is part of El Triunfo Biosphere Reserve in Chiapas. The methodological focus used mixed techniques, which included a survey, interviews, and indicative parameters of physical-chemical contamination. The results showed that coffee growers and representatives of local authorities are aware of the pollution caused by tributaries into La Suiza River. However, they would be willing to modify their humid benefit processes, specifically the humid benefit stage if they received technical support and assistance. They are aware that the reserve ecosystem services, including capture and water treatment, are necessary for coffee cultivation in the area.

Keywords Water · Honey · HBS · Perception

M. E. González-Ávila (✉)
Departamento de Desarrollo Urbano y Medio Ambiente, El Colegio de la Frontera Norte, Monterey Nuevo León, Mexico
e-mail: megamar@colef.mx

G. Eulselmar Vázquez-González
Independent Consultor, Monterrey, Nuevo León, Mexico

W. López-Báez
Centro de Investigación Regional Pacífico Sur, Instituto Nacional de Investigaciones Forestales Agrícolas y Pecuarias, Inifap, Chiapas, Mexico
e-mail: lopez.walter@inifap.gob.mx

A. Ortega-Rubio (ed.), *Socio-ecological Studies in Natural Protected Areas*,
https://doi.org/10.1007/978-3-030-47264-1_20

Abbreviations

CESMACH	Ecological Farmers of Sierra Madre de Chiapas
CONAGUA	National Water Commission
CONANP	National Natural Protected Areas Comissions
CONAFOR	National Forest Commission
CONAM	National Environment Commission
CONANP	National Natural Protected Areas Comissions
COD	Chemical oxygen demand
ECOSUR	El Colegio de la Frontera Sur
EPA	US Environmental Protection Agency
FONCET	El Triunfo Conservation Fund
GIAT	Intercommunitary Group of Territorial Action
GWF	Global Water Foundation
INMECAFE	Mexican Coffee Institute
INIFAP	National Institute of Forestry, Agricultural, and Livestock Research
LSMB	La Suiza Microbasin
PIACC	Integral plans for climate change adaptation
SAGARPA	Ministry of Agriculture, Livestock, Rural Development, Fishing, and Food
SEMARNAT	Ministry of the Environment and Natural Resources
SWRCB	State Water Resources Control Board
TNC	The Nature Conservancy
NPOs	Nonprofitable organizations

20.1 Introduction

In Mexico, the states who are coffee producers are in the central-south part of the territory: Chiapas, Veracruz, Puebla, and Oaxaca, which represent 94% of the national coffee production, 85% of the harvested surface, and 83% of the producers (Flores-Vichi 2015). Chiapas is considered the leading entity in coffee production because of the high yield per hectare that goes beyond any national export (Flores-Vichi 2015). At world level, it is the leading state in organic coffee production with an average of 18 million tons cultivated and processed annually by more than 60 thousand producers located in different regions of the entity (Mariscal 2011; Flores-Vichi 2015).

The region that stands out in Chiapas is Frailesca, located in La Suiza Microbasin (LSMB) in the municipality of Montecristo de Guerrero. The polygon that conforms all the basin is situated within the Sierra Madre (a mountain chain) with a surface of 6437 ha, where 82% of its extension is found within El Triunfo Biosphere Reserve (TriBioRe). The reserve is characterized by having a cloud forest that provides important ecosystem services, among those that stand out are water recharge and provision, mainly for the municipalities of Montecristo de Guerrero and Angel

Albino Corzo that are agricultural activity areas where coffee, corn, and legumes are mainly cultivated (López et al. 2014; Castro and López-Báez 2014).

Coffee cultivation is the greatest agricultural activity in this microbasin and the main source of economic income for resident families (Benedetti et al. 2012). Nevertheless, coffee production implies multiple activities that go from planting, renovating, and harvesting up to its humid benefit stage (HBS), which consists of detaching the pulp or skin of the mucilage (honey) of the coffee bean made by means of a mechanical depulper that requires high water volumes.

The humid benefit stage is the most used process in La Suiza Microbasin and other coffee cultivation areas because of water availability and the least time to obtain the coffee bean still in *parchment*,[1] which implies less cost when compared to other dry processes. It is important to highlight that the coffee bean in the humid benefit stage requires high water volumes. Hence, Bello-Mendoza et al. (1993) quoted that to produce 1 kg of coffee, approximately 20 L of water are required during washing and depulping until a clean coffee bean is obtained for sale. Other authors have reported that the volumes could reach up to 2000 and 300 L of water/kg of coffee produced, which depends on the type and conditions of the humid benefit stage.

Water consumption used at this stage in Las Suiza Microbasin would not seem exaggerated, considering that Chiapas has high precipitation that goes from *1200 to 4000 mm* (INEGI 2019) compared with coffee production regions in the central zone of Colombia where precipitation is from 2000 to 2500 mm (Gómez and Jaramillo 2017). However, the basic problem does not lie in the high consumption of water, but in the socio-environmental impacts associated with discharged honey and pulp,which increase the acidity of the soil, humidification, mineralization(Ángeles 2015) and alteration of the natural landscape (Romero et al. 2012). Besides the previous effects, honey and pulp discharged into water bodies also add up altering acidity values (pH), suspended solids, and chemical oxygen demand (COD) that generates serious disequilibrium in the receptor ecosystems and in general biological processes of water bodies (Matuk-Velazco et al. 1997); moreover, odor and change of water conferment favor the presence of insects that are vectors of different diseases for human beings and other organisms (Álvarez et al. 2011).

The socioenvironmental impacts previously mentioned have already been reported for La Suiza Microbasin by different authors (Castro and López 2014; López-Báez et al. 2014; Palacio-Herrera 2012); nevertheless, the problems still persist, such as lack of economic resources and technical training; obsolete equipment for the coffee-washing process (COMCAFE 2013); irregular distribution of the populations along the microbasin; lack of basic services of drinking water, sewage, and waste collection; and deficient health services (Benedetti et al. 2012; CONEVAL

[1] In México, coffee can also be classified as washed coffee, which is prepared by humid benefit; it constitutes the main type of coffee that our country produces and exports, estimating an average of 86% of the total green coffee produced during the last years (Fundación Produce and ITESM, 2003:14). Recent studies have indicated that the coffee-washing process has not changed due to the lack of technical support (Medina, et al. 2016).

2015). All these factors increase community vulnerability facing climate change and the consequences, such as landslides, fire, earthquakes, and hurricanes, just to mention some climate events that increase economic and environmental problems for these communities (Morales et al. 2010).

Thus, the objective of this study was to identify the affectation of honey wine discharge during humid benefit stage into the main river of La Suiza Microbasin, located in El Triunfo Biosphere Reserve. For this purpose, a mixed methodology was used, consisting of a documentary analysis, semi-structured survey applied to coffee growers, interviews to institutional actors related with the topic, and monitoring of physical-chemical indicative parameters, both in situ and water bodies of superficial waters close to the humid benefit stage.

The development and compliance of the objectives are shown in five sections of this chapter: (1) Introduction, (2) Description of the Study Area, (3) methodology explanation, (4) Analyses and Discussion, and finally (5) Conclusion.

20.2 Description of the Study Area

La Suiza is an exorheic microbasin that flows into La Angostura Dam; it is part of the sub-basin Yahuayita that belongs to Hydrological Region No. 30 called Alto Grijalva (Gutiérrez-Vizcaíno 2014). Such microbasin has an extension of 6083 ha with a perimeter of 37 km (Palacios-Herrera 2012) located within the municipality of Montecristo de Guerrero, and both are situated within El Triunfo Biosphere Reserve. Its surface represents 84%, that is, 5190 ha, of La Suiza Microbasin being part of the reserve (Palacios-Herrera 2012).

At hydrographic net level, La Suiza Microbasin comprises tributaries highly vulnerable to landslide effects and contamination due to the agrochemicals used in conventional agriculture, both in the high and medium parts of the basin besides the contamination problems caused by fecal matter of outsiders and the existing septic tanks (Gutiérrez-Vizcaíno 2014).

The type of soil of the basin is shallow mollic Leptosol, which is extremely gravelly and rocky with less than 20% of fine land, making it highly vulnerable to erosion. However, soils covered with mesophile forest allow high water infiltration in relation to those that have corn and coffee cultivations. Thus, the high zone of La Suiza Microbasin is where the greatest rainwater catchment takes place, supplying local populations and productive activities, of which coffee is one of them (López-Baéz et al. 2014; 2018).

On the other hand, different management strategies have been implemented in the microbasin at the level of integral plans of climate change adaptation (PIACC) where water quality has been assessed with the participation of the communities, whose objective is to reduce vulnerability of the natural systems and decrease the impact caused by anthropogenic activities through social participation (Castro and López 2014). In fact PIACC has the task of identifying the social actors to make synergy with researchers and governmental authorities, such as the Fondo de Conservación El Triunfo (FONCET), a global environmental organization called

The Nature Conservancy (TNC), Comisión Nacional de Areas Protegidas (CONANP), Global Water Foundation (GWF), and Instituto Nacional de Investigaciones Forestales, Agrícolas y Pecuarias (INIFAP); this last one has, among its competences, monitored superficial water quality necessary for agricultural activities and specially for coffee production (López-Báez et al. 2014; Castro and López-Báez 2014).

At social level La Suiza Microbasin has six communities, of which only five were included (Puerto Rico, Vista Alegre, Monte Virgen, Río Negro, and Toluca). The community of Zapata was not included because it did not accept participating in the study (Fig. 20.1). Such communities are considered in poverty conditions with low education levels and lack of basic public services, such as water, electricity, and health (CONEVAL 2015). Paradoxically, this zone provides the greatest coffee production and transformation in the basin (Benedetti et al. 2012), showing extreme poverty levels where coffee cultivation is more profitable than other grains, such as corn and beans. Furthermore, cultivating coffee has higher culture and identity value among the indigenous farmers that work in common land property (Sánchez-Suárez 2015), and only three large private plantations exist, which are La Suiza, El Olvido, and La Candelaria (Benedetti et al. 2012).

At economical level, coffee has the most profitable cultivation, but in the last years, its price in the national and international markets has dropped, which has led families to deforest greater surfaces of the biosphere reserve to use them as new cultivation areas, hoping to compensate the drop in coffee production and price in the international market. In addition, the price of coffee beans and change of land use have led to biodiversity losses and water contamination, placing not only family

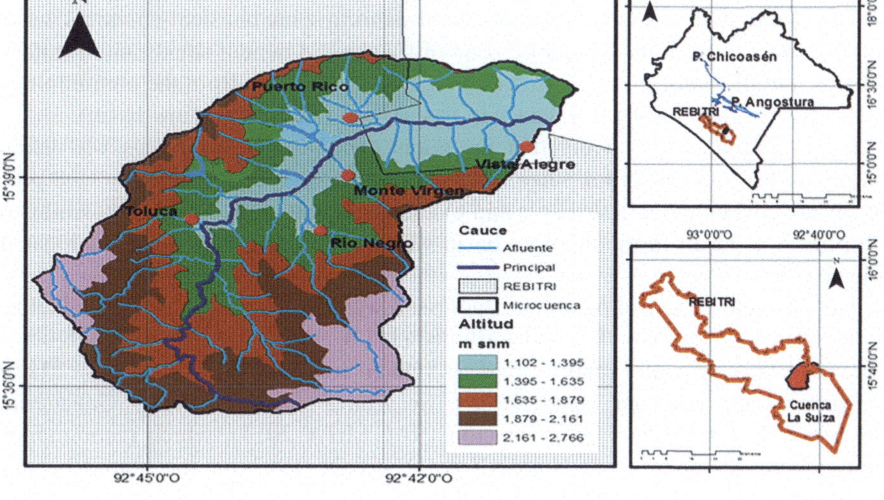

Fig. 20.1 Location of the communities in La Suiza Microbasin. (Source: Taken from López-Báez et al. 2014)

income of the population at risk but also a series of ecosystem services that La Suiza Biosphere Reserve has to offer (López-Báez and Reynoso-Santos 2016).

20.3 Materials and Methods

This study assessed the socioenvironmental affectation of the honey discharge from the humid benefit process by means of a mixed method that consisted of qualitative and quantitative assessments divided into two stages:

(a) Consulting: bibliographical review of the area of study and coffee production activity, selection of the community affluents to be monitored, and the humid benefit stage used by the community. The design of the questionnaire to be applied to the coffee growers was made considering five topics: (1) socioeconomic aspects (basic data of the respondents), (2) water use, (3) water quality of the main river, (4) characteristics of the humid benefit stage wastewater, and (5) water conservation in the basin. Sampling points were also established in La Suiza Microbasin, and an interview guide was designed for the key actors.
(b) Field: the questionnaire (n = 50) with 20 multiple choice questions were applied randomly to coffee growers older than 18 contacted within the 5 communities in study that belonged to the Grupo Intercomunitario de Accion Territorial (GIAT).

As to the interview guide, it was applied to seven key actors: the coordinator of Insituto Mexicano del Café (INMECAFE) in Chiapas, three coordinators of coffee production organizations, two coordinators of nongovernmental organizations (NGOs), and one coordinator of operating water manager, namely, Organismo de Cuenca Frontera Sur (OCFS).

The quantitative assessment comprised measuring some physical-chemical water parameters both from the discharge zone of five humid benefit stages and tributary streams into the main river, La Suiza. Such discharges come from the communities of Río Negro (high basin), Toluca (high basin), and Puerto Rico (low basin).

It is important to point out that the greatest number of samplings was taken mainly in the high part of La Suiza Microbasin, mainly because common land producers are concentrated in that area and to a lesser extent in the lower area, while no sampling was taken in the medium basin where the Zapata community is located (Fig. 20.2).

The physical-chemical parameters assessed were temperature (°C), pH, alkalinity (mg/L), hardness (mg/L), and turbidity (JTU), which according to Salguero-Zecena (1996) are water quality markers related to honey wine discharge. These parameters are assessed regularly by Secretaria de Agricultura, Ganaderia, Desarrollo Rural, Pesca y Alimentacion (SAGARPA 2014), which recommends applying the Mexican norm NOM-CCA-027-ECOL-1993 for the humid benefit stage in the basin (Gaceta Ecológica 1994).

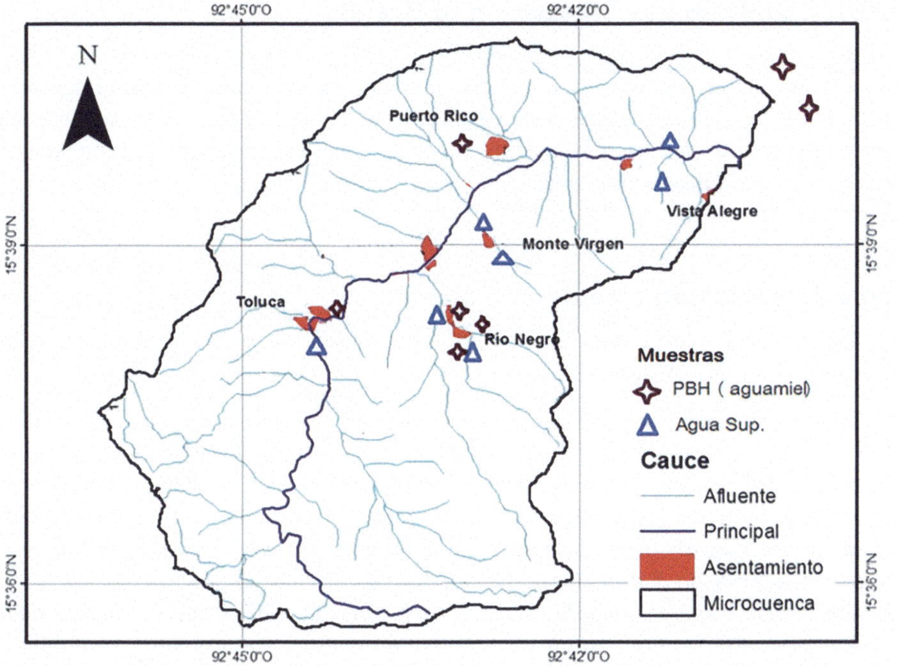

Fig. 20.2 Sampling points of superficial water and the humid benefit process (FBP). (Source: Adapted from Castro and López-Báez 2014)

These parameters were assessed by the *Global Water Watch* (GWW) method for water quality in rural environment and recommended by the US Environmental Protection Agency (US EPA) for their efficiency in measurements in situ. It is worth to mention that the analyses were referential to samplings that were taken from February to March (2016); according to those reported by Benedetti et al. (2012), it was within the dry period that allowed coffee bean drying.

Finally, with the information obtained from surveys, interviews, and the physical-chemical analyses, a database was made in Excel (version Windows 10). Results were shown in graphs and tables for analysis and discussion. Surveys were grouped in five categories (socioeconomic, water use, superficial water quality, humid benefit wastewater, water conservation) and interviews in water use, socioenvironmental effects, and institutional management. As to physical-chemical parameters, they were shown in two categories: honey wine and superficial water discharge.

20.4 Analyses and Discussion of Results

Results are shown in two sections: the first one shows qualitative findings that include the surveys applied to coffee growers and key social actors; the second one shows the results and discussion of the physical-chemical parameters of the area of study and coffee production sites.

20.4.1 Qualitative Analysis

(a) *Socioeconomic Aspects*

Of the 50 respondents, 76% were men (38) and 24% women (12), who according to that reported by Gutiérrez-Vizcaíno (2014) were specifically women who have a tight link with water because of their social role within their communities and their household (hygiene, food preparation, and cleaning). Nonetheless, female producers do not necessarily participate in decision-making processes in coffee production assemblies. They do participate in the sun-drying coffee bean process but not in the humid benefit stage as such. Hence, López-Baéz et al. (2014) quoted that female coffee producers did not participate directly on topics of water quality, availability, and consumption because of the role they had in the family, but common land females should be encouraged to participate in coffee production assemblies for decision-making and also be trained in monitoring water quality in La Suiza

Fig. 20.3 Assembly of coffee producers carried out in 2016

Microbasin because of their commitment in this topic. Thus, it is important to mention that such activities need to be planned with a gender perspective that would allow women equilibrium and equality among family functions, decision-making, and water quality monitoring. Otherwise, it would only increase their work load and responsibility, which would lead to widening the inequality breach in which female coffee producers live in relation to their male counterparts, including other women in Chiapas, Fig. 20.3.

In the case of male coffee growers, the survey detected they have the role of decision-makers within the coffee production assemblies besides being the food and labor activity providers, which not necessarily implies they are committed to water quality topics even if they know the importance of good water quality for their cultivations. Thus, the respondents suggested raising awareness in relation to water use and ecosystem services for the communities of La Suiza Microbasin.

The educational level of the respondents was 64% with primary education; 26% were illiterate; 6% had secondary education; and only 4% studied primary school. In general terms, 74% of the surveyed population knew how to read and write, which allowed them to participate in training projects to monitor the risks of climate change and promote the sustainable use of water and management programs in La microcuenca la Suiza as indicated by a TNC representative (2011).

The average age of coffee producers in the communities of Toluca, Rio Negro, Monte Virgen, Puerto Rico, and Vista Alegre was 44 years old, while it was less than 44 in the Zapata community; that is, the population groups are in productive age to cultivate coffee, which according to that reported by Escamilla et al. (2005) are within the age range of the coffee producers of the region, which is 49 years old with regional variations from 43 to 54 years of age.

With respect to years committed as coffee growers, it varied slightly by basin area; for example, those of the high basin had been dedicated to this activity for 17.6 years, while it was 17.1 years for those of the low basin and 15.5 years for those of the medium basin, which agrees with the period of coffee value increase and production capacity that Chiapas recorded from 2008 to 2010 (FIRA 2016).

As for coffee production technique, 96% of the respondents indicated that coffee production was their family subsistence type, and only 4% performed it individually. This coincides with that reported by Vargas-Venci (2007) and García-Benítez et al. (2015), who quoted that coffee production was a family matter because of the number of family members that represented the labor force for cultivation and other activities at home. This type of property was family-based in 100% of the cases.

As for type of coffee production (organic or conventional), 52% produced coffee with the conventional method and 20% organic; and 28% did not answer to this question. It is important to remember that in organic coffee production, strict controls in quality and environmental protection are required, which include a fertilizer-pesticide-free cultivation besides a certification as sustainable producer (Palomares et al. 2012). Even so, organic coffee production consumes 9 L of water that has to be of excellent quality (Vargas-Venci 2007) and where water from the depulper is not considered, which is why water consumption should be above 20 L.

As to land extension dedicated to coffee cultivation, the data obtained indicated a total of 116.76 ha for the cycle 2015–2016, in which the sowed surface in the low basin was 44.25 ha, followed by 41.05 ha in the medium basin and 31.46 ha in the high basin, which was the least sowed area (Fig. 20.4). In total, coffee production was 27657.5 kg of parchment or washed coffee, of which the medium basin was the most productive with 9976.25 kg of coffee, followed by the low basin with 9660 kg and the high basin with 8021.25 kg. In such productive cycle, 2060 men and 740 women participated, as it was the season that required greater physical effort, which was why men worked and women only participated in recollection, storage, and bean selection, but not in all the humid benefit stage, as quoted by Vargas-Venci (2007).

(b) *Water Use*

With respect to the water volumes used in the humid benefit stage and depulping, coffee producers ignored how much water they consumed per stage, so data from other authors were taken into account, indicating that water consumption depended on the infrastructure built. Thus, if the consumption of 40 L of water per kg (BH traditional) was considered, then 1,924,000 L (1924 m^3) of water would be consumed, which could vary according to the type and condition of the humid benefit stage and type of coffee (Table 20.1).

With respect to supply and shortage of the water used in the humid benefit stage, 48% of the respondents expressed they had never experienced shortage; 6% indicated sometimes; and only 2% referred to always; the rest of the respondents did not provide an answer to this question. As to the presence or absence of water, it seemed to be that this problem was not perceived by the coffee growers because the area traditionally has annual precipitation from 2000 to 3000 mm in the low part of El Triunfo Biosphere Reserve and from 2500 to 4500 mm in the high part, mainly from April to October (López-Báez and Reynoso-Santos 2016).

In the case of the perception of water quality deterioration causes in the basin, the responses were related to the geographic location of the persons surveyed. For

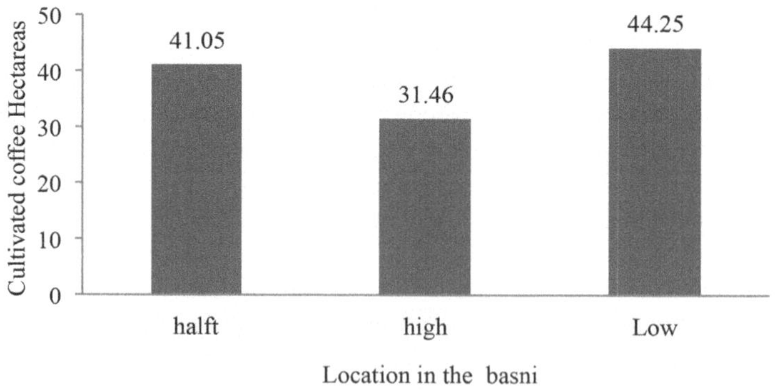

Fig. 20.4 Coffee production per hectare and site location in La Suiza Microbasin. Made with field data

Table 20.1 Water consumption per type of humid benefit stage

Water consumed by coffee (L/kg)	HBS type and coffe bean phase (country/city)	Author
20.5	Traditional (Mexico, Chiapas)	Bello et al. (1993)
40	Traditional (Colombia)	Arango et al. (1999)
24	Ecological (gold coffee) and traditional (Nicaragua)	Guerrero (2001)
10 to 13	Traditional (gold coffee) (Salvador)	Molina and Villatoro (2006)
20	Ecological (Colombia)	Cruz Palacios (2009)
40 to 50	Traditional (Colombia)	Cenicafe (2015)
10	Ecological (Colombia)	

Source: Authors' own creation

example, those located in the low basin referred to the inhabitants of the high and medium basins as the responsible for water quality deterioration.

On the other hand, in the question on water consumption used in the humid benefit stage, 48% of the respondents referred that the coffee-washing stage was the moment when more water was consumed; 14% indicated it was depulping; and 14% indicated that "milling" or *devanado* (selection and elimination of foreign material, density, and maturation degree of coffee cherries) was the moment when more water was consumed. The remaining respondents, 24%, did not know. However, different authors have quoted that water consumption depends on the physical condition and type of humid benefit (see Table 20.1).

With respect to the water-saving methods known, 34% of the producers did not know about the topic; 14% indicated they knew about the topic; and 52% did not know. As to the main actors that promote water-saving methods, Instituto Nacional de Investigaciones Forestales, Agrícolas y Pecuarias (INIFAP) is the main institution that coffee producers perceive as managing water saving and quality. However, other agencies also work on this topic, such as The Nature Conservancy (TNC) that has created seed and water fund for the sustainable management of Alto Grijalva, Sierra Madre, and coastal basin (López-Báez and Reynoso 2016) and is associated to public-private initiatives, such as civil associations.

Among the factors that hinder adopting any method for water saving by the producers, 38% indicated it was due to the lack of technical consultancy; 10% referred to the lack of organization; 20% expressed other reasons, while 24% indicated that it was due to lack of money; 8% referred to the lack of infrastructure; and the remaining respondents, 38%, did not know. It is important to highlight that in this point, the institutions that were regularly mentioned related to water-saving topic were INIFAP (belongs to Secretaría de Agricultura, Ganadería, Desarrollo Rural, Pesca y Alimentación) and in the health sector Secretaria de Salud, which has performed campaigns to prevent diseases associated to water or by the presence of insects that inhabit contaminated water. As for the Comisión Nacional del Agua and Secretaria del Medio Ambiente y Recursos Naturales, nothing was mentioned about their presence or intervention in any action competing with water quality in the basin.

Finally, when the question "Are any water conservation actions performed at the end of the humid benefit stage?" was asked, 24% of the respondents indicated that good water management was not performed; 12% indicated it was regular and that the humid benefit stage wastewater was sent to an infiltration pit instead of discharging it directly to a water body, while 6% indicated water conservation was performed; and the rest 58% did not respond. It is worth to mention in this point that although the infiltration wells are a way of removing and minimizing impacts of honey wine discharge, it does not imply that the problem disappears or much less neutralizes because it is a temporary mitigation method, which could be causing contamination to underwater and soil (Molina 1999). Vega-Montenegro (2012) reported that infiltration pits were not efficient and that its efficiency depended on soil and slope characteristics.

(c) *Superficial Water Quality*

In general terms, the respondents considered that the superficial water used at the humid benefit stage was of good quality despite the discharge they dump. In fact 96% of the producers considered they get good water quality at this stage and indicated not having had problems to this respect. Another 2% considered that water quality was bad and it could affect the humid benefit stage; and the rest, 2%, did not answer the question.

In relation to the type of problems detected by the honey wine discharge and with it the negative effects on water quality, 32% indicated none; 26% considered bad odor, 22% mosquitoes, and 10% health problems; and the other 10% indicated that it caused other problems without mentioning any of them (Fig. 20.5).

With respect to the question, "Do you consider that water from the washing and depulping processes have an effect on water quality of the river and streams?" 50%

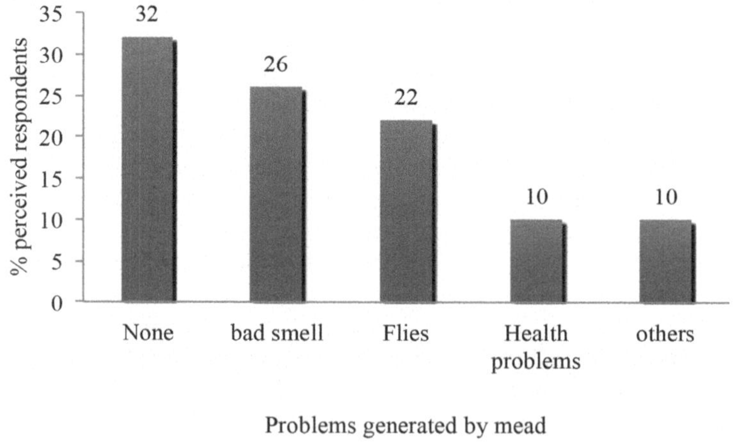

Fig. 20.5 Problem perception with the water for the humid benefit stage. Made with field data

indicated no; 26% knew of someone who had this problem; and 24% did not respond. This question helped to respond the next one about which communities got contaminated water: 14% indicated Puerto Rico (high basin); 22% mentioned Emiliano Zapata (high basin); and 64% indicated not knowing.

It is worth highlighting that the community Emiliano Zapata is the one that has caused the greatest change of land use in the forest areas, which is why they only have 10.59 ha with forest vegetation, necessary to recharge the aquifer, and the rest, 464.37 ha, is land dedicated to coffee cultivation (López-Báez and Reynoso-Santos 2016), which is considered as the second community that produces more coffee in La Suiza Microbasin, after the plantation Finca La Suiza.

In response to the question "Do you consider that bad use of water in washing and depulping stages has negative or positive effects on water quality of the rivers and streams?" 52% admitted that it had a negative effect, while 2% indicated it was positive, and 46% did not know. Undoubtedly, it was evident they had detected the problem, and they knew the causes. However, they did not consider they were the direct cause of contamination but indicated that other producers located in the high basin were the cause of contamination. This type of perception was quoted by Benes (2010), who mentioned that others were always causing environmental harm, that is, they were not personally responsible for their actions.

Finally, in the question "Has any of the laborers or family members experienced gastrointestinal, eyes, or skin diseases caused by drinking water from or bathing in the main river?" 52% indicated no health problems; 2% responded yes; and 46% did not answer. These data contrasts with that reported by López-Báez et al. (2014) who recommended installing a purifying water plant to reduce to 95% the cases reported on diarrheic diseases in the area (López-Báez and Reynoso 2016).

(d) *Wastewater Quality of the Humid Benefit Process*

The perception of the coffee growers on water quality of the depulping discharge discarded to La Suiza Microbasin indicated that 22% expressed the water was dirty, 14% detected solid waste (trash), 8% considered that other factors affected the river, and the rest 56% of the respondents did not answer. As to the destination of honey wine to decrease the harm caused, 34% dumped it into an infiltration pit; 12% let it

Table 20.2 Reports of environmental impacts caused by the humid benefit stages

Substage	Residue	Pulped	Environmental impact
Despulpado	Pulp	Solid	Alters chemical oxygen demand in receiving water bodies (García 2014)
	Pulping water	Liquid	It generates a bad smell, attracts insects by the amount of sugar, and alters the natural landscape (Salguero 1996)
Washed	Mucilage	Liquid	Contaminant capacity of 25% for any receiving body (Pineda et al. 2001)
	Mead	Liquid	It mainly alters water quality with an acidic pH and a high COD (García 2014)

Source: Taken from Vázquez et al. 2019

drain naturally; 4% used other removal system; and 50% did not answer. With respect to the pulping and honey wine discharge, several authors indicated that environmental impacts were significant, above all in the biochemical processes of the water bodies and soil (Table 20.2).

Under the previous context, INIFAP and TNC have recommended to encourage the majority of the coffee growers to dump honey wine and pulp discharge in filtering wells, which would attenuate the direct environmental impact to soil and close water bodies. The magnitude or type of environmental harm caused has not been assessed yet, so as to be able to prevent or mitigate it as required and, if so, avoid accumulative impacts. Even though water quality has been monitored, drainage and filtration flux or movements that reach La Suiza River have not been assessed to know of any effects on human, flora, and fauna health, much less in the potential affectations of the own agricultural activity of La Suiza Microbasin to propose the required mitigation measurements.

(e) *Water Conservation in La Suiza Microbasin*

In general terms, coffee growers showed disposition to participate in any water conservation project of La Suiza Microbasin, which was detected when 70% of the respondents indicated they were willing to participate in actions for water conservation of the basin, and only 30% did not respond to the question. To this respect, according to that reported by López-Baéz and Reynoso (2016), the communities of coffee growers knew, both by living experience and through the training implemented, the importance that water had in their lives and the productive activity of the microbasin, as well as the effects caused by climate change.

As to how willing they would be to participate in improving water quality, 34% responded with community or group work; 14% responded individually; and 52% would not participate with economic contribution. However, when they were asked about their economic participation to improve their equipment for honey wine and pulp treatment, 38% responded they would be willing to invest; 14% would not invest; and 50% did not answer. The only observation made was that they would invest provided they had the support or technical advice, so they would not lose their investment.

Finally, with respect to this topic, when coffee growers were asked about the institutional actors that helped them in water conservation, 28% indicated INIFAP, 16% other institutions that offer social support, 4% FONCET (El Triunfo Conservation Fund), and 2% CONAFOR (National Forest Commission). However, none of the respondents mentioned SEMARNAT or CONAGUA, which are the main institutions that keep an eye on wastewater discharges into the natural environment and that affect water bodies. This result reflected the degree of involvement that INIFAP has in both actions and technical assessment and training in these coffee production areas.

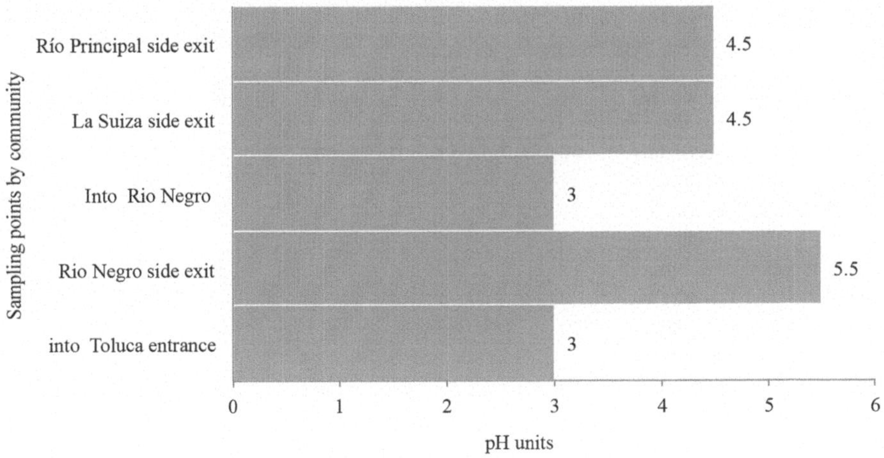

Fig. 20.6 Values of honey wine pH discharged in situ. (Source: Made from field work)

20.4.2 *Quantitative Analyses*

(a) *Honey Wine Discharge*

The pH measure of the honey discharged by the Río Negro community located in the upper basin (1CA-RN and 3CA-RN) showed a pH range of 3 to 5.5 (acid), outside of wahts is established by the Mexican standard (NOM-CCA-027-ECOL-1997) for the went discharge stage (Ecological Gazette 1994). While the temperature of the water discharged with honey was 8 ° C above 20 ° C for the main river water. Which contrasts with that reported by Castro and López (2014) as a pH = 7.2 and average temperature from 19 to 22.2 °C (Fig. 20.6). Nonetheless, the pH and temperature reported in the honey wine discharge by the Rio Negro community affects the biological activity of the water body and accelerates organic matter decomposition; at the same time dissolved oxygen decreases, according to that reported by the State Water Resources Control Board Meeting (SWRCB 2013).

On the other hand, Bailly et al. (1992) and Bello-Mendoza et al. (1993) referred that temperature over 7 °C in honey wine affected any biological process in the water used and discharged after washing. On the other hand, Gómez (2010) referred that a temperature increase in honey wine discharged in soil favors microorganism growth that affected local vegetation causing plant leaf loss and obstruction of functions and with it a state of lassitude in plants.

The production of honey wine discharged at high environmental temperatures causes affectations in local flora and fauna and human health risk (Alfaro and Rodríguez 1994). To this respect, coffee producers detected quality changes in water based on odor and color expressing them as "strong odor" and "stinky-dark" water, denoting it as landscape alteration and, in general, social environment.

Table 20.3 Quality parameters of superficial water stations in La Suiza Microbasin

Sampling points	Water body	T (°C)	pH	COD (ppm)	Alkalinity (mg/L)	Hardness (mg/L)	Turbidity (JTU)
Rio Negro side exit	Secondary stream	17	7.5	7.4–7.0	65	60	2
Into the Rio Negro	Secondary stream	17	7.5	6.8–6.8	80	70	2
Into the Monte Virgen	Secondary stream	21	7	7	30	50	2
Monte Virgen side exit	Secondary stream	18	7	7.2	40	50	2
La Suiza side exit	Secondary stream	20	6.5	6.8–6.6	85	90	2
Entrada a Toluca	Secondary stream	16	7	7.3	40	70	2
End the río principal	Principal river	20	7.5	7.6	65	70	2

Source: From field work

(b) *Discharge of Honey Wine in Superficial Currents*

Regarding the quality of the water from the effluents that transport honey and pulp that discharge the populations of Toluca and Puerto Rico. Its temperature and alkalinity were detected to be severely altered, because the contaminated water in the upper part of the basin is concentrated in the main river, this due to the hydrological behavior of the basin. Such fact was perceived by the respondents as a bad smell and mosquito presence.

The parameters of the secondary affluents assessed for this study (Table 20.3) did not show significant alteration and no parameters in relation to the Mexican norms NOM-127-SSA1–1994 (DOF 2000) and NOM-001-ECOL-1996 (DOF 1997) published in the official government bulletin. These results agree with the community monitoring promoted by INIFAP jointly with GWF in La Suiza Microbasin in 2013, which indicated that water did not have the quality for human consumption, above all due to the infiltration of the septic tanks from open-air waste discharge that the communities had done (Castro and López 2014).

The previous information had been identified by authors López-Báez et al. (2014), so it would be convenient to perform a systematic and exhaustive monitoring in the honey wine and pulp discharge before, during, and after harvest. Moreover, sampling stations should be relocated because they might be affected by the septic tanks and leachate from waste discharge, which could mask the contamination caused by the humid benefit stage.

Figure 20.7 shows pH values between stream discharges, with emphasis on the honey discharge, which caused a decrease in pH and, with it, the previously mentioned environmental affectations. However, because the monitoring performed in

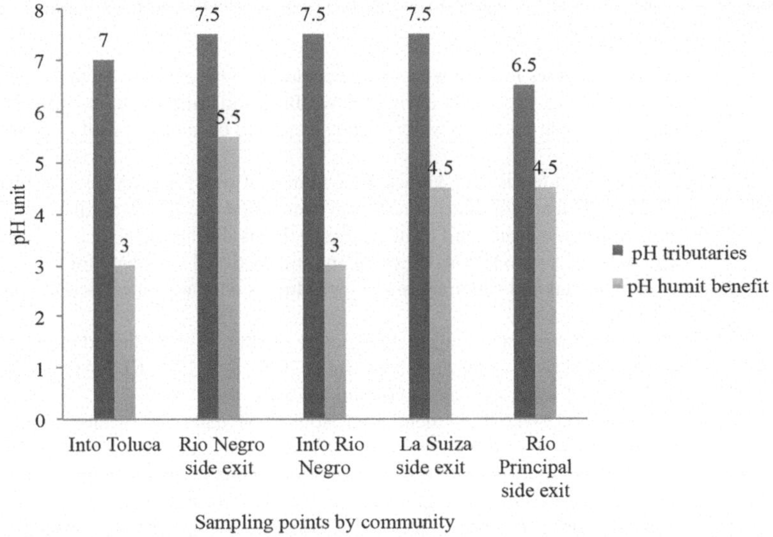

Fig. 20.7 Values of superficial pH and discharge areas of the humid benefit stage (Source: Made from field work (December–March 2016))

this study was timely, it is suggested that future studies should be performed medium to long term and by type of humid benefit.

(c) *Perception of the Respondents in Water Use, Management, and Quality*

Table 20.4 summarizes the main ideas detected during the interviews to key actors, among which they considered grass, coffee, corn, and banana as the plants that need abundant water. However, they clarified that in the case of coffee, it does not consume much water during cultivation per se, but it does during the humid benefit stage, specifically during washing and depulping, which is necessary to obtain the commercial coffee bean.

None of the respondents indicated knowing the exact or approximate water volumes used during the humid benefit stage and much less those that were discharged during the honey wine and pulp process.

As to the compliance of the official Mexican norms for water discharge of the humid benefit stage, the head of Organic Coffee Certification, CESMACH (Campesinos Ecológicos de la Sierra Madre de Chiapas), and the plantation Finca Triunfo Verde indicated that their members complied with the norms, without mentioning any, because all the communities that produced organic coffee must comply with international norms of no use of agrochemical, sustainable water use, providing other ecosystem services, etc., among other requirements to be certified as organic producers and thus be able to sell their product in European markets at a good price.

The Mexican norm NOM-CCA-027-ECOL-1993 (Gaceta Ecológica 1994) was not mentioned; they only referred to the NOM-001-SEMARNAT.1996 (SEMARNAT

Table 20.4 Perception summary of water discharge problems in the humid benefit stage

Topic	Key ideas
Water use	Everyone agrees that the wet coffee beneficiary phase is the one that consumes the most water, especially during the washing and pulping phase, with other large crops that consume water such as corn and bananas, as well as cattle graze None of the interviewees knows the volumes of water used during the FBH; Most of them are familiar with the environmental and health regulations applied to wastewater and drinking water that is discharged. The only ones who follow Mexican and international standards for environmental and water are organic producers who are supervised to comply with the standards. As for whether they know the agencies or institutions that should regulate the use and quality of water discharges by BHF, CONAGUA, CONAP, SAGARPA, and the state water agency and organizations like TNC and GWF were mentioned. Some have had relationships with institutions such as GWF, INIFAP and TNC, which they identify as institutions that work on water quality and help with problems in the basin in general; There is a spatial or physical limitation that defines their action in terms of water use and conservation for any of the interviewees, and all agreed that there is no adequate use of water in the FBH and in general in the basin.
Environmental impact	All respondents indicated that in the FBH the discharges of mead and pulp are causing environmental damage, loss of vegetation by deforesting the surface of the TriBioRe to continue growing coffee. They mentioned that discharges of mead and pulp cause acidity, dusting, increased organic matter, and effects on the biochemical processes of adjacent aquatic ecosystems. These damages are usually detected by bad smell and by the change in color of surface water bodies. There is no manual or guide for the management of the discharges, and the ones who count work rules are the producers of organic coffee, while the conventional ones only perform pits or infiltration wells, although it is not always done. None of the respondents has received or will invest money for water conservation, and only a technician from FONCET indicated that they can manage applications for this type of project, although the limitation is the lack of financing and the acceptance of new projects by community.
Water conservation	None of the respondents have managed finances for water conservation, and only one FONCET technician indicated that it can accompany a request to manage projects of this type, although the limitation is lack of financing and acceptance of new projects by the community. They suggest, as measures to improve the quality and use of water treatment plants, environmental education courses so that communities decrease water consumption at home and the FBH itself, until the use of alternative technology, wetlands, advice, and support for coffee growers.

1996) that established the maximum limits of contaminants permitted in wastewater discharge in water bodies and the NOM-127-SSA1–1994 (DOF 2000) that established the treatment and quality limits permitted to which water should be subjected for purification. Although both norms apply to water quality and use for human consumption and water bodies, none of the official authorities they represent have competence in monitoring or following up on water quality for the environment or human consumption in the basin.

The respondents indicated that governmental authorities from CONAGUA, SEMARNAT, CONAM, and SAGARPA should be mainly present with actions in monitoring water quality, use, and conservation, but that was not the case. On the other hand, those that performed more activities were INIFAP, TNC, El Colegio de la Frontera Sur (ECOSUR), and the German Agency for International Cooperation (GIZ) that have supported with outreach education projects and training at agricultural level. In general terms, the respondents agreed that conventional coffee growers have performed bad practices in the humid benefit stage due to the lack of training, technology, and economic support.

Generally, all the respondents indicated their concern for water contamination caused by honey wine and pulp discharge, considering it caused acidification of soils and water bodies, loss of habitat for some species, and thus harm to ecosystems and human health. It is worth to mention that although they do not have a guide or handbook for the sustainable use of water for the humid benefit stage, they think it is necessary to have this type of material to implement environmental courses for children of the coffee growers and to improve water quality of La Suiza Microbasin at a longer term.

Finally, when they were asked if they had performed any physical-chemical samplings to measure water quality, they admitted they had not but mentioned they had collaborated in this type of project and acknowledged that INIFAP had been successful in having the producers perform water quality samplings according to a climate change project promoted by GIZ. They considered this type of actions adequate besides working together with implementing technology, training, and environmental education that lead to water quality improvement in La Suiza Microbasin in favor of population health and El Triunfo Biosphere Reserve.

20.5 Conclusion

La Suiza Microbasin is well known for its productivity and the quality of coffee produced; nonetheless, the production of this coffee bean by the conventional humid benefit stage is causing environmental and social impacts that should be assessed in a timely manner and with a systematic follow-up. By means of surveys and interviews to key actors (coffee growers and representatives of several governmental and academic institutions), this study detected that water contamination caused by honey wine and pulp from the humid benefit stage goes directly into soil or affluents of La Suiza main river.

Furthermore, such key actors accepted not knowing, in fact, the type of impacts, harm magnitude, and affectation time on water quality on the basin altogether. Although they detected empirically that honey wine and pulp discharge caused harm to water mainly from the main river, describing it as “dirty water” and “stinky,” such fact was corroborated when physical-chemical parameters were monitored.

Among those that stood out were temperature and pH, which were out of range according to the NOM-CCA-027-ECOL-1993 and data reported by different authors for this area of La Suiza Microbasin. Such fact leads us to recommend performing studies in timely manner, systematically, and longer to make proposals for discharge prevention or mitigation of the humid benefit given along La Suiza Microbasin.

Other affectations associated to coffee production were socioeconomic and expressed as low profitability of the coffee cultivations, which implied that coffee growers intended to compensate this coffee profitability with greater cultivation surface, which would increase loss of the reserve area, as well as water catchment and treatment in La Suiza Microbasin.

20.6 Lessons Learned

An important point that was evident was the lack of coordination among the public agencies related with water quality of La Suiza Microbasin, where the producers and representatives of organizations only recognized INIFAP, GIZ, TNC, and ECOSUR actions in favor of water quality while CONAGUA, CONAM, and SEMARNAT had little or null presence in water matters.

Last but not least, this study detected that both coffee growers and representatives of institutions and organizations have known or detected the problem of water contamination and thus were willing to participate and find a solution, above all with group activities at the level of La Suiza Microbasin. At the individual level, they were willing to invest economically in technological improvements of the humid benefit stage, as long as they have support and technical advice. Although coffee producers did not admit causing contamination, they considered themselves part of the problem because they know the importance of maintaining a basin with good water quality required for consumption of the population and development of coffee cultivation, which is the main source of income in this Chiapas region.

Bibliography

Alfaro M. Rodríguez J (1994) Impacto ambiental del procesamiento del café en Costa Rica. Agronomía Costarricense 18(2). Ministerio de Agricultura y Ganadería de Costa Rica-Universidad de Costa Rica-Colegio de Ingenieros Agrónomos, Heredia, Costa Rica, pp 217–225. http://repositorio.uaaan.mx:8080/xmlui/bitstream/handle/123456789/8217/64185%20MARTHA%20CONCEPCI%C3%93N%20%C3%81NGEL%20CRUZ.pdf?sequence=1&isAllowed=y. Accessed 22 de january 2019

Álvarez J et al (2011) Evaluación de un sistema de tratamiento de aguas residuales del prebeneficiado de café (*Coffea arábica*) implementado en la comunidad Carmen Pampa provincia Nor Yungas. J Selva Andina Res Soc 2(1). Departamento de Investigación en Bioquímica and Microbiología, La Paz, Bolivia, pp 34–42

Ángeles CMC (2015) Diagnóstico participativo de la acidez en suelo cafetaleros en la Micro cuenca la Suiza, Municipio de Montecristo, Chiapas". Tesis de licenciatura, Universidad Autónoma Agraria Antonio Narro, Cintalapa, Chiapas, México

Arango R.M. (1999). El beneficio ecológico del café en Colombia. Cuaderno de Desarrollo rural , No. 42. (119 pag.) https://revistas.javeriana.edu.co/index.php/desarrolloRural/article/view/2370

Bailly H et al (1992) Proyectos de tratamientos de aguas residuales de beneficios húmedos. Café, Cacao y Thé, núm. 26, Association scientifique internationale du café, Paris, Francia, pp 129–141

Bello-Mendoza R et al (1993) Diagnóstico de la contaminación en las aguas residuales de los beneficios húmedos del café en el soconusco, Chiapas, México. XVI Simposio sobre la cafeticultura Latinoamericana, 25–29 de octubre, Instituto Interamericano de Cooperación para la Agricultura, Managua, Nicaragua

Benedetti A et al (2012) Evaluación integral de riesgo ante el cambio climático en la microcuenca La Suiza, municipio de Montecristo de Guerrero, Chiapas. Tesis de maestría, El Colegio de la frontera sur, San Cristóbal, Chiapas, México. http://aleph.ecosur.mx:8991/exlibris/aleph/a22_1/apache_media/RYQMMHMVP91TR9PTNF2MSYR2U1C3XI.pdf. Accessed 25 de Aug 2019

Benes M (2010) Percepciones ambientales de la calidad del agua superficial en la microcuenca del rio Fogótico, Chiapas. Frontera Norte 22(43), El Colegio de Frontera Norte, Tijuana, México, pp 129–158. http://www.scielo.org.mx/pdf/fn/v22n43/v22n43a6.pdf. Accessed 23 May 2018

Castro MI, López BW (2014) Monitoreo de la calidad del agua en cauces de La microcuenca La Suiza, Chiapas. Secretaria de Agricultura, ganadería, desarrollo rural, pesca y alimentación (Sagarpa, 2014). http://biblioteca.inifap.gob.mx:8080/xmlui/bitstream/handle/123456789/4174/010106303100069865_CIRPAS.pdf?sequence=1. Acccesed 28 May 2019

Centro Nacional de Investigaciones de Café (2011) Construyendo el modelo para la gestión integrada del recurso hídrico en la cafeticultura Colombiana, CENICAFE, Colombia

Comcafe, Comisión para el Desarrollo y Fomento del Café de Chiapas (2013) Programa Institucional de la Comisión para el Desarrollo y Fomento del Café de Chiapas. Gobierno del Estado de Chiapas, Tuxtla Gutiérrez, México. http://www.planeacion.chiapas.gob.mx/planeacion/ProgInst.%20COMCAF%C3%89/ProgInst%20COMCAFE.pdf. Accessed 12 de Aug 2019

Coneval (Consejo Nacional de Evaluación de la Política de Desarrollo Social) (2015) Evolución de las carencias sociales en Chiapas. Coneval, Ciudad de México, México. https://www.coneval.org.mx/coordinacion/entidades/Chiapas/Paginas/carencias-sociales20102015.aspx. Accessed 2 Jan 2018

Cruz Palacios D (2009) Beneficio Húmedo Ecológico. Centro de Investigaciones en Café Cedicafe – Anacafe. (8 pag.) https://www.originicaffe.it/doc/beneficiadoh%FAmedodlemarcruz.pdf

DOF (Diario Oficial de la Federación) (1997) Que establece los límites máximos permisibles de contaminantes en las descargas de aguas residuales en aguas y bienes nacionales. http://dof.gob.mx/nota_detalle.php?codigo=4863829&fecha=06/01/1997. Accessed 12 Feb 2018

DOF(Diario Oficial de la Federación) (2000) Modificación a la Norma oficial Mexicana NOM-127-SSA1–1994, Salud ambiental. Agua para uso y consumo humano. Límites permisibles de calidad y tratamiento a que debe someterse el agua para su potabilización. https://www.gob.mx/cms/uploads/attachment/file/110520/MODIFICACION_A_LA_NORMA_NOM_127_SSA1_1994_22_NOVIEMBRE_2000.pdf. Accessed 12 Feb 2018

Escamilla EP et al (2005) El agroecosistema de café en México. Rev. Manejo integrado de plagas y agroecología (Costa Rica.) No. 76, pp 5–16. http://www.sidalc.net/repdoc/A1853E/A1853E.PDF. Accessed 1 Aug 2019

Fira (Fideicomisos Instituidos en Relación con la Agricultura) (2016) Panorama agroalimentarios. Dirección de investigación y evaluación económica y sectorial. p 24. https://www.gob.mx/cms/

uploads/attachment/file/200636/Panorama_Agroalimentario_Caf__2016.pdf. Accessed 1 Jan 2019

Flores VF (2015) La Producción de café en México: ventana de oportunidades para el sector agrícola de Chiapas. ESPACIO I+D Innovación más Desarrollo IV(7), Universidad Autónoma de Nuevo León, San Nicolás. pp 175–793. http://www.espacioimasd.unach.mx/articulos/num7/pdf/produccion_cafe.pdf. consultado Accessed 20 Oct 2016

Gaceta Ecológica (1994) Nom-CCA-027-ECOL-1993.vol VI, no 28. http://www.paot.org.mx/centro/ine-semarnat/gacetas/GE28.pdf. Accessed 22 Agu 2019

García-Benítez E et al (2015) Caracterización de la producción y del comercio de café en el municipio de Cuetzalan, Puebla. Rev. Agricultura, sociedad y desarrollo, abril – junio 2015. https://www.colpos.mx/asyd/volumen12/numero2/asd-13-098.pdf. Accessed 14 Agu 2019

Gómez G (2010) Cultivo y beneficio del café. Geografía agrícola, núm. 45, Universidad Autónoma de Chapingo, Texcoco, México, pp 103–193

Gómez GL, Jaramillo Á (2017) El clima de la zona cafetalera y su relación con el café. http://biblioteca.cenicafe.org/bitstream/10778/713/5/5%20Clima%20zona%20cafetera.pdf. Accessed 11 May 2019

Guerrero-de León AA, Gerritsen PRW, Martínez-Rivera LM, Salcido-Ruíz S, Meza-Rodríguez D, Bustos-Santana HR (2001) Gobernanza y participación social en la gestión del agua en la microcuenca El Cangrejo, en el municipio de Autlán de Navarro, Jalisco, México. Economía Sociedad y Territorio

Gutiérrez-Vizcaíno M (2014) Pautas para una gestión integrada del agua con enfoque de género en la microcuenca La Suiza, Chiapas. México. Tesis de maestría, Centro Agronómico Tropical de Investigación y Enseñanza, Turrialba, Costa Rica. https://pdfs.semanticscholar.org/e3f8/67eaf0f1e78d529f688466f6211377838a82.pdf. Accessed 22 Oct 2019

INEGI (Instituto Nacional en Ecología Geografía e Informática) (2019) Cuéntame: información por entidad. http://cuentame.inegi.org.mx/monografias/informacion/chis/territorio/clima.aspx?tema=me&e=07. Accessed 2 May 2019

López-Báez W, Reynoso-Santos R (2016) Manejo integral del paisaje para la adaptación al cambio climático en la Sierra Madre de Chiapas, México. Revista Mexicana de Ciencias Agrícolas. Vol.7 Núm. 8 12 de noviembre – 31 de diciembre, pp 236–237. Accessed 17 Sept 2019

López-Báez W et al (2014) Servicio ambiental de la reserva de la biosfera El Triunfo, Chiapas, México. Ingeniería hidráulica y ambiental XXXV(3), Instituto Superior Politécnico José A. Echeverría. Cujae, La Habana, Cuba, pp 47–59. http://scielo.sld.cu/pdf/riha/v35n3/riha04314.pdf. Accessed 13 May 2018

López-Báez W et al (2018) Diagnostico de las propiedades en áreas de Bosque, y productivas de la Reserva de la Biosfera. Agroproductividad, vol 11, no 3, marzo 2018, pp 108–114. http://revista-agroproductividad.org/index.php/agroproductividad/article/download/225/168/. Accessed 4 June 2019

Mariscal Á (2011) El café orgánico de Chiapas crece a 'contracorriente' y sin incentivos. Expansión sección Nacional, 25 de enero, CNN Expansión, México, Ciudad de México, http://expansion.mx/nacional/2011/01/25/el-cafe-organico-de-chiapas-crece-a-contracorriente-y-sin-incentivos. Accessed 5 Feb 2019

Matuk-Velázco V et al (1997) Impacto biológico de los efluentes del beneficio húmedo del café. Cenicafé 48(4):234–252. Accessed 12 Sept 2019

Medina J et al (2016) Estudio del sistema de producción de café (*Coffea arabica L.*) en la región Frailesca, Chiapas. Ciencia UAT 10(2), Universidad autónoma de Chiapas, Facultad de Agronomía, Villa flores, Chiapas, México, pp 33–43

Molina GLV (1999) Gastos defensivos del beneficio de café en la zona de Nicaragua. http://orton.catie.ac.cr/repdoc/A0181e/A0181e.pdf. Accessed 22 July 2019

Molina A, Villatoro A (2006) Propuesta de tratamientos de aguas residuales en beneficios húmedos de café. http://ri.ues.edu.sv/4524/1/Propuesta%20de%20tratamientos%20de%20aguas%20residuales%20en%20beneficios%20h%C3%BAmedos%20de%20caf%C3%A9.pdf. Accessed 22 Agu 2019

Morales G et al (2010) Estrategia del sector cafetalero para la adaptación, mitigación y reducción de la vulnerabilidad ante el cambio climático en la Sierra Madre de Chiapas. Reporte técnico 94 pp. https://www.researchgate.net/publication/271826824_Estrategia_del_sector_cafetalero_para_la_adaptacion_mitigacion_y_reduccion_de_la_vulnerabilidad_ante_el_cambio_climatico_en_la_Sierra_Madre_de_Chiapas. Accessed 22 Jan 2019

Palacios-Herrera B (2012) Análisis participativo de la oferta, amenazas y estrategias de conservación de los servicios ecosistémicos (SE) en áreas prioritarias de la subcuenca La Suiza, Chiapas, México. Tesis de maestría, Centro Agronómico Tropical de Investigación y Enseñanza, Turrialba, Costa Rica. Accessed 12 Jan 2019

Palomares RJA et al (2012) Investigación del café orgánico en México, Edit. UNAM. http://www.cafeybiodiversidad.mx/archivos/cafe-organico-terminado.pdf, Accessed 22 May 2019

Romero HFO et al (2012) Caracterización de tres beneficios húmedos colectivos y uno industrializado de café con énfasis en una propuesta de mejora de un modelo de beneficio húmedo colectivo en la Unión de Cooperativa Agropecuaria del Norte UCANOR, Jinotega. Informe de Final, Universidad Nacional de Ingeniería de Nicaragua. En: http://ribuni.uni.edu.ni/1281/1/39805.pdf. Accessed 22 Julio 2019

Salguero-Zeceña ERT (1996) Valoración económica de la contaminación de las fuentes de agua por los desechos de la industria del beneficiado del café: el uso del concepto de costo defensivo. Tesis de maestría, Centro Agronómico Tropical de Investigación y de Enseñanza, Turrialba, Costa Rica http://orton.catie.ac.cr/repdoc/A0523e/A0523e.pdf. Accessed 22 Jun 2019

Sánchez-Suárez K (2015) Los pequeños cafeticultores de Chiapas: organización y resistencia frente al mercado. Tesis de Universidad de Ciencias y Artes de Chiapas, CESMECA. http://biblioteca.clacso.edu.ar/Mexico/cesmeca-unicach/20170419034553/pdf_655.pdf. Accessed 22 Jun 2019

Semarnat (Secretaría de Medio Ambiente y Recursos Naturales) (1996) Norma Oficial Mexicana NOM-001-SEMARNAT-1996. Con base en el acuerdo por el cual se reforma la nomenclatura de las normas oficiales mexicanas expedidas por la secretaría de medio ambiente y recursos naturales". Diario Oficial de la Federación, 23 de abril, Secretaría de Gobernación, Ciudad de México, México

SWRCB (State Water Resources Control Board) (2013) Folleto Informativo pH. https://www.waterboards.ca.gov/water_issues/programs/swamp/docs/cwt/guidance/3140sp.pdf. Accessed 22 Jun 2019

Vargas-Venci P (2007) Mujeres cafetaleras y producción de café orgánico en Chiapas. El cotidiano, Vol. 22, No. 142, pp 74–83. https://www.redalyc.org/pdf/325/32514209.pdf. Accessed 22 Mah 2018

Vázquez-González E et al (2019) Capítulo 7. Impactos socio-ambientales del manejo de aguas residuales del beneficiado húmedo del café: hacia la sostenibilidad hídrica en la microcuenca La Suiza, Chiapas.en e-libro: La Gestión hídrica en México: casos de estudio y propuestas de políticas públicas. (in procees e-book in El Colegio de la Frontera Norte)

Vega-Montenegro D (2012) Evaluación del transporte de nutrientes y materia orgánica a través del subsuelo y en el flujo de agua de la fosa de infiltración de un beneficio húmedo de café. Ubicado En la aldea San Antonio Chacayá, Santiago Atitlán, Sololá. http://biblioteca.usac.edu.gt/tesis/08/08_1265_Q.pdf. Accessed 22 Agu 2019

Part IV
Biodiversity Conservation Success from Socio-Ecological Approach

Chapter 21
Resistance of the Civil Society Against Mining Projects

Heidi L. Romero-Schmidt and Alfredo Ortega-Rubio

Abstract A very high proportion of concessions for open-pit mining projects overlap with Protected Natural Areas. Unfortunately, the lax environmental legislation favors this type of companies above the interest of local and regional communities to preserve their environment and health. Thus the reason why local communities, faced with the onslaught of the capital of these predatory companies, have organized themselves to challenge these social and ecological tragedies. This chapter analyzes the main methods of resistance used by local communities. Resistance to mining has evolved throughout history, mining conflicts have been originally dominated by strikes, protests, and demands, but in the last two decades, communities living in the areas surrounding mining projects oppose environmental issues and object to their lack of representation and participation in the decisions concerning its development. Its result a combination of local narratives, rights (clean water, human, indigenous) and environmental justice. This evolution have made it possible to make alliances among local groups and organizations, which has succeeded in increasing knowledge about these projects and giving them visibility by contributing to new strategies of social resistance, such as legal judgments, scientific collaborations, and public consultations to reject mining projects.

Keywords Mining · Social resistance · Socio-environmental struggles · Extractivism

H. L. Romero-Schmidt (✉)
Centro de Investigaciones Biológicas del Noroeste, La Paz, México
e-mail: heidi04@cibnor.mx

A. Ortega-Rubio
Red Nacional Áreas Naturales Protegidas RENANP-CONACYT, Centro de Investigaciones Biológicas del Noroeste CIBNOR-CONACYT, La Paz, Baja California Sur, Mexico
e-mail: aortega@cibnor.mx

A. Ortega-Rubio (ed.), *Socio-ecological Studies in Natural Protected Areas*,
https://doi.org/10.1007/978-3-030-47264-1_21

21.1 Introduction

More than 190 countries (ONU 2019) worldwide shelter from 60 to 70% of the total biodiversity of the planet, of which only 12 are considered megadiverse and Mexico is one of them (CECADESU 2006; Mittermeier and Goettsch 1992). Mexico occupies the fifth place among the countries with greater biological diversity after Brazil, Colombia, China, and Indonesia and hosts about 12% of global biodiversity (Fig. 21.1) (Llorente-Bousquets and Ocegueda 2008; Lopez-Rojas and Coutinho 2003; Yáñez 2007).

Though Mexico is a megadiverse country, a significant number of species are at risk of extinction due to the intensive destruction of their habitats because of agriculture, ranching, and extractive activities, such as mining besides pollution, among others (Fig. 21.2) (Lopez-Rojas and Coutiño 2003). The current ecological crisis, which exists not only in Mexico but also worldwide (Solano 2013), has brought, in consequence, the development of diverse ways of carefully protecting nature globally, of which the Protected Natural Areas (PNA) stand out.

Protected Natural Areas are representative terrestrial or aquatic regions of diverse ecosystems where human activities have not affected their original environment significantly (De la Maza 1999, CCFD 2010). Currently, the PNAs are one of the most important tools used to seek ecosystem conservation and the sustainable

Fig. 21.1 Mexican biodiversity. (Author: Fernanda Vega-Romero)

Fig. 21.2 Habitat destruction by activities like agriculture. (Author: Fernanda Vega-Romero)

development of the people who inhabit them. They are subjected to special protection, conservation, restoration, and development, according to categories established by law (De la Maza 1999; CCFD 2010), and in Mexico, a decree is enacted at municipal, state, or federal levels (Yáñez 2007).

Mexican government, until february 2017 as part of the environmental preservation policy, the Mexican government, through The National Commission of Protected Natural Areas (CONANP 2019), had decreed 182, whose area amounted to almost 22 million hectares (21 886 691 ha) - that is, 11.14% of the national terrestrial surface, not counting protected marine areas, and intended exclusively to safeguard the existing flora and fauna in the country (Martínez et al. 2018; CONANP 2019). PNAs are the most important instruments to achieve ecological balance and prevent biodiversity loss; however, its usefulness in Mexico is still questioned by more than 1900 mining concessions granted within these land areas (Martínez et al. 2018); 68% of the Mexican Protected Natural Areas have been affected by mining concessions, mostly for precious metal extraction (Martínez et al. 2018; SE 2017).

Comparing the hectares destined to protect the environment and those granted for mining, this study found that only 10% of the Mexican land territory has the goal of conserving biodiversity while 11% has been licensed to mining companies (Martínez et al. 2018). Apparently, the battle for the conservation of Mexican ecosystems is balanced in favor; however, in cross-checking data between the location of the Protected Natural Areas and that of the territory granted for mining, more than one third of the total PNAs have been affected by this type of concession; 61 Mexican PNAs have mining developments within their protected territory (Martínez et al. 2018; SE 2017).

Previously, mining was performed with spikes, shovels, hammers, and wedges to split the rock, digging galleries in caves, or grotto surrounded by vegetation. Current mining activity implies large-scale projects, specialized work, use of explosives, and heavy machinery to remove the entire surface and be able to access the subsoil resources. This technique called open pit erodes and deforests land, raids entire hills, forces fauna to migrate, and not only monopolizes available water but also pollutes it (Martínez et al. 2018; López and Eslava 2018). Furthermore, it requires skilled labor, which means that local inhabitants are rarely hired for this type of work.

The open pit technique is the most used and cheapest, which affects the environment with the most predatory production activities that exist in our country (Fig. 21.3) (López and Eslava 2018). In addition, highly polluting chemicals are used for leaching mineral, whose control is extremely difficult due to the lack of instruments and mechanisms to perform it and result in contaminating the environment and space where the activity is carried out. Moreover, if the leaching tanks burst, the risks of contamination increase, as it has happened in Sonora, Durango, Zacatecas, and Oaxaca where such accidents have produced incalculable damage among the population and environment (López and Eslava 2018).

Besides pollution, other problems exist. One of them is that the government grants mining concessions to private companies without taking into consideration the point of view of the populations inhabiting the place where the mineral is located and who will be mostly affected when the company begins exploration or exploitation activities (WGMHRLT 2014, Madrid and Avila 2019). Another aspect that has generated conflicts is the preparation of environmental impact manifests that companies show to meet the requirements that the law imposes, which are entrusted to

Fig. 21.3 Open pit mine (Author: Horjaraul from Pixabay)

specialized offices that usually make them according to the requirements of those who requested them. Mining companies already own affiliated concession titles or are related to large industrial emporiums – both national and foreign capital – enjoying great financial, political, or both powers; several of these groups are interrelated and also have interests in other sectors, such as telecommunication, commercial, or banking sectors (WGMHRLT 2014, Martínez et al. 2018). Paradoxically, mining companies participate in nine Advisory Councils of Protected Natural Areas, according to the 2016 report of the Chamber of the Mining Industry of Mexico (CAMIMEX). One of the achievements of this participation was to include "a sub-area of sustainable use to allow mining exploration and exploitation, in addition to continuing with the sub-area of special use" (Martínez et al. 2018).

21.2 Mining in Contemporary Mexico

The North American Free Trade Agreement (NAFTA) implementation allowed transnational companies to be interested in natural resources (Lomelín et al. 2007; Roblero and Hernández 2012; López and Eslava 2018). Besides the signing of NAFTA, the incidence of mining in the Mexican economy began to have importance in the 1990s where a reform of the Article 27 of the Constitution says, "Ownership of the lands and waters included within the limits of the national territory corresponds originally to the Nation, which has had and has the right to transmit ownership to individuals constituting private property" (Orden Juridico 2019).

Two important facts in the future of the mining industry stand out. The first one allowed changing the orientation of mining legislation. A new mining law was enacted on June 26, 1992, and it has been amended three times; (a) on December 24, 1996, it granted legal certainty to concessionaires and simplified the granting of permits for mining activity; (b) on April 28, 2005, it admitted to issue mining exploration and exploitation in a single title; and, (c) on June 26, 2006, it added rules for foreign investment (Gómez 2010; López 2011; Roblero and Hernández 2012). The second one established the guidelines for what must be approved, which was not only reduced to mining legislation but also included access to underground land where they found minerals and use of the land and water for processing minerals, preventing or remediating environmental pollution and foreign investment in this area.

For these reasons in recent years, Mexico has become one of the most coveted countries by big transnational mining companies because of the great amount of metal ores found in the country, but especially due to law flexibility, which includes tax benefits, lack of regulations, and a permissive legislation. This situation benefits multinational mining companies that use their economic power to buy political support and media needed to carry out their projects (Montenegro 2003; Fernandez 2010; Delgado 2011).

The majority of the mining concessions in Mexico have been granted to companies with foreign capital from 15 countries (SINEM 2019), most of which belong to Canada. Until this year (2019), the Mexican government had granted 921 concessions to Canadian miners in 26 of the 32 states of the Mexican Republic (SINEM 2019). Canada, in contrast to the green and ecological discourse that prevails within this nation, is the leading open pit mining country in the world. Canadian transnationals own 51% of the global mining capital (Godoy 2009), and Mexico is one of the countries in which mining activities are significant, allowing Canada to be the "mining champion." The Canadian magazine "The Corporate Ethics Monitor" in its July–August 1994 issue indicated that the Canadian mining companies had migrated to other countries for several important causes, including (a) tax incentives from the recipient country; (b) earnings from mineral deposits; (c) lower production costs; (d) less complicated and faster approval processes; and (e) less stringent environmental regulations or these did not apply.

In Latin America, the crisis of national economies was dangerously combined with aggressive market penetration strategies deployed by large mining corporations (Montenegro 2003). The mining industry enjoys a prominent role in today's global economy. Growth in consumption and production has escalated the need for energy and raw materials with use of resources reaching exceptionally high levels worldwide leading to a concomitant rise in the number of conflicts over resource extraction and waste disposal (Conde 2016; Aydin et al. 2017). Mining activities have generated various negative environmental and social impacts, including deforestation, biodiversity loss, high water consumption, groundwater contamination, and population migration (Urkidi 2010; Aydin et al. 2017).

All these impacts create discontent and conflict because the basic needs of some groups and their access to environmental resources and services are compromised, resulting in the loss of livelihoods, cultures, and worse of all even lives (Aydin et al. 2017). At the social level, various studies have shown that mining areas tend to have comparatively higher levels of economic inequity, AIDS, alcoholism, prostitution, and child labor, accentuating poverty and social conflict (Pegg 2006). Due to these factors, many affected communities tend to protest (Urkidi 2010).

Behind all the promises of employment, profits, technology, and development proclaimed as "benefits" by the mining industry, actually a greater enrichment of transnational corporations exists, while the greatest losses are distributed among peasant and indigenous people; in all rivers, estuaries, and forests, no matter how hard fauna and flora try to hide from the environmental disaster, they are irremediably affected by it day by day (Martínez et al. 2018).

In general, open pit mining alters the ways of life of the nearby towns where the projects are installed. These affectations bring contradictions with what the projects offer, so they generate mobilizations against mining activities (López and Eslava 2018). The principal causes of mining conflicts are by the state unilateral decisions when it grants concessions and mining licenses for the use of natural resources in their zones without any consideration to local populations and their rights. People are generally not informed about the state decision to grant rights to third parties on subsoil, which breaches human and settler rights to land and territory. Usually, the community will be merely informed of the decision by the company that it has been granted the concession to explore and extract resources from the subsoil of the lands in question. In many cases, the communities have no other option than to accept initiating negotiations between unequal actors regarding the use of their land while giving up their internationally recognized rights to their territories. If the community is not willing to reach an agreement, there is no efficient mechanism or procedure to bring its position to the attention of the state. It is on these grounds that the communities are now expressing their resistance against mining.

Moreover, the territories of communities and indigenous populations are neither abandoned nor unproductive: they serve traditional community usage or are used in intensive activities, such as agriculture, cattle raising, or tourism. Above all, these communities generally built a way of life and a vision of their own form of development based on the natural possibilities of the territory. Nonetheless, the concessions are granted even without evaluating the current use of land or the benefit that it generates for these communities.

Unfortunately, the figure of Protected Natural Areas has not been sufficient to stop open pit mining as seen not only in our country but also in Latin America, Africa, and elsewhere. For example:

Mexico. In San Luis Potosi, the Canadian company "First Majestic" operated a mine in the Huichol Wirikuta Sanctuary in the desert of San Luis Potosi, which is considered a protected natural area by UNESCO (Rodríguez-García 2011). Open pit mining has threatened the historic enclave of San Luis Potosi, Mexico, in an area declared in 1993 by the Mexican government as restoration and preservation of wildlife area. For 15 years associations and neighbors have confronted the Canadian "Minera San Xavier" without having won the legal battles to stop the project (Otramerica 2011).

Uruguay. Nationally, 44 Protected Natural Areas have been created to conserve biodiversity and environmental goods and services. However, they have been facing a serious problem with changing land use or mining activities within these areas (La Hora 2011).

Peru. Lima (Tierramérica) Mining concessions undermine Andean moors and mountain chain water springs in protected areas. More than 11% of the subsoil has been distributed to Peruvian mining concessions. The surface may have moors, springs, and even watershed protected areas. The mining maps and fragile ecosystems overlap in this country of great mineral richness (Salazar 2010).

Ecuador. Protected natural areas in Ecuador face critical situations in both administrative management controlling the use of resources by the surrounding towns and the presence of natural nonrenewable resource extraction (Andrade 2009). Gold mining with cyanide is obsolete in Europe and done in Costa Rica (Baltodano 2010). Mining companies have put pressure on protected areas. The mining companies were shocked by a "recommendation" approved by the World Conservation Congress, held in Amman in 2002, which demanded an end to oil extraction, mining, and gas in all protected areas in Categories I, II, III, and IV of the IUCN ("strict nature reserve," "wilderness," "national park," "natural monument," and "habitat management areas") (Bravo and Carrere 2004). The high environmental impacts, lack of participation, extra-local alliances, and distrust toward state and extractive companies tended to increase resistance (Conde and Le Billon 2017).

21.3 Social Movements Against Mining

All modifications to mining laws, political corruption, and transnational interests are articulators of organizations, groups, and individuals to carry out joint actions to protest against mining and protect natural resources where the main axes of resistance have been history, religion, identity, and risk threat (Fig. 21.4). When citizens and communities look at this situation, they realize that there is no other way to continue to exist than to fight.

In the last century, the resistance against mining was done through centralized and hierarchical organizations; now it is done directly by the affected communities, which are organized in their daily tasks and with their own resources without relying on a central organization (López and Eslava 2018).

Great mobilizations were neither seen in the style of the trade union and peasant struggles of the last century; they were less visible fights but sometimes more successful. The forms of struggle have varied depending on whether they were facing mining companies or the government that set the conditions for carrying out their plans. In recent decades, social movements against mining extraction have been transformed as the process of economic globalization progresses, and the methods of appropriation of the mineral have been modernized. Communities organized in regional, national, or international networks have been the most observed (López and Eslava 2018). More than social movements, societies are now facing movements that organizations create to become stronger (Zibechi 2017). For the purpose they pursue, the movements that oppose extraction mining are also known as

Fig. 21.4 Protest against mining. (Photo by Markus Spiske temporausch.com from Pexels)

socioenvironmental conflicts, that is, "problems linked to the control of natural resources and the territory they possess by the actors involved, divergent interests and values surrounding them in a great asymmetry of power"(Svampa 2012:19).

The emergence of a collective action is always the result of a tension that disturbs the balance of the social system. Tension produces widespread beliefs that mobilize action, that is, creates an identity and seeks to restore the balance of the system. The resistance of the ecological movements of indigenous people and peasants has been called "the environmentalism of the poor" (Roblero and Hernández 2012:84) because it has often been criminalized with threats, imprisonment, and murder of leaders. In Mexico, pointing out that "the poor are not always environmentalists but in environmental conflicts, they are often seen on the side of nature conversation" (Roblero and Hernández 2012:84) has been considered a dangerous activity. "Environmentalism of the poor" refers to conflicts caused by economic growth in the name of development and social inequality where the negative effects on environment by extracting resources and evacuating waste are suffered and paid by marginalized social groups (Roblero and Hernández 2012:84). These conflicts often take place at the borders of extraction in indigenous lands and refuge areas where no journalistic coverage occurs. Resistance is ecological even if it is not referred to as such (Martínez 2009).

Therefore, the artificial discourse of mining entrepreneurs to generate employment and "economic welfare" to the inhabitants of neighboring communities is not met; life quality of inhabitants is not improved; on the contrary, it destroys the existing social fabric, criminalizes social protest, and puts pressure on ecosystems. Historical experience, not only in Latin America but also in many other parts of the world, has demonstrated that development through mining, as an economic policy and alternative to surrounding communities, has failed to eliminate poverty. On the contrary it seems that the current results have been negative. Development implemented in a municipality by mining companies has resulted in a vicious circle. The promised infrastructure works were not fulfilled or partially fulfilled, and the mining activities had negative socioenvironmental effects that resulted in the emergence of anti-mining resistance, a phenomenon that has been growing radially, polarizing society. In this manner, anti-mining resistance arises in a context of socioenvironmental crisis as a community response manifested through the threat of dispossession, pollution, and destruction of the territory (Roblero and Hernández 2012).

During an exchange of experiences with organizations from Mexico, Guatemala, El Salvador, and Honduras, the process of territorial appropriation by mining companies was found to be similar in those countries; in many cases they were the same extractive companies with different names (Roblero and Hernández 2012). For example, among the most relevant strategies used by transnational corporations to appropriate the plots are the use of social programs as co-optation strategies; economic bribery toward more significant authorities and characters to intercede in their favor; the use of community organizational forms (monthly assemblies) as convincing instruments; a speech impregnated with religious, ecological, nationalist, progressive ideas; and taking advantage of ignorance about the mining issue, poverty, and unemployment. Of these actions, the ones that have most destroyed the social fabric and polarized society were the offer of infrastructure that they did not fulfill or partially comply with and the co-optation of government authorities through royalties (Roblero and Hernández 2012).

In Mexico, indigenous and peasant communities became aware of the problem, sought information, and discovered that they had rights that had been violated and began to fight for them. They organized themselves according to their own resources and capacities, began to weave alliances with other communities that faced the same problem, and sought external advice (López and Eslava 2018). At national level, two organizations set out to fight with the mining activity to respect the rights of peasants and their communities and indigenous people. One of them is the Mexican Network of People Affected by Mining (Red Mexicana de Afectados por la Minería, REMA) (López and Eslava 2018). Its constitution took place from June 19 and 21, 2008, in the community of Temacapulín, municipality of Cañadas de Obregón, State of Jalisco, within the framework of the First Meeting of the Mexican Network of People Affected by Mining (REMA), where hundreds of people from social, human, educational, indigenous rights, and peasant community organizations participated, as well as diverse groups from Chihuahua, Sonora, Nayarit, Jalisco, Oaxaca, Chiapas, Guerrero, State of Mexico, San Luis Potosí, Coahuila, Veracruz, and Mexico City (López and Eslava 2018). Among the objectives that were integrated by the Network were the following: raise awareness of the social and environmental effects of mining through the exchange of experiences and strategies for the defense of territory, water, natural resources, and communities; integrate the movement against mining and strengthen local struggles and movements for the resistance and defense of their rights; generate an organizational structure of the Network and develop joint strategies that strengthen alliances and coordinate with national and international networks against mining (López and Eslava 2018).

The other national organization is National Assembly of Environmentally Affected Persons (Asamblea Nacional de Afectados Ambientales ANAA), which emerged on August 31, 2008, at the Faculty of Economics of the Universidad Nacional Autónoma de México (UNAM), where several communities and popular organizations focused on defending natural resources, territory, and the rights of the people who decided to join forces to overcome the dispersion and invisibility of this type of struggle (López and Eslava 2018). Among the attendees were people from Jalisco, Guerrero, State of Mexico, Oaxaca, Tlaxcala, Puebla, Morelos, and Mexico City. The next day, ANAA members marched in front of the National Water Commission (CONAGUA) and the Ministry of Environment and Natural Resources (SEMARNAT) to demand a fair solution to the serious problems of water, land, and air pollution; dispossession of forests and minerals; and continuous deterioration of health, loss of flora and fauna, and harassment and repression of communities (López and Eslava 2018). The construction of national networks was important, but due to its nature, it could not meet the regional and local needs, so they decided to build fighting spaces, regionally and locally, to provide specific answers to specific problems (López and Eslava 2018). Diverse actors converged in these spaces, among them common land, agrarian communities and indigenous people, social and cultural organizations, nongovernmental organizations (NGOs), and environmentalists (López and Eslava 2018). As an example of this type of organizations is the Broad Opposing Front (Frente Amplio Opositor) in San Luis Potosí, which for a long time sustained the fight in defense of the hill of San Pedro, against the San Javier mining company (López and Eslava 2018).

Other examples of this type of organisms is the Wirikuta Tamatsima Wahaa Defense Front (Frente en Defensa de Wirikuta Tamatsima Wahaa), which proposed the defense of the sacred places of these indigenous people (López and Eslava 2018). Veracruz Assembly of Initiatives and Environmental Defense "La Vida," in Veracruz, fights against environmental pollution and the defense of water and rivers and Tiyat Tlali Council, of the Sierra Norte de Puebla, whose motto is "for the defense of life and territory" (López and Eslava 2018). Oaxaca Collective in defense of the territories and the Council of Agrarian Authorities focus on the Defense of the Mountain Territory and the Costa Chica de Guerrero; in Baja California Sur, civil organizations Environment and Society AC and Water is Worth more than Gold (López and Eslava 2018). Latin America Mining Conflicts Observatory (Observatorio de Conflictos Mineros de América Latina OCMAL) was founded in a meeting held in Oruro, Bolivia, in 2007. It was formed by various organizations (Table 21.1) that had been collaborating with each other since late 1990s to establish resistance strategies and alternatives to mining in Latin America. Its main objective is defending communities and populations that exercise their local activities, such as agriculture, livestock, forestry, fishing, tourism, housing, and culture, which are affected by mining impacts (OCMAL 2019).

21.4 Conclusions

Our country has natural and social richness seen little in other countries, which makes it one of the richest biodiversity countries. This natural richness works as the basis of many ecosystems and climates of the country, which allow the existence of diverse societies and cultures across the Mexican territory (López and Eslava 2018). This diversity is noticeable in everyday life from clothing, colloquial speech, and customs to economic forms used to meet their needs. However, with the emergence of neoliberalism, capital has taken over many spaces, disrupting inhabitants' lives, undergoing a profound change in the last three decades (López and Eslava 2018).

Mining activities seek to transform the property of the people, maintained as a common good for a long time into private property to introduce to the market and convert it into merchandise. Mining activity in Mexico, as in other Latin America latitudes and in general around the world, is not a synonym of development or progress; it does not generate well-being. By specializing in gold extraction, the consequences of its high-degree ecotoxicity and pollution represent a threat to life, traditions, and peacefulness of Mexican rural populations where mining enterprises are located.

Most works are executed under impunity and opacity. Many of them are already directed to certain companies and with bribery limitations for authorities of

Table 21.1 Observatorio de Conflictos Mineros de América Latina (OCMAL) (Latin America Mining Conflicts Observatory) members

Country	Organization name	Website
Argentina	Be.Pe Bienaventurados los pobres	http://bepe.org.ar/
Colombia	CENSAT Agua Viva	http://censat.org/ http://extractivismoencolombia.org/
Costa Rica	Unión Norte por la Vida-UNOVIDA	http://unovida.blogspot.cl/
Honduras	Centro Hondureño de Promoción para el Desarrollo Comunitario-CEHPRODEC	http://www.cehprodec.org/
Brazil	Justiça Nos Trilhos	http://www.justicanostrilhos.org/
El Salvador	Mesa Nacional Contra la Megaminería	http://esnomineria.blogspot.cl/
	Asociación de Desarrollo Económico Social-ADES	http://www.adessantamarta.sv/
	Centro de Investigación sobre Inversión y Comercio	https://www.facebook.com/CEICOMONG/
Nicaragua	Centro Humboldt	http://www.humboldt.org.ni/
Guatemala	Organizacion Diocesana y Parroquia San Miguel Ixtahuacan	No website
	Asociación CEIBA	http://ceibaguate.org/
Chile	Observatorio Latinoamèricano de Conflictos Ambientales	http://www.olca.cl/oca/index.htm
	Pastoral Salvaguarda de la Creación	No website
	Consejo Ciudadano Salamanca-OCAS	No website
	Comité de Defensa del Valle de Chuchiñi	No website
	Grupo Atacama Limpio	No website
	Concejo de Defensa del Valle del Huasco	http://consejodefensadelvallehuasco.jimdo.com/
Bolivia	Centro de Ecología y Pueblos Andinos-CEPA	http://www.cepaoruro.org/
	Centro de Documentación e Información Bolivia-CEDIB	http://www.cedib.org/
	Sociedad Potosina de Ecología	https://www.facebook.com/sociedadpotosinadeecologia.sope/
	Red Nacional de Mujeres en Defensa de la Madre Tierra	http://renamatbolivia.blogspot.cl/
	Colectivo CASA	http://www.colectivocasa.org.bo/
Peru	CooperAccion	http://cooperaccion.org.pe/main/index.php
	Grufides	http://www.grufides.org/
	Red Muqui	http://muqui.org/
	Fundación Ecuménica para el Desarrollo y la Paz-FEDEPAZ	http://www.fedepaz.org/
	Derechos Humanos Sin Frontera	http://derechosinfronteras.pe/
	VIMA	Sin sitio web

(continued)

Table 21.1 (continued)

Country	Organization name	Website
Ecuador	Acción Ecológica	http://www.accionecologica.org/
	Pastoral Shuar	No website
	Federación Interprovincial de Centros Shuar-FICSH	No website
	Coordinadora de Mujeres Intag	No website
	Frente de Mujeres en Defensa de la Vida y la Naturaleza	No website
	Comite de Defensa de la Naturaleza y la Vida	No website
	Fundacion Vientos de Vida	No website
Mexico	Observatorio de Conflictos Mineros de Zacatecas	https://www.facebook.com/OCMZacatecas/
Latin America	Red Latinoamericana de Defensoras de Derechos Sociales y Ambientales	http://www.redlatinoamericanademujeres.org/
	Iglesias y Minería	https://www.facebook.com/IglesiasyMineria/
	Alianza Centroamericana frente a la Minería Metálica	No website

https://www.ocmal.org/miembros-ocmal/

different levels, leaving disagreements for quality and for type of work since they are not mitigating the damage generated by mining activity nor generating a positive social impact on local population. This inequality situation has generated great discontent throughout the country. Social and environmental conflicts over mining have grown exponentially, mainly because most of projects are located in high vegetation territory (85%) and in Protected Natural Areas (FUNDAR 2018). Furthermore, because mega-mining has more visible and faster impact than traditional gallery mining through the destruction of hills and landscapes, water pollution, health damage, forced displacement, and devastation of whole villages are the main brands left by this new mining boom.

Given the inequality and grievances, affected people organize and protest, giving rise to the country's anti-mining and social movements. The population is looking for ways to build resistance, so mobilizations of different groups and expressions have been developed against mining extraction from different scales; struggles at local, regional, and national levels have taken place. To make it possible, adequate conditions are required. One of them has been law flexibility on the subject; the other one was the State reform at the service of mining companies, generating institutions that operated with the appropriate policies for this purpose. As a result, in the last two decades, we have seen emergence and resurgence of socioenvironmental movements that oppose open pit mining because it has not only destroyed the places where it settles but also caused negative impacts on lives of all the people that live in the surroundings.

With the socioeconomic characteristics of Latin America, social tolerance limits for plundering schemes are increasingly smaller, which has transformed environmental struggle in a class of several actors and with different languages and expressions (Hernández 2011). Currently, no exact number of mining conflicts in the country is known because mining extraction speed advances and increases conflicts. Besides, deep and alarming crisis of human rights in the field of environmental activism take place with violence scenarios (Svampa 2012). From 2010 to 2015, 33 murders were reported, while 11 extrajudicial executions were from June 2016 to May 2017 (Tornay 2018). 2018 closed with 73 murders of territory defenders according to the Defender report of Human Rights in Mexico of the Centro de Readaptacion Social (CERESO) Committee (2018).

In an environmental conflict, ecological, cultural, subsistence, and economic values of the populations are displayed. They are values that are expressed in different scales, and they are not measurable. Current mining legislation needs to be reformed, so that it complies with the constitutional principles and guarantees fundamental rights of indigenous people and communities, in addition to the agrarian nuclei. Just as foreign investment should regulate (Cardenas 2013), as long as the model (perspective) in mining has not changed, minerals will continue to be extracted causing territory deterioration. A new model is necessary where the inhabitants of natural resources may teach and continue the form of exploitation and sustainability that they have performed for years and centuries.

Lesson Learned

Resistance to mining have evolved, throughout history, mining conflicts have been originally dominated by strikes, protests, and demands, but in the last two decades, communities living in areas surrounding mining projects oppose environmental issues and object to their lack of representation and participation in the decisions concerning its development. Its result a combination of local narratives, rights (clean water, human, indigenous) and environmental justice. This has made it possible to make alliances between local groups and organizations, which have succeeded in increasing knowledge about these projects and giving them visibility by contributing to new strategies such as legal judgments, scientific collaborations, and public consultations to reject mining projects.

Current mining legislations needs to be reformed, in order both to comply with the constitutional principles and to guarantee the fundamental rights of indigenous people and local communities. If current model persist in mining activities by foreign companies, and minerals will continue to be extracted causing territorial and environmental deterioration, then the communities will follow their fight to preserve their environmental, ecological, cultural, subsistence, health, and economic values. Incommensurable values are much more important than the money obtained by the mining companies.

References

Andrade Mendoza K (2009) Las áreas naturales protegidas frente a la actividad hidrocarburífera. Las organizaciones ambientalistas y la gobernanza ambiental en el Ecuador. El caso del Parque Nacional Yasuní (Dossier) En: Letras verdes, Quito: FLACSO sede Ecuador. Programa de Estudios Socioambientales, (no. 3, abril 2009): pp. 14-16. ISSN: 1390-4280

Aydin CI, Ozkaynak B, Rodríguez-Labajos B, Yenilmez T (2017) Network effects in environmental justice struggles: An investigation of conflicts between mining companies and civil society organizations from a network perspective. PloS one 12(7):e0180494. https://doi.org/10.1371/journal.pone.0180494

Bravo E, Carrere R (Ed) (2004) Protected Areas. Protected Against Whom? Oilwatch and World Rainforest Movement. Montevideo, Uruguay 200 p.

Cárdenas Jaime (2013) La minería en México: Despojo a la Nación. Cuestiones Constitucionales. Revista Mexicana de Derecho Constitucional Núm. 28, enero-junio 2013.pp 35–74

CCFD Comité Cívico de Ford y sus Distribuidores (2010) Áreas Protegidas de México ¿Qué son y dónde están? Pagina web. http://www.apdm.com.mx/areas-protegidas/que-son-y-donde-estan Consultada el 10 de julio de 2019

CECADESU (Centro de Educación y Capacitación para el Desarrollo Sustentable) (2006) Cap. III. ¿Qué es la Biodiversidad? Cruzada Nacional por los Bosques y el Agua. Página Web SEMARNAT. http://cruzadabosquesagua.semarnat.gob.mx/iii.html

Comité Cerezo (2018). Defender los Derechos Humanos en México. Consultada el 15 de oct 2019. Pág. Web. https://www.comitecerezo.org/spip.php?article3063

CONANP. Comisión Nacional de Áreas Naturales Protegidas (2019) Pág. Web. Consultada el 10 de julio de 2019. http://sig.conanp.gob.mx/website/pagsig/datos_anp.htm

Marta C (2016) Resistance to Mining. A Review. Ecological Economics. Elsevier 132:80–90

Conde Marta, Philippe Le Billon (2017) Why do some communities resist mining projects while others do not?. Extr Ind Soc. Elsevier, 4(3): 681–697

De la Maza Elvira, Roberto (1999) Una historia de las áreas naturales protegidas en México. Gaceta Ecológica. Nueva Época, No. 51. Instituto Nacional de Ecología. http://www2.ine.gob.mx/publicaciones/gacetas/276/histanp.html

Delgado Ramos GC (2011) MINANDO MÉXICO A PRINCIPIOS DEL SIGLO XXI. Despojo, saqueo y criminalización social. Memoria 249. DICIEMBRE 2010 - ENERO 2011

Fernández Vega Carlos (2010) Pastel minero del país, entre pocas familias. Mexico S.A. La Jornada. http://cac08.wordpress.com/page/6/

FUNDAR (2018) Las actividades extractivas en México: minería e hidrocarburos hacia el fin del sexenio. FUNDAR Centro de Análisis e Investigación. Pág. Web. https://2018.fundar.org.mx/

Godoy E (2009) Daños a cielo abierto. 22 jul 2009. IPS Noticias. http://ipsnoticias.net/nota.asp?idnews=92784

Gómez Magdalena (2010) Saldos del centenario: las concesiones mineras, La Jornada, noviembre, 23, 2012.

Hernández Navarro L (2009) México: El nuevo saqueo: minería a cielo abierto. La Jornada. Opinión. 27 julio 2009. http://www.jornada.unam.mx/2009/07/21/index.php?section=opinion&article=017a1pol&partner=rss

Hernández Navarro L (2011) Wirikuta y la minería devastadora en México. 10/Febrero/11.http://www.ecoportal.net/Eco-Noticias/Wirikuta_y_la_mineria_devastadora_en_Mexico

La Hora (2011) Alerta frente a alteraciones en Áreas Protegidas. Diario Digital. Sábado, 15 de enero de 2011. http://www.lahora.com.ec/index.php/noticias/show/1101079083/-1/Alerta_frente_a_alteraciones_en_%C3%81reas_Protegidas.html

Llorente-Bousquets J, Ocegueda YS (2008) Estado del conocimiento de la biota. In: Capital natural de México, vol. I: Conocimiento actual de la biodiversidad. Conabio, México, pp 283–322

Lomelín J, de Quevedo XG, Luna E, Gutiérrez J, Salazar A, Almazán S (2007) "Minería mexicana 1960-2007". En: Revista Artes de México: Una visión de la minería. núm. 86. México. pp II–VII

López Bárcenas Francisco (2011) "Pueblos indígenas y megaproyectos en México. Las nuevas rutas del despojo. In: Los derechos de los pueblos indígenas a los recursos naturales y al territorio: conflictos y desafíos en América Latina / coord. Marco Aparicio Wilhelmi, 2011. pp 181–202. ISBN 978-84-9888-340-4

López Bárcenas Francisco y Mayra Montserrat Eslava Galicia (2018) Los movimientos sociales contra la minería extractivista. El Cotidiano 207 Revista de la División de Ciencias Sociales y Humanidades de la UAM – Azcapotzalco. Enero - Febrero 2018 Año 33. pp 121–129. ISSN: 1563-7417

López-Rojas J, Coutiño Ramos TA (2003) México un país Megadiverso. http://www.monografias.com/trabajos12/divmamif/divmamif.shtml

Madrid MR, Avila HM (2019) RIESGOS DE CORRUPCIÓN EN EL SECTOR MINERO: INFORME PERÚ. Un análisis de vulnerabilidades y riesgos de corrupción en el otorgamiento de licencias, permisos y contratos relacionados con el sector minero. PROÉTICA. Consejo Nacional para la Ética Pública (Capítulo Peruano de Transparency International). Lima, Perú. pp 70. https://www.proetica.org.pe/wp-content/uploads/2019/02/estudio-mineria-proetica.pdf

Martínez-Alier J (2009) El Ecologismo de los pobres. Barcelona, España. Editorial Icaria, 395 pp

Martínez Gisela, Erandi Flores Romero, Rocío Arias Puga, Daniel Hernández, Oliver Morales (2018). Belleza Concesionada. Pag web: http://www.ladata.mx/ (Consultada el 10 de julio de 2019)

Mittermeier, R., C. Goettsch (1992) La importancia de la diversidad biológica de México. en: Sarukhán J, Dirzo R (eds.) México ante los retos de la biodiversidad. CONABIO, México, pp 43–55

Montenegro RA (2003) Estudio Sobre el Impacto Ambiental y Sanitario de las Minas de Oro. El Caso Cordón Esquel. FUNAM (Fundación para la Defensa del Ambiente). Córdoba, Argentina. http://www.greenpeace.org/raw/content/argentina/contaminaci-n/miner-a-de-oro/estudio-sobre-el-impacto-ambie.pdf

OCMAL. Observatorio de Conflictos Mineros de América Latina (2019) Pág, web. https://www.ocmal.org Consultada 4 de octubre 2019

ONU (2019). https://www.un.org/es/about-un/index.html. Consultada 4 de sep 2019 11.30 am https://www.bbc.com/mundo/noticias-4815362127

Orden Jurídico (2019) Constitucion › articulos › www.ordenjuridico.gob.mx Pág Web, consultada el 4 de oct 2019

Otramerica (Redacción) (2011) La titánica pelea de San Luis de Potosí. Otramerica Periódico Digital. Sábado 30 de abril de 2011. http://otramerica.com/causas/la-titanica-pelea-de-san-luis-de-potosi/153

Pegg S (2006) Mining and poverty reduction: transforming rhetoric into reality. J Cleaner Prod 14(3-4):376–387

Roblero Morales Marín y Gerardo Porfirio Hernández Aguilar (2012) El despertar de la serpiente. La minería en la Sierra Madre de Chiapas. Revista de Geografía Agrícola núm. 48-49/75-88. Universisad Autónoma Chapingo, Mexico. https://www.academia.edu/25280094/Revista_de_Geograf%C3%ADa_Agr%C3%ADcola_no._48-49

Rodríguez García A (2011) La traición a los huicholes. Proceso – 10 Junio 2011. http://www.elmanana.com/XStatic/manana/template/notaimprimir.aspx?id=1169841

Salazar M (2010) Perú: Páramos peruanos en desamparo legal. Nota Periodística. Tierramérica. 29 de noviembre de 2010. http://www.paramo.org/content/per%C3%BA-p%C3%A1ramos-peruanos-en-desamparo-legal

SE. Secretaría de Economía (2017). https://www.gob.mx/se/acciones-y-programas/mineria

SINEM Sistema Integral sobre Economía Minera (2019) Minería de México: Proyectos mineros operados por compañías de capital extranjero. Dirección General de Desarrollo Minero https://www.sgm.gob.mx/extranjeras/Consulta_Minera.jsp. Consultada 29 de agosto de 2019 a las 2.32 pm

Solano Ortiz Lina Magali (2013) "Impactos Sociales De La Minería A Gran Escala En La Fase De Exploración Y El Rol De Las Defensoras De La Pachamama En La Resistencia

A Los Proyectos Río Blanco Y Quimsacocha". Tesis Previa A La Obtención Del Título De Magister En Sociología Y Desarrollo. Universidad De Cuenca Facultad De Ciencias Económicas Y Administrativas. Cuenca-Ecuador. Pp 150. http://dspace.ucuenca.edu.ec/bitstream/123456789/4701/1/TESIS.pdf

Svampa Maristella (2012) Consenso de los commodities y megaminería. In: Movimientos socioambientales en América Latina. Revista del OSAL Observatorio Social de América Latina. Año XIII N° 32 / publicación semestral / noviembre de 2012.Pág.15-39 http://biblioteca.clacso.edu.ar/clacso/osal/20120927103642/OSAL32.pdf

Svampa Maristella, Pablo Bertinat (eds) (2019) CONFLICTOS MINEROS EN AMÉRICA LATINA: EXTRACCIÓN, SAQUEO Y AGRESIÓN Estado de situación en 2018. Observatorio de Conflictos Mineros de América Latina, OCMAL www.ocmal.org. Marzo 2019

Tornay MMC (2018) Violencia e Impunidad en México en la Defensa de la Vida y el Territorio. Pueblos – Revista de Información y Debate. http://www.revistapueblos.org/blog/2018/01/25/violencia-e-impunidad-en-mexico-en-la-defensa-de-la-vida-y-el-territorio/

Urkidi Leire (2010) A glocal environmental movement against gold mining: Pascua–Lama in Chile. Ecol Econ 70(2): 219–227

WGMHRLT Working Group on Mining and Human Rights in Latin America (2014). "the impact of canadian mining in Latin America and Canada's responsibility: executive summary of the report submitted to the Inter-American commission on human rights", Working Group on Mining and Human Rights in Latin America. April 2014. Pp. 31 http://www.dplf.org/sites/default/files/report_canadian_mining_executive_summary.pdf

Yáñez Mondragón Carlos Francisco (2007) Las Áreas Naturales Protegidas en México, criterios para su determinación. Caso estudio: Sierra Tarahumara, Estado de Chihuahua. II Coloquio de Ingreso. Especialidad Geología. Academia de Ingeniería, A.C. México D.F. http://www.ai.org.mx/archivos/coloquios/2/Las%20areas%20naturales%20protegidas%20en%20Mexico.pdf

Zibechi R (2017) Movimientos sociales en América Latina. El « mundo otro » en movimiento. Ciudad de México: Bajo Tierra Ediciones: El Rebozo, 2017. ISBN: 9786079675134 6079675137

Chapter 22
Visions of the Future in the Oases of Baja California Sur, Mexico

Aurora Breceda, Alicia Tenza-Peral, Andrés Giménez-Casalduero, Micheline Cariño-Olvera, and Yven Echeverría-Ayala

Abstract The wetlands occupies less than 1% of the Baja California Peninsula and are patches with a mesic environment, habitat, and refuge of wild flora and fauna The wetlands were enclaves for the indigenous populations. However, it was not until the end of the seventeenth century, with the establishment of the Jesuit missions, were transformed in oases with a deep environmental and cultural changes, resembling as those of the arid belt of the northern hemisphere. These socio-ecosystems have subsisted for 300 years; they constitute sites of great biocultural value. However, it was until the arrived of the Jesuit missionaries, in the late seventeenth century, that the wetlands were transfomed into oases, with deep environmental and cultural changes, reseambing as those of the arid belt of the northern hemisphere. The recovery and revaluation of traditional socio-ecosystems can lead to a substantial change in the development model of Baja California Sur. In this chapter we summarize the results of the research of the Interdisciplinary Network for Integral and Sustainable Development of the Sudcalifornian Oases (RIDISOS) in the oasis of Comondú and present the results of visions of the future of the communities of San José and San Miguel de Comondú, which were analyzed during two

---

A. Breceda (✉)
Centro de Investigaciones Biológicas del Noroeste, La Paz, Baja California Sur, Mexico
e-mail: abreceda@cibnor.mx

A. Tenza-Peral
Universidad Miguel Hernández, Elche, Spain

Arizona State University, Tempe, AZ, USA
e-mail: atenza@umh.es; atenzape@asu.edu

A. Giménez-Casalduero
Universidad Miguel Hernández, Elche, Spain
e-mail: agimenez@umh.es

M. Cariño-Olvera
Universidad Autónoma de Baja California Sur, La Paz, Mexico

Y. Echeverría-Ayala
Independent Consultant on Climate Change Adaptation based on Ecosystems (AbE), Mexico City, Mexico

A. Ortega-Rubio (ed.), *Socio-ecological Studies in Natural Protected Areas*,
https://doi.org/10.1007/978-3-030-47264-1_22

workshops in these locations. In these workshops, the communities identified their main environmental units, assessed their current status, noted current community organization structures (COS) and existing gaps, and described the desired status of their communities for the future. The results of this activity were presented to the decision-makers (government authorities) and to the community, through written reports and a meeting on October 21, 2016, in San Miguel de Comondú, where agreements and commitments were signed between communities and authorities.

Keywords Traditional socio-ecosystems · Wetlands · Community participation · Comondú · RIDISOS

22.1 Wetlands of Baja California: Foundations of the Baja California Oases

In the Baja California Peninsula, Mexico, there are 184 wetlands, 171 of them located in the state of Baja California Sur (Maya et al. 1997). These ecosystems are characterized by the presence of water, whether by springs or subsurface water, allowing the development of dense vegetation which contrasts with the arid matrix around them. The surface of these wetlands represents only 1% of the peninsular land. However, these unique enclaves are patches with a mesic environment (more temperate), which constitutes the habitat and refuge for wild flora and fauna species, which find in their interior a more stable microclimate, and a source of water and food more constant than in the surrounding desert matrix. In these wetlands there are still plant species of tropical characteristics that can be relict of a vegetation that in the past had a greater distribution in the peninsula (CONANP 2006). These patches constitute stopover sites for migratory birds that stop at these sites on their trips to the south (Rodríguez-Estrella et al. 1997, 2005; Rubio et al. 1997).

The complex geological history of the peninsula, the diverse origins of its biota, and the evolutionary processes derived from its geographical and ecological isolation have developed a unique biodiversity, with a high number of endemic species, some of them endemic to a particular oasis such is the case of spiders (Jiménez et al. 2015). These ecosystems contain one of the densest forest biomass in the region, starting with the palm grove which have a complex vertical structure, with various overlapping strata, being the captures of CO_2, one of the environmental services they provide, as well as flood control, retention, and export of sediments and nutrients.

Within these ecosystems, soil saturation or proximity to the water permits the presence of different types of functional vegetation, which in turn creates different microenvironments, some of them highly fragile. In the water bodies and on its edges predominate a certain kind of vegetation like hydrophilic species such as the cattail (*Typha domingensis*), yerba del manso (*Anemopsis californica*), and reeds (*Phragmites* spp.). In the area that is occasionally flooded by water, palm trees are predominant, mainly from taco palm (*Washingtonia robusta)* and date (*Phoenix dactylifera*); and as we move away from the water body, the matrix is formed by dry scrub, and often there are an important number of mesquite (*Prosopis* spp.) and huisache (*Acacia farnesiana*) (Breceda and Pérez-Navarro 2013). Preliminary studies on the process of secondary succession in these ecosystems point to high resilience, especially in palm grove areas, in the face of relatively frequent disturbances such as floods associated with tropical cyclones and fires (Shiba-Reyes et al. 2019).

22.2 The Oases: Traditional Social-Ecological Systems of Baja California Sur

In the peninsula of Baja California, wetlands were fundamental enclaves for semi-nomadic hunter-gatherer-fishermen indigenous populations that used these spaces as central camps on their coast-to-coast journeys across the peninsula. However, it was not until the end of the seventeenth century, with the establishment of the Jesuit missions, when there was a deep environmental and cultural transformation (Cariño and Castillo 2017). The topology was modified by converting the sandy and rocky beds of streams and canyons into cultivated terraces. The hydrography was altered by channeling water for irrigation and building reservoirs. The biota was disrupted due to the introduction of numerous plant species – including date palms – and domestic animals, from many different regions of the world. From the eighteenth century, the typical cultural landscape of the arid belt of the northern hemisphere was formed on the peninsula, and the Baja California space began to resemble that of other oasis of the world such as the Spanish Levante, the Maghreb, the Middle East, North India, or West China (Cariño and De Grenade 2015).

The Baja Californian ranch culture, heiress and sentinel of the Baja California oases, is the result of the union between the bajacalifornian indigenous and the western Mediterranean culture and is characterized by austerity, self-sufficiency, and the varied and integral use of biotic diversity (Cariño 1996). In these oases, traditional irrigated agriculture has been successfully developed in the humid zone. This productive activity is complemented by extensive cattle raising in the surrounding dry land for the case of the mountain oases and with fishing for the case of the coastal oases (Breceda et al. 1997). Traditional agriculture is developed in terraces with stratified crops on three levels that take full advantage of the reduced fertile space and water. On the upper level, date palms (*Phoenix dactylifera*) and native palms (*Washingtonia robusta*) form a canopy that filters the sun's rays, reducing sunstroke and evaporation. The intermediate level is used to grow fruit trees, Mediterranean and tropical fruit trees. The lower level is dedicated to sowing legumes, cereals, and vegetables (Tenza et al. 2014). In the same way that in the wild flora, the different functional types of vegetation are distributed according to the saturation of the soils and the proximity to the water. The crops also follow a similar spatial organization, attending to their needs. Date palms are close to irrigation canals and in flood areas; sugarcane is also grown in this same flood zone. In the intermediate areas with better aeration of soil are fruit trees, and finally, on the terraces or *ancones* cereals, legumes and vegetables are grown. One key element of the cultural landscape of the oases are the irrigation systems, which include physical infrastructure for the channeling and distribution of water as ditches, channels, floodgates, waterwheels, dikes, dams, etc. and social infrastructure that entail the norms and rules for distributing and managing the irrigation water among the users. Thus, the oases are a coupled relationship between nature and society built in the wetlands of the arid regions by the societies of the deserts. They are a human construction in which the management of natural resources (water, fertile soil, and biodiversity) has been sustained for millennia, allowing the development of their populations and maintaining the ecosystem functionality of the wetlands that support them. However, the globalization threats their sustainability.

22.3 Regional Problem and Expectations of Change

Until the middle of the twentieth century, the oases of Baja California Sur were the main suppliers of raw materials to the urban centers of the federative entity, and even exported their products to other parts of the Mexican Republic. The economic and cultural development of the region was focused on these agroecosystems. However, since the mid-twentieth century, the modernization of the economy and its integration into globalization have decentralized the importance of these agroecosystems. The generalization of modern agriculture, with the imposition of the Green Revolution (since 1950) on valleys and agro-export production in greenhouses, has caused the decline of traditional irrigated agriculture. The growth of cities and tourist centers are now attraction poles for the impoverished and marginalized ranch population, who abandon their oases. Although regional development processes have had different effects on the oases of Baja California Sur, a generalized process has been the loss of traditional agrarian practices and the abandonment of primary activities. The oases best connected by roads and highways with other urban centers and tourist centers have experienced economic diversification with significant growth in the tertiary sector. In these cases, it is possible that the population has remained stable, or even increased. Nevertheless, in these oases, as is the case of the oasis of San José del Cabo, there have been irreversible losses of fertile soil due to the expansion of urban-tourist developments and contamination of water bodies. The oases of Todos Santos and Mulegé have also experienced a similar situation, but with a less massive urbanization. On the other hand, there are other oases that were more isolated from the regional development process, which is why they are ecologically better conserved, but socially disarticulated (Tenza et al. 2014; Cariño et al. 2016).

The regional development model of Baja California Sur has, roughly, encouraged the abandonment and deterioration of the oases, the sites with the greatest natural availability of water and of soil suitable for cultivation by agricultural production schemes highly dependent on external inputs (groundwater pumping, pesticides, and fertilizers) and land speculation with coastal tourism. The salinization of groundwater and soil generated by the overexploitation of aquifers in the great agricultural agro-industrial valleys of the region, uncontrolled urban growth, marginalization, and poverty on the periphery of Los Cabos and La Paz, and the deterioration of mangroves and others coastal ecosystems are some of the symptoms resulting from this development model.

If we consider the implications of this development model on biodiversity conservation, urban expansion and road infrastructure entail fragmentation and irreversible loss of natural habitat. In a fragmented habitat in which natural areas and human settlements are embedded in an agricultural-livestock matrix, the quality of that matrix condition compromises not only the biodiversity that persists in these anthropized spaces but also the migratory flows of individuals of species that move from some natural spaces to others. The quality, functionality, and permeability of this matrix are essential (Perfecto et al. 2010; Tenza et al. 2011). The large agricultural valleys of the state, characterized by extensions of monocultures, the use of pesticides and fertilizers, and experimentation with genetically modified organisms (GMOs), constitute a matrix of low quality and functionality, a little permeable, and can have serious consequences for the flora and wildlife.

The recovery and revaluation of traditional agroecosystems can lead to a substantial change in the development model of Baja California Sur. This could imply reestablishing a sustainable use of natural resources, capable of conserving biotic and cultural diversity, improving food security, and being competitive and resilient in the face of external drivers (e.g., climate, market prices). The revitalization and repopulation of rural areas could relieve some pressure on urban areas; and the creation of short market channels to encourage local consumption of regional products could improve and strengthen food security. It is a commitment to an endogenous development model that seeks to reduce the differences between urban and rural areas, improving the quality of life in both.

Each oasis has specific problems, so the actions aimed at the recovery and revitalization of each of these spaces must be done individually, avoiding "one-size-fits-all" approaches. It is for this reason, which we focus our study on one particular oasis, as a pilot case, the oasis of Comondú.

22.4 Pilot Case: The Oasis of Comondú

The Oasis of Los Comondú is one of the most representative and best preserved oasis of BCS. It has been a Ramsar site since 2008 and is currently part of a proposal for the creation of a new Biosphere Reserve. It is in the middle area of the mountain range La Giganta, within the municipality of Comondú, approximately 130 km north of the current municipal seat, Ciudad Constitución (Fig. 22.1). The wet area of the oasis, where traditional irrigation develops and the villages of San Miguel and San José de Comondú are located, is inside a canyon which it is width varies along

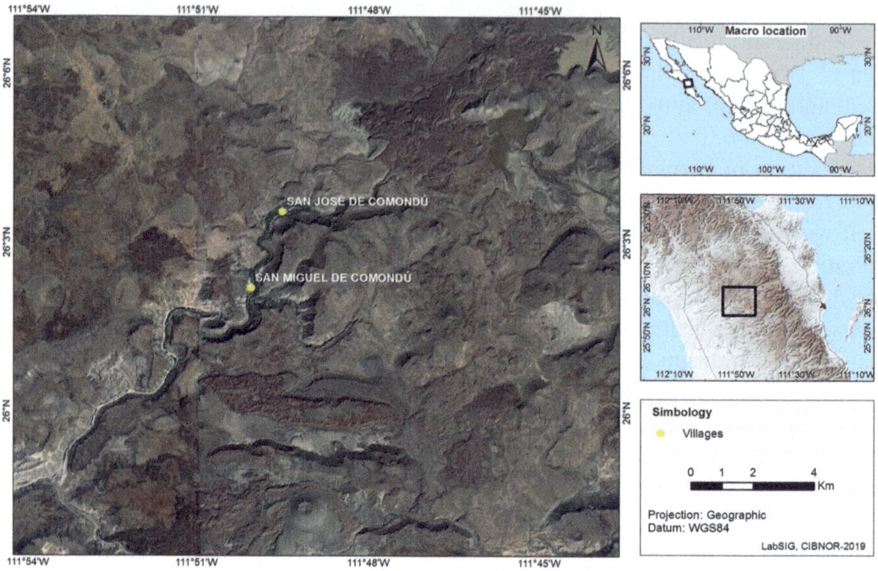

Fig. 22.1 Location of the oasis Comondú (San Miguel and San José de Comondú)

its route approximately between 50 and 500 meters and has a 16 km long although; only eight of them are recently used by the population. The area of the wetland is 88 hectares. However, to this surface it is necessary to add the surrounding dryland area, where the ranches and land of livestock use are located which are historically and currently interdependent of the wet area of the oasis. The dry land area reaches a total of 107,000 ha and includes the privately owned properties and the Comondú ejido (common land). The total population of the oasis was censored in 2010, counting 257 inhabitants, 109 of them in San José de Comondú, and 148 in San Miguel de Comondú.

In a study published in 1997, on the floristic composition and vegetation of eight South California oases, which did not include the oasis of Los Comondú, a total of 146 species were identified (Arriaga and Rodríguez-Estrella 1997). In a bibliographic review and recent sampling of the vegetation in the Comondú oasis, 124 species were registered, of which only 35% have been quoted in the other oases studied. In the same way, on a recent study of the vertebrate fauna, a total of 126 species were reported (Pérez-García et al. 2013), 31% of which are not mentioned in neighboring oasis and include 82 species of birds, 15 mammals, 3 amphibians, and 25 reptiles. Of these species, 13.5% are endemic (11 reptiles, 2 mammals, and 4 birds) and 30.5% (14 reptiles, 5 mammals, and 20 birds) are listed under some protection category in NOM-059-SEMARNAT-2010 (Pérez-García et al. 2013).

If we attend to its agrobiodiversity, the oasis of Los Comondú has one of the most diverse mixtures of perennial cultivated species of all the oases of the peninsula, including two types of cherimoya (*Annona reticulata* and *A. cherimola*), three varieties of Spanish tropical plums (*Spondias purpurea* red and yellow and *S. mombin*), old missionary oranges that can be reproduced by seed, royal lemon, grapefruit, various types of creoles mangoes and avocados, olive trees, grapevines and fig trees, guavas, pomegranates, green bananas, purple and white sugar cane, papayas, nisperos, and also new varieties of grafted mangoes and citrus (Valencian oranges, tangerines, and grapefruits). A total of 42 species of perennial crops belonging to 19 families have been described (De Grenade and Nabhan 2013). On the other hand, they also have a great diversity of annual species among cereals, legumes, and vegetables (e.g., cabbage, lettuce, coriander, garlic, onion, chili peppers, peas, beans, chickpeas, and corn).

Unlike other oases, such as the coastal ones, this has remained outside investments for the development of tourist complexity and secondary residences. Until 2011, the oasis Los Comondú was poorly connected to the transpeninsular highway, which links the most important urban centers of the entity that influences the economy and population dynamics of rural areas. Perhaps one of the consequences of this isolation has been the good conservation of biodiversity, both planned (crops) and associated (wild). However, the deterioration of the social net, the breakdown of local institutions, and the strong rural exodus that has lived in the last 70 years threaten to collapse this traditional agroecosystem with more than 300 years of history (Tenza et al. 2017, 2019).

The collapse and disappearance of this biocultural system would mean the irreversible loss of (1) a model of use and exploitation of sustainable territory, capable of producing food in sufficient quality and quantity and respecting biodiversity; (2) an important gene pool of local varieties adapted to the specific conditions of the region; and (3) the incalculable intangible heritage represented by traditional ecological knowledge transmitted and enriched generation after generation.

In recent years, the Baja California Sur government has paid more attention to the entity's oases, especially after the creation of a General Coordination of Sustainable Development within the Secretariat of Economic Promotion and Development (SPyDE). However, although actions have been taken, the limited deadlines available do not seem to have had great effects in solving the basic problems of these communities. Only through in-depth knowledge about the structure and functioning of these socioecological systems can policies and management measures be adapted to meet local needs and can make qualitative changes in their dynamics (Tenza et al. 2017, 2019).

The revitalization and participation of local communities is essential to empower the decision-making process on the actions (derived from policies and management measures) that can determine the future of their communities. That is why the Interdisciplinary Network for Integral and Sustainable Development of the Sudcalifornian Oases (called "RIDISOS" from the Spanish acronym) with the support of the German Cooperation Agency (GIZ), within the Biodiversity Governance project, created a plan that allowed to accompany the communities of Sam José and San Miguel de Comondú to express their vision of their future.

22.5 Visions of Future

In December 2015, two participatory workshops were held, one for each town. In these workshops, the communities identified their main environmental units, assessed their current status, noted current community organization structures (COS) and existing gaps, and, most important of all, described the desired status of their communities for the future. The results of this activity were presented to the decision-makers (government authorities) and to the community, through written reports and a meeting on October 21, 2016, in San Miguel de Comondú, where agreements and commitments were signed between communities and authorities (Figs. 22.2 a–b, 22.3, 22.4, and 22.5).

22.5.1 Environmental Units: State and Associated Community Organization Structures (COS)

Both communities distinguish two large environmental units: the ravine, which is the humid area where the villages, orchards, and palm groves are located and the surrounding rainfed area, where the ranches, mountains, hills, plains reliefs, and temporary watering holes, streams, and springs are located.

Among the mains activities associated with the area of the ravine are the traditional agriculture, palm leaf collection, obtaining materials for construction and handcrafts, the elaboration of fruit compote and missionary wine, as well as an incipient tourism.

In the mountains the main activities are animal husbandry (bovine and goat), farm animals milking, cheese producing and the sale of calves and goats; collection of medicinal and wild plants such as condiments, food, or firewood, hunting, and also incipient tourism activities (visit of cave paintings and petroglyphs).

All environmental subunits were evaluated, and the result was that almost all of them are considered in good condition. This means that these environmental units are healthy, and the main resources continue to be achieved with the same availability that previous generations had. However, some of the subunits were valued in a regular state, such as ranches, orchards, streams, palm groves, and villages. This means that these environmental units have lost their desired state of health and their natural resources have deteriorated, and this requires the intervention or human help to recover them.

The only existing community organization structure (COS) that is considered in good condition is the Ejido Comondú. In good condition means when the COS knows its responsibilities, has a work plan, meets periodically, makes decisions that represent the interests of the community, and implements all the actions required. The Ejido Comondú is mainly associated with the mountain area. The Events or Citizen Participation Committee (Comité de Eventos) is a COS associated with the villages but is considered in a regular-poor state. The "regular" state means that the

a

b

Fig. 22.2 a-b Workshops at San José and San Miguel de Comondú

Fig. 22.3 Women working at the workshop in San Miguel de Comondú

Fig. 22.4 Information meeting on the results of the workshops

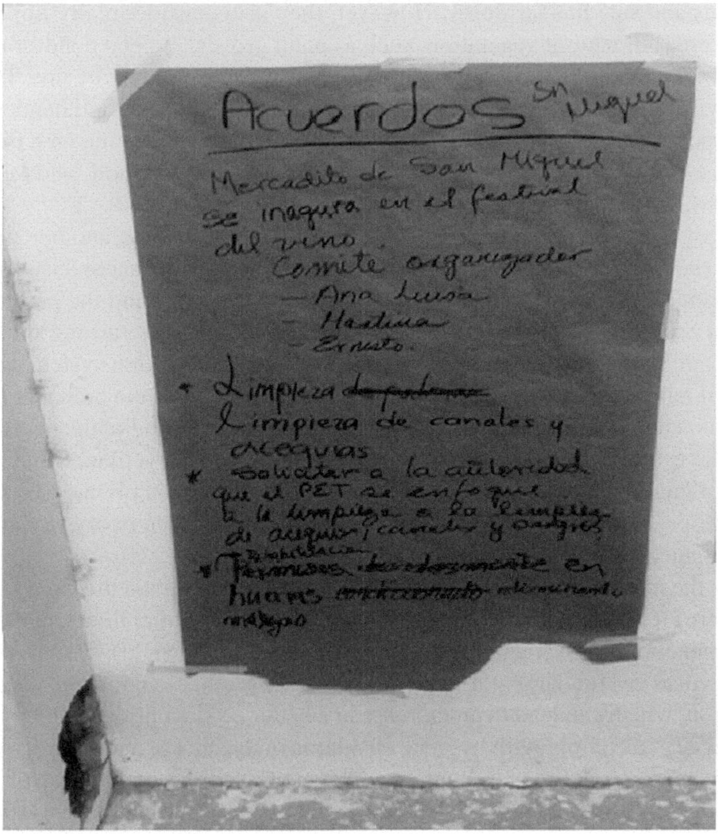

Fig. 22.5 Example of the community agreements

COS knows its responsibilities, has a work plan, and meets regularly. The “poor” state means that the COS only knows its responsibilities. The Temporary Employment Programs (PET) committees are the COS associated with palm trees and are considered in a regular state. Likewise, there are an Association for Livestock (Asociación Ganadera) in the area of the mountain but also classified in a regular state. There are gaps in community organization structures in orchards and streams.

22.5.2 Visions of the Future for the Oasis of Los Comondú

Both communities share a vision of the future for the oasis, and in some respects, although they maintain slight differences, they are complementary visions. In this report we will emphasize these similarities and complementarities.

The comondueños desire to reactivate and maintain part of the orchards currently abandoned and produce in sufficient quantity and quality to achieve the self-supply of families and sale the surpluses. However, they also consider really important to maintain areas of natural vegetation, such as palm groves, in good condition, which benefits biodiversity, but also allows its use and reduces the risk of fire. The palm grove is visualized not only as a source of construction materials and handcrafts but also as a potentially recreational area for camping. But this camping area must have infrastructure that does not cause a negative impact on water, soil, and vegetation (e.g., the waste management, drainage, etc.).

To recover the proper functioning of the area of the ravine, another important aspect is the correct channeling of water through ditches and secondary sprinklers (earth channels), existing channels (of stone and/or cement), and the periodic realization of "sangrías" to drain the excess water in the lands, returning them to the main stream. In this process of reactivation of orchards, irrigation system and maintenance of palm groves are necessary to organize the users of the orchards (owners, possessors, etc.), who know their responsibilities, manage, and make decisions. As well as the establishment of rules and regulations, and a surveillance and sanction system, to guarantee the proper functioning of the area of the ravine. The orchards must be respected in times of drought, denning the cattle to run freely, because it can cause damage to orchards in growth.

As for the mountain area, the communities aspire to maintain this area of pasture (agostadero) for cattle in good condition, considering its spatial limits, and continue with the genetic improvement of the cattle (to increase the yield). The organization of the breeders is a fundamental step to improve the production, market, and product distribution, which can lead to an increase in a better product pricing. In this process it is necessary to comply with hygiene standards in the cheese and dairy products as well as farming meat. The communities also expressed the importance of keeping water bodies in good condition and dredged. They also wish to improve management and relations with government authorities that facilitate them to obtain support, compensation for the loss of animals due to predation (and/or extreme weather events), and to solve existing problems with the current livestock registration system (aretado).

In relation to the state of the villages, the comondueños wish to maintain the current population and even the return of emigrants and new settlers. For this, the situation aims to have a greater and better amount of services that improve the life quality for the community. Among the services mentioned are middle education, high school and/or technical careers (tele-high school); constant health services (GP, medical clinic, ambulance); improvement of the drinking water distribution network; infrastructure for the artisanal product sales ("artisan's house"); restoration of old houses in town; and the restoration of the community development center (space to offer information to tourists).

Both communities agreed that for the attainment of these great objectives one of the key aspects to be addressed is land tenure and certainty in land ownership. While the community of San José was stricter in trying to obtain 100% of the legal certainty of land tenure by legal and administrative means, San Miguel community expressed the possibility to establish agreements for the use of land between resident, nonresident, and ejidatarios (landlords) possessors. It is a subject that we perceive as a very sensitive subject. A large part of the population accepts the possibility of establishing some kind of agreement.

22.6 Final Remarks

The maintenance of small-scale productive activities in arid regions is essential to meet the Sustainable Development Goals of the 2030 Agenda (SDG). The arid regions occupy 40% of the planet's surface, sustain 44% of world agriculture, and 50% of the livestock. Agro-livestock systems in arid regions are important for global sustainability, given their contribution to food security, as well as the economy and employment of their local populations, as it sustains more than two million people, half of them in poverty condition (IFAD 2016). The global change processes threaten the resilience and sustainability of these systems. Its disappearance reduces our possibilities as a global society to achieve sustainable development, irreversibly losing local institutions for the management of natural resources, ecosystem services, traditional ecological knowledge, management practices, and a large collection of species, varieties, and races, both vegetables and animals, adapted to local conditions.

The oases of the peninsula of Baja California are a sample of the biocultural relevance of traditional socio-ecosystems, as well as the threats and abandonment in which they are currently. The participation and organization of local communities is essential for their revival and resilience. With the development of this project, it was possible to identify visions of the future of the communities of a representative oasis in Baja California Sur. The visions of the future of the communities of San José and San Miguel de Comondú can be summarized in the desire to recover and improve the traditional productive activities, agriculture, and livestock, with the establishment of new marketing channels that improve income perceived by families. There is an incipient development of tourism; this activity can diversify the economy and improve the availability of economic and financial capital in the oasis, necessary for the growth of private services, and increase local employment. The increase in public services is another aspect that is part of their visions of the future, especially in matters of education, health, and/or communication, which would significantly improve the quality of life, and could be attractive aspects for the repopulation of the oasis. The community organization is presented as the "cornerstone" that must be present in any community development process. The request and management of external support is more efficient and equitable if there are community organization structures. The current state of land tenure is presented as one of the main obstacles in the community development process. It is important to explore the possible ways of solution, either by the legal-administrative route or by the establishment of agreements for the usufruct of the land between interested users. The results were presented to decision-makers and communities at a meeting on October 21, 2016, in San Miguel de Comondú, where agreements and commitments were signed between communities and authorities.

Acknowledgments The study was part of the activities developed into the Biodiversity Governance Project supported by the German Cooperation Agency (GIZ) and CONABIO. We acknowledge the collaboration of the MS Christian Silva for his help to carry out the workshops and meetings, as well to the Ing. Gil Ceseña who developed the map of the study oasis.

References

Arriaga L, Rodríguez-Estrella R (eds) (1997) Los oasis de la Península de Baja California. Centro de Investigaciones Biológicas del Noroeste. Publicación No 13, La Paz BCS, México

Breceda A, Pérez-Navarro JJ (2013) Vegetación. In: Cariño M, Breceda A, Ortega A and L Castorena (eds) Evocando el edén: Conocimiento valoración y problemática del oasis de Los Comondú, ISBN: 978-84-9888-491-3

Breceda A, Arriaga L, Coria R (1997) Características socioeconómicas y uso de los recursos naturales. In: Arriaga L, Rodríguez-Estrella R (eds) Los Oasis de la Península de Baja California SIMAC-CIBNOR, Publicación No 13, Centro de Investigaciones Biológicas del Noroeste, 292 pp

Cariño M (1996) Historia de las relaciones hombre/naturaleza en Baja California Sur 1500–1940 SEP-FOMES, UABCS, La Paz México

Cariño M, Castillo AL (2017) Oasis Sudcalifornianos: paisajes bioculturales con elevada capacidad adaptativa a la aridez y potencial para la construcción de la sustentabilidad local. Rev Front J Soc Technol Environ Sci 6(2):217–239

Cariño M, De Grenade R (2015) Oasisidad: una promesa de sustentabilidad para las zonas áridas. In: Cariño M, Castorena L (eds) Saberes para la sustentabilidad. Icaria Editorial, Barcelona, pp 109–142

Cariño M, Breceda A, Tenza A (2016) Los oasis: sistemas socioambientales tradicionales de la península de Baja California. In: Moreno-Calles AI, Casas A, Toledo VM, Vallejo-Ramos M (compilers) Etnoagroforestería en México. Universidad Autónoma Nacional de México, Escuela Nacional de Estudios Superiores Unidad de Morelia, Instituto de Investigaciones en Ecosistemas y Sustentabilidad, pp 149–169

CONANP (Comisión Nacional de Áreas Naturales Protegidas) (2006) Estudio Previo Justificativo para el Establecimiento del Área Natural Protegida: "Área de Protección de Flora y Fauna Oasis de Baja California Sur conformada en Archipiélago", México DF, pp 259

De Grenade R, Nabhan GP (2013) Agro-biodiversidad in-situ en el Oasis de Los Comondú. In: Cariño M, Breceda A, Ortega A, L Castorena (eds) Evocando el edén: Conocimiento valoración y problemática del oasis de Los Comondú. ISBN: 978-84-9888-491-3

Jiménez ML, Nieto-Castañeda IG, Correa-Ramírez MM, Palacios-Cardiel C (2015) Las arañas de los oasis de la región meridional de la península de Baja California, México. Rev Mexicana de Biodiversidad 86:319–331

Maya Y, Coria R and Domínguez R (1997) Caracterización de los oasis. In: Arriaga L, Rodríguez-Estrella R (eds) Los oasis de la Península de Baja California. Centro de Investigaciones Biológicas del Noroeste, Publicación No 13, La Paz, BCS, México

Pérez-García J, Botella-Robles F, Galina-Tessaro P, Arnaud G, Pérez-Ibarra I, Giménez A (2013) Fauna vertebrada. In: Cariño M, Breceda A, Ortega A, L Castorena (eds) Evocando el edén: Conocimiento valoración y problemática del oasis de Los Comondú, ISBN: 978-84-9888-491-3, pp 139–162

Perfecto I, Vandermeer J, Wright A (2010) Nature's matrix. Earthscan Ltd

Rodríguez-Estrella R, Rubio L and Pineda E (1997) Los oasis como parches atractivos para las aves residentes e invernantes In: Arriaga L and Rodríguez-Estrella R (eds) Los oasis de la Península de Baja California. Centro de Investigaciones Biológicas del Noroeste, Publicación No 13, La Paz, BCS, México, pp 157–195

Rodríguez-Estrella R, Blázquez M. C and Lobato J. M (2005) Avian communities of arroyos and desert oases in Baja California Sur: Implications for conservation. In: Cartron J.E, Ceballos G and Fleger R.S (eds) Biodiversity, ecosystems and conservation in Northern Mexico. Oxford University Press, NewYork, pp 334–353

Rubio L, Rodríguez-Estrella R, Pineda E (1997) Comparación del uso del hábitat por aves residentes e invernantes. In: Arriaga L, Rodríguez-Estrella R (eds) Los oasis de la Península de Baja California. Centro de Investigaciones Biológicas del Noroeste, Publicación No 13, La Paz, BCS, México, pp 221–248

Shiba-Reyes M, Breceda-Solís A,Troyo-Diéguez E, Martínez-Rincón R (2019) Efecto de ciclones tropicales sobre la cuenca baja y estero de San José del Cabo, BCS, México. In: Valverde (Presidencia) Simposio: Estrategias de conservación y uso sustentable del agua, suelo y agro-sistemas en cuencas deficitarias. VII Congreso Mexicano de Ecología. Juriquilla, Querétaro. México

Tenza A, Breceda A, Cariño M, Martínez-Fernández J, Giménez A (2014) Claves de sustentabilidad de agroecosistemas tradicionales basados en regadíos y ganadería extensiva en oasis de Baja California (México). In: Sanchis-Ibor C, Palau-Salvador G, Mangue Alférez I and Martínez-Sanmartín LP (eds) Irrigation Society and Landscape Tribute to Thomas F Glick, pp 791–804

Tenza A, Pérez I, Martínez-Fernández J, Giménez A (2017) Understanding the decline and resilience loss of a long-lived social-ecological system: insights from system dynamics. Ecol Soc 22(2):15

Tenza A, Martínez-Fernández J, Pérez-Ibarra I, Giménez A (2019) Sustainability of small-scale social-ecological systems in arid environments: trade-off and synergies of global and regional changes. Sustain Sci 14:791–807

Tenza-Peral A, García-Barrios L, Giménez A (2011) Agricultura y conservación en Latinoamérica en el siglo XXI: ¿Festejamos la "Transición Forestal" o construimos activamente "la matriz de la Naturaleza"? Interciencia 36(7):500–507

Chapter 23
The Challenge of the Science of Sustainability in Protected Natural Areas: The Case of the UMA "Wotoch Aayin" in the Ría Celestún Biosphere Reserve, Campeche

Mario del Roble Pensado-Leglise and Gerardo Rodríguez Quiroz

Abstract This study analyzes the current challenges in the science of sustainability, SS, in protected natural areas, particularly the case of a Unit of Environmental Management (UMA—Unidad de Manejo Ambiental) of the Moreletti crocodile in Ría Celestún, Campeche. SS is conceptualized as a holistic approach for identification and resolution of the main problems in sustainable systems with a global, social, and human scope. SS is not a basic or applied science; rather it distinguishes knowledge (scientific and non-scientific) to reach its objective to solve problems in the relationships between society and nature. The results of the analysis revealed intertwining of the activities of biodiversity conservation with sustainable practices in the territories parting from the approach of ecosystem services. This has been a key factor in obtaining advances in proper management of an intensive breeding farm associated with a scheme of productive diversification linked to development of both scientific-technological and social innovation, in which it is sought to conjugate scientific innovation with culture, traditional productive practices, and strengthening of local sociocultural identity of the mangroves to constitute a sustainable civil productive organization.

Keywords Sustainability · UMA · *Crocodylus moreletii* · Ría Celestún · Social innovation

M. R. Pensado-Leglise (✉)
Instituto Politécnico Nacional, CIIEMAD, Proy SIP 20196384, Ciudad de México, México
e-mail: mpensado@ipn.mx

G. Rodíguez Quiroz
Instituto Politécnico Nacional, CIIDIR Sinaloa, Proy SIP 20196690,
Guasave, Sinaloa, México
e-mail: grquiroz@ipn.mx

A. Ortega-Rubio (ed.), *Socio-ecological Studies in Natural Protected Areas*,
https://doi.org/10.1007/978-3-030-47264-1_23

23.1 The Science of Sustainability: Where Are We?

In 2001, the importance of a new interdisciplinary scientific effort centered in sustainability was put forward (Kates et al. 2001). Later, other collectives of authors backed the initiative, some with other criteria and arguments but equally concluding that it was necessary to generate a new way of doing science for the sustainability of the planet (Giannuzzo 2010; Takeuchi 2010; Bettencourt and Kaur 2011; Delgado and Rist 2016; Komiyama and Takeuchi 2011; Vilches and Gil-Pérez 2015; Díaz et al. 2015).

The emergence of this science is explained by the impossibility of finding a true solution to the problem of sustainability based on the same conventional scientific criteria, instruments, and paradigms that treat the objects of study in an unconnected way, ignoring their context (Wiek et al. 2015) and using linear, mechanist approaches that give partial isolated solutions that do not solve global problems.

It is not about denying interdisciplinary work indispensable for the solution of problems in the relationship between society and environment: from ecological economics, exact sciences, genomics, nanotechnologies, telematics, those related to artificial intelligence, and marine, atmospheric, and space sciences. However, maintaining these reductionist criteria and partializing scientific knowledge, together with ignorance of the perpetuation of the property rights of humanity's common global resources, promotes the battle to privatize the genomic and microbiomatic scientific fields as well as the disputes to strip the world's indigenous peoples of their cultural heritage in this era of knowledge (Cocho et al. 2017).

Current conventional science increasingly demonstrates its incapability to deal with problems generated in the process of globalization. Such is the failure of economic theory to prevent serious international financial crises, the ineffectiveness of the social and political sciences to decrease social inequality and poverty in the world, and the failure of agronomic sciences to solve the food problem and paradoxically contributing to impeding the functioning of a sustainable food safety system that global society requires. On the contrary, the new scientific praxis of sustainability (Sachs 1974; NRS 1999) lies in corresponding to the Anthropocene context (Steffen et al. 2011; Lewis and Maslin 2015; Waters et al. 2017) in harmony with solving problems of planetary sustainability (Vessuri 2016) of post-normal (Funtowics and Marchi 2009; Kønig et al. 2016), holistic (not reductionist) (Leff 2009), interdisciplinary (not partialized) character and centered on a systemic approach to solve complex environmental problems through nonlinear environmental solutions through methods of scientific approximation, not just deterministically (Hauge and Barwell 2016). Also, it is based on taking advantage of the wealth of scientific and non-scientific knowledge that comes forth in participative research (Santes and Pensado-Leglise 2017) with guidelines from a variety of methods, such as, for sustainable development with dynamic systems, the method of territorial analysis and, in economic-ecology (Pensado-Leglise 2011), collective social action (Beitl 2014) and agroecology and other alternative hybrid technologies in agronomy (Hausknost et al. 2017). The main result of these scientific efforts has solidified only through the union of political and citizen effort reached in the global accord on the 2030 Agenda and the 17 objectives of sustainable development, OSD, stipulated in the 2015 Paris Conference (ONU 2015; Hák et al. 2016) as well as a global wager for the future.

In this respect, this study on crocodiles, mangroves, and society addresses the ambit of two OSD: the first is Objective 14 "Conserve and sustainably use the oceans, seas and marine resources for sustainable development" and Objective 15 "Protect, reestablish and promote the sustainable use of terrestrial ecosystems ... and detain biodiversity loss" (ONU 2015; Robert et al. 2016). In this sense, the problem in the science of sustainability for this case study is how to solve the problem of threat of extinction of a species based on conservation and sustainable use of the natural space that corresponds to said species (Russi et al. 2013). In this sense, evaluating the experience described in this case study on the Unidad de Manejo Ambiental (UMA; Environmental Management Unit) "Wotoch Aayin" in the Ría Celestún Biosphere Reserve, Campeche, provides a magnificent example of the scope and limitations of the interaction of a new practice in scientific and non-scientific research in sustainability carried out by a small family cooperative.

23.2 Sustainable Science and Ecosystem Services in National Protected Areas

At the end of the twentieth Century, Costanza et al. (1997) proposed a new methodology to understand the contributions of "natural capital" and defined ecosystem services as those whose processes and ecosystem functions benefit people consciously or unconsciously, directly or indirectly (Costanza et al. 2017, p. 3). Later, "The Economics of Ecosystems & Biodiversity," TEEB (2010), the international institution of research in ecological economics defined ecosystem services as direct and indirect contributions of ecosystems to human well-being, enumerating four:

1. Supply services: for example, wild food, harvests, fresh water, and plant-based medicine
2. Regulation services: for example, filtration of pollutants through wetlands, climate regulation through carbon storage, and water cycle, pollination, and protection from natural disasters
3. Cultural services: for example, recreational activities, spiritual and esthetic values, education
4. Support services: for example, soil formation, photosynthesis, nutrient cycle (TEEB 2010, p. 9)

These conceptual and methodological advances have produced new experiences in public policy linked to conservation of biodiversity with sustainable use in countries such as Mexico (IPBES 2018; WWF 2018; L'humeau and Cordero 2012; Díaz et al. 2019). The result has been an enormous effort to make changes in scientific paradigms, evidenced with new discourse at the Convention on Biological Diversity, which, with its Strategic Plan for Biological Diversity 2011–2020 and the Aichi Goals for Biological Diversity, established the need to install public policies that simultaneously linked biodiversity conservation actions to sustainable development (CBD/COP10 2010) and to eradication of poverty (CBD/COP12 2014) relative to sustainable use of wild fauna and flora in plans and strategies for organisms cooperating in the development of pertinent social sectors (CBD/COP13 2016, p. 61).

Fischer and Eastwood (2016) framed part of this new scientific approach in the character of coproduction of the nature-society relationships. This implies considering evolution of these relationships as a process sustained in a dynamic system in which its context changes, leading to conceive the need for a form of economy (ecology) that corresponds to the configuration of a new state of social well-being and sustainability on the global level (Costanza et al. 2017).

23.2.1 The New Policy of Linking Biodiversity Conservation and Sustainable Development in Mexico

In the 1990s, the Mexican federal government developed a new stage of public policy aimed to protect some endangered or threatened species in which the territories of protected areas (initiated in the 1980s) containing biodiversity richness were defined by legal figures that involved landowners (ejidos, agrarian communities, and small private property) in adoption of protective measures for the protected areas and at the same time permitted some economic activities that did not negatively impact or put at risk the nuclear zones of those areas (Challenger et al. 2018).

In 2017, the area covered by natural protected areas was around 90.8 million hectares, of which one fourth was terrestrial and the rest was marine. Currently, the national system of protected areas in Mexico consists of 67 national parks, 40 areas of protected wild flora and fauna, 8 areas of protected natural resources, 5 national monuments, 18 sanctuaries, and 44 biosphere reserves (Brenner 2010).

In 1997, a new program for wildlife protection was established, Unidades de Manejo Ambiental (UMAs in Spanish; Environmental Management Units for conservation of wildlife), which authorized associations and private and social enterprises to use areas for sustainable exploitation of wild flora and fauna (INE-SEMARNAP 1997). Establishment of UMAs has been directed to link conservation goals and sustainable use of natural habitats and wild species through scientific research and technological transfer, ecological restoration, and reproduction, rescue, and reintroduction of species of wild flora and fauna or their protection, rehabilitation, and exhibition, as well as sustainable recreational and touristic activities and training of the population and communities that inhabit the protected areas (Cámara de Diputados 2000a). By 2017, there were 1722 UMAs in the country, of both extensive and intensive character, in around 38.5 million hectares, that is, an area equivalent to almost a fifth of the country's continental area (SEMARNAT 2018). All of these areas have a sustainable territorial management program in conjunction with the local population and sustainable use of the wild species in the UMAs.

It is clear that this implicates a change in logic that new models of social-institutional intervention bring (CONABIO 2014) based on the new interdisciplinary scientific research in the framework of the confluence of scientific and non-scientific knowledge (technological innovation with the wealth of local culture, indigenous productive practices, productive campesino diversification, etc.) in sustainable territorial development projects with social participation, all of this accompanied by a new legal-administrative normativity for natural protected areas and wildlife (Cámara de Diputados 2000b). However, the conflicts related to environmental problems, particularly loss of protected areas and biodiversity, have been increasing in recent years (Pensado-Leglise 2017).

23.3 Recovery of the Moreletti Crocodile and ANP Conservation: How Can Progress Be Sustained?

The problem of animal conservation varies depending on the type of animal and the number of individuals in its natural environment as well as on whether it is exploited commercially or as a food source for certain cultures. Conservation of some species in many places has taken place through commercial exploitation. One case is the Moreletti crocodile (*Crocodylus moreletii*), which was nearly extinct because of excessive and unregulated extraction. The conservation action implemented beginning in the 1970s has permitted its recovery, and it is currently no longer endangered.

All over the world, alternatives for conservation of the crocodile have been sought with the aim of sustainable use for the benefit of those communities that border the bodies of water where the crocodile lives and that see in this animal an opportunity to improve their quality of life as well as a commitment to conserve the species and its habitat (Barrios and Cremieux 2018).

Worldwide, there have been examples of successful management that has permitted the recovery and permanence of wild populations of different species of crocodiles. These examples include coordinated programs of sustainable extractive use in wild environments and protection of the habitat without affecting wild populations since they naturally compensate the population load and offspring mortality by promoting a higher rate of survival (Webb and Manolis 1992). This reinforces the theory that the populations subjected to sustainable use have greater growth than those that are not subject to use (Webb 2014).

There are other modalities of sustainable use of crocodiles in the world:

1. Direct extraction, or harvest, which consists of capture and extraction of specimens of commercial size from wild environments for commercial exploitation. Countries such as Venezuela and Papua New Guinea were pioneers in implementing this scheme.

2. "Farming," or extraction of eggs or offspring from the wild environment and taking them to a farm or nursery for incubation, development, and growth. This modality requires that specimens be later released to compensate those harvested for commercial purposes. The countries that have developed this scheme are the United States, Australia, and Argentina (Barrios and Cremieux 2018).
3. Extraction of problem specimens and/or trophy hunting. In this modality of use, a determined number of permits are granted to hunters and wild fauna controllers for hunting, and problem or very large specimens are removed for commercial use or are promoted as trophies. This practice has been successful in countries such as the United States and Australia, among others.

Other measures have contributed to recovery of crocodile populations, documented by the International Union for the Conservation of Nature (IUCN; Ross 2000), the CoPan project (Domínguez Laso 2006), and the Program of Moreletti crocodile (*Crocodylus moreletii*) monitoring in Mexico (Sánchez-Herrera et al. 2012, 2015; Rivera-Téllez et al. 2017).

In Mexico currently, the Moreletti crocodile is catalogued by the nom-059-semarnat 2010 as a species subject to special protection. At the international level, the IUCN has placed it at low risk/dependent on conservation and later in 2012, as one of lower concern. In 2010, the populations in Mexico and Belize were moved from Appendix I to II of the CITES with a zero quota for wild species for commercial ends. By 2012, based on a demographic analysis and projection (Sánchez-Herrera and Álvarez-Romero 2006), these populations were eliminated from the Endangered Species Act of the United States.

In Mexico, the Moreletti crocodile has been exploited under the scheme of closed-cycle breeding in captivity and, in the late 1990s, under the intensive modality of Unidades de Manejo Ambiental (UMA in Spanish, or Environmental Management Units) for wildlife conservation. The scheme consists of having a reproducing herd in captivity, fomenting their reproduction, collecting the eggs, and growing and developing the specimens to a commercial size in their natural environment. In this scheme, activities of environmental education and generation of knowledge of the species support to communities and social groups in management and co-existence with the crocodile are also contemplated in order to decrease and eradicate illegal commerce and thus indirectly contribute to conservation of wild populations and their habitat.

The plan of action for the conservation of this crocodile was published by the group of crocodile specialists of the IUCN. This plan points out that the species has a high potential for management supported by a program of sustainable use for the Moreletti crocodile in Mexico derived on the basis of information collected by national projects existing in locations where there are activities in UMAs for its care in captivity and later distribution between exploitation and release into the wild environment (Platt et al. 2010).

Since 2004, multidisciplinary work has been done by academic researchers, producers, NGOs, and government institutions such as CONABIO to generate robust information on the populational state of the Moreletti crocodile (*Crocodylus moreletii*) in the wild, and in 2010, a Mexico-Belize-Guatemala monitoring program of the crocodile was designed and implemented (Sánchez Herrera et al. 2011).

23.4 The Ría Celestún Biosphere Reserve

23.4.1 Mangroves in Mexico

Campeche is the state with the largest area of mangroves, and Mexico occupies fourth place in mangrove area in the world. Mangroves are found on the coasts and mouths of rivers that flow into the seas and other bodies of water. It is estimated that there is an area of 7801 km^2, 5% of the world total in 2010. Mangroves are one of the most productive ecosystems. Their ecological and economic importance lies in that they are coastal wetlands in which plant species adapted to flood environments, saltwater, and fresh water grow, like facultative halophyte plants (Valderrama-Landeros et al. 2017).

The main direct ecosystem services mangroves provide for the territory are formation of organic barriers of natural protection that mitigate the effects of erosion by wind and tides and attenuation of negative effects of catastrophic phenomena such as hurricanes and tsunamis by regulating water flows, thus serving as protection from high waves and controlling floods. Another important function is that of bodies that receive continental water; they intercept and process organic nutrients being carried from the continent toward the oceans. They are a biological filter that improves water quality and controls erosion by retaining sediments (Russi et al. 2013). Moreover, because they produce large amounts of nutrients, they contribute to growth and development of the populations of crustaceans, mollusks, and a large variety of fish. Mangroves also serve as habitat for sea birds, migratory and endemic. They provide spaces for reproduction of a diversity of insects and refuge for small mammals, reptiles, and crocodiles. For all these reasons, mangroves are a fundamental factor in biodiversity reproduction and conservation.

These buffer zones between terrestrial and marine ecosystems allow anthropogenic activities. Besides producing sea food for human consumption, they make other activities possible, such as exploiting plants to obtain natural dyes, medicine, wood, fibers, firewood, and charcoal. Moreover, they provide areas for rest and recreation, for enjoyment of natural panoramas, and for other activities related to ecotourism.

23.4.2 The Problem

The economic extractive modality has destroyed ecosystems and habitats of wild flora and fauna, demolishing species sustainability and generating profound ecological imbalances. In any case, if existing populations are to be recovered for conservation or for an economic activity, be it riparian or at a higher elevation, it is imperative to improve the quality of mangrove ecosystem services so that the fish, crustaceans, and mollusks can raise the repopulation rate of the local marine fauna that occurs in the mangroves.

Mangroves have suffered major losses. In 1981 there were 216.9 thousand ha, while in 2015 the area was 198.8 thousand ha, of which a little more than two thousand ha of mangrove is perturbed. Currently, around 90% of the total mangrove areas are under government protection as natural protected areas. This, however, has not stopped deforestation, desiccation of wetlands, and expansion of urban, touristic, and service areas in recent years (Vázquez-Lule et al. 2009).

In addition, global warming is elevating water temperature and sea level, causing erosion of coastal soil (SEMAAS 2015). In Campeche, where our study area is located, there are no up-to-date measurements, but much is compromised. The state's coast is 434 km, and in 2015 it was estimated that 79% was occupied by mangroves. The mangroves are the last line of defense for the population and the coast against the forces of the sea. One measure of mitigating climate change is conserving and restoring areas of mangroves and improving the quality of their ecosystem services and sustainable use of coastal flora and fauna, which provides better incomes, more jobs, and greater associative capacity for the population of the Campeche coastal region.

Inadequate management of federal terrestrial maritime zones and ignorance of land use regulations have led to the destruction of the mangroves. Examples of land use change affecting the mangroves are deforestation, pressure from urban, touristic and commercial expansion, and land transport infrastructure. These changes have directly affected trophic chains and exert pressure on fishing and riparian resources. Survival of protected species is shortened, and natural rates of repopulation are impeded, as well as the possibility of a variety of species to reach the ocean for their development. Areas for nesting and refuge for birds and other fauna are decreased, contributing to imbalances, which, if climate change worsens, will cause infestation of insect vectors of endemic diseases such as malaria (Vázquez-Lule et al. 2009).

23.5 Case Study: The Experience of the Environmental Management Unit "Wotoch Aayin SC de RL de CV"[1]

The case study was conducted with the methodology of participative rural evaluation (rapid). The precepts of introductory study were completed. Documentary and bibliographic information was reviewed. The community was sensitized, and a

[1]An initial version of the systemization of the experience was written by Dr. Pensado-Leglise, for the "Second exchange of innovations in public policy and practice of inclusive rural development linked to family agriculture in Latin America" held October 23 to 27, 2017, in the state of Campeche, organized by IICA with collaboration of IPN, INCA-Rural SAGARPA, and the GTD CONACYT network.

semistructured interview was designed for each type of participant (cooperative, families, fishermen, technicians, and authorities). Two general meetings (workshop type) were held with the members of the cooperative (Pretty and Vodouhê 1998).

23.5.1 The Rural Family Enterprise and Conservation of the Coastal Ecosystem

Wotoch Aayin is a family cooperative enterprise that operates a UMA (Management Unit for conservation of wildlife), and the main objective of their activity is rearing, conservation, and exploitation of the Moreletti crocodile in a context of sustainable use of the mangroves.

For some, the more than 20 years of effort of the family cooperative responsible for the functioning of the UMA may not be very impressive. Others might say that it is "a micro-business that has been lucky." But they are expressing their incapability of understanding that this small family cooperative is "turning things around" by carrying out a fundamental part of the process of constructing sustainable development in the state of Campeche and the country.

Over time, their strategy has addressed two transcendental aspects of the state's strategy in the face of climate change: (1) conservation of the mangroves through sustainable exploitation and diverse education activities and ecotourism services and (2) operation of the UMA, which has contributed to reducing the risk of extinction of the Moreletti crocodile. Twenty years ago, it was included in Appendix I of the CITES list of endangered species. In 2013, through the great effort of this type of small family cooperative, this species in Mexico was placed in Appendix II of species subject to controlled commercialization.

23.5.2 History of the Wotoch Aayin Cooperative

The cooperative emerged between two families, one of fishermen and the other of a biologist. They decided to create a crocodile farm on an island, Isla Arena, of the municipality of Calkini, which forms part of the Ría Celestún Biosphere Reserve (SEMARNAT 2000). They were able to request a piece of land for a crocodile farm through the municipal agency. This was granted in 2000; unlike local tourism projects, they did not request land near the beach, rather land that bordered the mangrove.

The proposal was to conciliate the need to protect the Moreletti crocodile and its habitat at the same time. It took them 7 years of red tape to obtain a permit to operate an Environmental Management Unit (UMA, the legal figure of an entrepreneurial unit for sustainable exploitation of wildlife) for rearing crocodiles. Meanwhile, they continued living by fishing and their jobs in services. In 2008, there was an opportunity to use a permit in force of an UMA operating in Yucatán. This permit allowed them to begin activities on Isla Arena. The National Commission for Natural Protected Areas (CONANP, in Spanish) provided 15 Moreletti crocodiles, 10 females and 5 males, and partly funded the protective fence. In 2009, CONANP granted support for installation of a "lookout-dock" that can be reached by walking through an "ecological path," which is long pontoon with a wooden handrail leading to a palm-roof palapa, which serves as a lookout and a dock (Fig. 23.1).

In 2010, the Secretariat of Agriculture, Livestock, Rural Development, Fishing and Food (SAGARPA), through the National Commission of Aquaculture and Fishing (CONAPESCA), provided funding for building the restaurant that specializes in crocodile filets, acquisition of kayaks, a generator, and other equipment. At that time, they only had a palapa, and so the funding of around 400 thousand pesos was very important in consolidating the diversification of incomes from the activities of their crocodile farm (Fig. 23.2).

In late 2012, the UMA permit from Yucatán expired, but the state delegation of the Secretariat of Environment and Natural Resources (SEMARNAT) authorized the UMA with its own permit. In 2013, the first exploitation rate was granted, with a vision to making this project sustainable and profitable (since it is family cooperative). Later, because of the seriousness and dedication of the associates of the cooperative, the project began to be well-regarded by some public institutions that had programs that promoted local economic, social, or environmental progress.

In the same year 2013, they obtained funding from the National Commission for Development of Indigenous Peoples (CDI) and the municipal government to construct rooms for visitors and to improve the installations. Later, they acquired a bank loan and constructed a small swimming pool for guests. With this, they completed the offer for a one- or two-day stay of visitors with interest in ecotourism and healthy living (food, housing, crocodile exhibit, ecological path, kayak ride, bird and wildlife observation, swimming pool, and souvenirs).

The proposal of the UMA was shared with other local families from the start. However, because they did not have direct economic government support or salaries as employees, there was no more interest in the activity. This is a serious problem. The clienteles' culture persists in part of the rural population, obstructing rural social initiatives for local development that require associative commitment.

In recent years, the State Secretariat of Tourism has promoted the crocodile farm, which has been received enthusiastically by tourists who visit Campeche. They swim on the beach, eat, visit the crocodile exhibit, and leave. Others have the option to stay in rustic hotels and eat dishes of fish and seafood in restaurants.

Family members growing at their parents' side in this foundational work have been integrated as active associates. Today, the family cooperative has 12 members and approximately 700 crocodiles, and Wotoch Aayin is a growing ecotourist center in the stage of consolidation. Its main objective is conservation of the Moreletii crocodile and the mangroves through sustainable use. The most significant role has been played by the members of the cooperative due to their dedication in planning, promotion and negotiation, development, and execution of the project.

Fig. 23.1 Crocodile in Wotoch Aayin facilities. (Photo of Mario del Roble Pensado-Leglise)

The public institutions that have made it possible are the following:

- SEMARNAT: permits
- SAGARPA and Comisión Nacional de Pesca: permits and acquisition of Kayaks
- CONANP: funding for the construction of the crocodile fence, a palapa and the ecological path
- CDI: promotion and improvement in infrastructure
- State Secretariat of Tourism: promotion and public information

Fig. 23.2 Wotoch Aayin view in the Ría Celestún Reserve. (Photo of Mario del Roble Pensado-Leglise)

The alliance with CONANP was particularly strategic because it permitted launching the project in 2008. The interests were common since CONANP (with SEMARNAT) was responsible for pushing CITES for authorization by presenting evidence of the success of the UMAs dedicated to rearing Moreletti crocodiles.

The objectives maintained by the project:

A. Productive: technological improvement in reproduction and sustainable exploitation of the Moreletti crocodile
B. Economic: generate income for the members of the cooperative and of the community by boosting sales of food, lodging, and recreational activities for sustainable tourism; sale of local handcrafts, buying secondary products of fishing and fishing by-products from local fishing families, and sale of goods derived from crocodiles (meat, skins, etc.)
C. Social: job creation and improvement of local incomes, elevating levels of social well-being; strengthening cultural identity based on sustainable use of the mangroves and petenes
D. Environment: conservation of the wild crocodiles and of the mangroves that form part of the Ría Celestún Biosphere Reserve (72 km from the city of Calkini)

The cooperative's president, Romualda Caridad Gómez Gómez, wife of the fisherman Carlos Rivero Leon, is responsible for the operation of the farm. She is cousin to the biologist Javier Omar Gomez Duarte, who is responsible for the technical aspect of the UMA. Associates who participate in the organizational structure of the cooperative include Javier Omar Gómez Duarte, Ninfa del Rosario Pool Gómez, José Eduardo Martin Pool Gómez, Juan de la Cruz Pool Gómez, and Carlos Raúl Rivero Gómez.

Their main achievement is the efficient operation of the farm because they feed the crocodiles nontarget fauna caught by fishermen who used to throw it away. Technical yields are stable (4-month gestation, 30–40 eggs per oviposition, 18–24 months to female size of 1.4 meters and 3–5 kg of filet per sacrificed animal).

The principal innovation is cooking dishes with crocodile meat filets and exploitation of the positive image of the mangrove for nature tourism, especially during the arrival of seasonal tropical birds, such as flamingoes, ibis, and cranes, and the abundance of fish, seafood, mollusks, snails, reptiles, and small mammals like the raccoon. For now, the main market is domestic and foreign ecological and nature tourism. The most important permanent publicity is through social networks and their participation in national forums, fairs, exhibitions, courses, and workshops.

23.5.3 Results of Adoption of These Innovations

1. Economic: Generation of incomes for the members of the cooperative, the personnel that works for the cooperative and for the community through acquisition of basic inputs

2. Sustainable productive use and conservation: Increase in the population of Moreletti crocodiles
3. Social: Jobs, economic spillover in the locality because it attracts tourists that acquire local inputs
4. Environmental: Preservation and cultural appreciation of the mangroves as the habitat of the wild crocodile, their valuable ecosystem as refuge of biodiversity, and importance in contributing to mitigating climate change

The cooperative Wotoch Aayin has demonstrated with its experience that a social enterprise in Mexico with goals of sustainable use and biodiversity conservation can be successful in alliance with institutions at three levels of government: federal, state, and municipal. The persistence and capacity of those in positions of responsibility to channel support was crucial to maintain the development of the UMA to transform it into an ecotourist center.

23.5.4 Replication and Current Challenges

It is indispensable to replicate this successful experience, but it requires associations that are responsible and committed and are not contingent on the culture that seeks government support to receive direct cash transfers to a lost fund. The design and establishment of public policies, coordinated at the three levels of government, are essential to stimulate projects of sustainable use and biodiversity conservation in the mangroves, which should be valued not only for their beauty and biodiversity richness but also as a source of human social and cultural development for the neighboring population.

The Wotoch Aayin cooperative needs to perfect its knowledge of management, administration, and finance. It also requires elevating the quality and efficiency of the farm's operation as well as of the ecotourist center accompanied by good environmental practices. It especially needs social innovation and a focus on economic-ecological productivity for the local population in fomenting the development of limited cultural, nature, and sustainable tourism.

23.5.5 Threats

One great weakness is the lack of insertion in the crocodile skin market. Given the scale economy, it is not feasible to obtain good prices in the domestic market. For this reason, it is necessary for the authorities to support this type of associations with a market study that would provide options for commercialization, although the products are handcrafts such as those they are currently making. (They sell wallets, belts, skins, and other souvenirs made of crocodile skins or by-products that are made in cottage maquiladoras.)

In order to promote public policies to protect and restore mangroves, the state government of Campeche must support research and technological transfer and innovation projects focused on mangrove biotic resources. Through needs analysis, a project to establish an economic cluster based on industrial ecology could be designed to process crocodile skins and by-products with the input of researchers, technicians, and, above all, enterprises specialized in the market of crocodile skins to give impulse to the profitability of the projects of Moreletti crocodile farm UMAs. A study is required to establish orderly industrial development grounded in industrial ecology engineering and in complete agreement with the federal and municipal governments and, especially, rigorous selection of associated groups for sustainable use and biodiversity conservation of the Campeche mangroves.

It is necessary to convince the local community with good de facto economic-ecological practices to propose, together with the municipal government, legal administrative norms that restrict overloads of tourists, and the overpopulation of local spaces due to the bad experiences of environmental degradation and ecocide that irresponsible promotion of massive tourism by local authorities has caused on the Yucatán peninsula.

Based on the experience of the crocodile farm, it is now necessary to diversify production through development of the science of sustainability so that local fishermen can establish cultures of shellfish in the marshes and mangroves of Isla Arena with sustainable practices and community participation. This should take place first with the aim of biodiversity conservation and later with projects of economic-ecological character for local sale as specialty goods. On the Campeche coast, there are species listed in nom-013-pesca-1994, such as the queen conch (*Strombus gigas*), the milk conch (*S. costatus*), the lightning whelk (*Busycon contrarium*), and the West Indian chank shell (*Xancus angulatus*), and in the nom-059-SEMARNAT-2001, species under special protection such as the flat tree oyster (*Isognomon alatus*) and the Carolina marsh clam (*Polymesoda caroliniana*) (Pech et al. 2010). Ecuador has had an important experience in sustainable use in this direction (Beitl and Gaibor 2018; Álvarez-Dagnino et al. 2017).

A sustainable waste management system and sewage water drainage that does not pollute the mangrove or the waters included in the federal land-sea zone is required for the local population and a limited population of cultural and nature tourism.

23.6 Conclusions

The Wotoch Aayin cooperative today is not just an intensive crocodile farm; it is also an ecotourist center in the growth and consolidation stage. Its main objective is to conserve the Moreletii crocodile and the mangroves through sustainable use. This allows the generation of a broad platform of proposals for scientific and technological research that can consolidate a model of sustainable use of the mangrove and biodiversity conservation. Success in management of the crocodile has permitted the recovery of endangered species with participation of the community.

A fundamental aspect is the interaction of the entire community of Isla Arena in the mangrove ecosystem economy through study, discussion, and adoption of sustainable productive activities that are not impaired by development of massive predatory tourism that threatens this site. An alternative mode is being signaled by the experience of the cooperative, which should broaden its productive diversification through cultivation of marsh and mangrove species for culture and nature tourism. The dynamism and the brief time in obtaining partial successes are achievements of having adopted the instruments and interdisciplinary analysis of the science of sustainability in natural protected areas, particularly as implemented in the case study of the Crocodile Moreletti UMA in Ría Celestún, Campeche.

In other words, SS is based on nature; it is a discipline with a holistic focus to identify and solve the main problems of sustainability in global, social, and human ambits. This has been key in achieving progress in management of the intensive farm associated with a scheme of productive diversification linked to development of scientific-technological and social innovation where scientific and cultural knowledge, traditional productive practices, and strengthening of local sociocultural identity of the mangrove are conjugated in the constitution of sustainable civil productive organizations.

References

Álvarez-Dagnino E, Santamaría-Miranda A, García-Ulloa M, Góngora-Gómez A (2017) Reproduction of Megapitaria squalida (Bivalvia: Veneridae) in the Southeast Gulf of California, Mexico. Rev Biol Trop 65(3):881–889

Barrios G, Cremieux J (Comp) (2018) Protocolo de rancheo para cocodrilo de pantano (Crocodylus moreletii) en México. Comisión Nacional para el Conocimiento y Uso de la Biodiversidad (CONABIO), México

Beitl C (2014) Adding environment to the collective action problem: individuals, civil society, and the mangrove-fishery commons in Ecuador. Rev World Dev 56:93–107

Beitl Ch, Gaibor N (2018) Rights-based Approaches in Ecuador's Fishery for Mangrove Cockles, paper, FAO Tenure and User Rights in Fisheries Seminar, Instituto Nacional de Pesca-Ecuador, The University of Maine

Bettencourt L, Kaur J (2011) Evolution and structure of sustainability. PNAS 108(49):19540–19545

Brenner L (2010) Gobernanza ambiental, actores sociales y conflictos en las Áreas Naturales Protegidas mexicanas. Rev Mex Sociol 72(2):283–310

Cámara de Diputados del H. Congreso de la Unión(2000a) Reglamento de la Ley General del Equilibrio Ecológico y la Protección al Ambiente en Materia de Áreas Naturales Protegidas, DOF 30-11-2000; (Última reforma publicada DOF 21-05-2014). http://www.diputados.gob.mx/LeyesBiblio/regley/Reg_LGEEPA_ANP.pdf

Cámara de Diputados del H. Congreso de la Unión (2000b) Ley General de Vida Silvestre DOF 3-07-2000 (modificada 6-04-2010; 6-06-2012 y última reforma publicada DOF 19-01-2018). http://www.diputados.gob.mx/LeyesBiblio/pdf/146_190118.pdf

CBD/COP10 (2010) El Plan Estratégico para la Diversidad Biológica 2011 2020 y las Metas de Aichi para la Diversidad Biológica. UNEP/CBD/COP/DEC/X/2, Nagoya, Japón, 18–29. https://www.cbd.int/doc/decisions/cop-10/cop-10-dec-02-es.pdf
CBD/COP12 (2014) Diversidad biológica para la erradicación de la pobreza y el desarrollo sostenible. UNEP/CBD/COP/DEC/XII/5, Pyeongchang, Repùblica de Corea
CBD/COP13 (2016) Informe de la conferencia de las partes en el Convenio sobre la Diversidad Biológica sobre su 13ª reunión (CBD/COP13), Cancún (México). https://www.cbd.int/doc/c/049d/59ca/66f0ef3b745e68ae7ed0a673/cop-13-25-es.pdf
Challenger A, Cordova A, Lazos E, Equihua M, Maass M (2018) La opinión experta evalúa la política ambiental mexicana: Hacia la gestión de socioecosistemas. Rev Gestión y Política Pública 32(2):431–473
Cocho-Gil G et al (2017) Ciencia, Humanismo y Sociedad. De los sistemas complejos a la imaginación heterodoxa, Mexico, CDMX: CopIt-arXives y EditoraC3
CONABIO (2014) Quinto Informe Nacional de México ante el Convenio sobre la Diversidad Biológica. CONABIO, México
Costanza R, d'Argec R, de Groot R, Farbere S, Grass M, Hannongh B, Limburg K, Naeem S, VO'Neil R, Paruelo J, Raskin R, Sutton P, van den Belto M (1997) The value of ecosystem services: putting the issues in perspective. Rev Ecol Econ 25(1):67–72
Costanza R, de Groot R, Braat L, Kubiszewski I, Fioramonti L, Sutton P, Farber S, Grasso M (2017) Twenty years of ecosystem services: how far have we come and how far do we still need to go? Rev Ecosys Serv 28:1–16
Delgado F, Rist S (2016) Las ciencias desde la perspectiva del diálogo de saberes, la transdisciplinariedad y el diálogo intercientífico. In: Delgado F, Rist S (eds) Ciencias, diálogo de saberes y transdisciplinariedad. CRIM-UNAM-AGRUCO-CLACSO, UMSS, Bolivia, pp 35–60
Díaz S, Demissew S, Carabias J et al (2015) The IPBES conceptual framework — connecting nature and people. [Intergovernmental Platform on Biodiversity and Ecosystem Services (IPBES)]. Rev Curr Opin Environ Sustain 14:1–16
Díaz S, Settele J, Brondízio E et al (2019) Summary for policymakers of the global assessment report on biodiversity and ecosystem services in Report of the Plenary of the Intergovernmental Science-Policy Platform on Biodiversity and Ecosystem Services on the work of its seventh session, Paris, IPBES, UNEP. https://www.ipbes.net/global-assessment-report-biodiversity-ecosystem-services
Domínguez-Laso, J. 2006. Determinación del estado de las poblaciones silvestres del cocodrilo de pantano (Crocodylus moreletii) en México y evaluación de su estatus en la CITES. Instituto de Historia Natural y Ecología. Informe final SNIB-CONABIO proyecto No. CS009. México D.F.
Fischer A, Eastwood A (2016) Coproduction of ecosystem services as human-nature interactions-an analytical framework. Rev Land Use Policy 52:41–50
Funtowics S, de Marchi B (2009) Ciencia Posnormal, complejidad reflexiva y sustentabilidad. In: Leff E (ed) La Complejidad ambiental. S. XXI Ed. /CIICH-PNUMA, México, pp 54–84
Giannuzzo AN (2010) Los estudios sobre el ambiente y la ciencia ambiental. Rev Scientiæ Studia 8(1):129–156
Hák T, Janoušková S, Moldan B (2016) Sustainable development goals: a need for relevant indicators. Ecol Indic 60:565–573. https://doi.org/10.1016/j.ecolind.2015.08.003
Hauge K, Barwell R (2016) Post-normal science and mathematics education in uncertaintimes: educating future citizens for extended peer communities. Rev Future. https://doi.org/10.1016/j.futures.2016.11.013
Hausknost D, Schriefl E, Lauk C, Kalt G (2017) A transition to which bioeconomy? An exploration of diverging techno-political choices. Rev Sustain 9(669):1–22
INE-SEMARNAP (1997) Programa de Conservación de la vida silvestre y diversificación productiva en el sector rural 1997–2000, Instituto Nacional de Ecología, Sría. De Medio Ambiente, Recursos Naturales y Pesca, México.
IPBES (2018) The IPBES regional assessment report on biodiversity and ecosystem services for the Americas. Rice J, Seixas CS, Zaccagnini ME, Bedoya-Gaitán M, Valderrama N (eds)

Secretariat of the Intergovernmental Science-Policy Platform on Biodiversity and Ecosystem Services, Bonn, Germany, p 656
Kates R, Clark W, Corell R et al (2001) Environment and development: sustainability science. Sustain Sci 292(5517):641–642
Komiyama H, Takeuchi K (2011) Sustainability science: building a new academic discipline. In: Komiyama H, Takeuchi K, Shiroyama H, Mino T (eds) Sustainability science: a multidisciplinary approach. United Nations University Press, Hong Kong
Kønig T, Børsen C, Emmeche C (2016) The ethos of post-normal science. Rev Futures. https://doi.org/10.1016/j.futures.2016.12.004
Leff E (2009) Pensar la complejidad ambiental. En: Leff E (Coord) La Complejidad ambiental. S. XXI Ed./CIICH-PNUMA, México, pp 1–18
Lewis S, Maslin M (2015) Defining the Anthropocene. Rev Nat 519(171). https://doi.org/10.1038/nature14258
L'humeau A, Cordero D (2012) Adaptación basada en Ecosistemas: una respuesta al cambio climático. UICN, Quito
NRS (1999) Our common journey: a transition toward sustainability. National Research Council policy division board on sustainable development. National Academy Press, Washington, DC
ONU (2015) Resolución aprobada por la Asamblea General el 25 de septiembre de 2015: "Transformar nuestro mundo: la Agenda 2030 para el Desarrollo Sostenible". Documento final de la cumbre de las Naciones Unidas para la aprobación de la agenda para el desarrollo después de 2015, Nueva York
Pech D, Ardisson P, Reguero M (2010) Moluscos marino-costeros. En: Villalobos-Zapata GJ, Mendoza-Vega J (Coords) La Biodiversidad en Campeche: Estudio de Estado. Comisión Nacional para el Conocimiento y Uso de la Biodiversidad (CONABIO), Gobierno del Estado de Campeche, Universidad Autónoma de Campeche, El Colegio de la Frontera Sur. México, pp 280–285
Pensado-Leglise M (2011) La necesidad de repensar la relación territorio y ambiente. Una introducción a la aproximación metodológica. En: Pensado Mario (Coord) Territorio y Ambiente: Aproximaciones Metodológicas, Ed. SXXI, ISBN 978-607-03-0350-0 México, pp 7–17
Pensado-Leglise, Mario del Roble (2017) Los juegos de poder, marco institucional y conflictos socio ambientales al nivel territorial en México. En: Pensado-Leglise MR, García Serrano LA (Coords) Los retos actuales de las ciencias ambientales y de la sustentabilidad en México. Altres Costa-Amic Editores, México, ISBN: 978-607-8518-09-8, pp 60–80
Platt SG, Sigler L, Rainwater TR (2010) Morelet's crocodile Crocodylus moreletii. In: Manolis SC, Stevenson C (eds) Crocodiles. Status survey and conservation action plan, 3rd edn. Crocodile Specialist Group, Darwin, pp 79–83
Pretty J, Vodouhê SD (1998) Using rapid or participatory rural appraisal. In: Swanson B, Bentz RP, Sofranko AJ Improving agricultural extension: a reference manual FAO, Roma
Rivera-Téllez E, López SeguraJáuregui G, Antaño Díaz LA, Benítez Díaz H (2017) Informe del Programa de Monitoreo del Cocodrilo de Pantano en México, temporadas 2014 a 2015 y análisis de tendencias del 2011 al 2015. CONABIO, México
Ross (2000) *Crocodylus moreletii*. In: IUCN 2012. IUCN Red List of Threatened Species. Version 2012.1
Robert C, Lorenzo F, Ida K (2016) The UN Sustainable Development Goals and the dynamics of well-being. Front Ecol Environ 14(2):59
Russi D, Brink T, Farmer A, Badura T, Coates D, Förster J, Kumar R, Davidson N (2013) The economics of ecosystems and biodiversity for water and wetlands. The Institute for European Environmental Policy, IEEP/Ramsar Secretariat, London and Brussels/Gland
Sachs I (1974) Ecodesarrollo: un aporte a la definición de estilos de desarrollo para América Latina. Estudios Internacionales 7(25):57–77
Sánchez-Herrera O, Álvarez-Romero JG (2006) Conservation Status of Morelet's crocodile (Crocodylus moreletii) in Mexico: a proposal for its reclassification in the U.S. Endangered

Species Act (esa). En: Crocodiles, proceedings of the 18th working meeting of the crocodile specialist group. IUCN/The World Conservation Union, Suiza y Cambridge, pp 255–265
Sánchez-Herrera O, Benítez Díaz H, García Naranjo Ortiz de la Huerta A, López Segura Jáuregui G (2011) Programa de Monitoreo del Cocodrilo de pantano (Crocodylus moreletii) México, Belice y Guatemala. CONABIO, México
Sánchez-Herrera O, López SeguraJáuregui G, García Naranjo Ortiz de la Huerta A, Benítez Díaz H (2012) Informe del Programa de Monitoreo del Cocodrilo de Pantano en México Temporada 2011. CONABIO, México
Sánchez-Herrera O, Rivera-Téllez E, López SeguraJáuregui G et al (2015) Informe del Programa de Monitoreo del Cocodrilo de Pantano en México, Temporadas 2011 a 2013. CONABIO, México
Santes V, Pensado-Leglise MR (2017) Presentación capítulo del libro. En: Pensado-Leglise MR, García Serrano LA (coords) "Los retos actuales de las ciencias ambientales y de la sustentabilidad en México", Altres Costa-Amic Editores, México, ISBN: 978-607-8518-09-8, pp 7–16
SEMAAS (2015) Programa Estatal ante el Cambio Climático 2030, Secretaría de Medio Ambiente y Aprovechamiento Sustentables, Gobierno C. del Estado de Campeche. San Francisco de Campeche, Campeche, México, marzo de 2015
SEMARNAT (2000) Programa de Manejo Reserva de la Biosfera Ría Celestún. SEMARNAT, México
SEMARNAT (2018) Las UMA'S, sitios dedicados a la Conservación de la Vida Silvestre. Artículo en el sitio electrónico: https://www.gob.mx/semarnat/articulos/las-uma-sitios-dedicados-a-la-conservacion-de-la-vida-silvestre
Steffen W, Persson A, Deutsch L, Zalasiewicz J, Williams M, Richardson K, Crumley C, Crutzen P, Folke C, Gordon L, Molina M, Ramanathan V, Rockstro J, Scheffer M, Schellnhuber HJ, Svedin U (2011) The Anthropocene: from global change to planetary stewardship. Ambio 40:739–761. https://doi.org/10.1007/s13280-011-0185-x
Takeuchi K (2010) Rebuilding the relationship between people and nature: the Satoyama Initiative. Ecol Res 25:891–897
TEEB (2010) La Economía de los Ecosistemas y la Biodiversidad: Incorporación de los aspectos económicos de la naturaleza. Una síntesis del enfoque, las conclusiones y las recomendaciones del estudio TEEB. Ginebra, Ed. TEEB, ONU-PNUMA
Valderrama-Landeros LH, Rodríguez-Zúñiga MT, Troche-Souza C, Velázquez-Salazar S, Villeda-Chávez E, Alcántara-Maya JA, Vázquez-Balderas B, Cruz-López MI, Ress R (2017) Manglares de México: actualización y exploración de los datos del sistema de monitoreo 1970/1980–2015. Comisión Nacional para el Conocimiento y Uso de la Biodiversidad, Ciudad de México, p 128
Vázquez-Lule A, Ríos-Saís G, Adame M (2009) Caracterización del sitio de manglar Celestún, en Comisión Nacional para el Conocimiento y Uso de la Biodiversidad (CONABIO). Sitios de manglar con relevancia biológica y con necesidades de rehabilitación ecológica Ría Celestún. CONABIO, México, D.F.. http://www.conabio.gob.mx/conocimiento/manglares/doctos/caracterizacion/PY59_Celestun_caracterizacion.pdf
Vessuri H (2016) La ciencia para el Desarrollo sostenible (Agenda 2030). Ed. Oficina. de Montevideo de UNESCO, AECI, Foro CILAC 2016
Vilches A, Gil-Pérez D (2015) Ciencia de la Sostenibilidad: ¿Una nueva disciplina o un nuevo enfoque para todas las disciplinas? Rev Iberoam Educ 69(1):39–60
Waters C et al (2017) The Anthropocene is functionally and stratigraphically distinct from the Holocene. Science 351(6269):aad2622-1. http://science.sciencemag.org/content/351/6269/aad262
Webb GJW (2014) Wildlife conservation: in the belly of the beast. Charles Darwin University Press, Northern Territory
Webb GJW, Manolis SC (1992) Monitoring saltwater crocodiles (Crocodylus porosus) in the Northern Territory of Australia. In: McCullough DR, Barrett RH (eds) Wildlife 2001: populations. Elsevier Applied Science, New York, pp 404–418

Wiek A, Harlow J, Melnick R, van der Leeuw S, Fukushi K, Takeuchi K, Farioli F, Yamba F, Blake A, Geiger C, Kutter R (2015) Sustainability science in action: a review of the state of the field through case studies on disaster recovery, bioenergy, and precautionary purchasing. Sustain Sci 10:17–31

WWF (2018) Living planet report – 2018: aiming higher. In: Grooten M, Almond REA (eds) WWF, Gland

Chapter 24
Results of Socio-ecosystem Institutional Management: Analysis of Two Protected Natural Areas of Central México

Cecilia L. Jiménez-Sierra, María Loraine Matias-Palafox, Joaquín Sosa-Ramírez, Victor Javier Arriola-Padilla, Daniel Torres-Orozco Jiménez, and Melany Aguilar-López

Abstract This chapter deals with historical and current data regarding the socio-ecosystems of two Protected Natural Areas (PNAs) of central México: El Chico National Park (ECHNP) and the Barranca de Metztitlán Biosphere Reserve (BMBR). Both PNAs are in the State of Hidalgo; despite covering different types of ecosystems and having different protection categories, they share the common objective of conserving the biological heritage. The achievements in maintenance of the biological elements and ecosystem functions were assessed based on bibliographic information. Socio-economic and cultural aspects were explored by conducting structured interviews to different population sectors. Visitor flow and administration problems of these PNAs were also obtained from interviews to visitors and reservation staff. This chapter discusses the challenges and success found in these PNAs with the information gathered.

C. L. Jiménez-Sierra (✉)
Universidad Autónoma Metropolitana – Unidad Iztapalapa, Ciudad de México, México

M. L. Matias-Palafox
Comisión Nacional para el Conocimiento y Uso de la Biodiversidad, Ciudad de México, México
e-mail: lmatias@conabio.gob.mx

J. Sosa-Ramírez
Universidad Autónoma de Aguascalientes, Aguascalientes, México
e-mail: jsosar@correo.uaa.mx

V. J. Arriola-Padilla
National Institute for Forestry, Agriculture and Livestock Research (INIFAP), Ciudad de México, México
e-mail: arriola.victor@inifap.gob.mx

D. Torres-Orozco Jiménez
Universidad Nacional Autónoma de México. Ciencias de la Tierra, Ciudad de México, México

M. Aguilar-López
ECOYDES A. C., Mineral de la Reforma, Hidalgo, México

A. Ortega-Rubio (ed.), *Socio-ecological Studies in Natural Protected Areas*,
https://doi.org/10.1007/978-3-030-47264-1_24

Keywords Protected Natural Areas · Biosphere Reserve · Socio-ecosystems Wildlife and Environmental Management Units · National Park

24.1 Introduction

México is one of the most diverse countries in the world in terms of biological and cultural richness. Protected Natural Areas (PNAs) are important to conserve natural spaces and promote sustainable development. They can be defined as sea or land extensions declared by the government to conduct biodiversity conservation in situ (CONANP 2012). However, the concept of PNA and its relevance have changed over time.

The first PNAs emerged at the end of the nineteenth century with the purpose of guaranteeing the conservation of great natural beauty sites, as well as to protect the watersheds of some important rivers (Torres-Orozco et al. 2015). The "Desierto de los Leones" was the first National Park (NP) decreed in 1917 by President Venustiano Carranza. However, the period when the greatest number of NPs was decreed was during the government of Lázaro Cárdenas (1934–1940) when nearly 30% of the Mexican territory obtained this status. Nowadays, NPs constitute the conservation category with the highest number of protection units ($N = 66$), covering 14113.19 Km^2. Throughout the 1960s of the last century, Man and the Biosphere Programme of the United Nations Educational, Scientific and Cultural Organization (UNESCO) proposed the creation of Biosphere Reserves (BR). México adopted this conservation figure to solve socio-environmental conflicts in regions with great biological and cultural richness. The BR system intended to reconcile the conservation of the ecosystems and their ecological processes together with the sustainable use and development of local people (Halffter 1984).

In 1970, "Complejo Lagunar Ojo de Liebre", located northwest of the State of Baja California Sur, became the first BR. The lagoon is not only the site of the largest commercial saltworks plant in the world but also the reproduction habitat and hibernation site of the grey whale (*Eschrichtius robustus*), the harbour seal (*Phoca vitulina*) and other marine mammals, such as the California sea lion (*Zalophus californianus*), the Northern sea elephant (*Mirounga angustirostris*) and the blue whale (*Balaenoptera musculus*). Nowadays, 45 BR have been decreed, covering an area of 777615.33 Km^2 and representing 85% of the protected national territory (CONABIO 2019).

In México, besides NP and BR, there are other conservation areas, such as (1) Flora and Fauna Protection Areas, (2) Natural Resource Protection Areas, (3) Sanctuaries and (4) Natural Monuments (General Law of Ecologic Equilibrium 2018). PNAs are regulated by the General Law of Ecological Equilibrium and Environment Protection (Ley General de Equilibrio Ecológico y Protección al

Ambiente), LGEEPA for its abbreviation in Spanish, and they are managed by the National Commission of Protected Natural Areas (Comisión Nacional de Areas Naturales Protegidas, CONANP) (CONANP 2012). Today, 182 federal PNAs have been established in the country, which cover a little more than 900 thousand ha. Each PNA should have a Management Plan or Program (MP) to achieve habitat conservation (Torres-Orozco et al. 2015). The proper development of the MP ensures the success of the functional ecosystem conservation to face the swirl of social and environmental dynamics (Jiménez-Sierra et al. 2018).

To promote conservation at regional or local level, the government has created "Wildlife and Environmental Management Units" (UMA, for its abbreviation in Spanish, Unidades de Manejo Ambiental para el Aprovechamiento Sustentable de la Vida Silvestre), which are areas or facilities voluntarily dedicated to the sustainable use of threatened wildlife populations. The UMA requires government approval and surveillance to ensure the sustainable use of exploited resources at risk (SEMARNAT 2009). Since the emergence of the UMA, certain indigenous common land/smallholder communities have been able to manage and use wildlife in nearly 26 million ha, with a financial movement of more than $250 million USD/year with a double benefit, to (1) contribute to repopulate wildlife in México and (2) improve economy of the population (Robles de Benito 2009).

In 2000, the first UMA was established in Hampolol (Campeche), while the "National Strategy for Wildlife" was unveiled. Today, UMA constitutes an alternative figure to allow the sustainable use of resources in lands of settlers (http://www.conabio.gob.mx). Historically, the rights of use of the resources had been limited through prohibitions, regulations and norms that caused social problems, discrimination and productive segregation that unleashed the illegal use of resources by the owners (Robles de Benito 2009). Nowadays, the relevance of UMA has increased because it is the only way to generate economic income for local people in many places. Therefore, it is possible to avoid problems, such as migration and family disintegration, which are some of the most important problems that affect rural society in México.

This chapter discusses an approach for two PNAs that are established in the central part of México (Hidalgo State). The first one is El Chico National Park (ECHNP), and the second one is Barranca de Metztitlán Biosphere Reserve (BMBR). Although both areas aim to conserve the ecosystems, their approaches differ. While the goal of an NP is conservation and implementation of recreational activities for the inhabitants, that of the BR is to seek the development of human communities using strategies of sustainable use of the resources. Thus, a brief description of these PNAs was considered, and some management aspects and perception of their inhabitants and visitors were included. This information was obtained through interviews to management staff, inhabitants and visitors, respectively. The goal of this research was to determine some of the success, as challenges and actions that are required to conserve the functionality of the socio-ecosystems of these selected PNAs.

24.2 Methodology

To describe some characteristics of the natural areas, the information of management programmes, technical data sheets and specialized literature were synthetized; interviews were also conducted to:

1. Management staff to know the problems related to PNA management;
2. Settlers to address economic aspects and perceptions about the importance of living in a PNA. In El Chico National Park (ECHNP) where the number of visitors was significant, a survey was also conducted to this sector to know their interest and opinion about the NP;
3. UMA owners in the Barranca de Metztitlán Biosphere Reserve (BMBR) to highlight if the establishment of PNAs have favoured the development of sustainable activities.

In the case of ECHNP, where an UMA is not found, small local entrepreneurs were interviewed. Based on the data collected, this study highlighted actions, success and challenges that are required to achieve the socio-environmental conservation and with it guarantee social welfare and biodiversity conservation.

24.3 Results

24.3.1 Historical Background and Biological Importance of the Protected Natural Areas Selected

24.3.1.1 El Chico National Park (ECHNP)

El Chico National Park (ECHNP) is located between the Sierra de Pachuca and Trans-Mexican Volcanic Belt. Its altitude ranges from 2320 to 3090 m asl (Melo and López 1993; Zavala 1995). The region has important rock formations, such as Las Ventanas, Peña Cercada and Peña del Cuervo, among others (CONANP 2005). In 1898, it was decreed as Forest Reserve and in 1982 as National Park. It covers an area of 2739.02 ha from which 369.6 ha were expropriated. The polygon of the ECHNP includes the municipalities of Mineral del Chico, Pachuca and Real del Monte (Melo-Gallegos and López-García 1993; Zavala 1995; CONANP 2005).

The weather at ECHNP is temperate sub-humid. The area belongs to the Floristic Province of the Meridional Mountains of the Mesoamerican Region. ECHNP has a great biological importance since it harbours one of the most beautiful relicts of oyamel (*Abies religiosa*) forest in the watershed of Valle de México. It also contains relict populations of oaks and pines, as well as pastures. (Fig. 24.1). The flora is valuable because it has six out of nine of the existing conifers in México (Medina and Rzedowsky 1981). In 1981, Medina and Rzedowsky reported 116 plant species in the main plant communities in the highlands of Sierra de Pachuca. Zavala (1995) reported the presence of 20 oak species (*Quercus* spp.) and a study of the phanerogamic flora

Fig. 24.1 Panoramic view of El Chico National Park (Hidalgo, México). (Author: CLJS)

including 530 species. The "red oyamel" (*Pseudotsuga macrolepis*), a species with limited distribution, and Mexican yew "romerillo" (*Taxus globosa*), both grow in the park and are subjected to special protection (Pr) by the Mexican government because their populations have been threatened due to their natural low regeneration and deforestation in the last decades (CONANP 2005).

The fauna in ECHNP is characteristic of the temperate humid areas. Most of the animal species have Nearctic affinity; small mammals found are mice, bats, moles, rats, rabbits and squirrels; medium-size mammals include armadillos, opossums, skunks, grey foxes, coyotes and ringtail raccoons; the most abundant mammals with a strong presence are the mouse (*Peromyscus difficilis*), rat (*Neotoma mexicana*) and rabbit (*Sylvilagus cunicularius*); other animals found are 48 bird species, 14 reptile species and 11 amphibian species (Camarillo and Casas 2001; CONANP 2005). Within the polygon of this NP, a section of the town Mineral El Chico (MECH) is included with an estimated population of 505 inhabitants (CONANP 2005). This town is made up of approximately 135 private houses, most of them with gable roofs and walls made of rock, block, adobe or bricks; they usually have a small orchard with fruit trees, such as pear, apple and plum trees (CONANP 2005). MECH has been named "Magic Town" (Pueblo Mágico) because of its historical and cultural richness (Fig. 24.2). In addition to MECH, seven other rural populations are directly related to the ECHNP. In this region, no specific ethnic group has been identified, but speakers of Nahuatl, Hñahñú and Zapoteco have been recorded. The population in this municipality is integrated by 51.74% women and 48.26% men (CONANP 2005).

24.3.1.2 Barranca de Metztitlán Biosphere Reserve (BMBR)

The Metztitlán Canyon is located at the central-east region of the State of Hidalgo. It is part of the Trans-Mexican Volcanic Belt and the Sierra Madre Oriental and belongs to the area of canyons shaped by tributaries of the Pánuco River, which

Fig. 24.2 Neoclassical church "La Purísima Concepción" at Mineral El Chico (Hidalgo, México). (Author: MLMP)

houses a great biological richness. The area was declared as a BR in November 2000 with an extension of 96,042 ha. It constitutes a biological corridor between the Nearctic vegetation from the northern zone of the country and the tropical vegetation located in the central Altiplano of México. The area has been catalogued as "priority conservation zone" (CONABIO 2000), and its water reservoirs (Metztitlán River and Metztitlán Lagoon) are important resources for the biota.

The topography of the BMBR is very rugged with altitudinal gradients ranging from 1000 to 2000 m asl (Fig. 24.3). The reserve covers partly the municipalities of Acatlán, Atotonilco el Grande, Eloxochitlán, Huasca de Ocampo, Metepec, San Juan Metztitlán, San Agustín Metzquititlán and Zacualtipán de Ángeles (CONANP 2003).

The weather in the BMBR is dry and warm with an average annual temperature from 18 to 22 °C, and its average annual precipitation is lower than 500 mm (CONANP 2003). The vegetation types are diverse, among them dry tropical forest, xerophilous shrub, sub-mountain shrub, grassland and riparian vegetation. The xerophilous shrub is the predominant vegetation considered as a Pleistocene refuge of Mexican desert biota since it shares characteristics of those in the States of Chihuahua and Sonora (CONANP 2003). Most of the flora belongs to the Altiplano Floristic Province, constituted mainly by southern and northern elements with a vast

Fig. 24.3 Panoramic view of Metztitlán Canyon, Hidalgo, México. Several species of cacti are found in the cliffs; in the bottom, "La Vega" is an area of fertile cultivated lands. (Author: CLJS)

number of endemic and native elements. Among the main plant families are Asteraceae, Cactaceae, Fabaceae and Poaceae (CONANP 2003).

The BMBR has been catalogued as one of the most important cactus areas in México (Fig. 24.4) (Sánchez-Mejorada 1978; CONANP 2003; Matias-Palafox and Jiménez-Sierra 2006) with approximately 70 cactus species, from which 11.42% are endemic and 15% are threatened. The locally called "viejito" or old man (*Cephalocereus senilis*) is an endemic cactus of the canyon that reaches more than 10 m high, and it has been selected as an emblematic species of the BMBR (Fig. 24.5).

Regarding fauna, the reserve is considered part of the biotic province Sierra Madre Oriental. Most species are from a Nearctic origin with desert and cold – humid affinities from the Sierra Madre Oriental. Alongside, the fauna merges with neotropical elements typical of the Huasteca area. Among the vertebrate species reported, 1.5% belong to fishes, 1.8% amphibians, (Order Anura: frogs and toads), 9.2% reptiles, 69.6% birds and 17.8% mammals (CONANP 2003). One of the most representative mammals recorded was bat species since the abrupt and diverse topography of the area facilitates the formation of caves and caverns where different species of bats find refuge, in addition to the abundance of columnar cacti that can be an important resource for them (CONANP 2003).

Fig. 24.4 Barrel cactus (*Echinocactus platyacanthus*) in the Barranca de Metztitlán Biosphere Reserve, Hidalgo, México. (Author: CLJS)

Human population in the municipalities of the reserve are constituted by 32,215 inhabitants from which 47% are men and 53% women. Nevertheless, just 25.38% of the total population from the eight municipalities are within the reserve. The municipalities with the largest area that belongs to the BMBR are Metztitlán and Eloxochitlán; the resident population within the reserve was estimated in 26,533 inhabitants (https://simec.conanp.gob.) (CONANP 2003; Jiménez-Sierra et al. 2017).

24.3.2 *Management of the Selected Protected National Areas*

24.3.2.1 The Management of the El Chico National Park (ECHNP)

The ECHNP was decreed in 1982, and the corresponding MP was published in 2005. The details about the park management were obtained in 2018 through interviews conducted to the reserve staff. The main office is in the city of Pachuca. The park has a visitor centre, which is a two-floor building with a meeting room and a small museum. It also has sanitary facilities, parking area and some cabins.

Management is conducted by CONANP staff (principal, assistant principal, three biologists and three technical assistants) and 13 park guards. The park is under

Fig. 24.5 *Cephalocereus senilis* "viejito" (old man) specimens, an endemic endangered species used as an emblematic species of the Biosphere Reserve Barranca de Metztitlán, México. (Author: CLJS)

co-management since the main administration is federal (CONANP), but the park guards are paid by the government of the State of Hidalgo. The annual economic budget of the park is precarious ($25,590 USD), which is used for the maintenance of the facilities. Even though the registration of the number of visitors to the park is partial, it is estimated to be around 2000 each year.

Most of the tourists come from the city of Pachuca about 25 km from the Park, but during weekends, people coming from México City, approximately 116 km away, are also frequent visitors. The highest tourism affluence occurs during the Easter break (March–April). The park entrance fee is currently $1.95 USD/day, but it is charged only to visitors who arrive at the visitor centre.

The average number of research projects recorded per year is around four, and at least two research theses are produced each year. Most researchers are from the Universidad Nacional Autónoma de México (UNAM), Universidad Autónoma del Estado de Hidalgo (UAEH) and Colegio de Postgraduados (COLPOS). The MP of the National Parks indicates that an advisory council formed by community representatives must be installed. However, in ECHNP, this council is still in process due to some conflicts that exist among the members since some common land owners claim about a fee for the right of access to the park and a greater support for forest conservation. Within the park, no UMA has been established, but there is an initiative for the creation of one of them for the cultivation of Mexican bay leaf (*Litsea glaucescens*), which is an endangered species.

Among the achievements of the park management that stand out are (a) proper collaboration between CONANP and State of Hidalgo offices, as well as the excellent disposition of the park staff; (b) appropriate signalling within the park area, which makes ECHNP the Mexican NP with the best signalling system; (c) good preservation of nearly 1500 ha of spruce and pine forests; (d) pest control; (e) reduction of illegal hunting (poaching); and (f) change in inhabitants' behaviour achieved through environmental education campaigns, which has allowed the settlers to be the guardians of their forests.

The main administrative challenges are to (a) improve resource management, (b) increase environmental education activities, (c) propose actions that promote the well-being of the settlers, (d) control harmful fauna and (e) monitor exotic species. The main initiatives implemented by the park management to increase the environmental conservation and social welfare have been (a) signalling foot paths, (b) developing diagnoses of ecosystem health, (c) creating interpretative trails and (d) controlling exotic species. Among the priority actions that are required to preserve the functionality of the socio-ecosystems are to (a) continue with forest pest monitoring and control and (b) complete the fauna and flora park inventory. To achieve it, the park has acquired 16 mammalian monitoring cameras.

24.3.2.2 Management of Barranca de Metztitlán Biosphere Reserve (BMBR)

The Barranca de Metztitlán Biosphere Reserve was decreed in 2000, and the corresponding MP was published in 2003. Interviews with the reserve staff helped to detail the information about the management of this PNA. The management department of the BR is in the city of Pachuca with another office in the town of Metztitlán. Management is conducted by CONANP staff (principal, assistant principal, two department managers, one administrator and two field technicians). All staff (except for the administrator) work in the field and in the office. The reserve has six park rangers (paid by the municipalities), and in 2019 three more young people from the "Youth Building the Future" government programme started to collaborate.

The annual budget of the reserve is precarious ($15,355 USD), which is mainly used for maintenance of the facilities. No accurate inventory of the visitors is available, but it is estimated to be close to 4000 people/year. This number can be considered small, taking into account that the city of Pachuca is about 67 km away from Acalome, the entrance to Metztitlán Canyon, and just 158 km away from México City. The entrance fee to this reserve is $1.95 USD/day as previously mentioned, but it has not been applied since no office is located at a strategic point. In the last years, the most important tourism has been the academic sector since students of different levels frequently visit the Reserve.

Regarding research activities, in the last 2 years, nearly 20 theses were developed with subjects concerning biological, socio-economic, historic and anthropologic aspects. Frequently, groups of different universities of the central part of México, such as UAEH, CP, Universidad Autonoma Chapingo (UACh), UNAM, and UAM, take study trips to this area.

Recently, the German Corporation for International Cooperation (GIZ) has been working here writing the new MP of the Reserve. Within the BMBR, five UMAs have been installed, dedicated to cultivation and sale of cacti, and soon an UMA for orchid cultivation is going to be established. The reserve managers have helped the UMA owners to obtain economic funds for the construction of infrastructure and greenhouses.

The achievements of the BMBR management are (a) excellent working attitude of CONANP staff; (b) empowerment of local people and their great disposition towards conservation; (c) environmental education programmes for inhabitants and visitors, which were implemented since the creation of the reserve; (d) proper management of the reserve that has achieved the respect and collaboration of the residents; and (e) willingness of residents to conserve their resources.

The challenges in management of the reserve are to (a) achieve greater efficiency in the distribution of the economic budget; (b) update the MP; (c) formulate regulations for the use of stone materials since to date they do not exist; (d) control pollution from the municipal water sewage of Zacualtipán, where textile companies are installed; and (e) regulate garbage dumps.

The main initiatives implemented by the reserve management to increase environmental conservation and social welfare have been (a) supporting settlers for installing an UMA; (b) installing the first UMA for orchid culture; (c) organizing workshops to prepare traditional candies (coconut sweets and traditional Mexican milk fudge "jamoncillo") to promote new economic alternatives; (d) implementing livestock-free areas to avoid foraging of cacti and land erosion; and (e) controlling forest pests in Tlaxco region, which affected from 80 to 100 ha of pine forest in 2017.

The actions required to preserve ecosystem functionality are to (a) complete the flora and fauna inventory, (b) control and monitor forest pests and (c) monitor invasive species, among others.

24.3.3 Perception of the Inhabitants About the Protected Natural Areas

24.3.3.1 Inhabitants from Mineral El Chico Municipality

Twenty-two interviews were randomly conducted to inhabitants from different localities in the municipality of Mineral El Chico, 41% of them were women and 59% men. Ages ranged from 16 to 83 years old with an average of 49. Their occupations were 14% domestic workers, 18% field workers (farmer), 27% construction

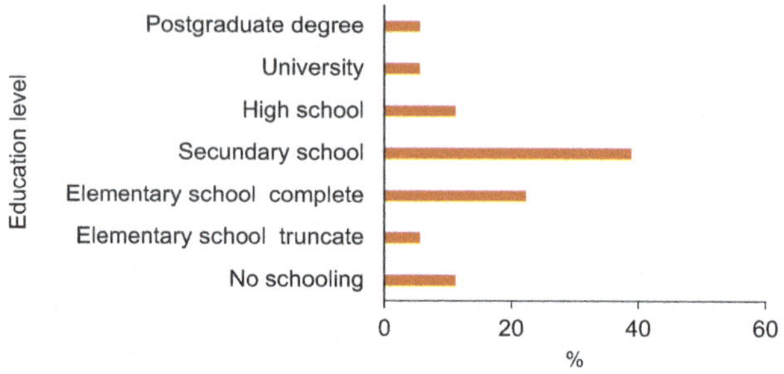

Fig. 24.6 Study level of interviewed inhabitants in the municipality of El Chico, Hidalgo, México (April, 2019)

workers (builders), 36% merchants and people dedicated to services and 5% students. The average workday time was of 09:20 h, and the daily salary ranged from $7.67 to 41.00 USD/day.

The study level of interviewed people was heterogeneous, but only 24% had a university degree (Fig. 24.6). The distance at which the schools were located and economic problems were the main causes by which the people dropped out of their studies. The number of schools in the municipality of El Chico has been recently raised, so now the people interviewed in the localities have elementary and high schools.

Even though several aspects of the houses were explored, the type of floor can give us an idea of the heterogeneous economic level of the inhabitants since some of them had dirt floor, others had wood and others are covered with tile, ceramic or stoneware.

Regarding the knowledge of the PNA concept, 89% of the population interviewed reported to know its meaning, and 64% was aware of living within or in the surroundings. Although numerous interviewees did not know exactly the date of the National Park creation, they had a rough idea that it was decreed a couple of decades ago. Approximately 90% of the people were certain that the NP was created to avoid plant and animal extraction and to control cutting down trees and wood extraction.

Regarding environmental protection measures, 59% stated they had been effective; 27% said the objectives had been partially achieved; and 14% said they had not been effective. Among the reasons why the optimal results had not been achieved, it was noted that the extraction of natural resources continued in some areas where surveillance was complicated.

In recent years, tourism in ECHNP and in MECH has increased due to the security and the good conservation status of the forest, as well as to the installation of small hotels and ecotourism agencies in the park surroundings (Estanzuela and Carboneras); 14% of the inhabitants reported that tourism had affected their lives; 21% considered it positive because of the increase in economic activities, while 79% said it had been detrimental because of the increase of waste (especially of non-organic waste like food packages and diapers, among others) and erosion, as well as loss of peacefulness due to the noise generated by rental vehicles (bicycles, motor bicycles and ATVs) which are very common especially during weekends. Regarding knowledge and use of medicinal plants, 77% of the people mentioned they used different medicinal plants from the forest, and 23% confirmed they did not use any of these resources.

24.3.3.2 Inhabitants from Barranca de Metztitlán Biosphere Reserve

Thirty-five inhabitants were randomly interviewed from different localities of the municipality of Metztitlán, of which 71% were women and 29% men. Ages ranged from 17 to 76 years old with an average age of 50. Among the occupations of the people interviewed, 46% were domestic workers, 26% were field workers, while 28% were employees. A workday was approximately 9 h, and the salary ranged from $7.73 to $10.3 USD/day.

Regarding the level of education, only 12% had finished high school or college level, and 37% did not finish elementary school. (Fig. 24.7). The distance at which the schools were and economic problems were the main causes by why the people dropped out of their studies. Nowadays the number of schools has increased since, in the case of the localities of interviewees, 55% have a centre for early education, 90% have elementary school, and 46% have high school.

Even though several aspects from housing condition were evaluated, the number of bulb lights/house can tell much about the economic status; after the inspection, the number of bulbs ranged from 1 to 14, with an average of 4.7 bulbs/house.

With respect to knowledge of the PNA concept, 80% of the people reported to know its meaning, and 80% was aware of living within or in the surroundings of a PNA. Only two people knew the exact year of the creation of the BMBR. Most of

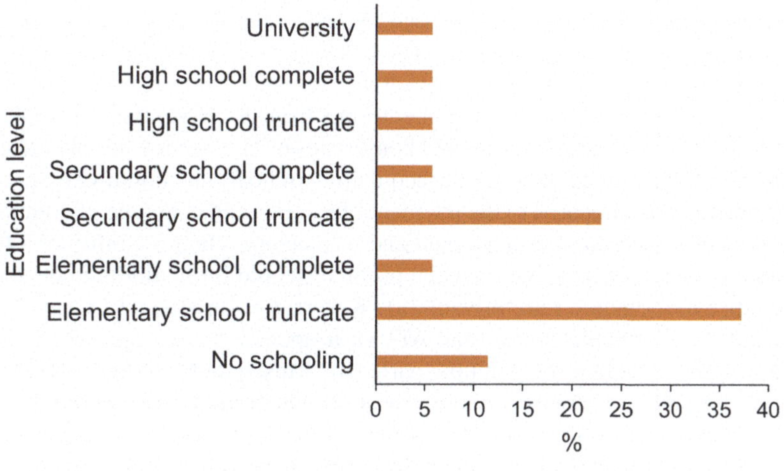

Fig. 24.7 Study level of interviewed inhabitants in the municipality of Metztitlán, Hidalgo, México (April, 2019)

those interviewed were certain that the reserve was created to avoid plant and animal extraction, and they did refer especially to cactus extraction and wildlife hunting.

Most of the inhabitants thought that the decree of the reserve was adequate and very important for them since preserving the environment and the reserve has favoured the arrival of tourism, thus increasing the economy within the region. Nevertheless, 40% declared that living in the reserve has not changed their lifestyle. The rest of the people expressed to be pleased with the implementation of measures that help to avoid the loss of natural resources, but some settlers showed discomfort to the fact that it is no longer permitted to cut trees because they need wood for their stoves. Most of those interviewed confirmed that the illegal extractions of wildlife were controlled, and 23% asserted to have witnessed illegal deforestation (cutting down trees or extracting cacti). Regarding the knowledge and use of medicinal plants, 91% of the people mentioned that they use different medicinal plants from the reserve.

24.3.4 Local Tourism Entrepreneurs and Ecotourism Agencies in El Chico National Park

Visitors' affluence in the town of Mineral del Chico (MECH) and in ECHNP has allowed the emergence of different local tourism and ecotourism agencies, promoting job creation for local people. This study reported the information obtained from the interviews to owners and workers of five agencies.

The first agency "Bien Ejidal La Estanzuela" includes 182 common land members who, at the beginning, were not certain if the programme could represent a financial benefit for them. The agency installed a recreational and camping zone, lodges and grills. This area is surrounded by forest and the dam La Estanzuela is nearby, where tourists can go fishing and take boat trips. This agency started 50 years ago because of the constant arrival of visitors looking for places where they could camp out. This agency has received mentoring from ECHNP staff to implement strategies for resource preservation and conservation. The agency has also received financial support from the State for the acquisition of boats and building lodges. The money generated is allocated to the payment of salaries and coverage of community needs, such as the creation of a hospital and a cemetery. The possibility of a future extension in the area exists, but financial problems could limit building new lodges.

The other four agencies are private. (a) "Recorridos Turísticos El Chico" offers hiking and tours in vehicles (bus and ATVs); it opened 2 years ago when visitors came asking for information about the park. (b) "Museo de Minería de Mineral del Chico" is a small and modest mining museum, which until now has not received financial support from the State; it was funded thanks to a resident's initiative in 2012. The museum comprises a collection of objects used in mining and old photographs gathered (Fig. 24.8). It is installed in a small space (one room) and supported by donations; five people depend on this income, of which four of them are women.

Fig. 24.8 Mineral del Chico museum established by a resident with the intention of extending the exhibition of objects used in local mining during the last century. (Author: CLJS)

(c) "Hotel Vagabundo de Media Noche" is a small hotel that opened 6 years ago; the owner was a former local resident who migrated to México City looking for work opportunities; after some years, he came back to the MECH and opened the hotel in his childhood home, which has a panoramic view to the forest. (d) "H-GO Adventures" offers activities, such as canyoning, canopy, zip line and climbing. The last interview was applied to an employee of this ecotourism agency. Ten people who work there are mostly relatives. All these agencies have helped to maintain local people in their communities and increase their interest in environmental conservation of the area.

24.3.5 *Wildlife and Environmental Management Units (UMA) in the Biosphere Reserve Barranca de Metztitlán (BMBR)*

The information about the UMAs was obtained from the interviews to the managers of three cactus UMAs (UMA Santa, UMA El Viejo Cactus (Fig. 24.9) and UMA Biznaga Cactus).

The establishment of UMAs in the reserve rose from the information and support given to settlers by the reserve and government staff. Since the reserve was founded, the staff of the reserve has organized workshops about how to cultivate cacti. For two of the UMAs, the reserve staff negotiated financial support for the construction of greenhouses, one palapa for talks, tables, toilets and cisterns. Unfortunately, economic resources have decreased in recent years, making it more difficult to obtain funding each time for the creation of new UMAs.

It is interesting to mention that people interviewed said they did not imagine that cacti could represent an economic support for them. Currently, the UMAs are profitable, and each one employs from 10 to 20 workers, integrating women in their work teams. Cacti produced are directly sold to the visitors, and today approximately 35 species of cacti have grown, as well as some grafted specimens that nowadays have high demand because of their colourful features (Fig. 24.10). All UMA owners said

Fig. 24.9 Interviews in the facilities of the Wildlife Environmental Unit (UMA) "El Viejo Cactus" (Acalome, Metztitlán, México). (Author: RJS)

Fig. 24.10 Cactus variety cultivated in the Wildlife and Environmental Unit (UMA) of the Barranca de Metztitlán Biosphere Reserve (BMBR). (Author: CLJS)

that the production of cacti had high demand, but they were limited by financial support, space and resources, such as water to continue growing. The owners declared that the creation of UMAs has led to the conservation of wild cacti, some of them endemic and endangered, such as *Cephalocereus senilis* (old man), *Mammillaria longimamma* and *Echinocactus platyacanthus,* among others. Moreover, the creation of UMAs has strengthened family bonds and favoured the creation of new job opportunities that have helped to decrease social problems, such as migration and family breakdown.

24.3.6 Perception of Visitors About El Chico National Park

Thirty-four visitors were randomly interviewed (just one member of each group or family), to know their opinion about the importance of PNAs and especially about ECHNP; 44% of the interviewees were women and 56% men. Ages ranged from 21 to 73 years old with an average age of 47. Visitors came from diverse places: 44% were from the State of Hidalgo (of which, 86% came from the city of Pachuca); 32% came from México City; and 3% came from other States (Colima, Coahuila, Nuevo León, Veracruz, among others). The visitors' education level is shown in Fig. 24.11 where it was evident that most of them had university level.

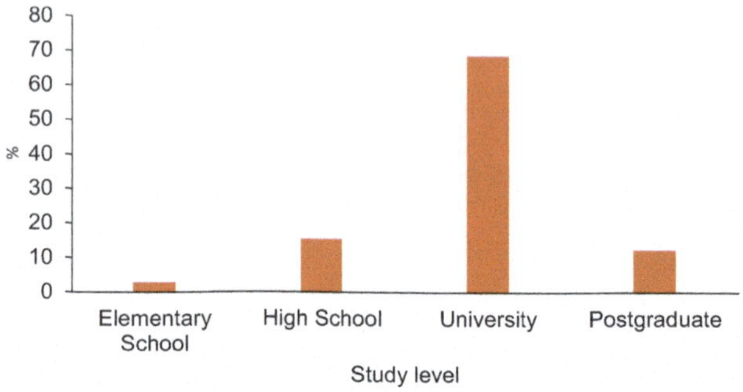

Fig. 24.11 Educational level from visitors interviewed at El Chico National Park (ECHNP)

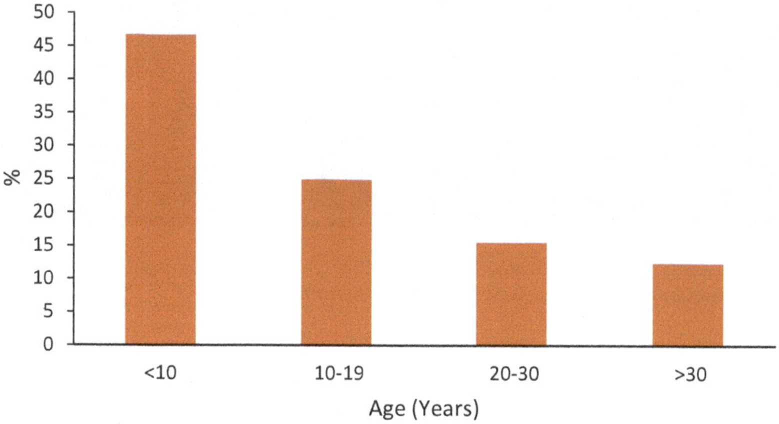

Fig. 24.12 Age at which visitors interviewed were for the first time in a National Park

Most of the visitor groups were families (36%) from six to ten members; some were non-familiar groups (26%); and 24% were couples. Most of the people interviewed declared to be interested in spending time in nature (40%), practicing hiking (32%) and having some recreational time in the area (9%). Less than 6% reported having a special interest in practicing sports like cycling, running or other sport activities. The age at which the interviewed people visited a NP for the first time was heterogeneous: 47% had made their visit to a NP before they were 10 years old, 29% when they were from 11 to 20 years old and 13% after they were 30 years old (Fig. 24.12).

The time people spent in the park is shown in Fig. 24.13 where in general visiting time was less than 24 h.

The visitors declared that they enjoyed visiting ECHNP because of the landscapes, trees, forest, weather and especially the peace it brings during their stay. All people interviewed said that the existence of PAs is very important. Regarding the question of who should pay for the protection of natural areas, 48% answered that

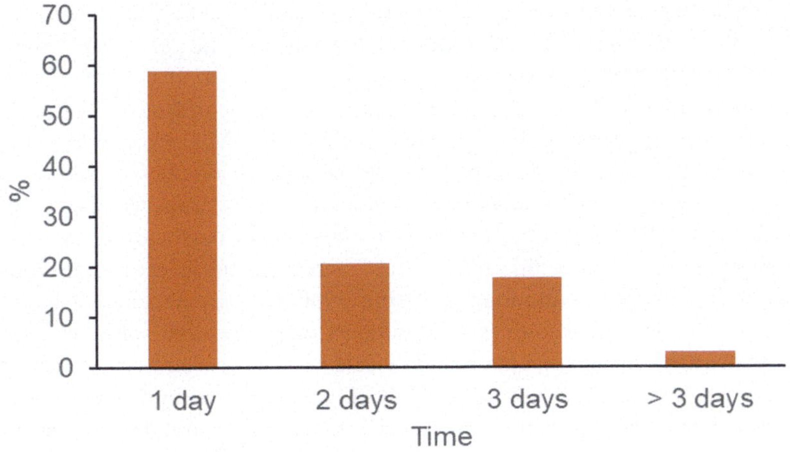

Fig. 24.13 Time spent at El Chico National Park by visitors that were interviewed

it was an obligation of the government; 9% said that was the citizens' obligation; and 42% said that it was a shared obligation. Nevertheless, concerning the entrance fee, only 21% confirmed to have paid it.

As to the management, success perceived by visitors of the ECHNP was (1) forest preservation; (2) maintenance, cleaning and garbage control; (3) signalling footpaths; (4) security; (5) organization and kindness of the park staff; (6) reducing the number of vehicles that transit within the park area; and (7) visitor centre infrastructure.

Among the visitors' suggestions to improve park management, they declared the need to improve signals, increase the number of rubbish containers, increase the surveillance and work with people to raise awareness of the importance of the park conservation; another suggestion was to limit the circulation of noisy vehicles around the park and in the town. However, 38% of the people interviewed had no suggestions.

With respect to the presence of wildlife in the park, 41% of the visitors reported that they had not seen animals; those who had seen them (59%) mentioned squirrels, snakes, frogs, chameleons, birds and insects. The plants that visitors could recognize at the park were fir, pines, oaks, cedars, fruit (apple, pear) trees, strawberries, ferns, bromeliads and some ornamental plants, but 15% of the visitors declared that they could not recognize the plants by their names.

24.4 Discussion and Conclusions

The two Natural Protected Areas selected for this study belong to a different conservation category and have different time of establishment and ecosystems, but both have reached their goals. The remarkable aspect of ECHNP is the conservation of relict forests of pine and fir trees. The work performed by the CONANP staff and guards is great, taking into consideration the low budget. The park staff should continue with outreach programmes of environmental education activities since the inhabitants have to be the promoters for the protection of their forests. The park staff should also start working with the residents to install UMAs related to the plants that occur in the forest. With respect to the increase in number of visitors, their impact and design strategies should be assessed to allow a proper management.

As to the BMBR, the achievements have also been outstanding. In less than 20 years after its creation, this reserve has increased the level of consciousness in local people about the importance of conservation. They have now become the protectors of their own resources. The support of the reserve staff for installing UMAs with cactus production has helped locals to develop a strategy of sustainable use of the resources, which have also allowed to slow down migration in the communities, at least in a small scale. The reserve staff should continue with the support for the creation of UMAs, not only for cacti but also for all the other species present at the reserve, such as magueys (*Agave*) and shrubs and trees like mesquite (*Prosopis* sp.) or copales (*Bursera* sp.). For both PNA, it would be advisable to have a registration office in a strategic place to control the entry of visitors and give them information about the objective of the PNA, the interesting points to visit, the activities that they can do during their visit and how they can help to the conservation of the PNA.

24.5 Lessons Learned

Ecosystem conservation involves several elements: local people, government, legislation, surveillance, researchers, science disseminators, teachers and visitors, among others. The goal of the PNAs is the conservation of the ecosystem functionality and to improve human well-being, which can only be achieved through the joint work of all the elements involved. In addition to the conservation that each PNA can achieve of its locality, it also plays an important role at regional level since it functions as a nucleus that influences the behaviour of the inhabitants of the surrounding areas, which can favour conservation of the socio-ecosystems at regional level.

Acknowledgements This study was part of the activities of the Red CONACYT Areas Naturales Protegidas. We thank the students of Ecology I (19-P) of the Universidad Autónoma Metropolitana-Iztapalapa, Lidia Ramírez H. (UAM-X), Irma G. Sánchez M., Víctor M. Martínez C. (UAA) and R. Jiménez S. for the help in making interviews, as well as those responsible for the UMAs: Guillermo Sánchez (UMA Santana), Raúl Balderrama (UMA El Viejo Cactus) and Edith Cinthya Bermeo López (UMA Biznaga Cactus).

References

Camarillo RJ, Casas AG (2001) Anfibios y Reptiles del Parque Nacional El Chico, Hidalgo, México. Anales del Instituto de Biología. Serie Zoología, Universidad Nacional Autónoma de México 72(1):105–123

CONABIO (2000) Regiones terrestres prioritarias de México. http://www.conabio.gob.mx/conocimiento/regionalizacion/doctos/Tacerca.html. Accessed 12 Sept 2019

CONABIO (2019) Áreas protegidas en México. https://www.biodiversidad.gob.mx/region/areasprot/enmexico.html. Accessed 12 sept 2019

CONANP (2003) Programa de Manejo Reserva de la Biosfera Barranca de Metztitlán. Comisión Nacional de Áreas Naturales Protegidas, SEMARNAT, D.F. México

CONANP (2005) Programa de conservación y manejo Parque Nacional El Chico. Comisión Nacional de Áreas Naturales Protegidas, SEMARNAT, D.F. México

CONANP (2012) Areas Naturales Protegidas. http://www.conanp.gob.mx/regionales/CONABIO. Accessed 21 Jan 2016

Halffter G (1984) Las reservas de la biosfera: conservación de la naturaleza para el hombre. Acta Zoológica Mexicana (ns) 5:4–48

Jiménez-Sierra CL, Torres-Orozco D, Martínez López JC, Toledo-Guzmán AD (2017) Una visión socioecositémica de las reservas naturales: la reserva de la biosfera barranca de Metztitlán como caso de estudio. In: Calderón-Contreras R (ed) Los Sistemas Socioecológicos y su resiliencia: Casos de estudio. Universidad Autónoma Metropolitana, Unidad Cuajimalpa, Gedisa, México, pp 138–158

Jiménez-Sierra CL, Torres-Orozco Jiménez D, Matias-Palafox ML (2018) Are current actions for conservation in México enough? Proximate and ultimate threats. In: Ortega-Rubio A (ed) Mexican natural resources management and biodiversity conservation: recent case studies. Springer, Cham, pp 53–71

Law of Ecologic Equilibrium (2018) Ley general del equilibrio ecológico y la protección al ambiente. Diario Oficial de la Federación el 28 de enero de 1988. http://www.diputados.gob.mx/LeyesBiblio/pdf/148_050618.pdf. Accessed 12 sept 2019

Matias-Palafox ML, Jiménez-Sierra CL (2006) Evaluación del Riesgo de Extinción (MER) de las poblaciones de Turbinicarpus horripilus (Lem.) Vác. John & Riha, cactácea endémica de la Barranca de Metztitlán. Cactáceas y Suculentas Mexicanas. 51(4):100–110

Medina JM, Rzedowsky J (1981) Guía botánico-forestal de la parte alta de la Sierra de Pachuca. En: Guías Botánicas de Excursiones en México. Soc Bot Méx IV:1–19

Melo-Gallegos C, López-García J (1993) Parque Nacional El Chico, marco geográfico-natural y propuesta de zonificación para su manejo operativo. Investigaciones Geográficas, boletín del Instituto de Geografía, UNAM 28:65–128

Robles de Benito R (2009) Las unidades de manejo para la conservación de vida silvestre y el Corredor Biológico Mesoamericano México. CONABIO, D.F. México

Sánchez Mejorada H (1978) Manual de campo de las cactáceas y suculentas de la Barranca de Metztitlán. Sociedad Mexicana de Cactología. México

SEMARNAT, CONAFOR (2009) Manual técnico para beneficiarios: Manejo de vida silvestre

Torres-Orozco D, Jiménez-Sierra CL, Sosa Ramírez J, Cortés-Calva P, Breceda A, Iñiguez LI, Ortega-Rubio A (2015) La importancia de las áreas naturales protegidas en nuestro país. In: Ortega-Rubio A (ed) Las Áreas Naturales Protegidas y la Investigación Científica en México. CIBNOR, UAY, UM de San Nicolás de Hidalgo, La Paz, pp 41–60

Zavala Chávez F (1995) Encinos Hidalguenses. División de Ciencias Forestales, Universidad Autónoma Chapingo, México

Part V
Community Well Living Improvement from Ecological Conservation

Chapter 25
Economic Valuation of Diving with Bull Sharks in Natural Conditions: A Recent Activity in Cabo Pulmo National Park, Gulf of California, Mexico

Carmen Pasos-Acuña, Marco Antonio Almendarez-Hernández, Edgar Mauricio Hoyos-Padilla, María Carmen Blázquez, and James Thomas Ketchum

Abstract Cabo Pulmo National Park (CPNP) is one of the main tourist destinations in the Gulf of California, especially for recreational diving with bull sharks (*Carcharhinus leucas*). This activity begun in the late 1990s after the National Park was created, the local community changed its main activity from traditional fishing to offer tourist facilities for observation of marine life; as a result, the economic benefits are now greater than ever before. However, these benefits have not been quantified or taken into account in the CPNP management; thus, the objective of this study was to estimate the specific economic value of diving with bull sharks by using a method of revealed preferences (travel cost) and calculating the consumer surplus (CS) of diving with bull sharks at CPNP as a tourist destination. We deducted the travel cost of each person per day through 250 on-site surveys directed to tourists who visited CPNP to dive with bull sharks. The economic analysis showed that the number of divers who travel specifically to CPNP for bull shark sighting represents

C. Pasos-Acuña · E. M. Hoyos-Padilla
Pelagios Kakunjá, A.C, La Paz, B.C.S, México
e-mail: mauricio@pelagioskakunja.org; james@pelagioskakunja.org

M. A. Almendarez-Hernández
Office of Intellectual Property and Technology Marketing and the Patent Center, Centro de Investigaciones Biológicas Noroeste (CIBNOR), La Paz, B.C.S, México
e-mail: malmendarez@cibnor.mx

M. C. Blázquez (✉)
Behavioural Ecology Lab. Environmental Planning and Conservation Department, Centro de Investigaciones Biológicas Noroeste (CIBNOR), La Paz, B.C.S, México
e-mail: blazquez@cibnor.mx

J. T. Ketchum
Pelagios Kakunjá, A.C, La Paz, B.C.S, México

Centro de Investigaciones Biológicas del Noroeste, La Paz, California Sur, México

A. Ortega-Rubio (ed.), *Socio-ecological Studies in Natural Protected Areas*,
https://doi.org/10.1007/978-3-030-47264-1_25

23% of the total tourists visiting the park. The recreational demand function with the econometric treatment of the Poisson model revealed a CS result of $694 USD per person per day with an average of a 4-day stay. This information was useful to promote improvements in management and conservation of bull sharks at CPNP.

Keywords Economy of the local resources · Diving with sharks · Marine protected areas · Tourism · Travel cost method · Data count

25.1 Introduction

In the 1970s, Cabo Pulmo, in the southern tip of the Baja California Peninsula (Mexico), was a small traditional fishing community; locals exploited the reef with traditional methods to extract fish for food or for ornamental trade. They also took tourists to sport fishing; hence they regularly caught several species of reef fishes, sharks, and other large predators as part of their support to the local economy. During the 1980s, excessive fishing pressure diminished these species, causing large reef species like grouper to decrease in number and size. By the early 1990s, the community observed a decrease in their resources, as well as economic instability. Cabo Pulmo villagers realized the need to seek for other alternative resource use, such as ecotourism activities and scuba diving implementation. The coral reef was declared a Natural Protected Area in 1995 and National Park in 2000, which occurred as a community request with the support of federal government, scientific community, and nongovernmental organizations (Benítez-Arce 2013). Ten years after its protection, Cabo Pulmo National Park (CPNP) showed important signs of recovery with a biomass increase of 463% (Aburto-Oropeza et al. 2011). Today, CPNP has become a symbol of conservation and a success story worldwide. Fishermen and villagers turned into proud guardians, conserving their resources and improving their quality of life, using the reef sustainably. CPNP has become one of the main diving sites in the Gulf of California, hosting an important number of tourists from around the world, similar to other places like Palau and the Galápagos where ecotourism has also been an alternative use of ecological services. In these places, the local community's perception of natural systems thankfully changed, as a response to the degradation of the ecosystem. Using ecotourism as a strategy, these countries promoted a sustainable use of resources, providing benefits to the ecosystems and the local communities (Pires et al. 2016).

Shark diving tourism has been increasing in the last few years in many places around the world, such as Australia, South Africa, Ecuador, Colombia, Mexico, Bahamas, etc. (Laroche et al. 2007; Cubero-Pardo et al. 2011; Cisneros-Montemayor et al. 2013; Huveneers et al. 2013; Apps et al. 2015). This activity also brought a wide range of economic, educational, and conservation benefits to local communities (Garrod and Wilson 2004; Vianna et al. 2012; Pires et al. 2016). It was estimated that approximately 590 thousand divers were interested in shark diving worldwide in 2013. This large number of divers, which may have increased in recent years, generated \$314 million USD per year (Cisneros-Montemayor et al. 2013). In some countries, such as Palau, the shark diving industry was estimated to have generated \$18 million USD in benefits per year, accounting for 8% of the gross domestic product of the country (Vianna et al. 2012).

Shark diving can provide an economic alternative to fishing, promoting a sustainable economic use of these animals, favoring their conservation. Understanding the socioeconomic aspects of this activity is the first step toward its sustainability and, ultimately, shark conservation (Torres et al. 2017). This study used the travel cost valuation method (TCVM) for its economic valuation in CPNP. This method is applied to the value of recreational services provided by nature when a person must move to a particular environment to see and enjoy it (Azqueta 2002). It is a method based on revealed preferences; it measures changes in people's well-being that results from changes in natural resources or environmental quality. The aforementioned is given by changes in transportation costs, associated with the target activity. The method is also based on the theoretical implicit cost valuation involved in access to a specific good or service, such as diving with bull sharks in this case. The cost of access to the marine park is \$2.00 USD, which is mainly for the park management and not for the conservation of the bull shark; however, the real expense to access the activity is actually much more because every visit by each user implies greater expenses generated by displacement. The function of the demand is related to the number of visits (demand amount), with the cost of displacement (price) (Azqueta 2002; Hernández-Trejo et al. 2012a).

CPNP is one of the main sites for shark diving in Mexico; the other sites are Guadalupe Island, Playa del Carmen, and Revillagigedo Archipelago, but CPNP is the only one in the country without using supplemental feeding (chumming) for recreational diving with bull sharks. This activity has been attracting divers to the park in the last years to enjoy this ecotourism experience (Boncheva et al. 2010). Hence, we generated information regarding the economic value of sharks as an ecotourism resource, which can be useful for better management to the park, as well as for shark conservation. This study (I) evaluates the bull shark diving activity in CPNP according to the travel cost and calculates the population segment that generated the consumer surplus (CS), (II) generates a database about the economic importance that implicates diving with bull sharks in CPNP, and (III) provides management tools for the design of public policies, focusing on maintaining and/or improving both the sustainable use of bull shark diving activity within CPNP and the associated urban infrastructure.

25.2 Methods

Data were collected at CPNP (Fig 25.1) located between 23° 22′ 30″ and 23° 30′ 00″ N and 109° 28′ 03″ and 109° 28′ 03″ W in the southwestern Gulf of California, Mexico. The National Park is a marine reserve with an area of 70 km^2 of which 99% corresponds to marine habitat surrounding the reef (CONANP 2006).

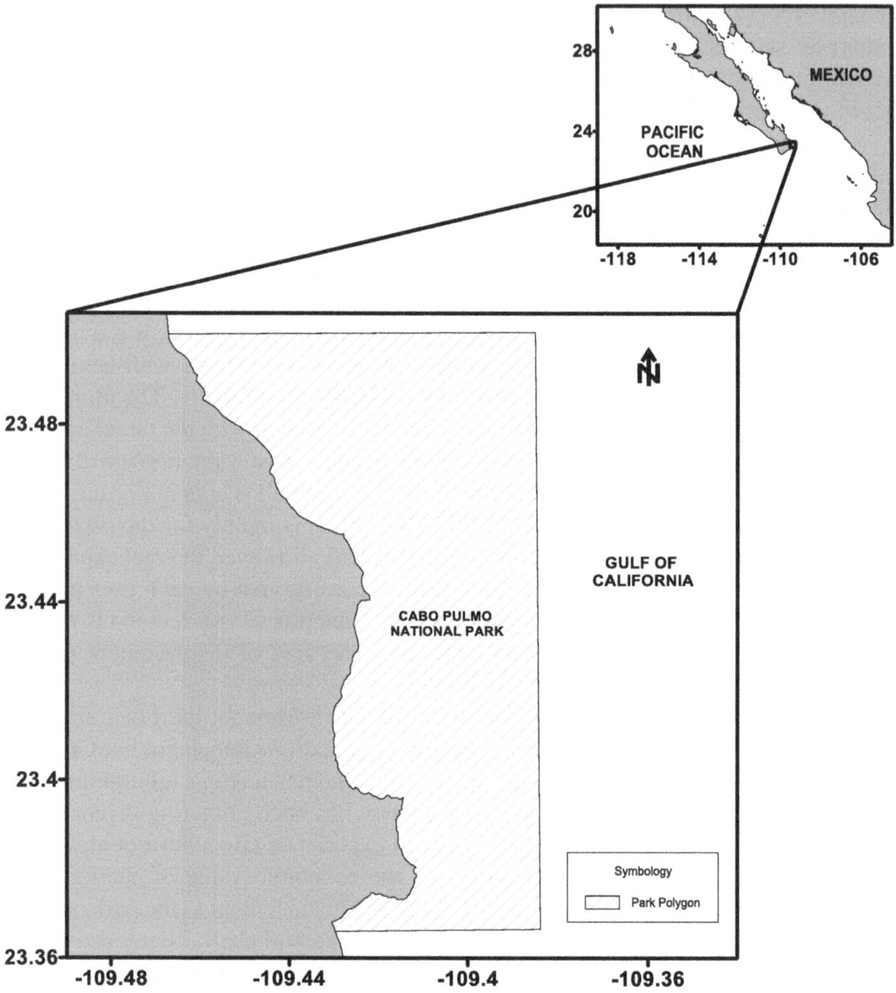

Fig. 25.1 Location of study site (Cabo Pulmo National Park), Baja California Sur, Mexico

25.2.1 Travel Cost Method (TCM)

The TCM allowed comparing the value of marketing activities with the value of outdoor recreation (Haab and McConnell 2002; Rolfe and Prayaga 2007). It is an indirect method of revealed preferences. It describes changes in the quality of recreational services that tourists experience by estimating a demand curve, to obtain a measure of well-being, such as CS (Turner et al. 1993).

Travel cost assessment has three classifications: (I) zone travel cost method, (II) individual cost method, and (III) random utility travel cost model (Prayaga et al. 2010). We used the individual cost method in this study. It establishes a regression between the number of visits (Tourkolias et al. 2015) and the number of days that each individual or interviewee spends in the site and the cost of travel, as well as other predictors (Mendes and Proenca 2011).

In a model of trips to a single site, such as CPNP, all individuals face the same site quality. In this case, we use the single-site model to measure the economic value of access to the site, but not the value of changes in the quality of the site. On the other hand, the travel cost serves in the model as a proxy for price, or the basis for welfare measurement. All demand models for recreation are essentially models of time allocation, and as such, they must meet the individual's time and income constraints (Haab and McConnell 2002).

The TCM has several assumptions: (1) Travel and time are variable for recreational trip price. (2) Travel time is neutral; it does not provide usefulness or disutility. (3) The unit of decision are trips of equal distance to the site of interest for each income; the wage rate represents the opportunity cost of time. (4) The only reason for the trip is to visit the interest site, simple purpose trips. (5) The amount consumed is the trip to the same site for all consumers (Azqueta 2002; Haab and McConnell 2002; Hernández-Trejo et al. 2012a).

The disadvantages or limitations of the model are:

1. Assuming that people would have the same perspective and respond in the same way to changes in travel cost and fare price increment to do the recreational activity (diving with sharks).
2. Defining and measuring the opportunity cost of time or the value of time spent traveling are problematic. If people enjoy the time spent traveling, then time would be a benefit, not a cost, so the result will be overestimated (Turner et al. 1993).
3. Having availability of substitute sites (alternative sites for bull shark diving) will affect the value. In our case, no other sites for diving with bull sharks in natural conditions (without chumming) are available in the area; therefore, this aspect did not apply.

On the other hand the advantages are (1) based on the real behavior of the tourist, revealed preference data; (2) high tourist participation; (3) cheap to be carried out; (4) easy interpretation; and (5) approximate to conventional methods that estimate economic value based on market prices (Nyangwe 2005).

25.2.2 Demand Model Construction

To model the demand for recreation, this study considered the two main restrictions: the individual's income and time allocation, where income was defined by:

$$\sum_{n}^{j=1} x_{ij} c_{ij} + z_i \leq y_i \tag{25.1}$$

where person i spends X_{ij} number of days at the site n, (in our case n = 1) and C_{ij} is the round-trip travel cost spent in order to enjoy the activity j (diving with bull sharks); the tourist also purchases souvenir Z_i at a price normalized to equal 1. At the same time, the tourist is not able to spend more than his actual income Y_i.

The time component was defined by:

$$\sum_{n}^{j=1} x_{ij} t_{ij} + h_i = T_i \quad (25.2)$$

where t_{ij} is the time to travel to the site j; h is the working hours; and T is the total time available.

Considering the time and income constraints in the equation, the demand for individual i for site 1 was:

$$x_{i1 = f_1 (c_{i1} + W_i t_1, \ldots c_{in} + W_i t_{in}, q_1 \ldots q_n, y_i^f} \quad (25.3)$$

where q_1 was the exogenous quality for the site; j, y_i^f means that income spent was a function of fixed income and the available salary; W_i is income after taxes; C_{ij} is travel costs; and t_{ij} is time.

The next component for TCM was to estimate the parameters associated with each trip determinant. The most common model in TCM analysis is the Poisson counting model, which is a regression model for counting data to specify the quantity demanded, expressed as number of visits (Haab and McConnell 2002) or days of stay per person (used for the first time by Mendes and Proenca 2011). This study expressed the demand as days of stay per person at CPNP. The Poisson model considers dependent variables with nonnegative values.

For this study, the Poisson model objective was to estimate changes in frequency of days of stay per visit of individuals to CPNP. For single-site models, as in this case, the general counting model was:

$$\Pr(x_i = n) = f(n, z_i \beta), n = 0, 1, 2 \ldots \quad (25.4)$$

For this model, the functional form of the expected demand was typically exponential where the demand variable x can take values from 0 to ∞. Density function of Poisson probability is given by:

$$\Pr(x_i = n) = \frac{e^{-\lambda_i \lambda_i^n}}{n!}, n = 0, 1, 2 \ldots \quad (25.5)$$

where λi is the mean and variance of the distribution, and it is required that $\lambda i > 0$ be specified as an exponential function $\lambda i = \exp(zi\beta)$. Zi is the vector of explanatory variables, which were eleven in this case:

1. Number of days of stay to the place j by person i
2. Displacement cost
3. Quality of the site: taking values from 1 to 5
4. Number of sharks sighted

5. Chumming implementation, taking value of 1 if the person was favorable to the possibly implementing chumming in CPNP and 0 if the person did not agree
6. Proximity to the sharks
7. Positive change of perception toward sharks, taking a value of 1 if there was a change in perception and 0 if not
8. First shark dive experience, taking a value of 1 if it was the first-time diving with sharks and 0 if it was not
9. Days of stay
10. Tourist age
11. Income

Once the estimates of the parameters associated with the *z* vector have been made, the measure that expresses changes in well-being can be calculated by accessing the recreational service (diving with bull sharks), known as consumer surplus or willing to pay (WTP).

$$\text{WTP acces} \int_{\infty}^{c^0} e^{b_0 + b_1 c}\, dC = \left[\frac{\mathrm{e}^{bo + b_1 c}}{b1} \right]_{c = c^0}^{c \to \infty} = -\frac{x}{b_1} \tag{25.6}$$

where WTP is willing to pay; C^0 is the actual trip cost; *C* is the trip cost; *x* is the average of days of the trip; and *b1* is the coefficient associated with the cost of travel.

The Poisson model makes some assumptions that if it is not fulfilled, it would lead to overdispersion or underdispersion (Cameron and Trivedi 2005): (1) The data set must follow the equality known as equidispersion, which implies the difficulty to obtain a good fit and make valid inferences. (2) A used database may contain excess values of zero, which is not consistent with the model. (3) Independence of the variables. (4) The consideration of a Poisson distribution with parameter λi as a deterministic function causes unobservable heterogeneity not to be captured.

25.3 Results

The surveys were applied from January 2016 to March 2017. The final sample was 250 people surveyed who dived with bull sharks at CPNP or that were interested in diving with bull sharks although they did not see them.

25.3.1 *Tourist Profile*

25.3.1.1 Gender

From the 250 interviewed, 94 (38%) were women and 154 were men (62%). The age varied between 18 and 70 years, of which the average age was 37 years and the most frequent was 30 years (Fig 25.2).

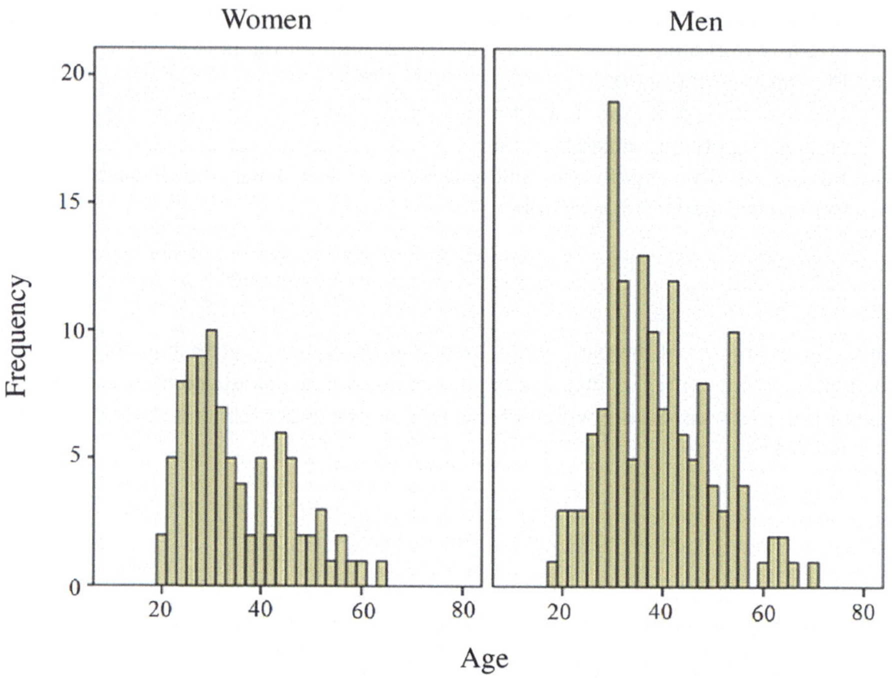

Fig. 25.2 Frequencies of sex and age of the tourists at Cabo Pulmo National Park, Baja California Sur

25.3.1.2 Nationality

From the 250 tourists interviewed, 105 were Mexicans (42%). Overall, the Americas accounted for 82% of the total tourist influx, while Europe and Asia accounted for only 12.4% and 2.8%, respectively (Fig. 25.3). In total, 115 tourists were foreigners, which corresponded to 46%, while 30 respondents did not answer the question.

25.3.1.3 Occupation

Occupation was divided into categories: self-employed, employed by others, student, professional, and retired (Table 25.1), of which the most frequent category of tourists interviewed in CPNP was professionals (46.4%) and the least frequent was retired (0.8%) in descending order.

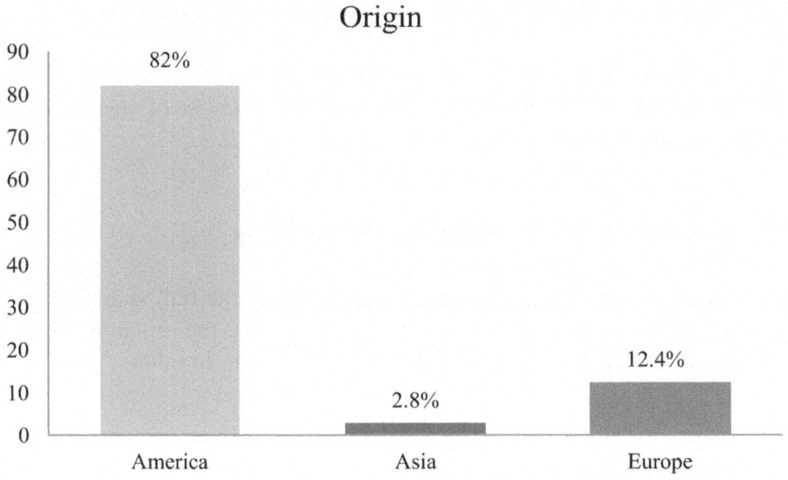

Fig. 25.3 Origin percentages of tourists interviewed in CPNP during 2016

Table 25.1 Frequencies of occupation of the tourists who dived with bull sharks at Cabo Pulmo National Park, Baja California Sur, Mexico, during the season January 2016–March 2017

Occupation	Frequency	Percentage	Valid percentage	Accumulated percentage
Did not answer	32	12.8	12.8	12.8
Self-employed	40	16	16	28.8
Employees	39	15.6	15.6	44.4
Student	21	8.4	8.4	52.8
Professional	116	46.4	46.4	99.2
Retired	2	0.8	0.8	100
Total	250	100	100	

25.3.2 *General Aspects of Diving with Bull Sharks at Cabo Pulmo National Park*

25.3.2.1 Assessment of General Diving Quality in Cabo Pulmo National Park Given by Tourists Who Visited the Park in 2016

From the people interviewed (n = 250), 62% rated diving in the park as excellent, 31% good, 6% regular, 1% poor, and 0% very poor (Fig 25.4). The average quality of diving in CPNP was rated 4.5 on a scale of 1–5 where the most frequent rating was 5.

25.3.2.2 Number of Shark Sightings per Person During the Dive

The average number of sharks sighted was 3 sharks per dive; the maximum number of sharks sighted was 15 (Table 25.2). However, the number of sharks seen most frequently was 0–2 sharks per dive.

25.3.2.3 Favorite Animal for Sightings During Dives

Eighty-four percent of the people surveyed pointed to the bull shark as the most attractive animal to see in such activity (Fig. 25.5), while the remaining 16% were more interested in observing turtles, corals, dolphins, fish, or others.

25.3.3 Demand for the Activity

25.3.3.1 First Shark Dive Encounter

From all the people interviewed, 32% responded that it was the first time they dived with sharks in their lives (Fig 25.6), while 68% had done it before.

25.3.3.2 Days of Stay

Regarding the stay of tourists in CPNP, the average number of visiting days was 4.23; standard deviation = 3.044; the lowest value was 1 and the highest was 20 days; the highest frequency was maintained from 1 to 5 days with mode of four (Fig 25.7).

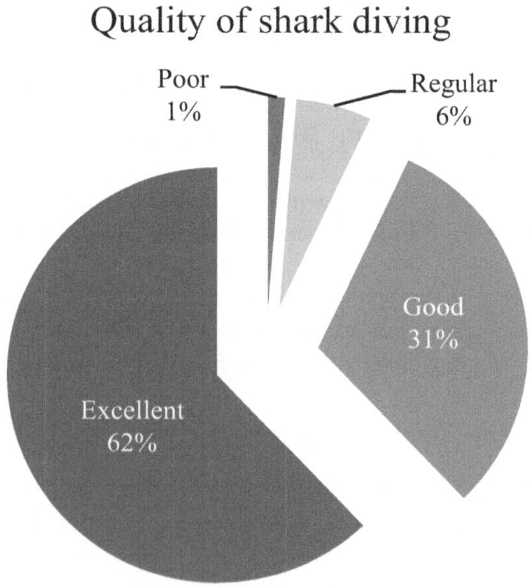

Fig. 25.4 Graph of diving quality in CPNP on a scale of 1, very poor; 2, poor; 3, regular; 4, good; and 5, excellent

Table 25.2 Central tendency measures of number of sharks sighted per person during the dive

N	246
Average	3.01
Median	2
Mode	0
Minimum	0
Maximum	15

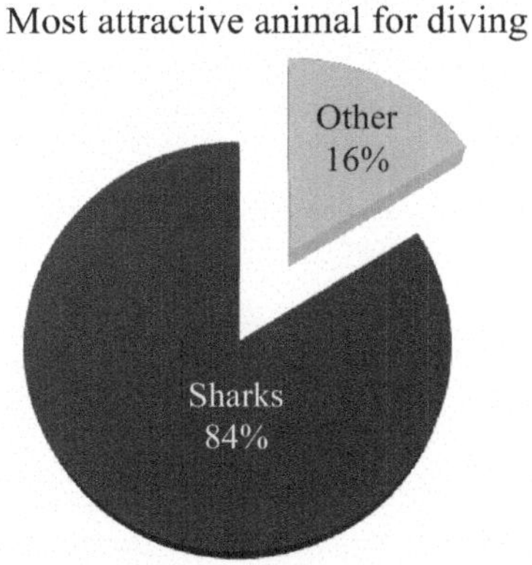

Fig. 25.5 The chart represents the most attractive animal in recreational diving n = 250

25.3.3.3 Diving with Sharks as the Main Reason to Visit CPNP

From the total number of divers interviewed, 22.8% traveled to CPNP with the main objective of diving with bull sharks (n = 51 tourists from 223), while the rest of the divers had another motivation to visit the park although they did dive with bull sharks (Fig. 25.8).

25.3.4 Other Aspects

25.3.4.1 Other Activities to Be Carried Out in PNPC

Almost forty-two percent of divers (99 divers out of n = 238) interviewed had not planned to carry out another activity inside the park; the remaining 58% (139 divers out of n = 238) responded to be interested in carrying out more activities within the park.

25.3.4.2 Support for Implementation of Regulations with Conservation Routes

Eighty-nine percent of divers (223 out of n = 248) interviewed answered to be willing to support implementation of regulations for bull shark conservation in CPNP. The other 11% were not interested in such implementations or in the possible conservation of the species.

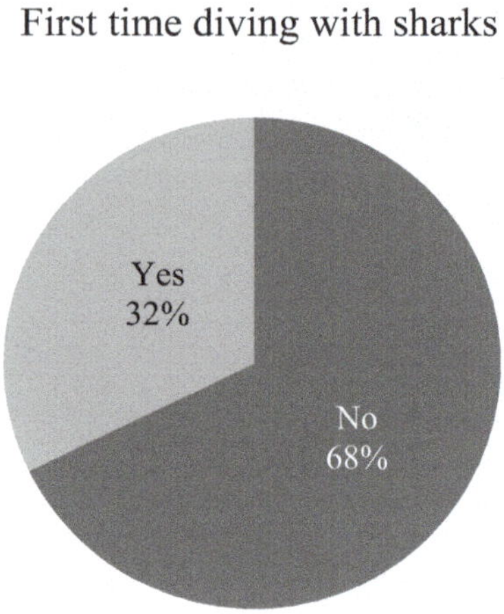

Fig. 25.6 Percentages of people interviewed who dived with sharks for the first-time n = 250

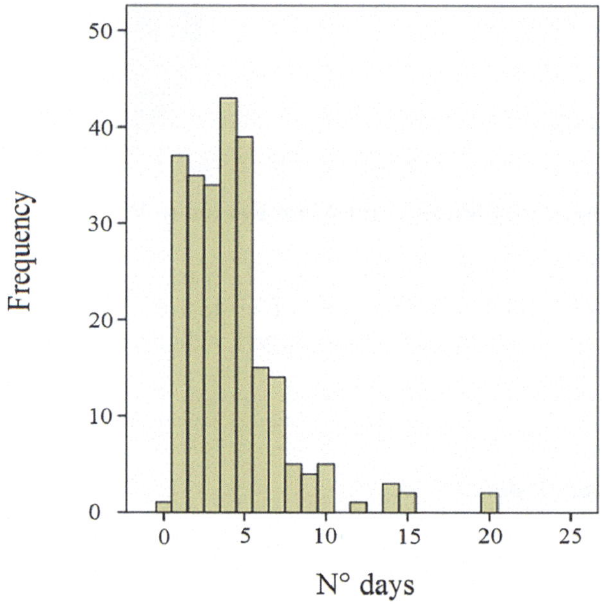

Fig. 25.7 Histogram of frequencies of number of days that the tourists spent in CPNP n = 240

25.3.4.3 Changing Tourist Perception of Sharks After the Bull Shark Diving Experience

Thirty-six percent of divers interviewed (86 out of n = 239) reported having changed their perception toward sharks after the diving experience with bull sharks in a positive way. The other 64% did not indicate any change.

However, from the divers who had experienced their first shark dive at CPNP with bull sharks, 55% of them (45 out of n = 82) reported having changed their perception in a positive way (Table 25.3).

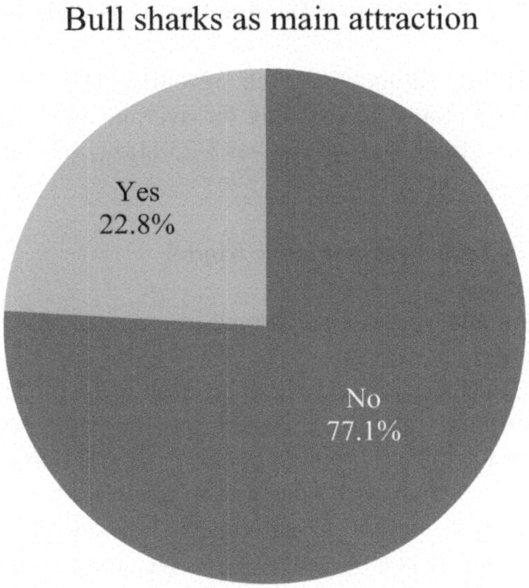

Fig. 25.8 Graph of percentages from tourists who traveled to CPNP with the main objective of diving with bull sharks

Table 25.3 Description in percentages of people who had a positive perception change after their first shark dive at CPNP

Perception changed					
Response		Frequency	Percent	Valid percentage	Cumulative percentage
Valid	No	37	45.1	45.1	45.1
	Yes	45	54.9	54.9	100.0
	Total	82	100.0	100.0	

25.3.4.4 Attitude Toward the Implementation of Chumming (Organic Attractant) During Dives with Bull Shark at CPNP

Seventy-two percent of the divers interviewed said they did not agree with the implementation of chumming to attract the sharks at CPNP even if implementing it could increase sighting chances (n = 244).

25.3.5 Economic Valuation by Cost of Travel Method (CTM)

Travel costs per person ranged from $600 to $2000 USD with an average of $1370; SD = 433 944. The most frequent value was $925, representing 37.2% of tourism, followed in descending scale for $2000 (20.2%), $1175 (14.3%), $1400 (12.6%), $1875 (9.4%), and finally $1625 (6.3) (Fig 25.9).

25.3.5.1 Annual Income Level per Person

The annual income levels of tourists (USD) varied from $25 800 USD to $120 000 USD (Table 25.4) where the average annual income was $67 599.55 USD and the most frequent amount was $ 25 800 USD (Fig. 25.10).

25.3.5.2 Econometric Model

An analysis was performed to assess the total cost of travel per person, using the program NLogit5.

The explanatory variables used in the model to calculate the value of the response variable (final travel cost per person) were (a) number of days of stay at CPNP; (b) cost per day; (c) annual income; (d) opinion toward using chumming; (e) first shark dive encounter; (f) perception quality toward shark diving experience; (g) number of shark sightings by diving; (h) distance from sharks during dives; (i) perception change to positive after shark dive; and (j) age of the tourist. The over dispersion test

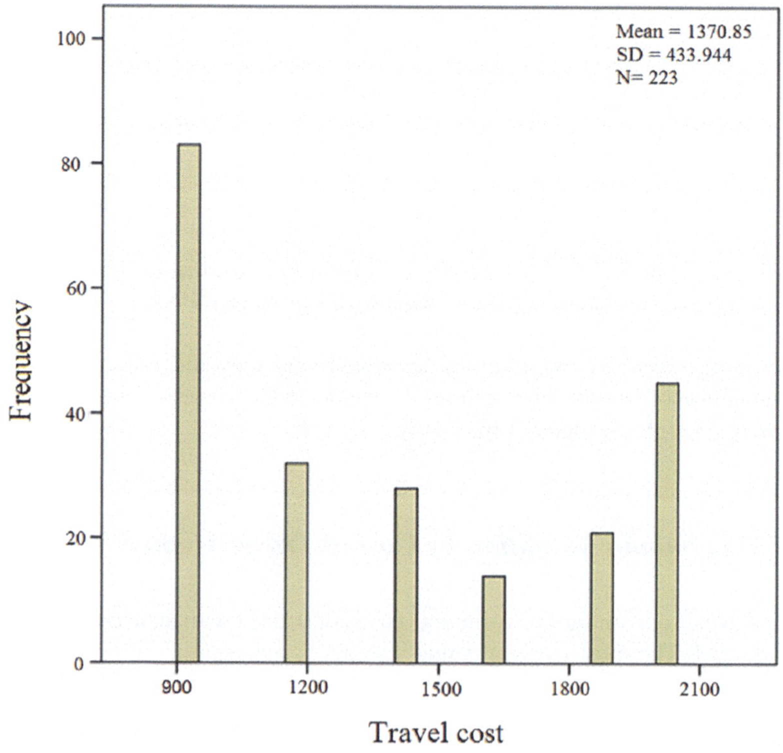

Fig. 25.9 Histogram of frequencies of travel costs to CPNP per person

Table 25.4 Description of variables used in the econometric model

Variable	Description
Days	N° days at CPNP
TC day	Cost of travel per day
Income	Annual income
Chumming	People's opinion toward supplemental feeding
First shark dive	First shark dive encounter
Quality	Perception quality toward shark diving experience
Shark number	Number of shark sightings by diving
Shark approach	Distance from sharks during dives
Perception change	Perception change to positive after shark dive
Age	Age of the tourist

values we obtained in our Poison model were ($g = \mu = 3.46$ and $g = \mu^2 = 3.37$). There are below the critical χ =3.84, for one degree of freedom (Table 25.5). Therefore, our study complied with the Poisson model assumptions. The model was significant with pseudo R-squared 0.217; significance level = $p < 0.01$; and AIC value = 881.4 (Table 25.6).

The model coefficients for the travel cost per day were −0.00144, standard error = 0.0001, and a probability value of Z = 0 (Table 25.7).

On the other hand, marginal effects gave a partial effect value = −0.00566 with a Z value = 0.0, which indicated that for every dollar, travel cost per day increased; the expected demand, quantified in the number of days that tourists spent in CPNP, decreased 0.0056 days (Table 25.8); in other words, the number of days of stay at CPNP change when the cost per day increases or if chumming is implemented. At the same time, tourists stayed longer when it was their first-time diving with sharks or when the number of sharks sighted increased.

The demand curve indicated that the number of days a person stayed at CPNP was negatively related to the cost of travel in the same way as with chumming implementation. Divers were willing to stay longer at CPNP to have their first shark experience with bull sharks and when the number of sharks sighted increased. Similarly, those with a higher income level were willing to stay for more days at CPNP.

25.3.5.3 Consumer Surplus or Willingness to Pay to Access the Activity

In the function of the cost valuation travel from January 2016 to March 2017 at CPNP, the consumer surplus was estimated at an average of $694 USD per person per day. Considering the average of a 4-day stay and the number of divers that has

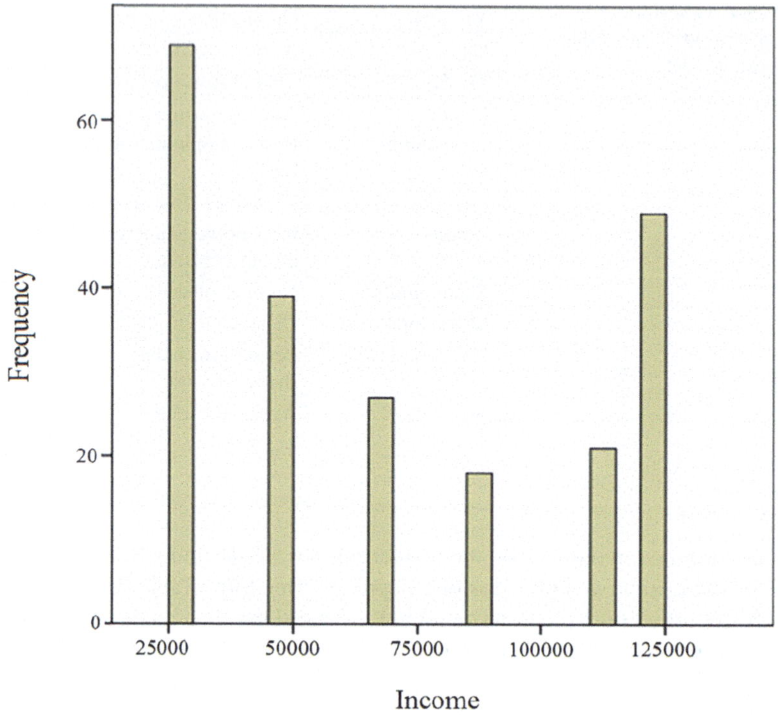

Fig. 25.10 Histogram of annual income (USD) frequencies per person (average = 67 599.55; SD = 37 919.631, n = 223)

Table 25.5 Econometric model values and significance level

Poisson Regression	
Dependent variable	DAYS
Log-likelihood function	−430.717
Restricted log-likelihood	−550.566
Significance level	0.000
McFadden pseudo R-squared (analog to the determination coefficient of minimum ordinary squares)	0.217
Chi-squared	239.697
Overdispersion test: g = mu (i)	3.460
Overdispersion test: g = mu (i) ^2	3.370

Table 25.6 Coefficients of the model variables

Days	Coefficients		Standard error	Prob.lzl>Z*
Constant	1.7301	***	0.0976	0.0
TC day	−0.0014	***	0.0001	0.0
Income	0.3723D-05	***	0.9769D-06	0.0001
Chumming	−0.0006	***	0.0001	0.0
First shark dive	0.0009	*	0.0005	0.0986
Quality	0.1596D-04		0.0002	0.9384
Shark number	0.0005	*	0.0003	0.0760
Shark approach	−0.8102D-04		0.0001	0.4235
Perception change	−0–7095D-04		0.0002	0.7489
Age	−0.0001		0.0002	0.3928

*p < 0.05; *** p < 0.001

Table 25.7 Marginal effects

Mean average conditional sample 3.9616			
Factor scale for marginal effects 3.9616			
Days	Partial effects	Standard error	Prob.lzl>Z*
Constant			
TC day	−0.00566	0.00042	0.0
Income	0.14625D-04	0.3585D-05	0.0
Chumming	−0.0025	0.00051	0.0
First shark dive	0.00361	.00218	0.00980
Quality	0.62702D-04	0.00081	0.9384
Shark number	0.00216	0.00122	0.0758
Shark approach	−0.00032	0.00040	0.4238
Perception change	−0–00028	0.00087	0.7488
Age	−0.00070	0.00082	0.3923

access to the park per year, specifically searching for shark dive encounter (n = 2864 tourists from 13 021 that have access to the park annually), the value of aggregate economic benefits by diving with bull sharks in CPNP was estimated at $7 950 464 USD per year.

25.4 Discussion

25.4.1 Tourist Characterization

The profile of the prevalent tourist visiting CPNP specifically to dive with bull sharks was as follows: men, from 30 to 60 years old, and an income equal to or greater than $25 000 USD per year. From the total of tourists surveyed, 42% were national tourists and 46% were foreigners, mostly from the USA. The number of foreign tourists remained regularly throughout the year, while national tourists

Table 25.8 Travel cost studies in the area

Author	Study	WTP (USD)
Hernández-Trejo et al. (2012a)	Recreation of the aquatic biodiversity of Espiritu Santo Archipelago National Park, BCS, Mexico	\$8 – 288
Hernández-Trejo et al. (2012b)	Recreation of sport fishing, Los Cabos, BCS, Mexico	\$22.27 – 70.55
Morales-Zarate et al. (2019)	Recreational beach ecosystem service at Los Cabos, BCS, Mexico	\$4 941
This study	Bull shark diving at Cabo Pulmo National Park, BCS, Mexico	\$694

increased during the holiday periods of summer and Easter. Our results show a change in tourism inflow compared with previous years; during 2003, most of the tourists came from the USA, while national tourism only accounted for 5% (National Institute of Ecology 2003). In other similar studies, the results show that when the distance is longer, travel cost increases, causing the number of visitors or demand to decrease (Rivera-Castañeda 2002; Labandeira et al. 2007,). However, this effect is less remarkable in CPNP because its main tourist influx comes from the USA, which is fairly close and cheaper compared to other countries (Rivera-Castañeda 2002). The average stay at CPNP was 4 days; this period was shorter than in other tourist diving places, which may be due to the easy access, just by the road, allowing tourists to change the days of stay freely. It is different to other diving sites located in oceanic islands where it is not possible for tourists to vary their time of stay so easily. Consequently, diving activity in CPNP becomes more accessible in economic terms.

The average shark sighted by tourists in CPNP was three per dive. During the high season, average sightings were about ten sharks and none during low season. This number is very low, compared to other places around the world, such as Galápagos, Malpelo, or Revillagigedo. In other sites like Aliwal Shoal, on the east coast of Africa, the observed number of sharks by diving was around 4, with 98% of tourists having shark sightings during their diving (Dicken and Hosking 2009). In CPNP bull sharks are in the area all year round (Pasos et al. unpublished data); however, the behavioral ecology of the individuals of the population influences group size, the rate of visits and the time they spend in sites looking for food or mates as well as their social interactions (Asúnsolo Rivera 2016). These factors are still not very well studied in the area, which make the possibility of encounters unpredictable.

In other places as Playa del Carmen (Mexico), where bull sharks are attracted by chumming, the number of animals that can be seen in high season is greater than 10; however, the comparison between both locations is not possible because they are a whole different experience. In fact, the method of attraction for shark sighting had a high level of rejection in tourists at CPNP where 72.1% of people did not agree to chumming implementation although it meant increasing the chances of sighting. Our results contrast with other studies in the Aliwal Shoal Marine Protected Area.

In Aliwal, 88.5% supported the use of chumming for a closer experience with tiger shark (*Galeocerdo cuvier*) where 44.3% would not visit the Protected Area again if they could not dive with tiger sharks (Dicken and Hosking 2009). Therefore, based on our results, we found that the ecosystem service of diving with bull sharks in CPNP attracts divers with the specific purpose of sighting sharks in the most natural conditions as possible. These divers came to CPNP looking for a different type of experience from that which happens with the use of chumming.

Public support for environmental problems can lead to important changes in conservation policies (O'Bryhim and Parsons 2015). Therefore, a better understanding of public attitudes and behavior toward sharks is necessary to encourage support for conservation initiatives (Acuña-Marrero et al. 2018). Most tourists (89%) who travel to CPNP were in favor of conservation. Tourists were willing to support the implementation of regulations in favor of bull shark conservation. This trend in the diving industry shows an attitude evolution toward sharks since 1974 when it was thought that human beings needed protection against sharks. In 1998 shark diving began to attract masses, and from 1999 to 2002, the perspective changed to the conception that sharks had to be protected (Dearden and Ziegler 2007). In general, large carnivores suffer from negative attitudes and low tolerance from humans (Crossley et al. 2014) although these can be compensated by the economic benefits perceived (Bruskotter and Wilson 2014). Consequently, shark-based ecotourism not only provides economic incentives for local communities; it can also promote more positive attitudes and behavioral responses to sharks in tourists (Acuña-Marrero et al. 2018). Finally, helping to understand the vulnerability of sharks and their ecological roles is essential to support conservation (Acuña-Marrero et al. 2018).

25.4.2 Demand

Sharks and their habitat represent an important form of natural capital with high potential to produce economic value (Cardenas-Torres 2006; Rodríguez-Dowdell et al. 2007). As a demand for diving with bull sharks, 84% of divers pointed to sharks as the most attractive animals for sightings, which shows a clear increase in the number of visitors attracted by shark encounters since 2010 where just 33% of divers were mainly interested in sharks (Boncheva et al. 2010). This increase has also been observed in other sites, such as Revillagigedo where sharks represent 27% of the preference for sightings (Ruiz-Sakamoto 2015). During our study, the number of tourists who traveled to CPNP, specifically to dive with bull sharks, represented 22.8% of its total tourists during 2016. This trend is comparable to the divers' percentage who visit other national parks, such as the Revillagigedo Archipelago, in which 23% of the people went specifically to dive with sharks (Ruiz-Sakamoto 2015). In Fernando de Noronha Archipelago, Brazil, 23% of the tourists interviewed became interested in shark diving after they arrived, demonstrating that this activity carries the potential to expand in the archipelago (Pires et al. 2016). Also in Playa del Carmen, Mexico, it is estimated that during 2012, 14 000 dives for bull shark sighting were scheduled (CONANP and PROCODES 2014). In another study using the TCM methodology, 143 tourists were interviewed in Los Cabos (70 miles south of CPNP); in that survey, 6.29% (9 of 143) of those tourists visited CPNP specifically for scuba diving where tourists were willing to pay for the entire trip just for the opportunity to dive in Mexico by flying to Los Cabos but without having Los Cabos as their final destination (Morales-Zarate et al. 2019).

During our study, 32% of the divers had their first shark dive in CPNP. Similarly, 36% of people changed their perception toward sharks in a positive way after the diving experience, which is a very significant result for the park because this change in perception might promote future visits and recommendation to carry out the activity again. We found that the element, reflected in our results, was the main factor for people to come to CPNP because of the recommendation from previous divers' experiences.

In this study, 62% of tourists rated diving in the park as excellent. Different authors have indicated that satisfaction can be associated with willingness of tourists to recommend the activity (Mustika 2011). In this sense, participants' motivation and satisfaction experience should be considered for the sustainable development of the industry (Dicken and Hosking 2009; Ziegler et al. 2012).

25.4.3 Economic Valuation by the Cost of Travel Method (MCV)

The recreational value of a species depends critically on the motives for targeting that particular species. Those species, which are sought mainly for recreational purposes, have a higher value than those targeted for eating purposes (Wheeler and Damania 2001). In the results obtained by the model, the variables cost of travel, first time diving with sharks, income, number of sharks sighted, and chumming were significant. The aforementioned coincides with the economic literature and other studies where travel distance, income, and length of stay were predictors for tourists' daily expenses (Pires et al. 2016), where increase in price has a negative effect on demand, so it gets a negative algebraic sign (Haab and McConnell 2002; Labandeira et al. 2007; Pires et al. 2016). Similarly, provisioning was highly rejected by tourists, which means that one of the special characteristics of bull shark diving at CPNP was the opportunity to meet sharks in their environment with the least possible disturbance and in the most natural way possible. Therefore, bull shark diving without the use of provisioning is one of the most important aspects for tourists.

Consumer surplus in CPNP was $694 USD per person per day, which ultimately resulted in a direct use value of $7 595 097 USD annually. These results were consistent with other studies in Mexico where this methodology has been used to assess ecosystem services (Table 25.8). In Mexico, the economic value of the manta ray (*Mobula birostris*) population at Revillagigedo Archipelago has been valued at $14 111 414 USD annually (Ruiz-Sakamoto 2015). In 2011 in global terms for Mexico, shark sighting was estimated to have generated USD 12 412 000 by 2013 only taking into account the cost of the sighting activity (Cisneros-Montemayor et al. 2013). However, the cost of sighting only increased 23% from 1999 to 2010 (Gallagher and Hammerschlag 2011).

In other places like Fernando de Noronha Archipelago in Brazil, the total recreational use was estimated at $90.2 million USD annually (TCM was used), where shark provided 4% of the total economic benefit within the tourism industry ($2.64 million USD) (Pires et al. 2016). Similarly various other countries have shown an increase in benefits from shark tourism, such as tiger shark diving at Aliwal Shoal Marine Reserve in Africa, which generates $1 055 139 USD annually (Dicken and Hosking 2009). Another case is in Seychelles where the potential value of whale shark (*Rhincodon typus*) estimate is up to $4.9 million USD for a 14-week season (Rowat and Engelhardt 2007), which has also been happening with other types of sightings, such as in Lovina, Bali, Indonesia where 37 000 tourists generate $9.5 million USD in dolphin sightings annually (Mustika 2011). At mid-Atlantic Azores, the estimated annual spill of shark diving in 2014 amounted to $2 244 890 USD; also about 70% of the participants were willing to pay an additional amount (up to $74 USD) to ensure that shark diving remains an option, and more than half (53%) would like to invest that amount in conservation (Torres et al. 2017). Thus, some countries have opted to drastically reduce shark fisheries and in some cases transformed completely to ecotourism activities; such is the case of Palau where shark diving is an important contributor to the economy, generating $18 million USD per year and representing approximately 8% of the country gross domestic product (Vianna et al. 2012). Another example is Galápagos where the tourism industry generated $418 million USD in 2006, of which $120.5 million USD correspond to marine tourism (Medina et al. 2012) and 38.9 from shark diving (Terán-Sevilla 2017). Since then, the total income from ecotourism (i.e., the gross national product income) increased by an estimated 78% from 1999 to 2005, placing Galapagos among the fastest-growing economies in the world (Taylor et al. 2006). In general terms, shark diving worldwide generates $314 million USD per year, directly supporting 10 000 jobs (Cisneros-Montemayor et al. 2013).The services from ecological systems and natural capital resources are critical to the operation of the earth life support. Using environmental resources for recreational purposes is likely to provide significant economic benefits to many levels of human societies (Pires et al. 2016), which contribute to human well-being in direct and indirect manners, therefore representing part of the total economic value of the planet (Costanza et al. 1997).

25.5 Conclusions

The economic benefits from the ecosystem service of diving with bull sharks in CPNP are positive and affect the economy of the community of Cabo Pulmo significantly. The TCM is a useful tool for assessing ecosystem services, which are based on revealed preferences. It allows us to assign an economic value to outdoor activities, such as diving with bull sharks. Through this method, the consumer surplus per visitor at CPNP was estimated at $694 USD per day with an average of 4 days per stay and $7 595 097 USD annually as added value of economic benefits. These values are consistent with those reported for other ecosystem services in the state.

Just as other shark species, bull sharks and their habitat represent an important form of natural capital with high potential to produce economic value. This study generated a baseline of information about the economic importance that implicates diving with bull sharks in CPNP. It demonstrated that from all the activities that can be carried out in CPNP, 22% of the tourists declared bull shark diving as the main activity. Furthermore, the importance of conservation and improvement in this ecosystem service, despite being a rather recent activity, is growing in popularity and attracting tourism from all over the world. Although the reef is the main attraction of the park, this result is relevant because diving with bull sharks in natural conditions has become an attraction that will increase in the next years, if proper management is carried out

Lesson learned: The travel cost assessment study provided a reference value, as a guide to make decisions in the design of proposals on the implementation of fees to access diving with the bull shark. At the same time, this study provided empirical information, useful in the development of public policies for bull shark conservation and management of CPNP.

Acknowledgments This study was developed with the economic support of the CONACYT scholarship number 589333 to C. Pasos-Acuña. The study was financed by Pelagios Kakunja, A.C., Paul M. Angell Family Foundation and Alianza WWW-Telmex Telcel. The authors thank Diana Fischer for English language edition, Abel Trejo Ramírez for image edition and the local community of Cabo Pulmo for their support during the surveys.

References

Aburto-Oropeza O, Erisman B, Galland GR et al (2011) Large recovery of fish biomass in a No-take Marine Reserve. PLoS One 6(8):e0023601. https://doi.org/10.1371/journal.pone.0023601

Acuña-Marrero D, Cruz-Modino R d l, Smith ANH et al (2018) Understanding human attitudes towards sharks to promote sustainable coexistence. Mar Policy 91:122–128. https://doi.org/10.1016/j.marpol.2018.02.018

Apps K, Lloyd D, Dimmock K (2015) Scuba diving with the grey nurse shark (*Carcharias taurus*): an application of the theory of planned behaviour to identify divers beliefs. Aquat Conserv 25:201–211. https://doi.org/10.1002/aqc.2430

Asúnsolo Rivera A (2016) Distribución, abundancia y riqueza de especies de tiburones del arrecife de Cabo Pulmo, B.C.S. México. Bachelor thesis, Universidad Autónoma de Baja California Sur. Mexico

Azqueta D (2002) Introducción a la Economía Ambiental, 2nd edn. McGraw-Hill Interamericana, Madrid

Benítez-Arce NE (2013) El proyecto cabo pulmo: análisis de caso sobre el turismo sustentable durante el periodo 2006–2012. Bachelor thesis, Universidad Nacional Autónoma De México. http://132.248.9.195/ptd2013/octubre/0703591/Index.html

Boncheva AI, López Fletes C, Ivanova Boncheva A (2010) Ictiofauna como recurso ecoturístico en Cabo Pulmo B.C.S. TuryDes 3(7):16. http://www.eumed.net/rev/turydes/07/lfb-resum.htm
Bruskotter JT, Wilson RS (2014) Determining where the wild things will be: using psychological theory to find tolerance for large carnivores. Conserv Lett 7(3):158–165. https://doi.org/10.1111/conl.12072
Cameron C, Trivedi P (2005) Microeconometrics. Methods and applications. Cambridge University Press, Cambridge. https://www.cambridge.org/9780521848053
Cardenas-Torres AN (2006) Valoración económica de la actividad recreativa con tiburón ballena y su relación con la calidad del hábitat en Bahía de los Ángeles, Baja California. Master's thesis, Universidad Autónoma De Baja California, Mexico. http://fcm.ens.uabc.mx/~enriquez/complementos/productos/Cardenas,N.2006TiburonBallena.pdf
Cisneros-Montemayor AM, Barnes-Mauthe M, Al-Abdulrazzak D et al (2013) Global economic value of shark ecotourism: implications for conservation. Oryx 47(03):381–388. https://doi.org/10.1017/S0030605312001718
CONANP (2006) Programa de Conservación y Manejo Parque Nacional Cabo Pulmo. México, D.F. https://simec.conanp.gob.mx/pdf_libro_pm/111_libro_pm.pdf
CONANP and PROCODES (2014) Manual de Buenas Prácticas para el buceo con tiburón toro (*Carcharhinus leucas*) en Playa del Carmen. http://www.savingoursharks.org/wp-content/uploads/2019/01/Manual-de-Buenas-Practicas-Para-el-Buceo-con-Tiburon-Toro.pdf
Costanza R, Arge R, Groot R et al (1997) The value of the world's ecosystem services and natural capital. Nature 387(May):253–260. https://doi.org/10.1038/387253a0
Crossley R, Collins CM, Sutton SG et al (2014) Public perception and understanding of shark attack mitigation measures in Australia. Hum Dimens Wildl 19(2):154–165. https://doi.org/10.1080/10871209.2014.844289
Cubero-Pardo P, Herrón P, González-Pérez F (2011) Shark reactions to scuba divers in two marine protected areas of the Eastern Tropical Pacific. Aquat Conserv 21(3):239–246. https://doi.org/10.1002/aqc.1189
Dearden P, Ziegler J (2007) Tourist interactions with sharks. In: Higham J, Luck M (eds) Marine wildlife and tourism management. CABI, Oxford, pp 66–89
Dicken ML, Hosking SG (2009) Socio-economic aspects of the tiger shark diving industry within the Aliwal Shoal Marine protected area, South Africa. Afr Mar Sci 31(2):227–232. https://doi.org/10.2989/AJMS.2009.31.2.10.882
Gallagher AJ, Hammerschlag N (2011) Global shark currency: the distribution, frequency, and economic value of shark ecotourism. Curr Issue Tour 14:797–812. https://doi.org/10.1080/13683500.2011.585227
Garrod B, Wilson JC (2004) Nature on the edge? Marine ecotourism in coastal peripheral areas. J Sustain Tour 12:95–120. https://doi.org/10.1080/09669580408667227
Haab TC, McConnell KE (2002) Valuing environmental and natural resources. Edward Elgar Publishing, Cheltenham. https://doi.org/10.4337/9781843765431
Hernández-Trejo V, Avilés-Polanco G, Almendarez-Hernández MA (2012a) Beneficios económicos de los servicios recreativos provistos por la biodiversidad acuática del Parque Nacional Archipiélago Espíritu Santo. Estudios Sociales 20(40):156–177. http://www.scielo.org.mx/pdf/estsoc/v20n40/v20n40a6.pdf
Hernández-Trejo VA, Ponce-Díaz G, Lluch-Belda D, et al (2012b) Economic benefits of sport fishing in Los Cabos, Mexico: is the relative abundance a determinant? WIT Trans Ecol Environ 161:165–176. https://doi.org/10.2495/ST120141
Huveneers C, Rogers PJ, Beckmann C et al (2013) The effects of cage-diving activities on the fine-scale swimming behaviour and space use of white sharks. Mar Biol 160(11):2863–2875. https://doi.org/10.1007/s00227-013-2277-6
Instituto Nacional de Ecología (2003) Demanda Turistica Del Parque Marino Cabo Pulmo. http://cabopulmovivo.org/portal/wp-content/uploads/2014/04/3.-cabo_pulmo.pdf
Labandeira X, León C, Vázquez MX (2007) Economía Ambiental. Pearson Educación. Madrid. https://doi.org/10.1007/s13398-014-0173-7.2

Laroche RK, Kock AA, Dill LM et al (2007) Effects of provisioning ecotourism activity on the behaviour of white sharks *Carcharodon carcharias*. Mar Ecol Prog Ser 338:199–209. https://doi.org/10.3354/meps338199

Medina C, Aravena C, Vásquez F (2012) Valoración económica de la conservación de tiburones en la Reserva Marina de Galápagos (No. WP34). Working Paper. Galápagos. https://www.researchgate.net/publication/291970999_Valoracion_economica_de_la_conservacion_de_tiburones_en_la_Reserva_Marina_de_Galapagos

Mendes I, Proenca I (2011) Measuring the social recreation per-day net benefit of the wildlife amenities of a National Park: a count-data travel-cost approach. Environ Manag 48:920–932

Morales-Zarate MV, Almendarez Hernández MA, Sánchez Brito I et al (2019) Economic valuation of recreational beach ecosystem service at Los Cabos, Baja California Sur (BCS), Mexico: an application of travel cost method. El Periplo Sustentable 36:23. https://www.researchgate.net/publication/334067691_ArticuloEconomic_valuation_of_recreational_beach_ecosystem_service_at_Los_Cabos_Baja_California_Sur_BCS_Mexico_An_application_of_Travel_Cost_Method

Mustika PLK (2011) Towards Sustainable Dolphin Watching Tourism in Lovina, Bali, Indonesia. Tourism. Phd thesis, James Cook University.New Zealand https://researchonline.jcu.edu.au/29750/1/29750_Mustika_2011_thesis.pdf

Nyangwe C (2005) An application of travel cost method in the valuation of recreational properties. Master's thesis, Nairobi University, Kenia.

O'Bryhim JR, Parsons ECM (2015) Increased knowledge about sharks increases public concern about their conservation. Mar Policy 56:43–47. https://doi.org/10.1016/J.MARPOL.2015.02.007

Pires NM, Garla RC, Carvalho AR (2016) The economic role of sharks in a major ecotourism archipelago in the western South Atlantic. Mar Policy 72:31–39. https://doi.org/10.1016/j.marpol.2016.06.016

Prayaga P, Rolfe J, Stoeckl N (2010) The value of recreational fishing in the Great Barrier Reef, Australia: a pooled revealed preference and contingent behaviour model. Mar Policy 34(2):244–225. https://doi.org/10.1016/j.marpol.2009.07.002

Rivera-Castañeda P (2002) Valoración económica del servicio ambiental reacreación en Bahía de los Angeles, Baja California. Master's thesis, Centro de Investigación Científica y de Educación Superior de Ensenada, Mexico http://biblioteca.cicese.mx/catalogos/tesis/ficha.php?id=15301

Rodríguez-Dowdell N, Enríquez-Andrade R, Cárdenas-Torres N (2007) Property rights-based management: whale shark ecotourism in Bahia de los Angeles, Mexico. Fish Res 84(1):119–127. https://doi.org/10.1016/j.fishres.2006.11.020

Rolfe J, Prayaga P (2007) Estimating values for recreational fishing at freshwater dams in Queensland. Aust J Agric Resour Econ 51:157–174. https://doi.org/10.1111/j.1467-8489.2007.00369.x

Rowat D, Engelhardt U (2007) Seychelles: a case study of community involvement in the development of whale shark ecotourism and its socio-economic impact. Fish Res 84(1):109–113. https://doi.org/10.1016/j.fishres.2006.11.018

Ruiz-Sakamoto A (2015) Estimación del valor económico total y catálogo de foto identificación de la manta gigante (*Manta birostris* Walbaum, 1792) en el Archipiélago Revillagigedo. Bachelor thesis, Universidad autónoma de Baja California Sur. Mexico

Taylor JE, Hardner J, Stewart M (2006) Ecotourism and economic growth in the Galapagos: an island economy-wide analysis (No. 06–001). Working Paper. California. https://escholarship.org/uc/item/3wc975kp

Terán-Sevilla PT (2017). Valoración económica del servicio estético y recreativo de los tiburones endémicos del Parque Nacional Galápagos para la actividad turística de buceo. Bachelor thesis, Pontificia Universidad Católica del Ecuador. Ecuador. http://repositorio.puce.edu.ec/bitstream/handle/22000/12677/TESIS27-01-2017.pdf?sequence=1&isAllowed=y

Torres P, Bolhão N, Tristão da Cunha R et al (2017) Dead or alive: the growing importance of shark diving in the Mid-Atlantic region. J Nat Conserv 36:20–28. https://doi.org/10.1016/j.jnc.2017.01.005

Tourkolias C, Skiada T, Mirasgedis S, Diakoulaki D (2015) Application of the travel cost method for the valuation of the Poseidon temple in Sounio, Greece. J Cult Herit 16:567–574

Turner K, Pearce D, Bateman I (1993) Environmental economics: an elementary introduction. Harvester Wheatsheaf, Nueva York/London

Vianna GMS, Meekan MG, Pannell DJ et al (2012) Socio-economic value and community benefits from shark-diving tourism in Palau: a sustainable use of reef shark populations. Biol Conserv 145(1):267–277. https://doi.org/10.1016/j.biocon.2011.11.022

Wheeler S, Damania R (2001) Valuing New Zealand recreational fishing and an assessment of the validity of the contingent valuation estimates. Aust J Agric Resour Econ 45(4):599–621

Ziegler J, Dearden P, Rollins R (2012) But are tourists satisfied? Importance-performance analysis of the whale shark tourism industry on Isla Holbox, Mexico. Tour Manag 33(3):692–701. https://doi.org/10.1016/j.tourman.2011.08.004

Chapter 26
Socio-ecological Effects of Government and Community Collaborative Work with Local Development in a Natural Protected Area

Elizabeth Olmos-Martínez, Juan Pedro Ibarra-Michel, and Mónica Velarde-Valdez

Abstract A successful example of collaborative work and smart use of natural resources in a Protected Natural Area (PNA) is an ecotourism cooperative in "El Verde Camacho," a small conservation site close to Mazatlán, México. This cooperative is formed by community members of the small village of "El Recreo," who have been involved in this project for nearly 7 years under the sponsorship and advisement of government officials. The result of this collaboration has been clear and positive, since it has generated socioeconomic development from the conservation, preservation, and management of the coastal wetland and endangered species, such as the olive ridley sea turtle.

Keywords Ecotourism · Community · Collaborative work · Local development

26.1 Introduction

Rural areas with natural ecosystems have been transformed into mosaics that contain a great diversity of land uses, including ecotourism. The challenge of conserving biodiversity in these heterogeneous landscapes has resulted in a growing interest in promoting and managing conservation initiatives run by the local population.

In that sense and according to Pinkus-Rendon and Pinkus-Rendón (2015), ecotourism is perceived with the ability to contribute to the generation of jobs for the populations that inhabit the natural reserves, together with the promotion of conservation and use of natural and cultural resources. The authors mention that it is the

---

E. Olmos-Martínez (✉) · J. P. Ibarra-Michel · M. Velarde-Valdez
Departamento Académico de Ciencias Económico Administrativas, Universidad Autónoma de Occidente, Mazatlán, Sinaloa, Mexico

A. Ortega-Rubio (ed.), *Socio-ecological Studies in Natural Protected Areas*,
https://doi.org/10.1007/978-3-030-47264-1_26

reason why a multiplication of Protected Natural Areas (PNA) was generated where ecotourism is practiced as a tool to achieve sustainable development.

According to the National Commission of Protected Natural Areas of Mexico (CONANP), ecotourism, unlike mass tourism, must be more sensitive and cause less negative effects on the environment and local culture while contributing to the conservation and sustainable development of the PNA and areas with other conservation modalities, as an economic alternative that benefits communities and local users (CONANP 2006). In this sense, government institutions have defined lines of action for the development of ecotourism projects in PNA through national and state programs, with a sustainable approach to responsible, equitable and beneficial use, which strengthens ecotourism (CONANP 2000).

Likewise, the World Tourism Organization (UNWTO) (2002) mentions that ecotourism must contribute so that the tourism sector as a whole is more sustainable, increasing the economic and social benefits for the host communities, actively favoring the conservation of resources natural and cultural integrity of the host communities and increasing the awareness of travelers regarding the conservation of natural and cultural heritage.

In addition, ecotourism is an activity that is based on the principles of stability in local and global development in its permanent evolution in the long term, in equity, in the distribution of its benefits, and also in the forecast and effective control of undesirable impacts on the economy, culture, natural and social environment of the recipient (Ayala 2006).

It is known that many of the community entrepreneurship projects in rural areas are the product of social or regional development programs created by the state in order to alleviate poverty (Bartra Vergés 2000). Against this background, social groups composed of farmers, housewives, or people without basic instruction, who have decided to collectively create and manage a business organization, are involved in projects (Fernández Aldecua et al. 2012).

On the other hand, one of the problems that hinders the consolidation of ecotourism activity is the low participation by the local community, so there is a need to strengthen mechanisms for the consolidation of community sovereignty through self-management and self-organization (Valencia 2008). Therefore, these organizations must acquire a vision of solidarity economy and cooperation, since there is the possibility that the population that inhabits the territory will gain control, generate benefits to their associates, and guarantee the maintenance of the resource system in the time (Fajardo Gómez 2010).

Given the above, the present research aims to examine the livelihoods and socio-ecological relations of the population of El Recreo, specifically the actors belonging to the El Verde Camacho Ecotourism Cooperative that develop their activity in the protected natural area of El Verde Camacho Sanctuary known for local development under a scheme of social and governmental cooperativism.

The methodology used is mixed (qualitative-quantitative), which allowed the description, analysis, interpretation, and argumentation of the information obtained on the livelihoods and socio-ecological relations of the population of El Recreo concerning ecotourism activity. Surveys were applied to cooperatives and collaborators. The information was reinforced by interviews with members of the steering committee and the associate technician from the administrative offices of the PNA, and a meeting was held with members of the El Recreo community.

The results of this research allow us to affirm that ecotourism that is developed by tourism cooperatives can have commitment toward conservation as well as an economic benefit. The basic aspect of socio-economic organizations can generate not only promote socially important programs but provide economic benefits to the participants that contribute to local development. The collaboration established by the Ecotourism Cooperative El Verde Camacho between the different levels of government and the participation of the inhabitants of the El Recreo community have created a network of collective action for the benefit of the community and conservation of the PNA, which has allowed this case to become a positive reference for collaborative work.

26.2 Background

The worldwide concern for conserving natural areas highlighted by their fragility and biodiversity has grown. For example, the areas of community management have taken great relevance at international level (Galvin and Haller 2008); thus the existence of local institutions that manage the access and use of natural resources, including for ecotourism activity, is of great relevance (Harris 2009). In the case of Mexico, this institution is the National Commission of Protected Natural Areas (CONANP).

Ecotourism stands out for responsible trips to natural areas that help conserve the environment and improve population conditions (TIES 2006b); In addition, it fosters environmentally sustainable local development and community participation in a way that takes advantage of social and environmental potential (Borrini-Feyerabend et al. 2004).

Community tourism and ecotourism privilege local populations and urge operating companies and agencies to take effective measures to integrate and share the benefits generated by tourism activities (Vargas del Río and Brenner 2013).

According to Drumm and Moore (2005), The International Ecotourism Society (TIES 2006a), and the European Commission (2002), ecotourism is considered as the fastest growing tourism segment; previous data shows that for more than 15 years, ecotourism has grown rapidly worldwide.

In addition to the above, according to Camacho-Ruiz et al. (2016), ecotourism is a market segment with the highest growth in the world and with the greatest potential to integrate the dimensions of sustainability since it is an activity that produces a minimum impact on the environment and relates aspects of environmental education aimed at conservation, understanding and appreciation for the environment.

In Mexico, ecotourism has experienced an annual growth rate of 25% in the last two decades, according to the National Commission for the Knowledge and Use of Biodiversity (2019). Also, according to SECTUR (2019) in recent years, the federal government has developed public policies to incorporate sustainability strategies into Mexico's tourism activity, including the creation of PNA (Protected Natural Areas), which has led to sustainable tourism. In these areas, growth has reached 257% in the period from 2012 to 2018.

It is not uncommon to find ecotourism projects in PNAs with high levels of biodiversity thanks to the involvement and sustainable management of the natural resources of the local population and with some market potential as they are located near established tourist centers (Vargas del Río 2010), such is the case of Ramos-García et al. (2017) who conducted a study on community management for ecotourism in the Tortuguero El Verde Camacho Sanctuary, which is 22 km from the destination of consolidated tourism of Mazatlan, Sinaloa, Mexico. For these types of projects, one of the strategies is that the local population voluntarily adheres to the concept of ecotourism by sharing interests such as protecting wildlife and obtaining economic benefits and supports to generate or improve the local tourism offer, according with Borrini-Feyerabend et al. (2004).

Some of the research that has been done locally to evaluate the benefits generated by ecotourism, both for conservation and for the complement of productive activities in rural communities, indicate positive results. Such is the case of the study conducted by Hernández et al. (2005) in the Lacandona jungle of Chiapas, in which ecotourism contributed complementary gains to those obtained in all its productive activities, for example, to the boatmen, employees, and owners of restaurants and cabins with 41%, 61% and 66% respectively. It also represented a new form of organization that allowed them to learn to manage resources collectively. The authors also refer to another successful example of community ecotourism, which has been developed in Capirona, a Quehua community in Ecuador, where its success was reflected in the diversification of the economy with little capital, with minimal ecological impact and control on total resources, together with the involvement of indigenous organizations for training, advice, and promotion.

In another study conducted in the Yucatan Peninsula, Santana et al. (2013) addressed the issue of life strategies of domestic groups and the economic contribution derived from ecotourism, and the results demonstrated that ecotourism is a complementary activity for daily per capita income. In the Yucatan study, the authors also mentioned that public policies are required that focus attention on the factors that influence the income of the population and local life strategies. Furthermore, they found that from the 1990s, government agencies and non-governmental organizations (NGOs) granted economic support through public policy for ecotourism projects in rural communities in the Mayan area (Brown 1999, cited in Santana et al. 2013), which helped to develop the implementation of projects that provided them with jobs and income, although it is known that the economic benefits have not yet reached the community so that public policy has failed to comply with the objective of making ecotourism a means to strengthen the standard of living of the population.

For its part, Fajardo (2010) conducted a study in the Sierra Nevada of El Cocuy and Güicán, Colombia, on collective action as a scenario for the development of ecotourism evaluating the capabilities of the cooperative and its contribution to the development of the activity. The results show that cooperativism and ecotourism are complementary and that one leads to the other, in addition, both imply social responsibility. Also, it was found that environmental education is a component of which the partners have appropriated and that they always try to implement in the provision of services. In this way, they handle the concept of environmental interpretation and put it into practice as a means to generate awareness in the visitor.

Continuing along the same lines, Fernández et al. (2012) conducted a study in Bahías de Huatulco, Oaxaca, Mexico, where they analyzed five social companies dedicated to alternative tourism, in rural contexts with economic and social objectives. The analysis has a business approach because the main problem faced by ecotourism cooperatives is the lack of knowledge and skills to function as companies and manage as entrepreneurs. The partners of these social enterprises come from economic activities of the agricultural sector, or from the unpaid sector (housewives, elderly people). Therefore, the lack of human capital or the lack of enrichment of this represents one of its greatest challenges to continue existing as social enterprises. However, in the specific case of Bahías de Huatulco, federal government authorities and some local NGOs have struggled to give it a sustainable destination image attractive to ecotourists. In that sense, Bahías de Huatulco has been recognized nationally as one of the success stories in the implementation of the Agenda 21 Local program in Mexico and has also been distinguished with the Green Globe environmental certification in 2005 (Manuel Aragón 2011).

However, there are not very encouraging experiences for the benefit of social development, such is the case of the study conducted by Pinkus-Rendon and Pinkus-Rendón (2015) on the reality of development from ecotourism in the Ría Celestún biosphere reserve, in the Yucatan Peninsula who, based on semi-structured interviews with ecotourism service providers, found that there are differences in the perception of the conservation of natural resources depending on the training received, which prevented them from carrying out and promoting activities that would not harm natural resources or make proper use of them when performing ecotourism activities, coupled with the lack of coordination between the various authorities, negatively influencing the sustainable management of the reserve and ecotourism cooperatives. The authors conclude that ecotourism can be a potential activity that provides better life opportunities for people in rural communities, but for that to happen, they must be more informed, sensitized, trained and, above all, committed to the development of their communities, in addition to having the required resources (labor, land, time, and economic funds) to carry out successful ecotourism.

Given the previous background, local tourism must propose a competitive offer that also meets the needs of tourists and contributes to the progress and well-being of its inhabitants (Morales 2006). According to Guerrero et al. (2008) to achieve economic and local development, an integrated quality management is necessary, that can offer a unique and original experience to the visitor while also satisfying the legitimate aspirations of the local population for living not only in a social, harmonious, and environmentally friendly development, but one that provides for their economic needs. The same authors mention that it is necessary to combine efforts with adequate management and administration of environmental, physical, human, and financial resources, through the creation of organizations that administer them, efficiently, for the benefit of the community. Therefore, from the institutional point of view, actors in charge of different functions must attend to plan, certify, accredit, and promote the activity of the sector.

26.3 Collaboration Strategies and Collective Action

The collaboration between different social actors and the creation of networks of collective action (lasting over time) allows to create opportunities for new interactions and therefore access to greater diversity of social and ecological resources; the above is essential to face threats, changes, and disturbances that the systems may face (Urquiza and Cadenas 2015; Espiner et al. 2017). Therefore, the expansion and consolidation of social networks, local, national, regional, and international, contribute to the increase of resilience by increasing links and diversity (Olsson et al. 2004).

Returning to the above, and including the concept of resilience, it can be understood that it is the relationship between a system and a specific environment based on its ability to react to the threats identified in it; in the case of partner systems-ecological, it can be indicated as a threat to both social and environmental disturbances. In that sense, studies on resilience in socio-ecological systems recognize that both ecological fragmentation and social fragmentation reduce the chances of system reaction, since isolated fragments lack connectivity decreasing access to social and ecological diversity (Urquiza and Cadenas 2015).

Furthermore, the relationship between local groups and organizations with greater territorial scope is also considered important, since the work strategies of these organizations tend to be more in line with local community conditions (Tompkins and Adger 2004). In addition, this allows for one to take advantage of the knowledge of the communities that are directly related to the environment (Espiner et al. 2017; Urquiza and Cadenas 2015). In this sense, relations between organizations and local actors can generate opportunities to access new resources or knowledge. Otherwise, socially excluded groups tend to be more vulnerable to various types of disturbances; this is because they lack social stabilizing links (such as access to services, support networks, government programs, etc.) (Cumming 2011).

In this context, the importance of vertical and horizontal collaboration is stressed, that is, the relationships, links and networks that are established to take advantage of common resources and knowledge, both among different actors in a community and with actors that are part of a higher level, but that have some kind of vertical relationship with the community (government offices, civil associations, universities, etc.) (Folke et al. 2005).

Likewise, it is necessary to emphasize that in the study of the environment-society relationship, the ability to act collectively is important. In that sense, collective action refers to coordination between individuals to achieve a common goal (Ostrom 1990). Access to natural resources requires coordination between individuals, mainly to achieve more sustainable and equitable access.

In relation to the above, the role of social enterprises and their collective capacity in search of a common goal requires a detailed review. In this regard, Peredo and McLean (2006) mention that social enterprises are a modality of social entrepreneurship; they are organizations created through a rational process based on social action with common expectations and goals. For these authors, the social adjective that entrepreneurship implies goals that go beyond individual interests. More akin to social entrepreneurship is practiced when a group of people have an objective of creating some kind of common good of social value.

The social value that is intended to be created in this type of enterprise consists of a series of collective social, cultural, and environmental benefits, in addition to the economic objectives. In this manner, a social enterprise of collective organizations is formed where a group of individuals and/or families manage these benefits based on the values of cooperation and solidarity, taking economic advantage of union bonds based on trust and reciprocity of their partners. In other words, they take advantage of social capital to create financial capital (Fernández Aldecua et al. 2012).

Therefore, in the ecotourism cooperatives, different scenarios can be generated in the receiving communities where a solid path can be formed for the improvement in the living conditions of the population, as well as paradoxical contradictions, depending on the external appropriation of resources by of tour operators, as well as the effort to meet tourist expectations even above local needs (Serrano-Barquín et al. 2010).

For what Peredo and Chrisman (2006) point out, cooperation, as the engine of collective action, is a valuable community resource to achieve local development. The authors mention that social capital and cooperation are, potentially, intangible assets that a community may or may not own but can only be achieved through its own historical processes.

Therefore, in villages where there is no minimum base of social capital or cooperative attitude, it is difficult to organize or manage a social enterprise successfully (Fernández Aldecua et al. 2012).

Collective action must then be understood as the sum of the dimensions generated by social interactions, which are particular in each case, and that help to solve in an organized way the problems derived from the shared use of some goods (Fajardo Gómez 2010).

In addition to the above, the dynamics of participatory processes are presented as a transversal component in all cooperative acts, so participation is an effective mechanism in the improvement of local development. If the interests and capacities of the inhabitants of a certain locality are taken into account, the development will become more effective, since it will be based on real needs, which contribute to improving the social and economic conditions that are closer to reality (Silva and Dávila 2002).

Finally, the UNWTO (2000) establishes that, in order to ensure tourism activities, it is necessary to encourage stable employment opportunities and guarantee the obtaining of income and social services for communities, strengthening collaboration between different sectors, through of cooperation models. Serrano-Barquín et al. (2010) point out that from the perspective of social phenomenon, tourism is linked to recreational activities that foster a dynamic relationship between tourists and local residents in order to foster understanding between social actors and their different cultures.

26.4 The Protected Natural Area of El Verde Camacho Sanctuary

The definition of a Protected Natural Areas (PNAs) is defined broadly as a section of land and/or sea for protection and conservation. In the case of Mexico, being a megadiverse country, it has a great variety of areas associated with species that need some protection regime (Ibáñez Pérez et al. 2019); such is the case of the El Verde Camacho Sanctuary, municipality of Mazatlan, Sinaloa, Mexico.

On October 29, 1986, El Verde Camacho Beach was declared, together with a group of beaches, reserve areas, and refuge sites for the protection, conservation, repopulation, development, and control of the various species of sea turtle; these included the places where they nest and spawn (DOF 1986). Subsequently, on July 16, 2002, PNA was declared under the category of Sanctuary as it presents adequate conditions of biodiversity, endemism, uniqueness, extent, and degree of conservation, which must be re-categorized according to their natural vocation and ecological importance, with the purpose of maintaining and developing institutional, human, and physical infrastructure capacities for its management and operation (DOF 2002).

Moreover, according to the Information Service on the online Ramsar site, the Playa Tortuguera El Verde Camacho was declared a Ramsar site on February 2, 2004, in accordance with the criteria established by the "Convention Relating to Wetlands of International Importance Especially as Habitat of Waterfowl" (Ramsar Convention, February 2, 1971), an international treaty of which Mexico is a party (Ramsar 2019). This wetland has white and red mangroves (Briseño Dueñas 2003; Briseño-Dueñas et al. 2002).

However, according to the representatives of the administrative office of the Sanctuary, to date, there is no management program for the study site; although they have different guiding instruments that mark the development of the PNA activities. This is the case of the Annual Operational Program (POA) which, for the year 2018 (and the previous ones), covers different conservation and management areas that allow them to administer and conserve the natural area, such as (1) surveillance (conservation and management of the biodiversity); (2) recovery of species at risk (conservation and management of biodiversity); (3) biological monitoring (conservation and management of biodiversity); (4) climate change (attention to the effects of climate change and reduction of gas emissions greenhouse effect); (5) conservation for development (conservation economics); (6) tourism and protected areas (conservation economics); (7) mainstreaming of public policies (strengthening of intersectoral coordination); (8) identity, communication, and dissemination (communication, education, culture, and social participation for conservation); and (9) participation (communication, education, culture, and social participation for conservation). Each of the themes has its objective, projects, activities, goals, and means of verification (SEMARNAT-CONANP 2018).

The study area borders to the northwest with Punta Gruesa, near the town of Marmol, to the southeast with Punta Cerritos, urban limit of the port of Mazatlan, to the west with the Pacific Ocean and to the east with marshes, estuaries, and the mouth of the River Quelite. It is located north of Mazatlan; has an approximate area of 6465 ha and 25 km in the coastal perimeter corresponding to 31% of the coastal extension of the municipality of Mazatlan, Sinaloa in Northwestern Mexico; and is the main habitat for the nesting of the golf turtle and for more than 20 thousand species of birds (Briseño Dueñas 2003; Briseño-Dueñas et al. 2002).

The area adjacent to the nesting beach has nine estuarine systems and marshes of variable size, which when temporarily connected to the sea provide a significant pulse of life force, such as nutrients to the adjacent coastal area. The largest coastal wetland is El Verde Camacho, fed by the seasonal flow of the Quelite River. This area provides a great amount of fresh, brackish, and marine water to the wetlands, which are used by many diverse species as a breeding area, as well for growth, shelter, and food, among which include resident and migratory birds, reptiles besides turtles, and other terrestrial mammals classified as endangered species.

The ichthyofauna diversity that migrates between coastal wetlands and a marine area is typically very high. For example, lobster, shrimp, octopus, oyster, marine mammals (sea lions and dolphins), and fish are some of the conspicuous species present in the marine area (Briseño Dueñas 2003; Briseño-Dueñas et al. 2002).

The core zone of the beach strip is delimited in the decree of the Sanctuary is typically outside federal property under the administration of the federal maritime-terrestrial zone. Outside the boundaries of the federal land maritime zone, land ownership corresponds to individuals. The main wetlands of Tortuguera El Verde Camacho Beach are concessioned to coastal fishing production cooperatives. The sanctuary is under the federal jurisdiction of the Ministry of Environment and Natural Resources (SEMARNAT), and the buffer zone is the jurisdiction of the municipality of Mazatlan (Briseño Dueñas 2003; Briseño-Dueñas et al. 2002).

26.5 Community of El Recreo and the Ecotourism Cooperative

The community of El Recreo is located in the municipality of Mazatlán, Sinaloa, Mexico. According to the Population and Housing Census conducted by the National Institute of Statistics and Geography (INEGI) for 2010,[1] it is reported to have a total population of 579 inhabitants, where 51% are women. The connectivity with the city of Mazatlan is by road-related dirt road; the main economic activity reported is agriculture, construction, and livestock; they have a drinking water network but not drainage; they also lack a street cleaning service, nor do they have paved streets.

[1] It is worth mentioning that the census carried out by the National Institute of Statistics and Geography (INEGI) is repeated over a period of 10 years, the closest to the present study is the one made in 2010, so the census is taken as a reference. However, an inter-census survey is carried out in the middle of the period, that is, for this period it was carried out in 2015, but with the limitation that it does not include data by location such as the census, it only includes data by state and municipality.

However, they have garbage collection and street lighting; it is also reported that they have a police force. Regarding stores in the community, only grocery stores are registered. Nor do they have a health center or a private doctor's office.

On the other hand, in the area of education, it is reported that they have schools that cover the basic levels of preschool, primary, and secondary. They also have a sports court (INEGI 2010). Likewise, the same source of information refers to that for the census year, 3% of the total population was illiterate, 36% were classified as economically active population, 35% were employed, and 7% were unemployed. Regarding health indicators, 62% of the total population is the right holder of the health service of the Mexican Social Security Institute (IMSS) or the Institute of Security and Social Services (ISSSTE); on the other hand, 40% have popular insurance. Likewise, the community of El Recreo registers 176 homes in total, of which 86% were registered as inhabited, 100% of them are private; the average of inhabitants by house is of 3.83; 84% of the houses have firm floor (different to the earth), 85% of them have electricity; and 84% have piped water.

According to information from the interviews, the Ecotourism Cooperative El Verde Camacho was founded in 2012, to date it has had two directives under the organization chart with President, Vice President, Treasurer, and members. The 17 founding members and the 13 current members are members of the EL Recreo community.

The ecotourism cooperative project arose from the need to organize due to the need to provide for a socioeconomic development that was effective for the Community of El Recreo, as well as the need to offer ecotourism products within El Verde Camacho Beach. Once the guidelines of the Sanctuary were declared, conservation and management gave different aspects among them the opportunity to offer organized ecotourism activity.

At the same time, in this area there has been a fishing cooperative that has been organized for this purpose for several years; from it the idea of registering an ecotourism cooperative emanated; some fishermen founded it by inviting other members of their families. In the current administration, some founding partners gave up and continued in the fishing activity, but the ecotourism cooperative resumed its strength with new members and with the valuable participation of women and youth in the community. Most of them are family members and have developed different projects for the benefit of the entire population.

In the same interviews, it is commented that the ecotourism cooperative has facilities for its operation in Playa Tortuguera El Verde Camacho; it has adequate infrastructure for monitoring work and field activities of ongoing research projects. The facilities consist of an incubation room for collected nests, a multipurpose room, a workshop for minor repairs, an energy generating plant, a kitchen-dining room, three rooms used as bedrooms (for members of the cooperative and some researchers) with sanitary services and showers, a cistern for rainwater collection, and a solar energy system, additional to the diesel generator that supports three air conditioners in the bedrooms and a refrigerator or refrigerator that works with domestic gas. The vehicles available are two vans, five ATVs, four kayaks, and an outboard motorboat, bicycles, for beach trips and beach entry and exit.

Environmental education programs are a component of the operational conservation program implemented at the El Verde Camacho Sanctuary. There is a visitor lounge for school children, community groups, and ecotourists. There is an interactive permanent showroom. In the sea turtle conservation and habitat restoration work, national and international volunteers are integrated in different tasks, including research (CONANP staff interview in June 2019).

In that context, the social, economic, and cultural value of the site has been determined to highlight its natural, landscape, and current and potential uses of these characteristics as a recreation site (in addition to the subsistence and commercialization extractive uses); such is the ecotourism case where activities such as sea turtle release, environmental education, hiking, kayaking, photography, beach, bird watching in the different habitats of the site, sport fishing, walks on the beach, and water sports (Briseño Dueñas 2003) stand out.

26.6 Materials and Methods

The present investigation is mixed (qualitative-quantitative), which allowed the description, analysis, interpretation, and argumentation of the information obtained on the livelihoods and socio-ecological relations of the population of El Recreo concerning ecotourism activity which is carried out in the PNA El Verde Camacho Sanctuary under a scheme of social and governmental cooperativism.

The information was collected from reality, allowing the development of an open method, based on the application of surveys to the members of the El Verde Camacho Ecotourism Cooperative and its main collaborators within the Community of El Recreo. Likewise, interviews were conducted with the President of the cooperative, as well as the Vice-president (former president) and the Associate Technician on behalf of the administrative offices of the PNA El Verde Camacho Sanctuary, which facilitated the contribution of information depending on the context and to support the study according to Tamayo and Tamayo (2003).

Given that the total number of current cooperative members is 13, plus 11 collaborators, it was decided to apply the survey questionnaire in census mode, so in total 24 surveys were applied. To this end, all the members were called to a meeting in the Community of El Recreo with the help of the President of the cooperative. This meeting was held in June 2019 at the facilities of the local Primary School. The meeting was attended by ten partners who were explained the purpose of the investigation and the survey was applied; for the rest of the partners and the main collaborators living in the same community, the survey was applied days after the meeting.

The survey includes qualitative and quantitative variables and was divided into five sections: (I) demographic data of the population, (II) social and economic welfare indicators, (III) the ecotourism cooperative (EL Verde Camacho Beach), (IV) conservation and management of the area natural protected (Playa El Verde Camacho), and (V) collaboration with government institutions. The maximum application time was 15 min. All the questions of the survey were closed, with several options of answer and/or dichotomous; some of the possible answers go in ranks and others in Likert scale with answers of ordinal level to know priorities, as recommended by Fernández de Pinedo (1982). Subsequently, the information was registered in a database for analysis and interpretation.

26.7 Results

26.7.1 Social and Economic Welfare

According to the results of the survey, the cooperative members, and the population of the Community of El Recreo that collaborate in the substantive activities, the group has a social profile with gender equity being 50% women and 50% men. Likewise, 46% of respondents are married, and 25% are single; the age of cooperatives 33% register an age between 36 and 45 years, 21% have more than 65 and 17% have between 18 and 25. Regarding the level of education of the respondents shows that 42% has a primary school level, with 33% secondary level, and 8% preparatory and undergraduate, with also 8% reporting having no formal schooling. Finally, the occupations data shows that 54% reported being a member of the cooperative, and thus they work locally, 38% said they are a housewife, and 8% reported being students; these last two groups are those who eventually collaborated with the cooperative in various jobs.

The data shows that in addition to the equity for carrying out the various activities in the cooperative, people over 65 and young people, including people who do not have educational instruction, as well as housewives seeking extra income, have an opportunity to integrate for the family, which indicates that collaborative work is inclusive and gives opportunity to people of all social characteristics.

Regarding the indicators of well-being, Table 26.1 shows that 42% are entitled to health services offered by the IMSS and 58% to popular health insurance. 100% reported that in their community, there are schools that cover the educational levels of kindergarten, primary, secondary, and high school. The monthly income per person was recorded, and we found that 75% of the cases had an income less than $ 4000 pesos, and the rest reports that their income varies between $ 4001 and $ 7000 pesos. In order to know what property they had access to, they were asked if they own a house (100%), have a gas stove (88%), have a refrigerator (83%), or have air conditioning equipment (58%), and also slightly more than half had a washing machine (54%). Less than 50% owned a car or a computer.

In addition to the above, another group of indicators was used to measure the degree of well-being of the inhabitants of the community that are related to the characteristics of the house they lived in (Table 26.1). In this case of the group studied, 92% reported that their homes have a concrete wall, only 8% report that it is of another material (among them it can be sheet and cardboard); likewise, 92% reported

Table 26.1 Welfare indicators of cooperative members and collaborators

Welfare characteristics	Percentage of the population
With health services	100
Monthly income per person	
Less than $ 4000.00	75
From $ 4001.00 to $ 7000.00	25
Property assets	
Own house 100	100
Gas stove 88	88
Refrigerator 83	83
Air conditioning equipment 58	58
Washing machine 54	54
Auto 17	17
Computer 8	8
Housing characteristics	
Cement wall	92
Wall of other material (sheet and cardboard)	8
Cement ceiling	92
Roof of other material (sheet and/or wood)	8
Municipal housing services	
Drinking water	100
Electricity	100
Garbage collection	100
Drainage and sewage	0
Number of inhabitants per dwelling	
5 people	42
4 people	21
3 people	8
2 people	16
1 person	13
Recreation and recreational activities	
Sports	38
Training and workshops	21
Recreational (visit to the beach and parks)	4
No activity	38

Own elaboration, with information from the surveys

that the roof of their house was made of cement, and like the previous case, 8% report that it was made of sheetrock and/or wood. Regarding the number of inhabitants per dwelling, most respondents commented that it amounts to five people (42%) and four people (21%). On the municipal services that the house has, it is reported that in 100% of cases, they have piped drinking water, electricity, and garbage collection. In contrast, 100% of cases reported not having drainage and sewer. Finally, respondents reported on recreational and recreational activities where 38% of them do some sport, and the same percentage reported not doing any activity.

26.7.2 The Ecotourism Cooperative as a Local Development Strategy

As previously mentioned, social action and collaborative work is an important tool to achieve common objectives that are related to economic, social, and environmental well-being. In the case of the Ecotourism Cooperative El Verde Camacho, it is necessary to know certain characteristics regarding networking, inclusion, and collaboration of the members and the local population of the community of El Recreo. This information should include data on family members who have also collaborated in the substantive actions of said cooperative. Table 26.2 shows that 83% of cooperatives and collaborators include their family in a range of 1–2 people, and to a lesser extent, it increases to involve up to more than 6 people of their family.

Regarding the time of participation in the cooperative, respondents show the following, 58% said they were collaborating in a time range of 1–3 years, 25% said they collaborated for more than 6 years, and 17% said they were collaborating for a range between 3 and 6 years. It is observed that cooperatives invite other relatives to join the work. There are also members who have collaborated for more than 5 years, which indicates that the linkage and collaboration has been beneficial for activities related to ecotourism and the conservation of the site as a fundamental element for nature tourism.

In the case of the responsibilities of the members of the cooperative and collaborators regarding substantive activities for the development of ecotourism and the conservation of the PNA, it was obtained that most of them carry out activities aimed at the release of turtles (83%), cleaning and maintenance of the beach (71%), maintenance work in the beach house (67%) (which they use to assist tourists in multiple activities), and activities related to the preparation of food for visitors (54%); on the other hand, below 50% of the responses, respondents report that they provide information and talks for tourists (46%), coordinate and manage ecotourism activities with the authorities (in this case with CONANP) (25%), coordinate and manage the resources of the ecotourism cooperative (17%), and plan and schedule the visits of tourists (13%) (Table 26.2). The foregoing shows that most cooperative members and collaborators have responsibilities in almost all ecotourism and conservation needs and/or activities, so they cover virtually all positions.

Table 26.2 Characteristics of collaboration between cooperative and community

Collaboration characteristics	Percentage of the population
Family members within the cooperative	
From 1 to 2	83
From 3 to 4	8
From 5 to 6	4
More than 6	4
Time of participation in the cooperative	
From 1 to 3 years	58
From 3 to 6 years old	17
More than 6 years	25
Responsibilities in the cooperative	
Turtle release	83
Cleaning and maintenance of the beach	71
Maintenance of the beach house	67
Food preparation	54
Information and talks for visitors	46
Coordination and management with authorities	25
Internal coordination and administration	17
Planning and agenda of visitors	13
Role of women and gender equality	
Women hold important positions within the cooperative	100
The voice of women is taken into account in decisions	75
The work is carried out equally between men and women	67
Women facilitate the training of diverse subjects	42

Own elaboration, with information from the surveys

On the other hand, all respondents reported that the role of women and gender equality are respected in the cooperative. All the members polled commented that women have important positions within the cooperative; the voice and vote of women is respected in decisions (75%); they also think that the jobs are performed equally well by both men and women (67%); and finally, they perceived that women facilitate training in diverse subjects (42%) (Table 26.2). In that sense, in the meeting with the cooperatives, the women were satisfied with their performance and work within the cooperative. Some woman commented that they had never participated before in any association, nor had worked in any private company, so the opportunity they were given in the cooperative not only helps them obtain extra income for their families, but helped them strengthen and grow personally.

On the other hand, according to the survey, the most common sites for the practice of ecotourism are the beach house, nesting sites, and the estuary with 96%,

75%, and 42% of perceptions, respectively. In that sense, in the interview with the authorities of CONANP (June 2019), it was commented that there are two ecotourism companies, Explora Caribe Tour and Pronatours, located in the tourist destination of Mazatlan that are responsible for promoting visits to the PNA El Verde Camacho in coordination with the president and vice president of the cooperative.

26.7.3 Visitor Reception

According to information from the interview with the authorities of CONANP and the visitation log of the Ecotourism Cooperative El Verde Camacho (June 2019), in the period from March 15, 2018 to April 25, 2019, a total of 932 people had visited the study area, of which 207 were young students from primary, secondary, and preparatory schools from different parts of the state of Sinaloa. These groups were received in coordination with CONANP staff in order to provide them with better management attention of their groups. In addition, a total of 693 adults (29 Nationals and 664 Foreigners (89% of the USA and 11% of Canada)) visited the area including with them were 32 children (all of them foreigners from the USA). The high season of visits was between the months of October and February, the time of sea turtle release, since this is the main tourist attraction of the site.

26.7.4 The Conservation and Management of Natural Resources in the PNA

The activities of conservation and management of natural resources are of fundamental importance to maintain the productive base and the essential ecological processes that guarantee the life of the different ecosystems. In that sense, regarding the activities and priority sites for conservation that are carried out in El Verde Camacho Beach, the following was found; 100% of the cases commented that activities related to the conservation of sites are carried out within the PNA turtle nesting sites, followed by 79% that stated that surveillance is carried out only in one of them. A total of 58% reported that two of the activities that are also carried out are landscape protection and solid waste management. In addition, 50% stated that activities related to the conservation of endemic species are also carried out, and the rest of the activities with responses of less than 50% are shown in Table 26.3. In this regard, it was observed that the bulk of activities were related to protection, surveillance, and conservation rather than maintenance, restoration, and reforestation. It should be mentioned that in an interview with CONANP staff, it was reported that, within the conservation and management strategies, the number of visitors was controlled to not exceed a maximum of 15 people at any one time according to the loading capacity of the site. CONANP staff also mentioned that they organized visitors according to educational and tourist groups for scheduled visits.

Table 26.3 Conservation and management of El Verde Camacho Beach

Management conservation and management in sites	Percentage of the population
Preservation and conservation activities	
Conservation of nesting sites	100
Surveillance	79
Landscape protection and solid waste management	58
Conservation of endemic species	50
Habitat rehabilitation	42
Mangrove reforestation	38
Maintenance and improvement of recreation sites	38
Signage of recreational sites	33
Elimination of invasive species (flora and fauna)	29
Soil protection and recovery	21
Delimitation of tourist trails	21
Coastal restoration	13
Environmental and natural resources conservation sites	
Turtle nesting area	79
Estero El Verde Camacho	75
Mangrove forests	54
Dunes	50
Shoreline	46
Tourist trails	42
Tropical deciduous forests	17

Own elaboration, with information from the surveys

Also, Table 26.3 shows the sites of environmental conservation and natural resources. For this case, the cooperatives mention that three of the most important sites for preservation and conservation activities are carried out where those of the turtle nesting area (79%), followed by El Verde Camacho estuary (75%), and mangrove forests (54%). These sites coincided with the areas where tourist activities are carried out, where the beach house (96%) is located, as well as the nesting sites (75%) and the estuary (42%).

On the other hand, in the context of collaboration and participatory actions between the community and government agencies, the participation of the members of the El Recreo community in the works and activities of preservation and conservation of the aforementioned sites was investigated in the first instance. In this regard, all the cooperatives mentioned that they are carried out by women and men from the El Recreo community, together with 83% of the responses that indicate that the children of that community are also involved and, in addition, 4% said that the population of other communities.

As already mentioned, collaborative work between social actors and government organizations is of vital importance for collective action to reach levels of sustainable management of natural resources, so participation was also investigated of the authorities, call government organizations of the three levels of government, in this regard the answers are as follows, 100% recognized the participation of federal government authorities, specifically CONANP, on the other hand, 33% said the authorities. Municipal level government agencies are involved in conservation and management activities, and 25% said they also involve state level government authorities. The foregoing indicates that the degree of involvement in the cooperative's work is of the whole family (adults and children) and makes it clear that the federal government is the body primarily involved in substantive work.

In this context, the level of commitment and participation in the conservation of the site (population and government authorities) was addressed; the answers try to rescue this level from a scale of greater to lesser commitment and awareness, in this regard, about the authorities; 46% mention that they get involved sometimes, and 38% report that they always get involved. In addition to the above, regarding community commitment, 79% mention that they are always involved (Table 26.4). In addition, they were asked if they consider that the conservation of El Verde Camacho Beach has improved thanks to the collaborative work between community and government; in that sense all the responses were positive and related to aspects such as there are more turtles, it is better the site and provide support and employment with conservation and surveillance programs, they pay us to conserve, there is no longer looting of species or logging of mangroves, the recovery of the sea turtle is observed and the government provides permits and resources.

Another important aspect is the certification of the clean beach in the Priority Conservation category for 6 km of beach that includes the Tortuguero El Verde Camacho Sanctuary, a work that has been achieved thanks to the coordinated work of authorities of the three levels of government and the active participation of the inhabitants of the community of El Recreo. For this purpose, sometimes the communities of Marmol (municipality of Mazatlan) and Llanitos (municipality of San Ignacio) are involved since they also have subsidies for days of cleaning and certification of beaches under the work programs of the PNA Meseta de Cacaxtla, located between the municipalities of San Ignacio and Mazatlán. This certification

Table 26.4 Level of commitment and conscientization of authorities and community

Actors	Level of involvement	Percentage of the population
Authorities	Always	38
	Almost always	17
	Sometimes	46
	Almost never	0
	Never	0
Community	Always	79
	Almost always	8
	Sometimes	13
	Almost never	0
	Never	0

Own elaboration, with information from the surveys

was achieved for the first time in 2012, and to date they have obtained it five consecutive times; recent work in the month of June validated them for the last certification given to them in September 2019 (interview with CONANP staff, June 2019).

26.7.5 Government Strategies for the Conservation of the PNA and Local Development

Local development is a dynamic process of expanding capacities (economic, cultural, social, and governmental) that allows working to improve the quality of life of all members of the population; the same goes for the conservation of natural resources in a territorial scenario. When it comes to local development based on ecotourism, development strategies must be based on territorial potential and natural, cultural, and social wealth. In view of the above, cooperativists were investigated about their collaboration and participation in federal government strategies, specifically with CONANP, related to the conservation of the PNA and that at the same time contribute to the local development of the community.

As a result of the above, 92% of the respondents stated that they had participated or collaborated in some government program, this group of people continued to be asked about the government programs where they had participated, and it was obtained that 95% of them have participated in the Conservation for Sustainable Development Program (PROCODES), followed by 64% who have participated in the Temporary Employment Program (PET) and 36% who have participated in the Social Inclusion Program (PROSPERA). In addition to the above, Table 26.5 shows other characteristics of the aforementioned programs; the first one refers to the number of times the study population has participated in them, where the response with the highest percentage is 1–2 times with 41% of cases. Secondly, and to highlight the influence and benefit that these government programs have left and the rapprochement with the community, on the number of beneficiaries, it was obtained that 64% say that they have benefited from 1 to 2 members of their family. Finally, to know the time in which the respondent has participated in these programs, 36% of them say they have participated more than 6 years. The above indicates that the most representative program is PROCODES, that the programs benefit more than one person from the families, and that the time in which people have participated in these programs is for several years, without a doubt indicating the benefit to the population.

In the same manner, the population of El Recreo (cooperatives and collaborators) was investigated about their participation in courses and workshops for the management and conservation of the PNA, as well as for the administration and management for the ecotourism cooperative, which, incidentally, they have also served as training for the personal improvement of the population involved.

In the first place, 88% responded affirmatively about their participation in management and conservation workshops, of which over 50% say they have taken courses in subjects such as turtles and their care (90%), environmental education (76%), inspection and surveillance (57%) (Table 26.6). Once again it highlights the importance of sea turtles both in the conservation of sites, the sites most visited by

Table 26.5 Strategies for local conservation and development

Participation in government programs	Percentage of the population
Number of times	
1–2 times	41
3–4 times	27
5–6 times	18
7–8 times	5
More than 8 times	9
Number of beneficiaries per family	
From 1 to 2 people	64
From 3 to 4 people	36
Time in which you participated	
1–3 years	36
From 4 to 6 years old	27
More than 6 years	36

Own elaboration, with information from the surveys

tourists and the training around it. It is also noted that the sites most visited by tourists are the beach house (96%), the nesting sites (75%), and the estuary (42%). Second, on the administration and management workshops for the ecotourism cooperative, 42% of the total respondents responded that they had taken one of them, of which Table 26.6 shows that all have taken workshops on gender equity and 50% on administration, and the rest of the topics are accounting, leadership, capacity development, group management, conflict resolution, and first aid.

Finally, in relation to the increase of well-being in the community from the collaborative work between the members of the cooperative (and its collaborators) and the federal government (CONANP), 88% of the respondents gave an affirmative answer. Because of this positive result of government and community collaboration, the following answers were also recorded: women who do not work have temporary income thanks to the cooperative and government programs; in addition thanks to the work provided, the economy of the community has increased, and this in turn has led to an increase in beach cleaning campaigns, resulting in less garbage and less looting of sea turtles. In summary, the work done here by the government with the community has led to families having personal and economic growth and a community that is family integrated.

26.8 Conclusions

The results of this research allowed us to affirm that ecotourism that is developed by tourism cooperatives that have a commitment toward conservation can benefit local communities. The basic aspect of social economy organizations can generate important social and economic benefits that can as shown by this study contribute to local development. The collaboration established by the Ecotourism

Table 26.6 Participation in training courses and workshops

Type of courses (subjects) and workshops	Percentage of the population
Management and conservation of the PNA	
Turtles and their care	90
Environmental education	76
Inspection and surveillance	57
Estuaries and conservation	48
Natural resources	38
Habitat	29
Solid waste	24
Biodiversity 19	19
Conservation and sustainable use of water	5
Environmental management and planning	5
Administration and management for the cooperative and ecotourism	
Gender equity	100
Administration	50
Accounting	40
Leadership	20
Capacity development	20
Group management	20
Conflict resolution	30
First aid	10

Own elaboration, with information from the surveys

Cooperative El Verde Camacho between the different levels of government and the participation of the inhabitants of the El Recreo community has created a network of collective action for the benefit of the community and conservation of the Protected Natural Area, which has allowed this case to become a positive reference for collaborative work.

The cooperative has had a positive impact on the well-being of the community, mainly due to the integration of inhabitants that are not members of the fishing cooperative. Thus, incorporating previously non-economically productive community members (at least not directly), such as older adults, youth, and housewives into the workforce, which contribute to the development of the locality. This is reflected in the better living conditions of the inhabitants, with incomes above the minimum wage and goods that are classified in the category of luxury for rural communities as is the case of the present study. Consequently, this has positively permeated in society those who have a positive and convenient perception that the work developed by the cooperative contributes to their well-being.

Regarding the strategies for local development promoted by the ecotourism cooperative, a strong link and collaboration was found between the partners and the local population, which has been favorable for the conservation of the PNA and for the contribution of tourism development while benefiting the entire community. Similarly, another strategy identified is responsibility and collaboration, which highlights activities such as turtle release, cleaning, and maintenance of beaches, among others. A gender strength is also identified, since women in the cooperative have important positions and responsibilities at the same level as men, and a broad respect for women's decisions is also identified, where women are perceived as important and relevant for the association.

Ecotourism if handled inappropriately can harm and generate negative impacts on PNAs; this type of tourism is not necessarily the solution when talking about the care and protection of natural resources. However, this research shows the positive aspects of ecotourism, among which are the conservation of nesting sites and endemic species, surveillance, landscape protection, solid waste management, as well as control of the carrying capacity for site visits. One of the most relevant actions related to the conservation of this area is the Clean Beach certification obtained in 2012, which has been endorsed five times and includes this year (2019).

In this sense, the close relationship of the cooperative with the Federal Government, specifically through CONANP, stands out as a fundamental element for the conservation of the PNAs. The commitment of government authorities to the population is highlighted, where it has positively permeated the mentality of the inhabitants of the community, since they openly attribute a strong relationship between conservation and collaborative work with the authorities. It highlights the participation of 92% of the population in various government subsidy programs such as PROCODES, PET, and PROSPERA. In summary, it was identified that the cooperative has contributed to local development, which has had a significant impact on improving the quality of life of the population's inhabitants. Likewise, the participation of CONANP has provided a fundamental support element for local social economy organization, conservation, restoration, as well as responsible use of natural resources.

26.8.1 Lessons Learned

This research study provides as its legacy a positive example of collaboration and goodwill between a local community and government. Both the community of El Recreo and the administrative and operational staff of the El Verde Camacho Sanctuary were always willing to provide the necessary information and had an attitude of interest in the research carried out. Specifically, the members of the cooperative were happy with their work, with their personal growth, and with the growth and social welfare of the entire community. It is worth highlighting that being a community with few inhabitants, almost all of them were family members, which implies a social and family union with common ends. The authors are especially grateful for all their attentive participation in this study.

References

Ayala H (2006) Contradicciones entre turismo, economía y ecología. Rev Turismo y Ecología 133(2):69–88

Bartra Vergés A (2000) Conciertos y desconciertos del desarrollo sustentable. Participación social y gestión en los programas regionales. En: Toledo C, Bartra A (Coords) Del círculo vicioso al círculo virtuoso. Cinco miradas al desarrollo sustentable de las regiones marginadas. Plaza y Valdés, PNUD, SEMARNAP, DFID, México, pp 57–116

Borrini-Feyerabend G, Kothari A, Oviedo G (2004) Indigenous and local communities and protected areas: towards equity and enhanced conservation. IUCN, Gland/Cambridge

Briseño Dueñas R (2003) Ficha informativa de los humedales Ramsar. Playa Tortuguera El Verde Camacho. Servicio de Información sobre Sitios Ramsar. Disponible en: https://rsis.ramsar.org/RISapp/files/RISrep/MX1349RIS.pdf?language=es [07/07/2019]

Briseño-Dueñas R, Guido-Sánchez S, Wong GP, Ríos Olmeda D (2002) Ordenamiento Ecológico y Plan de Manejo de la Franja Costera "Cerritos-Mármol", Mazatlán, Sinaloa, con énfasis en la Zona de Reserva para la Protección y Conservación de Tortugas Marinas "El Verde". Informe Final del Proyecto bajo convenio UNAM-CIAD

Camacho-Ruiz E, Carrillo-Reyes A, Rioja-Paradela TM, Espinoza-Medinilla EE (2016) Indicadores de sostenibilidad para el ecoturismo en México: estado actual. LiminaR 14(1):156–168. Recuperado en 16 de agosto de 2019, de http://www.scielo.org.mx/scielo.php?script=sci_arttext&pid=S1665-80272016000100011&lng=es&tlng=en

CONABIO (Comisión Nacional para el Conocimiento y Uso de la Biodiversidad) (2019) Sistemas productivos sostenibles y biodiversidad (SPSB), Ecoturismo. Disponible en: https://www.biodiversidad.gob.mx/corredor/SPSB/ecoturismo.html [16/08/19]

CONANP (Comisión Nacional de Áreas Naturales Protegidas) (2000) Programa de Manejo Complejo Insular del Espíritu Santo. Secretaría de Medio Ambiente y Recursos Naturales, México, p 194

CONANP (Comisión Nacional de Áreas Naturales Protegidas) (2006) Programa de turismo en áreas protegidas 2006–2012. Secretaría de Medio Ambiente y Recursos Naturales, México

Cumming GS (2011) Spatial resilience in case studies of SESs. In: Spatial resilience in social-ecological systems. Springer, Dordrecht, pp 205–229. Disponible en: https://doi.org/10.1007/978-94-007-0307-0_10

DOF (Diario Oficial de la Federación) (1986) Decreto por el que se determinan como zonas de reserva y sitios de refugio para la protección, conservación, repoblación, desarrollo y control, de las diversas especies de tortuga marina, los lugares en que anida y desova dicha especie. Gobierno de la República Mexicana, 29 de octubre de 1986

DOF (Diario Oficial de la Federación) (2002) ACUERDO por el que se determinan como áreas naturales protegidas, con la categoría de santuarios, a las zonas de reserva y sitios de refugio para la protección, conservación, repoblación, desarrollo y control de las diversas especies de tortuga marina, ubicadas en los estados de Chiapas, Guerrero, Jalisco, Michoacán, Oaxaca, Sinaloa, Tamaulipas y Yucatán, identificadas en el decreto publicado el 29 de octubre de 1986. Gobierno de la República Mexicana, 16 de julio de 2002

Drumm A, Moore A (2005) An introduction to ecotourism planning, vol 1. Arlington, The Nature Conservancy

Espiner S, Orchiston C, Higham J (2017) Resilience and sustainability: A complementary relationship? Towards a practical conceptual model for the sustainability–resilience nexus in tourism. J Sustain Tour 25(10):1385–1400

European Commission (2002) Using natural and cultural heritage to develop sustainable tourism in non–traditional tourist destinations. European Commission Tourism Unit, Luxemburgo

Fajardo Gómez CE (2010) La acción colectiva como escenario para el desarrollo del ecoturismo. Evaluación de las capacidades cooperativas de ASEGÜICOC y su contribución al desarrollo del ecoturismo en la Sierra Nevada de EL Cocuy y Güicán, Colombia. Revista Cooperativismo y Desarrollo 18(97):139–158

Fernández Aldecua MJ, Castillejos LB, Ramírez Luna JA (2012) Empresas sociales y ecoturismo en Bahías de Huatulco, México. Estudios y Perspectivas en Turismo 21:203–224
Fernández de Pinedo I (1982) Construcción de una escala de actitudes tipo Likert. Centro de investigación y asistencia técnica de Barcelona, España
Folke C, Hahn T, Olsson P, Norberg J (2005) Adaptive governance of social-ecological systems. Annu Rev Environ Resour 30:441–473
Galvin M, Haller T (2008) People, protected areas and global change: participatory conservation in Latin America, Africa, Asia and Europe. Swiss National Centre of Competence in Research North–South/University of Bern, Berna
Guerrero DE, Zambrano WR, Salazar Alonso MC, Vásquez Bernal JA, Sandoval Torres J, Posada Arrubla A (2008) Formulación de una cooperativa de servicios turísticos en los municipios de Sopó y La Calera, Cundinamarca (Colombia). Revista Actualidad & Divulgación científica 11(1):175–184
Harris M (2009) Introducción a la antropología general. Alianza, Madrid
Hernández R, Bello E, Montoya G, Estrada E (2005) Social adaptacion ecotourism in the Lacandon forest. Ann Tour Res 32:610–627
Ibáñez Pérez RM, Armenta L, Olmos Martínez E, Arizpe O (2019) Áreas naturales protegidas en Baja California Sur. In: Hernández Trejo VA, Ibáñez Pérez RM, y Valdivia Alcalá R (coord) Economía, manejo y conservación en áreas protegidas de México. Universidad Autónoma de Baja California Sur y Universidad Autónoma Chapingo, pp 15–38
INEGI (Instituto nacional de estadística y geografía) (2010) Censo de población y vivienda 2010, Resultados sobre infraestructura y características de las localidades con menos de 5 mil habitantes 2010. Disponible en: https://www.inegi.org.mx/app/tmp/scitel/default?ev=6 [08/07/2019]
Manuel Aragón M (2011) "Agentes difusores centrales y grado de difusión de la información en la red social Agenda 21 Local de Bahías de Huatulco, Oaxaca". Tesis de Licenciatura, Universidad del Mar, Huatulco
Morales H (2006) Turismo comunitario: una nueva alternativa de desarrollo indígena. Revista de Antropología Iberoamericana 1(2):249–264
Olsson P, Folke C, Hahn T (2004) Social-ecological transformation for ecosystem management: the development of adaptive co-management of a wetland landscape in Southern Sweden. Ecol Soc 9(4):2. https://doi.org/10.5751/ES-00683-090402
OMT (Organización Mundial de Turismo) (2000) Observaciones sobre el Turismo Internacional. El desarrollo sostenible. OMT
OMT (Organización Mundial de Turismo) (2002) Cumbre mundial del ecoturismo: Informe final, OMT
Ostrom E (1990) Governing the commons: the evolution of institutions for collective action. Cambridge University Press, Cambridge, MA
Peredo AM, Chrisman JJ (2006) Toward a theory of community-based enterprise. Acad Manag Rev 31(2):309–328
Peredo AM, McLean M (2006) Social entrepreneurship: a critical review of the concept. J World Bus 41(1):56–65
Pinkus-Rendón MJ, Pinkus-Rendón MA (2015) El ecoturismo: quimera o realidad de desarrollo en la Reserva de la biosfera Ría Celestún, México. Revista LiminaR. Estudios Sociales y Humanísticos XIII(1):69–80
Ramos-García J, Ibarra-Michel JP, Velarde-Valdez M (2017) Community-based ecotourism management: the case of a cooperative in Mazatlán, Sinaloa, Mexico. Czech J Tour 6(2):155–170. https://doi.org/10.1515/cjot-2017–0008
Ramsar (2019) Servicio de Información sobre Sitios Ramsar. Disponible en: https://rsis.ramsar.org/es/ris/1349?language=es [07/07/2019]
Santana R, Salvatierra Izaba B, Parra Vázquez MR, Arce Ibarra AM (2013) Aporte económico del ecoturismo a las estrategias de vida de grupos domésticos de la Península de Yucatán, México. Pasos, Revista de turismo y patrimonio cultural 11(1):185–204

SECTUR (Secretaría de Turismo) (2019) Comunicado de prensa: Turismo Sostenible Genera Crecimiento De 257% En Las Áreas Naturales Protegidas De México: Sectur. Disponible en: https://www.gob.mx/sectur/prensa/turismo-sostenible-genera-crecimiento-de-257-en-las-areas-naturales-protegidas-de-mexico-sectur [16/08/19]

SEMARTAT-CONANP (Secretaría del medio ambiente y de los recursos naturales-Comisión nacional de áreas naturales protegidas) (2018) Sistema general de programas operativos anuales, Noroeste y Alto Golfo de California, POA 2018, Playa El Verde Camacho. Dirección de Evaluación y Seguimiento. Subdirección de Planeación Operativa y Evaluación, 10p

Serrano-Barquín R, Pérez-Ramírez C, Manjarrez-Campos E, González-Melgarejo L (2010) Turismo armónico como alternativa sustentable para una comunidad en el estado de México. Estudios y Perspectivas en Turismo 19:970–993

Silva J, Dávila R (2002) Gestión y desarrollo: La experiencia de las cooperativas en Colombia. Pontificia Universidad Javeriana, Bogotá

Tamayo, Tamayo M (2003) El proceso de la Investigación Científica, 4ta. Ed. Limusa

TIES (The International Ecotourism Society) (2006a) Global ecotourism statistical fact sheet. Disponible en: http://www.ecotourism.org [05/06/2017]

TIES (The International Ecotourism Society) (2006b) What is ecotourism? Disponible en: http://www.ecotourism.org [05/05/2006]

Tompkins EL, Adger WN (2004) Does adaptive management of natural resources enhance resilience to climate change? Ecol Soc 9(2):10. https://doi.org/10.5751/ES-00667-090210

Urquiza GA, Cadenas H (2015) Sistemas socio-ecológicos: elementos teóricos y conceptuales para la discusión en torno a vulnerabilidad hídrica. L'Ordinaire des Amériques, 218. Disponible en: http://journals.openedition.org/orda/1774; https://doi.org/10.4000/orda.1774

Valencia D (2008) Aplicación del conocimiento ecológico local campesino en la actividad ecoturística. Estudio de caso: sector del Valle del río Lagunillas-Cabañas Kanwuara, zona de amortiguación occidental del Parque Nacional Natural El Cocuy, Güicán y/o Chita, municipios El Cocuy y Güicán-Boyacá. Tesis de posgrado. Bogotá, Pontificia Universidad Javeriana, Carrera de Ecología

Vargas del Río D (2010) Espacios comunitarios sujetos a conservación y turismo: retos para la gestión costera participativa. Tesis de doctorado, Barcelona, Universidad Politécnica de Cataluña

Vargas del Río D, Brenner L (2013) Ecoturismo comunitario y conservación ambiental: la experiencia de La Ventanilla, Oaxaca, México. Estudios sociales (Hermosillo, Son.) 21(41):31–63. Recuperado en 16 de agosto de 2019, de http://www.scielo.org.mx/scielo.php?script=sci_arttext&pid=S0188-45572013000100002&lng=es&tlng=es

Chapter 27
Integration of Resident Fisherfolk Communities in Marine Protected Areas by Social Micro-entrepreneurships of Mariculture: A Case Study at La Paz Bay, South Baja California, Mexico

Mario Monteforte-Sánchez

Abstract The global deterioration of vital ecosystems and natural resources has compelled to constantly redefine alternative futures. The creation of protected areas (PAs) is considered among the best prescriptions to rise environmental conservation insofar as multiple-objective established in the nations' policies allows resident communities to access productive opportunities through social micro-entrepreneurships (SMEs). Although implementing PAs assumes conversion-diversification processes designed for unique socio-ecologic timelines and components by case, it is recognized that SMEs of ecotourism, small-scale farms, and a rational exploitation of local natural resources are inherent vocational assets that improve PA profitability. The integration of rural/coastal groups in PA management by productive SMEs is analyzed in this chapter. We discuss the development and outcomes of this model applied in the real conditions of a marine-coastal PA (MPA) in La Paz Bay, Baja California Sur, Mexico, the "Balandra-Merito" coastal lagoon system. Here, the model comprises small-scale mariculture and coastal fisherfolk actors determined by the historical context of a case study. Using the narrative visualization approach, we review the extension (capacity-building) programs in which the roles of technologies and knowledge-holders signify decisive influences on the configuration of SMEs into or around protected areas. Prospects to make Balandra-Merito MPA more cost-effective are examined.

Keywords Coastal fisherfolks · Protected areas · Profitable conservation · Mariculture · Social micro-entrepreneurships

---

M. Monteforte-Sánchez (✉)
Centro de Investigaciones Biológicas del Noroeste (CIBNOR),
La Paz, Baja California Sur, México
e-mail: montefor04@cibnor.mx

A. Ortega-Rubio (ed.), *Socio-ecological Studies in Natural Protected Areas*,
https://doi.org/10.1007/978-3-030-47264-1_27

27.1 Framework of the Case Study

The global socioeconomic model has laid heavy footprints on almost every ecosystem and natural resource, with evidences of fast deterioration in the last 30 or 40 years at such an extended rate that great awareness about the present widespread crisis has been rising alerts to what it may come in the near future. Many summits, conventions, and other global reunions assembled principally from the 1980s and on keep searching for alternatives to regulate the consumption/waste model based on the overexploitation of natural resources, and thus the character of sustainable goals is in constant redefinition from shorter to longer intervals (2050 so far), in parallel to progressively complex socio-ecological scenarios. For example, 20 strategic goals, the Aichi targets, with biodiversity conservation as foremost aim, were supposed to be achieved by 2020. We are there now, and it seems the compromises comply just partially.

Because of many reasons, the social dimension in the first place, marine coastal areas have received systematic attention in finding better strategies for the management stewardship of fundamental ecosystems and natural resources, particularly in developing countries. For instance, the United Nations Ocean Conference (New York, June 2017)[1] set forth a factsheet that summarized the ocean's role in human societies, and vice versa, to 51 points classified in 14 sectors (main roles) where tropical coastal zones are highlighted in a substantial proportion: demography; sustainable livelihoods; food security and health; conditions of small-scale fisheries and coastal fisherfolks, similarly in rural-inland areas; effects of diverse forms of tourism and land-based activities such as urban, agriculture, industries, ports, etc.; status of mangroves, coral reefs, and coastal biodiversity in general; physical alterations on coastlines; threats from sea-level rise and extreme weather events; issues of displacement; and vulnerability at large. It is no surprise to observe the increasing global move to create marine-coastal protected areas (MPAs henceforth) and/or special measures on the utilization of vital ecosystems and key species.

There is, however, a situation of balance in the perception of the holistic effects that PAs generate in terms of advantages and disadvantages in real conditions that can be as specific and different as a single case study (Du et al. 2015; Oldekop et al. 2016; Le Gouvello et al. 2017). Although the IUCN categories system is the global standard for PA governance adopted in practically every nation owning these natural assets, the configuration of virtual actions (e.g., planning stage and forecasts) or adjustment of these actions to the current management framework (administrative, productive, extractions, research, capacity-building, etc.) should find alignments on a wide array of components (actors and their institutions, development trend adopted, cultural/psychological constellations, ecological-biogeographic contour, timeline construction, etc.) and of terminology in consequence, that vary at national and local scale (e.g., reserves, no-take zones, refuges, sanctuaries, parks, or a

[1] https://www.un.org/sustainabledevelopment/wp-content/uploads/2017/05/Ocean-fact-sheet-package.pdf Accessed 4 December 2019.

generic "conservation area"). Likewise, there are different interpretations about what is and how to apply multipurpose or multiple-use objectives in PA and dilemmas on conservation and production (e.g., Agardy et al. 2011; Bottema and Bush 2012; Oldekop et al. 2016; Du et al. 2015; Le Gouvello et al. 2017), which addresses to insights necessarily related to management and planning methods that connect in particular way to a "tropical" stereotype, the said Third World developing countries, more so if rural or coastal communities are involved (Agardy et al. 2011; Bennett and Dearden 2014a; Segi 2014; Du et al. 2015). In the practice, implementing PAs (what and how will be protected, by whom, reasons/aims, benefits and target recipients, and other criteria) entails that the "willingness," a common guideline in research surveys of perception/acceptance, should be more properly defined as the attempt to convince proud and distrustful people to appropriate a proposal and of skeptical decision-makers in facilitating the process (e.g., Perkins et al. 2002; Brugère et al. 2008; Awortwi 2012; Cundill et al. 2017).

Certainly the IUCN categories encourage community-based management through social micro-entrepreneurships (SMEs hereafter),[2] but lessons learned in real conditions reveal that the concept of multiple objective alluded to those categories delivers few and narrow windows to accommodate local communities with productive activities into or on the periphery of PAs. The initiatives aiming at the protection of a given area or resource have to deem the presence of resident communities whose livelihoods rely on the exploitation of those native resources aimed to protect (fisheries, wood, recollection, soils of artisanal mining, etc.), and/or they have farms (agriculture and terrestrial and/or aquatic commercial livestock), and/or work on ecotourism (freelance or employed by private enterprises),[3] or other income-producing activities derived from regional/local assets, even hunting and leisure/sport fishery. Evidence shows that tourism-oriented SMEs have been the preferred choice to realize profitable community-based actions in continental and coastal PAs. This image is the mainstream projection of (tropical) PAs in the tourism market. On the other hand, the IUCN system allows farming modalities into PAs if explicit benchmarks are fulfilled within two compartments: restoration-restocking and countless modalities of small-scale production.[4] Therefore, a direct

[2] Community-based social micro-entrepreneurship (SME). See Brugère et al. (2008), Berkes and Davidson-Hunt (2010), and Davidson-Hunt et al. (2012), among others.

[3] Community-based tourism (tourism-oriented SMEs) is inherent to low-impact sustainable tourism in many forms (ecotourism, geotourism, ethnic/cultural tourism, pearling tourism, adventure, gastronomic, health, etc.), and there are different definitions for each (see Boley et al. 2016). In this chapter the reference to ecotourism is used in generalized term, unless it is specified where necessary.

[4] It is common to find productive SMEs doing husbandry of local species into or around PAs with different degrees of protection. Examples of rustic cultivation technologies include medicinal, ornamental, and "organic" plants; honey; cheese and other milk by-products—goats, cattle; leather from animal breeding, even alligators; pet breeding (iguanas, turtles, aquarium fish, etc.); and diverse gastronomic products that include marine and freshwater aquaculture, besides nacre and pearl-bearing mollusks like naiads, abalone, and pearl oysters (Monteforte and Cariño 2011; Bennett and Dearden 2014b p. 98). Browse the Internet.

deduction that emerges from conveying profitable conservation and community-based SMEs of ecotourism and/or cultivation of target species (plant or animal) integrated to PAs management would lean to underscore biogeographic interactions which, in terms of the geographic distribution of PAs (e.g., type of ecosystem and species), are concentrated on certain kinds of biodiversity (e.g., Agardy et al. 2011; Monteforte and Cariño 2011; Hjalager and Johansen 2013; Slater et al. 2013; Hill 2017; Le Gouvello et al. 2017). In addition, the scope to implement ecotourism and/or farmer SME models in real conditions, rather than being theoretical, depends on strategies specially devised to meet socio-ecological and cultural profiles prevailing in a focus case, where multifaceted dynamics of ecosystem-based factors and case-based indicators vary even within a locality and over time (e.g., Bondad-Reantaso et al. 2009; Slater et al. 2013; Le Gouvello et al. 2017). Overarching distinctions linked to regional/local geographical features of marine coastal and continental PAs stand out in the first place. Although both domains equally share the sense of iconic value (cf. natural asset, emblematic belonging, or marketable attribute) ascribed to a set of ecosystems and species biogeographically determined therein, those such dynamics take place in totally different spaces; obviously the separation is blurry in most MPAs. It also applies to cultivation modalities with a set of commercial species that ultimately can propel gastronomic tourism (Hjalager and Johansen 2013; Slater et al. 2013) and/or added value from ornamental marine species, seaweeds, and pearl oysters (Ferse et al. 2012, Monteforte and Cariño 2013, 2018, Bennett and Dearden 2014b p. 98, Le Gouvello et al. 2017, among others).

Another point of view to integrate resident communities in their inland or marine PA concerns the premise that conversion-diversification processes should not beget tangible changes to entrenched livelihoods (Bondad-Reantaso et al. 2009; Slater et al. 2013; Bennett and Dearden 2014a; Cundill et al. 2017). In approach to MPAs, these processes have to focus in coastal fisherfolk groups often organized in cooperatives or similar associations, considering that the regulations in many MPAs may restrain artisanal fisheries as well as certain modalities of mariculture or commercial species. Resident fisherfolk represent one of the major conflictual issues in MPAs since they often feel the proposals about conservation and protection as a threat of privatization and exclusion (e.g., Cárdenas and Ostrom 2004; Leopold et al. 2013; Bennett and Dearden 2014a; Segi 2014; Bennett et al. 2015; Hill 2017; Woodhouse et al. 2018). Contrary cases do happen when the locals undertake the initiative to create their PA and regulate the extractions of resources and/or access to the site ecosystems. In any event, the participants into collective initiatives aimed to conservation programs should have to believe on it as a potential instrument to obtain some benefit (abstract, material, economic) from the trade-off (Cárdenas and Ostrom 2004; Awortwi 2012; Bennett and Dearden 2014a; Cundill et al. 2017). Taking in consideration the MPAs' own natural vocations to develop mariculture ventures, fisherfolk should become primary actors to organize the management policies for the area, whether planned or in course. The IUCN system advises a zoning approach (e.g., buffer and core or nucleus zones and a peripheral strip) as the best practice guide to organize multiple-objective PAs, each type of zone has different objectives, and some allow for greater utilization of resources than others.

In MPAs, this implies tridimensional zoning in order to cover the water column and the seabed. There are open-ocean PAs as well though not yet formal because industrial fisheries introduce a different outlook (Bennett et al. 2015; Briscoe et al. 2016) and some consider that islands are oceanic.

This chapter subscribes to coastal planning practices in rapidly changing sceneries where decision-makers at the moment almost always worsen the troubles already accumulated over time, often to alert states, or introduce new ones. Due to the growing pressure from development models at different scales, a question is reiterated in forums, workshops, and case studies: What to do with fisherfolk communities? Or rather, what could motivate them to participate in conversion/diversification processes related (or not) to MPAs?

With that in mind, a case study is examined alongside four main components that regularly concur in PAs where private or communitarian farms existed before and had to be adjusted (e.g., modify the technology and/or change to small-scale) or were closed or are part of the planned perspective. These components are the following: (1) availability and access to certified technologies (mariculture in the case) adequately designed for (2) target groups (fisherfolk herein) and for (3) the prevailing factors of governance on the PA case study (MPA, this study) that includes the attitude of different stakeholders (e.g., national/international agencies, academic institutes, civil organizations and their sponsors, commercial and entrepreneurial sectors, special interest groups, and other leading persons) and, finally, (4) a timeline review at conversion and diversification processes of target groups upon technology-based SMEs, like coastal fisherfolks with small-scale mariculture of which the literature contains hundreds, maybe thousands, of examples (many with gender orientation) investigated in tropical coasts (e.g., Brugère et al. 2001; Bondad-Reantaso et al. 2009; Slater et al. 2013; Ateweberhan et al. 2018), which comprise pearl oysters (see Monteforte and Cariño 2013) and ornamental marine species (Ferse et al. 2012).

The present study is located in La Paz Bay, Baja California Sur (BCS), México, into a coastal lagoon system locally known as "Balandra" which owns the label of MPA since 2012 as Area of Protection of Flora and Fauna (DOF 2012). Actually, the MPA extends southward to El Merito and Falsa Bay coastal lagoons and comprises the adjacent marine area (Fig. 27.1). This chapter addresses a narrative visualization approach (Ma et al. 2011; Dahlstrom 2014) to explain the historic role of the main factors that led to the current scenery in this MPA and their decisive influence in making La Paz an important hub of mariculture science and technology. This attribute should be applied in the implementation of fisherfolk-based mariculture SMEs as a profitable alternative to achieve better management of this site, which could be replicated in other MPAs that have good conditions to ensure this type of models be successful. Ultimately, the analysis would be useful elsewhere to motivate capacity-building programs and conservation policies by linking beneficiaries, professional experts, and stakeholders by effective actions in that direction.

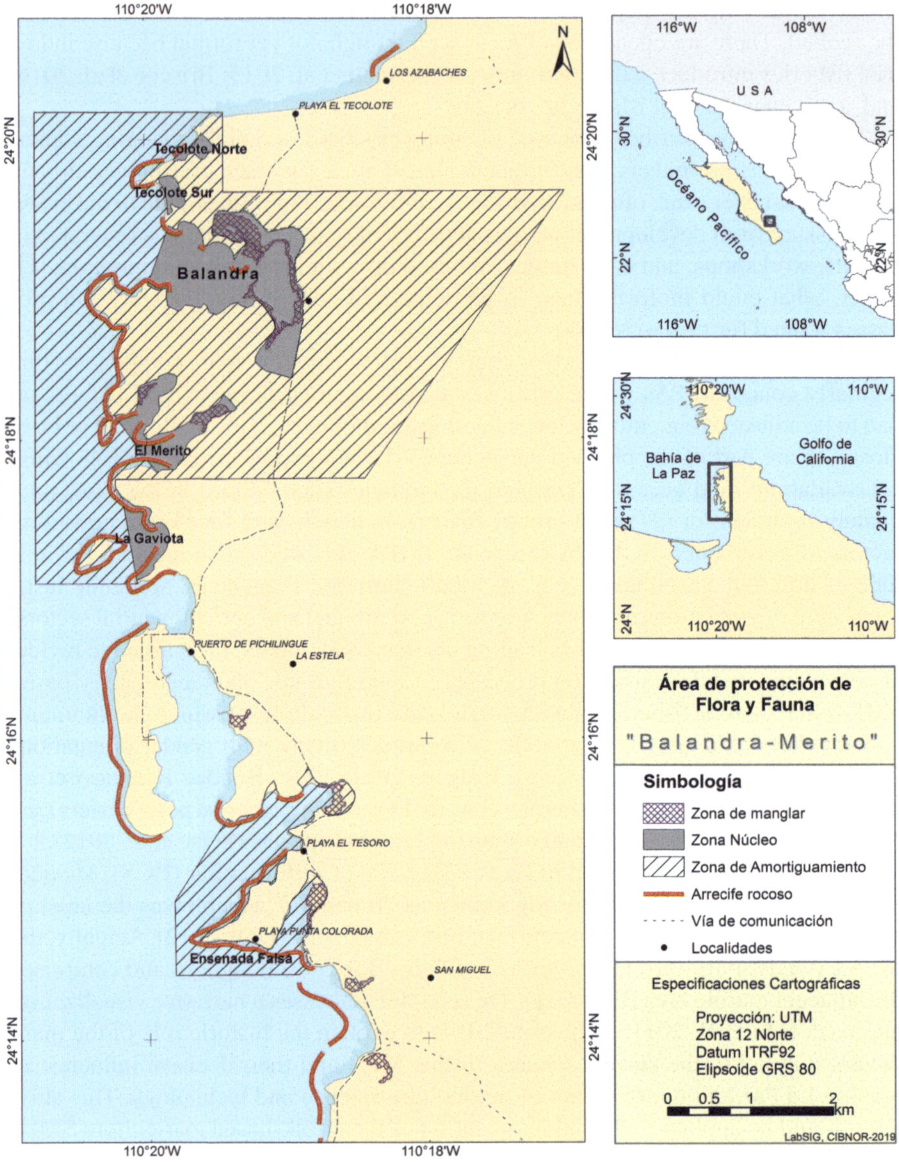

Fig. 27.1 Geographic location of the Marine Protected Area Balandra-Merito and delimitations of the zones at La Paz Bay, Baja California Sur, Mexico. Elaborated by the Laboratory of Geographic Information Systems and the Department of Extension and Scientific Diffusion (CIBNOR)

27.2 Description of the Study Area

Numerous legal instruments—laws, decrees, secretarial agreements, and official standards—regulate coastal and marine issues in Mexico (Fraga and Jesus 2008; Havard et al. 2015). Among the most important laws pertaining to the rights to exploit natural resources for any purpose is the General Law of Ecological Equilibrium and Environmental Protection (LGEEPA). It defines the tools of the national environment policy within the logic of a sustainable management of natural resources so that economic benefits could be obtained while preserving the ecosystem (Fraga and Jesus 2008). In addition to the LGEEPA, a considerable number of legal instruments focus in conservation and use of the biological diversity (Table 27.1). Likewise, La Paz Bay is a Marine Site Priority declared by the National Commission of Biodiversity (CONABIO); the Espiritu Santo archipelago is a National Park and belongs to the World Natural Heritage (UNESCO) stated for the Gulf of California islands, and Balandra-Merito and Falsa Bay coastal lagoons zones encompass an Area of Protection of Flora and Fauna (Fig. 27.1). These lagoons host wetland scrubs-mangrove ecosystems under Ramsar status; besides, the character of arid mangrove represents eminent importance as a relict ecosystem (Arreola-Lizárraga et al. 2004; Adame et al. 2018). The statement applies to the madreporian corals forming the typical rock-coralline ecosystem present along the BCS east coast, specially thriving in Loreto and Cabo Pulmo national parks. There are more labels related to the study area:

- Ensenada de La Paz (Fig. 27.2) is a Conservation Area for Birds (1998), and its wetland-mangrove ecosystem is Ramsar. There are initiatives to create in this singular Ensenada a refuge or sanctuary dedicated to dolphins that are common dwellers in the channels and deep areas there inside (deep, 6–10 m at most).
- Several resident and migrant marine species present in La Paz Bay are under special protection (NOM-059-SEMARNAT); therefore also their territories are fairly protected (turtles, birds, marine mammals, and some species with commercial value).
- There are social initiatives to protect La Paz Bay cultural heritage, like the remains of Indian groups, Pericúes and Guaycuras, in El Conchalito (south coast of Ensenada de La Paz, nearby the downtown), and the Compañia Criadora de Concha y Perla (CCCP) founded by Gaston Vives in Espiritu Santo island (Cariño 1998; Cariño and Monteforte 1999). Some local sectors wish the original CCCP installations be restored as a museum. Other initiatives aim to restrain transit (off-road vehicles), sand extraction, or residential buildings in sand dune ecosystems that are the dominant morphology of La Paz Bay continental coastline (Fig. 27.2).

Table 27.1 Main legal instruments that regulate conservation and use biological diversity in Mexico

Environmental regulations	Management/production sector
General Wildlife Law (LGVS) Federal Fisheries Law (LFP) General Law for Sustainable Fisheries and Aquaculture (LGPAS) General Law for Sustainable Forestry Development (LGDFS) Official Mexican Standard (NOM), for instance, NOM-059-Secretaría de Medio Ambiente y Recursos Naturales (SEMARNAT)-2001 lists threatened and endangered species and NOM-022-SEMARNAT-2003 regulates conservation, sustainable use, and restoration of coastal wetlands located in mangrove areas General Law of National Property (LGBN), which incorporates legislation on coastal areas (such as beaches), the 20 m federal maritime-terrestrial zone (ZOFEMAT) and reclaimed land Federal Tourism Law (LFT) regulates all tourism activities Federal Sea Law (LFM), the General Health Law (LGS), and the National Waters Law (LAN) control ocean pollution Ports Law (LP) and the Navigation Law (LN) regulate marine transportation, prohibiting all vessels from contaminating the country's waters	Secretariat of Agriculture, Livestock, Rural Development, Fisheries and Food (SAGARPA). Mainly intermedia of financial functions in regard of productive projects Nacional Commission of Fisheries and Aquaculture (CONAPESCA), attached to SAGARPA. Concessions, permits, liaison agents with the productive sector (cooperatives) Secretariat of Environment and Natural Resources (SEMARNAT). Coordinates environmental management in Mexico, e.g., evaluation of reports of environmental impact assessments, management of PAs, etc. National Commission of Protected Areas (CONANP), subscribed to SEMARNAT. Vigilance, management, administration of PAs (regional) Federal Attorney for Environmental Protection (PROFEPA). Law enforcement of environmental protection Other entities related to quality and health controls on the products (e.g., cultivation) and the environment Trend toward scientific-technologic parks to promote private investment (national and foreign) in the sector of primary production such as aquaculture. This trend involves technologies that usually are in the hands of academic institutions housing their park

Modified from Fraga and Jesus (2008) and Havard et al. (2015)

27.2.1 La Paz Bay Presentation

La Paz Bay and islands close by (Cerralvo, Espiritu Santo archipelago, San Francisquito and San José) are placed on the southeast of Baja California Sur (BCS) (Figs. 27.1 and 27.2) facing the Gulf of California. The Bay is about 83 km on the NW-SE axis and 35 km on the NE-SW axis (2600 km^2). The aerial view shows that landscapes of dunes and sandy beaches of smooth slope are the dominant landscape along the continental coast. There are discontinuous bands of rock-and-boulder beaches and rocky-coralline reefs (Fig. 27.2) better developed on the southeast coast from La Paz city to the boundary of Balandra, around Espíritu Santo archipelago, and on the north entrance of the Bay (El Portugués and Punta del Mechudo area). Rocky-coralline reefs represent biodiversity centers in the Bay; they stretch down like cliffs of different inclination on the range of 8–15 m of depth in general and to 25–30 m of depth in some places such as the Balandra-Merito MPA zone and west face of Espiritu Santo.

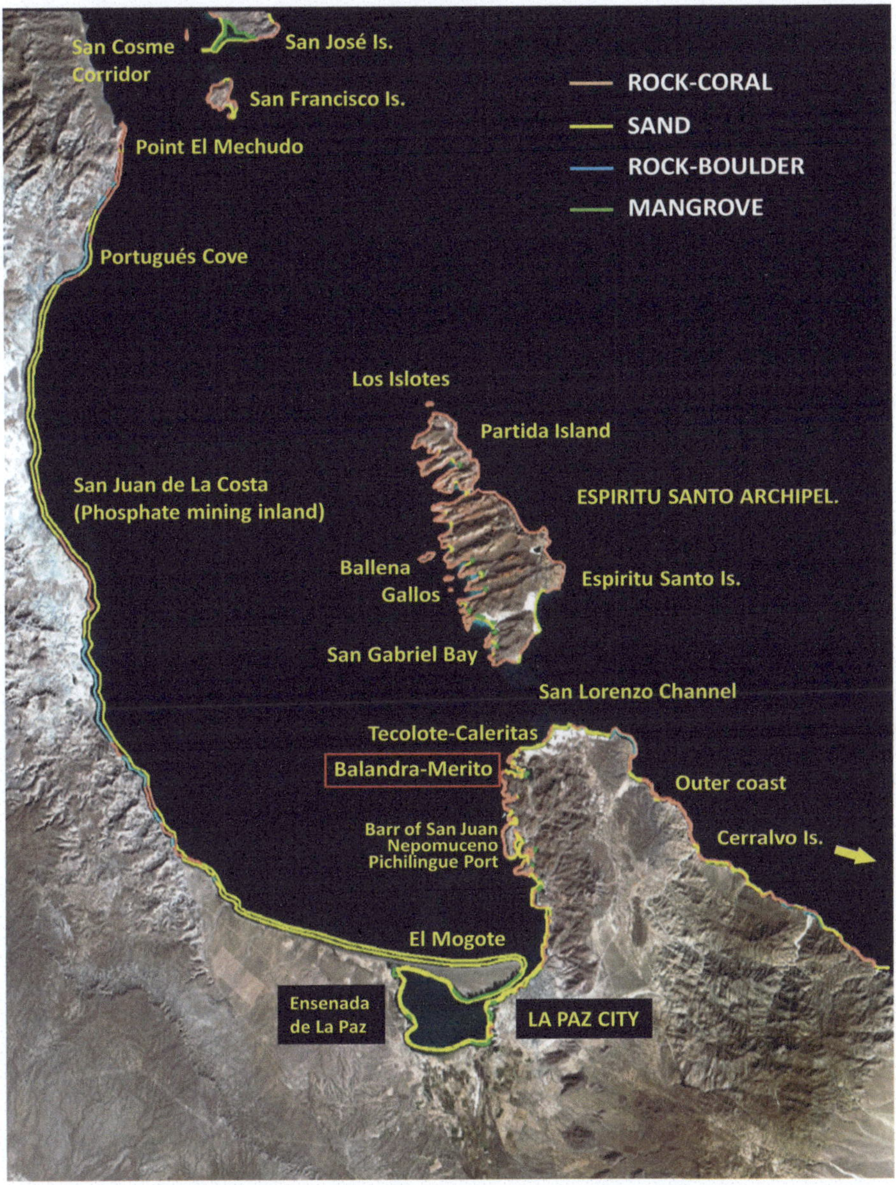

Fig. 27.2 Representative types of coastline around La Paz Bay and islands close by, with reference points mentioned in the text. (Modified from Google Earth. Author's work)

Small colonies of mangrove and wetland vegetation (some barely one or two square meters) are found into many small bays and coves or compose larger extensions in coastal lagoons (Fig. 27.2). Over time, a narrow sand bar called El Mogote enclosed a shallow coastal lagoon (Ensenada de La Paz) with marginal wetlands and branched channels barely 12–14 m of depth in the deepest points.

La Paz Bay contains the largest and deepest water body enclosure around the Gulf of California. The seafloor topography resembles an elongated funnel with its vertex, the Alfonso Basin, placed at 420 m of depth and a slope mounting to shallow flats of less than 20 m of depth along 60–65 km toward La Paz city coastline (Fig. 27.3). This shape was formed by water current regimes that alternate a dominant vector entering the northern mouth (San Cosme Corridor, 300–350 m depth) with tidal exchange through San Lorenzo channel, which is as shallow as 20 m of depth, while the outer border of the Bay rapidly falls to more than 2000 m of depth in a short distance toward the Gulf of California. Therefore, La Paz Bay behaves like a trap of water bodies and sediments that have been transported by the southward dominant current from as far as the Colorado River (Cadena-Cárdenas et al. 2009; Silverberg et al. 2014). This orientation and its position in respect of oceanic water bodies (Gulf of California and Pacific) impel a dominant yet moderate counterclockwise direction to La Paz Bay water body (Obeso-Nieblas et al. 2002, 2014; Silverberg et al. 2014), thereby the funneled shape of which about 95% of the seabed is sandy and 50% of it is at less than 50 m of depth (López-Cortés et al. 2012), and the hydrographic dynamics toward the southeast area during live tides that improve the primary productivity and nutrient contents (e.g., water exchange, flux of currents at wide depth range, upwellings, revolver-eddies, rip currents in the coves, events of water stratification, etc.) (Obeso-Nieblas et al. 2002, 2014; López-Cortés et al. 2012; Silverberg et al. 2014). However, the overall circulation and water exchange rate within the Bay is slow ("sluggish currents," as Silverberg et al. 2014 wrote down), which means that the water body tends to remain inside for some time. Actually, the maximum normal tidal height range fluctuates between 0 + 2 m and 0 + 2.5 m, in occasion to 0 + 3 m, and the maximum normal in-out water speed in San Lorenzo Channel (live tides) oscillates on 3–4 knots (about 6 km per hour). In steady tide conditions, you can cross with a kayak or swimming. It is approximately 6.5 km on the shortest distance.

La Paz is BCS' state capital city with a population of 280,000 habitants, roughly the 40% of the total state population (719,000). The cities of San José del Cabo and Cabo San Lucas (popularly gathered as Los Cabos) sum 240,000 habitants together. Other relevant concentration cities are Ciudad Insurgentes, Ciudad Constitución, Mulegé, Loreto, Santa Rosalía, Guerrero Negro, San Ignacio, and Todos Santos.

There are 20–24 fisherfolk cooperatives (Cooperative Society of Fishery Production, or SCPP) of different size (some SCPP are formed by 3 or 4 friends or relatives, others are gathered into two or three Federations, and there is an unstable number of fisherfolks registered as "free fishermen"—*pescadores libres*) operating from La Paz over a marine area as extended as their simple equipment (26–28 ft. panga and outboard engine up to 120 HP) lets them draw away the coast (Guzmán-Vizcarra and Ibáñez-Pérez 2018). In fact, the major fishery effort (principally fish) is concentrated outside the Bay, though some product is fished in the interior, mostly bivalves like Catarina scallop, pen-shell, two or three species of

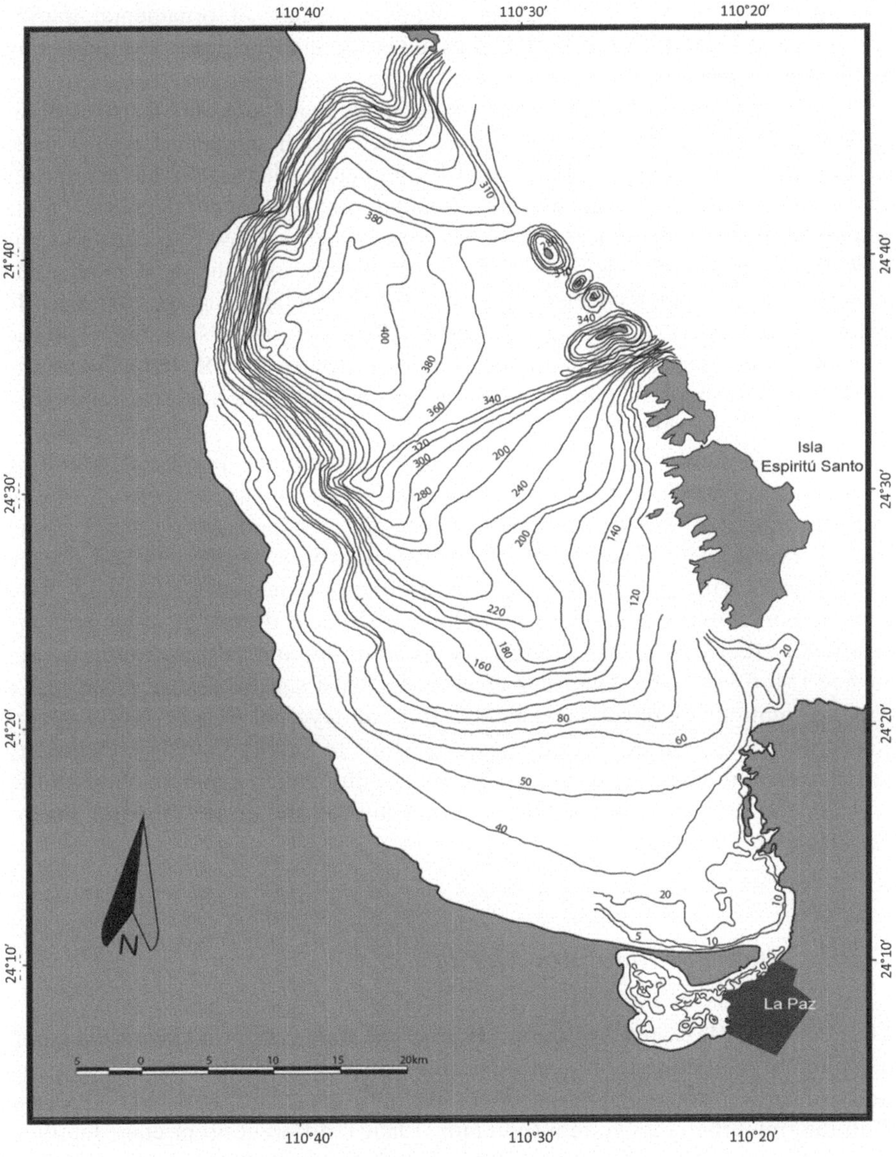

Fig. 27.3 Bathymetry (in meters) of La Paz Bay. (Template from Lechuga-Devéze (2012))

clam (CONAPESCA 2018), and an unknown capture of ornamental marine species, yet the traditional fisherfolk does not work on the activity. It is a different sector, furtive much of it.

Additional features of La Paz Bay worth to know point out a very active ecotourism that exploits the biogeographic regional assets (sea lion colonies, whale-shark, marine birds, snorkeling, camping, sport-fishing, etc.). There is private enterprises (some are micro, and some are larger habitually foreign) in one hand and short tours in touristic boats in the other. Almost any boat owner is allowed to carry out this activity as long as it meets a long series of requirements, like training as a tour guide included first aid and services quality (e.g., food, sanitary, garbage disposal). In addition, boats must comply to mandatory specifications (size/design, engine power, safety, comfort), and the requirements overall are stricter upon SCUBA diving and snorkeling. Therefore, this model may deter the average fisherfolk, or simply they cannot afford the technical requirements.

Another feature in the Bay is the presence of incipient (private) mariculture in operation: one shrimp farm (white-legged *Litopenaeus vannamei*), one of yellowtail *Seriola lalandi* in offshore cages and aquapods, and a pearl oyster farm of *Pteria sterna* (see Sect. 32.3). Importantly, urban sprawl and coastal megaprojects (the classical gated mini-cities of marina-golf-villas) have impacted on ecosystem services provoking inefficient drainage, water loss and insufficient recharge of the aquifer, and overmining and salinization in consequence (McEvoy 2014; Haeffner et al. 2018). We see excessive waste and a growing number of obsolete vehicles. Two smoky fuel-hungry power plants placed near La Paz must be fed by tanker ships to supply the city's electricity and also to both Cabos and villages in the middle, which generates high pollution and constant spills and accidents. Domestic gas and most of the fuel for the vehicles comes through the same way.

27.2.2 Status of Balandra-Merito

After the 1910 Mexican Revolution, slices of the Baja California Peninsula (short coastlines and broad terrain inland) became a sort custom gift to notable persons who had participated in the insurrection, like high-graded military and politicians. With the time, the *ejido* system (communal land), different proprietary families, and real-estate agencies got intermixed in land tenures (Gámez and Angeles 2010; Anderson 2017). Furthermore, several Presidents, top-placed functionaries of the Mexican government, and rich magnates had the customary practice of declaring "from-here-to-there" a personal (governmental?) possession. The aftermaths of these circumstances, combined with the characteristic aridity and isolation of the coastal zone around the Peninsula, the southern part in particular, have drawn striking divergences between large unproductive areas bounded to isolation against regions such as La Paz and "Los Cabos" that have been unable to match the increasing demands of the development model established in the region (Gámez 2008; Gámez and Angeles 2010).

The MPA of our case study did not linger to that trend of appropriation. Among the many hot spots related to real-estate speculation on touristic megaprojects in Mexico (Murray 2007; Monteforte and Cariño 2009; Gámez and Angeles 2010; Anderson 2017), this MPA has attained high relevance even in the international media. Balandra itself, in addition to housing one of the largest mangrove/wetland (relict) colony on the western coast of the Gulf of California (aside Mulegé and Concepción Bay), owns emblematic value of great significance for the locals (paceños). The proximity to La Paz City (10–12 km from downtown and there are public buses and paved road), and its shape of natural swimming pool with nice sandy beach, converts this area the favorite destination of families with children and anyone who wishes to spend a pleasant day in the beach. Balandra, the "jewel of BCS," is among the beautiful and worldwide famous beaches labeled by the Travelers Choice Awards 2019 of the TripAdvisor as second best beach in Mexico, only topped by Playa Norte, Mujeres Island in the Caribbean.

Truly said, the coastal zone at Merito lagoon and also other portions of the MPA do not display such as popular beaches; nevertheless, the exceptional ocean-front landscape, the proximity to La Paz, and consequently the high real-estate value are coveted objects. Long-time environmentalist movements have escalated to protect the area from the descendants of a former President who claim the ownership of 1500 Ha with adjacent coastline and pretend to develop another one of those classic megaprojects (Cariño et al. 2008). As expected, other persons and families dispute older land titles that still present overlapping perimeters anyway. Crisscrossed wire fences and grilles with brightly colored signs of "private property" (English and Spanish) coined to different companies or persons are the normal sightseeing on the way to Balandra, and in many coastal areas of BCS, as a matter of fact (Cariño et al. 2008; Monteforte and Cariño 2009). Despite the circumstances, local civil movements to protect Balandra, and Merito in extension, have achieved several successes, for example, Balandra enjoyed the label of Municipal Park between 1998 and 2002, not entirely official but rather as the result of those such movements that culminated in 2004 with a decree of MPA (Ecological Conservation Zone and of Interest to the Community) on nearly 2000 Ha and adjacent water (Cariño et al. 2008), and finally the present status of Area of Protection of Flora and Fauna since 2012 (DOF 2012) (Fig. 27.1). Hashtags #BalandraEsNuestra (Balandra is ours) are constant in social networks; eventually it triggers public manifestations as often as the promoters re-insist in re-commence their project, which is standby for the moment. Today, the Colectivo Balandra constitutes one of the primary sprockets of the local environmentalism movement centered in the area. To a certain extent, the decree of Espiritu Santo archipelago as National Park and its inscription in UNESCO's list of the World Natural Patrimony at Risk, and also the Ramsar designation for wetlands and the distinctive biodiversity of La Paz Bay, have been supportive arguments to incorporate Balandra-Merito in the same terms.

27.3 Influence of the MPA Balandra-Merito in the Mexican Mariculture

The marine coastal area in this MPA harbors one of the three best suitable areas in La Paz Bay[5] to install mariculture farms of bivalves and fish (Fig. 27.4) (Lechuga-Devéze 2012; Monteforte 2013). Table 27.2 compiles a chronologic resume (1939–2019) related to mariculture development in La Paz Bay, with emphasis in the MPA Balandra-Merito-Falsa Bay and based on landmark projects and pioneers in their respective institutions. We did not append the start interval of 1902–1914 in which the Compañia Criadora de Concha y Perla could have installed spat collection experiments for the mother-of-pearl *Pinctada mazatlanica* in the extant MPA (Cariño 1998). After the destruction of this enterprise in 1914, the fishery for pearl oyster shell (*P. mazatlanica*, primarily)[6] became so intensive that the wild stock reached high level of exhaustion leading the economic bonanza of La Paz to end by the middle 1930s and to a permanent ban on pearl oyster fisheries in 1940. In the meantime, there is no evidence of developments in mariculture science and technology[7] or active farms in Mexico whereas freshwater fish aquaculture received more attention (Morales-Díaz 2015; Cuéllar-Lugo et al. 2018). It was until 1939 when a private project focused in *P. mazatlanica* occurred in La Paz Bay, yet it did not work by cultivation techniques like Gaston Vives in his CCCP. Instead, they employed diving gear to extract wild specimens and gather them in culture artifacts to practice pearl induction surgery (Table 27.2). This project was the first of the three major fiascos (1979 and 1980) on the matter of pearl oysters, so far in the Gulf of California, because they aimed to an already depleted wild stock to produce cultured pearls. In addition, the three projects took place in the MPA and were assisted by (Japanese) specialists who were meeting *P. mazatlanica* for the first time. There is no evidence

[5]The single Balandra lagoon is unsuitable for mariculture installations. Falsa Bay is placed on the mouth of Pichilingue Port that lodges cargo ships, fishing boats, and large ferryboats. Although the Port capacity is small (amount/size of units and transit flow), this is a minus point to water quality for mariculture of edible species, filter-feeding mollusks in particular. The area between Diablo Point and the southern tip of San Juan Nepomuceno Barr is the second favorite in La Paz Bay. Number one is placed on the southeast of Espiritu Santo archipelago (Ballena islet to San Gabriel Bay), and the third is on the northwest of the Bay (Figs. 27.2 and 27.3). Other hierarchical criteria include technical and logistic variables, oceanography, depth, experimental data, etc. (Monteforte 2005, 2013; Lechuga-Devéze 2012).

[6]For centuries, and in every pearling region in the world, fisheries for pearl oysters aimed to nacre shells as the main objective. Species of the genus *Pinctada* (mother-of-pearl oyster) were preferred because of their flat-shaped shell that suited the enormous button-making and nacre inlay industry until the 1870s decade when the nacre was substituted by plastic.

[7]The first traces of mariculture research in Mexico appear by 1940, curiously most of them concentrated in the Gulf of California. For instance, the universities of Sonora and Sinaloa opened the line of marine sciences in that period (Morales-Díaz 2015; Cuéllar-Lugo et al. 2018). Also, the UNAM creates the BSc. in Hydrobiology in the Biology Faculty that established the basis for the Institute of Marine Sciences and Limnology in the 1960s and its campus at Mazatlán, Sinaloa, in the 1970s.

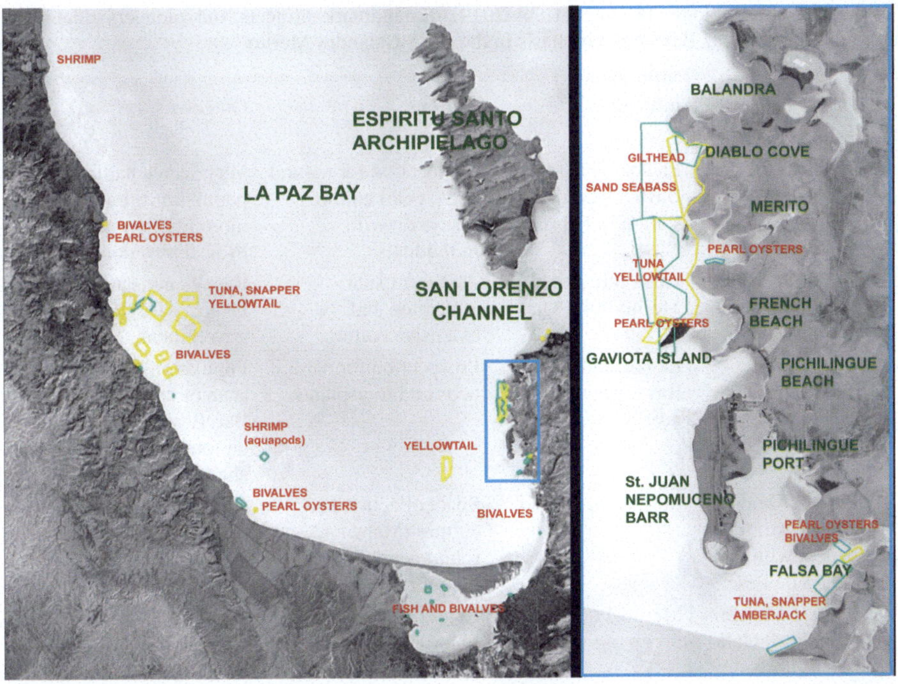

Fig. 27.4 Location of aquaculture licenses (Permit for Aquaculture Fomentation -PAF and Aquaculture Concession -CA) in La Paz Bay with zoom on Balandra-Merito-Falsa Bay area. Yellow polygons correspond to licenses in force; green ones are in process of renewal or have ceased their activity. (Modified from Google Earth with the KML/KMZ layers for PAFs and CAs available in Acuasesor SAGARPA/CONAPESCA. Updated to 2018)

known about round pearls resulting from their learning, and it said that they also tried the surgery on the winged pearl oyster *Pteria sterna*, the sister pearl oyster species of the region.

27.3.1 Timeline Overview

The MPA remained in standby from 1940 until the early 1960s when it started to acquire a significant role in research studies and commercial mariculture in Mexico. The pearling vision has persisted throughout new developments on other commercial native species of gourmet bivalves and fish that began by the 1970s and were diversified and perfected over time (Table 27.2). Much of the importance of La Paz as scientific and technologic center of mariculture in Mexico is linked to a variety experimental studies installed in the said MPA that constitute today a valuable body of systematic information and postgraduate thesis about the area and La Paz Bay. However, its suitable qualities for mariculture brought private entrepreneurs to

Table 27.2 Chronologic briefing (1939–2019) of landmark projects and pioneers related to mariculture in La Paz Bay, with emphasis in the MPA Balandra-Merito

Date/actors	MPA and/or influence	Actions	Outcomes
The pearl oysters sector			
1939. Y. Matsuii, Mexico-Japan agreement	La Paz and Loreto Bays. Operative station in Falsa Bay	Prospection for natural beds; pearl culture assays on wild individuals	Very scarce natural populations. Large mortality postsurgery. Project abandoned
1961–1962. Secretaría de Pesca	Falsa Bay and Merito Point Diablo	Assays of spat collection and extensive culture	Results not satisfactory. Project cancelled, change government actors
1962 A. Martínez CRIP, Secretaría de Pesca	La Paz and Loreto Bays, islands nearby	Prospection for natural beds and transplants	Populations in alarming state of exhaustion. Barely acceptable results on transplanting. Project abandoned
1969–1972 Denis George (Australia), agreement with the Secretaría de Pesca	Merito-Pt. Diablo	Spat collection and culture (Mabé and round pearls in wild individuals)	Great mortality and rejection postsurgery. Promising results on extensive culture
1970–1971 M. Díaz-Garcés and A. Gallo (Mexico); trained by D. George in 1969	Merito-Pt. Diablo	Extensive culture and Mabé implants	Good results, but the commercial initiative did not progress because of political rivalries. Project abandoned with a large economic loss
1979 Shoei Shirai and K. Sano (Japan), agreement with the Secretaría de Pesca	La Paz Bay (Falsa Bay)	Prospection for natural beds attempt to install a pearl farm (Falsa Bay); pearl culture in wild individuals	Deceiving results concerning abundance of pearl oysters. Great mortality and rejection postsurgery. Project abandoned
1976–1978 government project	La Paz Bay. Install in Falsa Bay and Merito-Pt. Diablo	Assay of extensive culture	Acceptable results. Project abandoned because of budget shortfall and administrative shifts
1979–1980. K. Sano and Yamamoto, agreement with private group in La Paz	Extraction of wild stock from La Paz Bay and islands. Installations at Merito-Pt. Diablo	Pearl culture in wild individuals	Scandalous failure. Ransack of natural beds. Expensive installations were abandoned and later pillaged
1981–1982 Jaime Singh (CRIP-BCS, Secretaría de Pesca, Mexico)	La Paz Bay. Installs. In Falsa Bay	Assays of extensive culture; pearl culture in wild individuals	Good results overall. High mortality postsurgery. No pearls

(continued)

Table 27.2 (continued)

Date/actors	MPA and/or influence	Actions	Outcomes
1987 Manuel Mazón (CRIP-BCS, Mexico)	Laboratory research	Hatchery studies in *P. mazatlanica*	Gonad conditioning and larval growth and survival were acceptable. No fixation. Project abandoned
1988–1998 Pearl Oysters Research Group (GOP) of CIBNOR	La Paz Bay and islands nearby. Station network, installations in La Paz Bay included Espiritu Santo Island. Surveys on the north coast up to Bay of Los Ángeles	Research on extensive culture and pearl production in *P. mazatlanica* and *Pt. sterna*. Restocking in La Paz Bay by repopulation cells of these species and other commercial bivalves studied there	Harvest high-quality Mabé pearls in April 1992. Certified technology of extensive culture in 1998. Punctual experiments continued until 2004 (certification of Icon Pearls). Influence in the NOM-058 for pearl oyster farms and pearl culture in Mexico (2013)
2002–2004 Ingeniería y Síntesis (private group assisted by GOP/ CIBNOR)	Merito-Pt. Diablo	Pilot commercial, extensive culture of *P. mazatlanica* and Mabé	Harvest of 1500 high-quality Mabé. The project did not continue
From 2001 UABCS (Perlas del Cortéz)	Gaviota Island	Extensive culture of *Pt. sterna*	Smaller than the farm at Guaymas. Harvest is only Mabé so far, jewelry and nacre-based cosmetic products
Gourmet and marine aquarium sectors			
Late 1970s. Araceli Avilés and group (CRIP La Paz)	Falsa Bay	Research on fish and bivalve culture	Technologic basis
From 1986. Carlos Cáceres. UABCS	La Paz Bay and Espiritu Santo Island. Station network. Campus in Pichilingue	Research on extensive and semi-extensive culture of bivalves and pearl oysters. Restocking of corals	Detailed timeline of environmental parameters and mariculture indicators on the species. Owner of Perlas del Cortéz farm (*Pteria sterna* extensive culture
From 1990 Arturo Tripp CETMAR	Ensenada de La Paz, Falsa Bay	Research on extensive and semi-extensive culture of scallop *Argopecten ventricosus* and pen-shell *Pinna rugosa* and *Atrina maura*	Detailed timeline of environmental parameters and mariculture indicators on the species

(continued)

Table 27.2 (continued)

Date/actors	MPA and/or influence	Actions	Outcomes
From 2007. Miguel Robles. Acuacultura Robles	Commercial bivalve seed production in laboratory. San Juan Nepomuceno Barr	Major regional supplier of seed of the common commercial native species (Catarina scallop, both pen shells, oysters included the introduced Japanese *Crassostrea gigas* and *Ostrea sikamea*, and mule-paw cockles *Anadara* spp.)	Most of the product is received by producers in the west coast of BCS although a small farmer has recently installed in Ensenada de La Paz. Isolated experiments in Merito-Point Diablo (CIBNOR's installation)
From 2002 CIBNOR. Different users of the Aquaculture Program	Merito-Point Diablo, Gaviota Island, and intermittent sites in Espiritu Santo Island, Ensenada de La Paz, and the Bay	Research on extensive and semi-extensive culture of commercial fish (Lutjanids, *Seriola*, *Totoaba*), edible bivalves and pearl oysters, and ornamental marine species. Monitoring	Certified technologies have ensured the commercial production of fish in La Paz Bay and mainly bivalves in other sites of the west BCS coast
From the 1990s. CRIP, CICIMAR, CETMAR, UABCS. Different users (e.g., student thesis, sponsored projects, etc.)	Merito-Point Diablo, Gaviota Island, and intermittent sites in Espiritu Santo Island, Ensenada de La Paz, and the Bay	Diverse experiments of short duration, principally on bivalves and fish (snappers, Lutjanidae, and seabasses, Serranidae)	Important assets of data and information
2010–2014 fisherfolk cooperatives (SCPP) assisted by Mario Monteforte	Merito-Point Diablo	Polyspecific extensive culture of gourmet bivalves; special devices tested for ornamental marine species	A capacity-building action; promising results; abandon due to issues of group dynamics and finances (see Monteforte and Cariño 2018)
From 2012 to 2013. CIBNOR creates the aquaculture park Biohelix (*)	La Paz Bay, Ensenada de La Paz.	Commercial culture of Lutjanids, *Seriola* spp., *Totoaba*, and shrimp *L. vannamei* and *Penaeus californiensis* (brown shrimp)	Promote private investment in aquaculture. Fisherfolk groups often do not fit in this kind of models
Hot spot cases			
2002–2004 private Mexican enterprise foreign partners.	North Gaviota Island	Yellowfin tuna ranching	Issues of logistic, provision, and financial nature led the project to fail. Moving to semi-extensive culture of *Seriola lalandi* with CIBNOR support

(continued)

Table 27.2 (continued)

Date/actors	MPA and/or influence	Actions	Outcomes
2005 private foreign enterprise BIOTECMAR	Merito-Point Diablo	Cage culture of introduced hybrids of striped bass *Morone saxatilis x M. chrysops*	Project failed. High mortality from invasion of local hydrozoans (*Apolemia, Obelia, Lytocarpus*) (Monteforte et al. Monteforte-Sánchez et al. 2014)
2006–2007. Same enterprise	Same site	Cage culture of introduced (normal) *Sparus aurata*	Project failed. Legal problems, strong opposition in the locality. Escapees were confirmed in amounts estimated between 20,000 and 60,000 fertile individuals (Monteforte et al. Monteforte-Sánchez et al. 2014)

(*) The idea of industrial technologic parks was originally based on business centers for electronics, multimedia, software, engines, biomedics, etc. The Monterrey Technologic Institute of Superior Studies (ITESM) was the entryway of this trend to Mexico. Agriculture and livestock parks spread out since the 1990s in nations attached to mono-specific production (Soares-Silva et al. 2016). Several aquaculture-based parks were created at that time, like the TALIARTE in the Grand Canaria Island, the Cawthron Park in New Zealand, the Büsum Park in Germany (focused mainly in aquaculture equipment), and Mikimoto Pearl Island in Japan, just to mention some.

install a tuna ranch in 2003 (Monteforte 2008) and cage culture of exotic hybrid striped bass *Morone saxatilis* × *M. chrysops* and Mediterranean gilthead seabream *Sparus aurata* in 2006–2007 (Monteforte-Sánchez et al. 2014), despite their well-known invasive potential, particularly the second one.

These projects are considered hot spots in the Mexican mariculture and had corollaries at larger scale (Table 27.2). Tuna ranching in La Paz Bay (the MPA) was first experienced beyond Ensenada area (Baja California, north State of the Peninsula) where the largest farm is placed since the early 1990s (Lechuga-Devéze 2012). Unfortunately several problems led the project to fail, principally due to the level of depletion of tuna wild stock and low captures the boat had to go farther off and then trawl the capture back, suffering high mortality as a result. Additional difficulties to obtain fresh sardine to feed the tuna delivered food shortage. Sardine is essential for the sport-fishery sector as bait; therefore it raised harsh disputes (Monteforte 2008; Lechuga-Devéze 2012).

In reference to offshore cultivation of non-modified and genetically modified marine fish, turning them into exotic species had not been tried commercially at that time aside the salmon industry in course of expansion. The introduction of foreign species in mariculture was, and continues, being restricted or precautionary principally in developed countries (Europe, United States, Canada, Australia). Cage culture of hybrid striped bass had not been tried but in the United States by leisure fishery in dams or controlled reservoirs and some experiments of cage culture in New York Bay, Atlantic coast, neither *S. aurata* away the Mediterranean, yet it had been already introduced in Aqaba, Israel, and Canarias Islands (Monteforte-Sánchez et al. 2014). In La Paz Bay the project with hybrid striped bass failed because the cages and the culture area got infested by aggressive hydrozoans (*Lytocarpus nuttingi*, *Apolemia uvaria*, *Obelia* spp.), locally known as "grullas," a pest for swimmers and snorkelers, that led to high mortality to the juvenile fish in the cages. This natural defense seems quite an uncommon event in fish mariculture (Monteforte-Sánchez et al. 2014).

The venture with *S. aurata* yielded satisfactory results, probably they used repellent paint on the cages, or the fish was immune. Nonetheless, some groups of the local society had been opposing to both projects, in particular the invasive potential of *S. aurata* motivated greater awareness. There were queries concerning the legal import of the fish fry through the La Paz international airport, and the name of "golden snapper" (*pargo dorado*) that the entrepreneur invented to disguise the species like a member of the regional snappers (Lutjanidae—*pargo, huachinango*) raised further suspicions. By July 2007 one escapee was caught way inside Ensenada de La Paz (Balart et al. 2009), and during the next weeks more specimens caught in the Bay were donated by fishermen. The situation generated a massive movement against BIOTECMAR, the enterprise proprietary of both projects that culminated in the closure, extract, and sacrifice of the fish and the enactment of an official survey to estimate the impact. Persons of academic institutions in La Paz (CIBNOR, UABCS and CICIMAR) and of environmental departments of the municipal government (PROFEPA, CONANP, SEMARNAT) carried out explorations (25 November to 5 December) on snorkeler, SCUBA, and fishing nets that resulted in 35 individuals observed free in the MPA area and 3 more captured with the nets in different parts of the Bay. In addition, contradictions on the amount of fish fry entered by the airport, the notification to the corresponding departments (SAGARPA/CONAPESCA) as to how many fish were placed alive in the cages, and the report of individuals extracted at the time of closure remind us that an amount between 20,000 and 60,000 healthy young adults, many of them aged enough to be fertile, escaped from the MPA 12 years ago (Monteforte-Sánchez et al. 2014). There has been some physical captures and reports in decreasing frequency to none in the last 4–5 years although once in a while news arrive from distant sites (e.g., Loreto, Cabo San Lucas, even Magdalena Bay and Guaymas). This is not trustable enough to surmise any forecast.

In synthesis, the contribution of La Paz Bay to the Mexican mariculture has been minimal in terms of commercial role but significant in terms of accumulated knowledge and influence at a larger scale. For example, La Paz has always been source of multidisciplinary information related to the world history of fisheries, commerce, and culture of nacre and pearls, and the MPA is reputed to be the cradle of the modern science and technology available for *P. sterna* and *P. mazatlanica* (Table 27.2).

In fact, much of the standards for extensive culture and pearl production in Mexico,[8] and used in recent ventures in Latin America, are assigned to results of research studies carried out in the MPA (see Monteforte 2005, 2013). However, diverse factors, mainly of oceanographic nature (e.g., lower temperature average, higher primary productivity, larger coastal lagoon systems), have directed the commercial mariculture of BCS to the Pacific coast aiming in Japanese oysters *Crassostrea gigas* and *C. sikamea* and a few white-legged shrimp *L. vannamei* farms. There are recent development in La Paz Bay (Fig. 27.4) with cage culture and aquapods with this shrimp at its sister species, the brown shrimp *Penaeus californiensis* and diverse fish (yellowtail *S. lalandi* and *S. rivoliana*, snappers *Lutjanus argentiventris* and *L. peru*, sand seabass *Paralabrax maculatofasciatus*, and *Totoaba macdonaldi*). Most of these projects belong to private transnational enterprises attached to Biohelix, the CIBNOR's aquaculture park (Table 27.2). Nevertheless, apart from the private pearl oyster farm at Gaviota Island rooted to the UABCS, commercial mariculture in the Bay accounts for just three or four pilot scale records of pearl oysters and gourmet bivalve farms of short duration, all of them placed in the MPA under study in this chapter.

27.3.2 Mariculture into the MPA Precint: What Works?

The mariculture chronology in La Paz Bay and the MPA reveals the lack of extension programs to promote integration of fisherfolk cooperatives (SCPPs) through mariculture SMEs, and in the other side there are private projects that monopolize propitious sites in a disarranged pattern (Fig. 27.4) commonly seen around the Gulf of California (Sonora, Sinaloa, and Nayarit) and other coastal sites in Mexico. Moreover, the information of each lease[9] could suggest that certified technology is available for all of the commercial species of fish and bivalves—pearl oysters too, inhabiting the Gulf of California (CONAPESCA 2018). Undoubtedly the modern technology should be able to ensure this to happen; even so, the reality does not reflect productive benefits from direct applications. Figure 27.5 serves to illustrate the statement on three cases that are ordinary in many areas with attributes for mari-

[8] NOM-058-SAG/PESC/SEMARNAT-2013 regulates the pearl oyster cultivation in Mexico: mother-of-pearl (*Pinctada mazatlanica*), nacre shell (*Pteria sterna*), Atlantic mother-of-pearl (*Pinctada imbricata*), and Atlantic winged pearl oyster (*Pteria colymbus*).

[9] SAGARPA and CONAPESCA regulate two leasing forms of productive aquaculture for private groups, cooperatives, or other kinds of entrepreneurial associations: a Permit for Aquaculture Fomentation (PAF) renewable by merits every 4 years (e.g., proofs of good work) that eventually can be leveled up to an Aquaculture Concession (CA) of 10 or 20 years, or it is directly conceded if the entrepreneurs fulfill technical and financial requirements. The average surface for the licenses is 10 Ha, up to 20 Ha in some cases of fish farms, amberjack and tuna, in particular. Detailed information of PAFs and CAs is of public access online through a tool called Acuasesor (https://acuasesor.conapesca.gob.mx/) which contains all about fisheries and aquaculture production in Mexico.

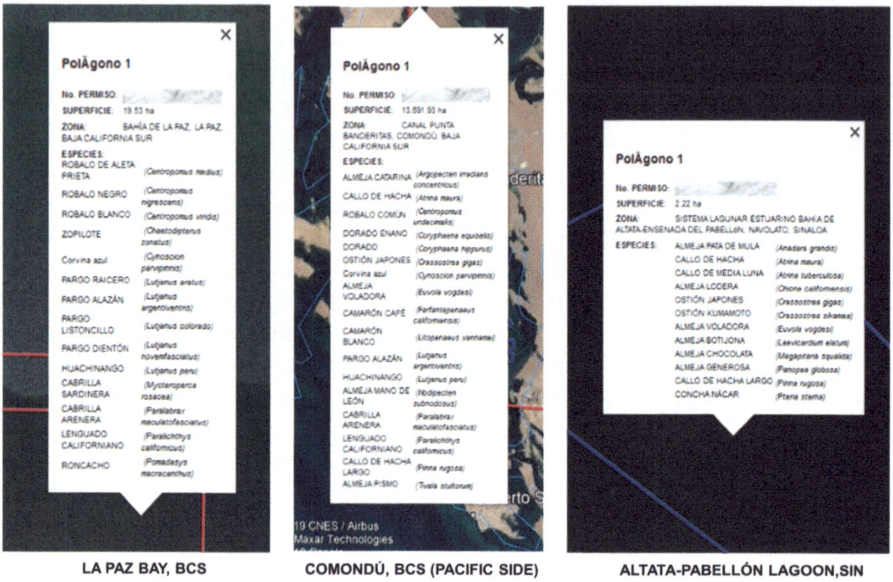

Fig. 27.5 The Acuasesor KML/KMZ layers deploy summary information on each license in aquatic areas in Mexico. The database contains nearly 2000 records in coastal marine areas. This figure presents three examples. (Modified from Google Earth with the KML/KMZ layers for PAFs and CAs available in Acuasesor SAGARPA/CONAPESCA. Updated to 2018)

culture in Mexico: La Paz Bay (the figure corresponds to the MPA in scrutiny here), and the coastal lagoons of Magdalena on the Pacific coast of BCS, and Pabellón-Altata, this one in Sinaloa, southeast of the Gulf of California. It is conceivable that the bioecology of bivalves would allow rustic less costly schemes than fish culture, particularly if laboratory generations are part of the equation; nevertheless, inferences on Fig. 27.5 may yield unlikely indicators. Perhaps some of the farms could have achieved small and sporadic harvests of some species for which the cultivation technology is sufficiently advanced or in process of certification (e.g., pilot harvests), so announcing a long list of species seems justifiable, or at least technically feasible. In reality, just a handful of PAF (Permit for Aquaculture Fomentation) or CA (Aquaculture Concession) sited in La Paz Bay have commercialized occasional pilot harvests during the last 38 years.[10] The current ventures installed in the study area (Fig. 27.4) announce around 25 species of fish and bivalves overall; this is likely true in terms technologic feasibility (some species still are in experimental

[10] The first Statistical Yearbook of Aquaculture and Fisheries was published in 1980 (SAGARPA/CONAPESCA) and has been continued more or less constantly every 2 years. Since then, the Mexican commercial mariculture is concentrated on the east coast of the Gulf of California and the Pacific coast of the Baja California Peninsula. White-legged shrimp and Japanese oysters are the flag production. Notably, the only two pearl oyster farms in Mexico are placed in Guaymas and La Paz. A third one in Cozumel Island is just starting, and there are more in prospect in the Gulf of California (La Paz, Loreto, and near Cabo San Lucas).

stage), but not in terms of commercial production. Furthermore, most of the licenses (10 Ha in average, up to 20 Ha in some cases of fish farm) belong to private groups well established in the market. Of course, there are professional specialists in aquaculture (e.g., ex-alumni and scientists of academic institutions) that create entrepreneurial entities. The increasing interest to develop mariculture in La Paz Bay is certainly novel though a rapid progress is expected; however, academic institutions display low participation except as a logistic need within the technologic park (e.g., Biohelix in La Paz or similar schemes like the CREMES and PIDEMAR in Kino Bay, Sonora), or source of information and solutions, such as diseases, locally made food, or support to obtain the licenses (PAF, CA) such as elaboration of technical documents and guidance in administrative procedures, even financial funds from the government. Fisherfolks are aware that turning into mariculture farmers would bring them better income; hence they often seek assistance in government departments like CONAPESCA or in the academic institutions. This is a global model that has produced both good and deficient results.

In Mexico, various obstacles prevail for the SCPPs, for example, they are insolvent to afford counterpart funds, a mandatory requirement in most government programs, while knowledge-holder scientists generally confer social extension activity less priority because it is underrated in academic evaluation criteria. This convergent factors marginalize small entrepreneurs and cooperatives considering that the installation and operation of a minimally profitable aquaculture venture often exceeds their financial, logistical, and technological capacity, to which we must add that for fisherfolk that live from their daily work, the waiting for dividends from the harvest may be untenable, not to mention the influence of complex group dynamics and conflicts into or among different groups. Table 27.2 includes one of these projects occurred in the MPA between 2010 and 2014. To present it in minimalistic form, the project attempted to join two SCPPs in developing polyspecific farming[11]of gourmet bivalves, pearl oysters, and ornamental marine species at pilot scale (Monteforte and Cariño 2018). Previous socioeconomic surveys in La Paz led us to contextualize this bipartite SME as a means to optimize scarce funds slowly gained and share the logistic management of the farm. We needed almost 2 years to acquire the elementary equipment and material to manufacture the cultivation systems and artifacts and install them in Point Diablo area, including a small temporal work camp on the Cove. Once everything settled and in function on the site, the next 7 months yielded promising results. The fisherfolk attained good training level, and nearly 40,000 specimens of gourmet bivalves distributed in nine species (Fig. 27.6) were growing to young adult stage, including 15,000 *P. sterna* specimens (Monteforte and Cariño 2018). A prototype device (inspired on the functional principle of a "portable" artificial reef) to cultivate ornamental marine species in extensive culture modality was in test process. Nevertheless, the SCCPPs showed disproportionate commitment to the project, likely because one partner owned better elements to

[11] Plans included extensive culture (wild spat collection and grow-out phase in the sea) and semi-extensive culture (based on the reception of laboratory-reared spat).

Fig. 27.6 Exhibit of some fieldwork activities in the mariculture project with two fisherfolk cooperatives (SCPP) in Diablo Point area (2010–2014). (**a**) Juveniles of commercial bivalves -pen shell, Catarina scallop, lion-paw scallop, winged nacre shell, and mule-hoof clam on the mesh board underneath. (**b**) and (**c**) fisherfolk of the 'good' SCPP in a spat harvest session. (**d**) A maintenance session of Nestier trays containing livestock of bivalve juveniles. (Photos in situ taken by the author (see text and Table 27.2))

accomplish the tasks than the other one (e.g., financial solvency—in their scale, infrastructure in La Paz such as workshop and storage area, two vehicles and two *panga* boats with 95 HP outboard engine, one air compressor diving gear, etc.). In particular, this SCPP (6–8 constant workers) was more convinced toward the project and therefore proved more enthusiastic participation (Fig. 27.6) even though the other SCPP had more than 20 members. However, something else beyond financial constraints and possible jealousies was instigating psychologic stress in the backgrounds (Monteforte and Cariño 2018). We discovered, too late (how did that come out?), that the senior presidents of the SCPPs had old feuds pendent when the level of hostilities was nearly uncontrollable. Within a couple of weeks, the installations were abandoned and vandalized, yet we were able to save about 12,000 specimens that were used for repopulation (throwing them overboard, basically).

At the bottom line, the double-partnership was a mistake. As Perkins et al. (2002) put it: "Psychological factors point to what motivates individuals to participate in particular settings and behaviors, how to maintain that participation, and how those motivations and behaviors interact with various setting and organizational characteristics to promote effective social capital." In addition, administrative issues were an impasse because aquaculture licenses are given to a single beneficiary and cover only its territory (polygon), even if contiguous projects are the same and/or directed by the same expert. This pyramidal policy amplifies the cost of licenses and the making of bulky files differentiated only by the name of the licensee and the polygon shape. Other administrative variables also had parallel impacts; for example, the cultivation site is part of Balandra-Merito MPA, and some of the targeted species are under some level of protection (e.g., *P. mazatlanica*, *Pt. sterna*, the lion paw scallop *Nodipecten subnudosus*, and various fish and invertebrates identified as ornamental). However, aquaculture permits had been issued in the MPA before the formal declaration, in particular the CIBNOR has carried out research studies in that location since 1988 (Table 27.2). Ultimately, these issues should have been a minor hindrance; we always stressed that repopulation would be carried out and that natural populations would not be touched. The farm itself was seen as a bank of larval dispersal. Nonetheless, the denouement of experience showed the laxity of the governmental bureaucracy and a clear manipulation of federal resources.

27.4 Final Reflections

Mariculture comes in many modalities (species, type of technologies, degree of intensification, etc.). Enterprises and rustic community-based small scale, there exists a wide range of practices, types, and scales that could provide opportunities for greater synergies within multiple-use MPAs. Multiple-use designs could also offer alternative livelihoods to develop sustainable financing opportunities for the MPA management (e.g., Brugère et al. 2008; Berkes and Davidson-Hunt 2010; Davidson-Hunt et al. 2012; Le Gouvello et al. 2017). The assessment of multiple-use in the case study seems predetermined by natural assets of the MPA related to ecotourisand to likelihood for mariculture SME models. Gourmet bivalves, pearl oysters, ornamental marine species, and pan-sized fish species (Lutjanidae and Serranidae), either in mono or polyspecific modality, may be advised as best conversion/diversification choice for the fisherfolk in La Paz Bay so that the MPA could acquire a socioeconomic role by re-activating its proven potential as productive site. It stands to reason that fisherfolks will be more likely to invest and stay rooted in places that are worth caring about, places with appealing local identity, an ambiance of belonging, and a sense of place. In corollary, it is clear that fisherfolk fit into distinct frameworks in accordance with species type and culture modalities that should ensure results within reasonable timeframes in order to prevent them from losing cohesion and interest towards a project. Since Balandra-Merito MPA has been submitted to thorough research studies on mariculture, and on environment and socioeconomic lines, a subsequent review of capacity-building actions might vindicate the fact that specific approaches are required in order to integrate target groups in collective projects without disrupting the sense of being local. Fostering a sense of place or nurturing local identity and authenticity begins by investigating what site-specific distinctive potentials determine a case study and how they could be used to raise community capital. This principle should guide to better understanding the profile of local and regional players (e.g., Emery and Flora 2006; UN 2011; Davidson-Hunt et al. 2012). Therefore, the livelihood of fisherfolk communities in the context of socio-ecological systems may be interpreted by behavioral and/or stance expressions of group dynamics toward the management of common-pools, such as altruism, self-interest, and conformism (Cárdenas and Ostrom 2004; Vélez et al. 2009); collaboration (e.g., willingness to engage in commitments, a sense of appropriation, solidarity, and collective participation) (Leopold et al. 2013; Bennett and Dearden 2014a, b; Barratt et al. 2015); and internal and external corruption (Tonoyan et al. 2010; Barratt et al. 2015). Moreover, heterogeneous perceptions of conservation and protection introduce broadened meanings to these expressions. Thus, questions arise when technology-based research is conducted on production processes such as those involved in aquaculture, in particular when matters related to gender and/or small communities, and/or isolation and marginalization in general, hinder collective/proactive actions.

It could be concluded by arguing that creating added-value in Balandra-Merito MPA is feasible if the civil society and knowledge-holders reconsider their capacity for empowerment in the management of this emblematic place. The acquisition of skills to develop a mariculture in SMEs is a first step, but not enough. What follows is to balance the terms with the use, management, and control of the sources and benefits related to the MPA. Advantages and disadvantages will have to be evaluated upon the expectation of ideal sceneries that are difficult to generate. Evidence indicates that every conversion/ diversification process has to face complex relationships between society and nature, with distinct expressions even among specific cases locally. In addition, perceptions of privatization, conservation, and production are deeply interwoven, as are political, socioeconomic, and long-term histories in which cultural and psychological factors and group dynamics engender crucial structures.

Acknowledgments Big thanks for the figures to Patricia González-Zamorano and Joaquín Rivera-Rosas of the Laboratory of Geographic Information Systems and also to Gerardo R. Hernández-García of the Department of Extension and Scientific Diffusion, both CIBNOR's. All the structural information in reference to the study site comes from published work either by the author, or extracted from academic databases, or can be consulted in government portals accessible to the public in internet. In the list of pioneers presented in Table 27.2 probably I forgot to add more than one, or nobody. My apologies. Finally, some may criticize the excessive auto-citation: I cannot help it; this MPA has been my professional fieldwork for nearly 30 years.

References

Adame MF, Najera E, Lovelock CE, Brown CJ (2018) Avoided emissions and conservation of scrub mangroves: potential for a Blue Carbon project in the Gulf of California, Mexico. Biol Lett 14. https://doi.org/10.1098/rsbl.2018.0400

Agardy T, Di Sciara GN, Christie P (2011) Mind the gap: addressing the shortcomings of marine protected areas through large scale marine spatial planning. Mar Policy 35(2):226–232

Anderson R (2017) Roads, value, and dispossession in Baja California Sur, Mexico. Eco Anthropol 4(1):7–21

Arreola-Lizárraga JA, Flores-Verdugo FJ, Ortega-Rubio A (2004) Structure and litterfall of an arid mangrove stand on the Gulf of California, Mexico. Aquat Bot 79(2):137–143

Ateweberhan M, Hudson J, Rougier A, Jiddawi NS, Msuya F, Stead SM, Harris A (2018) Community based aquaculture in the western Indian Ocean: challenges and opportunities for developing sustainable coastal livelihoods. Eco Soc 23(4):17. https://doi.org/10.5751/ES-10411-230417

Awortwi N (2012) The riddle of community development: factors influencing participation and management in twenty-nine African and Latin American communities. Community Dev J 48(1):89–104

Balart EF, Pérez-Urbiola JC, Campos-Dávila L, Monteforte M, & Ortega-Rubio A (2009) On the first record of a potentially harmful fish, Sparus aurata in the Gulf of California. Biological invasions, 11(3):547–550

Barratt C, Seeley J, Allison EH (2015) Lacking the means or the motivation? Exploring the experience of community-based resource management among fisherfolk on Lake Victoria, Uganda. Eur J Dev Res 27(2):257–272

Bennett NJ, Dearden P (2014a) Why local people do not support conservation? Community perceptions of marine protected area livelihood impacts, governance and management in Thailand. Mar Policy 44:107–116

Bennett NJ, Dearden P (2014b) From measuring outcomes to providing inputs: governance, management, and local development for more effective marine protected areas. Mar Policy 50:96–110

Bennett NJ, Govan H, Satterfield T (2015) Ocean grabbing. Mar Policy 57:61–68

Berkes F, Davidson-Hunt TJ (2010) Innovating through commons use: community-based enterprises. Int J Commons 4(1):4 pp

Boley B, Nickerson N, Bosak K (2016) Abstract: a critical examination exploring the differences between Geotourism and ecotourism. Travel and Tourism Research Association: Advancing Tourism Research Globally. University of Massachusetts Amherst. https://scholarworks.umass.edu/ttra/2009/Abstracts/1

Bondad-Reantaso MG, Bueno PB, Demaine H, Pongthanapanich T (2009) Development of an indicator system for measuring the contribution of small-scale aquaculture to sustainable rural development. In: Bondad- Reantaso MG, Prein M (eds) Measuring the contribution of small-scale aquaculture: an assessment. FAO Fisheries and Aquaculture Technical Paper 534, pp 161–179

Bottema MJ, Bush SR (2012) The durability of private sector-led marine conservation: a case study of two entrepreneurial marine protected areas in Indonesia. Ocean Coast Manag 61:38–48

Briscoe DK, Maxwell M, Kudela R, Crowder LB, Croll D (2016) Are we missing important areas in pelagic marine conservation? Redefining conservation hotspots in the ocean. Endanger Species Res 29(3):229–237

Brugère C, Felsing M, Kusakabe K, Kelkar G, Colls (2001) Women in aquaculture. Asia-Pacific economic cooperation, project FWG 03/99. Ed. University of Stirling, 61 pp

Brugère C, Holvoet K, Allison E (2008) Livelihood diversification in coastal and inland fishing communities: misconceptions, evidence and implications for fisheries management, Working paper, Sustainable Fisheries Livelihoods Programme (SFLP). FAO/DFID, Rome

Cadena-Cárdenas L, Méndez-Rodríguez L, Zenteno-Savín T, García-Hernández J, Acosta-Vargas B (2009) Heavy metal levels in marine mollusks from areas with, or without, mining activities along the Gulf of California, Mexico. Arch Environ Contam Toxicol 57(1):96–102

Cárdenas JC, Ostrom E (2004) What do people bring into the game? Experiments in the field about cooperation in the commons. Agric Syst 82:307–326

Cariño M (1998) El porvenir de la Baja California está en sus mare. In: Vida y legado de Gastón J. Vives Gourieux, el primer maricultor de América. Ed. H. Congreso del Estado de Baja California Sur, México

Cariño M and Monteforte M (1999) El primer emporio perlero sustentable del mundo: La Compañía Criadora de Concha y Perla de la Baja California, S.A. y perspectivas para México. Editorial UABCS, CONACULTA, FONCA. México D.F

Cariño M, Cordero-Sauceda NL, Zariñán J (2008) Procesos de conservación impulsados por la gestión de las ONG que trabajan en Baja California Sur. In: Cariño M, Monteforte M (Coords.) Del saqueo a la conservación. Historia ambiental contemporánea de Baja California Sur (1940–2003). Ed. UABCS, SEMARNAT, INE, CONACYT, pp 673–722

Cuéllar-Lugo MB, Hoyos AA, Juárez-Sánchez JP, Mendiola JLR, López FG (2018) Normative and institutional evolution of aquaculture in Mexico. Agricultura, Sociedad y Desarrollo 15(4):541–564

CONAPESCA (2018) Anuario Estadístico de Pesca y Acuacultura. SAGARPA/CONAPESCA https://acuasesor.conapesca.gob.mx/

Cundill G, Bezerra JC, De Vos A, Ntingana N (2017) Beyond benefit sharing: place attachment and the importance of access to protected areas for surrounding communities. Ecosyst Serv 28:140–148

Dahlstrom MF (2014) Using narratives and storytelling to communicate science with non-expert audiences. PNAS 111(4):13614–13620

Davidson-Hunt IJ, Turner KL, Mead ATP, Cabrera-Lopez J et al (2012) Biocultural design: a new conceptual framework for sustainable development in rural indigenous and local communities. SAPIENS 5(2):33–45

DOF [Diario Oficial de la Federación] 30/11/2012. Secretaría de Gobernación, México. http://dof.gob.mx/nota_detalle.php

Du W, Penabaz-Wiley SM, Njeru AM, Kinoshita I (2015) Models and approaches for integrating protected areas with their surroundings: a review of the literature. Sustainability 7(7):8151–8177

Emery M, Flora C (2006) Spiraling-up: mapping community transformation with community capitals framework. Community Dev 37(1):19–35

Ferse SC, Knittweis L, Krause G, Maddusila A, Glaser M (2012) Livelihoods of ornamental coral fishermen in South Sulawesi/Indonesia: implications for management. Coast Manag 40(5):525–555

Fraga J, Jesus A (2008) Coastal and marine protected areas in Mexico. Samudra monograph, International Collective in Support of Fisherworkers (ICSF), Chennai, 97 pp

Gámez AE (2008) Turismo y sustentabilidad a la vera de áreas naturales protegidas. In: AE Gámez (Ed.). Turismo y sustentabilidad en Cabo Pulmo, BCS, Pub. San Diego State University, pp 11–30

Gámez AE, Angeles M (2010) Borders within. Tourism growth, migration and regional polarization in Baja California Sur (Mexico). Borderl Stud 25(1):1–18

Guzmán-Vizcarra E, Ibáñez-Pérez RM (2018) Percepción de necesidades y cohesión social en comunidades pesqueras de México. Análisis de sociedades cooperativas de la Bahía de la Paz, BCS, México. Revista Contribuciones a las Ciencias Sociales (Abril 2018) https://www.eumed.net/rev/cccss/2018/04/sociedades-cooperativas-mexico.html. Accessed 12 Dec 2019

Haeffner M, Baggio JA, Galvin K (2018) Investigating environmental migration and other rural drought adaptation strategies in Baja California Sur, Mexico. Reg Environ Chang 18(5):1495–1507

Havard L, Brigand L, Cariño M (2015) Stakeholder participation in decision-making processes for marine and coastal protected areas: case studies of the south-western Gulf of California, Mexico. Ocean Coast Manag 116:116–131

Hill A (2017) Blue grabbing: reviewing marine conservation in Redang island marine park, Malaysia. Geoforum 79:97–100

Hjalager AM, Johansen PH (2013) Food tourism in protected areas–sustainability for producers, the environment and tourism? J Sustain Tour 21(3):417–433

Le Gouvello R, Hochart LE, Laffoley D, Simard F, Andrade C, Angel D et al (2017) Aquaculture and marine protected areas: potential opportunities and synergies. Aquat Conserv Mar Freshwat Ecosyst 27:138–150

Lechuga-Devéze CH (2012) Una piscicultura sustentable: Bahía de La Paz, Golfo de California, un caso de estudio. Editorial Académica Española, Starbrucken

Leopold M, Beckensteiner J, Kaltavara J, Raubani J, Caillon S (2013) Community-based management of near-shore fisheries in Vanuatu: what works? Mar Policy 42:167–176
López-Cortés D, Bustilllos-Guzmán J, Hernández-Sandoval F, Núñez-Vázquez EJ (2012) Ambiente Acuático. In: Lechuga-Devéze CH (ed) Una piscicultura sustentable: Bahía de La Paz, Golfo de California, un caso de estudio. Editorial Académica Española, Starbrucken, pp 31–62
Ma KL, Liao I, Frazier J, Hauser H, Kostis HN (2011) Scientific storytelling using visualization. IEEE Comput Graph Appl 32(1):12–19
McEvoy J (2014) Desalination and water security: the promise and perils of a technological fix to the water crisis in Baja California Sur, Mexico. Water Altern 7(3):518–541
Monteforte M (2005) Ecología, biología y cultivo extensivo de la madreperla de Calafia, Pinctada mazatlanica, y la concha nácar arcoíris, Pteria sterna, en Bahía de La Paz, Baja California Sur, México. Post-PhD thesis, Universidad de La Habana, Cuba
Monteforte M (2008) Modelos de Desarrollo Acuícola en Baja California Sur: ¿Saqueo o Alternativa Sustentable? In: M Cariño and M Monteforte (Coords) Del Saqueo a la Conservación. Historia Ambiental Contemporánea de Baja California Sur (1940-2003). Editorial UABCS, SEMARNAT, INE, CONACYT (México, D.F.), pp 337–364
Monteforte M (2013) Instalación y operación de granjas perleras. Un manual para Pinctada mazatlanica y Pteria sterna. Editorial Académica Española, Starbucken
Monteforte M, Cariño M (2009) El Mar de Cortés no existe. Biodiversitas 86:12–15
Monteforte M, Cariño M (2011) Maricultura: Oportunidad de conversión sustentable para la pesca ribereña. Ciencia, tecnología e innovación para el desarrollo de México. PCTI, 78. http://pcti.mx. Accessed 26 Feb 2018
Monteforte M, Cariño M (2013) Condiciones para el desarrollo de granjas perleras y producción de perlas: Estado del arte, potencial y perspectivas. Rev Biol Mar Oceanogr 48(1):1–16
Monteforte M, Cariño M (2018) A history of nacre and pearls in the Gulf of California. In: Price LL, Narchi NE (eds) Coastal heritage and cultural resilience. Springer, Cham, pp 79–112
Monteforte-Sánchez M, Balart E, Pérez-Urbiola JC, Ortega-Rubio A (2014) Rentabilidad empresarial vs. protección contra especies acuáticas invasoras: Dilema para el desarrollo de la acuacultura, caso Bahía de La Paz, Baja California Sur, México. In: Low-Pfeng AM, Quijón PA, Recagno EMP (eds) SEMARNAT, INECC, University of Prince Edwards, pp 567–602
Morales-Díaz A (2015) Historia de la Acuicultura en México. Peces, Crustáceos y Moluscos. AGT Editor, México D.F.
Murray G (2007) Constructing paradise: the impacts of big tourism in the Mexican coastal zone. Coast Manag 35(2–3):339–355
Obeso-Nieblas M, Gaviño RJH, Jiménez-Illescas AR, Rafael-Shirasago GB (2002) Simulación numérica de la circulación por marea y viento del noroeste y sur en la Bahía de La Paz, B.C.S. Oceánides 17(1):1–12
Obeso-Nieblas M, Gaviño-Rodríguez JH, Obeso-Huerta H, Muñoz-Casillas SI (2014) Variabilidad espacial termohalina, masas de agua y circulación geostrófica en Bahía de La Paz, Golfo de California. Rev Biol Mar Oceanogr 49(3):413–426
Oldekop JA, Holmes G, Harris WE, Evans KL (2016) A global assessment of the social and conservation outcomes of protected areas. Conserv Biol 30(1):133–141
Perkins DD, Hughey J, Speer PW (2002) Community psychology perspectives on social capital theory and community development practice. Community Dev Soc J 33(1):33–52
Segi S (2014) Protecting or pilfering? Neoliberal conservationist marine protected areas in the experience of coastal Granada, the Philippines. Hum Ecol 42(4):565–575
Silverberg N, Aguirre-Bahena F, Mucci A (2014) Time-series measurements of settling particulate matter in Alfonso Basin, La Paz Bay, southwestern gulf of California. Cont Shelf Res 84:169–187
Slater MJ, Mgaya YD, Mill AC, Rushton SP, Stead SM (2013) Effect of social and economic drivers on choosing aquaculture as a coastal livelihood. Ocean Coast Manag 73:22–30
Soares-Silva AL, das-Chagas-Teles Â, Machado-de-Aragão-Gomes I, y Leitão-Russo, S (2016) Ciencia y parques tecnológicos. Evidencia en América, Asia y Europa. Razón y Palabra, 20(2–93):473–482

Tonoyan V, Strohmeyer R, Habib M, Perlitz M (2010) Corruption and entrepreneurship: how formal and informal institutions shape small firm behavior in transition and mature market economies. Entrep Theory Pract 34(5):803–832
UN [United Nations University Institute of Advanced Studies Operating Unit Ishikawa/Kanazawa] (2011) Biological and cultural diversity in coastal communities: exploring the potential of Satoumi for implementing the ecosystem approach in the Japanese archipelago. Secretariat of the Convention on Biological Diversity, Montreal, Technical Series no 61
Vélez MA, Stranlund JK, James JM (2009) What motivates common pool resource users? Experimental evidence from the field. J Econ Behav Organ 70(3):485–489
Woodhouse E, Bedelian C, Dawson N, Barnes P (2018) Social impacts of protected areas: exploring evidence of trade-offs and synergies. In: Schreckenberg K, Mace G, Poudyal M (eds) Ecosystem services and poverty alleviation. Routledge, Taylor & Francis Pubs, London, pp 222–239

Chapter 28
Community Water Management and Conservation in Cabo Pulmo National Park (Baja California Sur, Mexico)

Alba E. Gámez, Juan Carlos Graciano, Hitandehui Tovar, and Miguel Palmeros

Abstract Understanding how social and economic aspects influence the use and management of environmental systems is key to developing long-term approaches to conservation in natural protected areas. Relying on official data about the Cabo Pulmo Aquifer and on-site observation, this chapter explores water management practices in the town established next to Cabo Pulmo National Park (Baja California Sur, Mexico), a region which has been set as an example of community organization for successful conservation. It also elaborates on the social implications that private and community types of water management have in a context of tourism exploitation of natural resources in the Park and the human right to water.

Keywords Community water management · Conservation · Cabo Pulmo National Park · Baja California Sur · Mexico

28.1 Introduction

This chapter explores water management in the community established off Cabo Pulmo National Park (Baja California Sur, Mexico), which has been set as an example of social organization for successful marine conservation (Cariño et al. 2008; CONANP-GIZ 2017) in the form of a natural protected area (NPA). The town of Cabo Pulmo is located on the Tropic of Cancer, 169 kilometers (km) away from La Paz, capital of the state, and 108 km by road and 57 km by a dirt road from San José

A. E. Gámez (✉) · J. C. Graciano
Department of Economics, Universidad Autónoma de Baja California Sur, La Paz, Baja California Sur, Mexico
e-mail: agamez@uabcs.mx; jgraciano@uabcs.mx

H. Tovar · M. Palmeros
Marine Conservation Program, Sociedad de Historia Natural Niparajá, La Paz, Baja California Sur, México
e-mail: hitatovar@niparaja.org; palmeros@niparaja.org

A. Ortega-Rubio (ed.), *Socio-ecological Studies in Natural Protected Areas*,
https://doi.org/10.1007/978-3-030-47264-1_28

del Cabo, one of the cities that form the Los Cabos Corridor, which constitutes one of the main sun-beach tourist destinations in Mexico.

Over a century, Cabo Pulmo had suffered intense overexploitation of its various resources due to disorderly commercial fishing. Poverty predominated in the community, and its inhabitants realized that fishing resources were running out, because each time they had to go further into the sea and often returned without fish (Castro 2014). In 1995, the government declared Cabo Pulmo a NPA due to the environmental value of its coral reef, the scientific research carried out by local academics, the support government as well as national and international nongovernmental agencies provided, and the organizational willingness local inhabitants displayed to achieve such an end. Scuba diving, snorkeling, and other aquatic tourism activities replaced fisheries, and community members of Cabo Pulmo even became involved in the conservation monitoring developed by the government, academics, and civil society organizations.

The Cabo Pulmo reef is home to a variety of species converging from the Pangeic, Californian, and Indo-Pacific biogeographic provinces, which makes it one of the areas of highest biological diversity in the Mexican Pacific Coast. It is the northernmost coral reef in the Americas, the only hard reef in the entire peninsula of Baja California, and one of the only three living reefs left in all of North America, with an estimated age of more than 20,000 years (Glynn and McIntyre 1977; SEGOB 1995; CONANP 2006).

25 years after the NPA decree, Cabo Pulmo (both community and Park) remains a site of interest for conservation agencies that, in many cases, keep a permanent office in the area. Regarding its ecological success, in a period of 14 years since its creation, biomass increased by more than 460%, which made it, at the time, the most successful marine reserve in the world (Aburto-Oropeza et al. 2011; Cosío 2015). Key to this has been the commitment to refrain from fishing within the NPA, even for family consumption purposes. The accompaniment of environmental authorities and about 20 civil society organizations (Niparaja 2019) has also been relevant, as well as the abovementioned national and foreign collaboration networks against proposals for the establishment of massive tourism projects (mega developments) in La Ribera, in the immediate vicinity of Cabo Pulmo (AIDA 2013; Cosío 2015).

Alternative tourism and lodging activities have flourished with the active participation of community members. A recent estimate valued in USD 3.73 million the direct benefit for Cabo Pulmo inhabitants, while overall visitor expenditures (8638 people a year) in the amount of USD 5.8 million remained in the region (CONANP-GIZ 2017). Other studies even account for nearly 19,500 registered tourists and USD 1.7 million but just for Scuba diving and snorkeling services in 2017 (Niparajá 2019). An economic and lifestyle transformation has occurred that is represented in the number of business and improvement of some services and life conditions.

At present, the fishing community became a tourist community, where 91 businesses operate providing tourism aquatic services, lodging, souvenirs, meals, and beverages (Niparajá 2019). However, this has not implied the elimination of conflict or challenges stemming from an increasingly socially diverse Cabopulmeño community, in spite of its still small size. Those conflicts and challenges are mostly related to land tenure, since property rights are not fully established, as well as to intracommunity power relations regarding the use and access to natural resources (Barrera 2018; Anderson 2015; Menares 2008). Contributing factors may be that the creation of the 71.11-square-kilometer (27.46 sq. mi) NPA was 99% marine, thus providing little guidance about the management and conservation of adjacent land to the Park; also, current tourism and urban growth was not clearly foreseeable at the time the process of formal conservation started. In any case, the massive growth of tourism and real-estate investment in the municipality of Los Cabos (a model which is expanding across the state) add pressure to natural resources and people in Cabo Pulmo (Graciano et al. 2019).

There is a clear social divide in Cabo Pulmo represented by the distribution and type of households. Original fishermen and their families share their space with a significant expatriate community, as well as with Mexicans from other regions who arrive searching for jobs and the rustic lifestyle developed in this rural town, which lacks paved roads, urban planning, connections to the grid, and health or educational services. On the one hand, 117 *locals* (Cabo Pulmo's original inhabitants and their subsequent offspring) plus some foreigners and other Mexicans long ago integrated into the community account for 35 scattered houses. On the other, The Resort, a gated community on the beach whose managers also offer dive and lodging packages to tourists, accommodates 154 expatriates mostly from the United States and Canada, who enjoy sun-powered water service and electricity, as well as other amenities. In addition, 56 foreigners lease properties to up to 215 occupants (Niparajá 2019). During the high season, from autumn to spring, Cabo Pulmo becomes a busy community, a situation that leaves official government population census data (generally gathered in May) far behind.

Understanding how social and economic aspects influence the use and management of environmental systems is key to developing long-term approaches to conservation, especially in natural protected areas. This is particularly so in cases of relatively isolated communities in water scarce regions, such as Cabo Pulmo, which are subject to a growing tourism interest. Relying on official data on the aquifer, on-site interaction with locals and observation in October and November 2019,[1] we analyze water management in a context of tourism exploitation of natural resources in the Park and the notion of human right to water. Systematic content analysis (López 2002) produced a water management diagram for Cabo Pulmo, later presented in the text, which gives a glance at the complex social consequences of the lack of clear water regulations and accountability, but also at the opportunities for the local community to include water in its processes of conservation awareness. This would imply debating rights and sustainability with an approach that goes beyond the environment.

[1] We are grateful to Sociedad de Historia Natural Niparajá, SC, for facilitating access to their documents and to the community of Cabo Pulmo. The analysis and views expressed in this text belong solely to us.

Water management in conditions of scarcity becomes more important especially in arid or semi-arid regions. Focusing on the case study on Cabo Pulmo allows analyzing the forms of water management in place when the State does not intervene and how this affects ecosystem and social dynamics. Although there have been multiple studies about the use and management of the Cabo Pulmo National Park resources for conservation, the problem of water supply and its forms of management have barely been addressed in depth, with possibly the exception of Niparajá, a civil society organization that later in 2020 is to present the results of an ongoing research on that matter and about which Tovar (2019) provides some insight.

This chapter is organized in four parts after this introduction. The first refers to the debate about water as a human right or commodity in order to contextualize inequality in water access in Cabo Pulmo. The second shows the extent of the overexploitation of the Cabo Pulmo aquifer, on which Cabo Pulmo's inhabitants depend. The third deals with the community's forms of water management in Cabo Pulmo. Finally, the fourth section offers some considerations about the consequences of private water management in Cabo Pulmo and the need to better address equal access to such a key public good.

28.2 Water: A Human Right or Commodity?

The declaration of Human Right to Water and Sanitation (HRWS) by the United Nations Organization in 2010 has led to new controversies about water management and uses, in a context of various environmental, economic, and social crises derived from improper water use, management, and appropriation. On the one hand, the United Nations (2002) proclaims that water for human consumption should override any other use; but, on the other, international organizations such as the Inter-American Development Bank argue that water supply must be market-led, under the idea that it is a scarce commodity. In addition, in this latter position, it is proposed that State administration of natural or common-use resources causes market failures, while the assignment of an economic value allows them to be profitably used and managed (Moreno 2012).

According to the World Water Association (GWP 2008), Integrated Water Resources Management (IWRM) is a process that promotes the coordinated management and development of water, soil, and other related resources; and its purpose is to maximize economic results and social welfare in an equitable manner, but without compromising the sustainability of vital ecosystems. This institution was founded in 1996 by the World Bank, the United Nations Development Program, and the Swedish Agency for International Development Cooperation (GWP 2017). According to its definition, IWRM is an empirical concept that has been under construction since its Global Conference of Mar del Plata in 1977. However, it was not until after Agenda 21 and the World Summit on Sustainable Development in 1992, in Rio de Janeiro, when the concept of IWRM was subject of deep debates that included its practical implications (GWP 2015).

Water management is provided by public institutions and by the market. The former refers to formal (state) and informal (tradition, ministries, and communal councils) institutions. Market water management includes all those actors that operate under a logic of economic gain. A debate surrounds which of the two is the most "efficient" way to manage water and guarantee HRWS. A current of thought states that water is so important that its allocation should not be left to the forces of supply and demand (Bakker 2007; Veraza 2007; Arrojo 2017). The other holds that the absence of prices leads to excessive consumption (Hanley et al. 2002; Moreno 2012). In either case, there are pros and cons. On the side of public institutions, equitable distributions can be achieved by preventing abuse, in addition to preserving sustainable community traditions; however, their distribution can be manipulated by powerful political or economic actors, thus depleting the resource, which would represent a State failure.

On the other side, the market can achieve efficient distribution and generate income to maintain the resource, but (like public institutions) it can be manipulated by powerful political or economic actors and generate failures that would deplete the resource (Ostrom 2000). Key to this is how the two alternatives can live together. Water requires balancing all forms of management in a creative way.

Benavides (2015) calls for analyzing market mechanisms in depth regarding water due to the social consequences of its operation. The market paradigm considers water scarcity as the result of the lack of proper free market water transfer and distribution: had the latter prevailed, water would be transferred to regions with scarcity, and the rise in prices would lead to conservation (Shiva 2003). It is important to underline that capital exploits and appropriates common goods: the common becomes private property and, as private, deprives or excludes other members of the productive community from the right to decide (Dussel 2014). Therefore, HRWS cannot be subject to market logics because it would cease to be a human right.

28.3 Water Availability in Cabo Pulmo: An Overexploited Aquifer

As mentioned above, Cabo Pulmo is located in the Mexican state of Baja California Sur's southeast. It borders the Gulf of California to the north, the Pacific Ocean to the south, the town of La Ribera to the northwest, while the city of San José del Cabo can be reached to the southwest (Fig. 28.1).

A desert climate prevails in the locality, according to the Köppen climate classification chart, modified by E. García for Mexico (1964). Although Cabo Pulmo does not have a weather-monitoring station, with data from nearby stations (Las Barrancas, Boca del Salado and San José del Cabo), CONAGUA estimated that average temperatures range between 18 °C and 23.7 °C. The summer rain regime varies from approximately 200 mm to 250 mm on average; however, during the months of August and September, temperatures can reach 40 °C (CONAGUA 2018).

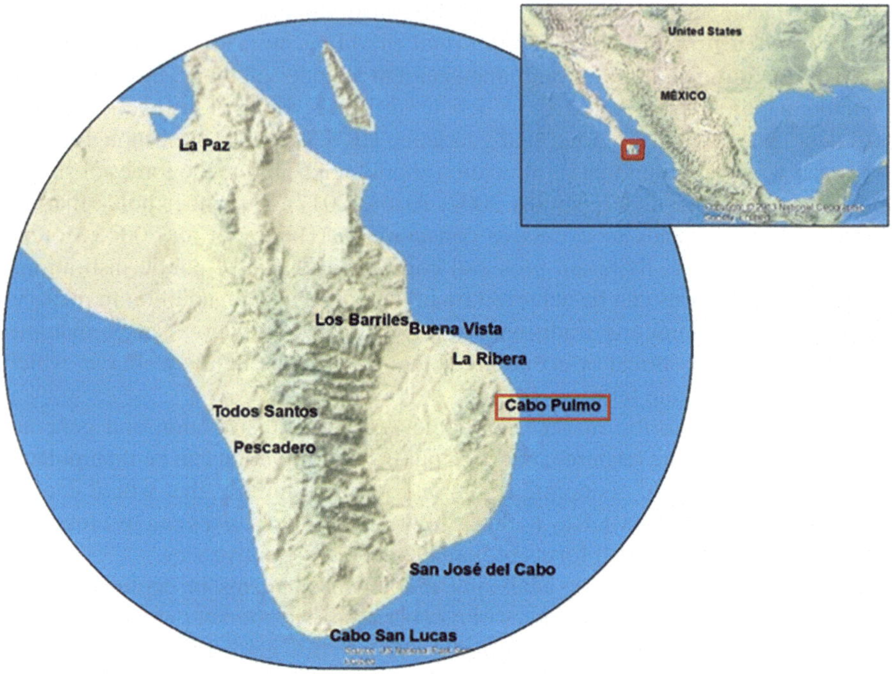

Fig. 28.1 Cabo Pulmo location. Juan Carlos Graciano, 2019

Clear or partly cloudy skies are observed during nearly 80% of the year; the tropical and cyclonic rainy season in the North Pacific goes from the second half of May to the second half of October, with different fluctuations due to El Niño and La Niña (INEGI 1996; Romero et al. 2007). However, almost half of the rains come from hurricanes and tropical storms, which tend to facilitate the recharge of aquifers (Wurl and Martínez 2006). Aquifers are the main source of water availability in the state. The area under study depends on the Cabo Pulmo aquifer, which is subject to the decree of ban zones, type III, that allows limited extractions for domestic, industrial, irrigation, and other uses. According to the Mexican Federal Law on Water Rights of 2013 (MFLWR), CONAGUA is obliged to make the availability of groundwater by aquifer public and update the data at least every 3 years.

The NOM-011-CONAGUA-2000 Official Mexican Standard (NOM) demands that the capacity of extraction or concession of an aquifer must be established according to the estimated annual average recharge, less the amount demanded by the ecosystems. However, some hydrology specialists point out that this norm is obsolete when calculating availability, given that it presents several inconsistencies and margins of error (Hernández-Morales and Wurl 2017). Article 22, Second Paragraph of MFLWR, demands that the extraction and use of water resources follow sustainability principles, but this is not always the case. The latest update on water availability in the Cabo Pulmo aquifer published by CONAGUA (2018) estimates an annual recharge of 2.2 Mm^3 and a deficit of 1.76 Mm^3 (Table 28.1).

Table 28.1 Cabo Pulmo aquifer. Average annual water availability

Code	Aquifer	Recharge	DNCOM	VCAS	VAPTYR	DAS	DEFICIT
0318	Cabo Pulmo	2.20	2.00	0.888570	1.080500	0.00	−1.769070

Abbreviations in Spanish: *DNCOM* compromised natural discharge, *VCAS* groundwater licensed volume, *VAPTYR* volume of groundwater extraction pending titration or registration in the Public Registry of Water Rights, *DAS* average annual groundwater availability
Source: Developed with data from CONAGUA (2018)

Table 28.2 Concession volumes by type of use for the Cabo Pulmo aquifer

Type of use	Number of concessions	Licensed volumes m^3/year
Agricultural	1	10,000
Multiple	9	40,660
Domestic	6	11,350
Livestock	31	70,500
Service	15	756,110
Total	62	888,620

Source: Estimations developed from PRWR data corresponding to 2019

Each water concession title must be duly accredited in the Public Registry of Water Rights (PRWR) to show the name of the concessionaire, the water volume the title covers, the type of use, the name of the aquifer to which it belongs, and its geographical location. However, CONAGUA recognizes that there are 1.08 Mm^3 pending or in the process of titling. It should be noted that this volume is greater than what is already registered, which denotes an over-concession and therefore an overexploitation of the aquifer. Similarly, 43 uses are reported but without specifying their specific types; in contrast, PRWR reports 62 concessions, mostly for *Livestock* use but, in terms of volume, *Services* use[2] accounts for 85% (Table 28.2).

The town of Cabo Pulmo has at least six wells: three are in the hands of real-estate companies, each with a water concession permit; one is managed by the local community; and two more belong to local residents; although whether they have a water title is unknown. Official data show water over-concession in the context of clearly limited water resources. Improper handling of concessions sharpens the crisis of water resources and ecosystems associated with the basin. In other words, the overexploitation of water resources in arid conditions produces less availability as well as a qualitative and quantitative water loss.

[2]The use of water for Services is defined by the NWL as the use of national water for services other than those indicated in Sections XVI to XXV of Article number 2 of that Law (which are agricultural, agroindustrial, domestic, aquaculture, industrial, ecological conservation, livestock, urban public, and others). As noted, the law is unclear about what *Services use* entails.

28.4 Private and Community Water Management in Cabo Pulmo

In Mexico, municipal governments have the constitutional mandate to supply water to urban and conurbation areas. However, an informal private market tends to operate in places where water services do not arrive or do so poorly; less frequently, social or community management becomes another means of water supply. In the community of Cabo Pulmo, these two latter "atypical" forms for water management coexist: the first or main one (due to its volume and operating capacity) rests on a private actor (Promotora Pelicano SA de CV), which is constituted as a Mexican company with direct foreign investment of US capital according to the National Registry of Foreign Investment (2019). Given the inability of the municipal government to serve this small remote community, this real-estate company operates under a profit logic and has at its disposal a captive market composed entirely of expatriates of high purchasing power. The second form is managed by the local community, whose members are organized to directly use a communal well or are supplied by a modest water system.

Promotora Pelicano SA de CV owns two water concessions registered in the PRWR, which provide it with 10,000 and 15,000 m^3/year, respectively, for the *Services* use that allows it to profit from groundwater extraction. With these concessions, it sells water exclusively to the foreign community living in The Resort, a gated community in Cabo Pulmo, through a private hydraulic network supported by hydraulic equipment and a solar cell system, as well as storage tanks that allow it to supply almost uninterrupted service to just over 50 residences (Fig. 28.2). The monthly water service fee ranges from USD 250 to USD 1000 which, according to the residents themselves, are deposited in a bank account abroad.

For their part, local community members depend on a well that was initially a CONAGUA exploration site. The Mexican local community, through its own

Fig. 28.2 Promotora Pelicano SA de CV's hydraulic infrastructure

Fig. 28.3 The community's water infrastructure

means and with governmental aid, managed to install a pumping equipment powered by solar panels (Fig. 28.3); it has a smaller capacity storage tank (10,000 liters), which allows water to be distributed at least twice a week to about 20 families and about 30 small businesses through the abovementioned independent water network (Tovar 2019). Both families and businesses pay a monthly fee of 200 pesos (about USD 20).

Some businesses are not directly connected to the network, but they usually gather water directly from the community's tank (Fig. 28.4), which often leads to decompensations in the fragile community water network, forcing local residents to use considerable hours of their time (and resources) to store and/or carry the water they need for their various activities. This opens the door to another form of informal water market through the sale of water in pipes, which complements water provision (at a higher price) when demand exceeds the limited local supply.

Considering the expansion of tourism, new actors begin to gradually show up, causing disputes over land and water use in Cabo Pulmo. Such is the case of Cragar Desarrollos, SA de CV, which in 2007 was accused of destroying houses and performing acts of intimidation against the inhabitants of Cabo Pulmo (Progreso 2007). This real-estate developer has a concession of 500 m^3 per year for *Services* use in Cabo Pulmo. Another firm, Hansa Baja Investments, SA de CV, proposed in 2008 a huge tourism/residential project in an area adjacent to the Cabo Pulmo National Park that would have severely impacted the local ecosystem. This was denounced by international organizations, among them Greenpeace (Magaña and Guzmán 2015), and received such opposition that the project was canceled by presidential decree. The successful process of resistance importantly rested on a sense of sustainability on the part of local inhabitants (Anderson 2015). Yet, Costa Palmas, another megaproject located in the nearby La Ribera (previously Cabo Riviera), has sparked old debates about the desirability of changes caused by tourism (Niparajá 2011) and the likely adverse impacts their establishment will have on water access.

Fig. 28.4 Water supply for business use from the community's well

28.5 Social Implications of Water Management in Cabo Pulmo

Private and community forms of management have diverse implications that are important to analyze (Fig. 28.5) in a community so small and totally dependent on tourism activity such as Cabo Pulmo. In the first place, these forms of water management show the municipal government is not meeting the responsibilities or using the powers granted by the Mexican Constitution in Article 115, which could be translated even as a loss of sovereignty, nor fulfilling international commitments to guarantee access to water as a human right.

Privatizing a common good becomes what David Harvey calls "accumulation by dispossession" (Harvey 2004), that is, the process by which those who lack the means to pay are excluded from the enjoyment of such a good. Therefore, HRWS is defined in terms of income. Regarding Cabo Pulmo, a flexible water policy encourages the definition of scarcity in terms of the resource's physical limitation; scarcity, thus, becomes purely a technical issue and a source of opportunism that produces social unrest among those who are disadvantaged.

On the other hand, the provision of community service in Cabo Pulmo lacks sufficient infrastructure to guarantee equitable and quality access to water. As mentioned above, current water distribution patterns reflect on a poor-quality service to which the population has to allocate an important part of its time. This issue is exacerbated by some community members who go directly to the source of water supply mainly for commercial activities and not for domestic use. In both cases, a

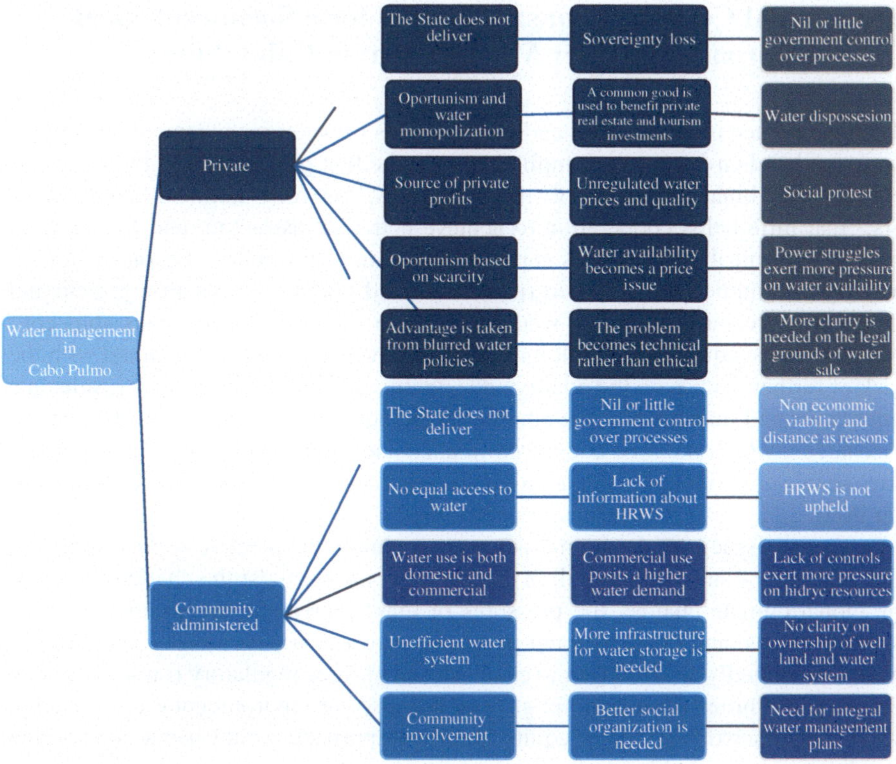

Fig. 28.5 Social implications of water management in Cabo Pulmo

comprehensive water management plan for the community is much needed. The increasing coexistence of these forms of distribution has been accompanied by a gradual process of appropriation by water dispossession. Water and its current forms of distribution have rested on a discourse that normalizes the status quo and hides the social interactions and power dynamics that turn water appropriation into a commodity.

Figure 28.5 represents a diagram of the various implications involving both forms of management. Private management highlights the monopolization of water services and the appropriation of a common good converted into merchandise to seek a profit, and expatriate residents are forced to pay high water prices. On its side, community water management shows unequal access to water that prevents the fulfillment of the human right to water, making local community members the most vulnerable social group insofar as they are forced to dedicate long hours to water provision, purification, and distribution. Both forms of management also express the municipal government's inability to perform its constitutional responsibility insofar as it has little or no control over the forms of management. This contributes to disputes over water between power groups and to make the human right to water a price issue.

28.6 Final Considerations: The Need for a Socio-ecological Approach to Water Management in Cabo Pulmo

The coexistence of community and private forms of water management has important social and environmental implications. In the first place, in Cabo Pulmo, it has encouraged a binary discourse of "they against us," between "nationals and foreigners," that little helps cooperation to achieve better infrastructure and financing to increase technical water access capacity for all in Cabo Pulmo. Secondly, tourist activity has implied an increased (permanent and floating) population growth and therefore a greater demand for water under both forms of management, thus contributing to the aquifer overexploitation. Third, overexploitation combined with the lack of further studies on the quality and quantity of water in the aquifer results in a qualitative and quantitative degradation of the aquifer's "committed water natural discharge", which must be conserved in adequate limits to prevent a negative and irreversible environmental impact to the fragile ecosystem of the Cabo Pulmo aquifer itself.

A central issue in Cabo Pulmo is how to maintain the positive results of collective action for conservation both in the Park and town amidst the challenges posed by global climate change, the pressures of mass tourism in the vicinity, internal social tensions, and economic and population growth in the coastal Cabopulmeño area. Undoubtedly, it is relevant to establish and enforce regulatory frameworks that favor opportunities for economic growth, but also for environmental conservation within a framework of social equity. Such an approach would mean considering regulations on the Park's adjoining area, where residential and tourist constructions and developments are carried out if terrestrial resources, among them potable water towers, are to be conserved.

In this regard, after so many years of community commitment to the Park's conservation, as well as some government and civil society intervention to favor locals' welfare, a lesson to be learned may be to extend to the terrestrial zone the positive results to protect Cabo Pulmo National Park's marine biodiversity and the ecosystem services it provides; that is, to explicitly recognize in actions the links between terrestrial and marine biodiversity. Nowadays, with commendable exceptions, attention mostly focuses on the NPA environmental conservation and, although social cohesion in Cabo Pulmo has proven strong against external impacts, much is needed to deal with internal social imbalances within the community as a whole. It is not appropriate to ignore the negative effects of human and tourist settlements and relationships on community life, as well as the current rate of water extraction from aquifers on biodiversity. Tending to this could help in solving the problems associated with water management for human and economic uses, thus contributing to stronger sustainability.

Ignoring social tensions does little favor to either conservation goals or the continuation of economic profits: in Cabo Pulmo, a community effort to deal with land tenure and water rights is in place aided by civil society organization intervention, but the absence of municipal government responsibilities must also be addressed.

As an anticipated response to what can be expected from the increase in environmental pressures that global climate change and tourism trends imply, tending to social inequalities in access to key natural resources, such as water, is required for the conservation of this NPA's biodiversity and the community members' full enjoyment of their human rights.

References

Aburto-Oropeza O, Erisman B, Galland GR, Mascareñas-Osorio I, Sala E, Ezcurra E (2011) Large recovery of fish biomass in a no-take marine reserve. PLoS One (August 12). https://doi.org/10.1371/journal.pone.0023601

AIDA. Asociación Interamericana para la Defensa del Ambiente (2013) Coalición internacional denuncia incumplimiento de la legislación ambiental en la autorización de cuatro megaproyectos turísticos en el Golfo de California, April 11. http://www.aida-americas.org/es/coalici%C3%B3n-internacional-denuncia-incumplimiento-de-la-legislaci%C3%B3n-ambiental-en-la-autorizaci%C3%B3n-de

Anderson R (2015) Sustainability, ideology, and the politics of development in Cabo Pulmo, Baja California Sur, Mexico. J Pol Ecol 22(1). https://doi.org/10.2458/v22i1.21107

Arrojo Agudo P (2017) Privatización del agua y corrupción, Dominio Publicoes, 22 March. http://blogs.publico.es/dominiopublico/19468/privatizacion-del-agua-y-corrupcion/

Bakker K (2007) The "commons" versus the "commodity": Alter-globalization, anti-privatization and the human right to water in the global south. Antipode J Compilation:430–455

Barrera Osuna FA (2018) Los conflictos socioterritoriales y ambientales en Cabo Pulmo, Baja California Sur: el dilema entre la conservación y el turismo, Masters dissertation, Colegio de la Frontera Norte, Tijuana, B. C

Benavides JM (2015) Agua en América Latina. Abundancia en medio de la escasez. Bogota: Banco Interamericano de Desarrollo-Universidad de los Andes de Colombia

Cariño M, Valle E, Monteforte M, Arizpe C, Aceves J (2008) La creación del área natural protegida: actores, procesos y retos. Gámez AE (ed.) Turismo y sustentabilidad en Cabo Pulmo, Baja California Sur. San Diego: SDSU

Castro J (2014) Historia de Cabo Pulmo. Amigos para la Conservación de Cabo Pulmo. http://www.oceanhealthindex.org/news/Historia_de_Cabo_Pulmo

CONAGUA. Comisión Nacional del Agua (2018) Actualización de la Disponibilidad media anual en el acuífero de Cabo Pulmo (0318). Diario Oficial de la Federación, 20 April 2015

CONANP. Comisión Nacional de Áreas Naturales Protegidas (2006) Programa de Conservación y Manejo Parque Nacional Cabo Pulmo. CONANP, México

CONANP-GIZ (2017) Valoración de los Servicios Ecosistémicos del Parque Nacional Cabo Pulmo. Ciudad de México. Proyecto de Valoración de Servicios Ecosistémicos de Áreas Naturales Protegidas Federales de México: una herramienta innovadora para el financiamiento de biodiversidad y cambio climático (EcoValor MX). CONANP-GIZ: Ciudad de México. https://www.conservation-strategy.org/sites/default/files/field-file/CONANP_GIZ_17_-_valoracion_cabo_pulmo.pdf

Cosío J (2015) La exitosa recuperación ecológica de Cabo Pulmo, La Paz, Baja California Sur. Agencia Informativa Conacyt, 3 December. http://www.conacytprensa.mx/index.php/ciencia/ambiente/4311-arque-nacional-cabo-pulmo-referente-de-la-conservacion-en-mexico

Dussel E (2014) 16 tesis sobre Economía Política: Interpretación Filosófica. Siglo XXI, Mexico

Glynn PW, Macintyre IG (1977) Growth rate and age of coral reefs on the Pacific coast of Panama. D.L. Taylor (ed.) Proceedings of Third International Coral Reef Symposium Vol. 2: Geology. Miami: Rosenstiel School of Marine and Atmospheric Science, pp 251–259

Godínez C, Gámez AE, Arizpe O, Aburto O, Reyes Bonilla H (2015, coords.). Simposio sobre los 20 años de Cabo Pulmo. La Paz, BCS: CONANP
GPD. Global Water Partnership (2017) History. http://www.gwp.org/en/About/who/History/
Graciano JC, Ángeles M, Gámez AE (2019) A critical geography approach to land and water use in the tourism economy in Los Cabos, Baja California Sur, Mexico. J Land Use Sci, https://doi.org/10.1080/1747423X.2019.1699613
GWP. Global Water Partnership (2008) Principios de Gestión Integrada de los Recursos Hídricos. Bases para el desarrollo de planes nacionales, Global Water Partnership. http://www.gwp.org/globalassets/global/gwp-sam_files/publicaciones/sobre-girh/2008-principios-de-girh-base-para-el-desarrollo-de-planes-nacionales.pdf
Hanley N, Shogren JF, White B (2002) Environmental economics in theory and practice. Palgrave Macmillan, Nueva York
Harvey D (2004) The 'new' imperialism: accumulation by dispossession. The Socialist Register, 4. https://socialistregister.com/index.php/srv/article/view/5811/2707
Hernández-Morales P, Wurl J (2017) Hydrogeochemical characterization of the thermal springs in northeastern of Los Cabos Block, Baja California Sur, México. Environmental Science and Pollution Research 24 (15):13184-13202. https://doi.org/10.1007/s11356-016-8087-x
INEGI. Instituto Nacional de Geografía e Informática (1996) Estudio Hidrológico. Estado de Baja California Sur. Edición 1996. Aguascalientes: INEGI
López Noguero F (2002) El análisis de contenido como método de investigación. XXI Rev de Educ 4:167–179
Masse Magana MV, Guzmán Hernandez C (2015) The state and the touristic megaprojects: the case of Cabo Pulmo, Baja California Sur, Mexico. Teoría y Praxis 11(18):101–129
Menares B (2008) Construcción y usos de los espacios sociales en la localidad de Cabo Pulmo. Gámez AE (ed.) Turismo y sustentabilidad en Cabo Pulmo, Baja California Sur. San Diego: SDSU, 277–292
Moreno SR (2012) Incentivos económicos para la conservación. Un marco conceptual. Lima: Sociedad Peruana de Derecho Ambiental
Niparajá (2011) Alcances del proyecto y aspectos de biodiversidad. Campaña de Orgullo para la Conservación de los Recursos Naturales de La Ribera. La Paz, BCS: Sociedad de Historia Natural Niparajá, A.C.,
Niparajá (2019) Síntesis de resultados del diagnóstico social de la comunidad de Cabo Pulmo. La Paz, BCS: Niparajá
Ostrom E (2000) Diseños complejos para manejos complejos. Gaceta Ecológica http://www.redalyc.org/articulo.oa?id=53905404
Proceso (2007) Inmobiliaria, acusada de destruir casas e intimidar a una comunidad de pescadores, en Los Cabos. https://www.proceso.com.mx/208919/inmobilaria-acusada-de-destruir-casas-e-intimidar-a-una-comunidad-de-pescadores-en-los-cabos
Romero-Vadillo E, Zaytsev O, Morales-Pérez R (2007) Tropical cyclone statistics in the northeastern Pacific. Atmósfera 20(2):197–213. http://www.scielo.org.mx/scielo.php?script=sci_arttext&pid=S0187-62362007000200005&lng=es&tlng=en
Sauvage A, Gámez AE (2013) Introducción. Los usos del patrimonio cultural en Sudcalifornia. A. Sauvage y A. E. Gámez (eds.). La Paz, BCS: Instituto Sudcaliforniano de Cultura, 7-18
SEGOB. Secretaría de Gobernación (1995) Decreto por el que se declara área natural protegida con el carácter de Parque Marino Nacional, la zona conocida como Cabo Pulmo, ubicada frente a las costas del Municipio de Los Cabos, B.C.S., con una superficie de 7,111-01-00 hectáreas. Diario Oficial de la Federación, Tomo DI, 5. http://dof.gob.mx/nota_detalle.php?codigo=4875024andfecha=06/06/1995
Shiva V (2003) Las Guerras del Agua. Privatización, Contaminación y Lucro. Mexico: Siglo Veintiuno Editores
Tovar H (2019) Diagnóstico Social de Cabo Pulmo. Sociedad de Historia Natural Niparaja, AC, La Paz

United Nations (2002) Organización de las Naciones Unidas (2002) Observación General No. 15. El derecho al agua (artículos 11 y 12 del Pacto Internacional de Derechos Económicos, Sociales y Culturales). Nueva York: Naciones Unidas. http://www.solidaritat.ub.edu/observatori/general/docugral/ONU_comentariogeneralagua.pdf

Vanderplank S, Wilder BT, Ezcurra E (2014) Descubriendo la biodiversidad terrestre en la región de Cabo Pulmo. Botanical Research Institute of Texas. http://nextgensd.com/wp-content/uploads/2014/04/Cabo-Pulmo-Report_Final.pdf

Veraza J (2007) Economía y política del agua. El agua que te vendo primero te la robé. Mexico, Editorial Itaca

Wurl J, Martínez Gutiérrez G (2006) El efecto de ciclones tropicales sobre el clima en la cuenca de Santiago, Baja California Sur, México. III Simposio Internacional en Ingeniería y Ciencias para la Sustentabilidad Ambiental y Semana del Ambiente, 5–6 June, Mexico, DF

Part VI
Governance Changes from Sociecological Approach

Chapter 29
Walls of Water, Socioecological Perspectives of Governance Changes in a Protected Natural Area of Mexico

Ricardo V. Santes-Álvarez

Abstract For more than 100 years, Islas Marias, a 600,000 ha archipelago located in the Mexican Pacific Ocean, hosted a federal detention centre that was acknowledged as one of the most infamous jails devised to punish illegal or anti-regime conduct. At the same time, the natural beauties and biodiversity of the area have been acknowledged by the scientific community, up to the point that, as of 2000, the archipelago acquired the status of protected natural area in the category of biosphere reserve. Unfortunately, the rich biota also means high economic revenue; hence, it has long been a motive for illegal exploitation. Quite recently the government announced the closure of the ancient jail; the built infrastructure is to be converted into an environmental education and culture centre. This new development will modify the governability of the Islas Marias Biosphere Reserve, overall due to relaxation of surveillance around the site by federal authorities once the prison compound is closed and the remaining inmates have been transported to other detention centre. If administrative and management measures are not adequately planned, it is expected that threats to nature will increase. This chapter will examine options to protect the reserve from such perils.

Keywords Protected natural area · Islas Marias Biosphere Reserve · Environmental governance · Mexico

29.1 Introduction

Protected natural areas are spread widely in a variety of countries that aim to achieve the goal of sustainability. Conceived as environmental policy artefacts to preserve and conserve areas containing valuable natural assets, their governance is oriented

R. V. Santes-Álvarez (✉)
Department of Urban Studies and Environment, El Colegio de la Frontera Norte, Tijuana, Baja California, Mexico
e-mail: rsantes@colef.mx

A. Ortega-Rubio (ed.), *Socio-ecological Studies in Natural Protected Areas*,
https://doi.org/10.1007/978-3-030-47264-1_29

primarily by criteria that are biological in nature; for that reason, social factors occurring inside their designated limits and certainly those happening at the outside are given lesser attention or are frankly disregarded. Nevertheless, those territories are in no way untouched by socio-economic pressures, either interior or exterior.

Biosphere reserves are types of protected areas that presumably better reconcile the objectives of natural resource protection and conservation with those of social well-being and economic gain. In the developing world, their importance is praised mainly because most of them are designated for territories where human communities already live. Consequently, the biosphere reserve model allows implementing actions to harmonize environmental and socio-economic interests. The reality, though, is not as plain as the discourse. With regard to decision-making about biosphere reserves, social groups and interests are hardly taken into account; that may be the cause why such regions fall short of their alleged objectives. Hence their adequacy as policy instruments towards sustainable development is brought into question.

Since it is considered a mega-diverse country—that is to say, hosting more than two thirds out of all terrestrial, freshwater, and marine planetary biodiversity—Mexico included the biosphere reserve model in its protected natural area (PNA) system since the late 1970s. Currently, Mexico possesses a number of 44 biosphere reserves that amount to nearly 70% of all nominated areas in its national territory.

Islas Marias, a more than 600,000 ha archipelago comprising four islands located in front of the coast of the state of Nayarit in the Mexican Pacific Ocean, has plenty of natural beauty and unique biodiversity—rich in terrestrial vegetation, birds, and marine species of fishes, crustaceans, and the like. Over time, the site has drawn the attention of the scientific community to the extent that, as of 2000, it acquired the status of PNA in the category of biosphere reserve. It is important to note that, different from others of its kind in the country and certainly from most of those around the world, Islas Marias has shared since its creation both territory and governance with a prison complex that for more than 100 years functioned on one of the isles. It is noteworthy that the government announced recently the shutting down of the prison; thus, after the speedy transportation of the inmates to other facilities, the conversion to the built infrastructure started: while keeping its status as a biosphere reserve, it has become a cultural, environmental, and research centre.

It is reasonable to assume the administration and management of the entire reserved area are going to change. In light of the new institutional accommodations, in which the closure of the detention centre took on the duties of a security sector, it might be argued that the new dynamic will contribute to the relaxation of surveillance over the archipelago; hence, ecological dangers will increase. This interpretation is not far from reality, for the natural resources of the region have long inspired illegal exploitation, mainly from fishermen and from foreign tourists as well, who are attracted by sport fishing, and raises the need to reflect on the viability of Islas Marias as a successful PNA and propose alternatives to harmonize the new administration and management approach with the objective of sustainability. It seems relevant, therefore, to examine the manner and extent to which the governance of the reserve should be modified.

The following section of the chapter offers an overview of biosphere reserves, emphasizing the case of Mexico. The third section describes the main biological and ecological features and sociopolitical aspects of the Islas Marias archipelago. It follows an exposition of the feasible changes that in a short time will be made to the governance of the area and their link to the goal of sustainability. The latter serves to confront the author's argument, that is to say, that after the archipelago modifies its longstanding governing ways because of the prison system shutting down, pressures on the biosphere reserve will increase due to reduction of strict surveillance and administrative moves. In the last part of the chapter, it is proposed that governance can be improved and all sorts of threats can be prevented by means of administrative and management measures that are adequately planned, but mainly through an alternative approach that gives strong weight to interaction among actors from inside and outside the protected area in decision-making.

29.2 Biosphere Reserves: An Overview

Biosphere reserve (BR) is a type of PNA comprising terrestrial, marine, and coastal ecosystems that holds the objectives of protection and conservation of natural resources and cultural possessions for sustainable use by encouraging their rational use. Being primarily concerned with species richness, endemism, and threatened and menaced species (SSP-Semarnat 2007) but also with adequate management, BR covers extended surfaces of the planet. They are nominated by national governments and remain under the sovereign jurisdiction of the states where they are located. UNESCO defines them as sites "for testing interdisciplinary approaches to understanding and managing changes and interactions between social and ecological systems, including conflict prevention and management of biodiversity" (UNESCO 2019a).

According to the World Database on Protected Areas, as of June 2018, there are on the planet 238,563 designated PNAs of all types, covering a total surface of 46,414,431 km^2, most of them located on land (UNEP-WCMC 2018). Regarding the BR type, a recent counting by UNESCO (2019a) reported a figure of 701 distributed over 124 countries, including 21 transboundary sites. In Latin America and the Caribbean, there are 130 BR spread over 21 countries (UNESCO 2019b).

Mexico defines PNA as parts of the territory where the nation exerts sovereignty and jurisdiction in which original environments have not been altered significantly by human action or places where the existing conditions and functions demand preservation and restoration. In legal terms, those zones are attached to the regime prescribed by the country's environmental law, the General Law of Ecological Equilibrium and Environmental Protection (LGEEPA), and the complementary ordinances (Gobierno de México 2018). The LGEEPA acknowledges six types of federal PNA: biosphere reserve, national park, natural monument, natural resource protection area, flora and fauna protection area, and sanctuary (Gobierno de México 2018). Table 29.1 shows the list of PNA in Mexico in accordance with number, category, total area, and per cent of coverage.

Table 29.1 Federal PNA of Mexico by management category (ha) in 2019

No. of PNA	Category	Total area (ha)	Per cent
44	Biosphere reserve	62,952,750.50	69.30
40	Flora and fauna protection area	6,996,864.12	7.70
8	Natural resource protection area	4,503,345.23	4.96
67	National park	16,220,099.30	17.86
18	Sanctuary	150,193.29	0.17
5	Natural monument	16,269.11	0.02
Total: 182		90,839,521.55	100

Source: Gob. de México and Semarnat-Conanp (2019)

BR is formally identified as relevant biogeographic environments representing one or more ecosystems inhabited by species representing national diversity, including those that are classified as endemics, menaced, or in danger of extinction (Gobierno de México 2018). From Table 29.1, it is worth noting the important role of BR not only due to their greater physical dimensions compared to other PNA but also to the expected function that they are presumed to play with respect to the objectives of sustainable development.

At the regional and subcontinental level, Mexico maintains a principal position before 20 other countries in Latin America and the Caribbean that possess BR. Data from UNESCO as of July 2018 revealed that Mexico possessed the greatest share: 32.3% or 42 out of a total of 130, followed by Argentina (11.54% or 15) and Chile (7.69% or 10) (UNESCO 2019b). In light of the fact that the number of Mexican BR increased to 44 in 2019, it is expected that the subcontinental gap increased too.

BRs are organized spatially by means of two major physical delimitations: the nucleus and the buffer zones. The first refers to regions that are better conserved or remain unaltered, constituting or hosting ecosystems and natural phenomena (flora and/or fauna species) under some risk category; the latter according to an official norm is related to environmental protection and biodiversity care, better known as NOM-059-SEMARNAT-2001. The second zone is considered to protecting the nucleus zone from impacts of human activities that put the evolution of ecosystems' natural processes within the reserve at risk. It is classified as the arrangement of areas in which productive activities that are compatible with the conservation of natural resources are allowed, always in compliance with the law and the soil and water restrictions set by the specific creation decree (SSP-Semarnat 2007).

29.3 Islas Marias

The Islas Marias archipelago is located in the Pacific Ocean, in the dry tropical zone of Mexico in front of the coast of Nayarit state, 176 km from the port of Mazatlan, Sinaloa, 132 km from San Blas, Nayarit (SSP-Semarnat 2007; Santos and Negrete 2010). It is composed of four islands, namely, Maria Madre, Maria Magdalena,

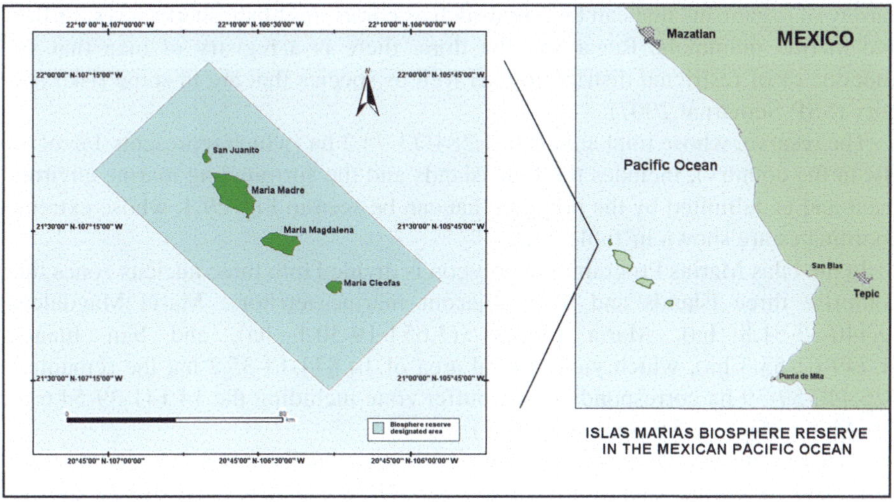

Fig. 29.1 Islas Marias Biosphere Reserve in the Mexican Pacific Ocean. (Source: Own elaboration)

Maria Cleofas, and San Juanito (Fig. 29.1), with rugged topography whose altitude varies from sea level to 620 m (SSP-Semarnat 2007).

29.3.1 Ecological Importance

The archipelago has been catalogued as a Conservation Priority Area and also as an Important Bird Area. In addition, in November 2000 it was designated in the category of biosphere reserve after a technical study revealed that the islands' ecosystems remained significantly unaltered (Gobierno de México 2000; SSP-Semarnat 2007; Santos and Negrete 2010; Senado de México 2010; Semarnat 2011).

As highlighted in the Islas Marias Biosphere Reserve Conservation and Management Program (hereafter Islas Marias Program), published in 2007, which is the governing document for all bioecological activities, the site is a habitat for a number of fragile ecosystems whose richness in species of flora and wild fauna manifests in the jungle of the terrestrial environment, the crassicaule thicket, the reefs, coasts, and pelagic environments found in the surrounding sea. Likewise, the islands are regarded as relicts of the biota of the Mexican dry tropics that have remained isolated from the continent for more than 8 million years and that currently functions as a rich reservoir of wildlife species endemic to Mexico (SSP-Semarnat 2007).

The Islas Marias islands are an important nesting and feeding site for large colonies of seabirds. Marine fauna is also represented by dozens of species, including endemic and endangered. In the sea there is high productivity that is reflected in the variety of organisms that can be found such as corals, reef fish, sharks, rays, turtles, and marine mammals. Regarding the flora, there is a registry of taxa that are endemic or of restricted distribution, as well as species that are in some risk category (SSP-Semarnat 2007).

The reserve, whose total area is 641,284-73-74.2 ha (which represents 1% of all BR in the country), includes the four islands and the surrounding marine environment and is delimited by the polygon that can be seen in Fig. 29.1, whose extreme coordinates are shown in Table 29.2.

In the Islas Marias Program, the polygon is divided into three nucleus zones that comprise three islands and their adjacent marine territory: Maria Magdalena (9,440-32-51.8 ha), Maria Cleofas (3,654-19-30.1 ha), and San Juanito (1,749-62-53.3 ha), which yield a total area of 14,844-14-35.2 ha; the remaining 626,440-59-39 ha corresponds to the buffer zone including the 14,141-39-54.6 ha Isla Maria Madre (SSP-Semarnat 2007).

As noted, the islands are habitats for fragile ecosystems rich in flora and fauna of biological, scientific, and cultural relevance. However, it is clear that in forests, reefs, coasts, and pelagic surroundings there is biodiversity of great economic importance (Semarnat 2011). That is why intrusion has occurred for a long time, essentially from coastal fishermen from San Blas and nearby towns on the Nayarit coast, who cross the boundaries of the BR in search of species of fishes, crustaceans, and other benthic fauna, but also of foreign fishermen who pursue pelagic species of high commercial and sporting value such as marlin (SSP-Semarnat 2007; Semar 2009, 2011). In their ambition, they intrude illegally on the reserve, decimating biodiversity and even threatening or endangering species. As expressed in the Islas Marias Program, the main problem in the seas surrounding the islands, which is under federal surveillance, has to do with illegal fishing (SSP-Semarnat 2007). The disturbing fact is that biodiversity in alleged protected territory is always in peril.

It is argued that poaching decreases due to the presence of the Marine Secretariat (SEMAR) personnel, who use speedy vehicles for the task of protecting the reserve from illegal activities. Also, in the period of the BRs' existence, tons of marine products, fishing gear, and smaller vessels have been seized, and people performing illegal actions have been secured. However, the external threat has not ended. It is the case that Maria Cleofas island, the furthest from the site where SEMAR has its advanced naval station, that is, Puerto Balleto on Maria Madre island, has been a refuge for poachers because its location makes their small vessels difficult to detect (SSP-Semarnat 2007).

Table 29.2 Extreme coordinates of Islas Marias Biosphere Reserve

Vertex	Latitude	Longitude
1	22°04′ 00″ N	106°40′ 00″ W
2	21°20′ 00″ N	105°54′ 00″ W
3	20°58′ 00″ N	106°16′ 00″ W
4	21°42′ 00″ N	107°03′ 00″ W

Source: Conservation and Management Program (SSP-Semarnat 2007)

In the opinion of the authorities, poaching in the area reflects the conditions of scarcity of the fishery of the coastal zone within the so-called Golden Triangle, a marine fishing set of relatively less deep and more productive waters whose vertices comprise Punta Mita, Islas Marias, and Mazatlan (Fig. 29.1) but whose productivity decreases because of not only an asymmetric catch capacity among fishermen's organizations but also an irrational fishing pattern that lacks a vision of sustainability. In that context, the element that encourages fishermen to make incursions into the prohibited zones of the BR is the certainty of finding a variety of species that is already difficult to find in the coastal seas of the continent, given the existing overexploitation (SSP-Semarnat 2007).

29.3.2 A Prison Colony

In accordance with the Political Constitution of the United Mexican States, the archipelago is set among the territorial seas of the country and is taken into account as a national quality, since the land possession is federal in its completeness (SSP-Semarnat 2007). The previous legal premise is a fundamental criterion when acknowledging the manner in which the realm of the reserve has been managed since its beginning. Islas Marias BR has operated under federal administration, not only in its natural setting but also the peculiar social activities developing there.

Since 1905, by decree of President Porfirio Díaz, the archipelago was marked for the establishment of a federal criminal colony since 1908, starting operations on the largest of the islands, Maria Madre, and was administered by the Ministry of the Interior, although many years later, from 2000 onwards, by the Ministry of Public Security or SSP (SSP-Semarnat 2007; Santos and Negrete 2010). Then in April 2010, the colony became a guardhouse compound, integrating several prisons (Semarnat 2011). In addition, since its publication in 1939, the statute of the Islas Marias Jail stipulated the power of the federal executive to prepare for the exploitation of the natural resources of the archipelago (SSP-Semarnat 2007).

The Islas Marias human settlement system consists of several camps. In 2011, six of these housed the same number of detention centres of the complex (each formally known as the Federal Centre for Social Re-adaptation or Cefereso); these are Cefereso Feminine of Minimum Security Zacatal, Cefereso Feminine Rehilete, Cefereso of Minimal Security Aserradero, Cefereso Morelos, Cefereso Bugambilias, and Cefereso of Maximum Security Laguna del Toro.

The anthropogenic activities carried out for so long inside the reserve have impacted the natural system. On the one hand, the regular ones, or those related to daily life in the prisons, due to the fact that the inmates' population as well as the administrative, custody, and provisioning and maintenance personnel became the main impact agents when carrying out agricultural and livestock work, aquaculture, and logging of timber resources for handicrafts, among others. On the one hand, those related to daily life in the prisons, as agricultural and livestock works, aquaculture, and logging of timber resources for handicrafts, carried out by the inmates' population as well as the administrative and maintenance personnel; on the other hand, all kinds of illegal actions, as it is the case with poaching.

Maria Madre is the only one inhabited formally. Together with the place of confinement and the inherent infrastructure, there is on the island the SEMAR advanced naval station, as well as an airport, and from there the monitoring and logistical support work is deployed throughout the biosphere reserve. It is from Puerto Balleto and the airport where official communication with the continent is established. Therefore, the entry or exit from the island of any person had to be done under the supervision of naval personnel stationed on-site.

29.3.3 The Conventional Governance Model

The jail complex remained in operation until February 2019, when the new federal government decreed its definitive closure and transferred all inmates, formally named persons deprived of liberty, to other prisons. It is understandable that the administration and management model, commanded by federal security forces of the Mexican State, extended to the remaining islands of the archipelago and the entire reserve. That is, while the Maria Madre island was presented as the most guarded both by SSP and SEMAR personnel, the Maria Magdalena, Maria Cleofas, and San Juanito islands were protected by members of SEMAR. Of course, these units have been accompanied by personnel of the National Commission of Natural Protected Areas (CONANP), under the Ministry of Environment and Natural Resources (SEMARNAT) whose responsibility has been limited to ensuring the good state of the natural resources of the region.

It is argued that the BR was intended to promote training for administrative and security personnel, as well as persons deprived of liberty and visitors in different productive fields in the knowledge and care of the environment (SSP-Semarnat 2007; Santos and Negrete 2010). Such work contributed to the vision of sustainability, although it should be stressed that the approach was limited, since it was only aimed at promoting that vision within the protected territory.

Regarding human rights and social reintegration, some critics point out that the prison system was a role model for a place where surveillance and attention to inmates were governed under high standards of respect for human rights and where the objective of re-entry into society of offenders was privileged once they regained their freedom (Villanueva 2019). However, others point out that in time, the positive images of the penal system of the islands have been much less prevalent than the negative ones (Pérez 2012).

All in all, it cannot be denied that the prison complex and reserve have been the axis of conventional governance contributing to the care and preservation of the natural resources of the area. It is relevant, therefore, to discuss the reasons that led the Mexican government to close the detention centre—an action that, as argued here, creates uncertainty about the viability of the ecosystem.

29.3.4 Reasons to Close the Jail

Much has been written about the prison system in Islas Marias since its beginnings in the early twentieth century. Although some refer to it as a positive punitive model, at least in some periods of its existence, the majority opinion is that it was one of the most infamous jails devised to punish illegal conduct and anti-regime expression. Five reasons support the closing of the site: size of the detained population, allotment availability in the prisons, budgeting, dubious social reintegration, and environmental protection. Let's take a look at each of them.

29.3.4.1 Size of the Incarcerated Population

A principal reason during the last years of activity of the complex was that the population of inmates was decreasing, so that the facilities were emptied more and more. As can be seen in the graph in Fig. 29.2, according to data from the National Human Rights Commission or CNDH (2018), in the 2012–2018 period, the population decreased significantly, with a maximum in 2012 of almost 2800 inmates in the Morelos and Laguna del Toro prisons, up to a maximum of 213 and 200 in Morelos and Aserradero, respectively, during 2018. In addition, the Rehilete prison, designed to house the female inmate population, ceased operations in 2016.

29.3.4.2 Allotment Availability in the Prisons

A second aspect is the decrease in the availability of space for the inmates. Figure 29.3 shows that during the period between 2012 and 2018, the accommodation capacity of the compound's penitentiaries gradually decreased. Data from the CNDH (2018) reveal that the worst cases were the penitentiaries of Bugambilias and Morelos, whose accommodation capacities went from 1015 and 2888, respectively, in 2012, to levels of 127 and 700, respectively, in 2018. The figures are strong enough; while Bugambilias had a decrease of 87.5% in availability during this period, Morelos had 75.8%.

It is possible that such decreases were due to the deterioration of the prisons' infrastructure, but this explanation contradicts investment data from the period precisely to rehabilitate and increase it. By 2014, the government through the Ministry of Finance announced a programme to build prisoner stays and increase accommodation capacity, increasing housing infrastructure for employees as well as rehabilitating the airport and the Puerto Balleto dock. In September of that year, it was formally announced that the programme had progress of 92% (SHCP 2014).

From the above, the material deterioration in Maria Madre island is due not only to regular human activity but also to the impact of meteorological phenomena. The damage caused by the most recent Hurricane Willa in October 2018 was of such a magnitude that it left much of the existing infrastructure in the camps in a deplorable condition. The pier of Puerto Balleto, for instance, suffered severe damage.

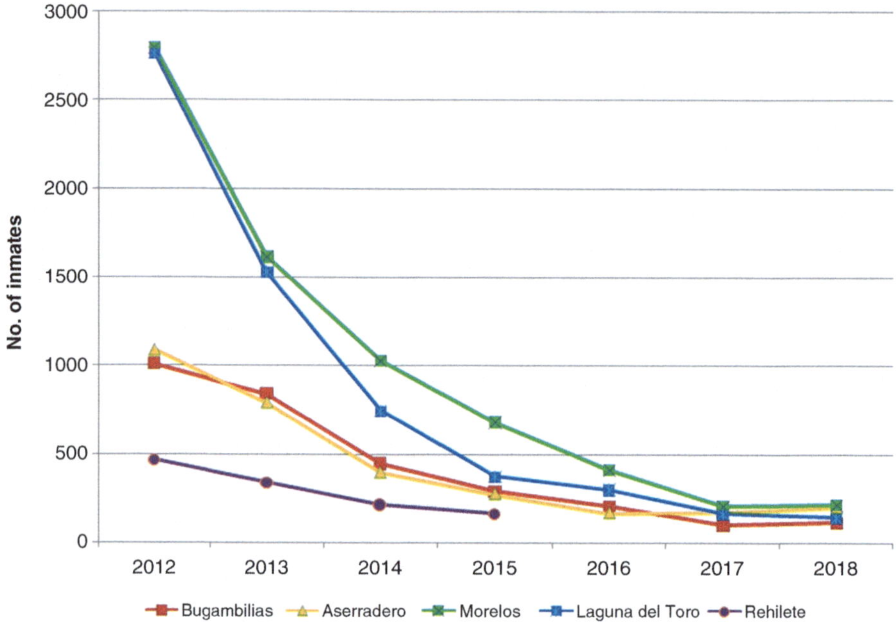

Fig. 29.2 Inmate population in prisons of Islas Marias penitentiary complex (2012–2018). (Source: Own elaboration based on data from CNDH (2018))

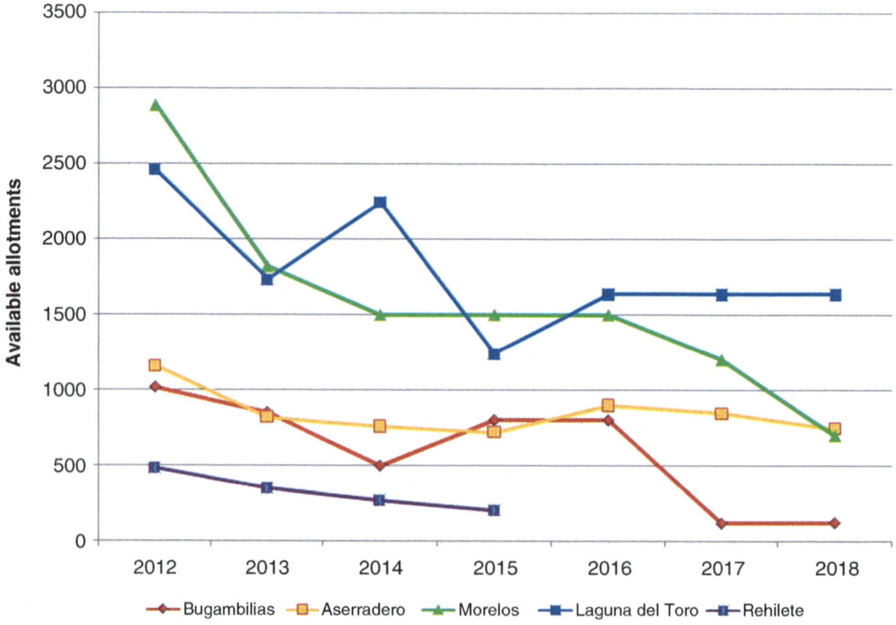

Fig. 29.3 Allotment availability in Islas Marias penitentiary complex (2012–2018). (Source: Own elaboration based on data from CNDH (2018))

29.3.4.3 Budgeting

An equally relevant and determinative factor in the closure of the prison compound was the annual budget, which is estimated to be 1,053,025.00 Mexican pesos per person/year (approximately US$ 55,422.00 per person/year). This amount was distributed for feeding and maintenance of inmates and administrative and custody personnel in the prisons, transportation of food and people from the continent, maintenance of school infrastructure, health clinic, court offices, workshops, aquaculture farm, and so on, also assignment of surveillance personnel of the perimeter of the complex, as well as of the other islands and the entire BR.

It is inevitable to weigh the national policy that the government of President Andres Manuel López Obrador has implemented since December 2018, in which he clearly puts forward the austere exercise of the treasury. By that logic, it is understandable that the burdensome allocation of resources for the maintenance of prisons will come into question. In effect, the decree that formalized the removal of the Cefereso compound of Islas Marias from the federal jail system indicates that "it is burdensome and financially unfeasible to maintain an obsolete detention regime" with high costs in the maintenance and payment of services of labour on-site (Gobierno de México 2019).

29.3.4.4 Dubious Social Reintegration

The decree of disincorporation of the Cefereso complex of Islas Marias warns that the achievement of social reintegration of persons deprived of their liberty demands that they maintain a periodic family life (Gobierno de México 2019); however, in the Maria Madre island, this was not possible in most cases, due to the distance between the prisons and the continent. It was not feasible that all the relatives of the inmates could have been able to come from different and very distant points of the country to the port of Mazatlan to the sea and move to Puerto Balleto. For this reason, despite the existence of a semi-freed detention regime for some (basically, those considered to be of low risk), it is argued that prisoners were not attracted to the idea of serving out their sentence on that site. We would question the viability of a convict to effectively re-enter society when his sentence is served on an island with no other company but that of the guards and peers and at a distance of hundreds of kilometres from the society that condemned him.

29.3.4.5 Environmental Protection

Possibly one of the reasons for the closure of the prison complex that is less appreciated by those interested in prison security issues is the preservation and conservation of biodiversity in the Islas Marias BR. But the presidential decree takes this factor seriously (Gobierno de México 2019), making it clear that the new vocation of the infrastructure and activity to develop in Isla Madre and the entire area of the reserve will contribute to complement the main objective of sustainability. The federal executive considered that after the completion of the island as a place of reclusion, an immediate transformation would take place in the centre for the promotion of education, culture, and research, taking advantage of the installed infrastructure and carrying out reconstruction or ad hoc remodelling.

29.4 Changes in the Archipelago

The closure of the Islas Marias prison complex put an end to what President López Obrador himself called a "history of punishment, torture, and repression over more than a century". With this speech, he evoked the reputation of the site for more than a century, where torture practices, confinement for political reasons, and all kinds of human rights violations were committed. However, some considered the role of the islands compound as an archetype of a prison centre, although for prisoners of low danger only, where the objective of social reintegration was privileged, through coexistence in semi-freedom, as in family, and attending to the respect of human rights, to the extent that they rejected the news of the closure. However, it was clear that there were reasons of greater weight, as stated above, to arrive at the final decision.

As of March 8, when a total of 584 convicts left the complex to be transferred to a federal detention centre on the continent, the place was available to start a different project: the so-called Jose Revueltas Walls of Water Environmental Education and Culture Centre, which, according to the government's directive, will promote environmental education, tourism, and research in the reserve. Consolidation of the new project is going to take time because, as noted, the existent infrastructure has deteriorated to a great degree; therefore, the premises are in need of repair and recondition. According to those in charge of the plan, in the foreseeable future, Jose Revueltas Walls of Water will receive an estimated 100,000 children on a yearly basis.

29.4.1 Administration and Management

In light of administrative changes, the prospects are not promising. Even with the alleged surveillance deployed by SEMAR personnel in the four islands and in the vast marine territory of the BR, it is feasible that the risks of further illegal interference will continue and most likely increase. Under the previous protocol, those responsible for the coordination of the criminal complex were the Ministry of Public Security (SSP) and SEMAR, who guarded the prisoners and protected employees and property on-site (SSP-Semarnat 2007), but with the disappearance of the prisons, the surveillance deployed by the SSP (now the Secretary of Public Safety and Protection) ceased to exist.

In reality, although the decree of disincorporation stipulates that it will be the SEMARNAT that is responsible for administering and preserving the Islas Marias BR and ensuring that the actions carried out within its territorial limits conform to its purpose of protecting the natural area (Gobierno de México 2019), SEMAR will be in charge of monitoring and protecting both the islands and the marine surroundings from external threats. In this regard, it is worth noting that Semar faces a great challenge. Despite the fact that on the Maria Madre island maintains its advanced naval station and in the continental port of San Blas holds a military zone (Semar 2011), the evidence shows that it has not had enough human resources and infrastructure to contain the illegal activity of fishermen, both from San Blas and neighbouring communities and those from the powerful fishing centre of Mazatlan.

Illegal fishing in the reserve is closely related to the scarce catch that can be made in the Golden Triangle, so the natural consequence is that fishermen enter the BR, where they know they will have guaranteed catches (SSP-Semarnat 2007). The closest continental point to the archipelago with Maria Cleofas island is Punta Mita (95 km), and coincidentally Maria Cleofas is the furthest from the Puerto Balleto naval station. It is not strange therefore that that island is the point of arrival of the poachers, where they even establish camps.

In terms of the objectives of the Islas Marias BR, the existence of a detention centre might be judged to be positive as long as it meant the placement of rigorous surveillance, both at the prison site and its surroundings. Apparently there was a certain degree of guarantee of the protection of terrestrial and marine biodiversity in the reserve.

While it is true that the conception and purpose of the entire system of reserves is oriented essentially by biological criteria such as abundance and distribution of fauna and flora, endemic species or endangered species, and it seems that everything that remains "outside" is irrelevant (Toledo 2005), this position does not reflect the reality, at least not the reality of Islas Marias. Indeed, the illegal exploiters of nature can allege that they maintain threatened by the biodiversity of Islas Marias BR because they themselves see compelled to act as outlaws, for in the places where they live the fishing activity is precarious or none at all; nevertheless, they would hardly recognize that that situation is due to over-exploitation of natural resources they themselves have engaged in for so long.

Bearing this in mind, it is clear that any attempt to protect the reserve must take into account a number of social and economical factors occurring outside its limits. Governments at the federal and local levels must take the inhabitants of places like San Blas, Punta Mita, and nearby communities on the coast of Nayarit and Sinaloa into an integrated plan that includes a variety of income alternatives in addition to fishery. Currently, at least at the federal level, the government of López Obrador is implementing a set of measures to alleviate social problems such as unemployment and lack of opportunity. The local, state, and municipal authorities must do more to resolve and satisfy most of the socio-economic needs of the people.

29.4.2 Challenges Ahead

It has been noted that the administration of the protected natural areas in Mexico is not a simple task, since the fact of managing ecosystems as diverse as those existing in the country demands a greater investment of resources, be they human, legal, financial, or political. However, it is clear that the financing aspect is of major importance, since PNA usually lack the necessary surveillance. In this context, activities carried out on the Maria Madre island from March 2019 onwards will demand a governance model of application extended to the entire BR. As noted, SEMAR and SEMARNAT are responsible for monitoring, protecting, and administering the reserve, but new actors are incorporated into this dynamic such as the Secretariats of Public Education, Culture, and Tourism, which will play a leading role in the success of the Jose Revueltas Walls of Water Centre.

The challenge of combating illegal activities of external agents is added that of coordinating the efforts of different sectors of government around the centre's goals. About the first aspect, some have already proposed substantive improvements for protected natural areas in general, which could be applied to Islas Marias. An alternative to the lack of vigilance "could be a concession scheme for NGOs experienced in conservation", in addition to the involvement of the population (González-Ocampo et al. 2015, p. 37). It also stresses the importance of establishing intersectoral agreements of environmental authorities and federal forces to prevent the export of fishery products from illegal activity to the international market, additionally exercising better local control of fisheries through close collaboration between fishermen and operators of protected areas (Ortiz-Gallarza et al. 2015).

Those observations are very pertinent, since they point out the need for an alternative model of governance in the PNA and the adjacent environs, which involve both formal or conventional actors and alternatives through ingenious arrangements that, taking into account the context of each problem, reach broad agreements and commitments regarding the fulfilment of the major goal of sustainability.

As for the second aspect, intersectoral coordination is also invoked here so that cultural, educational, and tourist activities are conducted with strict respect for biodiversity and all resources of the reserve. Knowing that changes in the administrative arrangements have just begun, it is advisable to wait to know the way in which the different governmental instances will organize around the objective.

29.5 Lessons Learned, Hopes for the Sustainability of Islas Marias

Caring for the Islas Marias BR has never been a simple task, much less efficient and satisfactory. In the reserve, biodiversity has been permanently threatened by forces that perceive rich natural assets as opportunities for achieving major economic profits. By means of any excuse, intruders decimate threatened or endangered species without any consideration, well aware of the low risk of being punished for their illegal behaviour. If the government decides to change this state of affairs, it must be to reform the way in which PNA as Islas Marias BR is governed.

As in other fields of the public sphere, the conventional way of governing in which traditional efficacy schemes are privileged by a public administration guided by standards, results, and money (Kjær 2004), in addition to focusing on "who decides", has to give way to alternative ways of addressing problems. Along this line, Prats i Català has called for verifying the quality of the interactions at different levels of government and between them and business and society organizations. The transition from the structure-function approach to the quality of the interaction entails a major change in the management of public affairs because the interaction demands conflict management and consensus building (Prats i Català 2005).

In a system of interactions, it is expected that shared responsibility becomes unavoidable; it is the quota that must be paid by those who have a voice and see their interests crystallized, modified, or impeded in terms of policy actions (Santes-Álvarez 2009). In that sense, the articulation and harmony with which the parties face the issues is the core of successful co-responsibility.

The approach to co-responsible governance implies an institutional reform that repositions stakeholders interested in decision-making. Unlike the action-submission condition (where the government and influencing groups decide, while the governed passively grants), the change translates into shared commitments. Therefore, in a region that is both fragile and conflicted, such as the Islas Marias archipelago, shared responsibility is the best way to face the problems that jeopardize its viability. In fact, in a system of diverse interests, confronted and in frank conflict such as those of who intends to protect and conserve natural resources and those of who insist on exploiting them to obtain economic benefits, it is urgent to devise better governance schemes that allow managing conflicts, building consensus, and harmonizing activities towards the ultimate goal of sustainability.

There is no doubt that the Mexican state must remain a guarantor of the safety and well-being of society and the environment, but it is clear that economic and social actors must assume a significant share of responsibility towards that goal. The vision of governance for the benefit of the Islas Marias Biosphere Reserve must transcend its designated limits and venture where the threats are created—that is, the economic centres dedicated to fishing, as well as public and private structures that have been erected around this economic activity. A comprehensive governance scheme for Islas Marias must embrace the economic, social, political, cultural, and environmental spheres in a vision of broad and long-term existence. The sustainability of resources and certainly the viability of the reserve as a whole do not admit any less ambitious alternative.

References

CNDH (Comisión Nacional de Derechos Humanos) (2018) Diagnóstico Nacional de Supervisión Penitenciaria 2018. http://www.cndh.org.mx/sites/all/doc/sistemas/DNSP/DNSP_2018.pdf

Gobierno de México (2000) Decreto por el que se declara área natural protegida, con el carácter de reserva de la biosfera, el archipiélago conocido como Islas Marías, ubicado en el mar territorial mexicano del Océano Pacífico, con una superficie total de 641,284–73-74.2 hectáreas. Diario Oficial de la Federación, 27 de noviembre. https://simec.conanp.gob.mx/pdf_decretos/23_decreto.pdf

Gobierno de México (2018) Ley General del Equilibrio Ecológico y la Protección al Ambiente. Diario Oficial de la Federación. 5 de junio

Gobierno de México (2019) Decreto por el que se desincorporan del Sistema Federal Penitenciario los Centros Federales de Readaptación Social que se indican, ubicados en el Complejo Penitenciario Islas Marías. Diario Oficial de la Federación, 8 de marzo

Gobierno de México, Semarnat-Conanp (2019) Listado de Áreas Naturales Protegidas de México. Junio. http://sig.conanp.gob.mx/website/pagsig/listanp/

González-Ocampo H, Rodríguez-Quiroz G, Ortega-Rubio A (2015) Una revisión panorámica de las áreas naturales protegidas de México. In: Ortega–Rubio A, Pinkus-Rendón MJ, Espitia-Moreno IC (eds) Las Áreas Naturales Protegidas y la Investigación Científica en México. Centro de Investigaciones Biológicas del Noroeste S. C., La Paz B. C. S., Universidad Autónoma de Yucatán, Mérida, Yucatán y Universidad Michoacana de San Nicolás de Hidalgo, Morelia, pp 19–40. 572 pp

Kjær M (2004) Governance. Polity Press, Cambridge

Ortiz-Gallarza SM, Uribe-Osorio F, Ortega-Rubio A (2015) Caracterización, diagnóstico y manejo de las Áreas Marinas Protegidas de México: propuesta metodológica. In: Ortega–Rubio A, Pinkus-Rendón MJ, EspitiaMoreno IC (eds) Las Áreas Naturales Protegidas y la Investigación Científica en México. Centro de Investigaciones Biológicas del Noroeste S. C., La Paz B. C. S., Universidad Autónoma de Yucatán, Mérida, Yucatán y Universidad Michoacana de San Nicolás de Hidalgo, Morelia, pp 85–128. 572 pp

Pérez AL (2012) Ceferesos, otro caos de García Luna. Contralínea, 1 de febrero. https://www.contralinea.com.mx/archivo-revista/2012/02/01/ceferesos-otro-caos-de-garcia-luna/

Prats I, Català J (2005) De la burocracia al management, del management a la gobernanza. Instituto Nacional de Administración Pública, Madrid

Santes-Álvarez R (2009) Replanteando la interacción gobierno-sociedad: lecciones de la gobernación ambiental en la Franja México-Estados Unidos. Eumednet/U. de Málaga. http://www.eumed.net/libros-gratis/2009c/603/index.htm

Santos K, Negrete G (2010) Las islas marías: hacia su conservación y manejo sustentable. In: Carabias J, J Sarukhán, J de la Maza and C Galindo (coord.) Patrimonio natural de México. Cien casos de éxito. Comisión Nacional para el Conocimiento y Uso de la Biodiversidad. México. https://www.biodiversidad.gob.mx/pais/cien_casos/pdf/cap55.pdf

Semar (Secretaría de Marina) (2009) En distintos operativos la armada de México evita el saqueo de recursos marítimos. Comunicado de Prensa 059/2009. http://2006-2012.semar.gob.mx/sala-prensa/prensa-2012/328-comunicado-de-prensa-059-2009.html

Semar (Secretaría de Marina) (2011) Personal de la armada de México realiza diversos apoyos en la mar. Comunicado de prensa 160/2011. http://2006-2012.semar.gob.mx/sala-prensa/comunicados-2011/1601-comunicado-de-prensa-160-2011.html

Semarnat (Secretaría de Medio Ambiente y Recursos Naturales) (2011) Acuerdo por el que se da a conocer el resumen del Programa de Manejo de la Reserva de la Biosfera Islas Marías. Diario Oficial de la Federación, 10 de junio. https://simec.conanp.gob.mx/pdf_pcym/23_DOF.pdf

Senado de México (2010) Dictamen de las comisiones unidas de Gobernación y Estudios Legislativos a la iniciativa con proyecto de decreto por el que se reforma el Artículo 4° del Estatuto de las Islas Marías. Gaceta del senado. http://www.senado.gob.mx/64/gaceta_del_senado/documento/27723

SHCP (Secretaría de Hacienda y Crédito Público) (2014) Informes sobre la Situación Económica, las Finanzas Públicas y la Deuda Pública. Tercer trimestre. https://www.finanzaspublicas.hacienda.gob.mx/work/models/Finanzas_Publicas/docs/congreso/infotrim/2014/iiit/04afp/itanfp23_201403.pdf

SSP-Semarnat (Secretaría de Seguridad Pública-Semarnat) (2007) Programa de Conservación y Manejo Reserva de la Biósfera Islas Marías. https://simec.conanp.gob.mx/pdf_libro_pm/23_libro_pm.pdf

Toledo V (2005) Repensar la conservación: ¿áreas naturales protegidas o estrategia bioregional? Gaceta Ecológica, núm. 77, octubre-diciembre, pp. 67-83

UNEP-WCMC (2018) 2018 United Nations List of Protected Areas. Supplement on protected area management effectiveness. UNEP-WCMC, Cambridge

UNESCO (2019a) Ecological sciences for sustainable development. Biosphere reserves –learning sites for sustainable development. http://www.unesco.org/new/en/natural-sciences/environment/ecological-sciences/biosphere-reserves/
UNESCO (2019b) Ecological sciences for sustainable development. Latin America and the Caribbean: 130 biosphere reserves in 21 countries. http://www.unesco.org/new/en/natural-sciences/environment/ecological-sciences/biosphere-reserves/latin-america-and-the-caribbean/
Villanueva R (2019) Por qué fue un error cerrar Islas Marías. El Mundo del Abogado, 1 de abril. https://elmundodelabogado.com/index.php?option=com_zoo&task=item&item_id=1379

Chapter 30
A Socioeconomic Assessment for Creating Successful Resource Management Policies for Protecting the Champotón River in Campeche

Lorraine A. Williams-Beck and Evelia Rivera-Arriaga

Abstract Estimating the viability to propose municipal-level natural protected areas in Campeche, Mexico, represents not only an opportunity to promote and reconcile long-term ecosystem integrity, but also delivers copious data sets to study and judge sustainable use resources practices through addressing historic and modern socioeconomic production strategies. One such study located in the Champoton River drainage system also provides indicators that help to identify grassroots support for designing cultural systems conservation and natural community sustainable management strategies among distinct local-, municipal-, and state-level social sectors. Those markers offer a myriad of options to local authorities to propose specific political policies and action agendas to strengthen social well-being programs and, at the same time, pose parameters for co-responsibility and participation strategies to fortify government and community-based civic organizations. The populace support-based bedrock for these programs resides on recognizing plausible grassroots indicators, describing real-time economic production strategies, and designing or modifying other use practices so as to provide a more ample means to feasibly sustain local communities residing in the proposed protected area. Designing new cooperative environments and social networks within these local community, municipal, and state systems provides critical feedback mechanisms that enhance these collaborative policies, particularly when this study reveals differential use strategies among community sectors that exploit distinct resource niches and other groups that have customary or foreign cultural baggage traditions for resource use and/or proposing mitigating mechanisms for curbing misuse strategies.

L. A. Williams-Beck
Centro de Investigaciones Históricas y Sociales, Universidad Autónoma de Campeche, Campeche, Mexico
e-mail: lwilliam@uacam.mx

E. Rivera-Arriaga (✉)
Instituto de Ecología, Pesquerías y Oceanografía del Golfo de México, Universidad Autónoma de Campeche, Campeche, Mexico
e-mail: evrivera@uacam.mx

A. Ortega-Rubio (ed.), *Socio-ecological Studies in Natural Protected Areas*,
https://doi.org/10.1007/978-3-030-47264-1_30

Keywords Champoton River · Environmental policy · Local protected area
Coastal management

30.1 Introduction

In theory, designing local natural protected area public policies that strive to combine social and diverse levels of governmental, institutional, nongovernmental operators and private initiative or local business actors would present a unique opportunity to promote and reconcile long-term ecosystem integrity with economic development schemes. This hypothetical notion stresses such programs might constitute a better way to integrally manage both cultural and ecological resources as an inalienable unit by incorporating people that are social activists for family and community betterment, that are historically cognizant about their shared heritage resources, and that could use both kinds of assets to provide new options to economically support their families living in rural communities that lack readily available employment or income sources and heightened social welfare programs. In addition, if said regions comprise long-term cultural traditions that practiced relatively sustainable resource exploitation patterns through time, fomenting such political programs also might inevitably lead to greater sustainable use resources practices; provide innovative means to partner social, governmental, and institutional sectors; design new economic development possibilities; and promote rural community social welfare.

These working principles drove our initiative to create an informed assessment of how to design a novel kind of multiple stakeholder public conservation and sustainable development policy for a rural area located along the Champoton River drainage (Fig. 30.1). A first step in building this enterprise required knowing both the available environmental resources and cultural heritage inventories strategies, to be able to estimate both biological and social assets' potentials in the area. Combined resource portfolios provide qualitative and quantitative indicators that would either emanate from local community initiatives or help craft programs to build grassroots support for creating cultural systems conservation and natural community sustainable management strategies among distinct local-, municipal-, and state-level governmental, nongovernmental, private, and social sectors. Those markers would then offer a myriad of options to local authorities to propose specific political policies and action agendas to strengthen social well-being programs and, at the same time, model parameters to structure co-responsibility and participation strategies to fortify government and community-based civic organizations.

A popular support-based bedrock for these combined management programs resides on identifying those social factors involved with individual and collective activism patterns that promote family and community well-being, modern and historic resource conservation and use, and environmentally sound economic production strategies, as well as designing or tweaking other exploitation practices so as to provide more ample means from which to feasibly sustain local communities

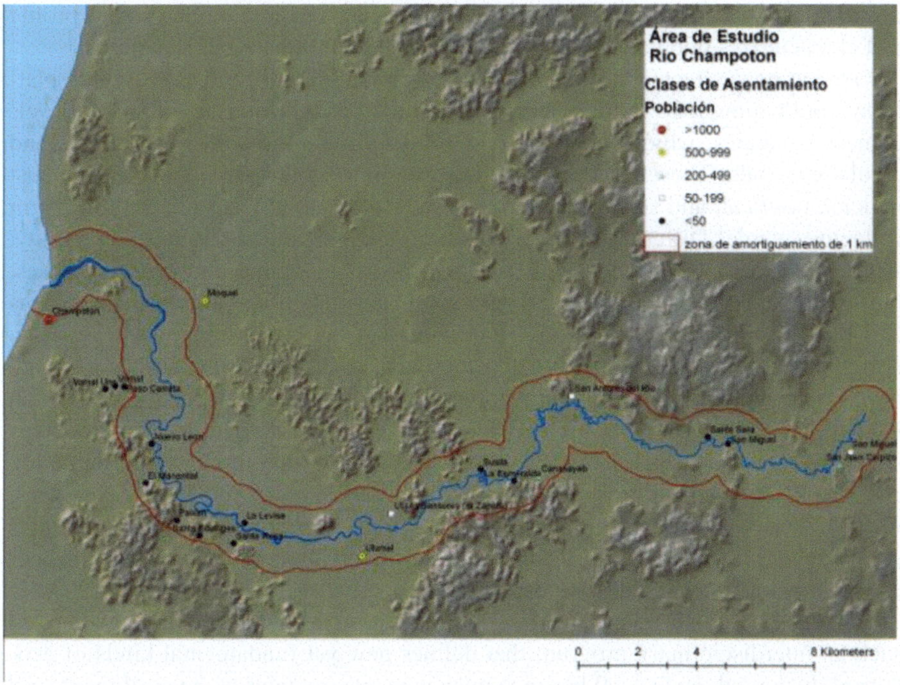

Fig. 30.1 Municipal seat and rural inland communities located along both Champoton 1 and Champoton 2 aquifer recharge zones

residing in the proposed protected area. Achieving a critical balance between actual, proposed, and potentially realistic rational use programs between local and state governments with community input is a key factor in successful management agendas proposed in other areas (Paz Salinas 2008; Durand and Jiménez 2010; Riemann et al. 2011; Valle Rodríguez 2006). Creating new cooperative environments and social networks within individual households, communities, and local, state, and federal government systems might also generate critical feedback mechanisms that would enhance those collaborative policies, particularly when a study reveals that certain community sectors exploit distinct resource niches and other groups possess foreign cultural baggage traditions for approaching resource use or suggesting mitigating mechanisms to curb detected misuse strategies.

Unique ecological, cultural, and historical substrates combine within the Champoton River drainage area to offer a robust case-study opportunity from which to design sustainable use policy programs that could successfully incorporate those fundamental qualities. Singular feedback factors that drive our research also could help to solve possible negative conundrums by helping to define economic production as well as individual and collective social activism profiles. These communities and historic settlements are designed to rationally or groundlessly appropriate and utilize localized cultural and natural assets, found within a finite set of biologically diverse floral and faunal resources. Those patterns provide a unique social profile that characterizes populations located along this exceptional river system.

Generating such a territorial data matrix from an interdisciplinary research platform would allow for a more holistic approach to promote conservation and development strategies, which share the possibility of attaining well-rounded use and regulatory goals. These strategies at the same time would provide a more robust judicial, political, and social basis from which to essentially sustain the long-term collective whole. Different sectors, including local social actors and community, state, and federal governmental as well as nongovernmental and private entities, must come together to establish proper use, management, and administration policies for defending integral cultural and natural resources' programs. In order to design this kind of holistic coordinated approach, a feasible goal would be to create new civic and institutional structural components to achieve these lofty goals, which include active community participation and pre-establishing co-responsibility steps among all actors to reach them. Promoting new kinds of community participation and social co-responsibility among local, municipal, state, and federal agencies, which also should incorporate higher education and research institutions, nongovernmental organizations, local business, and other private sectors, provides the necessary impetus to move these kinds of larger cooperative initiatives forward.

Innovative regional growth and development opportunities could emanate from such an interdisciplinary program that defines new yet fundamental kinds of ecological, historical, and social heritage management strategies as *integral territorial and productive instruments that specifically are designed for local and discrete regionally based components*. In this chapter we will discuss building such a program proposal that melds social profiles, economic and productive strategies, and resource use from data recovered through a tailor-made research design that sampled two distinct residential populations. One study group comprises nine inland communities located along the Champoton River drainage area. The second, more-focused survey polled two particular collective economic sectors that reside in the municipal seat on the river's estuary, which earn their living within a multifaceted environmental niche: shallow-water ocean products, freshwater fish species, and ecotourism opportunities.

While the states of Campeche, Yucatan, and Quintana Roo comprise shared yet locally unique social trajectories within the Yucatan Peninsula region which collectively boasts 25 natural protected areas to date, Campeche is the clear leader in this public lands conservation initiative and claims second place nationally (CONANP 2019). Their united profile in this particular case study constitutes ecological species and natural community reserve niches that serve as a fundamental means to assure not only biodiversity conservation but also ecological goods and services, in addition to long-term cultural heritage preservation activities. As an active part of the United Nations Conference on Environment and Development Congress in 1992, international programs paired with diverse governmental entities to work toward conserving biodiversity. Diverse optics coalesced from this meeting to help create natural protected areas (NPAs), which today comprise essential instruments for promoting a platform from which to achieve environmental and community-well-being synergies, in addition to their main goal to protect strategic natural and socially valuable assets considered as local, regional, and national identity heritage resources (Villalobos 2000; PNUMA 2003).

Numerous state-level statutes provide information to construct an ideal legal structure within which to base declaratory criteria for creating protected heritage resource reserves. The *Campeche State Law for Ecological Equilibrium and Environmental Protection* (1994) fosters a notion for considering natural protected areas as those areas which contain and preserve essential natural resources that underpin environmental balance and stability to help promote overall local social well-being. Chap. I, Section I, entitled "Categories, Declarations, and Administration of Natural Protected Areas" (1994), provides a detailed explanation of this legislative framework and establishes all essential content. Articles 57, 58, and 59, and portions of Article 62, propose precise parameters for managing water, soil, flora, and fauna resources. Within those legal charters, three separate and combined statutes are applicable to the Champoton River case: urban parks; recommended ecological conservation areas; and special flora, wildlife, and aquatic species protection zones (1994: 31–32).

A first methodological step to address this ambitious endeavor is to compile complete water, soil, and natural species content, as well as ancient, historic, and modern cultural features and site inventories. We achieved this goal through systematic field reconnaissance programs, census and bibliographic research, and opinion polling strategies that provided the basis for compiling abiotic, flora, fauna, and sociocultural diversity inventories for the projected protected area. Collecting precise facts regarding social activism, community and individual socioeconomic subsistence strategies, and resource use policies practiced within that spatial context adds key informational subsets required to assess short- and long-term protected heritage resource feasibility. Assessing access to basic public services, such as potable water, residential construction components (cement floors, domestic wastewater tanks), community solid waste management strategies, minimal yet basic primary or emergency care health facilities, and primary-level education services available to suburban and rural enclaves throughout the river drainage, is another critical information to be included. A combined field resource and published records database is the essential component to understanding the area's natural and social universe potential. Analyzing these data pinpoint modern consumption tendencies and resource exploitation strategies, which, in turn, help establish buffer and major impact zones detected within the protected area. The complete heritage resource universe, once compiled, serves as the crucial platform from which to create local and state monitoring systems that enable pertinent protection policy designs for sustainable development and rational use of endemic species within those particular geographic confines.

Our study also assessed immediate community location and social activism profiles, and then probable diachronic resource use strategies from ancient to modern times, to develop indicator baselines for drafting proper integral resource management plans to be carried out by diverse local governmental, nongovernmental, private initiative, and social sectors. Our goal was to provide integrated options to key decision-makers so as to propose actions that would strengthen community welfare while at the same time help to create new social participation and co-responsibility action programs among those diverse stakeholders residing along the Champoton River. Building such an integrated network like this also would encourage crafting other sustainable development programs among local, state, and federal governmental agencies that could effectively harmonize natural, cultural, historic, and human resources management policies. Integrating academic actors and nongovernmental cooperative groups provides another key conservation balance factor to private sector extractive activities and other areal development-related initiatives. Micro-regional development opportunities derived from such an integral resource study simultaneously would create an integrated land-use instrument for local planning and area conservation. Incorporating individual actors and local community participation, identified as direct resource river basin users in this schematic management plan, should promote long-term micro-regional cultural and ecological assets conservation, protection, and sustainable use in theory. While the state of Campeche currently contains seven other natural protected areas administered by federal governmental programs, some in conjunction with the State Ministry of Environmental Affairs, proposing an innovative micro-regional agenda to be managed by local stakeholders in close coordination with local community, municipal, state, and federal governmental agencies, could constitute an emerging integral heritage resource management trend for crafting cutting-edge public policy.

30.2 Macro- and Micro-regional Considerations for the Champoton River

Free-flowing surface freshwater sources comprise a unique exception to the generalized hydrological norm within the Yucatan Peninsula. Those 141,523 km^2 that constitute the political geography of Campeche, Quintana Roo, and Yucatan, located within Mexico's southeastern tropical confines between 17° 49′ 00″ and 21° 36′ 00″ North and 86° 45′ 00″ and 91° 20′ 00″ West, display an immense, relatively flat-relief limestone shelf. This leveled plane continues relatively unabated, except the northwest-southeast transversal limestone tuft Puuc hill country and the southern Campeche Chenes "mountainous" region that both sustain roughly 100–120 m vertical elevations. Hydrological Region 7 encompasses a vast yet porous area including the Mexican states of Yucatán, Quintana Roo, and most of Campeche, and excluding just the southwestern Palizada municipal area. This federal watershed management unit, whose centralized office is located in Merida, Yucatán, contains 124 municipal governing entities that cover some 137,800 km^2.

The Champoton River's east-west channel roughly flows through a 48 linear kilometer distance, 35 of which are hydrographically navigable waters beginning inland at Canasayab. This inland community resides along the watercourse's eastern bank near the late nineteenth- and early twentieth-century San Dimas ex-hacienda remains from which a small-gage rail system departed to the vast, tropical hardwood rainforests located to the south-southeast in what is today's Calakmul Biosphere Reserve (Lundell 1933). Champoton waterway's geographic area incorporates some 710 km^2 (Vega et al. 2011), which for federal management reasons designates two local hydrological watersheds that abut the Gulf of Mexico. Champoton 1 waterway stretches from the spring source to Canasayab (259 km^2) and Champoton 2's (391 km^2) from that rural community toward its estuary on the Gulf of Mexico. Both are sandwiched between the corresponding North Yucatan Region 32, South Grijalva-Usumacinta Region 30, and the East Yucatan Region 33. (Fig. 30.2).

Emerging from some 13 freshwater springs located at 120 m above sea level in the immediate area's easternmost elevated point near the community of San Miguel (Fig. 30.3), the river flows in westerly, northwesterly and southwesterly directions and links an unknown number of interconnected aqueous sources that comprised a vast subterranean peninsular hydrographic system, whose as yet undetermined points of origin flow beneath the West Yucatan Hydrological Region 31. This average 4-m-deep channel meanders rather swiftly along a softly decreasing slope that incorporates numerous springs along either shore within the first 10 km distance from the rural town Canasayab. While curiously classified in the literature as a "river," the slight topographical variation from source to outlet, within which no

Fig. 30.2 Sharp bend in the Champoton River near the "island" feature. The channel meanders some 12 km in a roughly west northwestwardly direction from this point toward the estuary

Fig. 30.3 One of the 13 underground water sources from which springs the Champoton River, near the rural community San Miguel

riverine tributaries merge into the main channel along its 50–80-m-width course toward the estuary, mockingly betrays the literal definition when arriving at the broad estuary at Champoton city and municipal seat located on the Gulf of Mexico.

Five equally diverse forest communities grace the river's shores and lateral buffer zones within the immediate region. One of the most exploited woodland communities through time, the medium perennifolia niche, shows differential forest succession characteristics for spiny palms (*Bactris balanoidea*), Mayan cherry (*Lonchocarpus castilloi*), el granadillo (*Platymiscium yucatanum*), flowering hibiscus tree (*Hampea trilobata*), and tropical cottonwood (*Ficus obtusifolia*). Chicle latex (*Manilkara zapota*); mahogany (*Swietenia macrophylla*); black olive or *pucté* in Mayan (*Bucida buceras*); ramón (*Brosimum alicastrum*); guanacaste (*Enterolobium cyclocarpum*); black creosote (*Metopium brownei*); silk cotton, *ceiba*, or *ya'ax ché* in Mayan (*Ceiba pentandra*); and fiddlewood or *ya'axnik* in Mayan (*Vitex gaumeri*) characterize this waterway's medium semi-deciduous forest content. Alternating among secondary and fallow cultivated areas are lowland wetland species such as logwood (*Haematoxylum campechianum*) (Fig. 30.4), and white creosote or *chintok* trees (*Sebastiana adenophora*).

And finally, thick red, black, white, and buttonwood mangrove forests blanket both riverine shores for almost its entire length and lend safe haven to abundant waterfowl and fish varieties (53 different species), in addition to manatees, crocodiles, and a wide range of smaller reptile groups (López-López and Sedeño 2011). When compared with other statewide faunal micro-regional totals, the Champoton River's biodiversity sustains a remarkable 74% of the amphibian, 52% of the reptilian, 25% of the birds, and 74% of other mammal species (Villalobos-Zapata y Mendoza 2010; Ceballos et al. 2002; Ceballos and Gisselle 2005; León 2006; Cedeño et al. 2010; Calderón et al. 2010; Escalona-Segura et al. 2010; Retana et al. 2010). Those collective factors pose a unique case study to propose innovative integral heritage resource management strategies.

30.3 Demographics, Land Use, and Socioeconomic Survey Design

While an intradisciplinary heritage resource research program can generate a territorial matrix from which to describe ecosystem communities and environmental diversity, knowing the social fingerprint and use approaches of those assets is an essential component for designing optimal management plans. Requisite information includes local demographics, existing community infrastructures and programs, social population profiles, estimating social activism potentials, and describing historic and modern human land-use and resource exploitation potentials, as well as current economic productive strategies that harness or misuse those natural resources which are crucial factors to consider for designing more holistic real-time conservation and sustainable development strategies. Such holistic policies promote

Fig. 30.4 A logwood (*Haematoxylum campechianum*) forest near San Miguel and the river source

well-rounded use and regulatory goals and provide a more robust basis from which to essentially estimate measures geared toward nurturing the collective whole.

Nine communities with greater than 50 people reside along the Champoton River's 1 km north-south lateral buffer zone and within this 710 km^2 linear east-west proposed protected area (Fig. 30.5). Champoton city and neighboring Paraíso

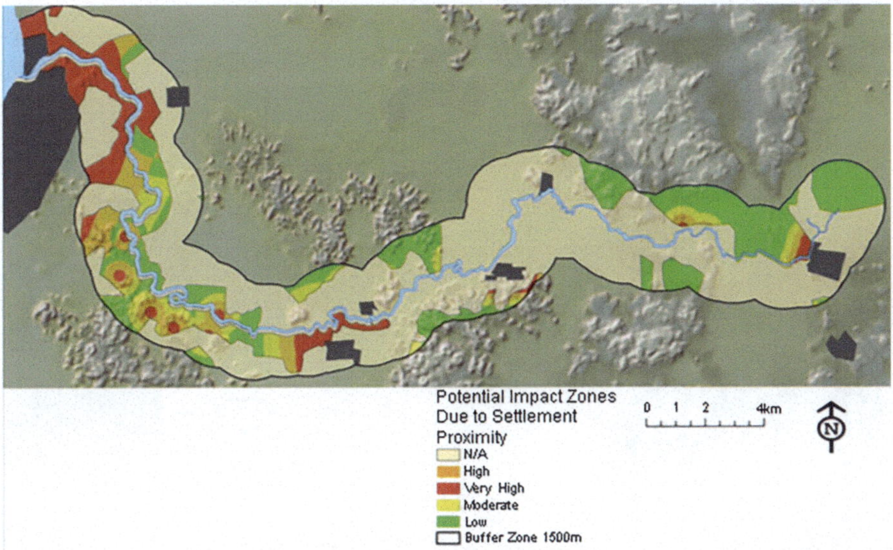

Fig. 30.5 Communities and settlements along the Champoton River display different potential impact zones due to settlement proximity. Red-color-coded areas illustrate places and communities with greater impact

register by far the largest urban municipal seat and linked suburban district within the immediate area, with a total of 27,500 people residing at the river estuary. Easternmost San Miguel, located near the watercourse's 13 springs' point of origin, contains less than 100 houses with yards and/or with adjacent plots for animal husbandry or cultivated gardens and reflects the most common settlement pattern for the river's rural population enclaves.

As the largest inland riverside place in the immediate area with 753 occupants and second oldest community following Champoton's founding in 1545, Ulumal's abandoned historic district located a scant kilometer west-southwest of the modern town contains multiple ex-hacienda residence and industrial structures in addition to the first convent established in 1608 (López Cogolludo 1955; García Bernal 1978; Gonzalez Cicero 1978; Williams-Beck 2017a). Generalized census data note most of the river's inhabitants are communal land tenure residents and subsistence farmers, and many people have house lot gardens with adjacent private or family or community collective livestock pastures to complete their productive strategy profiles (Fig. 30.6). While none of these rural river towns possessed outfield irrigation systems, community hospitals, or resident populations with individual or family having access to medical insurance, domestic telephone or Internet service, and indoor plumbing connected to private or community septic tanks, external company or local job opportunities beyond subsistence agriculture or small animal husbandry activities are also nonexistent. Some places have local primary schools, most people have access to electricity and public water systems, and almost all have pervasive televisions with some small cable satellite dishes in their homes. Studies show many

Fig. 30.6 An atypical residence for the upriver area, but common among Mayan rural communities, crafted from woven wood, covered with wattle and daub spackle, and crowned by a perishable, huano palm (*Sabal* spp.) thatched roof

of these connectivity features, such as telecommunications and Internet facilities, as well as the presence of guilds, individual and family-based economic initiatives, or collective service programs that promote family or community development, are key indicators for rural growth success (Marquez et al. 2006).

In order to fine-tune generalized census inventories, given the low population figures for towns within the region, we developed two extensive questionnaires (Babbie 1999) to randomly sample 8% of both the total rural inland residents and two strategic productive collective associations among the municipal seat population at the river estuary (Fig. 30.7).

The inland research design included specific topics using closed, multiple choice, and open question formats, based upon birth origins; length of residence in the area; linguistic affinities; cultural traditions celebrated; household solid refuse disposal and indoor hygiene; food preparation and storage strategies; secular memberships in craft production guilds or community-based service associations; religious activities; governmental assistance program affiliations; principal subsistence economical pursuits; the kinds of crops planted/harvested; kinds and quantity of domestic husbandry species raised; other specific economic activities such as aquaculture, hunting, or tourist guide services; traditional group hunting customs; and wood fuel and plant gathering techniques that exploit natural resources, in addition to a basic, overall understanding of local history and knowing about where ancient and historic cultural heritage assets are found and what potential those cultural resources might have for them personally or for their communities. Certain issues such as those

Fig. 30.7 Champoton River estuary urban-suburban dynamics, with the municipal seat located on the lower right and suburban Paraíso placed on the upper left

involved with Mayan cultural tradition observance, customary habits for resource exploitation and land fallow uses, and other cultural heritage celebrations include particularly repetitive elements to detect and discern local analogous or incompatible trends and tweak possible anomalies in respondent's answers. In total 67 men and 66 women responded to 133 questionnaires with 59 different queries. Forty-seven percent of those surveyed report their economic activity as primarily agricultural, with 5% noting animal husbandry as complementary, with one teacher, four students, and three workers having participated in the survey. One-hundred ten respondents are cooperative land tenure associates, and 23 are private community residents. We included no cattle ranches or other private initiatives or productive agro-company conglomerates in the upriver survey.

The municipal seat urban and suburban survey addressed a systematic 10% gender-specific focus group random sample of three fishing coops, whose associates include private boat owners and contracted journeymen, as well as members of a single river fishing/environmental tourism collective business. The questionnaire design stressed closed multiple choice, open answer, and complementary redundant questions regarding birth origins; length of residence in the area; linguistic affinities; cultural traditions celebrated; household and work-related solid refuse disposal and indoor hygiene parameters; food preparation and storage strategies; community-based association and religious activities memberships; governmental assistance program affiliations; principal subsistence economical pursuits; other specific economic activities such as aquaculture, hunting, or tourist guide services; fish procurement techniques; common species caught; and places where those activities occur. Supplemental productive or visitation actions that exploit natural resources, in addition to a basic, overall understanding of local history and knowing about where ancient and historic cultural heritage assets are found and what potential benefit those cultural resources might have for them personally or for their communities, completed the survey. Certain topics involved with Mayan cultural tradition observance, customary habits for marine or riverine resource exploitation, and other cultural heritage celebrations include particularly repetitive elements to discern analogous or incompatible trends as those mentioned for the upriver survey. Of the total 31 people that replied to 61 survey questions, 22 claim fishing as their sole means of economic activity, 2 complement this socioeconomic activity with limited agricultural production during seasonally prohibited species capture, 2 people are boat owners and fishermen contractors, 1 respondent works at a marine and riverine species processing plant, and 2 men are ecotourism guides.

30.4 Survey Results and Discussion

A specific detail the census data did not reveal is that while 93% of the urban and suburban municipal seat population residing at the estuary claims indigenous Mayan descent, the upriver communities represent non-Mayan people that tend not to speak other native languages. Due to a governmental colonization program to repopulate certain abandoned but arable lands within Mexico executed by the federal Land Reform and Agricultural Development Secretariats during the mid- to late 1940s (de Regil 2001a, b; Eckstein et al. 1978; Ellis and Porter-Bolland 2008; Kay 1998; Signet 2010; Walsh 2000), people from diverse ethnic origins emigrated from the central Highlands and the states of Michoacan, Veracruz, and Oaxaca to resettle in bygone towns and create new ones along the waterway. Although the Champoton River excels in the academic literature as a significant thoroughfare through time, particularly during pre-Columbian and early to late colonial periods, relatively few historic records discuss its diachronic significance (Williams-Beck 2011, 2017a, b; Williams-Beck et al. 2013; Williams-Beck and Anaya Hernández 2016; García Bernal 1978; Gerhard 1991). While 85% of respondents voiced approval of promoting local archeological and historical tourism, most upriver residents feign knowledge of where these potential assets are located (Fig. 30.8). This indifference leads many to disregard a deeper understanding of their own recolonization significance within local historical traditions or how their presence is connected to that trajectory. Few survey individuals mentioned how to conserve and protect cultural heritage resources, and at least 15% of the population considered ancient ruins would be best utilized in building construction fill or road repair activities.

Social dichotomy and disjunction between Mayan descendants historically living in the municipal seat and adjoining suburban neighborhood at the estuary, and

Fig. 30.8 Upriver regional hub at Ulumal, a sixteenth- to nineteenth-century logwood and precious tropical hardwoods ex-hacienda remains, whose church and associated ruined convent were founded in 1608, according to López Cogolludo (1955)

other non-Mayan people settling in repopulated areas, presented a unique problem to be solved regarding potentially differential resource use and socioeconomic development strategies. This particular circumstance also required designing a more focused resource base and socioeconomic pursuit questionnaire for the urban and suburban sectors at the estuary to develop whether traditional socioeconomic patterns for resource use exist or define other customary actions that stress shallow depth marine and riverine fishing exploitation, coastal ecology, and land-use strategies.

The heterogeneous cultural nature of inland communities entailed creating a divergent questionnaire design that delved into markedly different issues and how those people addressed their particular socioeconomic solvency requirements. Unlike their urban and suburban counterparts that use propane gas stoves for food preparation, most inland river dwellers cook on three-stone enclosed hearths or rudimentary yet slightly more efficient stoves, all wood-fueled. Ten to fifteen kilograms of firewood constitute the daily quantity consumed among upriver households for nutritive sustenance. None of these food preparation methods is particularly healthy or resource-savvy, given modern fuel gathering techniques that overtly deplete local species by commissioning weekly pickup truckloads from certain entrepreneurial residents, who employ local tree harvesting as their full-time economic pursuit. This practice sharply contrasts with a traditional Mayan sustainable resource use practice which stresses an individually, tumpline-carried daily roll of wood for domestic needs. A logical result of these ill-conceived selective use and exploitative strategies has had dire consequences for overall forest ecology policies and species biodiversity indicators. Resource deforestation in areas immediately adjacent to local community confines and others more distant from town centers but accessible by river or vehicle is a particularly salient trend inland.

Upriver solid residual and domestic hygienic refuse disposal trends also note troubling ecological tendencies. Almost all surveyed indicated residential refuse burning to rid domestic use areas of trash, with no organic remains recycling activities detected in communities or at domiciles during survey activities. If households raised cattle, sheep, or porcine species close to or within their house lots, many organic residuals could be effectively recycled via these activities. However, the lack of organic composting at the domestic level or collective areas within communities not directly adjacent to the river or the separation of animal or domestic human waste products could become a significant public health issue if current disposal inclinations continue.

Despite upriver cultural diversity trends when compared with the suburban and urban municipal sectors populations, poverty indices gleaned from census data for rural districts are considered from medium to highly marginal in character. These data also suggest a plausible relationship between economic impoverishment and other social hardship issues such as the paucity of healthcare options due to negligible medical services, lower scholastic development possibilities beyond primary education, available environmental education programs to quell disparate resource depredation, as well as additional sources of viable income other than domestic subsistence production agricultural and/or animal husbandry modes. Survey results note that while at least 39% of the local resident population hails from the area (many respondents are under the age of 40 years), few people gather or use traditional medicinal plants or have knowledge of those curative biocultural resources available in their vicinity. Our opinion poll data also revealed few individual or family artisanal activities for procuring alternative sources of income or adherence to certain localized cultural traditions and/or celebrations other than generic holidays such as "Mother's" or "Father's Day", young women's 15-year-old debutant parties, or other kinds of traditional festivities not initiated among children who attend rural basic education programs within their immediate community.

Urban and suburban Champoton respondents unfortunately share with their inland counterparts a marked lack of knowledge regarding solid waste disposal and domestic septic tank designs for inadequate liquid waste discharges. An immense sugarcane production and processing plant and sown cane fields wedged between collective community or individual agricultural parcels along the river and within the watershed produce additional ecological contamination concerns through chemical byproducts presence and agrochemical runoff. In addition to industrial waste from this agro-business, inadequate urban and suburban as well as rural community-level infrastructures for fish and marine products processing plants, wastewater treatment facilities, and substandard domestic septic tank construction systems also directly impact the river and estuary environment. We detected few individual or community-based upriver or municipal social or environmental protection programs, or even additional budding ecotourism business models or other low-impact economic initiatives which, if present, could serve as springboards to enhancing social activism profiles for heritage resource grassroots participation and protected areas support.

Perhaps the most egregious conclusion derived from the upriver questionnaire which the opinion survey did not address but emerged from direct personnel observation involves domesticated fauna presence in house lots and rural community township areas. Despite being countryside enclaves that in traditional Maya-speaking populations generally contain mix breed canine species, almost all inland towns had abundant quantities of specialized hunting dog varieties which illustrate stark contrasts to traditional peninsular rural counterparts. Clandestine commercial sport hunting and sport fishing activities appear to be alternative socioeconomic development pursuits that left unchecked could seriously undermine biodiversity and resource conservation programs.

30.5 Lessons Learned: Threading the Needle – Final Considerations for Surmounting Social, Legal, and Political Hurdles to Create Localized Protected Area Management

Our study identified key indicators for developing public policy initiatives through decentralization processes to create public participation programs that would not only provide positive sustainable environmental impact but could also enhance economic growth possibilities for rural municipal populations in Campeche. We have detected several areas that need to be addressed and fortified in order to foster requisite popular support for such an ambitious multi-stakeholder enterprise.

In order to understand human population-riverine socioeconomic and ecological relationships, we divided the Champoton River universe in micro-regions, which are characterized by similar elements such as the origins of their communities' populations, practical real-time resources used, and the environmental impact suffered (See Fig. 30.5). We only considered the communities located in a 1 km radius from the river: Champotón, Canasayab, Villa Guadalupe or San Dimas, San Juan Carpizo, El Zapote, Ulumal, San Antonio del Río, Moquel, and San Miguel. That 1 km distance between water and suburban and urban areas provides critical data to assess immediate river impacts. Therefore, this geographic micro-region constitutes the socioeconomic dynamic of the Champoton River, as the product of the interaction between people and the natural environment. We include Villa Guadalupe in this study because of its location within access thoroughfares to other communities close to the river's origins, and this geographic proximity also creates impacts from other communities placed along the river banks.

Additional hurdles to declaring natural protected areas as conservation and sustainable use-strategy programs require collating four distinct political bodies with corresponding legal frameworks and jurisdictional imperatives into a unified whole: (a) federal and state legislation protecting all cultural historic manifestations, rivers and all natural water sources, and threatened flora and fauna species; (b) precise territorial planning with natural resources sustainable use policies; and (c) promoting notions of popular acceptance and stewardship to protect and rationally use crucially important integral heritage resource areas.

To date the biological and cultural assets identified within this watershed area have yet to tap into federal- or state-mandated or federal- or state-funded protection programs to create a more robust basis for conservation, sustainable use, and public policy platforms from which heightened social well-being objectives for both inland and coastal residents could be proposed. Natural protected areas (NPAs) comprise terrestrial or aquatic geographies which represent diverse ecosystems with pristine characteristics that generate, recognize, and value biocultural goods and services. Presidential decree currently creates NPAs. Federal Ecological Equilibrium and Environmental Protection Laws determine those activities that can be undertaken within established NPAs. In addition to those authorized parameters, other legal instruments such as guidelines, individual areal resource management plans, and land planning programs must join forces to carefully craft each NPA's particular conservation and sustainable use profile. NPAs also adhere to special protection, conservation, restoration, and development regimes according to their individual nature and salient characteristics. The National Commission for NPAs currently is the administrative entity which oversees and manages those areas in Mexico (Arellano Guillermo et al. 2008).

Federal legislation exercises strict prerogatives over protecting certain cultural traditions and crafted masonry components as key roles to constructing national, regional, or local identities. While manifestations such as all archeological sites, historic and artistic monuments, and still-entombed in situ paleontological remains persist under federal mandate (Ley Federal del Patrimonio... 1971, Ley Federal sobre Monumentos...1972), the National Institute of Anthropology and History can designate special management permits to legally constituted organizations, such as local *ejido* communal land associations whose constituent members collectively authorize in sworn legal affidavits conservation and participation in managing and protecting these assets.

Similar conditions, instruments, and principles pertain to water administration, conservation, and use policies in Mexico through the National Water Act (2004) administered by the National Water Commission (*Comisión Nacional del Agua*, CONAGUA in Spanish). States, municipalities, individual users, and organized social associations can oversee water concerns in each of the country's 13 watersheds. All regional development, growth, and productive activities intensity depend on available water sources. Because poverty indices and food security are tied to regional economic and public security dynamics, access to clean water displays deep regional contrasts. Aquifers in Mexico receive an average annual rainfall recharge of 78 km^3 that counteracts an extraction of 28 km^3 (CONAGUA 2008,

2010). Conserving and protecting biodiversity becomes a critical factor in successful clean water management policies (Kauffer Michael and Villanueva Aguilar 2011).

As the micro-region's socioeconomic hub, Champoton city is where the political, administrative, and economic powers of the municipality reside. At the same time, the metropolitan area fluvial channel and overland routes provide access to the micro-region constituent communities. Each rural town inserts itself into local, state, and federal government program initiatives through their common land grant or *ejido* authorities and via those regulated service and aid programs within which the corresponding collective groups or individual initiatives are inscribed. However, rural political prerogatives also must be incorporated into state and federal political and economic agendas. Interdigitating those three or four realms, when diverse governmental policy programs and their temporal implementation durations vary from 3 (local)- to 6-year (state and federal) time periods do not coincide, and rotation among local and national political faction constituencies do not tend to overlap with one another. These convoluted dynamics complicate matters and may clash rather than coalesce around shared initiatives that could cultivate additional hurdles. All initiatives require specific means to broker consensus among differing spheres of influence that may not mutually coincide.

Upriver rural enclaves in the Champoton River drainage resemble a cloak-of-many-colors population from different corners of the Mexican universe. As previously mentioned their presence in this micro-region resides within a federal and state government colonization program implemented from the 1940s to the 1960s. While their influence creates the diverse social and cultural contexts manifested as a national average culture, the people inhabiting this rural district have yet to adopt traditional Mayan resource use and customs more adapted to local ecological parameters.

Whereas within the socioeconomic context for regional growth and development opportunities could emanate from such an interdisciplinary program as the one we outline, that defines new yet fundamental kinds of ecological, historical, and social heritage integral management strategies, free enterprise among these particular constituents tends to foster the mother of invention and create certain business opportunities that might offer viable means, viewed from an entrepreneurial rather than environmentally sustainable lens, from which to complement small-scale family agricultural production strategies. For example, some enterprising individuals residing between Ulumal and Canasayab clandestinely installed tilapia aquaculture tanks on the river channel's south side shore. Tilapia's rapid reproduction cycles and predatory nature create additional problems. This exceedingly aggressive exotic species tends to displace other prized riverine varieties. River bass varieties, as the supremely popular fillet menu choice in most local restaurants, are now more critically endangered in the immediate area due to overexploitation or unintended exotic species competition. Timely access to environmental or sustainable growth information could effectively channel resource abuse to creative more ecologically balanced economic initiatives.

While the 2009–2015 municipal government expressed keen interest in crafting a protected area to defend this unique freshwater resource, the successive local and state governmental programs have not mirrored or upheld those efforts. Champoton hydric resources, represented by the river drainage and springs and underground fluvial sources, currently remain insufficient to supply water demands required by natural ecosystems, agriculture crops, and people. Protecting the Champoton River, and the archaeological, cultural, and bioecological assets associated with it, requires pondering the entire hydrological watershed as one integral management component. The watershed concept is conceived here as a particular kind of hydrography-orography in which water runoff is stored and flows to a receptor water body through individual or collective hydrology channels. Champoton's regional, environmental, and social complexity compels careful consideration and attention in order to better understand how natural and social contexts interact.

Another weakness detected for building a complex multiple stakeholder integral resources platform requires developing social activism programs, whose objectives aim to increase individual, family, or local community collective civic participation. A localized management strategy would have to develop significant policies to environmentally educate local populations and to cultivate and stimulate political inclusion strategies that could substantiate the inherent internal intercommunity cohesion feedback factors necessary for proposing the Champoton River drainage as a candidate for presidential decree.

Regarding political concerns, establishing a municipal-level NPA policy road map begins with the local government council which must first approve specific proposals. From there this decision-making entity then authorizes the mayor of Champoton to declare the reserve. However, achieving a coherently derived strategy to produce such an outcome requires considerable measures and additional actions for guaranteeing such an endeavor's success. A localized management strategy such as one proposed here demands developing significant policies to educate local populations with which to cultivate and stimulate political inclusion strategies that could substantiate the inherent internal intercommunity cohesion feedback factors necessary for proposing this particular river drainage as a candidate for presidential decree. Protecting this unique cultural, ecological, and fluvial component could provide win-win policies through key social and economic benefits for local constituents: delivering clean water security, natural and cultural resource stability, heightened pollution and waste disposal regulations, promoting recreation and environmental education areas, guaranteeing the conservation and protection of cultural and archeological assets, providing enhanced heritage resource appreciation, and endowing carefully planned and managed sustainable productive activities such as apiculture and ecotourism. Much work remains to thread the needle in order to formulate these kinds of local programs and to create with them possible sustainable options for balanced resource conservation, social wellbeing, and economic development.

References

Arellano Guillermo A, Fraga J, Robles de Benito R (2008) Áreas Naturales Protegidas y Descentralización en la Península de Yucatán. In: Fraga J et al (eds) Descentralización Costera y Manejo Ambiental: Gobernanza costera en México. Plaza y Valdés Editorial, México, pp 121–139

Babbie E (1999) The basics of social research. Wadsworth Publishing Company, Belmont

Calderón Mandujano RR, Cedeño Vázquez JR, Perera Trejo EE, Retana Guascón OG, Corbala Bermejo JA (2010) Reptiles, 332–336 pp. In: Villalobos-Zapata GJ y Mendoza Vega J (Coord) La biodiversidad en Campeche: Estudio de Estado. Comisión Nacional para el Conocimiento y Uso de la Biodiversidad (CONABIO), Gob. Estado de Campeche, Universidad Autónoma de Campeche, El Colegio de la Frontera Sur, México 730 p

Ceballos G, Gisselle O (2005) Los mamíferos silvestres de México. Fondo de Cultura económica, Comisión Nacional para el Conocimiento y Uso de la Biodiversidad (CONABIO), México DF 986 pp

Ceballos G, Chávez C, Rivera A, Manterola C (2002) Tamaño poblacional y conservación del jaguar (Panthera onca) en la Reserva de la Biosfera Calakmul, Campeche, México. pp 403–481. En: Medellín RA, Chetkiewicz C, Rabinowitz A, Redford KH, Robinson JG, Sanderson E, y Taber A (eds) Jaguares en el nuevo milenio: una evaluación de su estado, detección de prioridades y recomendaciones para la conservación de los jaguares en América. Universidad Nacional Autónoma de México-Wildlife Conservation Society, Mexico

Cedeño Vázquez JR, Calderón Mandujano RR, Perera Trejo EE, Retana Guascón OG (2010) Anfibios. 326–331 pp. In: Villalobos-Zapata GJ, y Mendoza Vega J (Coord) In La biodiversidad en Campeche: Estudio de Estado. Comisión Nacional para el Conocimiento y Uso de la Biodiversidad (CONABIO), Gob. Estado de Campeche, Universidad Autónoma de Campeche, El Colegio de la Frontera Sur, México 730 pp

CONAGUA (Comisión Nacional de Aguas) (2008) Estadísticas del agua en México 2008 sobre sequía, México, CONAGUA, capítulo 2

CONAGUA (Comisión Nacional de Aguas) (2010) Estadísticas del agua en México 2010, México, CONAGUA

CONANP (Comisión Nacional de Áreas Naturales Protegidas) (2019) Listado de Áreas Naturales Protegidas de México. http://sig.conanp.gob.mx/website/pagsig/listanp/. Consulted on the 7 of November, 2019)

De Regil AJ (2001a) The case of Mexico I: neo-colonialism and truncated revolution. Global Economic Development – A TLWNSI Issue Essay Series (12):1–23

De Regil AJ (2001b) The case of Mexico II: globalization and destitution. Global Economic Development – A TLWNSI Issue Essay Series (13):1–18

Durand L, Jiménez J (2010) Sobre áreas naturales protegidas y la construcción de no-lugares. Notas para México. Revista Líder vol 16 año 12 pp 59–72. http://ceder.ulagos.cl/lider/images/numeros/16/[LIDERVol16A%C3%B1o12-2010-ISSN-0717-0165]4.-Sobre%C3%A1reasnaturalesprotegidas.pdf. Consulted on October the 30th, 2019

Eckstein GD, Horton D, Carroll T (1978) Land reform in Latin America: Bolivia, Chile, Mexico, Peru, and Venezuela. World Bank Staff Working Paper No. 2, pp 202, April

Ellis EA, Porter-Bolland L (2008) Is community-based forest management more effective than protected areas? A comparison of land use/land cover change in two neighboring study areas of the Central Yucatan Peninsula. Mexico For Ecol Manag 256:1971–1983

Escalona Segura G, Salgado Ortíz J, Vargas Soriano J, Vargas Contreras JA (2010) Aves, 350–357. In: Villalobos-Zapata GJ, y Mendoza Vega J (Coord) La biodiversidad en Campeche: Estudio de Estado. Comisión Nacional para el Conocimiento y Uso de la Biodiversidad (CONABIO), Gob. Estado de Campeche, Universidad Autónoma de Campeche, El Colegio de la Frontera Sur, México, 730 p

García Bernal MC (1978) Población y encomienda en Yucatán bajo los Austrias. Archivo General de Indias, Sevilla

Gerhard P (1991) La frontera sureste de la Nueva España, UNAM: IG-IIH, México
González Cicero SM (1978) Perspectiva religiosa en Yucatán 1517–1571. Yucatán, los franciscanos y el primer obispo fray Francisco de Toral. El Colegio de México, México
Kauffer Michael EF, Villanueva Aguilar CL (2011) Retos de la gestión de una cuenca construida: la península de Yucatán en México. AquaLAC 3(2):81–91
Kay C (1998) Latin America's agrarian reform: lights and shadows. FAO Land Reform Land Settl Coop 2:8–31
León PN (2006) Aprovechamiento de la fauna silvestre en una comunidad aledaña a la Reserva de la Biosfera Los Petenes, Campeche. Unpublished Masters' thesis. Centro de Investigaciones y de Estadísticas Avanzadas del Instituto Politécnico Nacional. Unidad Mérida, Yuc., 108 pp
Ley de Aguas Nacionales (1992). http://www.diputados.gob.mx/LeyesBiblio/pdf/16_240316.pdf
Ley Estatal del Equilibrio Ecológico y Protección al Ambiente de Campeche (1994). https://legislacion.congresocam.gob.mx/index.php/etiquetas-x-materia/328-ley-del-equilibrio-ecologico-y-proteccion-al-ambiente-del-estado-de-campeche-1
Ley Federal del Patrimonio Cultural de la Nación (1971) Secretaría de Educación Pública, México
Ley Federal sobre Monumentos y Zonas Arqueológicos, Artísticos, Históricos y Paleontológicos (1972) Secretaría de Educación Pública, México
López Cogolludo D (1955) Historia de Yucatán (3 Volúmenes). Talleres Gráficos del Gobierno del Estado, Campeche
López-López E, Sedeño J (2011) El Río Champotón, área de alta diversidad y riqueza íctica. FOMIZ Campeche, enero-marzo: 39–43 p
Lundell CL (1933) Archeological discoveries in the Maya area. Proc Am Philos Soc 72:147–179
Márquez J, González L, Oreggioni W, Pastorini M (2006) Investigación – acción – participativa en el desarrollo del cooperativismo rural. Paper presented in the symposium La intercooperación del concepto a la práctica, Universidad de la República de Uruguay, November, 22 pp
Paz Salinas MF (2008) De áreas naturales protegidas y participación: convergencias y divergencias en la construcción del interés público. Nueva Antropología 21(68) http://www.scielo.org.mx/scielo.php?script=sci_arttext&pid=S0185-06362008000100004. Consulted on October the 30th 2019
PNUMA (Programa de Naciones Unidas de Medio Ambiente) (2003) GEO América Latina y el Caribe. Perspectivas del medio ambiente 2003. Oficina Regional para América Latina y el Caribe- Programa de las Naciones Unidas para el Medio Ambiente, México, D.F
Reglamento de la Ley de Aguas Nacionales (2004). http://www.conagua.gob.mx/CONAGUA07/Publicaciones/Publicaciones/SGAA-37-12.pdf
Retana Guascón OG, Weber M, Guzmán D (2010) Mamíferos terrestres. 372–377. In: Villalobos-Zapata GJ y Mendoza Vega J (Coord) La biodiversidad en Campeche: Estudio de Estado. La biodiversidad en Campeche: Estudio de Estado. Comisión Nacional para el Conocimiento y Uso de la Biodiversidad (CONABIO), Gob. Estado de Campeche, Universidad Autónoma de Campeche, El Colegio de la Frontera Sur, México 730 pp
Riemann H, Santos-Álvarez RV, Pombo A (2011) El papel de las áreas naturales protegidas en el desarrollo local. El caso de la península de Baja California. Gestión y Política Pública XX(1):141–172. http://www.gestionypoliticapublica.cide.edu/num_anteriores/Vol.XX_No.I_1ersem/04_Hugo-Riemann_y_co-autores.pdf. Consulted on October the 30th, 2019
Signet WD (2010) Grading a revolution: 100 years of Mexican land reform. Law Bus Rev Am 16(3):481–530
Valle Rodríguez S (2006) Las Áreas Naturales Protegidas de México. Un ejemplo de propuesta de gestión de un área protegida y plan de manejo en "La Sierra de Monte Escobedo" (Zacatecas, México). Unpublished doctoral dissertation, Instituto de Ciencia y Tecnología Ambiental, Universidad Autónoma de Barcelona, 233 pp. https://www.tdx.cat/bitstream/handle/10803/5809/svr1de1.pdf. Consulted on October the 30th, 2019
Vega Serratos E, Martínez Reyes A, Posada Vanegas G, Ruiz Martínez G, Nava Fuentes JC, Pulido Yah A (2011) Estudio hidrológico de la cuenca del río Champotón. Jaina 22(1):5–30

Villalobos I (2000) Áreas Naturales Protegidas: Instrumento Estratégico para la Conservación de la Biodiversidad. Gaceta Ecológica 54:24–34

Villalobos-Zapata GJ, Mendoza Vega J (Coord) (2010) La biodiversidad en Campeche: Estudio de Estado. Comisión Nacional para el Conocimiento y Uso de la Biodiversidad (CONABIO), Gob. Estado de Campeche, Universidad Autónoma de Campeche, El Colegio de la Frontera Sur, México 730 pp

Walsh C (2000) Migration, repatriation, and colonization in Mexico, 1911–1940. Demobilizing the revolution: migration, repatriation, and colonization in Mexico 1911–1940. The Center for Comparative Immigration Studies, University of California, San Diego, pp 1–26

Williams-Beck LA (2011) Rivers of ritual and power in the Northwestern Maya lowlands. In Ecology, power, and religion in Maya Landscapes, editado por C Isendahl y B Liljefors Persson, pp 69–90. Mälmo Högskola, Mälmo, Suecia y Verlag Anton Saurwein, Markt Schwab

Williams-Beck LA (2017a) El paisaje sagrado de Cochi[s]tan, Campeche. In La memoria de los nombres: la toponimia en la conformación histórica del territorio. De Mesoamérica a México, coordinado por Karine Lefebvre y Carlos Paredes. Instituto de Geografía Ambiental, UNAM, Morelia, pp 99–123

Williams-Beck LA (2017b) La provincia Cochi[s]tan. In Los investigadores de la cultura maya XXIX:325–345. San Francisco de Campeche: Universidad Autónoma de Campeche, Difusión Cultural

Williams-Beck LA, Anaya Hernández A (2016) Relaciones sociopolíticas de la provincia Cochistan con sus vecinos del área Peten Campechana y guatemalteca. Memorias del XXIX Simposio de Investigaciones Arqueológicas en Guatemala, Guatemala Ministry of Culture, Museo Popol Vuh

Williams-Beck LA, Liljefors-Persson B, Anaya Hernández A (2013) Back to the future for predicting the past: Cuchcabal – Batabil – Cuchteel and may ritual political structures across archaeological landscapes. In: Ethnohistoric texts, and through cosmological time, pp 251–278. In: Jaroslaw Zralka, Wieslaw Koszkul (eds) Contributions to new world archaeology. Jagiellonian University and Polish Academy of Arts and Sciences, Kracow

Chapter 31
Socio-ecological Approach of Two Fishery Resources in the Centla Wetland Biosphere Reserve

Everardo Barba-Macías, Juan Juárez-Flores, Cinthia Trinidad-Ocaña, Alberto de Jesús Sánchez-Martínez, and Manuel Mendoza-Carranza

Abstract Natural Protected Areas (NPAs) are very important among other attributes because of their high biodiversity, in addition to the ecosystem services they provide. The Centla Wetland Biosphere Reserve (CWBR) in Tabasco, Mexico, is one of the most extensive wetlands of Mesoamerica, where the two rivers converge with the greatest flow to the Gulf of Mexico, the Grijalva and Usumacinta Rivers. The CWBR presents several problems caused by anthropogenic activities. One of these is fishing; in this reserve, there is a problem caused by the numeric dominance of the invasive non-native genus of the armored catfish (*Pterygoplichthys* spp.) even with a minimal economic importance. In addition, the local and regional overfishing of the blue land crab (*Cardisoma guanhumi*) has affected its reproductive cycle. Results of research projects on the environmental and socioeconomic situation of both resources are exposed here as well the perception of this situation by fishermen in the area. The situation of each resource is exposed, on the one hand how to control and reduce the damage caused by the armored catfish and on the other hand how to regulate the exploitation of the blue land crab through the proposal of legal mechanisms such as the closure and regulation of catches.

Keywords Wetlands · Armored catfish · Blue land crab · Fisheries

E. Barba-Macías (✉) · J. Juárez-Flores · C. Trinidad-Ocaña · M. Mendoza-Carranza
Sustainable Management of Basin and Coastal Zones, Sustainability Sciences Department, El Colegio de la Frontera Sur, Villahermosa, Tabasco, Mexico
e-mail: ebarba@ecosur.mx

A. de Jesús Sánchez-Martínez
Diagnostic and Management of Tropical wetlands laboratory, División Académica de Ciencias Biológicas, Universidad Juárez Autónoma de Tabasco, Villahermosa, Tabasco, Mexico

A. Ortega-Rubio (ed.), *Socio-ecological Studies in Natural Protected Areas*,
https://doi.org/10.1007/978-3-030-47264-1_31

31.1 Introduction

Protected Natural Areas (PNAs) Natural Protected Areas (NPAs) are an instrument intended to conserve areas with high biological diversity by integrating the human populations residing in them (Bennett and Dearden 2014; Gurney et al. 2014), enabling the residents' sustainable use of biota, and improving their well-being (Adams et al. 2004; Bennett and Dearden 2014). Paradoxically, PNAs are subject to alterations incentivized by poorly targeted government programs, created to meet the population's needs and fight poverty (Galindo et al. 2006). These activities benefit some sectors while reflecting loss of opportunities and natural resources for others (Guerra-Martínez and Ochoa-Gaona 2008).

Tabasco (Mexico) has one of the most important wetlands in Mesoamerica, the Centla Wetland Biosphere Reserve (CWBR), which has a rich biodiversity and is considered the most important energy storage unit in North America due to its high primary productivity (Toledo 2003). It lies at the confluence of the two most abundant rivers in Mexico, the Grijalva and Usumacinta, and represents the largest wetland area in Mexico, with a great diversity of plants that makes it one of the top seven most diverse areas worldwide (Ramírez-Martínez et al. 2015). Given the importance of the area where the CWBR is located, which is considered a PNA, it hosts a wide variety of aquatic habitats (coastal, rivers, lagoons, streams, and flood areas) with a predominance of marshy wetlands (Barba et al. 2006), which support a wide and varied source of fishery resources (Mendoza-Carranza et al. 2013). In addition, it has a diverse array of plants consisting of 737 species, 1.35% of which are in some category of vulnerability (Guadarrama and Ortíz 2000), and 546 animal species, 24.3% of which have some level of vulnerability due to habitat destruction or direct extraction (IREBIT 1994).

However, the activities carried out there (e.g., oil industry activity, fishing, disorderly hunting, livestock activity, deforestation, among others), as well as the multiple human settlements with limited municipal services, make the CWBR a highly threatened and vulnerable area; therefore, there must be no delay in reconsidering the conservation programs, notably those for protection of documented biodiversity. The main economic activities in the CWBR are fishing, livestock, and agriculture (Romero et al. 2000; Godoy et al. 2017). The latter two are among the least incompatible activities in the CWBR; however, they receive important subsidies from government programs for cultivation, thereby accelerating the land-use change that has contributed to the loss of vegetation and natural habitats (Challenger and Dirzo 2009; Barba et al. 2014).

Along with these two activities, there is also oil industry activity, which has caused pollution, channel construction, and erosion (Flores 2006). Oil activity in the state of Tabasco has been key to regional and national growth (Cato 2009), but it also represents marked alterations to ecosystems (wetlands, lagoon, and river systems) and socioeconomic systems (fishery resources) of great significance for the state economy; therefore, the government is currently interested in the socio-environmental heritage of the state (Hernández-Santana et al. 2008). From 1980 to

2005, the oil boom accelerated population growth by 45% and caused a 10% decrease in the harvested area, as economic resources were directed toward the oil sector (Pinkus and Contreras 2012). In addition to the aforementioned activities, habitat fragmentation caused by continuous fires in the CWBR has had a great impact on the loss of medium semi-evergreen forest area and has represented an increase in grassland areas (Guerra-Martínez and Ochoa-Gaona 2008).

Fishing is one of the predominant activities in coastal and river communities (Villalba 2006, Mendoza-Carranza et al. 2013) and has increased enormously in the last 50 years (Hernández and Kempton 2003). In addition to this, De la Torre-Valdez and Sandoval-Godoy (2015) mention that communities with a strong dependence on riparian fishing have generally been disadvantaged by various factors, namely, the randomness of resources, overexploitation, and the loss of habitats themselves as caused by high-impact extractive activities, such as those of the oil industry (Godoy et al. 2017). Therefore, the OECD (2014) has mentioned that Mexico needs to reform the fishing sector by taking advantage of biological resources while also caring for the environment.

Aquatic fauna has also been impacted by the actions of government programs involving the introduction of non-native species as the tilapia and the herbivorous carp. However, the impact has been lower, since these fish are exploited for commercialization or personal consumption. In addition, there are unexploited introduced species that have increased in distribution and abundance, such as the armored catfish (*Pterygoplichthys* spp.) (Fig. 31.1), which is to date unexploited by the fishing community, as it lacks commercial value, but it has become a severe problem for riparian fishery, causing economic losses for fishermen (Barba 2010; Barba et al. 2013) and their working equipment (fishing gear and motors) (Wakida-Kusunoki et al. 2007).

This invasive species is highly adaptable to the tropical environmental conditions of aquatic systems and can live in systems with hypoxic conditions, as it has the ability to breathe atmospheric air (Armbruster 1998; Hoover et al. 2004; Armbruster and Page 2006; Mendoza et al. 2007) There is a known latent risk that its

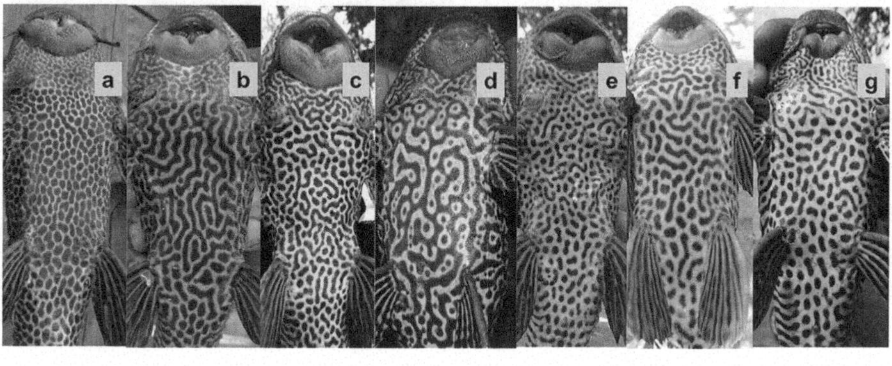

Fig. 31.1 Different patterns of abdominal pigmentation of the armored catfish: (**a**) *P. pardalis*, (**b**) *P. disjunctivus*, (**c, d, e, f, g**) not determined

introduction will cause native species to disappear (Cambray 2003) and result in various ecosystem disturbances such as habitat alteration, competition, and predation (Leppäkoski et al. 2013; Govinda-Das 2010).

In contrast with the introduced armored catfish with very low commercial importance, there is a resource of great importance within the CWBR, the blue land crab (*Cardisoma guanhumi*). This crab is a native species exploited for local and regional consumption throughout the year, with a highly cultural importance; its populations have been impacted by several anthropogenic activities (agriculture, livestock, oil extraction) and overexploitation due to catches intended for commercialization and direct consumption, (Fig. 31.2). Despite being a resource of great importance, it has been subject to little study (Rejane and Oshiro 2002). Although this species is not considered endangered, its populations are rapidly decreasing due to its uncontrolled exploitation and to the degradation and modification of its natural habitats (Firmo et al. 2012).

It is important to approach and compare both resources from a social and environmental perspective and to provide an adequate dimension to the problems caused by the inhabitants with the overexploitation and commercialization of the blue land crab and the introduction of an invasive species (Pterygoplichthys spp.).

Fig. 31.2 *Cardisoma guanhumi* blue land crab in its natural habitat (**a, d**) and capture for sale on the roadsides (**c, b**)

31.2 Study Area or Physical and Ecological Context

The CWBR represents one of the largest wetland areas (Barba et al. 2006) and is considered by the Ramsar Convention to be among the wetlands of greatest international importance (RAMSAR 1995; DOF 1992). This area contains two of the most abundantly flowing rivers in Mexico (the Grijalva and the Usumacinta), with a joint drainage basin containing 28% of Mexico's overall water area (Mendoza-Carranza et al. 2010). The CWBR basin is the second after that of the Mississippi River in contributing freshwater and nutrients to the Gulf of Mexico and the seventh in discharge worldwide (Yáñez-Arancibia and Sánchez-Gil 1988). The CWBR covers 302,706 ha (17°57′–18°39′ N and 92°06′–92°45′ W) (Bautista-Jiménez et al. 2000; Barba et al. 2015). The climate is warm and humid with an average annual rainfall of 1573 mm and an average annual temperature of 26.6 °C (CONAGUA 2000); rainfall is highest in October and November (Corella et al. 2001).

31.2.1 *Methods for Distribution and Abundance of the Armored Catfish (an Invasive Species)*

Samples were taken during the rainy season (September–November 2015) in ten sampling locations (El Viento, El Coco, Salsipuedes, El Cometa, and San Pedro lagoons and their waterways) (Fig. 31.3). In each location, the physicochemical variables of the water column were recorded, and monofilament nets were put in place with mesh measuring 9 cm between knots, a length of 25 m, and a depth of 3 m, with three repetitions in each location ($n = 90$) for 3 h at night time. Once the organisms were collected (armored catfish), they were counted, measured, and identified by sex. Likewise, the accompanying fauna was quantified, identified, weighed, and measured in order to determine the structure of the populations and the effect of the presence of the armored catfish on the populations of the commercial and native species of the reserve.

31.2.2 *Methods for Evaluation of the Population Dynamics of the Blue Land Crab* (**Cardisoma guanhumi**)

Seven sampling locations were established in preserved mangrove areas: four sites beside the San Pedro and San Pablo River (Basurero, Huarache, Perico, and San Juan) 50 meters away from the coastline, perpendicular to the river, and three sites in areas adjacent to the beach (Bateria 1, Bateria 2, and Bateria 3) where a quadrant was established for each location, 200 m away from the coastline, in order to carry out the mangrove structure. Additionally, three transects per quadrant were prepared from the coastline, each separated 15 m from each other. For each transect, the blue

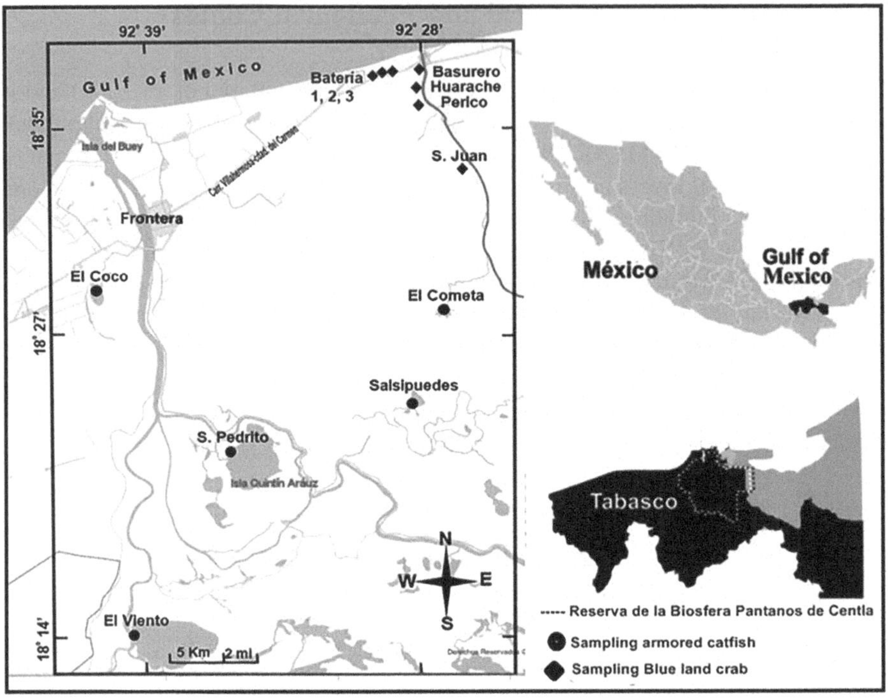

Fig. 31.3 Study area and sampling sites

land crab's burrows located within a 5-meter area on each side, and along the same line perpendicular to the coastline, were counted and measured (Fig. 31.3), in the dry and rainy seasons (June to November 2015). It should be noted that the blue land crab is a semiterrestrial species, and its distribution was therefore evaluated in the semiaquatic habitats on the bank of the San Pedro River.

31.2.3 Environmental Characterization

A micro-topographic survey was carried out at each site to obtain the terrain profiles from the coastline to the point of establishment of the quadrant where the mangrove structure was evaluated, along with the vegetation and its distribution along the transect, starting at the coastline until passing the monitoring point of each site. The length of the transect depends on the distance from the coastline to the monitoring point. Mangrove species were identified in accordance with Agraz-Hernández et al. (2006), and the vegetation associated with the mangrove was identified based on Novelo (2006). The structural attributes of the mangrove were determined using the technique proposed by Valdéz (2002) in sampling units (SU) with an area of 50 × 50 m. For each mangrove species, the abundance of adult trees with a diameter at breast height (DBH) ≥ 2.5 cm was recorded, as well as the height, length, and crown width of 20% of the adult trees in order to determine coverage (Corella et al. 2001). Additionally, the contribution to mangrove biomass production was determined, for which five circular traps were set beneath the canopy (Moreno-Casasola and Warner 2009) to quantify the contribution of each component on a monthly basis: leaves, flowers, fruits, and stipules per species of mangrove. The production of fallen leaves was expressed monthly as g. m^{-2} .$month^{-1}$.

Sediment samples were collected with a core (0.0033 m2). It was determined the texture with the Bouyoucos method (Klute 1986), pH at a 1:2 ratio in water through electrometry, organic matter (Walkley and Black 1934), bulk density, and soil moisture content (Moreno-Casasola and Warner 2009). In each location, the hydroperiod (underground water fluctuation) was determined by following the criteria proposed by Peralta et al. (2009), installing two piezometers per location and burying them at depths of 0.5 m (interstitial water) and 1.5 m (groundwater) for each sample to determine the variables of salinity (psu), redox potential (Eh mV), temperature (°C), and pH using a Hanna HI9828 multiparameter probe.

31.2.4 Biological Characterization: Estimation of the Population Parameters of the Blue Land Crab (Run)

For the direct estimation of the relative abundance of crabs, random stratified quadrants were plotted in each sampling unit (SU) with a variable or constant length in which visual censuses of the crab burrows were conducted to establish a potential relative abundance. Measurements of the galleries were taken with a tape measure to obtain the representative sizes of the crabs, measure reproductive populations, and determine reproductive potential. Six sites on the highway will be selected in order to measure the caught crabs (at least 30 organisms per site) for sale and in order to estimate their abundance, weight (g), size (total length and width), and sex ratio.

31.2.5 Socio-environmental Perception

Due to the socio-environmental problems caused by the increasing abundance of armored catfish, and diminution of abundance of blue land crab, participatory workshops were held with autonomous fishermen and fishing cooperatives located at the sampling points in order to gather relevant information from other areas or types of habitats with which the presence of the armored catfish may be associated, as well as the fluctuation of its daily and temporary abundance. This is in addition to considering the possibility of proposing, together with the fishermen, other possible alternatives for the use of the armored catfish that would also represent economic revenue for fishermen and their families and conservation strategies for blue land crab.

31.3 Results

31.3.1 Armored Catfish: Environmental Characterization and Abundance

In the channel habitat, the temperature registered a minimum average value in the locality of El Cometa and the maximum in San Pedro, as well as the pH; the lowest concentration of dissolved oxygen (DO) was recorded in El Viento and the highest in San Pedro; salinity recorded the minimum value in El Viento and the maximum in El Coco; the depth recorded very similar values in all locations, while transparency showed the minimum in El Viento and the maximum in Salsipuedes. Regarding the sampling points made in the lagoons, the physicochemical variables showed the following behavior: the temperature and the pH showed the minimum in El Cometa and the maximum in San Pedro; DO had lowest concentration in Salsipuedes and highest in San Pedro. Salinity recorded the minimum in Salsipuedes and the maximum in El Cometa; the lowest depth was recorded in El Cometa and the largest in El Coco and Salsipuedes; the minimum transparency values corresponded to the towns of El Coco and El Viento, while the maximum corresponded to San Pedro (Table 31.1).

Regarding sediments, clays predominate at most sampling sites. The locations with the highest percentage of organic matter (OM) were the channel and lagoon of El Viento (15.7 and 11.7%, respectively). In turn, the lagoon of El Coco had the lower percentage of organic matter at 3.9% (Table 31.2).

A total of 765 armored catfish specimens were collected, weighing a total of 185.57 kg during the 3 months of sampling. The location with the greatest abundance was El Viento (203 individuals), while El Cometa had the lowest (106). In El Coco, El Cometa, and San Pedro lagoons, more organisms were captured in the channels than in the lagoons; conversely, in El Viento and Salsipuedes, the results were the opposite (Fig. 31.4).

The regression analysis revealed that males of armored catfish are longer and heavier than females (Fig. 31.5). Isomorphic measurements revealed that the most common sizes for male specimens of *Pterygoplichthys* spp. are between 27 and 35 cm; females are between 27 and 32 cm. No variation was observed in the sizes, habitat, and sex recorded during the sampling (Fig. 31.6).

Sex ratios evidently favored to males, with the highest male proportion in Salsipuedes and the lowest in the lagoon of El Cometa; the highest ratio for females was recorded in El Viento and the lowest in El Cometa. In addition to males and females, a third group of organisms of indeterminable sex was recorded; the largest

Table 31.1 Average values (standard deviation) of the physicochemical variables of the water column

	Locality	Temperature (°C)	pH	OD (mg/L)	Conductivity (mScm^{-1})	Salinity (psu)	ORP	Depth (m)	Transparency (m)
Channel	El Coco	28.41 (± 2.37)	6.98 (±0.91)	4.65 (±2.06)	2571.33 (±3205)	1.36 (±1.75)	107.07 (±19.95)	1.00 (±0.35)	0.30 (±0.001)
	Salsipuedes	28.57 (± 1.84)	7.30 (±0.36)	5.44 (±1.16)	494.00 (±77.97)	0.23 (±0.04)	106.97 (±17.04)	1.07 (±0.06)	0.63 (±0.29)
	El Cometa	27.77 (±3.34)	6.98 (±1.06)	5.14 (±1.41)	2538.67 ± (±2390)	1.33 (±1.29)	102.93 (±34.37)	0.60 (±0.001)	0.40 (±0.17)
	El Viento	29.45 (±1.18)	7.07 (±0.36)	4.24 (±1.17)	508.67 (±98.82)	0.23 (±0.05)	101.73 (±14.06)	1.00 (±0.17)	0.23 (±0.06)
	San Pedro	30.05 (±1.16)	7.72 (±0.21)	6.82 (±0.12)	919.33 (±370.50)	0.45 (±0.19)	101.50 (±2.90)	0.67 (±0.12)	0.50 (±0.17)
Lagoon	El Coco	28.96 (±2.57)	7.21 (±0.56)	4.84 (±1.52)	645.61 (±511.64)	1.44 (±1.70)	101.40 (±20.66)	1.13 (±0.64)	0.37 (±0.06)
	Salsipuedes	28.65 (±1.73)	7.14 (±1.24)	4.39 (±5.05)	424.67 (±69.58)	0.20 (±0.03)	110.07 (±20.51)	1.13 (±0.12)	0.80 (±0.17)
	El Cometa	28.09 (±3.11)	7.00 (±0.85)	4.57 (±1.03)	279.35 (±239.65)	2.40 (±3.82)	89.13 (±19.72)	0.70 (±0.00)	0.70 (±0.00)
	El Viento	29.79 (±1.00)	7.22 (±0.45)	5.41 (±1.58)	332.83 (±235)	0.25 (±0.06)	111.07 (±5.38)	1.00(±0.00)	0.37 (±0.06)
	San Pedro	31.75 (±2.15)	7.82 (±0.04)	6.99 (±0.40)	979.67 (±389.64)	0.48 (±0.20)	101.20 (±3.86)	1.07 (±0.06)	0.87 (±0.40)

Table 31.2 Percentages of the different parts that make up the soils collected for each sampling site during the month of September

Site	Habitat	O. M.	Granulometric fraction (%)			Texture
			Clay	Silt	Sand	
El Coco	Channel	5.5	62	16	22	Clay
Salsipuedes		6.6	58	33	9	Clay
El Cometa		4.4	60	27	13	Clay
El Viento		15.7	48	33	19	Clay
San Pedro		24.5	34	14	52	Sandy clay loam
El Coco	Lagoon	3.9	28	16	56	Sandy clay loam
Salsipuedes		9.8	50	35	15	Clay
El Cometa		3.2	18	8	74	Sandy loam
El Viento		11.7	32	59	9	Silty clay loam
San Pedro		10.1	28	37	35	Clay loam

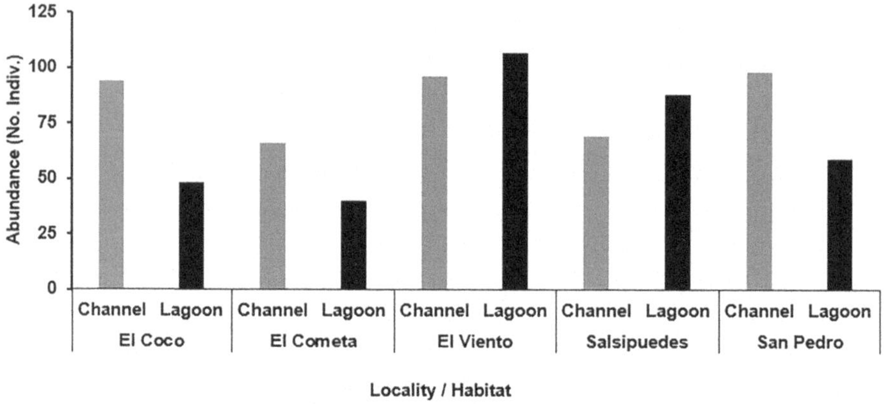

Fig. 31.4 Number of armored catfish caught for each location and habitat

number of these was identified in San Pedro (20 organisms) average size, while less than 10 organisms were recorded in the other locations (Fig. 31.7).

To determine the socioeconomic impact of the armored catfish on fishery activity, the abundance of other species collected while monitoring the armored catfish was also recorded. These species are of commercial, local, and regional importance; some of these were introduced, such as the grass carp (*Ctenopharyngodon idella*), the common carp (*Cyprinus carpio*), and the Nile tilapia (*Oreochromis niloticus*), and others are native, such as the Bay snook (*Petenia splendida*), the Mexican mojarra (*Mayaheros urophthalmus*), the redhead cichlid (*Vieja melanura*), and the tropical gar (*Atractosteus tropicus*). There were also species that have no commercial value, such as Mexican river gizzard shad (*Dorosoma anale*). Salsipuedes had the greatest species richness with 14 species, followed by El Viento (10) and San Pedro (3). In terms of environment, the channel had greater richness, with a record of 16 species, while 11 species were caught in the lagoon.

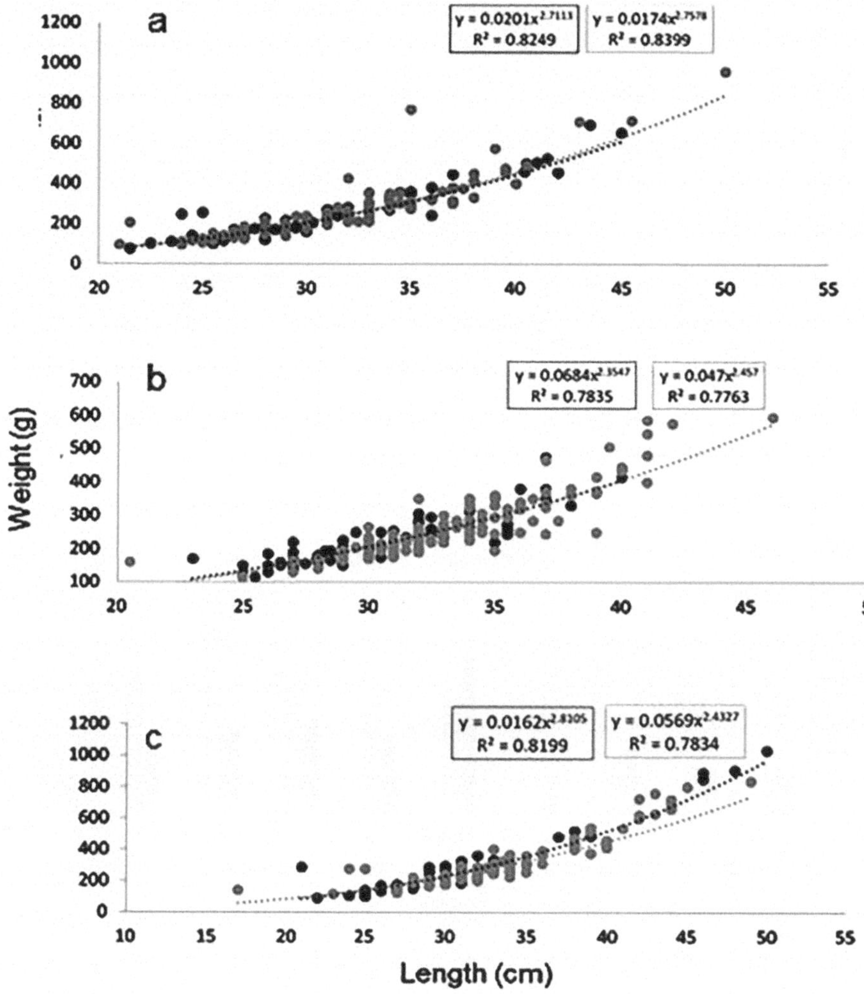

Fig. 31.5 Length-to-weight ratio for males (gray color points and lines) and females (black color points and lines). (**a**) September, (**b**) October, and (**c**) November

The greatest abundance of other fish (accompanying fauna) was recorded in the locations of Salsipuedes and El Viento, followed by San Pedro, while the lowest abundances occurred in El Cometa and El Coco. The weight of these species followed the same pattern as abundance. It is important to mention that the highest abundance was generally recorded in channel environments (Fig. 31.8).

The armored catfish evidently far exceeds the weight of the catch with respect to fish of commercial and economic importance at four of the sampling points, allowing us to infer that abundance must have the same tendency. It also bears mentioning that, in the location of El Cometa, armored catfish catches still represent no threat. This relates to two factors, the first of which is that most fishermen do not use gill

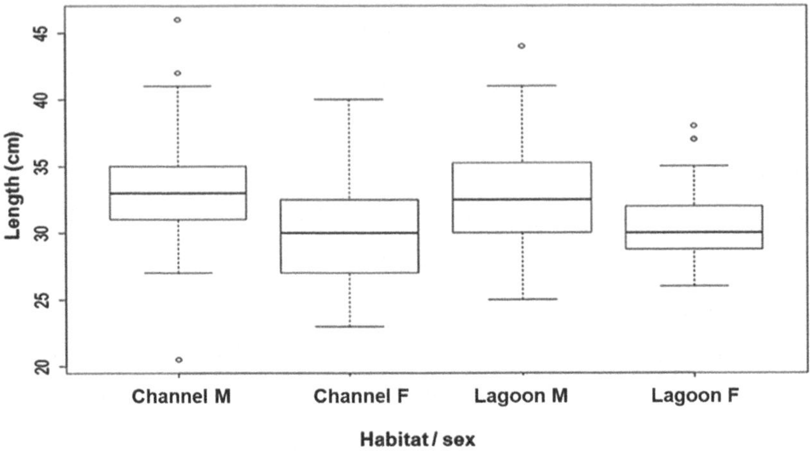

Fig. 31.6 Relationship between the length (cm) and the sex (*M* male, *H* female) of the armored catfish in habitat (channel and lagoon)

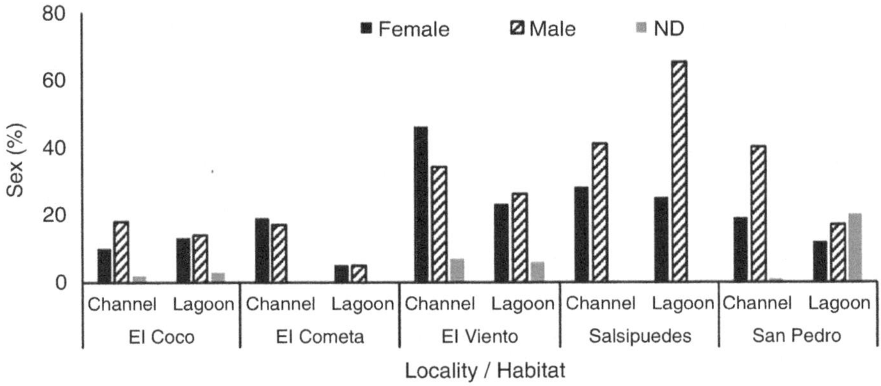

Fig. 31.7 Armored catfish sex proportion by location and habitat of armored catfish

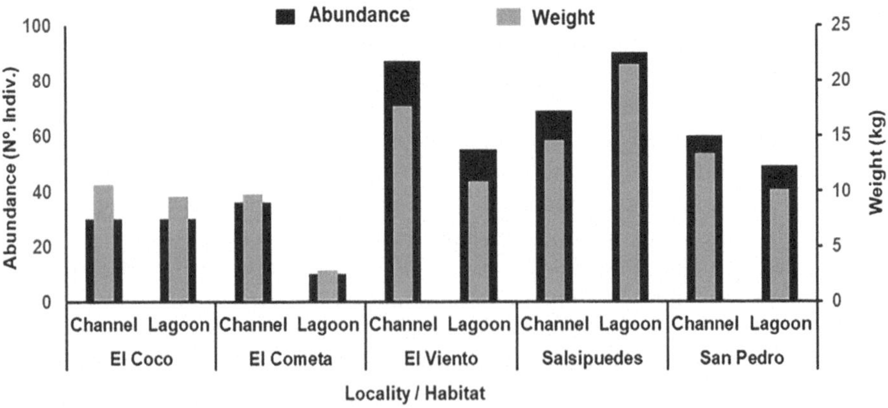

Fig. 31.8 Abundance and biomass of accompanying fauna by habitat type

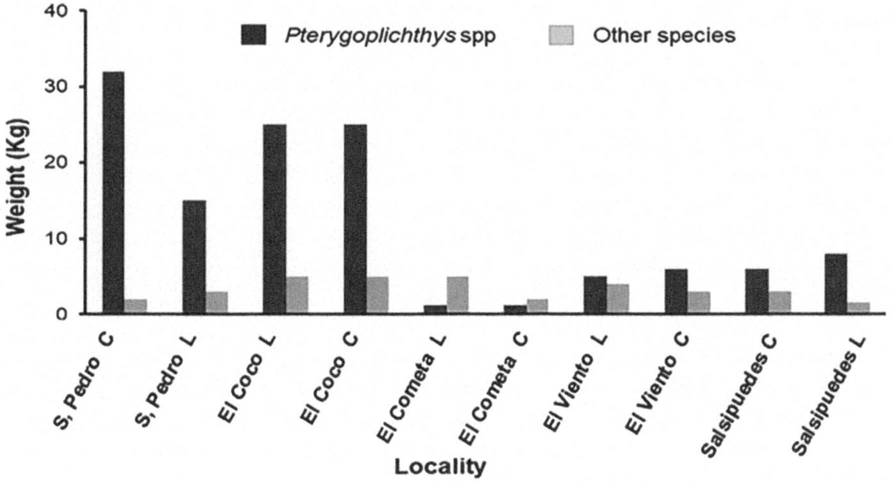

Fig. 31.9 Comparison of weights caught between armored catfish and other native species

nets and the second of which is that water salinity remains high for longer periods and represents a natural barrier for armored catfish (Fig. 31.9).

The impact of the armored catfish can be observed by confirming its constant presence on the working days of the fishermen who use gill nets, since the species was present in 100% of fishing nets in the sampling processes carried out by the fishermen of the cooperative and the autonomous fishermen in the location of Salsipuedes (38 work days). The armored catfish represented the highest volume with an average of 79 kg per working day, a value well above that represented by the rest of the commercial species, thereby underscoring the negative impact caused by the armored catfish on catches of other commercially important species (Fig. 31.10). hardhead sea catfish (*Ariopsis felis*), bigmouth sleeper (*Gobiomorus dormitor*), cichlids (Cichlidae), tropical gar (*Atractosteus tropicus*), armored catfish (*Pterygoplichthys* sp.), Bay snook (*Petenia splendida*), Nile tilapia (*Oreochromis niloticus*)

31.3.2 Blue Land Crab: Environmental Characterization and Abundance

Microphotographic profile and quantification of burrows. The San Juan site exhibited a pronounced relief, and the monitoring unit was slightly high with respect to the river level; therefore, in flood seasons, this area does not receive any surface contribution from the river. However, a contribution is received from sub-surface water, unlike the Huarache and Perico sites, which do receive surface flooding. The latter of these sites exhibited a uniform relief, without slope variations and with an

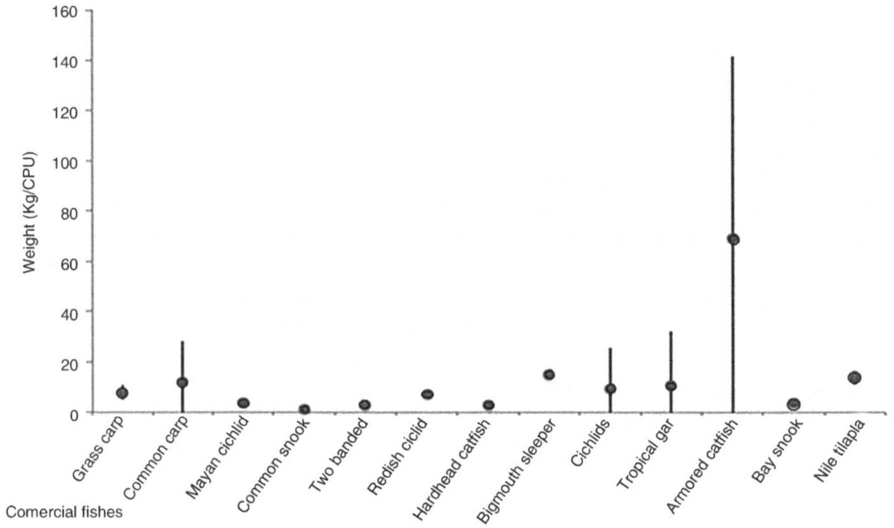

Fig. 31.10 Average catch (kg) per fishing trip of the different species exploited in the location of Salsipuedes. Grass carp (*Ctenopharyngodon idella*), common carp (*Cyprinus carpio*), Mayan cichlid (*Cichlasoma urophthalmus*), common snook (*Centropomus undecimalis*), two-banded reddish cichlid (*Paraneetroplus* sp.)

abundant mangrove cover consisting of *Rhizophora mangle* and *Laguncularia racemosa* and ultimately of mixed mangroves consisting of *R. mangle*, *L. racemosa*, and *Avicennia germinans*. Burrows have a greater presence as of 75 m from the coastline in the area with the presence of the three types of mangroves. For the Huarache site, the land has a slight slope within the first 40 m from the shore of the river. The amplitude of each profile was estimated to be about 50 m from the monitoring site where the piezometers are installed. The distribution of the galleries starts 55 m from the river's shore, and another portion is scattered 90 m away. The mangrove vegetation's composition is red and white mangrove (Fig. 31.11).

The locations adjacent to the shore (Bateria 1, 2, and 3) had the greatest relief variation, which was recorded as up to 300 cm from sea level to the sampling area; the transects for the burrow census were located approximately 350 m from the shore in an area containing induced pasture. Burrows were distributed uniformly throughout the entire transect in the pasture area and decreased in mangrove areas (Fig. 31.12).

The mangrove was composed of *R. mangle*, *A. germinans*, and *L. racemosa* at the sites of the San Pedro and San Pablo Rivers (SPySP), and of *A. germinans*, and *L. racemosa* at the Bateria sites. A basal area with a density of 151,640 m^2/ha was recorded. The maximum value was recorded in Basurero (41.14 m^2/ha), while the minimum value was recorded in Perico (6.73 m^2/ha) (Table 31.3).

The maximum average mangrove height was recorded at the sites of Perico, Huarache, and Basurero (21.3 ± 1.4 m), followed by that of San Juan (17.6 ± 1.2 m),

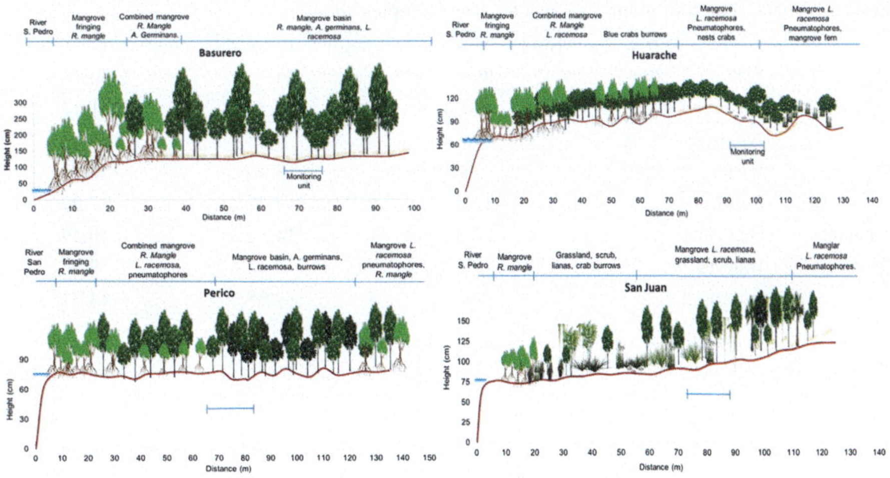

Fig. 31.11 Microtopography of soil profiles and distribution in the monitoring unit of the San Pedro and San Pablo River sites

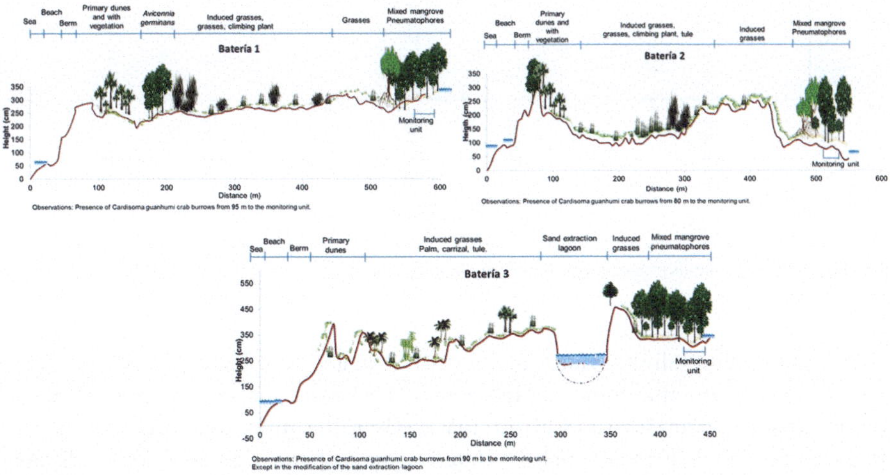

Fig. 31.12 Microtopography of soil profiles and distribution in the monitoring unit of the coastal zone sites

with the minimum average recorded at Bateria 1, 2, and 3 (15.3 ± 1.15 m) (Fig. 31.13).

At the sites of SPySP, the maximum leaf contribution values were recorded for *L. racemosa*, followed by the leaves and fruit of *R. mangle*; for the Bateria sites in the shore area, the maximum leaf contribution values were recorded for *A. germinans* (Fig. 31.14).

Table 31.3 Basal area of mangrove species from samples sites

	Site	*R. mangle* (m^2/ha)	*A. germinans* (m^2/ha)	*L. racemosa* (m^2/ha)	Total (m^2/ha)
SPySP	Basurero	9.22	10.34	21.58	**41.14**
	Huarache	1.48	3.64	6.35	**11.47**
	Perico	0.59	2.65	3.49	**6.73**
	San Juan		14.53	13.29	**27.82**
Coastal zone	Batería 1		15.47	14.82	**30.29**
	Batería 2		12.06	7.42	**19.48**
	Batería 3		4.86	9.85	**14.71**

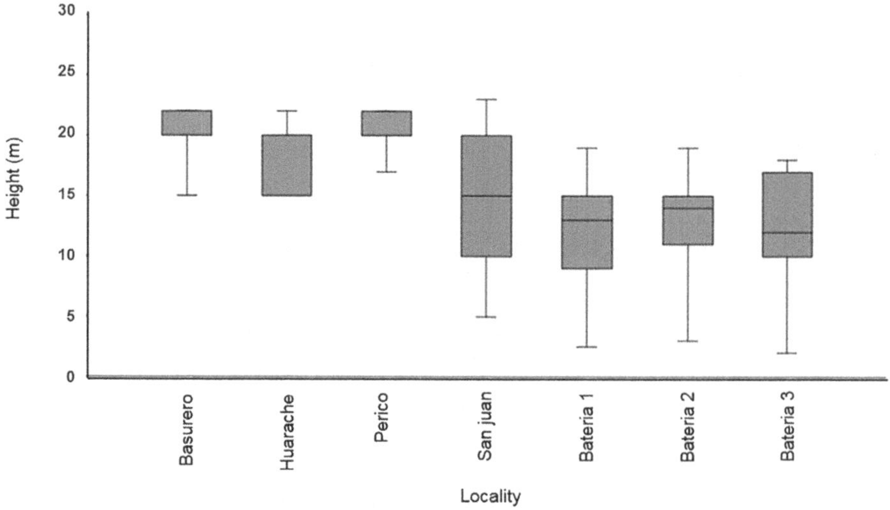

Fig. 31.13 Mangrove average heights of San Pedro, Centla

The maximum salinity value in interstitial water was identified at Bateria 1 (39.39 psu) and the minimum in Huarache (21.74 psu); the maximum ORP value was recorded in San Juan (48.60 Eh.mV), while the minimum was recorded in Basurero (−374.28 Eh.mV). pH exhibited a range from 7.14 to 7.51 at all sites except San Juan, where a value of 6.84 was recorded. The maximum temperature value was recorded at Bateria 2 (30.93 °C), and the minimum was recorded in San Juan (28.02 °C). The maximum dissolved oxygen value was recorded at Bateria 3 (3.59 mg/l), and the minimum was recorded in Basurero (0.78 mg/l); the maximum conductivity value was recorded at Bateria 1 (55.09 mS/cm), and the minimum was recorded in Huarache (34.54 mS/cm) (Table 31.4).

The sediment texture exhibited a higher sand content at the shoreline sites of Bateria 1, 2, and 3 (69%, 77.4%, and 54.9%). The maximum silt content was found at the shoreline sites of Basurero (38%) and the minimum in the shoreline site of Bateria 2 (8.4%). The maximum clay content occurred in San Juan (58.8%) and the

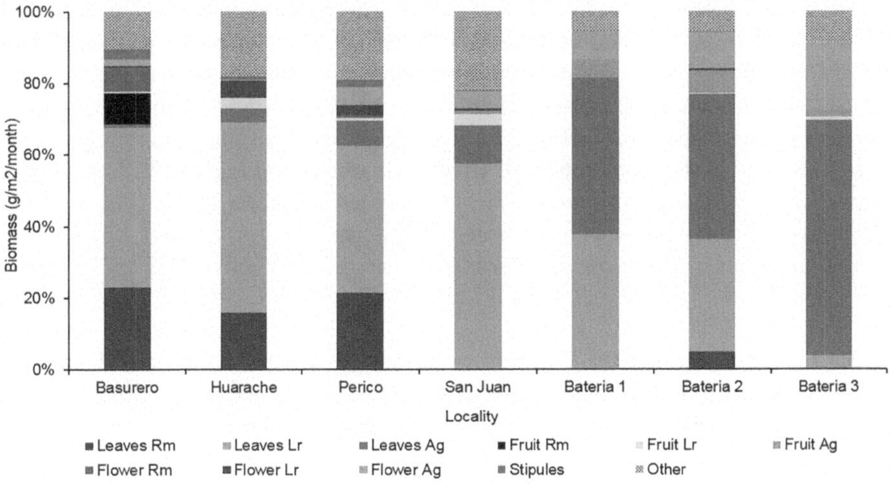

Fig. 31.14 Litter contribution at the sites of San Pedro, Centla

Table 31.4 Physicochemical parameters of water during the study in Centla Wetland Biosphere Reserve, Tabasco

Zone	Site	pH	°C	DO (mg/l)	(mS/cm)	Sal (psu)	ORP (mV)
SPySP	Basurero	7,45	28,53	0,73	43,88	28,20	−374,28
	Huarache	7,29	29,02	0,88	34,54	21,74	−337,27
	Perico	7,41	28,97	0,77	45,66	29,47	−361,40
	San Juan	6,84	28,02	2,68	39,84	25,34	−48,60
Coastal zone	Batería 1	7,14	28,58	1,28	55,09	36,39	−5,00
	Batería 2	7,35	30,93	1,70	53,12	34,92	−9,10
	Batería 3	7,51	29,91	3,59	47,83	31,05	−8,78

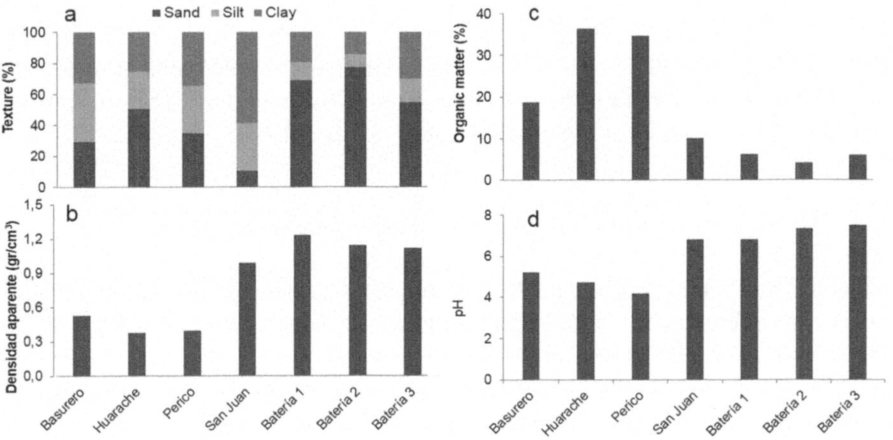

Fig. 31.15 Physical chemical in sediment. (**a**) Texture, (**b**) bulk density, (**c**) organic matter, and (**d**) pH

minimum at Bateria 2 (14.3%) (Fig. 31.15a). The maximum values for bulk soil density were found at sites of Bateria 1, 2, and 3 (1.23 g/cm^3, 1.14 g/cm^3, and 1.11 g/cm^3, respectively), and the minimum values were identified at the sites of the SPySP Rivers, namely, Basurero, Huarache, and Perico (0.52 g/cm^3, 0.37 g/cm^3, and 0.39 g/cm^3, respectively) (Fig. 31.15b). The maximum values of organic matter were presented at the sites of SPySP Rivers, Huarache and Perico (36% and 35%, respectively), while the minimum value was founded at Bateria 2 (4%) (Fig. 31.15c). The maximum pH values were recorded at Bateria 2 and 3 (7.5 and 7.3) and the minimum values in Huarache and Perico (4.7 and 4.2, respectively) (Fig. 31.15d).

31.3.2.1 Estimation of the Population Parameters of the Blue Land Crab (Count and Measurement of Burrows) The maximum values of relative abundance of burrow of the blue land crab, was presented in Huarache with 487 burrows,with a more dramatic increase in the first 25 m established, since the coastline that moved inland contained 199 burrows, with a lateral coastal extension of 140 m; the San Juan stream had the lowest abundance, with a total of 32 burrows in its entire sampling unit

The site where the maximum density values were identified was in Basurero (1.05 ind/m^2), followed by Huarache (0.99 ind/m^2), with the minimum value found at Bateria (1, 2, and 3) with 0.04 ind/m^2, 0.02 ind/m^2, and 0.01 ind/m^2, respectively. These are the sites closest to the shore (Fig. 31.16).

The maximum burrow values were recorded at the sites of SPySP (Huarache and Perico), with average lengths of 11–20 cm at each site. In Basurero, no burrows were identified. Regarding the sites in the shore areas, the minimum frequency values were recorded with average lengths of 8–32 cm (Fig. 31.17).

Individual sizes have been estimated directly and indirectly. The indirect way involved measuring the entrance of the galleries, and the direct way involved manually catching individuals only during the full moon in the month of September (known as a run). The regression analysis revealed a positive relationship between length and width for all San Pedro sites. The isomorphic measurements revealed the most common sizes for crabs to be between 10 and 20 cm for the SPySP sites, while they ranged from 5 to 25 cm for the Bateria sites (Fig. 31.18).

31.3.2.2 Relationship of Crab Size and Weight During the Run

100 crabs were caught during the 24-hour run; 88% of the individuals caught were females at different stages (75% adult females, 12 ovigerous females and 2 juveniles), while only 12% of total crabs caught were males.

The sizes obtained in the run are not related to the sizes of the galleries obtained at the different sampling sites. Therefore, it is estimated that, during the run, most of the crabs larger than 10 cm are captured for sale, or if not all crabs follow the

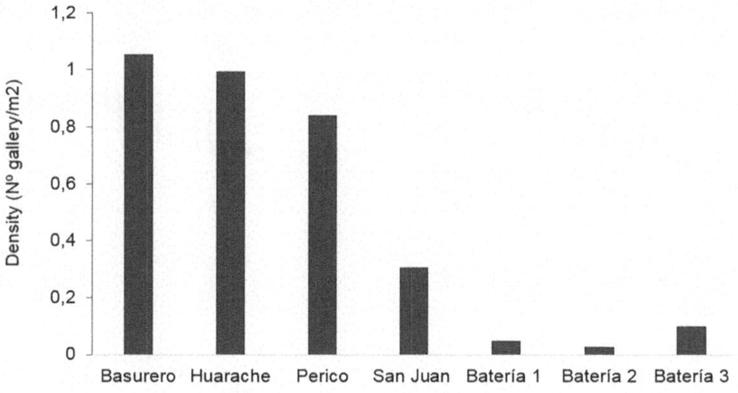

Fig. 31.16 Gallery density from samples sites, San Pedro, Centla

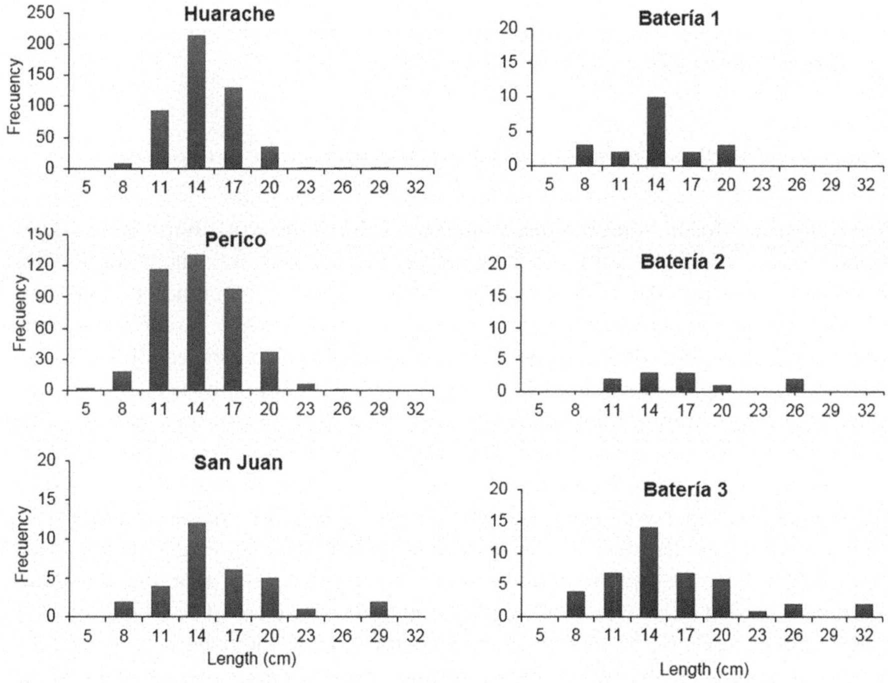

Fig. 31.17 Histograms of frequency of the sizes of crab burrows *Cardisoma guanhumi* of the different sampling units

same route; fishermen mention observing that, during the run, crabs go or cast themselves into the river to avoid predation and then depart on the current to the beach or delta to the sea to continue their route (Fig. 31.19).

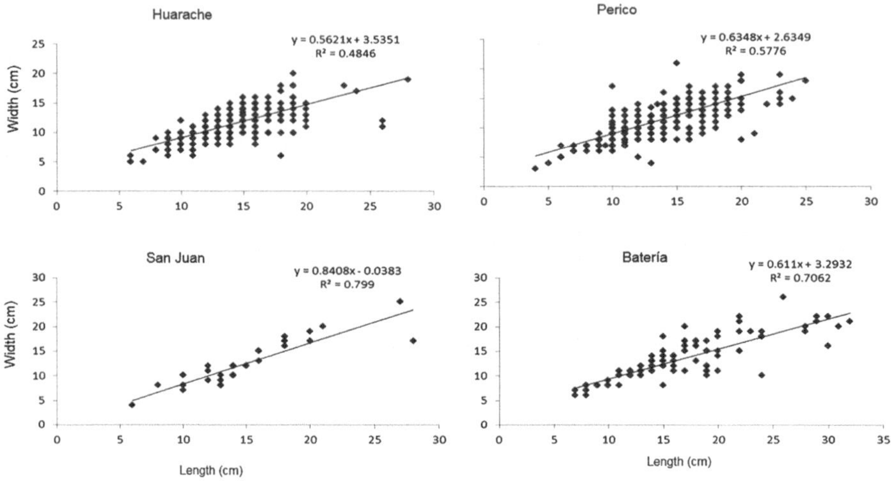

Fig. 31.18 Linear regressions of the length and width ratio of the *Cardisoma guanhumi* crab galleries in each sampling unit

31.3.3 Armored Catfish: Socio-environmental Perception

Regarding the information obtained from the results of the workshops, it was determined that, from the fisherman's perspective, the lagoons are where the greatest abundance of armored catfish is caught with respect to other aquatic systems. Additionally, they mentioned that the armored catfish is strongly associated with the presence of grasslands, docks, piles, drains, carcasses, etc., but especially fishermen associate the greatest abundance of armored catfish with grasslands. They also mentioned that another important factor for the abundance of armored catfish is the vegetation in the area; for example, whenever the common water hyacinth (*Eichhornia crassipes*) is found, the abundance of the armored catfish is higher, but it also decreases in areas with southern cattail (*Typha domingensis*), panicgrass (*Panicum* sp.), and Brown's Indian rosewood (*Dalbergia brownei*). They further mention that, because aquatic systems are closer to the coastal area, the abundance of armored catfish decreases considerably due to the presence of brackish water.

Regarding the daily fluctuation of the armored catfish, the fishermen agreed that the armored catfish's abundance in the morning until midafternoon is minimal and increases at dusk, with the greatest abundance of catches in the early morning. They provisionally mentioned that the dry season exhibits the greatest abundance, which decreases with rainfall and northerlies, when abundance is lowest (in a normal year). In turn, during a dry year, abundance is high throughout most of the year. During a rainy year, abundance decreases due to greater availability of habitats and greater dispersion of organisms over a larger area.

Fishermen have been affected economically by the presence of the armored catfish, since, before it appeared, gill nets had a useful life of up to 6 months, which is

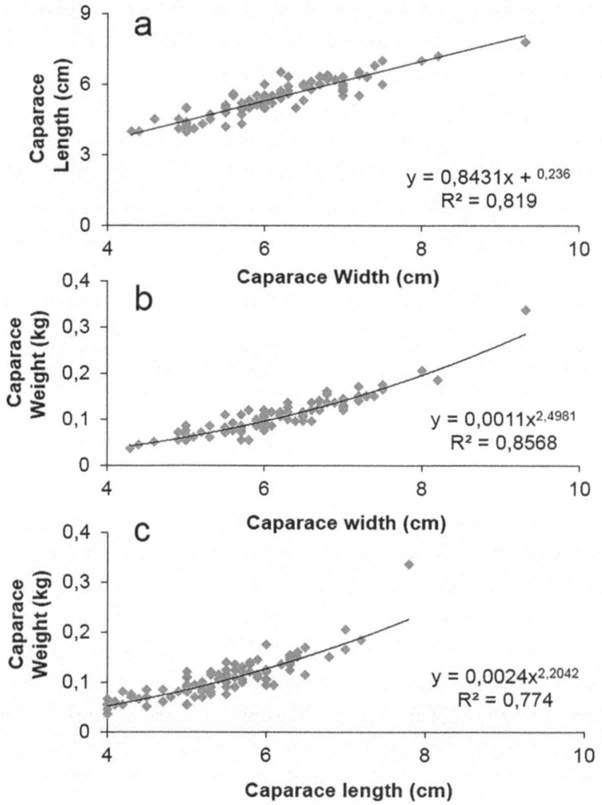

Fig. 31.19 Relationship of the measures of the shell and weight of the blue land crab *Cardisoma guanhumi*. (**a**) Shell width-length (**b**) Shell weight-width, (**c**) Shell weight-length

now 2 months. As this fish has scales that form a shell of spiny dermal shields, it becomes very tightly tangled in the nets, and, when removed, it tears and affects the net's effectiveness in catching commercially important species, generating a negative economic impact. Additionally, larger specimens can damage motor propellers through accidental impact. Taken together, such damages represent losses greater than $35,000.00 mnx annually, making fishing a low-return activity.

All the fishermen agree that, since the appearance of the armored catfish in the CWBR, all species of commercial importance have drastically decreased in abundance, such as the tropical gar (*Atractosteus tropicus*), the gafftopsail catfish (*Bagre marinus*), the hardhead sea catfish (*Ariopsis felis*), the flathead grey mullet (*Mugil cephalus*), the bigmouth sleeper (*Gobiomorus dormitor*), the fat sleeper (*Dormitator maculatus*), the common snook (*Centropomus undecimalis*), and cichlids like the Bay snook (*Petenia splendida*), the quetzal cichlid (*Vieja melanura*), the Nile tilapia (*Oreochromis niloticus*), the jaguar cichlid (*Parachromis managuensis*), and the Mexican mojarra (*Mayaheros urophthalmus*).

It is worth highlighting that fishermen are aware that overfishing is another one of the possible factors influencing the decreased catch rate, in addition to the increased number of fishermen and permit holders.

The workshops concluded with proposals from the residents of the reserve, on alternatives to exploit the armored catfish that have already been employed at other sites and may be viable to be implemented in the communities of the CWBR, including activities such as the use of fish skin for leather products (wallets, belts, sandals, etc.); preparation of meal for feed and food supplements for backyard animals such as chickens, pigs, and livestock; as well as preparation of organic fertilizers and composts to grow vegetables, habanero peppers, tomatillos, and chives, which would allow for additional revenue from the use of the fish.

31.3.4 Blue Land Crab Socio-environmental Perception

The activities carried out during the workshop made it possible to discover what the villagers know about the distribution of the blue land crab in their community, its habitat type, and the variation of its populations over different timescales, along with addressing economic and fishery issues relating to the resource, considering threats and alternatives for its conservation. The distribution of the blue crab in CWBR is recorded along the entire San Pedro River, toward the Tembladeras Channel and the Frontera, Tabasco-Ciudad del Carmen, and Campeche federal highway (Fig. 31.3). They also noted that, in the locations of Bateria and Nuevo Centla, during the rainy season, large quantities of blue land crab migrate from the mangroves to the sea, and they said that this is related to the species' reproductive process, specifically for the females to lay their eggs in the sand.

The inhabitants of the areas mentioned that crabs prefer habitats such as panicgrass (*Panicum* sp.), the white mangrove (*Laguncularia racemosa*), and the black mangrove (*Avicennia germinans*), where the greatest abundances occur in sampling survey of panicgrass. They are also found in wetlands with southern cattail (*Typha domingensis*) and on beach shores, but in lower abundances. During their breeding season, they prefer the edges of lagoons or swamps. In clay loam soil, they only appear when they go out to sea and pass through dry unflooded areas.

The fishermen noted that, between 1986 and 1995, during the blue land crab's reproductive migration process toward the marine zone, its abundance was very high and it could be seen throughout the banks of the San Pedro River and on the roadsides. From 1996 to 2005, an accelerated decline in the abundance of the species was first noticed, in connection with intensified commercialization. Likewise, they mentioned that, since 2010, there has been a notable decrease in the abundance of the blue land crab due to its high demand, which led fishermen to refrain from waiting for the run season, instead of extracting them from their burrows to sell throughout the year.

The start of the reproductive migration, when they leave their burrows naturally, consists of the nights of the full moon from July to September, which is the rainy season. For the locals, the full moon and heavy rains are highly important to the reproductive process due to their influence on tide conditions. Since, according to the inhabitants, these factors stimulate the start of the reproductive migration process. If these factors are delayed, the blue land crab's reproductive cycle is delayed too. The full moon is highly important to the reproductive process due to its influence on tide conditions.

31.3.4.1 Economic Importance of the Blue Land Crab

The variation in the sale of crabs from 1997 to 2015 was analyzed, precisely because the inhabitants were only able to define their price fluctuation in that specific period. Like all resources, the price of this species is subject to periodic fluctuations according to its abundance, albeit with a sustained increase over the years.

Between 1997 and 2000, crabs were about 14–15 centimeters wide, and each specimen was sold for eight Mexican pesos. They are currently sold in bundles of four crabs at $100.00 MXN, or $25.00 MXN per crab. In the past, the blue land crab was only caught during its reproductive migration; it was very abundant, and relatively high quantities of crabs could be caught in a short time during the reproductive migration season, making its sale highly profitable. Some people have now started to catch crabs regularly throughout the year, but even when the price is higher, there are few specimens for sale, since catching crabs is usually only meant to supplement other activities.

The blue land crab used to be bought not only in the surrounding communities, but people came to buy from other municipalities or took them elsewhere for sale, as the species' abundance allowed for this geographical range. Currently, the sales chain is local, and only the people who catch the blue land crab throughout the year sell it. It may be said that the economic importance of the resource has decreased due to the low abundance of the crab, although its price has increased. At present, this species is still caught mainly for personal consumption.

31.3.4.2 Threats to and Impacts on the Blue Land Crab Population

Indiscriminate catches, especially of ovigerous females. One of the greatest concerns of the attendees is catching throughout the year. In this type of catch, the organisms are removed from their burrows by various methods, ranging from the use of shovels, fire, and water to the use of carbide gas.

Fires. Fires are directly connected with a profound deterioration of the habitat and therefore represent impact on all the organisms living in the affected areas, including the blue land crab.

Dredging and contamination. Although attendees could not identify a correlation between dredging and low abundance of the blue land crab, it is very possible that changes in hydrology can affect this species, as well as contamination from various sources. Mangrove cutting is directly related to habitat loss and the resulting impacts.

Another threat to the species, is the Villahermosa-Frontera-Ciudad del Carmen federal highway, which runs parallel to the shoreline, representing a geographical accident involving crabs that entails a high mortality rate. This highway slows down migration and is an ideal location for the inhabitants to catch the crabs during their reproductive migration. In addition to this, the crabs' risk of being run over and killed is very high and constant almost all time of day.

In relation to these threats, the inhabitants mentioned the following proposals for the conservation and sustainable management of the blue land crab.

Implement a season that covers the breeding period (July to September). Prohibit the catching of females at any stage of their life cycle. Establish a minimum size for organisms to be caught. Organize volunteer groups who will perform nightly inspection rounds during the migration season around roads, mangroves, and beaches to help specimens reach the breeding grounds safely. Prepare flyers aimed at drivers to prevent the killing of crabs when these are crossing streets and roads to reach the sea. Encourage the population surrounding the mangrove area to create farm pens for the species, thereby creating the opportunity for an extra source of family income. Hold a crab festival to promote conservation and responsible consumption of this resource. The residents of the community of San Pedro were highly interested in the subject, as they say that they disagree with what some members of nearby communities are doing, as they poach the crab throughout the year, thereby depleting the resource.

The gradual disappearance of the crab impacts the food chain, as they are scavengers that feed on dead organisms and thereby cleanse the bodies of water. Despite inaction by the authorities and the paucity of scientific studies on the species, people need to increase their awareness regarding the conservation of this species. The blue land crab is a species that can live between 4 and 6 years, requiring several molts to reach maturity, which makes them susceptible to threats and risks during their growth. In addition to this, their reproductive migratory behavior is another point of risk that makes them more susceptible to be caught, thereby reducing the number of specimens that successfully reach the beach to lay eggs.

31.3.5 Discussion

The highest recorded number of armored catfish matches the figures reported by Sánchez et al. (2015), who found high abundances of armored catfish in the low basins of the Grijalva and Usumacinta Rivers. Furthermore, Wakida-Kusunoki et al. (2007) mention that these high densities may affect the food chain, as these fish feed on the same resources that serve as a basic food source for other species (Hossain et al. 2008). They also affect other species that are not aquatic (i.e., birds) that catch the armored catfish, which, once ingested, extends its spiny fins and may pierce the bird's esophagus and cause death by suffocation (Wakida-Kusunoki et al. 2007).

The results obtained indicate that males are longer and heavier than females, while Govinda-Das (2010) and Gomes et al. (2011) mention that females are larger than males. The results obtained may be due to intrinsic (genetic) factors but may also be related to habitat and time of year (Gómez and Guzmán 1998).

Another correlation observed was that specimens caught in channels generally were larger and heavier, unlike organisms observed in lagoon systems. Barba-Macías and Cano-Salgado (2014) report similar conditions in their work carried out in the basin of the Usumacinta River in Tabasco, noting that this may be due to the conditions of temperature, sediments, and organic matter, which provide raw material with which they build their shelters.

Regarding the sex ratio, discrepancies with Govinda-Das (2010), Duarte et al. (2007), and Gomes et al. (2011) were found, as they report a higher ratio of females to males, while male specimens were more abundant according to the sampling processes in this study.

31.3.5.1 Socio-ecological Approach of Resources

We highlight the armored catfish's high adaptive ability based on its reproductive, dietary, and physiological characteristics that allow it to inhabit bodies of water with low oxygen levels. Within its ecology, it is very important to highlight that, although this species is classified as freshwater, it was able to live in estuarine environments and was found in locations with salinities above 6 psu (Capps et al. 2011). Because there seems to be no marked preference for a particular habitat, the species can exploit almost all types of environments within the CWBR.

In the case of the CWBR, and generally throughout the Grijalva-Usumacinta basin, this species is almost present in the area since 15 years ago, and even though the year of its introduction is debatable, there is communication suggesting that it was 1999, and Barba (2005) subsequently recorded juvenile organisms in the rivers and streams of Teapa and in the Usumacinta River (Tres Brazos) in 2003.

As per the previous scenario, it is important to consider that any technological proposal or innovation must be "thought out" and directed toward the fishermen in order to be carried out (Ortiz 1985). For this purpose, it is necessary to analyze the fishing cooperatives, i.e., how they operate in their environment, how they work, and with whom they form relationships, that is, their organizational capacity, which will allow us to observe not only whether they accept and adopt the proposal but which cooperative has the physical tools (nets, boats, etc.) and social tools (openness and willingness to engage in new ways of working, establishing relationships with institutions to form a united front against poaching, species invasion, globalization, among others) to ensure the success of the innovation.

Different efforts are currently being made to take advantage of the species, both by research carried out by the academy on catches and manufacture of flour with experimental formulas for fattening juveniles and juveniles of tilapia (Cano-Salgado et al. 2012; Filigrana 2016), there are government proposals to support the construction of a filleting plant for sale (CONANP 2016), together with these there have been competitions to prepare food and dishes based on this fish, as well as the inclusion of dishes in restaurants. However, it is necessary to disseminate the possibilities of exploitation of the species and raise awareness in the community for its feasibility of incorporation into current production systems.

Abetter understanding of the attributes of each species treated in the present study, as well as the socio-environmental problems they generate, first species armored catfish as species introduced in Mexico, originally from South America, it can be found in all kinds of aquatic habitats (rivers, lagoons, streams, swamps). Sincethis species has a highly negative impact on the habits and customs of the residents, social importance is reduced. However, some communities have organized around the problem in search of alternatives for use and control of the species. Although these species do have not any cultural and positive economic importance, some residents of the CWBR have begun to incorporate the image of these resources into their customs, knowledge, and culture; however, there is a nascent interest in its use. Its high abundances and its armored, spiny structure cause damage to netting. Increased catches of this species have resulted in a decrease in catches of commercial fish and, therefore, important economic losses. Currently this fish has different uses by people such as: direct consumption, inputs for the manufacture of agricultural food (flour and silage), use of skins for crafts, biogas production (Barba-Macías et al. 2017).

Moreover, blue land crab is a native species distributed along the Atlantic Coast of the Americas from Southeast Florida to Brazil, which inhabit the coastal area up to 5 km inland, occupying various wetland and flood areas. This species have a highly positive commercial importance, as the catching process is community- and family-based, with the crabs being caught during their reproductive migration; this process characterizes the region and brings the inhabitants together socially, providing a high cultural importance, as it is a resource that characterizes a region and is strongly rooted in the customs and food culture of the coastal area of Tabasco. Fisheries have a positive commercial importance: a bundle of four specimens can cost up to $120 mxn. However, its population decline has resulted in a very low catch; therefore, economic benefits have decreased. Despite the ecological importance in maintaining and recycling nutrient flows in mangrove areas and other flood areas, their populations are decreasing ever more sharply (Barba-Macías et al. 2017).

Acknowledgments The study was part of the activities developed with the economic and personal support of the National Commission for Natural Protected Areas (CONANP), to authorities and fishermen of Salsipuedes, etc. Additional support was provided by the Network for the Knowledge of Coastal Resources in Southeastern Mexico (RECORECOS) Red Temática de Investigación CONACYT Project **293368** and National Network of Protected areas (RENANP) Red Temática de Investigación CONACYT Project **269540**.

References

Adams WM, Aveling R, Brockington D et al (2004) Biodiversity conservation and the eradication of poverty. Science 306(5699):1146–1149. https://doi.org/10.1126/science.1097920

Agraz-Hernández CM, Noriega-Trejo R, López-Portillo J et al (2006) Guía de Campo: Identificación de manglares en México. Universidad Autónoma de Campeche, México

Armbruster JW (1998) Review of the loricariid catfish genus Aphanotorulus and redescription of A. unicolor (Teleostei: Siluriformes). Ichthyol Explor Freshw 8(3):253–262

Armbruster JW, Page LM (2006) Redescription of Pterygoplichthys punctatus and description of a new species of *Pterygoplichthys* (Siluriformes: Loricariidae). Neotrop Ichthyol 4:401–409

Barba E (2005) Valor del hábitat: distribución de peces en humedales de Tabasco. Ecofronteras 25:9–11

Barba E (2010) Situación actual de los recursos acuáticos en Tabasco: impacto económico y social de los plecos. In: Koleff P, González AI, Born-Schmidt G (eds) Estrategia nacional sobre especies invasoras en México, prevención, control y erradicación. Comisión Nacional para el Conocimiento y Uso de la Biodiversidad, Comisión Nacional de Áreas Protegidas, Secretaría de Medio Ambiente y Recursos Naturales, México, 58p

Barba E, Cano-Salgado MP (2014) Abundancia del plecos (*Pterygoplichthys pardalis*) en sistemas lagunares y ribereños de la cuenca del Usumacinta, Balancán, Tabasco, México. In: Low Pfeng AM, Quijon PA, Peters EM (eds) Especies invasoras acuáticas: casos de estudio en ecosistemas de México. Secretaría de Medio Ambiente y Recursos Naturales (Semarnat) Instituto Nacional de Ecología y Cambio Climático (INECC) University of Prince Edward Island (UPEI), Mexico, pp 293–311

Barba E, Rangel J, Ramos R (2006) Clasificación de los humedales de Tabasco mediante sistemas de información geográfica. Universidad y Ciencia 22(002):101–110

Barba E, Magaña M, Juárez-Flores J (2013) Nuevos registros del loricárido *Pterygoplichthys pardalis* en las cuencas de Carmen-Pajonal Machona, Grijalva y Usumacinta, en Tabasco. In: Low Pfeng AM, Quijon PA, Peters EM (eds) Especies invasoras acuáticas: casos de estudio en ecosistemas de México. Secretaría de Medio Ambiente y Recursos Naturales (Semarnat) Instituto Nacional de Ecología y Cambio Climático (INECC) University of Prince Edward Island (UPEI), Mexico, pp 233–251

Barba E, Valadez-Cruz F, Pinkus-Rendón MA, Pinkus Rendón MJ (2014) Revisión de la problemática socioambiental de la Reserva de la Biósfera Pantanos de Centla, Tabasco. Investigación y Ciencia de la Universidad Autónoma de Aguascalientes. 60:50–57

Barba E, Valadez-Cruz F, Pinkus-Rendón M et al (2015) Reserva de la Biosfera Pantanos de Centla: aspectos socio-ambientales prioritarios. In: Ortega-Rubio A, Pinkus-Rendón M, Espitia-Moreno C (eds) La Investigación Científica y las Áreas Naturales Protegidas en México. Centro de Yucatán y Universidad Michoacana de San Nicolás de Hidalgo, México, pp 11–29

Barba-Macias E, Mendoza-Carranza M, Trinidad-Ocaña C et al (2017) Contrastes en el manejo del cangrejo azul y el pez diablo: perspectiva de los pobladores de la Reserva de la Biósfera Pantanos de Centla. Tabasco, El Colegio de la Frontera Sur, 47 pp

Bautista-Jiménez A, García-Muñiz A, Pérez-Alejandro P (2000) Caracterización de la Reserva de la Biosfera Pantanos de Centla. Universidad y Ciencia 15:15–28

Bennett J, Dearden P (2014) Why local people do not support conservation: community perceptions of marine protected area livelihood impacts, governance and management in Thailand. Mar Policy 44:107–116. https://doi.org/10.1016/j.marpol.2013.08.017

Cambray J (2003) Impact on indigenous species biodiversity caused by the globalization of alien recreational freshwater fisheries. Hydrobiologia 500(1–3):217–230. https://doi.org/10.1023/A:1024648719995

Cano-Salgado MP, Bello E, Barba E (2012) Innovación socialy capacidad de organización de las cooperativas pesquerasen el municipio de Balancán, Tabasco, México. Centro de investigación en alimentacion y Desarrollo a.c. Estudios Sociales 20(39):65–98

Capps KA, Nico LG, Mendoza-Carranza M et al (2011) Salinity tolerance of non-native suckermouth armoured catfish (Loricariidae: Pterygoplichthys) in South-Eastern Mexico: implications for invasion and dispersal. *Aquatic Conservation: Marine and Freshwater Ecosystems* 21, 528–540. Available at: doi:https://doi.org/10.1002/aqc.121

Cato J (2009) Gulf of Mexico origin, waters, and biota. Texas A&M University Press, Mexico, 136 pp

Challenger A, Dirzo R (2009) Factores de cambio y estado de la biodiversidad. In: Dirzo R, González R, March I (eds) Capital natural de México: Estado de conservación y tendencias de cambio, vol II. Comisión Nacional para el Conocimiento y Uso de la Biodiversidad, México, pp 37–73

CONAGUA (Comisión Nacional del Agua) (2000) Recopilación de datos meteorológicos del estado de Tabasco. Informe del gobierno estatal. Villahermosa, Tabasco, p 677

Corella JF, Valdez-Hernández JI, Cetina AV et al. (2001) Estructura forestal de un bosque de mangles en el noroeste del estado de Tabasco, México. Ciencia Forestal en México 26(90):73–102

De La Torre-Valdez HC, Sandoval-Godoy SA (2015) Resiliencia socio-ecológica de las comunidades ribereñas en la zona Kino-Tastiota del Golfo de California. Ciencias Pesqueras 23(1):53–57

DOF (Diario Oficial de la Federación) (1992) Decreto de la Reserva de la Biosfera Pantanos de Centla. México, Poder Ejecutivo Federal. http://www.dof.gob.mx

Duarte S, Araújo FG, Sales A et al (2007) Morphology of gonads, maturity and spawning season of *Loricariichthys spixii* in a subtropical reservoir. Braz Arch Biol Technol 50:1019–1032

Filigrana G (2016) Uso de la harina de pez diablo (Pterygoplichthys spp) en la alimentación de tilapia nilótica (Oreochromis niloticus) El Colegio de la Frontera Sur 52pp

Firmo A, Tognella M, Silva S et al (2012) Capture and commercialization of blue land crabs ("guaiamum") Cardisoma guanhumi (Latreille, 1825) along the coast of Bahia state, Brazil: an ethnoecological approach. J Ethnobiol Ethnomed 8(12):1–12. https://doi.org/10.1186/1746-4269-8-12

Flores JM (2006) Chontales de Tabasco. México Comisión Nacional para el Desarrollo de los Pueblos Indígenas (CDI):52p

Galindo A, Gama L, Salcedo M et al (2006) Programa de ordenamiento ecológico del estado de Tabasco. Secretaría de Desarrollo Social y Protección al Ambiente, Gobierno del Estado de Tabasco, 115p

Godoy AE, Ramírez MA, Martínez MC (2017) Factores que afectan al desarrollo local en México y Argentina. In: Piotrowska Z, Zabecki K (eds) En: Lisocka-Jaegermann B. La vulnerabilidad socioeconómica y ambiental en el contexto local y regional. Universidad de Varsovia, México, pp 149–159

Gomes ID, Araújo FG, Uehara W et al (2011) Reproductive biology of the armoured catfish *Loricariichthys castaneus* (Castelnau, 1855) in Lajes reservoir, southeastern Brazil. J Appl Ichthyol 27:1322–1331

Gómez G, Guzmán R (1998) Relación longitud peso y talla de madurez de la petota (*Umbrina coroides*), en el norte del estado de Sucre, Venezuela. Zootec Trop 16:267–276

Govinda-Das H (2010) Biología reproductiva de la especie exótica invasora Pterygoplichthys pardalis (Siluriformes: Loricariidae) en los humedales de La Libertad (sitio RAMSAR No. 79), río Usumacinta, Chiapas, México. Tesis de Maestría, El Colegio de la Frontera Sur (ECOSUR), Unidad Villahermosa, México

Guadarrama MA, Ortiz G (2000) Análisis de la flora de la Reserva de la Biósfera Pantanos de Centla, Tabasco, México. Universidad y Ciencia 15(30):67–104

Guerra-Martinez V, Ochoa-Gaona S (2008) Evaluación del programa de manejo de la Reserva de la Biósfera Pantanos de Centla en Tabasco, México. Universidad y Ciencia 24(2):135–146

Gurney GG, Cinner J, Ban NC et al (2014) Poverty and protected areas: an evaluation of a marine integrated conservation and development project in Indonesia. Glob Environ Chang 26:98–107. https://doi.org/10.1016/j.gloenvcha.2014.04.003

Hernández A, Kempton W (2003) Changes in fisheries management in Mexico: effects of increasing scientific input and public participation. Ocean Coast Manag 46(6–7):507–526. https://doi.org/10.1016/S0964-5691(03)00032-2

Hernández-Santana JR, Ortiz-Perez MA, Méndez-Linares AP et al (2008) Morfodinámica de la linea de costa del estado de Tabasco, México: tendencias desde la segunda mitad del siglo XX hasta el presente. Investigaciones Geográficas 65:7–21

Hoover JJ, Killgore KJ, Cofrancesco AF (2004) Suckermouth catfishes: threats to aquatic ecosystems of the United States? Aquatic nuisance species research program. Engineers Research and Development Center, Vicksburg, MS. ANSRP Bulletin 4:1–13

Hossain MY, Rahman MM, Ahmed ZF, Ohtomi J et al (2008) First record of the south American sailfin catfish *Pterygoplichthys multiradiatus* in Bangladesh. J Appl Ichthyol 24:718–720. https://doi.org/10.1111/j.1439-0426.2008.01117.x

IREBIT (Instituto de Recursos Bióticos de Tabasco) (1994) Programa de manejo Reserva de la Biósfera de los Pantanos de Centla. Tabasco, México, p 106

Klute A (1986) Methods of soils analysis. Part 1. Physical and mineralogical methods, 2nd edn. EUA, Madison, p 1188

Leppäkoski E, Gollasch S, Olenin S (2013) Invasive aquatic species of Europe. Distribution, Impacts and Management. Springer Netherlands, Dordrecht, p 584. https://doi.org/10.1007/978-94-015-9956-6

Mendoza R, Contreras S, Ramirez C, Koleff P et al (2007) Los peces diablo. Biodiversitas 70:1–5

Mendoza-Carranza M, Hoeinghaus DJ, García AM et al (2010) Aquatic food webs in mangrove and seagrass habitats of Centla Wetland, a Biosphere Reserve in Southeastern Mexico. Neotrop Ichthyol 8(1):171–178

Mendoza-Carranza M, Arévalo-Frías W, Inda-Díaz E (2013) Common pool resources dilemmas in tropical inland small-scale fisheries. Ocean Coast Manag 82:119–126. https://doi.org/10.1016/j.ocecoaman.2013.06.004

Moreno-Casasola P, Warner B (2009) Breviario para describir, observar y manejar humedales. In: Serie Costa Sustentable No. 1. Xalapa Ver. México: RAMSAR Instituto de Ecología A.C., CONANP, US Fish and Wildlife Service US State Department

Novelo AR (2006) Plantas acuáticas de la Reserva de la Biósfera Pantanos de Centla Universidad Autónoma de México. Universidad Autónoma de México, México, p 260

OCDE (Organización para la Cooperación y el Desarrollo Económicos) (2014) Evaluaciones de la OCDE sobre el desempeño ambiental: México 2013. OECD Publishing, Paris. https://doi.org/10.1787/9789264189188-es

Ortiz de Montellano A (1985) Tecnologías pesqueras en el trópico húmedo de México. Centro de Ecodesarrollo, Litográfica ingramex, 146p

Peralta PL, Infante DM, Moreno-Casasola P (2009) Construcción e instalación de Piezómetros. In: Moreno-Casasola P, Warner BG (eds) Breviario para describir, observar y manejar humedales Xalapa, Veracruz, México: Serie Costa Sustentable No. 1. RAMSAR, Instituto de Ecología, A.C., CONANP, US Fish and Wildlife Service, US State Department, pp 17–30

Pinkus-Rendón M, Contreras-Sánchez A (2012) Impacto socioambiental de la industria petrolera en Tabasco: el caso de la Chontalpa. Limina R 10(2):7–16

Ramírez-Martínez C, Naranjo E, Caspeta JM et al (2015) Ecosistemas acuáticos. In: Carabias J, de la Maza J, Cadena R (eds) Conservación y desarrollo sustentable en la Selva Lacandona. 25 años de actividades y experiencias México D.F. Natura y Ecosistemas Mexicanos, México, pp 193–207

Rejane S, Oshiro Y (2002) Aspectos da reprodução do caranguejo guaiamum, *Cardisoma guanhumi* Latreille (Crustacea, Decapoda, Gecarcinidae) da Baía de Sepetiba, Rio de Janeiro, Brasil. Revta bras Zool 19(Supl. 1):71–78

Romero GJ, Garcia MA, Bautista JA et al (2000) Caracterización de la Reserva de la Biosfera Pantanos de Centla. Universidad y Ciencia 15(30):15–20

Sánchez AJ, Florido R, Álvarez-Pliego N et al (2015) Distribución de Pterygoplichthys spp. (Siluriformes: Loricariidae) en la cuenca baja de los ríos Grijalva-Usumacinta. Revista Mexicana de Biodiversidad 86(4):1099–1102. https://doi.org/10.1016/j.rmb.2015.06.016

Toledo A (2003) Ríos, costas, mares. Hacia un análisis integrado de las regiones hidrológicas de México, INE-SEMARNAT, México, México, 117p

Valdéz JI (2002) Aprovechamiento forestal de manglares, en el estado de Nayarit, costa pacífica de México. Madera y Bosques (Nùmero especial) 1:129–145

Villalba A (2006) Hacia la sustentabilidad de las pesquerías ribereñas del Golfo de California. In: Salas S, Cabrera MA, Ramos J (eds) Memorias Primera Conferencia de Pesquerías Costeras en América Latina y el Caribe. Evaluando Manejando y Balanceando Acciones (Coastfish 2004). CINVESTAV Unidad Mérida, Centro EPOMEX, pp 208–219

Wakida-Kusunoki A, Ruiz C, Amador-del Ángel E (2007) Amazon sailfin catfish, Pterygoplichthys pardalis (Castelnau, 1855) (Loricariidae), another exotic species established in southeastern Mexico. Southwest Nat 52:141–144

Walkley A, Black LA (1934) An examination of the method for determining soil organic matter, and proposed modification of the chromic acid tritation method. Soil Sci 37:29–38

Yáñez-Arancibia A, Sánchez-Gil P (1988) Ecología de los Ecosistemas Costeros del Sur del Golfo de México: La Región de la Laguna de Términos. Caracterización Ambiental de la Sonda de Campeche frente a la Laguna de Términos. In: Yáñez-Arancibia A, Day-Jr JW (eds) Instituto de Ciencias del Mar y Limnología (pág. 518). D.F. México: Universidad Nacional Autónoma de México. Coastal Ecology Institute, L.S.U. Editorial Universitari

Chapter 32
Ecotourism as a Means to Promote Community Inclusion and Nature Conservation: The Case Study of Maya Ka'an

Gonzalo Merediz-Alonso

Abstract Sian Ka'an Biosphere Reserve is one of Mexico's top World Heritage Sites. Since the 1990s, local NGOs, communities, and government have been using ecotourism to promote sustainable development among local people, while leading nature conservation in and around the reserve. Sian Ka'an is now the region's most important ecotourism destination. Since 2014 Sian Ka'an and its surroundings are in what is now called Maya Ka'an: Mexico's new ecotourism destination. Maya Ka'an will spread Sian Ka'an's success in the region, reducing pressure to the reserve while benefiting Mayan communities, showing that there are profitable non-massive tourism models that are environmentally and socially viable. Maya Ka'an represents an opportunity for Quintana Roo and Mexico to design new governance models based on policies oriented to more income from tourism by adding value to the destination rather than promoting the arrival of more visitors every year. Maya Ka'an is also an option for local Mayan people to receive more significant and direct benefits other than only job opportunities far from their communities.

Keywords Maya Ka'an · Ecotourism · Tourism sustainability · Sian Ka'an · Quintana Roo

G. Merediz-Alonso (✉)
Amigos de Sian Ka'an A.C., Cancún, Quintana Roo, Mexico
e-mail: gmerediz@amigosdesiankaan.org

A. Ortega-Rubio (ed.), *Socio-ecological Studies in Natural Protected Areas*,
https://doi.org/10.1007/978-3-030-47264-1_32

32.1 Introduction

The Yucatan Peninsula, in southeastern Mexico, is a region rich in cultural and natural resources. Its Caribbean coast, in the state of Quintana Roo, is part of the second largest coral reef in the world and the second largest tropical forest of America, after the Amazon. It also has the largest wetlands in Mesoamerica and the largest underground rivers on Earth. Given this environmental relevance, in 1986, the Mexican government declared 528,000 ha of tropical forest, wetlands, and marine ecosystems as the Sian Ka'an Biosphere Reserve under the Mexican legislation (SEDUE 1986) and the Man and the Biosphere Programme (UNESCO 2018). In December 1987, Sian Ka'an was also declared as a World Heritage Site (UNESCO 1988). This biosphere reserve is in the central coast of Quintana Roo, Mexico (Fig. 32.1).

Sian Ka'an was established about a decade after the declaration of Quintana Roo as a new Mexican state and the creation of the major tourism resort in Latin America: Cancun. That was the first time when a state and a tourism destination were created with the balance of a pole dedicated to nature conservation and sustainable tourism and a pole focused on massive tourism. Benefiting from this particular situation, Amigos de Sian Ka'an, a local NGO working in Quintana Roo since 1986, proposed to local stakeholders and authorities the creation of a new tourism destination in and around Sian Ka'an, based on local communities and the natural and cultural wealth of the region: Maya Ka'an. Since 2014, Maya Ka'an is 1 of the 11 destinations offered from Mexico to the world in the Mexican Caribbean (Fig. 32.1).

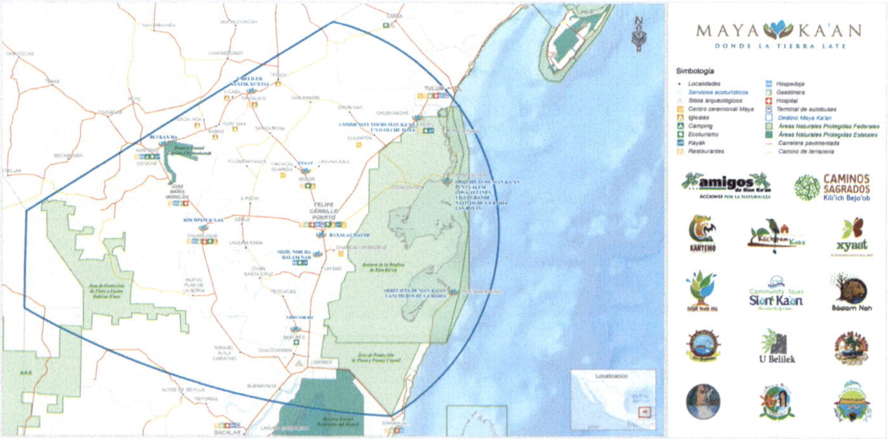

Fig. 32.1 Location of the Sian Ka'an Biosphere Reserve (green large polygon in the east side) and Maya Ka'an tourism destination (blue polygon) in Quintana Roo, Mexico. Other polygons represent federal (light green) and state (dark green) protected areas. (Map: Amigos de Sian Ka'an)

32.2 A History of Ecotourism Evolution in Sian Ka'an

Six months after the establishment of Sian Ka'an in 1986, Amigos de Sian Ka'an was created as the first environmental, nonprofit organization of the region. Although Amigos de Sian Ka'an's work today covers all the state of Quintana Roo, its first conservation approach was focused on Sian Ka'an. The concept of a biosphere reserve seeks a balance between ecosystem and species protection and the well-being of local people. Thus, Amigos de Sian Ka'an started working with the communities within and around Sian Ka'an immediately after its foundation, promoting sustainable lobster fishing and good agriculture practices.

In the late 1980s, Amigos de Sian Ka'an organized informal excursions to Sian Ka'an to show donors, journalists, and scientists the conservation work in the area and its ecological values. Based on that experience, in 1990, Amigos de Sian Ka'an designed a boat and swimming tour in the wetlands of Sian Ka'an and created the institutional and logistic frame to operate it. The project had several goals: (a) offer an opportunity for tourists and local people to visit a pristine protected area, (b) raise funds for Sian Ka'an's conservation, (c) generate income opportunities for local communities, (d) increase environmental awareness among visitors, and (e) support the authorities in supervising and monitoring the reserve (Barrios-Martínez-Rojas 1991). During the first year of operation, the tour to Sian Ka'an received a total of 642 visitors (Bezaury-Creel 1991, Fig. 32.2) a seed that, as we will see, evolved through the years, as an entire tourism sustainability initiative.

In the late 1990s, some members of the local communities in Sian Ka'an (Chunyaxché and Punta Allen) were interested in the success of the Amigos de Sian Ka'an's tour and requested formal permission to also operate it. The Mexican authorities granted their authorization, and, by 1997, Amigos de Sian Ka'an in alliance with

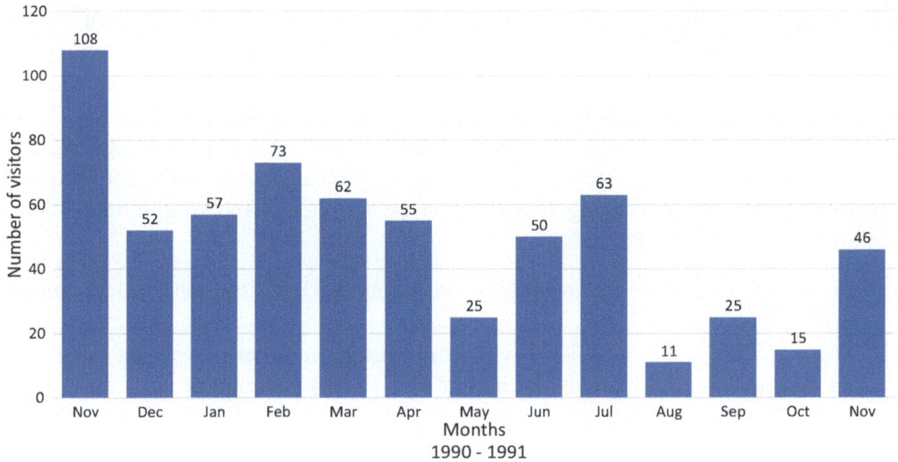

Fig. 32.2 Number of visitors to the Sian Ka'an Biosphere Reserve, Maya Ka'an, Quintana Roo, Mexico, from November 1990 to November 1991. (Bezaury-Creel 1991)

RARE Center for Tropical Conservation organized the first training program for nature tourism guides in Sian Ka'an (Iglesias 1999). Some of the trainees continue leading ecotourism activities in Sian Ka'an over 20 years later.

Amigos de Sian Ka'an continued operating the tour from Cancun until 2002 when it was fully transfered to the local cooperatives already formally established and operating. During the time Amigos de Sian Ka'an oversaw the ecotourism in Sian Ka'an, using two boats allowing up to 9 passengers per boat, 3–5 times a week, it was possible to strongly position Sian Ka'an in the international tourism market. Amigos de Sian Ka'an implementation of the tour also opened the possibilities for local communities to adopt ecotourism as a significant source of income and, as this chapter will show, start a long-term process that is looking for local sustainability through new tourism models. Because of this pioneer work promoting ecotourism in Sian Ka'an, Amigos de Sian Ka'an received the Condé Nast Traveler Ecotourism Award in 1996.

Today, ecotourism in Sian Ka'an is managed directly by 10 local cooperatives based on the Mexican environmental law that allows the use of natural resources within biosphere reserves only by the local people (SEMARNAT 2018). The main cooperative operating in Sian Ka'an works under the "Community Tours" brand has been so successful that it received the Equator Prize, granted by the Equator Initiative (UNDP 2012).

32.3 Ecotourism in Sian Ka'an Today

Given the institutional and community efforts during the 1990s and early 2000s, Sian Ka'an Biosphere Reserve became one of the top ecotourism destinations in Mexico. In 2017, for example, it received 5.6% of the visitors registered for all the federal protected areas in the country (CONANP 2018) (CONANP unpublished Sian Ka'an's visitation data). Between 1997 and 2018, Sian Ka'an received a total of 1,638,184 visitors, that is, an average of 74,463 visitors per year. However, the absolute numbers jumped from 15,189 in 1997 to 180,129 tourists in 2018 (Fig. 32.3) (CONANP unpublished Sian Ka'an's visitation data). The average annual increase is 7,506 visitors per year. However, there are four clear phases in the evolution of tourism in Sian Ka'an during that 22-year period (CONANP unpublished Sian Ka'an's visitation data, Fig. 32.3):

(a) From 1997 to 2005 with an average visitation of 33,402 tourists, and an annual average increase of 3,084.
(b) During 2006, 2007, and 2008 when the average number of visitors passed to 84,704 per year, with an annual increase of 19,994 passengers.

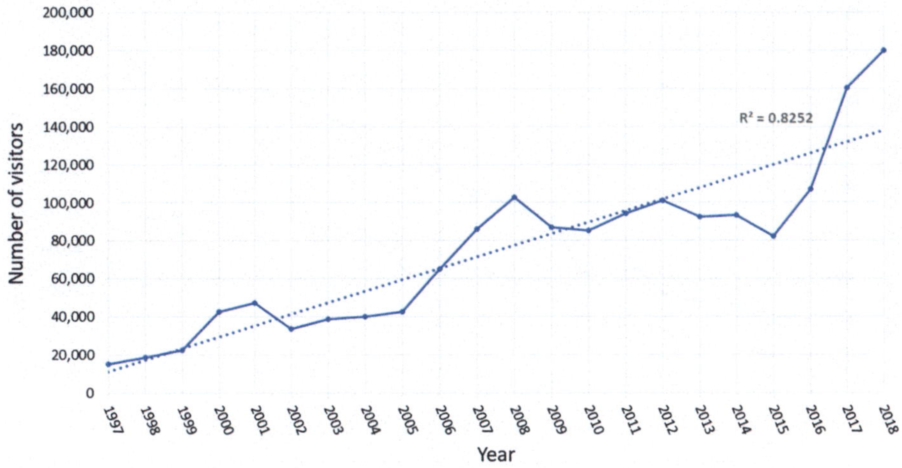

Fig. 32.3 Annual number of visitors to the Sian Ka'an Biosphere Reserve, Maya Ka'an, Quintana Roo, Mexico, from 1997 to 2018. (CONANP unpublished Sian Ka'an visitation data)

(c) In 2009[1] there was a drop in the number of tourists and the beginning of a 7-year period of stability with an average visitation of 90,836 people per year and a negative average increase of −2,902 visitors annually.

(d) An important period of growth during 2016, 2017, and 2018 when, in average, 149,202 people visited Sian Ka'an every year, that is, an annual increase of 32,570 visitors in those 3 years.

32.4 The Conditions for a New Tourism Model

In 1969 Mexico decided to expand tourism as a main source of foreign currency and focused its attention in the Caribbean coast of the Yucatan Peninsula: the federal territory of Quintana Roo. Cancun, an island inhabited by a few people at a seasonal basis, was chosen as the center of a new tourism development and was founded as a city and resort in 1970 (Martí 1985). In October 1974, the territory of Quintana Roo received the status of state (SEGOB 1974), totally transforming its political, social, and economic dynamics. In 1970, Quintana Roo's population was 88,150 people (Secretaría de Industria y Comercio 1972); by 2010 it was 1,501,562 people (INEGI

[1] During 2009 Mexico was affected by the pandemic H1N1 influenza outbreak (Novel Swine-Origin Influenza A (H1N1) Virus Investigation Team 2009). National tourism industry was significantly affected by a reduction in the tourists that traveled to Mexico (Monterrubio 2010), thus probably affecting the flow of tourists to Sian Ka'an.

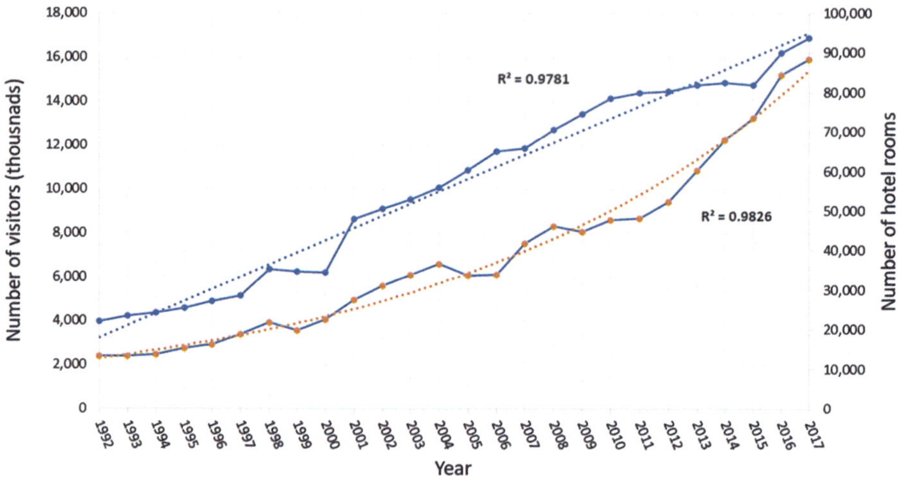

Fig. 32.4 Annual number of hotel rooms built (line with blue dots) and annual number of visitors (line with brown dots) in Quintana Roo, México, between 1992 and 2017. (SECTUR 2018b)

2016a), representing a 1,703% growth in 45 years or 37.8% per year. Such population increase is explained by the rapid development of the tourism industry not only in Cancun but also in Riviera Maya, another destination created in 1999 that expands along 120 Km of coast between Cancun and Tulum.

From 1992 to 2017, the number of hotel rooms available grew from 21,944 to 93,909, an average of 2,879 rooms per year (SECTUR 2018b). In the same period, the number of visitors to Quintana Roo grew exponentially from 2,383,036 to 15,926,071, a 668% increase in 25 years (Fig. 32.4) (SECTUR 2018b). Focusing the analysis in a 10-year frame, from 2007 to 2017, numbers of tourist in Quintana Roo grew 211%, while income for tourism increased only in 162% (SECTUR 2018b; SEDETUR 2018). Each additional visitor to Quintana Roo represents a higher economic, environmental, and social cost; however, the income generated by tourism is probably not compensating those costs.

Quintana Roo's average income from tourism between 2007 and 2017represented 47% of the tourism income of all Mexico: US$6.98 billion (SEDETUR 2018; SECTUR 2018a). At the same time, the unemployment rate was 4.06% in average[2] (INEGI 2019). Nonetheless, in the same period, 33.2% of the population was under poverty conditions, representing 493,700 people in average (CONEVAL 2019a). The absolute number of poor people changed from 428,100 in 2008 to 474,800 in 2018 (CONEVAL 2019b). In other words, hotel room construction and constant growth in visitors to Quintana Roo have not been able to significantly reduce the poverty condition for almost half a million people through the years.

Environmentally, tourism development and its related population growth have also an impact. Between 2001 and 2013, deforestation affected 333,240 ha of tropi-

[2]In the first quarter of 2019, Quintana Roo had the 13th lowest unemployment rate of the 32 Mexican states: 2.9% (INEGI 2019).

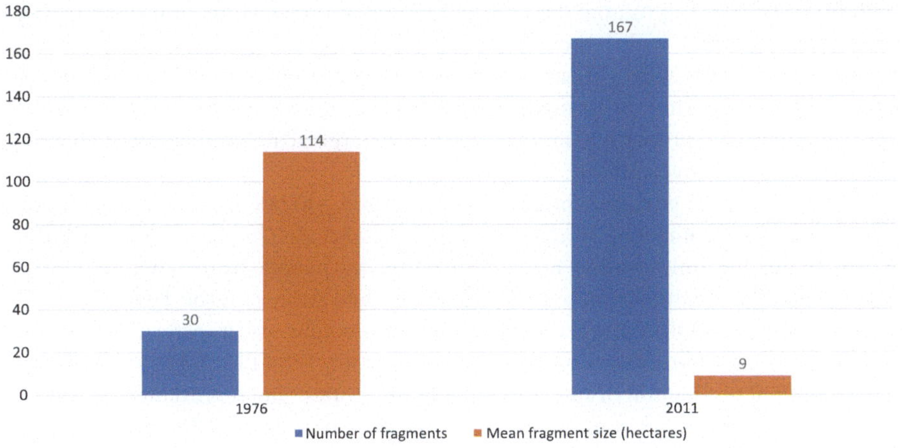

Fig. 32.5 Number of mangrove fragments (blue column) and mean fragment size in hectares (brown column) in the northern coast of Quintana Roo, Mexico, in 1976 (left side) and 2011 (right side). (Reza 2011)

cal forest in Quintana Roo (Ellis 2017). Mangrove coverage in the northern coast of Quintana Roo in 1976, before the expansion of coastal development, was 3,429 ha. By 2011, it was 1,569 ha, that is, a 54.2% reduction (Reza 2011). Mangrove deforestation effects are worsened by its fragmentation: in 1976 there were 30 fragments with an average size of 114 ha. In 2011, the number of fragments reached 167 with 9 ha each in average (Reza 2011, Fig. 32.5).

In the case of coral reefs in northern Quintana Roo, between 1998 and 2011, there was an average 50% loss of alive coral cover (Franquesa 2009). Elimination of coastal mangrove and coral reef generates an increased vulnerability to climate change, storm tides, and hurricanes for ecosystems and infrastructure, therefore, putting tourism industry in peril (Silva et al. 2014).

While the northern coast of Quintana Roo has a hectic dynamic of tourism, economic growth, employment generation, social inequalities, and environmental degradation, the center of the state has a different conformation. The population is slightly above 119,000 people of which 91% are indigenous (INEGI 2016a), mostly Yucatecan Maya. Economy is more precarious: the proportion of the population in poverty conditions increased from 68.6% in 2010 to 75.7% in 2015 (CONEVAL 2015). Main activities in the region are self-consumption agriculture, low-scale commercial agriculture, tropical timber harvest, honey production, handicraft manufacture, fishing, and ecotourism. The average annual income per person is US$2,452. Local gross production represents only 0.83% of the total for Quintana Roo (INEGI 2016b).

Today, local economy and governance is in a transitional period. In 1902, the federal authorities created the territory of Quintana Roo in an attempt to develop governance conditions in the region after 57 years of armed conflict with the local Mayan people (Macías-Richards 1999). For three decades the government tried, with little success, to promote the production of natural chewing gum (*chicle*), logwood (*palo de tinte*), and timber, through private concessions (Keyes 1998). It was until the late 1930s when President Lázaro Cárdenas institutionalized the governance in Quintana Roo by formalizing the land tenure for the Mayans by the creation of communal properties (*ejidos*) to promote large-scale production of chewing gum and timber (Keyes 1998).

In the 1950s, the government gave in concession 540,000 ha for timber production to a private company, *Maderas Industrializadas de Quintana Roo* (MIQRO) incentivizing the immigration and colonization by people from other regions of Mexico and the creation of new *ejidos* (Dachary and Arnaiz-Burne 1990). In 1982, after the end of MIQRO's concession, timber production was transferred to the *ejidos*, through the *Plan Piloto Forestal* (PPF or Forestry Pilot Plan). The PPF allowed the communal production of raw timber and railway ties, and represented up to 49% of the income for local communities during the 1980s and 1990s (Keyes 1998). However, in 1992 the Mexican Constitution was reformed allowing the privatization of *ejido* lands (Olivera-Lozano 2005). Since then, several *ejidos* have been subdivided, as it is the example case of *ejido* Tulum (RAN 2019, Fig. 32.6); a communal property with 20,692 ha is now fragmented in hundreds of small properties of tropical forest although the subdivision of forested *ejido* lands is forbidden by Mexican law (Ley Agraria 2018).

Today land use and economic activity are shifting from traditional, self-sustaining, or commercial agriculture to real estate speculation, land urbanization, and tourism. Therefore, governance needs to focus more in the regulation of land

Fig. 32.6 Example of communal land subdivision into private properties in *ejido* Tulum (light green grid polygon), Quintana Roo, Mexico. Other non-grid polygons represent other *ejidos*. (RAN 2019)

use, the preservation of ecosystem services, and the generation of new tourism models.

Even if the economic conditions in central Quintana Roo, now called Maya Ka'an, are poor, local communities are culturally rich. Today's Mayan culture is vibrant and evolving to current social and political conditions but deeply rooted in their ancient heritage. Old Mayan ruins are surrounded by villages with their traditional temples and corn fields (milpas), while many of their inhabitants are connected to the Internet and social media. This cultural wealth is surrounded by the exuberant Maya Forest, the Mesoamerican Coral Reef, and the most extensive underground river network in the planet, exposed through thousands of sinkholes called *cenotes*. Part of these natural treasures is in five protected areas declared by the federal and state governments in the region: Bala'an K'aax Flora and Fauna Protection Area (SEMARNAT 2005), Chichankanab Lagoon System State Reserve (SEDUMA 2011), Uaymil Flora and Fauna Protection Area (SEDESOL 1994), Sian Ka'an Reefs Biosphere Reserve (SEMARNAP 1998), and Sian Ka'an Biosphere Reserve (SEDUE 1986), covering a total of 780,583 ha. The relevance of these protected areas, and particularly of Sian Ka'an, is internationally recognized by its World Heritage Site status (Sect. 32.1). Sian Ka'an, Bala'an K'aax, and Chichankanab are also recognized as wetlands of international importance under the Ramsar Convention (RAMSAR 2019). Although protected areas are commonly seen by the public as regulatory instruments that stop development and ban human activity, the reality is that Mexican legislation provides numerous social benefits for the inhabitants of protected areas, making them tools for local sustainable development (SEMARNAT 2018). Therefore, protected areas should be considered as governance mechanisms to improve life quality, protect natural resources and ecosystem services, and generate opportunities for sustainable tourism. For instance, some sectors in the local society in Felipe Carrillo Puerto, the main city in Maya Ka'an, perceive Sian Ka'an Biosphere Reserve as an obstacle for local development, since the "progress model" they have is Cancun. However, the presence of a World Heritage Site is a unique opportunity to generate competitive tourism models where nature conservation provides the occasion for social progress and economic growth.

In central Quintana Roo, nature, culture, and people are well connected with the rest of the state, Mexico, and the world. Good-quality paved roads can easily take people from central Mexico, Belize, Riviera Maya, and Cancun. The connectivity with Cancun is important since it is the second busiest airport in Mexico with 25,202,000 passengers in 2018 (ASUR 2019) and the first in terms of international passenger number with 12,499,637 in 2016 (SCT 2018). The potential to receive more visitors is enormous, but society and government do not take advantage of such privilege, compared to other ecotourism destinations.

Fig. 32.7 Schematic representation of the elements or "equation" that facilitate the creation of Maya Ka'an as an ecotourism destination

32.5 The Creation of Maya Ka'an

Quintana Roo's social and environmental conditions described above are elements of an equation to allow new governance schemes and innovative paradigms for tourism development. The result of such equation is the proposal of Amigos de Sian Ka'an to promote the center of Quintana Roo as a low-density, community-based, ecotourism destination branded as Maya Ka'an (Solano 2014, Fig. 32.7).

Maya Ka'an covers more than 1 million hectares of marine and terrestrial ecosystems, comprising 3 municipalities (Felipe Carrillo Puerto, José María Morelos, and Tulum), more than 32 villages and towns (Fig. 32.1), and over 29 community tourism companies or cooperatives operating. Maya Ka'an was conceived by Amigos de Sian Ka'an to reduce development pressure on local protected areas, especially Sian Ka'an, while creating development opportunities for local communities. The goal is to have a tourism destination based on the natural, historical, and cultural attractions existing there. Maya Ka'an is also conceived to benefit directly local communities through the creation not only of jobs but also of business opportunities for the Mayan people as well (Solano 2014; Ferreyra-Acosta and Merediz-Alonso 2015).

Maya Ka'an, as a new type of tourism destination, is a governance tool to diversify Quintana Roo's tourism offer, mostly based on the "sun and beach" concept, and therefore to open new markets. It is also an opportunity to plan regional development to preserve some of the traditional Mayan production systems (*milpa*) and promote a profitable tourism development based on a low construction density and its own carrying capacity.

The development of Maya Ka'an first as a concept, then as a project, and finally as a destination has been a long process. In 2007, Amigos de Sian Ka'an conceived the idea of promoting the area in the vicinity of Sian Ka'an as an ecotourism destination in response to the vision of some local stakeholders that the presence of Sian Ka'an was opposed to the economic prosperity of the area. In 2010, with the support of UNESCO and The Nature Conservancy, Amigos de Sian Ka'an organized a working trip to Costa Rica, with a select group of local leaders. The goal was to study the experience of a country, with the size of Quintana Roo, that was a world champion in developing tourism around the conservation of protected areas. The trip was an eye-opening event that motivated some of the participants to see both Sian Ka'an and tourism development from a new perspective.

Short after the Costa Rica trip, Amigos de Sian Ka'an obtained funding from the Multilateral Investment Fund (MIF) of the Inter-American Development Bank (IDB), the WWF-Fundación Carlos Slim Alliance, and other sources to diversify the tourism offer of Riviera Maya based on the natural resources in and around the Sian Ka'an Biosphere Reserve. The project had four main objectives (Solano 2014):

1. Structure the alternative sustainable tourism destination Sian Ka'an (so called at the beginning of the project, today Maya Ka'an), promoting a private-public management scheme, based on sustainable tourism products and services, and a defined brand and image
2. Provide small- and medium-sized local entrepreneurs with access to knowledge and financial tools for the design and development of sustainable tourism services
3. Implement an access mechanism to promotional and market channels, generating a catalogue of existing and new tourist products and services, placed on web marketing and booking platforms
4. Consolidate and disseminate experiences and results through the systematization of project processes, developing dissemination tools for other similar areas of the Mexican Caribbean

The project generated several outcomes to support the design of a new destination. The supply and demand surveys indicated that 60.7% of the potential tourist for Maya Ka'an was from the United States, 34.5% from Europe and Canada, and 4.8 from other places. The profile of the potential customer was single, independent, professional women, and men, 18- to 35-year old, with more interest in hiking, nature contemplation, and local product purchasing. The potential tourists were interested in receiving safety, cleanliness, preserved nature, and proper services (Sierra-Magaña 2013a).

The supply surveys studied 32 locations of potential tourism interest. Of those, 15 had elements to attract visitors in the short term, offering archaeological sites, local gastronomy, landscape, lakes, swimming, hiking, bicycling, history, culture, monuments, wildlife observation, bird-watching, organized tours, established cooperatives, trained guides, etc. Some special examples were the floatation tour in the canals of Sian Ka'an (Sect. 32.2 & 32.3), bat-eating snake observation in the caves of Kantemó, a visit to the caste war in Tihosuco, and many more (Sierra-Magaña 2013b).

Another important step during the project was the design of the destination brand. After various scouting trips, deep desk research, interviews, and participatory workshops, different designs for three possible brands were defined: Sayab, Sian Ka'an, and Maya Ka'an (Alchemia 2014). A market study to evaluate the potential of the selected brands showed that 65% of surveyed tourists in Cancun International Airport preferred the name Maya Ka'an over Sayab or Sian Ka'an. Of those people, 68% preferred the graphic logo that now represents the destination brand. Maya Ka'an is easy to remember and has a harmonious sound, it is simple, and its colors represent land, nature, and sea (AD Consultores 2014). An advisory board formed with representatives of different stakeholder sectors made the final decision of adopting Maya Ka'an as de destination's brand.

Maya Ka'an was created with clear sustainability principles, based on the Global Sustainable Tourism Council (GSTC) criteria (GSTC 2019), that should be followed by all the stakeholders in the destination. Those principles look for a balance between economic income, environmental conservation, and the generation of benefits for local communities, which own and manage the natural resources. In general, the sustainability criteria are based on four main aspects (Vanegas 2013):

Sociocultural: Consider the social and cultural benefits to the community, and the stakeholder's ability to contribute to intercultural understanding and tolerance, minimizing negative impacts from tourism on local culture and social structures.

Environment: Promote the conservation and well use of biodiversity and natural resources within and outside protected areas, increasing the potential for nature tourism and local people well-being.

Economic: Ensure healthy economic returns at the enterprise, community, and destination levels, promoting access to business and employment opportunities for local people. Economy must be based on fair trade, prices, and salaries for all.

Management: There should be effective destination and business management practices, based on the law, looking for competitiveness, health, and safety.

Having sustainability criteria and principles is one thing. Another, and more important, step is to apply them at all levels in the destination. Based on the criteria for Maya Ka'an, Amigos de Sian Ka'an developed a series of manuals to help local tourism operators to incorporate and implement good sustainability practices in small hotel construction (Ángel et al. 2014a), operation of small hotels (Ángel et al. 2014b), recreational aquatic activities (Ángel et al. 2014c), and ecotourism practices (Ángel et al. 2014d).

To test the good practice manuals, a baseline analysis of sustainability compliance by different tourism companies and cooperatives was performed in 2014. It was an assisted self-evaluation by managers and leaders of small hotels and aquatic tour operators in Tulum and Felipe Carrillo Puerto municipalities, both within Maya Ka'an. A total of 17 basic sustainability criteria were measured in 12 small hotels. The results show an average compliance of 47.38%. The hotel with the higher sustainability levels had a 66% compliance. The hotel with the lower performance had 28.4%. Five hotels were under the average performance (Valle 2014, Fig. 32.8a). An average sustainability performance of almost 50% in the hotels operating in Maya Ka'an is a strong basis that facilitates a long-term improvement.

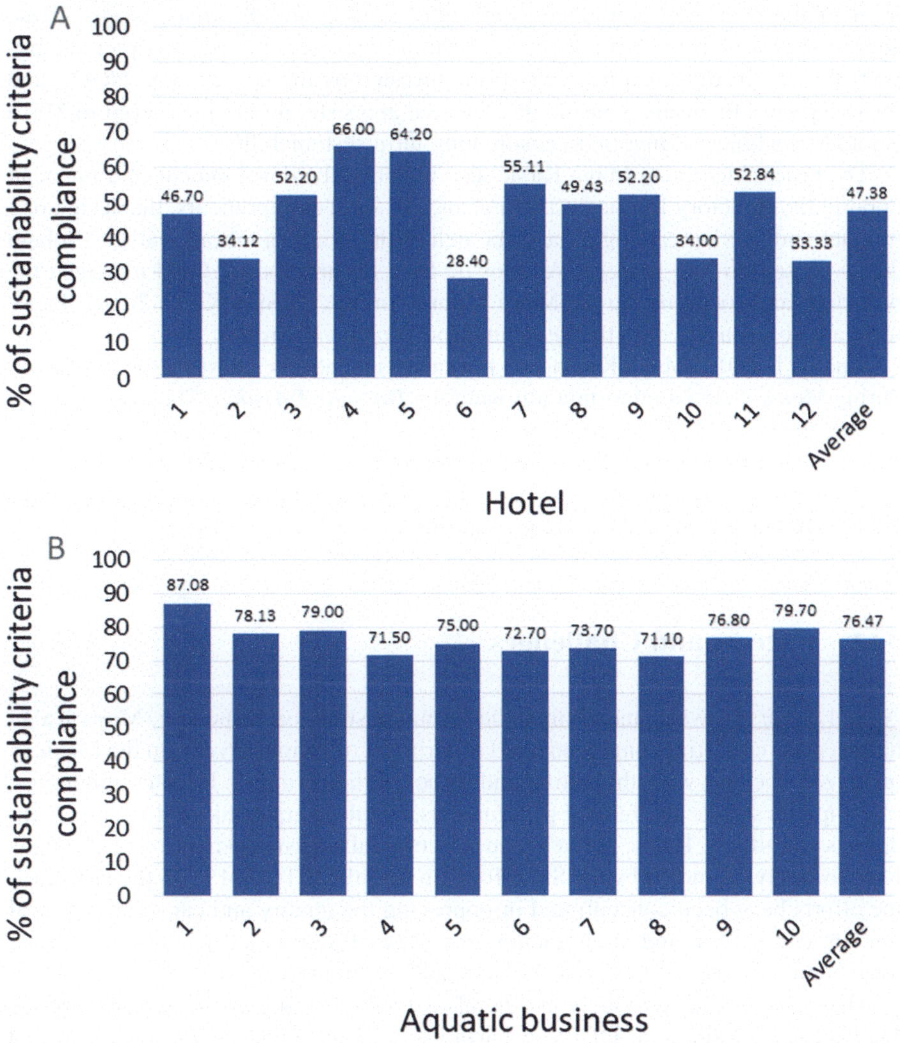

Fig. 32.8 Compliance proportion to the sustainability criteria defined for Maya Ka'an, Quintana Roo, Mexico, by small hotels (Graph **A**) and aquatic tour operators (Graph **B**). (Valle 2014; González-Haro et al. 2014)

In the case of aquatic operators, 10 businesses working in Sian Ka'an Biosphere Reserve participated in the evaluation of 115 criteria. The average sustainability performance was 76.47% of the evaluated criteria by organization. The higher value was 87.08% and the lower one was 71.10%, showing more uniformity in the criteria compliance among operators (González-Haro et al. 2014, Fig. 32.8b). It is interesting to compare the hotel and the aquatic operators' performance. In the hotel case, most of the evaluated companies were outside of protected areas, while all the aquatic operators provide their services within a protected area. This pattern suggests that the institutional training efforts invested by the government, NGOs, and the companies themselves inside of protected areas should be transferred to Maya Ka'an as a whole destination to ensure long-term sustainability.

The process to create Maya Ka'an also included the development of a financial opportunity directory for the destination and the local entrepreneurs, the design of a product and service catalogue, and the definition of a promotion strategy (Solano 2014). Based on that strategy, Amigos de Sian Ka'an produced and managed the web page and social media for Maya Ka'an (Amigos de Sian Ka'an 2014). Those tools are now managed by the Quintana Roo Tourism Board.

Finally, in 2014, Maya Ka'an was formally launched as a destination in Cancun during Mexico's most important tourism fair: *Tianguis Turístico*. The event was led by the state and national tourism secretaries, the Inter-American Development Bank, and Amigos de Sian Ka'an. It had the presence of almost 200 representatives from the press, government, tour operators, NGOs, and private sector from Maya Ka'an, Quintana Roo, and different countries.

32.6 Success and Challenges

After 12 years of work and 5 since the formal destination launching, Maya Ka'an has been an interesting, unprecedented experiment of an initiative from the Mexican organized society, with the active and direct participation of local communities, municipalities, state and federal governments, a multilateral bank, and various foundations and NGOs. Today, Maya Ka'an is an official tourism destination of Quintana Roo, formally promoted by the State Tourism Board (CPTQ 2019). In the last years, the efforts have been concentrated in improving the quality and capacities of local tourism companies, and strengthening the Maya Ka'an brand positioning among local communities, political stakeholders, and different markets.

There are 29 tour providers and hotels actively offering services as part of the Maya Ka'an destination. Of those businesses, 25 are engaged in applying good sustainability practices in their day to day operation, while 25 tour guides are already certified in ecotourism. Tourism expenditure in Maya Ka'an during 2018 is estimated to be over 30 million Mexican pesos (more than US$1.5 million at November 2019 exchange rate), generated by over 180,000 visitors (Amigos de Sian Ka'an and CONANP unpublished data).

Nine of the rural ecotourism cooperatives in Maya Ka'an decided to organize themselves in an alliance: Maya Zone Community Tourism Network (RTCZM, in Spanish). To strengthen the capacities and sustainability of those cooperatives, the National Commission for the Knowledge and Use of Biodiversity (CONABIO, in Spanish) in coordination with Amigos de Sian Ka'an provided financial support to the RTCZM, so it improved its promotional and administrative strategies, allowing the installation of clean water and energy facilities, while directly participating in biodiversity conservation actions and monitoring. This effort has effectively supported the coordination among local cooperatives in their work to attract visitors and compete with national and global markets.

The philanthropic investment to make possible the creation of Maya Ka'an and the direct benefit for local people generated a social return of US$0.54 for each US$1.00 invested during the 2011–2014 period (Sheehan 2014). For the period between 2015 and 2018, the social return on investment (SROI) was US$0.48 for each philanthropic dollar (Carrillo-García et al. 2018). That is, between 2011 and 2018, the creation of Maya Ka'an as a tourism destination has generated an average SROI of US$0.51 per dollar invested.

In other words, Maya Ka'an has generated initial concrete and significant benefits for some sectors of the local population. These results are an important basis that shows the potential that Maya Ka'an represents as a new model of tourism directly benefitting local people beyond the mere creation of jobs. For this new model to work as a means for sustainability and poverty reduction, it needs to be assumed as a governance tool by most sectors of the local society.

Further steps are needed to address the challenges that Maya Ka'an as destination and governance instrument represents. The national Secretary of Tourism (SECTUR), Amigos de Sian Ka'an, and other NGOs will implement, through the United Nations Development Programme (UNDP), a project to promote tourism sustainability in Mexico, financed by the Global Environmental Facility (GEF). One of the focal sites for the project implementation will be Maya Ka'an. The project will facilitate and leverage the efforts and investments of different institutions to address some of the crucial actions required to strengthen Maya Ka'an and its governance:

1. Develop a destination master plan that formalizes, with the participation of local stakeholders, the tourism model for Maya Ka'an, the strategic actions required to have a sustainable development of the region, and the financial plan for the destination management.
2. Declare Maya Ka'an as a Sustainable Tourism Development Zone, a legal figure that allows SECTUR to promote in specific regions "tourism sustainability with a green economy approach, … improving well-being and social equity, while significantly reducing environmental and ecological risks, driving low-carbon tourism development and harmonizing economic and environmental models to consolidate tourism as a source of quality jobs, investment, training and development to improve people's standard of living" (SECTUR 2017).
3. Establish a Destination Management Organization (DMO) as a formal space to coordinate actions and consolidate tourism as a relevant economic engine in the region. The DMO will have five objectives (Santaolaya 2017):

    (a) Facilitate an adequate flow of visitors, through the permanent promotion of the destination, based on its carrying capacity, and the need to respect the cultural and natural

(b) Safeguard the application and encouragement of Maya Ka'an's sustainability principles, guidelines, and a code of ethics that should be followed by any entity using the brand
(c) Promote conservation and increase the value of the natural and cultural heritage of the products and services offered in Maya Ka'an through innovation, technology, and professionality
(d) Implement sustainability, quality, and competitiveness plans and programs, with verifiable goals and indicators for the destination
(e) Ensure tourist's safety and satisfaction, through the quality verification of products and services provided, following national and international indicators, standards, and certifications

4. Contribute to the design of the municipal zoning plans in Maya Ka'an to ensure their concordance with the destination master plan, looking to promote low hotel density and the protection of natural and cultural resources that are key for the long-term tourism success for the area. Both the zoning and the master plans should orient investments according to the sustainability principles of Maya Ka'an, inducing business opportunities for local people.
5. Strengthen the operational and management capacities of the protected areas in the region and increase the participation of communities in the preservation of biodiversity in all the ecosystems in Maya Ka'an.
6. Increase training programs for new and existing businesses operating in Maya Ka'an to ensure good and uniform service quality and sustainability. Training should include administrative, fiscal, and financial components to promote financially healthy business initiatives.
7. Improve the promotion and marketing strategies for Maya Ka'an to reduce the risks of attracting massive tourism, considering the destination's carrying capacity.
8. Implement a sustainability strategy at the destination and the community level to align tourism development with housing improvement, adopt adequate water management and clean energy technologies, link local production and local tourism industry, implement environmental education actions, etc.[3]

All the actions presented above should consider contributing to the generation of a model for a new tourism paradigm in Mexico. Currently Mexico is far from being efficient and sustainable in terms of tourism. Although the country received 39,291,000 visitors in 2017, ranking 6th globally, it received an income of US$22,467,000,000 ranking only in the 15th place globally (UNWTO 2019; The World Bank 2019). In the same year, Mexico ranked 22 in Travel and Tourism Competitiveness Index (Crotti and Misrahi 2017). Even worse, Mexico occupied the 116th competitiveness place in terms of environmental sustainability and number 133rd on threats to fauna species (Crotti and Misrahi 2017, Sect. 32.4).

[3] This strategy is already being designed and applied by Quintana Roo's secretaries of social development, tourism, and ecology and environment, with the participation of Amigos de Sian Ka'an, among other institutions.

Maya Ka'an is an opportunity to start changing the tourism paradigms and policies that generate high costs and prevent better benefits from tourism. Maya Ka'an needs to focus on increasing the added value of nature and cultural attractions and therefore increase the income from tourists, while not incentivizing the onset of higher and higher numbers of hotel rooms built and visitors arriving. In other words, Mexico and Maya Ka'an tourism policies need to change the goals from increasing the destination position in the number of visitors ranking to improving the position in the income ranking. In that way, the cost (economic, social, and environmental) from each visitor will be reduced, and the destination's net benefit will increase.

This approach may contribute to improve the governance in central Quintana Roo, reduce poverty level on the local population, increase Mexico's tourism competitiveness, and access new markets.

32.7 Conclusions

Tourism has been an engine for Mexican economy for decades, and Quintana Roo is a fundamental part of that engine. However, Mexico has not obtained the most benefits from tourism as it could. The reasons for that are various: policies focused on bringing more tourists rather than increasing expenditure per visitor, emphasis on low-quality job generation rather than entrepreneurial opportunities for all, destruction of natural resources, economic inequalities, and many others.

The presence of Quintana Roo in the global tourism market, its connectivity, and its richness in terms of culture and nature provide an excellent opportunity to define

Fig. 32.9 Panoramic view of the Muyil archaeological zone and lagoon in Sian Ka'an Biosphere Reserve, Maya Ka'an, Quintana Roo, Mexico. (Photo: Amigos de Sian Ka'an)

new tourism paradigms that, in the medium and long term, can generate much more benefits than costs at all levels: nation, destination, local communities, and people.

The implementation of that new paradigm is the proposal of Amigos de Sian Ka'an by positioning the center of Quintana Roo as an ecotourism destination under the Maya Ka'an brand. It is also an opportunity to preserve ancient Mayan culture and high biodiversity sites, including the Sian Ka'an World Heritage Site (Fig. 32.9).

Maya Ka'an is not only a tourism brand but an opportunity to explore new ways of sustainable development based on tourism. The challenge is extensive, but it is already addressed in the field by authorities, communities, NGOs, and funders. Maya Ka'an is also a concept more demanded by the global markets. Quintana Roo was born as a state balancing tourism development (Cancun) and nature conservation (Sian Ka'an). Now it is the moment to take such a balance to the next level.

References

AD Consultores (2014) Proyecto de la nueva marca-destino de Sian Ka'an. Amigos de Sian Ka'an A.C., Cancún

Alchemia (2014) Taller de inmersión de marca: Maya Ka'an. Amigos de Sian Ka'an A.C., Cancún

Amigos de Sian Ka'an (2014) Maya Ka'an: Donde la Tierra Late. Retrieved from http://www.mayakaan.travel/

Ángel A, Valle F, Ávila O (2014a) Buenas Practicas sustentables para la planeación, diseño y construcción de pequeños desarrollos turísticos. Amigos de Sian Ka'an A.C., Cancún. Retrieved from http://mayakaan.travel/Guias/Guia_PDC/index.html

Ángel A, Valle F, Ávila O (2014b) Buenas prácticas sustentables de operaciones para pequeños hoteles. Amigos de Sian Ka'an A.C., Cancún. Retrieved from http://mayakaan.travel/Guias/Guia_Operacion_Hotelera/index.html

Ángel A, Valle F, Ávila O (2014c) Buenas prácticas sustentables para actividades acuáticas. Amigos de Sian Ka'an A.C., Cancún. Retrieved from http://mayakaan.travel/Guias/Guia_Actividades_Acuaticas/index.html

Ángel A, Valle F, Ávila O (2014d) Buenas prácticas sustentables para el ecoturismo. Amigos de Sian Ka'an A.C., Cancún. Retrieved from http://mayakaan.travel/Guias/Guia_Ecoturismo/index.html

ASUR (2019) ASUR Airports investor relations presentation July. Retrieved from ASUR: chrome-extension://oemmndcbldboiebfnladdacbdfmadadm/ http://www.asur.com.mx/assets/files/es/inversionistas/presentaciones/presentaciones/ASUR-Airports-Investor-Relations-Presentation-July-2019.pdf

Barrios-Martínez-Rojas C (1991) A day in Sian Ka'an. Boletín Amigos de Sian Ka'an 9:6–8

Bezaury-Creel J (1991) Conservation and tourism. Boletín Amigos de Sian Ka'an 9:3–5

Carrillo-García M, Quintero-Gómez M, Tello-Leyva M (2018) Documento de análisis del retorno social de la inversión del destino Maya Ka'an. Amigos de Sian Ka'an A.C., Cancún

CONANP (2018) Marco estratégico de turismo sustentable en áreas protegidas de México. Secretaría de Medio Ambiente y Recursos Naturales, México

CONEVAL (2015) Anexo estadístico - Pobreza municipal. Retrieved from CONEVAL: coneval.org.mx/Medicion/Paginas/AE_pobreza_municipal.aspx

CONEVAL (2019a) Pobreza 2018. Retrieved from CONEVAL: https://www.coneval.org.mx/Medicion/MP/Paginas/Pobreza-2018.aspx

CONEVAL (2019b) Pobreza 2018/Quintana Roo. Retrieved from CONEVAL/Página principal: https://www.coneval.org.mx/coordinacion/entidades/QuintanaRoo/Paginas/Pobreza_2018.aspx

CPTQ (2019) Marketing. Retrieved from Consejo de Promoción Turística de Quintana Roo: http://cptq.mx/marketing/

Crotti R, Misrahi T (2017) The travel & tourism competitiveness report 2017. World Economic Forum, Geneva

Dachary A, Arnaiz-Burne S (1990) Quintana Roo: sociedad, economía, política y cultura. Universidad Nacional Autónoma de México, México

Ellis ER (2017) Deforestation processes in the state of Quintana Roo, Mexico: the role of land use and community forestry. Trop Conserv Sci 10:1–12. https://doi.org/10.1177/1940082917697259

Ferreyra-Acosta V, Merediz-Alonso G (2015) El ecoturimso como motor de desarrollo: el caso innovador de Maya Ka'an, Quintana Roo. Cuadernos de Turismo 5:84–92

Franquesa A (2009) Riviera Maya reef health. A preliminary report 2008–2009. Amigos de Sian Ka'an A.C., Cancún

González-Haro A, Patiño-González E, Vásques J (2014) Implementation of technical assistance and capacity building for adopting sustainable best practices, for snorkel and diving tourism companies. Amigos de Sian Ka'an A.C., Cancún

GSTC (2019) GSTC criteria. Retrieved from Global Sustainable Tourism Council (GSTC): https://www.gstcouncil.org/gstc-criteria/

Iglesias O (1999) Cursos de capacitación para guías naturalistas. Boletín Amigos de Sian Ka'an 18:7–9

INEGI (2016a) Panorama demográfico de Quintana Roo 2015. Instituto Nacional de Estadística y Geografía, México

INEGI (2016b) Censos económicos 2014: Quintana Roo. Instituto Nacional de Estadística y Geografía, México

INEGI (2019) Indicadores por entidad federativa. Retrieved from INEGI: https://www.inegi.org.mx/app/estatal/?ag=07000023#grafica

Keyes M (1998) Cultura forestal en Quintana Roo, México; observaciones y perspectivas. Madera y Bosques 4(1):3–13

Ley Agraria (2018) Artículo 59. México

Macías-Richards C (1999) El territorio de Quintana Roo. Tentativas de colonización y control militar en la selva maya (1888–1902). Hist Mex 1(193):5–54

Martí F (1985) Cancún. Fantasía de banqueros. México, Editorial Uno

Monterrubio J (2010) Short-term economic impacts of influenza A (H1N1) and government reaction on the Mexican tourism industry: an analysis of the media. Int J Tour Policy 3(1):1–15. https://doi.org/10.1504/IJTP.2010.031599

Novel Swine-Origin Influenza A (H1N1) Virus Investigation Team (2009) Emergence of a novel swine-origin influenza A (H1N1) virus in humans. N Engl J Med 360:2605–2615. https://doi.org/10.1056/NEJMoa0903810

Olivera-Lozano G (2005) La reforma al artículo 27 constitucional y la incorporación de las tierras ejidales al mercado legal de suelo urbano en México. Revista Electrónica de Geografía y Ciencias Sociales IX(194):33. Retrieved from http://www.ub.edu/geocrit/sn/sn-194-33.htm

RAMSAR (2019) Mexico. Retrieved from Ramsar: https://rsis.ramsar.org/ris-search/?f[0]=regionCountry_en_ss%3AMexico&pagetab=1

RAN (2019) Datos geográficos de las zonas de tierra parceladas por estado. Retrieved from Datos abiertos: http://datos.ran.gob.mx/filescd/dgcat/ran_da_dgcat_poligonos_zona_parceladas_qroo.kml

Reza M (2011) Proyecto de monitoreo del manglar en el municipio de Solidaridad (Riviera Maya) Quintana Roo, México. Amigos de Sian Ka'an A.C. Unpublished Analysis, Cancun

Santaolaya W (2017) Organismo gestor del destino Maya Ka'an. Amigos de Sian Ka'an A.C., Cancún

SCT (2018) Infraestructura aeroportuaria. Retrieved from SCT: https://datos.gob.mx/busca/dataset/infraestructura-aeroportuaria/resource/42322d35-4c64-4b55-8e43-7c8be3fbc071?inner_span=True
Secretaría de Industria y Comercio (1972) IX Censo General de Población 1970. Resumen General Abreviado. Dirección General de Estadística, México
SECTUR (2017) Acciones y programas. Retrieved from Secretaría de Turismo: https://www.gob.mx/sectur/acciones-y-programas/ordenamiento-turistico-sustentable
SECTUR (2018a) DATATUR, Compendio estadístico. Retrieved from Secretaría de Turismo: https://www.datatur.sectur.gob.mx/SitePages/CompendioEstadistico.aspx
SECTUR (2018b) DATATUR, Quintana Roo. Retrieved from Secretaría de Turismo: https://www.datatur.sectur.gob.mx/ITxEF/ITxEF_QROO.aspx
SEDESOL (1994) Decreto por el que se declara como anp, con el carácter de área de protección de flora y fauna, la región conocida como Uaymil, con 89,118-15-35.5 has, ubicada en los mpios. de Felipe Carrillo Pto. y Othón P. Blanco, Q. Roo. (2a publicación). Diario Oficial de la Federación CDXCIV(17):10–13
SEDETUR (2018) Indicaddores turísticos. Retrieved from Secretaría de Turismo de Quintana Roo: https://qroo.gob.mx/sedetur/indicadores-turisticos
SEDUE (1986) Decreto por el que declara como área que requiere la protección, mejoramiento, conservación y restauración de sus condiciones ambientales la superficie denominada Reserva de la Biósfera Sian Ka'an. Diario Oficial de la Federación CCCXCIV(13):26–31
SEDUMA (2011) Decreto mediante el cual se declara como área natural protegida la región denominada Sistema Lagunar Chchankanab con la categoría de Reserva Estatal, Ubicada en el Municipio José Ma. Morelos, Quintana Roo. Periódico Oficial del Estado de Quintana Roo II(32 Extraordinario):10–15
SEGOB (1974) Decreto por el que se reforma el artículo 43 y demás relativos de la Constitución Política de los Estados Unidos Mexicanos. Diario Oficial de la Federación CCCXXVI(26):1–7
SEMARNAP (1998) Decreto por el que se declara área natural protegida, con el carácter de reserva de la biosfera, la región denominada Arrecifes de Sian Ka'an, con una superficie de 39,927-15-84 hectáreas. Diario Oficial de la Federación DXXXIII(1):50–54
SEMARNAT (2005) Decreto por el que se declara área natural protegida, con la categoría de área de protección de flora y fauna, la región conocida como Bala'an K'aax, ubicada en los municipios de Othón P. Blanco y José Ma. Morelos, en el estado de Quintana Roo. Diario Oficial de la Federación DCXX(2):40–46
SEMARNAT (2018) Ley general del equilibrio ecológico y la protección al ambiente. Diario Oficial de la Federación Junio 5:38
Sheehan A (2014) Evaluación final. Pronóstico SROI: Proyecto: Diversificación de la oferta turística de la Riviera Maya con base en el aprovechamiento sustentable de los atractivos de la Reserva de la Biosfera Sian Ka'an. Amigos de Sian Ka'an A.C., Cancún
Sierra-Magaña C (2013a) Caracterización de la demanda. Diversificación de la oferta turística de la Riviera Maya con base en el aprovechamiento sustentable de los atractivos alrededor de la Reserva de la Biosfera Sian Ka'an. Amigos de Sian Ka'an A.C., Cancún
Sierra-Magaña C (2013b) Caracterización de la marca-destino. Diversificación de la oferta turística de la Riviera Maya con base en el aprovechamiento sustentable de los atractivos alrededor de la Reserva de la Biosfera Sian Ka'an. Amigos de Sian Ka'an A.C., Cancún
Silva H, Rosas G, Secaira F, Meller T, Mendoza M (2014) Sobrevive al cambio climático. Catálogo de buenas prácticas para reducir los impactos del clima en la zona costera de Quintana Roo, México. Instituto Tecnológico de Chetumal, The Nature Conservancy, Amigos de Sian Ka'an, Chetumal
Solano L (2014) Diversificación de la oferta turística de la Riviera Maya con base en el aprovechamiento sustentable de los atractivos de la Reserva de la Biosfera Sian Ka'an. Amigos de Sian Ka'an A.C., Cancún
The World Bank (2019) International tourism, receipts (current US$). Retrieved from The World Bank: https://data.worldbank.org/indicator/ST.INT.RCPT.CD?most_recent_value_desc=true
UNDP (2012) Community tours Sian Ka'an, Mexico, Equator initiative case study series. UNDP, Mexico, pp 1–8

UNESCO (1988) Report of the World Heritage Committee. 11th Session. UNESCO, Paris

UNESCO (2018) Ecological sciences for sustainable development. Retrieved from Latin America and the Caribbean: 130 biosphere reserves in 21 countries: http://www.unesco.org/new/en/natural-sciences/environment/ecological-sciences/biosphere-reserves/latin-america-and-the-caribbean/

UNWTO (2019) International tourism highlights. World Tourism Organization, Madrid. https://doi.org/10.18111/9789284421152

Valle F (2014) Implementación de los criterios de sustentabilidad de operaciones turísticas en los pequeños hoteles de: Felipe Carrillo Puerto, Tulum, Akumal y Punta Allen. Amigos de Sian Ka'an A.C., Cancún

Vanegas M (2013) Consultoría para el establecimiento de estándares que rijan el uso y la aplicación de la marca del destino y su relación con criterios de sustentabilidad y calidad. Amigos de Sian Ka'an A.C., Cancún

Chapter 33
Effective Management of the National Park Espíritu Santo, Through the Governance, Planning, and Design of an Integral Strategy for Los Islotes

Claudia J. Hernández-Camacho, Irma González-López, Lili Pelayo-González, David Aurioles-Gamboa, Elia López-Greene, and Martha P. Rosas-Hernández

Abstract The Parque Nacional Zona Marina del Archipiélago de Espíritu Santo (PNZMAES; Espíritu Santo Archipelago Marine National Park) is unique for its scenic beauty and for housing the only California sea lion (CSL) breeding colony in the Gulf of California with a healthy growing population. The region is economically relevant to artisanal fisheries and tourism service providers, resulting in CSL entanglements and an increase in tourist visitation to the colony. Few years ago, the PNZMAES management plan implemented conservation strategies to protect the different areas essential to the distinct life cycle phases of species of conservation interest like the CSL. In this study, all demographic data and ecological and environmental variables available for this colony were compiled in order to assess its population's growth trend and evaluate how diet and the entanglement rate have affected CSL abundance and the number of births. The colony continues to grow, and none of the variables analyzed had an effect on population parameters. The rookery is important for the conservation of the species in Mexico as it may become

---

C. J. Hernández-Camacho (✉) · L. Pelayo-González
Centro Interdisciplinario de Ciencias Marinas, Instituto Politécnico Nacional, La Paz, Baja California Sur, México
e-mail: jcamacho@ipn.mx

I. González-López · E. López-Greene
Comisión Nacional de Áreas Naturales Protegidas, Secretaría de Medio Ambiente y Recursos Naturales, La Paz, Baja California Sur, México
e-mail: irma.gonzalez@conanp.gob.mx; elia.lopez@conanp.gob.mx

D. Aurioles-Gamboa
Facultad de Ingeniería Ambiental, Universidad Popular Autónoma del Estado de Puebla, Puebla, México

M. P. Rosas-Hernández
Centro de Investigación Científica y de Educación Superior de Ensenada, Unidad La Paz, La Paz, Baja California Sur, México
e-mail: mrosas@cicese.mx

A. Ortega-Rubio (ed.), *Socio-ecological Studies in Natural Protected Areas*,
https://doi.org/10.1007/978-3-030-47264-1_33

a core population from which historic and current breeding colonies in the Gulf of California at serious risk of disappearing may be recolonized.

Keywords California sea lion · Population growth · Diet · Trophic niche · Entanglement

33.1 Introduction

The Parque Nacional Zona Marina del Archipiélago de Espíritu Santo (PNZMAES; Espíritu Santo Archipelago Marine National Park) is a locus of particular ecological and economic significance in the Gulf of California, Mexico. The marine national park harbors a breeding colony of California sea lions (CSLs; *Zalophus californianus*) (Fig. 33.1), a protected otariid species; the CSL population at this colony is in good health and growing (Szteren et al. 2006).

As an international tourist destination located in the La Paz municipality, the PNZMAES receives more than 50,000 visitors a year; this scenic and biologically diverse region is also of primary importance to local fisheries. Specifically, 75% of the tourists traveling to La Paz visit the Complejo Insular Espíritu Santo-La Partida (Espíritu Santo-La Partida Island Complex) to participate in the many activities offered by local tourism service providers (Fig. 33.2). This has led to a considerable increase in visitation relative to other Protected Natural Areas also located in

Fig. 33.1 Los Islotes California sea lion breeding colony

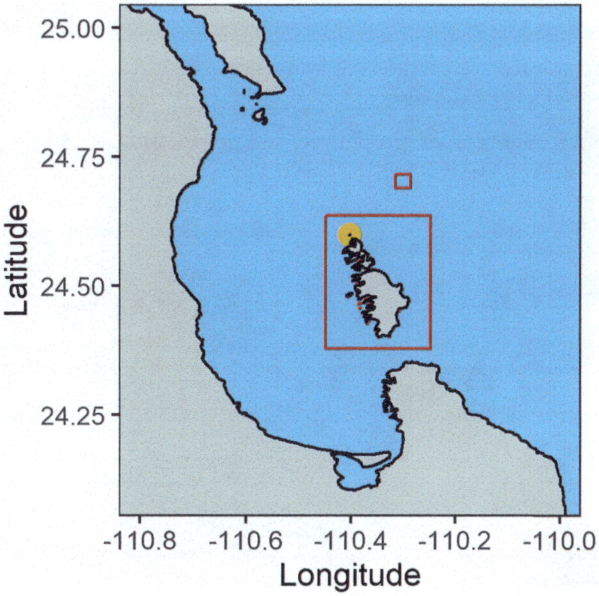

Fig. 33.2 Location of the Los Islotes California sea lion breeding colony (shaded area) in the northern part of the Espíritu Santo-La Partida Island Complex. The red boxes represent the two polygons established with the creation of the PNZMAES: the small one includes El Bajo and the large one includes Espiritu Santo and Los Islotes

popular tourist areas. A variety of problems have arisen that affect the natural conditions of the different species and ecosystems found in the PNZMAES.

In particular, the potential impact on the Los Islotes CSL breeding colony is high considering the number of visitors registered daily as this is the main site for swimming and diving within the Island Complex (Fig. 33.3). Moreover, CSLs interact regularly with artisanal fisheries during their feeding trips at sea during which they sometimes become entangled in fishing nets which may result in death by suffocation (Aurioles-Gamboa et al. 2003).

The Mexican government protected this area and the other islands of the Gulf of California through a conservation effort consisting of the establishment of Protected Natural Areas known as Islands and Protected Areas of the Gulf of California or Áreas de Protección de Flora y Fauna Islas del Golfo de California (APFF) (CONANP 2000b). The designation included the establishment of general policies and guidelines for the conservation and use of APFFs as a whole and the development of specific management programs for each island or archipelago tailored to the particular characteristics of each. The Programa de Manejo del Complejo Insular de Espíritu Santo (Management Program for the Espíritu Santo Island Complex) was published as a complement to the Programa de Manejo General del APFF Islas del Golfo de California BCS (General Management Program for APFFs and Islands of the Gulf of California BCS) by the National Commission on Protected Natural Areas or Comisión Nacional de Áreas Naturales Protegidas (CONANP 2000a).

Fig. 33.3 Tourist activity in the Los Islotes breeding colony

The Espíritu Santo Island Complex Management Program identified the need for a comprehensive management model to effectively and sustainably regulate the human activities occurring in and around the Archipelago as well as in the adjacent marine ecosystems. In conjunction with public consultation with the actors and sectors with relevant interests in the APFFs organized by the Sociedad de Historia Natural Niparajá, A.C. (Niparajá Natural History Society Civil Association), the APFF Director of the Islands of the Gulf of California BCS reviewed a study prepared by the Centro Interdisciplinario de Ciencias Marinas (CICIMAR; Interdisciplinary Center for Marine Sciences) of the Instituto Politécnico Nacional (IPN; National Polytechnic Institute); as a result, the area was designated as the Parque Nacional Zona Marina del Archipiélago de Espíritu Santo (PNZMAES) in the Official Gazette of the Federation on May 10, 2007.

Prior to the establishment of the PNZMAES, the activities carried out in the marine area adjacent to the island complex such as commercial and sport fishing as well as tourist activities (diving, snorkeling, and tourist routes) lacked regulation and zoning; both mechanisms have since been implemented in order to control and manage human activities in the area. One of the most significant problems affecting the Los Islotes CSL breeding colony is the use of harmful fishing gear, such as *chinchorros* or nets in which especially juveniles and breeding female CSLs in the vicinity of Los Islotes may become entangled (Aurioles-Gamboa et al. 2003). Prior to the decree, the area lacked regulations on snorkeling, swimming, and diving activities near the CSLs at Los Islotes; moreover, there existed a need for an anchoring system to reduce boat traffic and competition for space around the island.

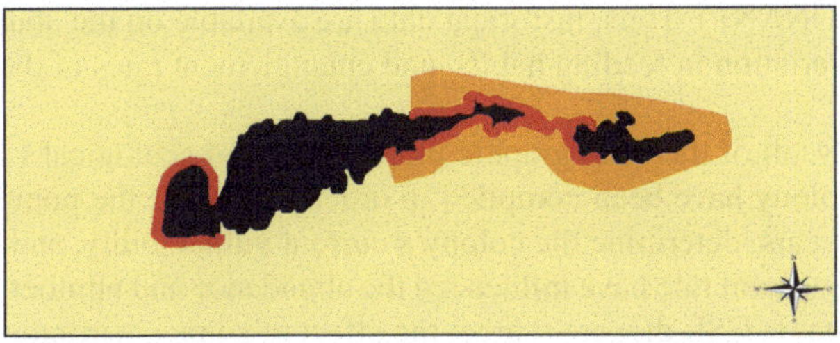

Fig. 33.4 Zoning of the Los Islotes rookery following the Espíritu Santo National Park management plan (CONANP 2014). The restricted zone wherein all activities are prohibited (red) extends 10 m offshore from the coastline in order to protect the critical breeding sites of California sea lions and reduces the effects of tourism in addition to protecting tourists from possible attacks by California sea lions defending their breeding territory. The public use area (orange) extends 10–40 m offshore immediately in front of the critical areas for California sea lion reproduction. The limit is marked by mooring buoys installed by the PNZMAES Directorate in order to protect the sites where California sea lions reproduce, in addition to reducing the effect of the use of anchors on the rocky reef

With the publication of the PNZMAES Conservation and Management Program (CONANP 2014), two broad polygons were established, one of which included Espíritu Santo and Los Islotes (Fig. 33.2); extractive activities such as commercial fishing, sports, and aquaculture are prohibited from three core areas in the second larger polygon (San Gabriel bay, Punta Lobos-Barra La Bonanza, and Los Islotes). These core zones were established in order to maintain or increase the abundance of species like CSLs as well as to protect the areas essential to the different phases of their life cycle. In particular, the management of the colony was achieved through subzonification and the installation of a mooring buoy system to facilitate regulation of use (Fig. 33.4).

In response to the increase in the frequency of attacks (bites) reported by tourists during the 2017 reproductive season, CONANP in collaboration with the Government of Baja California Sur, the Secretaría de Medio Ambiente y Recursos Naturales (SEMARNAT; Secretariat of Environment and Natural Resources), the Procuraduría Federal de Protección al Ambiente (PROFEPA; Federal Attorney General for Environmental Protection), and the Capitanía de Puerto (Port Capitaincy) of the Secretaría de la Marina (SEMAR; Secretariat of the Navy) restricted swimming and free and autonomous diving activities around Los Islotes during the CSL breeding season from June 1 to August 31 each year.

The CSL colony at Los Islotes has been monitored continuously since the 1980s by CICIMAR's "Burney J. Le Boeuf" Pinniped Ecology Laboratory and in recent years as part of the Protected Natural Areas Management Program or Programa de Manejo de Áreas Naturales Protegidas (PROMANP-CONANP), the Biological Monitoring Program in Protected Natural Areas or Programa de Monitoreo Biológico en Áreas Naturales Protegidas (PROMOBI-CONANP), and the Program for the Protection and Restoration of Ecosystems and Species at Risk or Programa para la Protección y Restauración de Ecosistemas y Especies en Riesgo

(PROREST-CONANP). Thus, historical data are available on the abundance, number of births, variation in feeding habits, and entanglement rates of the CSLs at Los Islotes.

In this study, all of the demographic information and ecological variables available for this colony have been compiled in order to analyze the population growth trend over 30 years, determine the colony's current vulnerability, and evaluate how diet and entanglement rate have influenced the abundance and number of births over time. Since data on CSL diet are scarce, the effect of some environmental variables (sea surface temperature and chlorophyll-*a* concentration) was assessed in conjunction with isotopic data from pup fur as proxy indicators of prey abundance and availability. These data were interpreted in the context of the abundance and number of sea lion births in order to assess the role of these factors in population dynamics .

33.2 Methodology

33.2.1 Population Trend and Abundance

A count-based Population Viability Analysis (PVA) was performed using a simple stochastic diffusion approximation model (Dennis et al. 1991); the latter uses linear regression parameters to estimate the stochastic population growth rate (λ). The model also provides the median time to extinction and the probability of the population declining to threshold abundances or the cumulative quasi-extinction probability (Dennis et al. 1991). A scenario of quasi-extinction thresholds occurring when the population falls below 100 individuals was evaluated based on the lowest historical population size recorded for Los Islotes (n = 169) and historical reports of CSL colonies in the Gulf of California (González-Suarez et al. 2006).

The probability of CSLs going extinct in 10, 20, and 100 years was determined based on the unconditional cumulative extinction probability and its bootstrapped 95% CI (Morris and Doak 2003). Whether the Los Islotes CSL population met the International Union for the Conservation of Nature (IUCN) quantitative analysis criteria were assessed for the following classifications: Critically Endangered (i.e., the probability of extinction is ≥ 50% within 10 years or three generations, whichever is longer), Endangered (i.e., the probability of extinction is ≥ 20% within 20 years or five generations, whichever is longer), or Vulnerable (i.e., the probability of extinction is ≥ 10% within 100 years) (IUCN 2012). Generation time is the average age of parents in the population or 10 years for the CSL based on the species life table (Hernández-Camacho 2001). All analyses were performed using the popbio library (Stubben and Milligan 2007) in R (Team 2019).

The PVA models assumed that (1) all individuals were identical and count data represent the true number of individuals in a population (i.e., observation error is minimal); (2) the patterns of population growth and fluctuations will be the same in the future (i.e., the mean and variance of the population growth remain constant); (3) environmental conditions are uncorrelated from one year to the next; and (4) there are no catastrophes or bonanzas (Dennis et al. 1991; Morris and Doak 2003).

Historical Los Islotes population counts were made continuously during the breeding season (June–July) from 1978 to 2019 by researchers of the CICIMAR-IPN Pinniped Ecology Laboratory and as part of the following federal programs: the Protected Natural Areas Management Program (PROMANP-CONANP), the Biological Monitoring Program in Protected Natural Areas (PROMOBI-CONANP), and the Program for the Protection and Restoration of Ecosystems and Species at Risk (PROREST-CONANP) (Gallo-Reynoso et al. 2011, 2014, 2015; Aurioles-Gamboa et al. 2012, 2013; Hernández-Camacho et al. 2016). The breeding season was chosen because it is when most animals of all age- and sex-classes are present at the colony, including newborn pups (Fig. 33.5).

The maximum number of births occurs from the end of June to mid-July; during this period, <6% of young die; therefore the pup count was considered the total number of births (Berkson and De Master 1985; Hernández-Camacho 2001). Pups are difficult to identify onshore as they are sometimes found behind or below large rocks or other animals; thus, they were counted on land at the rookery. A correction factor of 28% was applied to account for the number of females at sea during a given survey or the number of females onshore that were not counted due to substrate or other factors obstructing the view (Hernández-Camacho et al. 2015).

This correction factor is the difference between the number of females observed and those expected based on a stochastic demographic model structured by age and sex and using the survival and fecundity rates of female CSLs from Los Islotes and San Miguel Island in the Channel Islands of California, USA (Hernández-Camacho et al. 2008a, b; Melin 2002). Based on the demographic model, 14% of females are

Fig. 33.5 Group of pups observed during one of the counts to determine the number of births during the reproductive season at Los Islotes

lost during a survey (95% CI: 0–28%); to obtain a more conservative estimate, the upper confidence limit value was used to correct the counts.

33.2.2 Entanglement Rate

The entanglement rate at Los Islotes has been monitored over the last decade as an indicator of the interaction between CSLs and local fisheries (Aurioles-Gamboa et al. 2003, 2011). In the field, only CSLs with recent entanglement marks are recorded to avoid overestimation (Fig. 33.6). The entanglement rate is the ratio of the number of entangled individuals as a percentage of the total population. Data are available for 1996–2018, except for 1997, 1999, and 2004. A Spearman correlation was performed to assess the relationship between the entanglement rate and abundance. The number of births was not used because pups depend exclusively on milk and thus do not undertake foraging trips offshore.

33.2.3 Diet

The Los Islotes CSL diet has been studied for more than three decades but not consecutively or during the same months over several years. Thus, this study employed data only from research carried out during the reproductive seasons 1990, 1993, 2002, 2012-2013,2015, and 2019; the Prey Importance Index (PII) was used to determine the preferred prey items (García-Rodriguez 1995; Zavaleta-Lizárraga

Fig. 33.6 Female California sea lion with the remains of a fishing net entangled around her neck at the Los Islotes breeding colony

2003; García-Rodriguez and Aurioles-Gamboa 2004; Porras-Peters et al. 2008; Aurioles-Gamboa et al. 2012; Gallo-Reynoso et al. 2015; Hernández-Camacho et al. 2016). Species considered main prey had a PII ≥10 as recorded in the historical data on prey items observed in the CSL diet at this colony (García-Rodriguez and Aurioles-Gamboa 2004).

A Principal Components Analysis (PCA) was performed on the PII values of the main prey items to assess whether the diet composition varied over time. All analyses were performed using the factoextra package (Kassambara and Mundt 2017) in R (Team 2019).

33.2.4 Trophic Niche Width

The trophic niche width was evaluated based on the $\delta^{15}N$ and $\delta^{13}C$ from the 1999–2017 breeding seasons. Two- to three-month-old CSL pups depend on their mother's milk for nourishment; thus, their fur reflects the average $\delta^{15}N$ and $\delta^{13}C$ composition of the prey ingested by their mothers and the feeding area exploited by those adult females. The pup values are corrected to adult female values by applying a mother-to-pup fractionation value of ~2.1‰ (Porras-Peters et al. 2008; Aurioles-Gamboa et al. 2009).

A wider isotopic niche was assumed to be associated with unfavorable environmental conditions; that is, animals would have to move farther away from the colony in search of areas with more abundant prey in order to feed (Rosas-Hernández et al. 2019). Adult female CSLs feed as close to their colonies as possible because they have to return to nurse their young during the lactation period, which can last more than a year (Newsome et al. 2006; Villegas-Amtmann et al. 2012).

Fur samples were removed from the backs of pups using scissors. The samples were rinsed with distilled water to eliminate salt and sand residues, dried at room temperature, and later washed in a 1:1 solution of chloroform/methanol to remove lipids, after which they were dried again at room temperature for 12 h to allow all solvents to volatilize. The dried samples were cut with sterilized scissors and homogenized in an agate mortar to obtain weights of 1.0 ± 0.2 mg using a Mettler-Toledo MX5 micro balance (precision = 1 µg). The fur samples were placed in tin capsules for stable isotope analysis. Determination of the isotopic ratios of C and N through the combustion of CO_2 and N_2 was undertaken at the Stable Isotope Facility of the University of California at Davis, USA. The samples were analyzed using a PDZ Europa ANCA-GSL elemental analyzer interfaced to a PDZ Europa 20-20 isotope ratio mass spectrometer (PDZ Europe 20-20, Sercon Ltd) with an accuracy of ± 0.2‰ for both isotopes. The isotopic ratios are expressed as delta (δ): $\delta^{15}N$ or $\delta^{13}C = 1000 * [(R_{sam}/R_{std}) - 1]$, where R_{sam} and R_{std} are the ratios of $^{15}N/^{14}N$ or $^{13}C/^{12}C$ of the sample and the standard, respectively. The standards were Vienna-Pee Dee Belemnite limestone (V-PDB) for carbon and atmospheric N_2 for nitrogen. Units are expressed as parts per thousand (‰).

The standard ellipse areas corrected for small sample size (SEA_C) were used to obtain the isotopic niche or trophic niche width (potential foraging areas) of the female CSLs from each year from 1999 to 2017. These ellipses were obtained from the pup fur isotope values using the SIBER package (Jackson et al. 2019). This method is a Bayesian version of Layman's metrics (Layman et al. 2007); unlike Euclidean methods (e.g., convex hull), this technique incorporates uncertainties like sample biases and small sample sizes into the metrics of the niches (Jackson et al. 2011). Based on the Markov chain Monte Carlo (MCMC) simulation, this approach assigns uncertainty measures for constructing ellipse parameters in a manner similar to a bootstrap. One-way ANOVA was performed to identify differences between paired groups between years. All analyses were carried out using R (Team 2019).

33.2.5 Environmental Variables

Diet information for some years was scarce; thus, environmental variables were used as proxy indicators of prey availability. Sea surface temperature (SST) modulates the abundance and distribution of prey, while the chlorophyll-*a* concentration (CHL-*a*) is an indicator of prey availability (Nevárez-Martínez et al. 2001; Jaud et al. 2012). Anomalies in SST and CHL-*a* were identified in the area surrounding Los Islotes using satellite images for 1995–2018. Female CSLs feed around the colony at an average maximum distance of 60 km; thus, satellite information was extracted from a polygon with a 60 km radius (Kuhn 2006; Villegas-Amtmann 2009; Rosas-Hernández et al. 2018).

The daily satellite images of SST were obtained from the National Oceanic and Atmospheric Administration (NOAA) and the National Centers for Environmental Information (NCEI); the images are the result of interpolations and extrapolations with SST data taken both via Advanced Very High Resolution Radiometer (AVHRR) and in situ (e.g., boats and buoys). The images are level four (L4) with a resolution of 0.25° per pixel and are available from 1981 to the present. CHL-*a* information was obtained from the Copernicus Marine Environment Monitoring Service (Copernicus-CMEMS). The L4 daily satellite images were downloaded with a resolution of 4 × 4 km per pixel; the images are constructed from observations taken by several sensors (OLCI-A, MERIS, AquaMODIS, VIIRS, and SeaWiFS) and are available for 1997–2018. Annual anomalies were calculated from July to June of the following year (i.e., from the breeding season until the female gives birth the following year) based on the daily values of the environmental variables. Spearman correlations were carried out on these variables and the CSL population parameters.

33.2.6 El Niño Southern Oscillation (ENSO)

The monthly values of the Multivariate ENSO Index (MEI) for the period from 1995 to 2018 were obtained and annual averages (July to June of the following year) were calculated in order to evaluate their relationship with the CSL population parameters using Spearman correlations.

33.2.7 Generalized Linear Models (GLM)

Generalized linear models were constructed to explore how the different variables (entanglement rate, SST, CHL-*a*, MEI, and SEAc) influenced the CSL abundance and number of births. Models with different variables (three scenarios) were developed because some data were only available for some years. The global models were used to construct more parsimonious models that included uncorrelated variables; the relationships between these variables and the population parameters were justified in a biological context. The Akaike Information Criteria for small samples (AICc), the delta AICc (ΔAICc), the Akaike weight (***w***), and the explained deviance or pseudo R^2 (Zuur et al. 2009) were used to determine the level of support for each model. All analyses were performed using the MuMIn package (Barton and Barton 2019) in R (Team 2019).

33.3 Results

33.3.1 Population Trend and Abundance

The Los Islotes CSL colony has increased at an annual growth rate of 2% ($\lambda = 1.02$; CI 0.98–1.06) reaching ca. 600 individuals in 2019 (Fig. 33.7). Despite two significant declines in 2015 and 2019, the population size has remained relatively stable (between 569 and 659 individuals) since 2011. The colony has almost tripled in size from the first population size recorded in 1978. The number of births has increased at a slightly faster rate ($\lambda = 1.03$; CI 0.98–1.08), and, unlike the total population trend, the number of pups remained stable from 2001 to 2009 and from 2011 to 2018. In 2019, the number of pups decreased significantly. The Los Islotes colony does not meet the IUCN's quantitative analysis criteria for being considered Critically Endangered, Endangered, or Vulnerable (IUCN 2012) and is thus of Least Concern. However, the probability of extinction at the upper confidence limit meets the IUCN's quantitative criteria for being classified as Endangered.

33.3.2 Entanglement Rate

The entanglement rate varied from 7% to 10% between 1996 and 2003, with a dramatic decline in 2005. As of that year, the entanglement rate has remained <2%. The relationship between abundance and the entanglement rate at Los Islotes was inverse and significant ($r = -0.71$ and $p > 0.05$; Fig. 33.8).

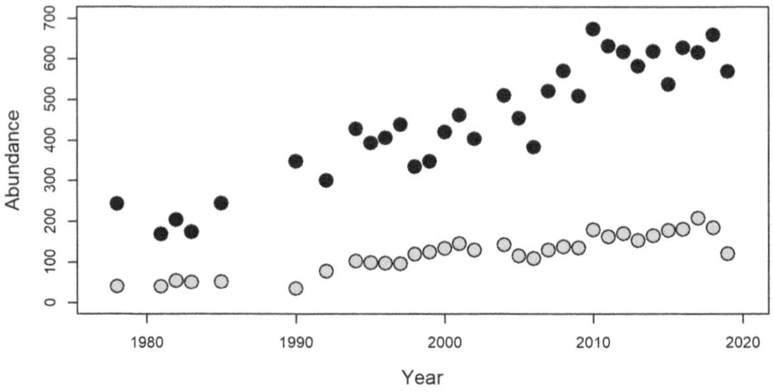

Fig. 33.7 Numbers of pups born (gray circles) and corrected total population size (including newborns) (black circles) for California sea lions on Los Islotes breeding colony

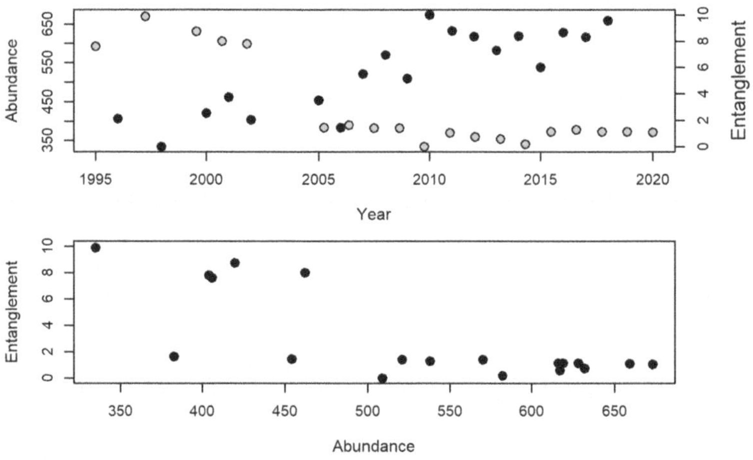

Fig. 33.8 Relationship between California sea lion abundance and entanglement rate at Los Islotes

33.3.3 *Diet*

The CSL diet during the breeding season at Los Islotes consisted of 83 species, of which 18 were main prey and only 6 appeared more than once (Table 33.1). Based on the PCA, the first three components explained 83% of the variance (40, 29, and 14%, respectively). The 1990 and 1993 diets were distinguished from other years for including unusually high proportions of *P. notatus* and *P. eos* (Fig. 33.9).

Table 33.1 Historical values of the Prey Importance Index for California sea lions at Los Islotes during the reproductive season.

Code	Species	1990	1993	2002	2012	2013	2015	2019
1	*Abraliopsis affinis*	0.0	0.0	18.2	0.0	0.0	0.0	0.0
2	*Aulopus sp1*	12.0	0.0	0.0	0.0	0.0	0.0	10.8
3	*Benthosema panamense*	0.0	0.0	0.0	0.0	0.0	17.0	0.0
4	*Engraulis mordax*	0.0	0.0	15.3	0.0	0.0	0.0	14.7
5	*Haemulopsis sp1*	16.3	0.0	0.0	0.0	0.0	0.0	0.0
6	*Merluccius productus*	0.0	0.0	0.0	0.0	17.0	10.0	0.0
7	*Myctophidae sp1*	0.0	0.0	0.0	0.0	0.0	0.0	16.8
8	*Ophidion scrippsae*	0.0	0.0	0.0	0.0	0.0	25.0	0.0
9	*Paralabrax nebulifer*	0.0	0.0	0.0	0.0	10.0	0.0	0.0
10	*Paralabrax sp1*	0.0	0.0	0.0	15.0	0.0	0.0	0.0
11	*Porichthys notatus*	67.3	0.0	0.0	0.0	0.0	0.0	0.0
12	*Prionotus sp1*	27.9	0.0	0.0	0.0	0.0	0.0	0.0
13	*Pronotogrammus eos*	0.0	53.0	0.0	0.0	0.0	0.0	0.0
14	*Pronotogrammus multifasciatus*	17.4	25.1	0.0	0.0	0.0	0.0	0.0
15	*Sardinops sagax*	0.0	0.0	0.0	0.0	0.0	15.0	10.0
16	*Serranus aequidens*	14.0	14.2	20.8	20.0	26.0	0.0	17.2
17	*Synodus sp1*	0.0	0.0	11.5	0.0	0.0	0.0	0.0
18	*Uroconger varidens*	0.0	28.0	0.0	0.0	0.0	0.0	0.0

33.3.4 Trophic Niche Width

The mean ± SD isotopic values obtained for CSL pup fur sampled over 19 years were 15.12 ± 0.3 ‰ for $\delta^{13}C$ and 21.81 ± 0.8 ‰ for $\delta^{15}N$ (Table 33.2). ANOVA revealed significant differences in both isotopes ($\delta^{13}C$: $F_{1, 18} = 8.063$, $p < 0.000$; $\delta^{15}N$: $F_{1, 18} = 41.983$, $p < 0.000$), observed in 1999, 2001, 2003, 2004, and 2007 (post-hoc Tukey HSD multiple comparison tests). The niche width also was significantly different between years; however, this pattern is not related to population size ($r = -0.12$, $p > 0.05$) or number of births ($r = 0.13$ and $p > 0.05$; Fig. 33.10).

33.3.5 Environmental Variables

No significant relationships were found between SST anomalies and abundance ($r = 0.26$, $p > 0.05$) or number of births ($r = 0.39$, $p > 0.05$). There were no obvious trends or decreases in abundance or number of births when the SST anomalies were positive (Fig. 33.11).

There was no relationship between CHL-*a* anomalies and abundance ($r = -0.34$, $p > 0.05$) or number of births ($r = -0.38$, $p > 0.05$). Abundance and number of births did not decrease when the CHL-*a* anomalies were positive. There were no obvious trends or decreases in the analyzed variables (Fig. 33.12).

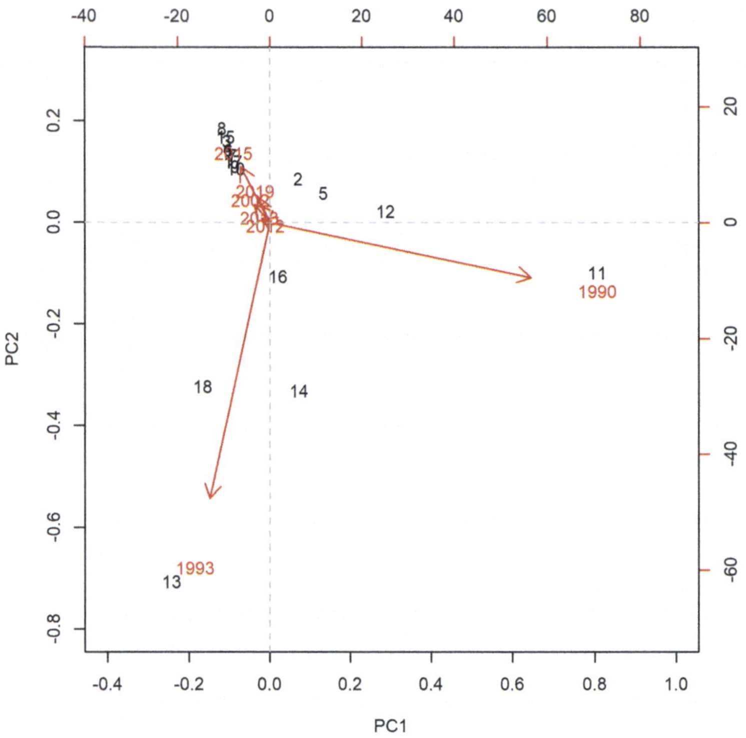

Fig. 33.9 Principal Component Analysis of the historical diet data for Los Islotes California sea lions. The first two components explain 40% and 29% of the variance, respectively

33.3.6 El Niño Southern Oscillation (ENSO)

The study period included both moderate (2002–2003 and 2009–2010) and strong (1997–1998 and 2014–2016) El Niño events; however, no significant decreases in abundance or number of births were observed during these episodes (Fig. 33.13). This pattern is confirmed by the lack of significant relationships between the MEI values and abundance (r = −0.06 and $p > 0.05$) or number of births (r = −0.10 and $p > 0.05$).

33.3.7 Generalized Linear Models

Eleven models of three scenarios were constructed for each population parameter (abundance and number of births). Most models for both parameters (except P3, P4, and P5) did not fit the data as they included values of explained deviance <50%. Models P3 and P4 presented higher values of explained deviance; however, it was

Table 33.2 $\delta^{13}C$ and $\delta^{15}N$ values (mean ± SD) for California sea lion pups from Los Islotes and summary statistics used in the SIBER analysis for each year

Year	$\delta^{13}C$ (‰)			$\delta^{15}N$ (‰)			TA	SEA(‰^{2})	SEAc (‰^{2})
	mean		SD	mean		SD			
1999	−14.7	±	0.5	20.7	±	0.4	1.355	0.586	0.645
2000	−15.1	±	0.5	20.0	±	0.6	1.790	0.877	0.986
2001	−14.7	±	0.6	20.5	±	0.1	0.285	0.190	0.221
2002	−15.4	±	0.3	21.5	±	0.3	0.605	0.288	0.324
2003	−14.6	±	0.6	21.5	±	0.4	1.540	0.611	0.649
2004	−14.7	±	0.1	21.4	±	0.5	0.170	0.175	0.219
2005	−15.1	±	0.1	21.7	±	0.3	0.115	0.059	0.064
2006	−15.0	±	0.3	21.7	±	0.4	0.825	0.399	0.430
2007	−14.7	±	0.3	21.9	±	0.4	0.755	0.309	0.328
2008	−15.0	±	0.2	22.1	±	0.5	0.920	0.539	0.599
2009	−15.1	±	0.4	22.1	±	0.2	0.495	0.208	0.231
2010	−15.2	±	0.3	23.3	±	0.3	0.425	0.179	0.195
2011	−15.3	±	0.2	21.9	±	0.3	0.565	0.204	0.220
2012	−15.6	±	0.2	22.8	±	0.3	0.787	0.331	0.357
2013	−15.3	±	0.4	22.4	±	0.3	0.506	0.254	0.285
2014	−15.5	±	0.4	22.6	±	0.2	0.663	0.373	0.420
2015	−15.4	±	0.5	22.5	±	0.4	1.400	0.605	0.652
2016	−15.4	±	0.6	21.6	±	0.5	1.810	0.839	0.904
2017	−15.5	±	0.5	22.2	±	0.4	1.295	0.597	0.643

TA Total area of the convex hull, *SEA* Standard ellipse area, SEA_C Corrected standard ellipse area

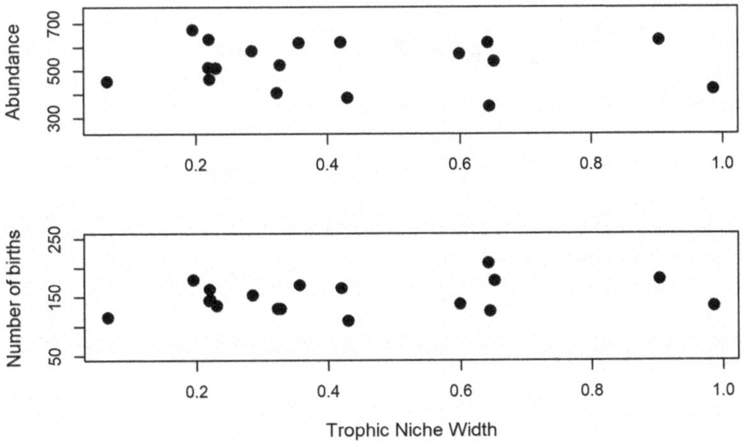

Fig. 33.10 Relationship between California sea lion population parameters and trophic niche width at Los Islotes

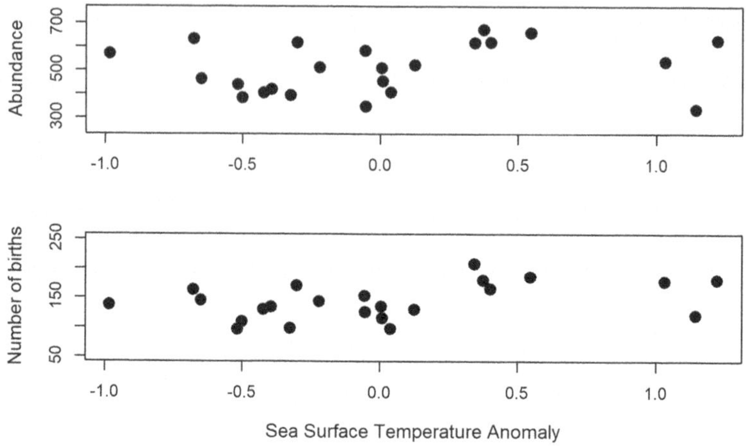

Fig. 33.11 Relationship between California sea lion population parameters and sea surface temperature anomalies around Los Islotes

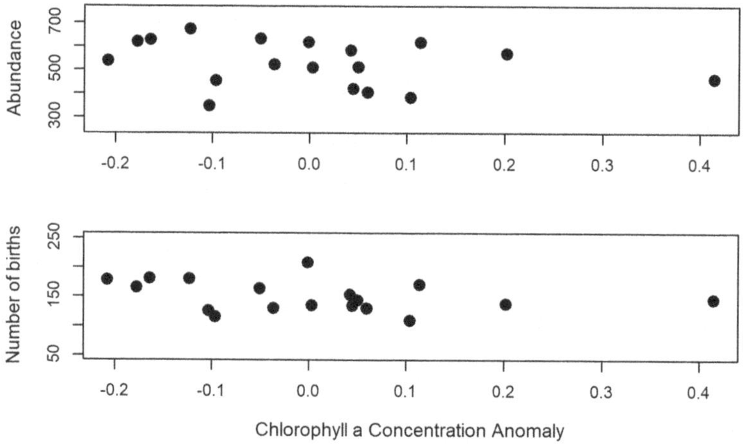

Fig. 33.12 Relationship between California sea lion population parameters and chlorophyll-*a* concentration anomalies around Los Islotes

the entanglement rate (P5) that explained the highest proportion of the variance (Table 33.3, Fig. 33.8).

33.4 Discussion

The Los Islotes CSL reproductive colony located within the PNZMAES is the only rookery in the Gulf of California with a positive growth tendency over the last three decades (Szteren et al. 2006; Gallo-Reynoso et al. 2011, 2014, 2015;

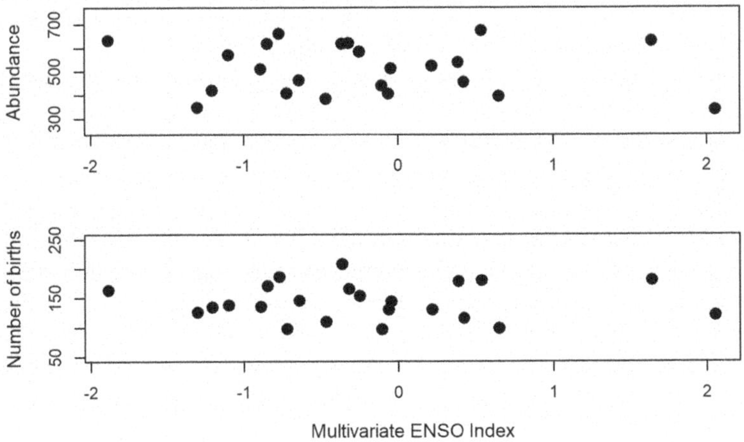

Fig. 33.13 Relationship between California sea lion population parameters and Multivariate ENSO Index (MEI) values at Los Islotes

Table 33.3 Generalized linear models of the population parameters of California sea lions at Los Islotes and the environmental variables

Parameter	Scenario	Model	Variables	AICc	ΔAICc	Weight (*w*)	Deviance (%)
Abundance	I	P1	~SST	667.9	0.00	0.997	3.837
		P2	~MEI	679.7	11.68	0.003	1.454
	II	P3	~SST+Entanglement	322.7	0.00	0.437	58.300
		P5	~Entanglement	323.1	0.40	0.357	57.460le
		P4	~MEI+Entanglement	324.2	1.51	0.206	57.910
	III	P6	~MEI+CHL-a+SEAc	441.0	0.00	0.499	10.186
		P7	~MEI+CHL-a	442.6	1.64	0.219	10.185
		P9	~CHL-a+SEAc	442.7	1.73	0.210	8.577
		P10	~CHL-a	444.9	3.99	0.068	6.941
		P8	~MEI+SEAc	451.0	10.05	0.003	5.949
		P11	~SEAc	465.1	24.14	0.000	0.577
Number of births	I	B1	~SST	283.8	0.00	1.000	17.041
		B2	~MEI	309.0	25.15	0.000	0.110
	II	B3	~SST+Entanglement	206.2	0.00	0.982	36.118
		B5	~Entanglement	214.8	8.59	0.013	25.511
		B4	~MEI+Entanglement	216.8	10.59	0.005	26.302
	III	B10	~CHL-a	198.2	0.00	0.284	8.941
		B9	~CHL-a+SEAc	198.5	0.30	0.245	12.328
		B8	~MEI+SEAc	199.4	1.17	0.158	11.197
		B7	~MEI+CHL-a	199.7	1.46	0.137	10.826
		B6	~MEI+CHL-a+SEAc	200.3	2.15	0.097	14.285
		B11	~ SEAc	200.7	2.53	0.080	5.669

P total population, *B* number of births

Aurioles-Gamboa et al. 2012, 2013; Hernández-Camacho et al. 2016) as well as being the most visited by tourism (Rubio-Castañeda 2018).

The implementation of conservation and use guidelines for the APFFs and islands of the Gulf of California (CONANP 2000a, b) and subsequent designation of the region as the PNZMAES (CONANP 2014) may have contributed to increasing the abundance of CSLs as the Los Islotes colony is one of the healthiest in the Gulf of California. The population at Los Islotes has increased 13% since the establishment of the PNZMAES, and although the colony has remained stable in recent years, we believe it has not stopped growing and that animals are moving to other nearby islands. In 2013, we recorded an increase in the number of CSLs that regularly occupy islands in the Bahía de La Paz and surrounding areas; moreover, new resting colonies have formed recently in areas previously occupied by CSLs.

Tourism and coastal fishing are the two main threats to the CSL colony contemplated by the PNZMAES Conservation and Management Program. In terms of tourism, the primary actions included the restriction of free and autonomous swimming and diving activities during the reproductive season, the subzonification of the area surrounding Los Islotes, the installation of a mooring buoy system to regulate activities and the tourist carrying capacity as well such as continuous surveillance and supervision by the Secretaría de Medio Ambiente y Recursos Naturales (SEMARNAT; Secretariat of Environment and Natural Resources), the Procuraduría Federal de Protección al Ambiente (PROFEPA; Federal Attorney General for Environmental Protection), and the Capitanía de Puerto (Port Capitaincy) of the Secretaría de la Marina (SEMAR; Secretariat of the Navy). These actions have contributed to ensuring that tourism activities do not affect behaviors (e.g., nursing, confrontations between males over reproductive territories) critical to the survival and reproductive success of the CSLs at this colony (Fig. 33.14) (Labrada-Martagón et al. 2005).

With regard to artisanal fishing, the restriction in the use of *chinchorros* nets in the immediate vicinity of CSLs contributed to a significant reduction in the entanglement rate from 10% to <2% over a 2-year period (Elorriaga-Verplancken 2005). Prior to the establishment of the PNZMAES, the Los Islotes rookery had the highest entanglement rate of all the reproductive colonies in the Gulf of California (0–2.24%) (Harcourt et al. 1994; Zavala-González and Mellink 1997).

The entangled CSLs observed at the Los Islotes colony are only one indicator of the degree of interaction between CSLs and coastal fisheries. An unknown percentage of CSLs cannot escape the nets and die from suffocation by immersion or indirectly due to injuries caused during their attempted escape, especially juveniles (Aurioles-Gamboa et al. 2003). Although the entanglement rate did not have a significant negative effect on the colony's population growth in this study, an increase in the number of juveniles dying after becoming entangled could affect the colony's viability as the population growth rate is more sensitive to changes in juvenile survival (Underwood et al. 2008). Thus, it is essential that fishing activities in the area continue to be regulated; in addition, a review of the PNZMAES Conservation and Management Program should be undertaken in order to take into consideration the changing distribution of CSLs within the Bahía de La Paz and surrounding areas.

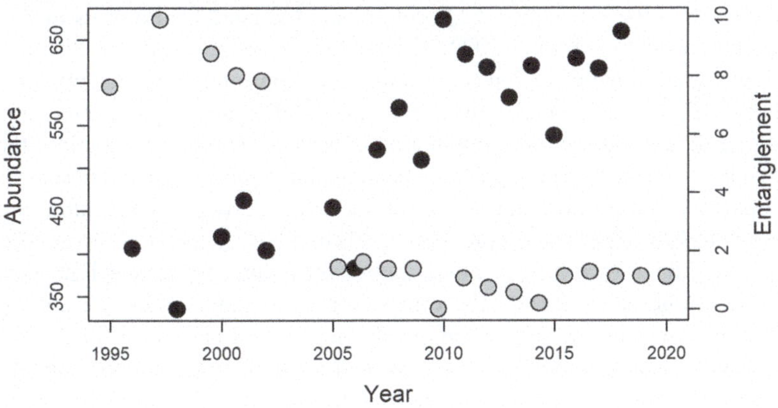

• •

Fig. 33.14 Female California sea lion nursing her pup during the breeding season at Los Islotes

The growth of the Los Islotes CSL colony and the good health of its individuals (Luque and Aurioles-Gamboa 2002; Hernández-Camacho et al. 2008b) not only depend on the effectiveness of the conservation actions and strategies implemented as part of local management programs. Prey availability, abundance, and quality are some of the main modulators of population dynamics. The rapid growth of the Los Islotes rookery suggests that prey have been available and were apparently of sufficient quality to support the individuals at the colony, especially the adult females, the most abundant age class and the individuals that feed in the area surrounding the island as they have to return to nurse their young during the year-long lactation period (Newsome et al. 2006). The effect of prey availability on the colony was evaluated by assessing the impact on the abundance and number of births of some environmental variables (SST, CHL-*a*) as well as analyzing isotopic indicators of prey abundance and availability. Our results provide the framework for predicting the potential impact of changes in seasonal and annual environmental conditions due to climate change.

Our models indicate that, thus far, neither environmental factors (e.g., SEA_C, SST, CHL-*a*) nor the entanglement rate have negatively affected the colony's growth rate or the number of births during the breeding season. Although not included in this analysis, information on diet suggests that the most important prey items in the CSL diet have remained constant over time and thus have consistently been available to animals at this colony. The prey items consumed by CSLs from Los Islotes are not of commercial value in the area (Aurioles-Gamboa et al. 2003). The only species of commercial value identified in CSL scat samples was the spotted sand bass (*Paralabrax maculofasciatus*); however, its relative importance was <5% (Aurioles-Gamboa et al. 2003). The low overlap between the main resources of the artisanal fishery and the Los Islotes CSL diet suggests that the animals accidentally become entangled during their nocturnal feeding trips when the nylon nets are more difficult to detect (Aurioles-Gamboa et al. 2003).

The Los Islotes CSL diet has been studied for decades; indeed, it is the colony for which the greatest amount of information is available. However, there are no systematic data on diet that would permit evaluation of changes over time. The CSL diet at this colony includes approximately 80 species, but only 18 of them are considered main prey. Most species persist in the diet over time but with a variable Prey Importance Index, which also reflects the sampling effort (Aurioles-Gamboa et al. 2003; Porras-Peters et al. 2008).

The lack of systematic information on diet did not permit analysis of the relationship with the CSL population parameters; however, the colony has grown and the body condition of pups has been reported as favorable over the last 15 years (Hernández-Camacho et al. 2016). Thus, the quality and availability of prey around Los Islotes has been adequate, preventing CSLs from experiencing nutritional stress, which in turn would impact their reproductive success and survival (Trites and Donnelly 2003; Soto et al. 2004).

The isotopic niche width was significantly different between years although no clear pattern was evident. This fluctuation may reflect local environmental variability as the characteristics of the immediate surroundings determine the amount and distribution of resources; thus, prey diversity and distribution change over time (Rosas-Hernández et al. 2019). However, the Los Islotes CSL diet did not present significant differences between years; indeed, the historical diet data revealed 18 main prey items that have been preferentially consumed over time. Thus, some other factor drives the variation in the CSL trophic niche width over time. The isotopic N data show a range of 4.6‰, which could reflect consumption of high trophic level prey (Kelly 2000; Newsome et al. 2010); however, the N also presents an annual increase associated with the expansion of the oxygen minimum zone recorded in the eastern tropical Pacific, which is most intense along the coast of Mexico (Aurioles-Gamboa et al. 2017).

In contrast, the minor variation in $\delta^{13}C$ over time suggests that CSL feeding areas did not change, regardless of the environmental conditions affecting the area. Initially, we assumed that the differences found in population counts could be associated with a change in feeding habitat ($\delta^{13}C$) or preferred prey items ($\delta^{15}N$); however, correlations were not significant (Fig. 33.10). Although the environmental conditions were anomalous during some years, this apparently did not affect the feeding behavior exhibited by adult female CSLs and thus did not influence their survival or reproductive success.

The analysis of the environmental variables confirmed the patterns revealed by the diet and isotopic niche width assessments. Neither the SST nor the CHL-*a* anomalies were significantly associated with abundance or number of births. The ENSO phenomenon did not exert a negative influence on the growth of the colony, in contrast to previous years (Aurioles-Gamboa and Le Boeuf 1991). The effects of ENSO within the Gulf of California are heterogeneous and particularly weak on the peninsular coast where Los Islotes is located (Herrera-Cervantes et al. 2010). In contrast, CSL colonies in the Pacific Ocean are more directly impacted by ENSO and significant decreases in both survival and number of births have been reported (Elorriaga-Verplancken et al. 2016; Laake et al. 2018).

Changes in CSL abundance at the Los Islotes colony also may be caused by disease and predation; however, it is difficult to prove either convincingly. The available health information is scarce due to the difficulty of capturing animals of different age-classes for the types of biological samples (e.g., blood, saliva, urine, muscle, etc.) required. In other studies, the presence of some pathogenic *Leptospira* serovars has been reported as a potential threat to the health of these animals (Godínez et al. 1999; Acevedo-Whitehouse et al. 2003; Avalos-Téllez et al. 2016). However, further research is necessary to determine its impact on the survival and reproduction of CSLs at this colony.

Natural predators of CSLs include the great white shark (*Carcharodon carcharias*), the bull shark (*Carcharhinus leucas*), and the killer whale (*Orcinus orca*) in different areas of its distribution both in the Gulf of California and at the colonies on the Pacific coast of California and the Baja California peninsula (Aurioles-Gamboa and Sinsel 1988; Klimley et al. 1992; Long et al. 1998). Over-exploitation of different Carcharhinidae shark species in the Bahía de La Paz region and in general in the Gulf of California in the late 1970s to early 1980s caused shark populations to decline (Sala et al. 2004; Galván-Magaña et al. 2010), as expected, shark attacks are not a major cause of mortality for CSLs in this region. Over the last 20 years of research at this colony, we have only observed one adult female with a shark bite on the side of her body, and we have never observed killer whales near the colony or feeding on CSLs in the vicinity. With the reduction in shark populations in the southern Gulf of California, high trophic predators were eliminated not only for CSLs but for lower trophic level bony fishes as well. This allowed their populations to increase and may have resulted in greater prey availability for CSLs. Since the main prey of the CSLs at this colony are not of commercial value, there is no information about changes in their abundance over time or how they respond to changes in SST and CHL-*a*.

The CSL population at Los Islotes has developed under favorable conditions for their diet, with low presence of natural predators and where fishing and tourism activities have not negatively affected their survival or reproductive success. This colony represents a success case exemplifying the achievement of the conservation and sustainable use of the significant natural resources of the PNZMAES. In recognition of the proper management of the PNZMAES, the marine national park was included in the Green List of Protected and Conserved Areas of the IUCN in 2018. Los Islotes is of considerable value for the conservation of this species in Mexico as it is the healthiest of the Gulf of California colonies. Under favorable conditions, the rookery might become a core population from which historic and current breeding colonies in the Gulf of California that are at serious risk of disappearing in the next 20 years might be recolonized (Szteren et al. 2006).

Acknowledgments This research was supported by funding provided by the National Polytechnic Institute (IPN; Instituto Politécnico Nacional: SIP 20100911, SIP 20120069, SIP 20130402, SIP 20140111, SIP 20150347, SIP 20160055, SIP 20170489, SIP 20180709, SIP 20195945), the National Committee on Science and Technology (CONACYT; Consejo Nacional de Ciencia y Tecnología: #132415), and National Commission on Protected Natural Areas (CONANP; Comisión Nacional de Áreas Naturales Protegidas: PROMOBI 2011, 2012, 2014, 2015; PROMANP 2016; PROREST 2019). This study was conducted under the following research permits approved by the Secretariat of Environment and Natural Resources (SEMARNAT; Secretaría de Medio Ambiente y Recursos Naturales): 240996-213-03, DOO 750.-4172/97, DOO 750.-4443/98, NUM/SGPA/DGVS 04311, 04160, 05325, 03269, 03018/08, 02012/11, 01847/12, 02209/13 02402/14, 03999/15, 07208/16, 08817/17, 08817/18, and 7395/19. We thank Luis Miguel Burciaga Cifuentes and Armando Martínez Castro for their collaboration in the fieldwork. We also thank Dr. K. Sullivan for translating the Spanish text.

References

Acevedo-Whitehouse K, De la Cueva H, Gulland FM et al (2003) Evidence of *Leptospira interrogans* infection in California sea lion pups from the Gulf of California. J Wildl Dis 39(1):145–151

Aurioles-Gamboa D, Le Boeuf BJ (1991) Effects of the El Niño 1982–83 on California sea lions in Mexico. In: Trillmich F, Ono KA (eds) Pinnipeds and El Niño. Springer, Berlin, pp 112–118

Aurioles-Gamboa D, Sinsel F (1988) Mortality of California sea lion pups at Los Islotes, Baja California Sur, Mexico. J Mammal 69(1):180–183

Aurioles-Gamboa D, García-Rodríguez FJ, Ramírez-Rodríguez M et al (2003) Interacción entre el lobo marino de California y la pesquería artesanal en la Bahía de La Paz, Golfo de California, México. Cienc Mar 29(3):357–370

Aurioles-Gamboa D, Newsome SD, Salazar-Pico S et al (2009) Stable isotope differences between sea lions (*Zalophus*) from the Gulf of California and Galapagos Islands. J Mammal 90(6):1410–1420

Aurioles-Gamboa D, Godínez-Reyes DC, Hernández-Camacho CJ et al (2011) Taller de análisis del estado de la población de lobo marino de California *Zalophus californianus* en México. Final report. CICIMAR, CONANP, INE. La Paz, B.C.S., Mexico. November 25 and 26 2010

Aurioles-Gamboa D, Gallo-Reynoso JP, Hernández-Camacho CJ et al (2012) Monitoreo de la población del lobo marino de California (*Zalophus californianus*) en colonias reproductivas del Golfo de California y estudio de la interacción con las pesquerías artesanales; tres casos de estudio. Final report. PROMOBI. Comisión Nacional de Áreas Naturales Protegidas

Aurioles-Gamboa D, Gallo-Reynoso JP, Hernández-Camacho CJ et al (2013) Monitoreo de la población del lobo marino de California (*Zalophus californianus*); análisis de la variación interanual del nicho isotópico y refuerzo a promotores comunitarios de Bahía Kino (RB Isla San Pedro Mártir) e Isla San Jorge (APFF Islas del Golfo de California). Final report. PROMOBI. Comisión Nacional de Áreas Naturales Protegidas

Aurioles-Gamboa D, Silverberg N, Aguirre-Bahena F (2017) Possible relation between enrichment of δ 15 N in the top predator *Zalophus californianus* and the expansion of the oxygen minimum zone. Mar Biol 164(8):157. https://doi.org/10.1007/s00227-017-3189-7

Avalos-Téllez R, Carrillo-Casas EM, Atilano-López D et al (2016) Pathogenic Leptospira serovars in free-living sea lions in the Gulf of California and along the Baja California coast of Mexico. J Wildl Dis 52(2):199–208. https://doi.org/10.7589/2015-06-133

Barton K, Barton MK (2019) Package 'MuMIn.' R package version 1(6)

Berkson JM, De Master DP (1985) Use of pup counts in indexing population changes in pinnipeds. Can J Fish Aquat Sci 42(5):873–879

Comisión Nacional de Áreas Naturales Protegidas (CONANP) (2000a) Programa de Manejo Complejo Insular del Espíritu Santo. CONANP, México

Comisión Nacional de Áreas Naturales Protegidas (CONANP) (2000b) Programa de Manejo del APFF Islas del Golfo de California. CONANP, México

Comisión Nacional de Áreas Naturales Protegidas (CONANP) (2014) Programa de Manejo del Parque Nacional Zona Marina del Archipiélago de Espíritu Santo. CONANP, México

Dennis B, Munholland PL, Scott JM (1991) Estimation of growth and extinction parameters for endangered species. Ecol Monogr 61(2):115–143

Elorriaga-Verplancken FR (2005) Incidencia y efectos del enmalle sobre el lobo marino de California (*Zalophus californianus californianus*, Lesson 1828) en Los Islotes, B.C.S., México. Master thesis. CICIMAR-IPN. La Paz, B.C.S., México

Elorriaga-Verplancken FR, Sierra-Rodríguez GE, Rosales-Nanduca H et al (2016) Impact of the 2015 El Niño-southern oscillation on the abundance and foraging habits of Guadalupe fur seals and California sea lions from the San Benito Archipelago, Mexico. PLoS One 11(5):1–16. https://doi.org/10.1371/journal.pone.0155034

Gallo-Reynoso JP, Aurioles-Gamboa D, Hernández-Camacho CJ et al (2011) Monitoreo de la población y condición de salud del lobo marino de California (*Zalophus californianus californianus*) en las colonias de reproducción del Golfo de California. Final report. PROMOBI. Comisión Nacional de Áreas Naturales Protegidas

Gallo-Reynoso JP, Aurioles-Gamboa D, Hernández-Camacho CJ et al (2014) Monitoreo de la población del lobo marino de California (*Zalophus californianus*) en las colonias de reproducción del Golfo de California; Análisis de la variación interanual del nicho isotópico fase II y participación de promotores comunitarios de Bahía Kino (RB Isla San Pedro Mártir) y Ejido Campodónico (APFF Isla San Jorge). Final report PROMOBI. Comisión Nacional de Áreas Naturales Protegidas

Gallo-Reynoso JP, Aurioles-Gamboa D, Hernández-Camacho CJ et al (2015) Monitoreo de la población del lobo marino de California (*Zalophus californianus*) en las colonias de reproducción del Golfo de California; islas del Golfo de California en Sonora, Sinaloa y la Reserva de la Biosfera de San Pedro. Final report. PROMOBI. Comisión Nacional de Áreas Naturales Protegidas

Galván-Magaña F, Hoyos-Padilla EM, Navarro-Serment CJ et al (2010) Records of White shark, *Carcharodon carcharias*, in the Gulf of California. Mar Biodivers Rec 3. https://doi.org/10.1017/S1755267210000977

García-Rodriguez FJ (1995) Ecología alimentaria del lobo marino de California, *Zalophus californianus californianus*, en Los Islotes, B.C.S., México. Bachelor tesis. UABCS. La Paz, B.C.S., México

García-Rodríguez FJ, Aurioles-Gamboa D (2004) Spatial and temporal variation in the diet of the California sea lion (*Zalophus californianus*) in the Gulf of California, Mexico. Fish Bull 102(1):47–62

Godínez CR, De Romillo BZ, Aurioles-Gamboa D et al (1999) Antibodies against *Leptospira interrogans* in California sea lion pups from Gulf of California. J Wildl Dis 35(1):108–111

González-Suarez M, Mc Cluney K, Aurioles-Gamboa D et al (2006) Incorporating uncertainty in spatial structure for viability predictions: a case study of California Sea Lions. Anim Conserv 9(2):219–227. https://doi.org/10.1111/j.1469-1795.2006.00022.x

Harcourt R, Aurioles-Gamboa D, Sanchez J (1994) Entanglement of California sea lions at Los Islotes, Baja California Sur, México. Mar Mammal Sci 10(1):122–125

Hernández-Camacho CJ (2001) Tabla de vida del lobo marino de California *Zalophus californianus californianus* en la lobera Los Islotes, BCS, México. Master thesis. CICIMAR-IPN. La Paz, B.C.S., México

Hernández-Camacho CJ, Aurioles-Gamboa D, Gerber LR (2008a) Age-specific birth rates of California sea lions (*Zalophus californianus*) in the Gulf of California, Mexico. Mar Mammal Sci 24(3):664–676. https://doi.org/10.1111/j.1748-7692.2008.00199.x

Hernández-Camacho CJ, Aurioles-Gamboa D, Laake J et al (2008b) Survival rates of the California sea lion, *Zalophus californianus*, in Mexico. J Mammal 89(4):1059–1066. https://doi.org/10.1644/07-MAMM-A-404.1

Hernández-Camacho CJ, Bakker VJ, Aurioles-Gamboa D et al (2015) The use of surrogate data in demographic population viability analysis: a case study of California sea lions. PLoS One 10(9). https://doi.org/10.1371/journal.pone.0139158

Hernández-Camacho CJ, Aurioles-Gamboa D, Valenzuela-Molina MY et al (2016) Monitoreo de las colonias reproductivas del lobo marino de California en las Islas del Golfo de California. Final report. PROMANP. Comisión Nacional de Áreas Naturales Protegidas

Herrera-Cervantes H, Lluch-Cota SE, Lluch-Cota DB et al (2010) ENSO influence on satellite-derived chlorophyll trends in the Gulf of California. Atmosfera 23(3):253–262

IUCN (2012) IUCN Red List categories and criteria: version 3.1, 2nd edn. IUCN, Gland. https://eee.iucnredlist.org/resources/categories-and-criteria

Jackson AL, Inger R, Parnell AC et al (2011) Comparing isotopic niche widths among and within communities: SIBER–Stable Isotope Bayesian Ellipses in R. J Anim Ecol 80(3):595–602. https://doi.org/10.1111/j.1365-2656.2011.01806.x

Jackson AL, Parnell A, Jackson MA (2019) Package 'SIBER': stable isotope bayesian ellipses in R. R package version 2(4)

Jaud T, Dragon AC, Garcia JV et al (2012) Relationship between chlorophyll a concentration, light attenuation and diving depth of the southern elephant seal *Mirounga leonina*. PLoS One 7(10). https://doi.org/10.1371/journal.pone.0047444

Kassambara A, Mundt F (2017) Package 'factoextra'. Extract and visualize the results of multivariate data analyses, 76

Kelly JF (2000) Stable isotopes of carbon and nitrogen in the study of avian and mammalian trophic ecology. Can J Zool 78(1):1–27

Klimley AP, Anderson SD, Pyle P et al (1992) Spatiotemporal patterns of white shark (*Carcharodon carcharias*) predation at the South Farallon Islands, California. Copeia:680–690

Kuhn CE (2006) Measuring at sea feeding to understand the foraging behavior of pinnipeds. Doctoral dissertation, University of California, Santa Cruz, United States

Laake JL, Lowry MS, Delong RL et al (2018) Population growth and status of California sea lions. J Wildl Manag 82(3):583–595. https://doi.org/10.1002/jwmg.21405

Labrada-Martagón V, Aurioles-Gamboa D, Martínez-Díaz SF (2005) Natural and human disturbance in a rookery of the California sea lion (*Zalophus californianus californianus*) in the Gulf of California, Mexico. LAJAM 4(2):175–185. https://doi.org/10.5597/lajam00080

Layman CA, Arrington DA, Montaña CG et al (2007) Can stable isotope ratios provide for community-wide measures of trophic structure? Ecology 88(1):42–48

Long DJ, Hanni KD, Pyle P et al (1998) White shark predation on four pinniped species in Central California Waters: geographic and temporal patterns inferred from Wounded Carcasses. In: Klimley AP, Ainley DG (eds) Great white sharks: the biology of *Carcharodon carcharias*. Academic, San Diego, pp 263–274

Luque SP, Aurioles-Gamboa D (2002) Estimation of body volume and body density in California sea lion pups. J Mar Biol Assoc UK 82(6):1019–1022

Melin SR (2002) The foraging ecology and reproduction of the California sea lion (*Zalophus californianus*). Ph.D. thesis, University of Minnesota, Minneapolis, MN

Morris WF, Doak DF (2003) Quantitative conservation biology: theory and practice of population viability analysis. Sinauer Associates, Sunderland

Nevárez-Martínez MO, Lluch-Belda D, Cisneros-Mata MA et al (2001) Distribution and abundance of the Pacific sardine (*Sardinops sagax*) in the Gulf of California and their relation with the environment. Prog Oceanogr 49(1-4):565–580. https://doi.org/10.1016/S0079-6611(01)00041-6

Newsome SD, Koch PL, Etnier MA et al (2006) Using carbon and nitrogen isotope values to investigate maternal strategies in northeast Pacific otariids. Mar Mammal Sci 22(3):556–572. https://doi.org/10.1111/j.1748-7692.2006.00043.x

Newsome SD, Clementz MT, Koch PL (2010) Using stable isotope biogeochemistry to study marine mammal ecology. Mar Mammal Sci 26(3):509–572. https://doi.org/10.1111/j.1748-7692.2009.00354.x

Porras-Peters H, Aurioles-Gamboa D, Cruz-Escalona VH et al (2008) Trophic level and overlap of sea lions (*Zalophus californianus*) in the Gulf of California, Mexico. Mar Mammal Sci 24(3):554–576. https://doi.org/10.1111/j.1748-7692.2008.00197.x

Rosas-Hernández MP, Aurioles-Gamboa D, Hernández-Camacho CJ (2018) Differences in the Isotopic Niche and Trophic position of Female California Sea Lions (*Zalophus californianus*) in distinct oceanographic conditions. Aquat Mamm 44(4):374–388. https://doi.org/10.1578/AM.44.4.2018.374

Rosas-Hernández MP, Aurioles-Gamboa D, Hernández-Camacho CJ (2019) Specialized foraging habits of adult female California sea lions *Zalophus californianus*. Mar Mammal Sci 35(4):1463–1488. https://doi.org/10.1111/mms.12603

Rubio-Castañeda N (2018) Capacidad de carga turística en la lobera de Los Islotes del Parque Nacional Archipiélago Espíritu Santo, BCS, México. Master thesis. CICIMAR-IPN. La Paz, B.C.S., México

Sala E, Aburto-Oropeza O, Reza M et al (2004) Fishing down coastal food webs in the Gulf of California. Fisheries 29(3):19–25. https://doi.org/10.1577/1548-8446(2004)29[19:FDCFWI]2.0.CO;2

Soto KH, Trites AW, Arias-Schreiber M (2004) The effects of prey availability on pup mortality and the timing of birth of South American sea lions (*Otaria flavescens*) in Peru. J Zool 264(4):419–428. https://doi.org/10.1017/S0952836904005965

Stubben CJ, Milligan BG (2007) Estimating and analyzing demographic models using the popbio package in R. J Stat Softw 22(11):1–23

Szteren D, Aurioles-Gamboa D, Gerber LR (2006) Population status and trends of the California sea lion (*Zalophus californianus californianus*) in the Gulf of California, Mexico. In: Trites AW, Atkinson SK, DeMaster DP et al (eds) Sea lions of the world. Alaska Sea Grant College Program, Fairbanks, pp 369–384

Team RC (2019) A language and environment for statistical computing. R Foundation for Statistical Computing; 2012, Vienna, Austria. https://www.R-project.org

Trites AW, Donnelly CP (2003) The decline of Steller sea lions *Eumetopias jubatus* in Alaska: a review of the nutritional stress hypothesis. Mammal Rev 33(1):3–28

Underwood JG, Hernández-Camacho CJ, Aurioles-Gamboa D et al (2008) Estimating sustainable bycatch rates for California sea lion populations in the Gulf of California. Conserv Biol 22(3):701–710. https://doi.org/10.1111/j.1523-1739.2008.00919.x

Villegas-Amtmann S (2009) Seasonal variation in the physiological ecology of the Galapagos (*Zalophus wollebaeki*) and California (*Zalophus californianus*) sea lions. Ph.D. thesis, University of California, Santa Cruz

Villegas-Amtmann S, Atkinson S, Paras-Garcia A et al (2012) Seasonal variation in blood and muscle oxygen stores attributed to diving behavior, environmental temperature and pregnancy in a marine predator, the California sea lion. Comp Biochem Physiol A Mol Integr Physiol 162(4):413–420. https://doi.org/10.1016/j.cbpa.2012.04.019

Zavala-González A, Mellink E (1997) Entanglement of California sea lions, *Zalophus californianus californianus*, in fishing gear in the central-northern part of the Gulf of California, Mexico. Fish Bull 95(1):180–184

Zavaleta-Lizárraga L (2003) Variaciones geográficas en morfometría craneal en machos adultos de lobo marino de california (*Zalophus californianus*) en México. Master thesis, CICIMAR-IPN. La Paz, B.C.S., México

Zuur A, Ieno EN, Walker N et al (2009) Mixed effects models and extensions in ecology with R. Springer, New York

Chapter 34
Analysis of a Socio-ecological System: Coastal Zone of the Yaqui Indigenous Community (NW México)

José Alfredo Arreola-Lizárraga, Gustavo Padilla-Arredondo, Luz María Cruz-García, Thelma Michelle Ruiz-Ruiz, and Elia Inés Polanco-Mizquez

Abstract In Northwestern Mexico, in the lower part of the Yaqui River basin that forms part of the territory of the Yaqui indigenous community, constitutes a coastal zone designated as a Ramsar site: Las Guásimas-Lobos, characterized by a deltaic plain with coastal lagoons bordered by mangroves. This chapter presents an analysis of the natural capital and ecosystem services of this coastal zone, as well as the social welfare of the Yaqui community, explaining the prevailing conditions, drivers of change, and challenges in this unique socio-ecological system.

Keywords Yaquis · Coastal ecosystem · Natural capital · Welfare

34.1 Introduction

Since its emergence in the 1970s, the term socio-ecological has become a dominant way of conceptualizing nature-society interactions (Stojanovic et al. 2016).

An ecosystem is any area of nature that includes living organisms and nonliving elements that interact through the exchange of materials and energy (Odum 1959). A socio-ecological system is defined as a spatially delimited region that contains an ecosystem and a social system that interact with each other (Tett et al. 2013). Both the social system and the ecological system are equally important and function as

---

J. A. Arreola-Lizárraga (✉) · G. Padilla-Arredondo · T. M. Ruiz-Ruiz
Centro de Investigaciones Biológicas del Noroeste SC, Guaymas, Sonora, México
e-mail: aarreola04@cibnor.mx; gpadilla04@cibnor.mx; tmruiz@cibnor.mx

L. M. Cruz-García
Universidad Autónoma de Baja California Sur, La Paz, BCS, México
e-mail: lmcruz@uabcs.mx

E. I. Polanco-Mizquez
Centro Intercultural de los Desiertos y Océanos, Pto. Peñasco, Sonora, México
e-mail: elia@cedo.org

A. Ortega-Rubio (ed.), *Socio-ecological Studies in Natural Protected Areas*,
https://doi.org/10.1007/978-3-030-47264-1_34

coupled, interdependent, and coevolutive systems, so that human actions affect ecological systems, ecological factors affect human well-being, and humans in turn respond to these factors (Berkes 2015).

The determination of the elements of socio-ecological systems can be based on recognizing four closely related components: natural capital, ecosystem services, human well-being, and drivers of change (Guerry et al. 2015; Brattland et al. 2019). Natural capital is the stock that generates a flow of useful natural goods and services or natural income over time. Ecosystem services are the benefits that society obtains directly or indirectly from the functions of ecosystems (Costanza et al. 1997; Costanza et al. 2014). Human well-being is based on basic human needs and focused on intangible values that give meaning to life, such as good social relations and well-preserved ecosystems (Aguado et al. 2012). The drivers of natural change and/or human activities affect socio-ecological systems in a complex and unpredictable way (Levin 1999).

In socio-ecological systems of the coastal zone, Tett et al. (2013) provided evidence that management for its sustainability can be generalized in conflict resolution in the biophysical scenario, need for human well-being, and establishment of social norms and institutional rules.

On the east coast of the Gulf of California (NW Mexico), in the lower part of the Yaqui River basin, a coastal area characterized by a deltaic plain with coastal lagoons is bordered by mangrove vegetation, which is part of the territory of the Yaqui indigenous community (Fig. 34.1). This coastal area and the Yaqui community are conceived as a socio-ecological system. In particular, in the delta plain, the agriculture of grains such as wheat, corn, and safflower is developed, as well as of vegetables, which are sown in an area of ~20,000 hectares, and in the coastal lagoons, artisanal fishing is developed where Yaquis fishermen capture shrimp, crab, and various species of fish and also develop small-scale oyster culture (Arreola-Lizárraga et al. 2019).

The Yaquis are the Mexican indigenous people who have been dispersed on more occasions and, paradoxically, are those who currently enjoy greater recognition of their autonomy and territoriality (Lerma-Rodríguez 2011). The Yaquis recognize the natural world, where animals and plants meet; however, they conceive a strong relationship between the people and the natural world, since the beings that inhabit each domain enter the space of the other and thus establish complex links between both groups and spaces. By establishing these close ties, the natural world becomes the natural sacred universe of the Yaquis (Moctezuma-Zamarrón 2007).

This chapter presents an analysis of the coastal area of the Yaqui territory, as well as a synthesis of the cultural, demographic, and social welfare aspects of the Yaquis, explaining the dynamics of interaction, the prevailing conditions, and the drivers of change in this unique socio-ecological system.

Fig. 34.1 Localization of the Yaqui territory showing traditional towns, fishing villages, and Ramsar site Guásimas-Lobos

34.2 Land and Coastal Zone Yaqui

34.2.1 Characteristics of the Yaqui Territory

The Yaqui territory has 3 zones differentiated by their particular geophysical characteristics: a mountain area, the Sierra del Bacatete; the valley where the irrigated lands are located, with 8 main towns and more than 50 rancherías; and a coastal zone where the fishing villages Las Guásimas and Bahía de Lobos are located (Olavarría 2000).

The Yaqui territory has an extension of 6054 km^2 and is located in the south of the state of Sonora (Fig. 34.1), between the Yaqui River basins (72,000 km^2) and that of the Mátape River (8004 km^2). Basically, this territory is characterized by the Sierras and Sonoran plain subprovince, which covers 62% of the territory, and by the coastal plain and delta of Sonora and Sinaloa subprovince, which covers 36%. The first consists of orogenic blocks with mountainous fragments at a lower altimetric position, which are presented as plateaus and hills composed of basaltic lavas and granite outcrops; the Sierra Bacatete constitutes the main tectonic block, and on the sides are the alluvial valleys with well-developed fertile soils for agricultural use. In the second subprovince mentioned, the dominant topological form is a wide coastal plain with deltaic fans, coastal lagoons with mangroves, and floodplains with salt flats. In addition, there are sandy beaches, stabilized coastal dunes on the coast, and small coastal spits. A minimum portion of the Sierras and Valles del Norte subprovince (2%) is located northeast of the territory (Padilla-Arredondo and Pedrin-Aviles 2019). Thus, the dominant geomorphological forms can be grouped in valleys, mountains, and terraces conditioned by the tectonic of the extensional regime that occurred in the area since the Oligocene. The altitude range in this territory varies from 1013 meters above sea level to the coast, with an average close to 145 meters above sea level, showing the dominance of the valleys and plains in front of the elevated areas. The slopes are relatively low (5.4 ± 9.4%) and are higher in the highlands and smaller in the coastal plains and valleys, so the runoff decreases in the flat and low areas of the basin, where water mainly infiltrates (Garatuza-Payan 2019).

The Yaqui territory has three climate units: very dry warm, dry hot, and dry semi-warm. The very dry climate has an influence from the coastal area toward the center of the territory. The average annual rainfall varies from 239 to 458 mm year^{-1}. The coastal zone receives less rainfall, 200–300 mm year^{-1}, than the mountainous area where rainfall is 300–400 mm year^{-1}. The average air temperature throughout the year is 24 °C; however, summer temperatures average 38 °C and in winter 8 °C. The maximum temperature recorded is 50 °C, while the historical minimum is −7 °C. The warmest month historically is the month of July, and the coldest month is January (Rojas-Robles and Robles-Morua 2019).

In the coastal area of the Yaqui territory, the dynamics of coastal processes is highly dependent on tide and wind, while water density plays an important role in mesoscale or geostrophic processes. The bathymetric contours of this coastal zone refract the incident waves generating a particular behavior in each of the environments: coast, channels of entrance to lagoons, and lagoon bodies. The tidal range is 1 m; the elevation of the average sea level is greater in summer and lower in winter. The influence of the seasonal wind pattern is of greater magnitude in winter (Burrola-Sánchez and Rosales-Grano 2019).

34.2.2 Water Use and Water Quality in the Yaqui Territory

The Yaquis have historically used river water for their cultural and economic survival, with access to water being the main problem they have faced (Padilla-Ramos and Moctezuma-Zamarrón 2017). This has worsened in recent decades because the flow of water from rivers and streams has decreased, for different reasons, and is expected to decrease further due to the effects of climate change (Garatuza-Payan 2019).

In the lower Yaqui River basin, agricultural activities and human settlements have been developed with a fundamental dependence on the Yaqui River water captured at the Álvaro Obregón dam (El Oviáchic). The lower Yaqui River basin can be divided into two particular areas: (1) the Yaqui Valley which represents the 040 irrigation district with ~220,000 ha and ~ 350,000 hab, and receives 92% of the dam's water and (2) the deltaic plain within the boundaries of the Yaqui territory where the 018 irrigation district is located with ~20,000 ha, ~ 25,000 hab. The Yaqui community receives 8% of the dam's water (Luna-Escalante 2007).

In the Yaqui territory, the diversion channel Chiculi receives water directly from the Álvaro Obregón dam and is subsequently distributed in irrigation district 018; finally, the agricultural return waters from the irrigation are conducted through a collector drain whose final destination is the Algodones lagoon. Lobos lagoon receives agricultural and urban wastewater through collector No. 2 from the Yaqui Valley (Ruiz-Ruiz et al. 2017). The Yaqui Valley has about 370,000 inhabitants with an economy based mainly on agricultural, shrimp, livestock, and fishing activities, as well as the support sectors for its development, and occupies 92% of the water of the Álvaro Obregón dam. There is a notable difference in the pressure of land use by urban and agricultural development, greater in the Yaqui Valley than in the part corresponding to the Yaqui territory (Arreola-Lizárraga et al. 2019).

A synthesis of the use of water and the quality of surface, groundwater, and coastal lagoons in the Yaqui territory is presented using a conceptual model (Fig. 34.2).

The surface water of the Yaqui territory has as its source the Yaqui River, and prior to its distribution and use by the Yaqui community, it is stored in the Álvaro

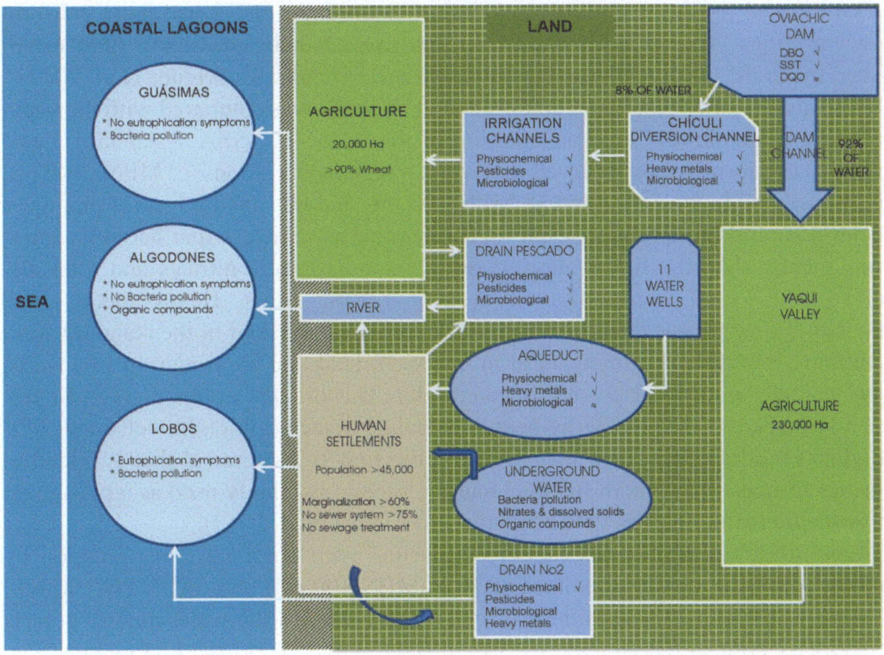

Fig. 34.2 Conceptual model of the water use and water quality in the Yaqui territory. (Sources: Luna-Escalante 2007; Arreola-Lizárraga et al. 2014; Méndez-Rodríguez et al. 2019; Reynaga-Franco et al. 2018)

Obregón dam. The water quality of this dam through an annual cycle (2012–2014) was good based on the indicators of total biochemical oxygen demand and suspended solids. The water quality indicated by the chemical oxygen demand was good for most of the year and reached pollution thresholds at the beginning of spring. The water from the diversion channel Chiculi and three irrigation channels, in terms of environmental, physical-chemical, heavy metals, microbiological, and pesticide parameters, were within the limits established by the ecological criteria for agricultural irrigation. The water at two sites of the Yaqui River evaluated in terms of environmental, physical-chemical, heavy metals, microbiological, and pesticide parameters was within the limits established by the ecological criteria for protection of aquatic life. The water in three sites of the Yaqui-Guaymas aqueduct, in terms of environmental, physical-chemical, heavy metal, and pesticide parameters, was within the limits established by both the ecological criteria for source of drinking water supply and Standard 127-SSA-1-1994 for human use and consumption. However, total and fecal coliform bacteria were recorded with values of 1.8–11 MPN/100 mL, which according to the standard should be absent (Méndez-Rodríguez et al. 2019).

Water quality in the agricultural collector drain No. 2 of the Yaqui Valley (discharge in the Lobos lagoon) and the El Pescado drain in the Colonias Yaquis irrigation district (discharge in the Algodones lagoon) based on environmental, physical-chemical parameters, and heavy metals was within the limits of NOM-001-ECOL. Based on the microbiological parameters, drain No. 2 had fecal coliform concentrations (16,000 MPN / 100 ml) that exceeded the maximum limit established by NOM-001-ECOL (1000 MPN/100 ml) (Méndez-Rodríguez et al. 2019).

The quality of groundwater determined in 16 wells was compared with the water criteria for human use and consumption of the NOM-127-SSA-1 standard. Fecal and total coliforms were observed in 8 wells with concentrations <2 MPN / 100 mL and 8 wells with concentrations>8 MPN/100 mL and the standard states that these bacteria should be absent. The physical-chemical parameters that had concentrations greater than the limits allowed by the standard were nitrates and total dissolved solids and were observed in 50% of the wells. Heavy metals with concentrations greater than the permissible limits established in the standard were cadmium, mercury, and manganese in one well, lead and aluminum in two wells, iron in three wells, and arsenic in five wells. The organic compounds most frequently detected were bis(2ethylhexyl) phthalate detected in eight wells and di(2-ethylhexyl) adipate detected in five wells; the first is a chemical that is added to plastics to make them more flexible, and the second is mainly used as a plasticizer in synthetic resins such as PVC (Méndez-Rodríguez et al. 2019).

Las Guásimas and Lobos lagoons had sites with bacterial contamination by enterococci with maximum values of 1011 MPN/100 mL and 691 MPN/100 mL, respectively. These concentrations of enterococci exceed the water quality criteria for primary contact recreational use established by the World Health Organization. Based on the trophic status index (TRIX), the Las Guásimas and Algodones lagoons showed no symptoms of eutrophication, and the Lobos lagoon showed symptoms of eutrophication (Arreola-Lizárraga et al. 2014; Reynaga-Franco et al. 2018).

34.2.3 The Coastal Area of the Territory and Its Importance of Conservation

The coastal area of the Yaqui territory has ~96 km of coastline and is characterized by a wide coastal plain with delta fans, mangrove coastal lagoons, floodplains with salt flats, sandy beaches, stabilized coastal dunes, and coastal cords (Mendoza-Cantú 1997). The coastal lagoons are associated with the deltaic prism of the Yaqui River and originated by differential terrigenous sedimentation and by intradeltaic and marginal depression (Lankford 1977). The most important lagoons due to their size are Lobos (102 km^2), Las Guásimas (37 km^2), and Algodones (4 km^2) (Arreola-Lizárraga et al. 2019). According to the criteria of Kjerfve and Magill (1989), they qualify as "restricted"-type lagoons, considering that they are permanently communicated with the sea, have a well-defined tidal circulation, influenced by seasonal winds, and are well mixed vertically. These bodies of water have flushing times that vary from 2 to 15 days depending on the lagoon system and the season of the year (Padilla-Arredondo et al. 2002; Valenzuela-Siu et al. 2007; Arreola-Lizárraga et al. 2015; Medina-Galván et al. 2018). Water temperature variations have minimum values of 13–14 °C and maximum values of 32–33 °C throughout the year. They are euhaline lagoons with annual salinity variation of 36–42 ups explained by the precipitation-evaporation balance, the good exchange of water with the ocean, and the low income of freshwater due to rains (Valenzuela-Siu et al. 2007; Ruiz-Ruiz et al. 2017). The dominant vegetation adjacent to the lagoon systems is halophyte: *Salicornia* spp. and *Batis* spp. Mangroves *Rhizophora mangle*, *Avicennia germinans*, *Laguncularia racemosa*, and *Conocarpus erectus* (INEGI 2010b) are under special protection by NOM-059-SEMARNAT-2010 (DOF 2010).

There are several elements that highlight the importance of conservation of the coastal area of the Yaqui territory, formed by the Guásimas-Lobos lagoon complex: (1) the Lobos barrier island that separates the lagoon of the same name from the adjacent sea is part of the area of protection of flora and fauna islands of the Gulf of California, Sonora; (2) the coastal zone is considered priority marine region and priority hydrological region (SEMARNAT 2007); (3) the Guásimas-Lobos lagoon complex is a Ramsar site (Ramsar 2008); (4) it is also a priority wetland for shorebirds in Mexico (SEMARNAT 2008) and priority wetland for migratory waterfowl in Mexico (Carrera and de la Fuente 2003); and (5) Bahía de Lobos was designated as site of regional importance and is part of the Western Hemisphere Shorebird Reserve Network (WHSRN 2018)). Bird census records corroborated its importance as a wintering site on the Pacific migratory route with 20,000 shorebirds, highlighting the presence of species such as the snowy plover (*Charadrius nivosus*), the western sandpiper (*Calidris mauri*), the red knot (*Calidris canutus roselaari*), and the marbled godwit (*Limosa fedoa*), among others. Likewise, due to its ecological and social importance, Bahía de Lobos was considered a candidate for a protected natural area in the category of area of protection of flora and fauna (Arreola-Lizárraga 1994).

34.3 Yaqui Community

34.3.1 Cultural Aspects

The history of the Yaquis (self-appointed *Yoremes*) is characterized by its struggle to conserve its territory. Since the arrival of the Spaniards in the Yaqui River, its inhabitants defended the right to conserve their living space. When the first Spanish expedition arrived in 1533 on the banks of the mighty Yaqui River, a Yaqui chief, accompanied by hundreds of warriors, drew a line on the ground and let Europeans know that if they crossed it they would be repelled militarily. As soon as they crossed the line, a battle was initiated in which the Spaniards were defeated. It would take a hundred years to try to penetrate the Yaqui territory again, but no longer with weapons but through Jesuit priests (Moctezuma-Zamarrón 2007).

History has shown the roots of the Yaquis to its territory and has exposed different forms of sociopolitical organization, strategic alliances, and war processes to defend it (Lerma-Rodríguez 2014). The Yaquis have proved resistant to the powerful interests of capital and state violence. For example, the defense of the Yaqui River has been the subject of many conflicts. Significant dimensions of the Yaqui culture that relate to water have been compromised, but their struggle for river water as an integral part of their territory, as well as their historical organization, has served the Yaquis to keep their farmland irrigated and keep water for consumption (Padilla-Ramos and Moctezuma-Zamarrón 2017).

The Yaqui tribe is organized around the eight traditional villages: Belén, Huirivis, Rahum, Pótam, Vícam Pueblo, Tórim, Loma de Bácum, and Loma de Guamúchil. Each represents a political, military, religious, and ritual unity. Vícam town is the head of the tribe. There are five groups that make up the political-religious organization, composed first of all by a group of civil authorities and complemented by the representation of the council of elders, through the Pueblo Mayor. The Yaqui religion is presented, as a result of the conquest, as a complex that juxtaposes native beliefs and practices with Catholics, without contradiction between them or subordination of one over another (Moctezuma-Zamarrón 2007).

Due to the damming of the Yaqui River, the aspect of the riverbed where most of the villages are located is exhibited as a long and deep strip on which a rugged pasture lives; only the irrigation canals temporarily enjoy the liquid, around which huge poplar trees grow. On the other hand, although the Yaqui community is not based on a culture settling in the sea, a part of it unfolds in the coastal strip of the territory. Although the sea is close to the villages, most of the Yaqui population does not attend frequently. They tend to remain in their localities, and only those who work in fishing come, mainly in the period of shrimp capture (Lerma-Rodríguez 2011).

The feast of the Virgen del Carmen in the fishing village of La Guásimas on the banks of the Yaqui territory and the feast of San Juan Bautista, in the main headwaters (in the center of the eight towns), show a relationship that can only be understood under a *Bawe* interpretation: Yaqui water. With these two ritual processes, a circle of space closure is created and at the same time an intersection between the different places that make up the territory and the significant elements of the Yaqui culture (Lerma-Rodríguez 2011).

34.3.2 Demography and Social Welfare

The Sonoran Yaqui population (~ 40,000 members) lives mainly in what is known as the Yaqui traditional territory. Approximately 8000 Yaquis also live in Arizona, USA. Between both populations there are constant exchanges of cultural, social, and material relations (Lerma-Rodríguez 2014).

Based on the last population and housing census, INEGI 2010, the Yaqui community consists of several settlements distributed in five municipalities in southern Sonora: Guaymas, Empalme, San Ignacio Río Muerto, Bácum, and Cajeme. The Yaqui population is 41,306 inhabitants. Of the total of localities, only 25 have more than 100 inhabitants (Fig. 34.3).

The Yaqui communities had a population growth of 0.9% in the period 2000–2010, being greater than one in most of the largest towns (Table 34.1).

The Yaqui community have high and very high marginalization ~94% (CONAPO 2010), and in most cases, the level of marginalization has increased over time (Table 34.2).

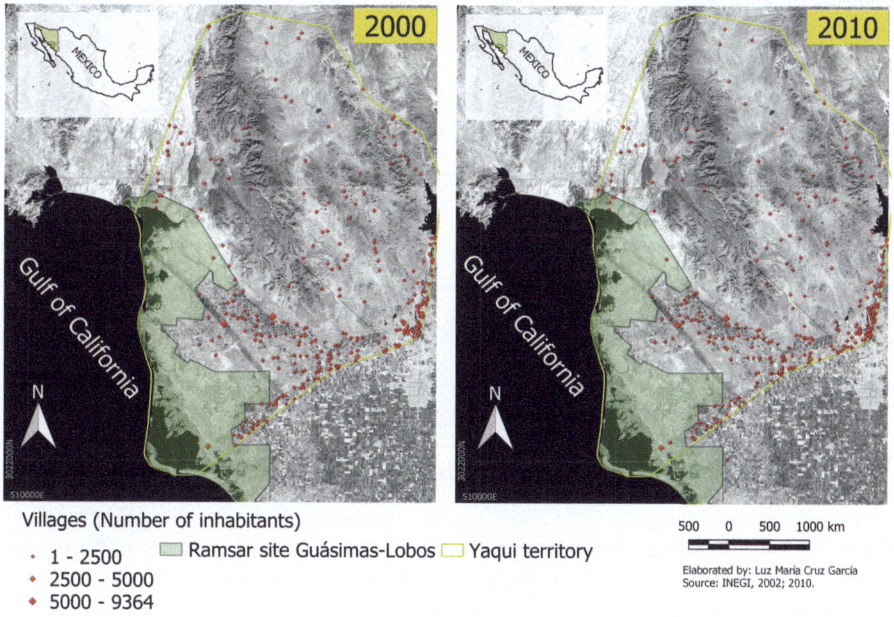

Fig. 34.3 A comparative census between 2000 and 2010 of total population in localities of the Yaqui territory. (Sources: INEGI 2000, 2010)

Table 34.1 Population growth in Yaqui localities and growth rate from 2000 to 2010. Sources: INEGI (2000, 2005, 2010)

Locality	Census 2000	Census 2005	Census 2010	Growth rate 2000–2010
Agua Caliente	63	–	15	−14.4
Babojori	144	145	169	1.6
Bahía de Lobos	2768	2879	2867	0.4
Baugo (Guásimas)	186	274	285	4.3
Casa Azul	101	112	131	2.6
Casas Blancas	364	454	366	0.1
Chumampaco	213	187	177	−1.9
Cócorit	8198	7953	7752	−0.6
Compuertas	214	233	245	1.4
Coracepe	74	84	73	−0.1
El Castillo	321	301	414	2.5
El Pescado	57	59	43	−2.8
Estación Oroz (Oroz)	316	388	468	3.9
Guásimas (de Belem)	1390	1629	1804	2.6
Guasimitas	231	209	190	−2.0
Huiribis	254	293	342	3.0
Lencho	173	179	170	−0.2
Loma de Bácum	1251	1300	1503	1.8
Loma de Guamuchil	1113	1124	1135	0.2
Los Limones	38	42	44	1.5
Pimienta	41	90	65	4.6
Pitahaya (Belem)	203	263	285	3.4
Pótam	5524	5782	6417	1.5
Rahum	224	280	272	1.9
San José de Bácum	3884	3882	4318	1.1
Tajimaroa	274	296	359	2.7
Tetabiate	414	450	492	1.7
Tórim	632	755	771	2.0
Urbalejo	19	7	11	−5.5
Vícam (Switch)	8332	8578	9364	1.2
Vícam Pueblo	603	709	759	2.3
Total	37,619	38,937	41,306	0.9

34.3.3 Productive Activities in the Coastal Zone

The main activities that sustain the economy of the Yaqui tribe are agriculture and fishing, but livestock, oyster culture, and logging such as mezquite are also developed.

Table 34.2 Index and degree of marginalization of Yaqui localities in 2005 and 2010

Locality	2005		2010	
	Marginalization index	Marginalization degree	Marginalization index	Marginalization degree
Agua Caliente	1.003	Very high	1.171	Very high
Babojori	0.744	Very high	0.656	High
Bahía de Lobos	−0.779	Medium	−0.612	High
Baugo (Guásimas)	−0.007	High	0.507	High
Casa Azul	0.463	High	0.726	Very high
Casas Blancas	0.778	Very high	0.809	Very high
Chumampaco	−0.737	Medium	−0.472	High
Cócorit	−1.380	Very low	−1.293	Low
Compuertas	0.451	High	0.741	Very high
Coracepe	0.960	Very high	1.513	Very high
El Castillo	−0.712	Medium	−0.571	High
El Pescado	0.86894	Very high	0.682	High
Estación Oroz (Oroz)	−0.026	High	0.332	High
Guásimas (de Belem)	−0.627	High	−0.162	High
Guasimitas	1.298	Very high	1.654	Very high
Huiribis	−0.103	High	0.054	High
Lencho	−0.219	High	0.173	High
Loma de Bácum	0.105	High	0.149	High
Loma de Guamuchil	−0.568	High	−0.449	High
Los Limones	1.102	Very high	1.290	Very high
Pimienta	1.805	Very high	1.909	Very high
Pitahaya (Belem)	0.441	High	0.648	High
Pótam	−0.289	High	−0.183	High
Rahum	0.188	High	0.183	High
San José de Bácum	−1.429	Very low	−1.281	Low
Tajimaroa	−0.612	High	−0.587	High
Tetabiate	−0.235	High	−0.419	High
Tórim	−0.108	High	−0.061	High
Urbalejo	−0.064	High	0.153	High
Vícam (Switch)	−0.798	High	−0.564	High
Vícam pueblo	1.030	Very high	0.811	Very high

Sources: CONAPO (2005, 2010)

34.3.4 Farming

The Yaqui tribe has practiced agriculture since before the arrival of the Spaniards. Currently, wheat, corn, sesame, and soybeans are the main crops farmed in the region, as well as beans, cotton, chickpea, flaxseed, safflower, sorghum, and alfalfa.

The cultivation land of the Yaqui tribe includes irrigation district 018 "Colonias Yaquis." Currently, the planting area of the Yaquis is approximately 20,000 ha, increasing the rent of land to individuals to about 90%, with problems in the efficient use and management of water, and increasing the monoculture of wheat that in 2015–2016 occupied almost 18,500 ha (Minjares-Lugo and Félix-Sánchez 2019).

34.3.5 Fishing

For the Yaqui community, fishing is a very important activity, so much so that the increase in fishing is the origin of the fishing villages of Las Guásimas and Bahía de Lobos.

The coastline of the Yaqui territory is made up of coastal lagoons, which are very productive and are breeding and development sites of commercially important marine species such as blue shrimp, brown shrimp, crabs, mullet fish, and jellyfish, which represent the main species exploited by the Yaquis.

Fishing by the Yaqui community is well organized; in 1958 the Yaquis Communities Fishing Production Cooperative was formed in the towns of Las Guásimas and Lobos, which was favored by a presidential resolution (DOF12 / 18/73) in that fishing exclusivity occurs in the coastal waters of its territory, and in this way the coastal water bodies of Las Guásimas, Las Cruces, Los Algodones, Camapochi, El Siuti, La Luna, and Lobos become exclusive to the Yaqui ethnic group, thus forming an area of 30,000 ha where they conduct their artisanal fishing (Figueroa 1994).

The main species that are fished are the blue shrimp (*Litopenaeus stylirostris*), the cannonball jellyfish (*Stomolophus meleagris*), and the crab (*Callinectes arcuatus* and *C. bellicosus*). The most important volumes of production (catches) of jellyfish occur in the vicinity of the Las Guásimas lagoon. Shrimp and crab have a greater fishing area, within the lagoon systems and on the coastal front of the Yaqui territory. The main fishing areas are Las Guásimas, Los Algodones, and Bahía de Lobos (Alcántara-Razo et al. 2019).

The towns of Las Guásimas and Lobos depend directly on fishing, with shrimp fishing being the most important due to the economic spill it represents, followed by the cannonball jellyfish due to the high fishing volumes and finally the crab. The Yaquis fishermans fish the three species in an alternate manner; the shrimp catch is from September to December, the jellyfish between March and May, and the crab from July to December. Shrimp and crab fishing production activates the economy of the towns of Las Guásimas and Lobos since their capture and commercialization in the main cities such as Guaymas, Empalme, and Ciudad Obregón (Alcántara-Razo et al. 2019).

34.3.6 Aquaculture

The shrimp farming and oyster culture potential for the Yaqui region is technically feasible; however the lack of economic, technical, and administrative capital has not allowed its development.

Historically, Yaqui community have participated in aquaculture activities. For example, in Las Guásimas, one of the first semi-intensive farming shrimp of Sonora (Ramos, 1989) was built, and its operation during 1993 demonstrated its technical and economic feasibility in at least two crop cycles (ITSON 1994). Technical feasibility of shrimp culture in floating cages was also demonstrated (Casillas-Hernández and Ibarra-Gámez 1996). However, this did not lead to the implementation of this economic activity by the community. Casillas-Hernández et al. (2019) consider that the reactivation of new shrimp farming projects for the Yaqui community should be raised taking into account previous experiences. A social enterprise scheme can be productive and successful as long as there is economic and technical capital, in addition to the cohesion and discipline of its members.

The oyster culture is another feasible activity to develop in the Yaqui territory, especially in the lagoons of Las Guásimas, Algodones, and Lobos, where geographical, water, and environmental characteristics suggest appropriate conditions for the cultivation of bivalves, in particular oysters *C. corteziensis* and *C. gigas*. Experiments in Las Guásimas using oyster boxes for *C. corteziensis* and *C. gigas* showed good growth (Chávez-Villalba et al. 2005, 2008; Castillo-Durán et al. 2010). In both Las Guasimas and Lobos, crops have been carried out with *C. gigas* cultivation boxes at the family level. Oyster cultivation could help the Yaqui community prosper with an increase in their income, and their consumption also would possibly improve their nutrition (Chávez-Villalba 2019).

34.4 Analysis

The conceptual framework for analyzing socio-ecological systems (SES) proposed by Ostrom (2009) exhibits the relationships between four first-level central subsystems of an SES that affect each other, as well as linked social, economic, and political environments and ecosystems. With this approach, the socio-ecological system of the coastal area of Yaqui territory is conceived, and the four subsystems correspond to the following:

1. Resource system corresponds to the Ramsar Guásimas-Lobos site, it represents the coastal area of the Yaqui territory, and it has a specific delimitation, where the two fishing villages of the Yaqui community are included: Las Guásimas and Bahía de Lobos.
2. Resource units correspond to mangrove, water body, sandy beaches, as well as species that constitute fishery resources and are of interest for conservation.

3. Governance system corresponds to regulations on issues of conservation of environments and species, as well as management of fishery and aquaculture resources, the Yaquis sociopolitical organization, and the set of government institutions with interference in issues of social and cultural development, conservation, and use of resources.
4. Users, the Yaquis, who are the exclusive usufructuaries of their territory and resources.

The drivers of change of this socio-ecological system correspond to (1) pollution by nitrogen and phosphorus and other substances contributed by wastewater, (2) the exploitation of fishery resources, and (3) climate change. The human activities that take place adjacent to the Guasimas-Lobos lagoon complex, but outside the Yaqui territory, correspond to irrigation agriculture, extensive livestock, shrimp farming, and salt extraction. The activities that take place inside the coastal lagoons correspond to small-scale fishing and oyster farming.

Based on the existing information, the effects of these drivers of change are summarized below for each SES subsystem:

Resource system. The Ramsar Guasimas-Lobos site exhibits a good environmental condition with a stable trend. It is notable that within Yaqui territory the pressure for land use is less than ~90% than observed outside the territory. The shrimp farming development (1000 ha) adjacent to coastal lagoons, but outside the Yaqui territory, has little influence on water bodies (Arreola-Lizárraga et al. 2019). However, the influence of human activities on the quality and flows of water in the coastal zone is manifested with changes in water quality by the contribution of nutrients from wastewater in the Lobos lagoon, which by nutrient enrichment exhibits symptoms of eutrophication that are expressed through increases in macroalgae biomass (Partida-Flores et al. 2019), presence of toxic or harmful phytoplankton species and microalgae blooms (Ruiz-Ruiz et al. 2017), as well as accumulation of some trace metals in sediments (Vargas-González et al. 2017). Considering that the discharge is in a specific area of the Lobos lagoon, the whole of the lagoon complex generally qualifies with good environmental condition (Arreola-Lizárraga et al. 2019). However, it should be considered that in this region the effects of climate change are mainly associated with the increase in sea surface temperature. In particular, it is estimated that by 2050 the average sea surface temperature will increase 0.63 °C, and this implies a potential imbalance in the composition and ecological function of ecosystems through gradual changes in species abundances, and therefore ecological monitoring will be important (Ayala-Bocos et al. 2015).

Resource units. Habitats such as mangroves, coastal dunes, and sandy beaches have a good state of conservation because the influence of agricultural development (20,000 ha) of the coastal plain of the Yaqui territory is relatively low and without direct influence on these environments. In particular, the mangroves of the Lobos lagoon maintain their development with an increase in coverage from 5782 ha in 1973 to 5906 ha in 2010 (Rodríguez-Zuñiga et al. 2013).

The Guásimas-Lobos lagoon complex is characterized by being a habitat of various species, some of which constitute fishery resources such as blue shrimp

(*Litopenaeus stylirostris*) and brown shrimp (*Farfantepenaeus californiensis*) (Arreola-Lizárraga et al. 2004) and blue crab (*Callinectes arcuatus*) and warrior crab (*Callinectes bellicosus*) (Arreola-Lizárraga et al. 2003). Inside the lagoons, fishing is an activity predominantly focused on the shrimp, crab, and cannonball jellyfish (*Stomolophus meleagris*) resources. There is evidence that environmental variability is more important than fishing pressure in the behavior of catches of these three resources (Alcántara-Razo et al. 2019), and therefore the monitoring of the behavior of the most important fishery resources of the lagoon complex must be considered. The anthropic impact of wastewater has an area of influence that represents <5% of the total area of the lagoon complex that requires restoration measures, but in most cases, environmental conditions conducive to the development of species that constitute fishery and aquaculture resources prevail.

In the winters of 2013 and 2014 (Carmona and Danemann 2013, 2014) and in winter-spring of 2017, 72 species of water birds were recorded together (Carmona and Danemann 2017). The ducks with 10 species, seagulls and the like with 11 species, pelicans and the like with 5 species, and herons and similar with 11 species stood out for their abundance. These 37 species accumulated more than 95% of total non-shorebird records. This evidence is an indicator of the good state of biodiversity in the lagoon complex (Arreola-Lizárraga et al. 2019).

Governance system. Five key elements correspond to the following: (1) it is a Ramsar site; (2) it is a territory awarded by presidential decree to the Yaquis, and they are the exclusive usufructuary of the resources; (3) for the most important fishery resources, there are management programs that apply regionally, shrimp, crab, and jellyfish; (4) Lobos lagoon belongs to the hemispheric network of shorebirds; and (5) the Yaquis have a traditional government that governs decision-making. However, there is an integrated functioning of this governance system, and the conception of a socio-ecological system would support in this regard.

Users. The fact that the Yaquis are exclusive users of the fishing resources of the coastal water bodies of their territory represents an opportunity for fisheries and aquaculture management, in the sense that conflicts are theoretically minimized. However, in reality, an internal division prevails in the Yaqui tribe that has its origins in the struggle for the recognition and respect of its territory and the rescue and conservation of its resources and its exploitation for its own benefit. This division is permeated by the particular economic and political interests that mainly benefit people outside the tribe (external investors) and that have blocked an autonomous process of the Yaquis to decide their own development, prevailing those interests that represent the fundamental cause of this division (Clavero 2019).

Based on the above, the ecosystem is in good condition and represents a valuable natural capital, but this has not translated into social welfare. In general the Yaqui community exhibits poverty and marginalization. The overall good state of natural capital represents an opportunity for development with social welfare. So in this socio-ecological system corresponding to the coastal area of the Yaqui territory, the most important challenge is to direct the ecosystem services toward the improvement of social welfare conditions for the Yaqui community.

Regarding this, the National Development Plan 2019–2024 of Mexico establishes in its strategies: “Consolidate the right to self-determination and autonomy of indigenous and Afro-Mexican peoples, and their coordination with the three orders of government; prioritize welfare policies and programs that aim to safeguard the rights of girls, boys, youth, women, older adults, people with disabilities, indigenous and Afro-Mexican communities and communities; and articulate policies and programs aimed at sustainable production to promote the supply of food at affordable prices, preferably in locations of high marginalization, indigenous and Afro-Mexican communities and communities, and with high levels of violence.” It also establishes: “develop in a sustainable and inclusive way the agricultural and aquaculture-fisheries sectors in rural territories, and in indigenous and Afro-Mexican villages and communities.”

To implement the above, it is necessary to promote the development of indigenous communities, through training, implementation of productive projects, and marketing of products, in accordance with their culture, values, and traditional knowledge. In addition, consider guiding public action so that it is in the indigenous community where the development planning process is carried out with an active, conscious, and congruent participation with your own identity (Clavero 2019).

34.5 Conclusions

The socio-ecological system of the coastal area of the Yaqui territory requires starting a medium- and long-term process aimed at strengthening the functioning of the governance system, investing in infrastructure to support fisheries and aquaculture production processes and training members of the Yaqui community to integrate their traditional knowledge in the best use of coastal ecosystem services. All this is oriented to generate the conditions that allow to reduce the social lag, improve the living conditions, and preserve the cultural values of the Yaquis. The process must be adopted by the Yaquis, supported by government institutions, and guided by scientists committed to their society.

Acknowledgments David Urias prepared Fig. 34.2, and Refugio López organized water quality databases. This chapter was part of the project for ecology and management of coastal ecosystems developed at the Centro de Investigaciones Biológicas del Noroeste, S.C.

References

Aguado M, Calvo D, Dessal C et al (2012) La necesidad de repensar el bienestar humano en un mundo cambiante. Papeles 119:49–76

Alcántara-Razo E, Padilla-Serrato JG, Aragón-Noriega AE (2019) La Comunidad Yaqui y su importancia en la producción Pesquera. pp. 233–260. In: Arreola-Lizárraga JA, Garatuza-Payán J, Yépez González EA Robles-Morúa A Capital Natural y Bienestar Social de la Comunidad Yaqui. Instituto Tecnológico de Sonora, Cd. Obregón, Sonora, México

Arreola-Lizárraga JA (1994) Bahía de Lobos, Sonora, un ecosistema costero candidato para el sistema de áreas naturales protegidas de México. Ecológica 3(2):19–24

Arreola-Lizárraga JA, Hernández-Moreno LG, Hernández-Vázquez S et al (2003) Ecology of *Callinectes arcuatus* and *C. bellicosus* (Decapoda: Portunidae) in a coastal lagoon of Northwest Mexico. Crustaceana 76(6):651–664. https://doi.org/10.1163/156854003322381487

Arreola-Lizárraga JA, Aragón-Noriega EA, Hernández-Moreno LG et al (2004) Co-occurence, habitat use and abundance of shrimps *Farfantepenaeus californiensis* and *Litopenaeus stylirostris* (Crustacea: Decapoda: Penaeidae) in a coastal lagoon on the Gulf of California. In: Hendrickx ME (ed) Contributions to the study of East Pacific Crustaceans 3. Instituto de Ciencias del Mar y Limnología, UNAM, México City, pp 143–151

Arreola-Lizárraga JA, Padilla-Arredondo G, Méndez-Rodríguez LC, et al (2014) Diagnóstico de la calidad del agua superficial y subterránea en las comunidades Yaquis del Estado de Sonora. Technical Report CIBNOR, SC, Guaymas, México

Arreola-Lizárraga JA, Padilla-Arredondo G, Garatuza-Payán J (2015) Flujos de nutrientes y metabolismo neto del estero Algodones (eurihalino, subtropical) en el Golfo de California. In: Paz-Pellat F, Wong-González J, Torres-Alamilla R (eds) Estado Actual del Conocimiento del Ciclo del Carbono y sus interacciones en México: Síntesis a 2015. Programa Mexicano del Carbono, Serie: Síntesis Nacionales, México, pp 495–501

Arreola-Lizárraga JA, Cruz-García LM, Ruiz-Ruiz TM et al (2019) Ecosistemas costeros: condición y tendencia ambiental del complejo lagunar Guásimas-Lobos. In: Arreola-Lizárraga JA, Garatuza-Payán J, Yépez González EA, Robles-Morúa A (eds) Capital Natural y Bienestar Social de la Comunidad Yaqui. Instituto Tecnológico de Sonora, Cd. Obregón, Sonora, México, pp 105–134

Ayala-Bocos A, Reyes-Bonilla H, Calderón-Aguilera L et al (2015) Proyección de cambios en la temperatura superficial del mar del Golfo de California y efectos sobre la abundancia y distribución de especies arrecifales. REVMAR 8(1):29–40. https://doi.org/10.15359/revmar.8-1.2

Berkes F (2015) Coasts for people: interdisciplinary approaches to coastal and marine resource management. Routledge Taylor & Francis Group, New York

Brattland C, Eythórsson E, Weines J et al (2019) Social–ecological timelines to explore human adaptation to coastal change. Ambio 48:1516–1529. https://doi.org/10.1007/s13280-018-1129-5

Burrola-Sánchez MS, Rosales-Grano P (2019) Descripción oceanográfica de la zona costera del Territorio Yaqui. In: Arreola-Lizárraga JA, Garatuza-Payán J, Yépez González EA, Robles-Morúa A (eds) Capital Natural y Bienestar Social de la Comunidad Yaqui. Instituto Tecnológico de Sonora, Cd. Obregón, Sonora, México, pp 69–85

Carmona R, Danemann G (2013) Monitoreo de aves playeras migratorias en 11 sitios prioritarios del Noroeste de México y reconocimiento invernal de cinco sitios adicionales. Technical Report Comisión Nacional de Áreas Naturales Protegidas, México

Carmona R, Danemann G (2014) Monitoreo de aves playeras migratorias en 11 sitios prioritarios del Noroeste de México y reconocimiento invernal de cinco sitios Adicionales. Technical Report Comisión Nacional de Áreas Naturales Protegidas, México

Carmona R, Danemann G (2017) Monitoreo de aves playeras migratorias en 11 sitios prioritarios del Noroeste de México y reconocimiento invernal de cinco sitios adicionales. Technical Report Comisión Nacional de Áreas Naturales Protegidas, México

Carrera E, de la Fuente G (2003) Inventario y Clasificación de Humedales en México. Ducks Unlimited de México, A.C

Casillas-Hernández R, Ibarra-Gámez JC (1996) Observaciones preliminares sobre el crecimiento y la producción del camarón blanco *Penaeus vannamei* en jaulas flotantes (Bahía de Guásimas, Sonora, México). Oceanologia 3(11):51–59

Casillas-Hernández R, Villa-Ibarra M, Magallón-Barajas F (2019) Potencial acuícola: Un desafío para el desarrollo social y sostenible de las Comunidades Yaquis. In: Arreola-Lizárraga JA, Garatuza-Payán J, Yépez González EA, Robles-Morúa A (eds) Capital Natural y Bienestar Social de la Comunidad Yaqui. Instituto Tecnológico de Sonora, Cd. Obregón, Sonora, México, pp 208–232

Castillo-Durán A, Chávez-Villalba J, Arreola-Lizárraga JA, Barraza-Guardado R (2010) Comparative growth, condition, and survival of juvenile oysters *Crassostrea gigas* and *C. corteziensis* cultivated in summer and winter. Cienc Mar 36:29–39. https://doi.org/10.7773/cm.v36i1.1623

Chávez-Villalba J (2019) Potencial ostrícola. In: Arreola-Lizárraga JA, Garatuza-Payán J, Yépez González EA, Robles-Morúa A (eds) Capital Natural y Bienestar Social de la Comunidad Yaqui. Instituto Tecnológico de Sonora, Cd. Obregón, Sonora, México, pp 185–207

Chávez-Villalba J, López-Tapia MR, Mazón-Suástegui JM et al (2005) Growth of the oyster Crassostrea corteziensis (Hertlein, 1951) in Sonora, Mexico. Aquac Res 36:1337–1344. https://doi.org/10.1111/j.1365-2109.2005.01345.x

Chávez-Villalba J, Hernández-Ibarra A, López-Tapia MR, Mazón-Suástegui JM (2008) Prospective culture of the Cortez oyster *Crassostrea corteziensis* from northwestern Mexico: growth, gametogenic activity, and condition index. J Shellfish Res 27:711–720. https://doi.org/10.1111/j.1365-2109.2005.01345.x

Clavero R (2019) Contexto y estrategia para el desarrollo económico y social de la tribu Yaqui. In: Arreola-Lizárraga JA, Garatuza-Payán J, Yépez González EA, Robles-Morúa A (eds) Capital Natural y Bienestar Social de la Comunidad Yaqui. Instituto Tecnológico de Sonora, Cd. Obregón, Sonora, México, pp 366–383

CONAPO (2005) Índice de marginación. Consejo Nacional de Población http://www.conapo.gob.mx/es/CONAPO/Indice_de_marginacion_a_nivel_localidad_2005

CONAPO (2010) Índice de marginación. Consejo Nacional de Población http://www.conapo.gob.mx/en/CONAPO/Indice_de_Marginacion_por_Localidad_2010

Costanza R, d'Arge R, de Groot R et al (1997) The value of the world's ecosystem services and natural capital. Nature 387:253–260. https://doi.org/10.1038/387253a0

Costanza R, de Groot R, Sutton P et al (2014) Changes in the global value of ecosystem services. Glob Environ Chang 26:152–158

DOF (2010) Norma Oficial Mexicana NOM-059-SEMARNAT-2010, Protección Ambiental: Especies nativas de México de flora y fauna silvestres, Categorías de riesgo y especificaciones para su inclusión, exclusión o cambio, Lista de especies en riesgo. Diario Oficial de la Federación, 30 diciembre 2010, México, DF

Figueroa A (1994) Por la tierra y por los santos: identidad y persistencia cultural entre Yaquis y Mayos. Consejo Nacional para la Cultura y las Artes, Serie Culturas Populares de México, México D.F

Garatuza-Payan J (2019) Hidrología del Territorio Yaqui. In: Arreola-Lizárraga JA, Garatuza-Payán J, Yépez González EA, Robles-Morúa A (eds) Capital Natural y Bienestar Social de la Comunidad Yaqui. Instituto Tecnológico de Sonora, Cd. Obregón, Sonora, México, pp 31–44

Guerry AD, Polasky S, Lubchenco J et al (2015) Natural capital and ecosystem services informing decisions: from promise to practice. PNAS 112(24):7348–7355. https://doi.org/10.1073/pnas.1503751112

INEGI (2000) Censo General de Población y Vivienda. Instituto Nacional de Estadística, Geografía e Informatica. https://www.inegi.org.mx/programas/ccpv/2000/

INEGI (2005) Censo General de Población y Vivienda. Instituto Nacional de Estadística, Geografía e Informatica. https://www.inegi.org.mx/programas/ccpv/2005/

INEGI (2010) Censo General de Población y Vivienda. Instituto Nacional de Estadística, Geografía e Informatica. https://www.inegi.org.mx/programas/ccpv/2010/

INEGI (2010b) Instituto Nacional de Estadística y Geografía. Conjunto de datos vectoriales de uso de suelo y vegetación serie IV Escala 1:250 000. https://www.inegi.org.mx/app/mapas/

ITSON (1994) Informe técnico sobre el cultivo semi-intensivo del camarón blanco *Penaeus vannamei*, en la granja de Guásimas. Technical Report Instituto Tecnológico de Sonora, Cd. Obregón, Sonora, México

Kjerfve BK, Magill E (1989) Geographic and hydrographic characteristics of shallow coastal lagoons. Mar Geol 88:187–199. https://doi.org/10.1016/0025-3227(89)90097-2

Lankford RR (1977) Coastal lagoons of Mexico: their origin and classification. In: M Wiley (ed), Estuarine processes, New York, pp 182–215
Lerma-Rodríguez E (2011) El nido heredado: estudio sobre cosmovisión, espacio y ciclo ritual de la Tribu Yaqui. Dissertation. Universidad Nacional Autónoma de México
Lerma-Rodríguez E (2014) Considerations about the ethnographic investigation of the Yaqui community. Des Cult 2(2):41–62
Levin SA (1999) Fragile dominion: complexity and the commons. Perseus Books, Reading
Luna-Escalante G (2007) Derechos, usos y gestión del agua en el Territorio Yaqui. Dissertation, Colegio de Sonora, Hermosillo, Sonora, México
Medina-Galván J, Padilla-Arredondo G, Arreola-Lizárraga JA (2018) Flujos de nutrientes y metabolismo neto del estero Siuti (eurihalino, subtropical) en el Golfo de California. In: Paz F, Velázquez A, Rojo M (eds) Estado Actual del Conocimiento del Ciclo del Carbono y sus Interacciones en México: Síntesis a 2018. Serie Síntesis Nacionales, Programa Mexicano del Carbono, México, pp 302–308
Méndez-Rodríguez LC, Acosta-Vargas B, Mendoza-Salgado R et al (2019) Abastecimiento y calidad del agua superficial y subterránea. In: Arreola-Lizárraga JA, Garatuza-Payán J, Yépez González EA, Robles-Morúa A (eds) Capital Natural y Bienestar Social de la Comunidad Yaqui. Instituto Tecnológico de Sonora, Cd. Obregón, Sonora, México, pp 135–160
Mendoza-Cantú ME (1997) Regionalización geomorfológica y de paisaje de la Zona Costera entre Guaymas y Agiabampo, Sonora, México. Dissertation, Instituto Tecnológico y de Estudios Superiores de Monterrey, Guaymas, Sonora, México
Minjares-Lugo JL, Félix-Sánchez JA (2019) Desarrollo de la Agricultura en la Tribu Yaqui. In: Arreola-Lizárraga JA, Garatuza-Payán J, Yépez González EA, Robles-Morúa A (eds) Capital Natural y Bienestar Social de la Comunidad Yaqui. Instituto Tecnológico de Sonora, Cd. Obregón, Sonora, México, pp 162–184
Moctezuma-Zamarrón JL (2007) Yaquis: Pueblos Indígenas del México Contemporáneo. Comisión Nacional para el Desarrollo de los Pueblos Indígenas, México, D.F
Odum EP (1959) Fundamentals of ecology. Saunders Company, Philadelphia
Olavarría ME (2000) Contexto calendárico Yaqui. Alteridades 10(20):35–57
Ostrom E (2009) A general framework for analyzing sustainability of social-ecological systems. Science 325:419–422. https://doi.org/10.1126/science.1172133
Padilla-Arredondo G, Pedrin-Aviles S (2019) Geología del Territorio Yaqui. In: Arreola-Lizárraga JA, Garatuza-Payán J, Yépez González EA, Robles-Morúa A (eds) Capital Natural y Bienestar Social de la Comunidad Yaqui. Instituto Tecnológico de Sonora, Cd. Obregón, Sonora, México, pp 45–68
Padilla-Arredondo G, Arreola-Lizárraga JA, Lechuga-Devéze C (2002) Las Guásimas coastal lagoon, Sonora, México. In: Camacho-Ibar VF, Dupra V, Marshall-Crossland JI, Wulff F, Smith SV, Crossland CJ (eds) Estuarine systems of the Latin American region and estuarine systems of the artic region: carbon, nitrogen and phosphorus fluxes, LOICZ R&S-23, pp 11–14
Padilla-Ramos R, Moctezuma-Zamarrón JL (2017) The Yaquis, a historical struggle for water. Water Hist 9:29–43. https://doi.org/10.1007/s12685-017-0194-1
Partida-Flores JM, Ruiz-Ruiz TM, Ochoa-Meza AR, Piñón-Gimate A, Arreola-Lizárraga JA (2019) Biomasas de fitoplancton y macroalgas como indicadores biológicos de impacto ambiental de las aguas residuales vertidas en Bahía de Lobos. In: Arreola-Lizárraga JA, Garatuza-Payán J, Yépez González EA, Robles-Morúa A (eds) Capital Natural y Bienestar Social de la Comunidad Yaqui. Instituto Tecnológico de Sonora, Cd. Obregón, Sonora, México, pp 281–308
Ramos JE (1989) Historia de un coctel. Cuadernos de divulgación No. 1, Universidad de Sonora, Hermosillo, Sonora, México
Reynaga-Franco F, Arreola-Lizárraga JA, Padilla-Arredondo G et al (2018) Estado trófico de tres lagunas costeras subtropicales del Golfo de California. In: Paz F, Velázquez A, Rojo M (eds) Estado Actual del Conocimiento del Ciclo del Carbono y sus Interacciones en México: Síntesis a 2018. Serie Síntesis Nacionales, Programa Mexicano del Carbono, México, pp 352–357

Rodríguez-Zuñiga MT, Troche-Souza C, Vázquez-Lule AD et al (2013) Manglares de México: Extensión, distribución y monitoreo. Comisión Nacional para el Conocimiento y Uso de la Biodiversidad, México DF

Rojas-Robles NE, Robles-Morua A (2019) La variabilidad espacial y temporal del clima en las Comunidades Yaquis. In: Arreola-Lizárraga JA, Garatuza-Payán J, Yépez González EA, Robles-Morúa A (eds) Capital Natural y Bienestar Social de la Comunidad Yaqui. Instituto Tecnológico de Sonora, Cd. Obregón, Sonora, México, pp 12–30

Ruiz-Ruiz TM, Arreola-Lizárraga JA, Morquecho L et al (2017) Detecting eutrophication symptoms in a subtropical semi-arid coastal lagoon by means of three different methods. Wetlands 37:1105–1118. https://doi.org/10.1007/s13157-017-0944-y

SEMARNAT (2007) Análisis de vacíos y omisiones en conservación de la biodiversidad marina de México: océanos, costas e islas. Secretaria de Medio Ambiente y Recursos Naturales, México

SEMARNAT (2008) Estrategia para para la Conservación y Manejo de las aves playeras y su hábitat en México. Dirección General de Vida Silvestre. Secretaria de Medio Ambiente y Recursos Naturales, México

Stojanovic T, McNae H, Tett P et al (2016) The "social" aspect of social-ecological systems: a critique of analytical frameworks and findings from a multisite study of coastal sustainability. Ecol Soc 21(3):15. https://doi.org/10.5751/ES-08633-210315

Tett P, Sandberg A, Mette A et al (2013) Perspectives of social and ecological systems. In: Moksness E, Dahl E, Støttrup J (eds) Global challenges in integrated coastal zone management. Wiley-Blackwell, West Sussex, pp 229–243

Valenzuela-Siu M, Arreola-Lizárraga JA, Sánchez-Carrillo S et al (2007) Flujos de nutrientes y metabolismo neto de la laguna costera Lobos, México. Hidrobiológica 17(3):193–202

Vargas-González HH, Arreola-Lizárraga JA, García-Hernández J et al (2017) Calidad de sedimentos asociada a actividades antrópicas en lagunas costeras semi-áridas subtropicales de la costa central este del Golfo de California. Rev Int Contam Ambient 33:7–22

WHSRN (2018). https://whsrn.org/es/whsrn_sites/bahia-de-lobos/

Chapter 35
Natural Protected Areas vs Integrated Watershed Management: People Participation Analysis in México

Clara Margarita Tinoco-Navarro, Raúl Francisco Pineda-López, Óscar Orlando Parra-Barrientos, and Roberto Urrutia-Pérez

Abstract Since its beginnings as a conservation strategy, natural protected areas (NPAs) in Mexico have been mostly managed in a top-down governmental approach. These management plans have had poor to moderate involvement with the population and little respect for their lifestyle and their relationships with the environment. Since people from natural reserves have been left aside, they do not always agree with the rules of the protected areas, and their involvement depends in many cases on external stakeholders. In this chapter, the importance of reassessing these conventional processes and strategies for conservation is presented in order to develop new and more inclusive approaches through continuous dialogue, incorporating institutional, scientific, and traditional knowledge. It is expected that the recovery of biodiversity and its management surpasses the conservation efforts of NPA traditional approach, but it must reflect the outcomes of integrative practices from watershed management.

Keywords Governance · Participation · Watershed management · Adaptation · Climate change

C. M. Tinoco-Navarro (✉) · Ó. O. Parra-Barrientos
University of Concepción, Concepción, Chile
e-mail: oparra@udec.cl

R. F. Pineda-López
Autonomous University of Querétaro, Santiago de Querétaro, Mexico

R. Urrutia-Pérez
Water Research Center for Agriculture and Mining (CRHIAM), University of Concepción, Concepción, Chile
e-mail: rurrutia@udec.cl

A. Ortega-Rubio (ed.), *Socio-ecological Studies in Natural Protected Areas*,
https://doi.org/10.1007/978-3-030-47264-1_35

35.1 Introduction

The current environmental crisis has positioned forest conservation and biodiversity as a strategy for mitigation and adaptation to climate change (CC), with emphasis on the conservation of water and the participation of local people. This has diversified community forest management schemes, community watershed management, establishment of natural protected areas (NPAs), and voluntary conservation areas (Porter-Bolland et al. 2012; UNEP 2012; de la Cruz-Hernández et al. 2016). In Mexico, it has been documented that environmental, forestry, and water management policies are poorly articulated, contradictory, and showing a lack of continuity and hinder the achievement of management and protection objectives for natural heritage (Merino and Segura 2007; Bautista 2007; de la Cruz-Hernández et al. 2016). The intention of this work is to compare some strategic processes to analyze social participation between two territorial management models: natural protected areas and integrated watershed management (IWM).

The central research question of this paper is which approach (integrated watershed management or natural protected areas) is more promising for conserving and managing Mexico's natural heritage in the context of global climate change. To answer this question, an analysis of social participation is presented, considering the strategic processes: the socio-ecosystemic vision, the territorial scales from local to regional, the expected and observed impact on the rehabilitation and conservation of the natural heritage, the legal framework available for the development of territorial management models, their governance mechanisms, and the possibility of establishing practice and learning communities.

35.2 Social Participation for the Management of Socio-ecosystems

Socio-ecosystems (SES) are complex adaptive systems composed of social agents (actors and their governance system) and ecological agents (natural resources and their processes) interacting at multiple spatial and temporal scales. These interactions include the management of natural resources, ecosystem services, and the impacts of individuals or groups that live in or make use of them. Each social group gives an identity to the SES through their practices, and they define the rules of resource use according to the sociocultural and economic value they have for the society. These, in return, will be influenced by the social, political, economic, and ecological context where the actors, conflicts, potential benefits, and costs are located (McGinnis and Ostrom 2014; Cusva 2014).

The growing pressures on ecosystems are reflected in degradation and land use change for competitive uses, such as livestock production and crops for human food, among others (Magrin et al. 2014). These are fundamental factors of environmental degradation, forest and biodiversity loss that, ultimately, generates an alteration and reduction of ecosystem services (Miranda et al. 2017; CR2 2019). Currently, there is a greater demand for decision-making, association, and influences on public affairs by nongovernment actors. However, a greater coordination by a wide range of public actors and territorial competences is required in open and participative processes to legitimize decision-making, resolve conflicts, and generate solid consensual commitments. These are key aspects to improve strategic planning for territory management (El Colegio de México 2012; Berggruen and Gardels 2013; Arriagada et al. 2018).

Some examples of strategic planning are the PADDD process (protected area downgrading, downsizing, and degazettement) with identified patterns of alteration (forestry, mining, oil and gas, industrial agriculture, industrialization, infrastructure, rural settlements, subsistence, degradation, land claims, sovereignty of displacement, refugee accommodation, and conservation planning), where the main causes correspond to the lack of management in the use of the territory (Mascia et al. 2014). This has been observed in changes in the decrees of NPAs, whether they are in the intention or realization of environmental protection policy or in its reduction of surface area and loss of vegetation cover. This is the case of Nevado de Toluca in the State of Mexico (de la Cruz-Hernández et al. 2016), Parque Nacional el Cimatario, El Tángano ecological protection zone in Querétaro, the Monarch Butterfly Biosphere Reserve, and the Cañada de los Dinamos in Magdalena Contreras (Fig. 35.1), among others. This trend toward changes in NPAs has generated a great deal of public opinion among sectors that consider that changing the category puts the conservation of environmental values at risk. Other sectors ensure that the changes will allow expansion in management option; those who consider that the changes are generated by constant pressure on land create speculation particularly in those natural protected areas located near cities with a high growth rate. Different

Fig. 35.1 Cañada de los Dinamos in Magdalena Contreras, México. (Author: Clara Tinoco)

studies document socio-environmental damage due to the lack of linkage and compatibility between conservation policies and natural resource use policies (Fernández et al. 2002; Campos et al. 2012; Tlapa et al. 2020).

In the Flora and Fauna Protection Area of Cuatro Ciénegas, Coahuila, there was discontent among the population due to the prohibition of traditional uses of natural resources and in the establishment of the natural protected area (NPA) –due to the lack of consensus in the decree and elaboration of the management program and a negative association regarding conservation with the restriction – perceiving that the ecosystem is favored more than the people (Ortiz and Romo 2016). Brenner and San German (2012) analyzed conflicts between actors in the Monarch Butterfly Biosphere Reserve, caused by a low social participation in the elaboration of the management program, resistance to logging constraints, and the establishment of parallel governance structures that resulted in the exclusion of non-communal landholders (no ejidatarios).

In the El Vizcaino Biosphere Reserve, the legal framework and functioning do not allow democratic and binding decision-making. Therefore, it emphasizes that the management processes of the NPAs should take advantage of local governance structures and strengthen capacities to avoid the exclusion of some sectors, reduce conflicts, take advantage of local initiatives, and promote the acceptance of environmental policies based on nonexclusive processes that integrate local perception and effective participation mechanisms.

In this regard, Monterrubio-Solís (2019), in an analysis of the formalization process of voluntary natural areas for conservation (VNAC) in the Chimalapas, mentioned the presence of unequal relationships, and a lack of articulation between the actors involved, along with the participation and monitoring in this certification, was dependent of the interpretation and discretion of institutional arrangements designed for their management. It also mentions the problem of simplification and homogenization of deliberation spaces for decision-making, diminishing the legitimacy of the processes and generating more conflicts, so that the communities "absorb" the social cost of the conservation of the territory.

Monroy et al. (2015), in a study on the NPA relationship in Oaxaca, indicated that it is necessary to incorporate regional or local conservation initiatives, in order to conserve greater biodiversity and have a greater representation. It emphasizes that in these cases, social and political conflicts over the control and use of resources are mitigated, since the conservation of these areas is an initiative of the communities themselves. Organized participation is encouraged since no distinction is made between biological, economic, or social conservation objectives, and other ecosystems are conserved that would not be necessarily protected in payment for environmental services (PES) schemes or other initiatives supported by government agencies.

Botana et al. (2019), analyzing the case of Cofre de Perote in Veracruz, emphasize that the production of spaces conceived for conservation is generally produced by experts and other external actors, including new modalities of management. In them, the characteristic of being restrictive with the use of natural resources as classical conservationist policies is maintained, and unequal relations of political, economic, and environmental power remain. This shelters a passive participation, while places where people live create their affections, and their history is fragmented or suppressed.

Finally, if it is considered that the NPAs in Mexico belong to communities in more than 85% of their total territorial coverage, it is determinant to recognize that the fulfillment of the decrees and management programs depends on the active participation of the communities (CONANP 2013). It is also important to mention that in the country there are about 150 communal protected areas, which correspond to indigenous and peasant community areas that do not have an official certification as PNA but are selected by the communities according to their uses and customs as conservation spaces (Monroy et al. 2015). This, as well as community-based land use planning, is an example of organized social participation for the management and conservation of ecosystems, where the local governance over the territory goes beyond institutional rules and norms or external management programs but rather represents important strategies that can reinforce the conservation and management of biodiversity from a local perspective. Therefore, the challenge is to move from participation and empowerment as mere enunciative concepts in management modalities, to generate the spaces where they implement them (Botana et al. 2019). This represents a transversal cultural transformation of officials, funders, management teams, and promoters, as well as for the communities.

The need to relearn how to interpret and manage the territory from other, more systemic perspectives has generated interest in integrated watershed management (IWM), a territorial management strategy that seeks to promote integrated care of the socio-ecosystem based on local participation and the articulation of stakeholders. This is done with the objective of establishing greater effectiveness in planning between conservation and the use of natural resources, risk prevention, promotion of agreements, and organized production, creating synergies among stakeholders to maintain ecosystem services, value local practices, and reduce vulnerability (Pineda and Ríos 2013; Pineda et al. 2016a, b; Tinoco et al. 2017).

In Mexico, watershed management projects are found between two configurations: the governmental one (regional vision, watersheds as providers of water, water for productive development, and cities) and the nongovernment actors, such as social and academic organizations (local, ecosystems as generators and users, capacity building, and community development). Some authors mention as advantages of IWM projects the local development perspective and the use of differentiated scales, runoff units, and micro-watersheds with a greater possibility to manage more inclusive decision-making, monitoring, and evaluation processes (Cotler and Pineda 2008; Luna 2014).

However, just like the NPAs, the watershed approach also faces different barriers to its effective implementation: lack of regulatory and institutional transversality, lack of information, or inequity in its access by social actors generate disconnect between institutional, normative, and informed participation in decision-making (Ortiz and Romo 2016).

There are some participation mechanisms determined from legislation that in practice have not been effective. One example is the case of some basin councils

(Ortiz and Romo 2016), which in several cases in the country are of a consultative nature and do not have legal powers, so their functioning depends on the capacities, commitment, and will of social actors (El Colegio de México et al. 2012; Ortiz and Romo 2016). This makes it clear that local community empowerment must be accompanied by political will, institutional support, and effective regulations. The co-creation and democratization of knowledge and spaces for effective participation and the strengthening of communication channels are strategies required to establish the appropriate environment for the generation, development, and implementation of effective and coordinated actions. These arrangements respond effectively to the management of the territory, taking into account its multiple dimensions and the new vulnerability scenarios of climate change (Aldunce et al. 2014; Arriagada et al. 2018; CR2 2019).

35.3 The Analysis of Social Participation in the Legal Framework

In Mexico, the General Law of Ecological Equilibrium and Environmental Protection (LGEEPA) supports the establishment, protection, and preservation of NPAs and ecological restoration zones (from federal to local administrative jurisdiction, according to the category of protection). This law mentions that the state is "responsible of guaranteeing the co-responsibility of the people, individually or collectively, in the preservation and restoration of the ecological balance and the protection of the environment." The state, as the coordinating entity, has a responsibility to promote social participation – according to the provisions of the legislation – of the different social groups in the elaboration of programs related to these public policy instruments (Articles 1 section VII, 18, 47 and 65), as well as in the establishment, administration, and management of the NPAs, specifying the participation of both inhabitants and owners, indigenous peoples and social organizations, and even specifying the participation of women as an essential part of sustainable development (Article 15, section XV) (DOF 2018).

This participation is differentiated and holds specific criteria related to the process of establishment, administration, and implementation of the management program. An example of this is the subdivision of the NPA according to biological, physical, and socioeconomic criteria where a territorial delimitation of the activities is made. Those divisions are I) core zones with exclusive preservation and education objectives with prohibitions for social use and participation and II) buffer zones with the function of orienting the activities of sustainable use in preservation zones, traditional use, and sustainable exploitation where socioeconomic and cultural aspects are considered that are carried out by the inhabitants to satisfy their needs, with a focus on long-term conservation of the ecosystems (Article 48) (DOF 2018).

In order to relate the government and society, within the National Commission of Natural Protected Areas, the participation of social representatives from *ejidos*, communities, and holders is considered for sessions where a matter of interest to the social group is discussed (Article 56 BIS). The National Commission of Natural Protected Areas (CONANP) has actions to strengthen social and institutional participation in some NPAs, and with the Federal Attorney for Environmental Protection (PROFEPA), participatory social monitoring committees have been integrated (Halffter 2011). In the case of declarations, social participation is established mainly as the state's obligation to solicit the opinion of different social sectors, including local governments, public administration agencies, academic and research institutions, public and private social organizations, indigenous peoples, and interested people. It also considers the social capacity to promote the establishment of NPAs in their territories when it comes to areas destined for the preservation, protection, and restoration of biodiversity, which are held to territorial zoning and use limitations (Articles 58 and 59). The management of the NPA is designated by the government as the coordinating body for the formulation, execution, and evaluation of the management programs (Article 65), which must specify the form of organization and the mechanisms for the participation of the population and institutions involved in the program (Articles 66 section III and 78), as well as in the declarations of ecological restoration zones and their corresponding programs (Article 78 BIS) (DOF 2018).

In these terms, although the current legislation clearly establishes the obligation of the authorities to promote the social participation of the different sectors interested in a specific territory, in the case of the NPAs, it implies a passive type of participation, since it is mostly delimited and directed based on the request of the authorities at certain specific moments of the process of declaration, establishment, execution, and evaluation of the management program. The protection objectives are then directed toward the maintenance of ecological components and processes, without giving relevance to the preservation and recovery of the structure of the ecosystems in order to recover their function for the benefit of both the communities living in the PNA and those who benefit from the ecosystem services generated in these territories.

In watershed management, the LGEEPA, Article 53, addresses watersheds and water protection as a justification for their preservation within the declaration of natural resource protection areas, emphasizing that they must be in forest lands or lands of preferential forest aptitude. Particularly considered are the protection areas of rivers, lakes, lagoons, springs, and bodies of water recognized as national waters, intended to supply water for the service of the population and where sustainable use can be made. Other laws such as the National Waters Law, the Sustainable Forestry Development Law, the Sustainable Rural Development Law, and the Climate Change Law propose river basins and micro-watersheds as territorial units for the operation of their guidelines. The Sustainable Forestry Development Law (Article 125) indicates institutional and social concurrence for the purposes of restoration and conservation of banks, watercourses, and recharge areas. It indicates that the owners and possessors must be previously heard by SEMARNAT, as well as the technical opinion of the councils, CONAGUA, and CONANP, the participation of indigenous peoples and communities, and the full and effective participation in the elaboration and execution of forest programs, as well as their knowledge of culture and traditions (DOF 2018). The General Law on Climate Change clearly establishes the powers of the state and local authorities to implement adaptation actions in watershed management policies and programs (Articles 9, 10, 11 and 27) (DOF 2018), whereas the Sustainable Rural Development Law establishes that the formulation of programs at the municipal and regional or watershed level must include the participation of authorities, population, and producers located in the watershed: the formation of local councils and interstate regional councils (Article 24) (DOF 2019).

The National Waters Law (Ley de Aguas Nacionales – LAN) defines water management as a process based on a set of principles, standards, policies, instruments, assets, resources, rights, responsibilities, and powers through which the state, users, and civil organizations coordinate to promote and implement actions geared to controlling and managing watersheds and aquifers; regulating the exploitation, use, or development of water; and preserving and sustaining it in terms of quantity and quality. This requires the creation of organizations and mechanisms for social participation in the management of river basins, such as river basin councils. These are collegiate bodies, with a mixed composition and for consultation. Their importance lies in the fact that they bring together the three levels of government (national, state, and municipal), users, individuals, and civil society organizations in decision-making and the assumption of commitments by these sectors and the National Water Commission (CONAGUA) (Articles 13 and 13 BIS) (DOF 2014). Although water management began as a centralist administration through federal institutions, later it has sought to move toward integrated watershed management, with the watershed as the unit of analysis, water as the guiding axis, and user participation as the elemental aspect for the conservation of ecosystems (Ortiz and Romo 2016).

The LAN consolidated social participation in the watersheds (Articles 5, 7 BIS, 9, 10 and 14 BIS) with the obligation for organizations to establish consultative councils that involve stakeholders in the decision-making processes (Article 12). However, currently the regulations of the act are still pending, which makes their implementation difficult (Wilder 2010; OECD 2012). Part of the functions of watershed councils is to promote the conservation of water and the vital ecosystems linked to it. By decree, the federal executive can declare the water reserve for environmental use, ecological conservation, and human consumption, in the respective aquifer, watershed, or hydrological region, in a regulated process in which the CONAGUA, the basin organizations, and the basin council participate, including the respective technical study and public consultation.

In summary, with the analysis of the aspects of social participation considered in regard of NPA and watershed management (Table 35.1), a key point to recognize is that, based on the management programs of the NPAs or IWMs, the legislation grants freedom of action to determine adequate and pertinent mechanisms of social participation, which generates two scenarios:

(a) The first scenario is the initial model of national parks in Mexico, replicating the US model of a classic vision where the use of natural resources in NPAs is excluded or restricted. In this model, the conservation perspective maintains that it is necessary to avoid the presence of human beings and the use of natural resources through surveillance, accessed restrictions, and the vigilance of the

Table 35.1 Regulations regarding social participation considered in NPA and watershed management (Elaborated with information of DOF 2018; LAN 1992, 2016; CONAGUA 2018a, c)

Social participation	ANP	Watershed management
Regulations and law	General Law of Ecological Equilibrium and Environmental Protection (LGEEPA)	National Waters Law (LAN)
Modalities of social participation	SEMARNAT asks for the opinion of different social sectors in the declarations and programs of the PNA	The CONAGUA carries out public consultation on water reserve decrees
	It establishes the capacity of any owner to promote the establishment of NPAs in their territory for the purpose of biodiversity conservation. Indigenous and peasant communities can establish conservation areas legally and voluntarily	Participation in basin councils and auxiliary organizations Participation in basin management programs
Social participation organizations	National Commission of Natural Protected Areas	Basin councils Auxiliary bodies (basin committees, groundwater committees, clean beaches committees)
Functioning of social participation organizations/mechanisms	The National Commission of Natural Protected Areas promotes the participation of social representatives from ejidos, communities, and landowners in the PNA to participate in the sessions in the case of having a matter of interest	The basin councils are responsible for channeling social participation in water management in the basin: 1. Agreeing on priorities for water use 2. To participate in the definition of general objectives and criteria for the formulation of water management programs in the basin 3. Promote coordination and complementarity of investments 4. To contribute to the sanitation of basins and bodies receiving wastewater 5. Contribute to the economic, environmental, and social valuation of water 6. Supporting the financing of regional water management 7. To know the information and documentation concerning the availability in quantity and quality, the uses of water, and the registered rights 8. To promote the efficient and sustainable use of water, to encourage the reuse and circulation of water 9. Promoting the establishment of subsidiary bodies
		Its powers include the establishment of working and social participation mechanisms for the operation of the council

authorities (Alcorn 2005). Therefore, the institutional work teams maintain vertical policies, where social participation is passive and not very functional, since only citizens are asked for consultations and opinions or the different social groups are considered as receivers of information generated by the "experts." This promotes the continuity of vertical conservationist policies, with social participation in information and consultation, which can lead to conservation strategies that are not very significant for the population, the possible generation of conflicts between the population of the communities and the governments, and the difficulty of maintaining and continuing with the conservation objectives.

(b) In a second scenario, the perspectives of sustainable co-management or conservation based on the communities, with the integrated watershed management (IWM), propose that the local population participates in the planning, management, and sustainable exploitation (Alcorn 2005). Therefore, the potential of functional social participation can be detonated if the declarations are the result of binding processes among the actors with a more integrated vision, the construction of transversal social processes aimed at the conservation and management of ecosystems, and the maintenance of ecosystem services by and for the communities.

35.4 Territory and Climate Change

In order to understand the environmental dynamics and transformations of the territory and to generate better conditions for social participation and articulation among actors, the concept of scale is a crucial analytical tool, as these are related and condition each other. In the study of the dynamics of the establishment of natural protected areas, Nardi and Mahmoud (2017) analyze the NPAs as spaces of "exclusion," unprotected spaces around the protected areas, and the mutual affectations between them, establishing that the processes of conservation and exploitation of natural resources need the incorporation of the multiscale vision to understand the territorial-environmental conflicts and the conservation of biodiversity.

In watershed management, the different types of scale can be glimpsed due to the socio-ecosystemic approach adopted: size (basin, sub-basin, micro-basin) and level (functional areas), which in turn allow other types of scale to be analyzed: network (public policies) and relationship (basin actors and their influence). Although this approach considers planning from the basin, management in the sub-basin, and implementation at micro-basin level, it's important to recognize that when the scale of analysis is changed, these same elements acquire a different but related importance, making conflicts and negotiation needs visible. This is relevant if we consider the new environmental challenges, such as climate change, where the effects on localities and regions are strongly differentiated in ecological and social terms but are generalized in public policies and management instruments at the global, regional, and state level. In order for these to be effective, they will require more integrated visions that allow, on the first hand, for large-scale strategies that are flexible enough to be implemented in conjunction with specific and locally relevant adaptation and mitigation actions and relationships.

The integrated watershed management (IWM) approach allows to face some impacts of climate change (CC) in an operational way, since it considers in an integral way the interrelations among the social, economic, and ecological components within a given territory (both in past, present, and future scenarios of the watershed), building participatory management programs where strategies strengthen adaptation to possible changes and uncertainty, since they are based on the strengthening of local capacities (CECADESU 2015). According to the Mexican Institute of Water Technology (IMTA), the main elements that can suffer impacts due to the effects of CC are water resources. This has variations in rain capture, runoff and infiltration, flood control, availability, and storage of water, which have implications for food, economic, and water security of populations worldwide. In Mexico, 43 watersheds (23% of the territory) require urgent measures to allow the recovery of the ecological-hydrological function, since, without substantial improvements in current use practices, the CC will exacerbate the pressure on the watersheds and their ecosystems, both because of the greater demand for water and because of changes in productive activities. The loss of water availability and quality is one of the impacts that will generate greater social and economic vulnerability, especially in the north and center of the country. According to the national climate change program, 1385 municipalities and 27 million inhabitants in Mexico are at risk of disaster due to the increase and intensification of climate events such as floods, agricultural droughts, heat waves, decreased precipitation, and increased temperature (Cotler 2010; CONAGUA 2017; 2018a, b).

Due to its potential to incorporate eco-hydrological functions and externalities resulting from economic activities and the social dynamics of the territory, IWM has been identified as a key element for adaptation to climate change. Reversing environmental degradation is a key element for adaptation, considering that global changes affect biodiversity and ecosystem functions, altering the provision of water and other environmental services that are fundamental for sustaining human activities and the quality of life of the population. In addition, it has been observed that the deterioration of the supply and quality of water resources can cause social conflicts and increase migration, vulnerability, and risk.

Adaptation and mitigation, although at a global level, constitute political strategies to deal with the problem of climate change; at the level of the communities, they are decision-making options, with a functional character, as an emerging response of the communities when they face their situation of vulnerability to the phenomenon. In this sense, transdisciplinary and inter-institutional research on the supply and demand of hydrological environmental services (their benefits, perceptions and interaction of stakeholders for their maintenance and recognition) is necessary, as well as the preparation of professional groups and the education and promotion of environmental awareness, which promote technical support for the development of local initiatives, where communication and education are the main gears to strengthen the participatory processes in the basin.

Problems such as climate change and the loss of biodiversity have spread as global phenomena that leave the population in a situation of uncertainty and inopportunity. Likewise, local perceptions define the interpretations of the phenomenon, its impacts, risks, vulnerabilities, and the criteria for adaptation and mitigation, so the key is to foster the capacity to respond to uncertainty. The IWM approach is based on strengthening the capacities of local actors in the basin, allowing a continuous dialogue that provides the right environment and the necessary tools for visualizing the impacts and risks of climate change in an integrated manner in the socio-ecosystem, thus promoting adaptation proposals from local lifestyles. This systemic analysis also allows for the recognition of possible future impacts and the causes that increase the vulnerability of the socio-ecosystem, determining individual, collective, and sectorial responsibilities, as well as differentiated adaptation (Pineda et al. 2016a, b; Tinoco 2017).

35.5 Communities of Practice: Some Watershed Management Models

IWM has been a process whose implementation and results are variable depending on the scale and context of application. Below are three models, their social participation mechanisms, and the link with biodiversity conservation and protected areas: 1) CRCC micro-watershed management based on the establishment of good practices and strengthening of local capacities; 2) WWF integrated management based on a process of knowledge exchange through models of sustainable water use (SWU), ecological flow, and water reserves; and 3) C6 coastal watershed management and climate change built on an integrated vision for improvements in governance for watershed management and strengthening of NPA.

35.5.1 Integrated Micro-watershed Management Model of Regional Watershed Training Center (RWTC)-Autonomous University of Querétaro (AUQ)

The integrated micro-watershed management model at RWTC was developed at the Autonomous University of Querétaro (AUQ), at local scales to promote a culture of conservation and natural resource management with an emphasis on water. It was implemented with a training circuit in four demonstration micro-watersheds (La Joya, Buenavista, and San Pedro Huimilpan in Querétaro state, Mesa de Escalante in Guanajuato state), which will allow the dissemination of best practices in watershed management from an integrated vision built on the knowledge dialogue, recognizing the role of communities, researchers, and organizations integrated in a pedagogical model based on learning by doing (Carvajal and Gilio 2015; Tinoco

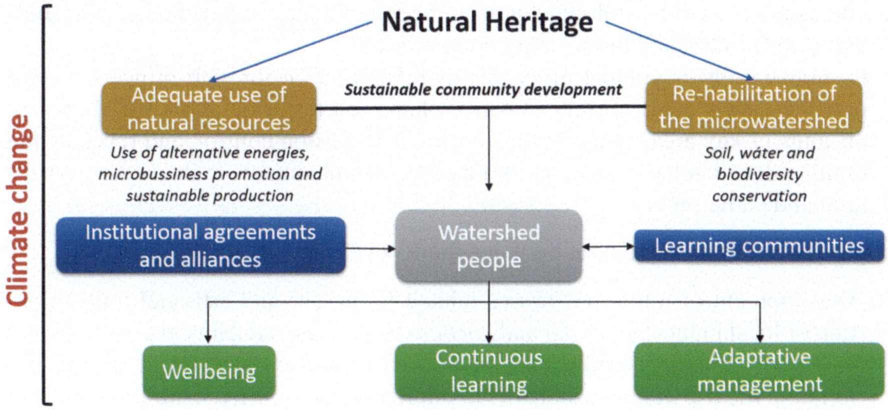

Fig. 35.2 Micro-watershed management model of the Regional Watershed Training Center (RWTC) at Autonomous University of Querétaro (AUQ) (own elaboration)

and González 2015). It seeks community and educational development toward a culture of sustainability based on collective learning and seven guiding processes (Fig. 35.2):

1. Soil conservation and management
2. Water conservation and management
3. Conservation and management of biodiversity
4. Promotion of alternative energy use
5. Promotion of sustainable production
6. Promotion of micro-businesses
7. Promotion of community development

The first three processes (1–3) seek to rehabilitate and conserve the structure and function of the watershed in an integral manner, while trying to reduce erosion and rehabilitate the formation of the soil and recover the green cover (which will improve the capture and increase the quality of the water that runs off in the micro-watershed). Those implementations are focused on the recovery of the natural heritage as a basis for establishing socioeconomic development. With the following three process (4–6), the aim is to promote actions that favor a system of conservation by producing and making use of the resources of the territory to have good quality ecological housing, seeking water and food security, and increasing economic benefits within a context of local adaptation to climate change.

The last process (7) seeks to establish an adequate balance between rehabilitation-conservation strategies and those of use and production, in order to jointly establish new forms of community development that move toward sustainability. The management of natural heritage by rehabilitation is promoted through:

1. Integrated management of runoff units agreed with the watershed people through various practices (structural, management, and vegetative) adapted to the spe-

cific conditions of hydrological units to be rehabilitated and named as SWBIOC units: soil, water, and biodiversity conservation.
2. Demonstration of gradual processes of rehabilitation through projects or best practices, such as the nursery of native plants and territorial work based on runoff units or key areas (using gully control, soil erosion control, and revegetation of hill sides) that have been incorporated by the inhabitants in their own agricultural and cattle plots.

In relation to the use of natural heritage, the RWTC approach is achieved by:

(a) Development of micro-businesses related to the use and artisanal transformation of local plants like *nopal* and *xoconoxtle* (leaves and fruits of *Opuntia* spp.) and medicinal and useful plants, the nursery of native plants for landscape restoration, the use of house orchards to improve food security, natural enrichment of traditional foods like corn tortillas, and the production of worm castings.
(b) Continuous training in various aspects of community life, like care of natural heritage, community organization, and the preparation of other micro-businesses as products that can boost the local economy. At the same time, they must ensure adequate food for the entire population, particularly strengthening vulnerable groups (women, elderly, and children).

In the La Joya and Buenavista micro-watersheds, the participative planning processes allowed agreements to be reached on the use of the territory based on the participatory analysis of the situation of the micro-watershed before, at present, and in the future. They identified priority problems, conservation values to be recovered and participatory geography to jointly define the runoff units, the current uses of the watershed functional areas, and selecting the best practices to be carried out in different runoff areas (Pineda et al. 2016a, b). The micro-watersheds of Mesa de Escalante (22 years), San Pedro Huimilpan (+50 years), and Buenavista (16 years) had management processes and were integrated into the RCWM into a watershed management and training route, where more than 7000 visitors had been trained in the last 7 years (Figs. 35.2, 35.3 and 35.4).

Various types of inter-institutional agreements and alliances were built during this project:

1. The creation of a CRCC operating group of academics, civil organizations and inhabitants, and foundations as a bridge for communication and operation
2. Installation of an advisory council composed of representatives of communities, agencies, academic, and government institutions
3. Participatory establishment of demonstration areas and livestock exclusion zones for restoration, seeking respect, consensus, and adequate management of runoff units
4. Creation of a network of local promoters and a permanent training program through a multipurpose (research, networking, training) and multilevel (preschool, undergraduate, graduate, farmer training) training route including the four participating micro-watersheds (Pineda et al. 2016a, b)

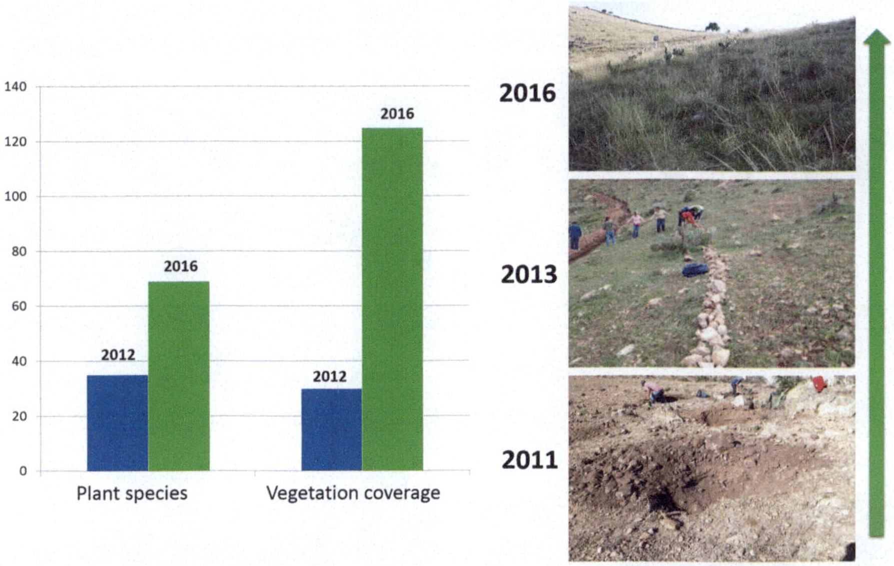

Fig. 35.3 Results of the plant cover rehabilitation on a SWBIOC unit at Puerto del Viento in La Joya micro-watershed. (Source: Pineda et al. 2016a, b)

Although the model proposes an intervention by seven processes, the best practices that integrate them respond to aspects of attention to climate change (Table 35.2).

Some lessons derived from this experience are summarized in aspects where social participation and dialogue are structural to establish trust as a main value in the relationship between operative group and watershed people (Pineda et al. 2014).

(a) Good decisions within the management process require a scientific basis for their selection and operation, a democratization of knowledge, and the development of a common language.
(b) A link of trust with the community and recognition of the value of teamwork is necessary to promote meaningful learning and knowledge dialogues.
(c) Conservation of natural heritage depends on the resolution of basic problems (food, water, money) most felt by the watershed people.
(d) In the communities there are strong capacities for the transmission of traditional knowledge, the acquisition of knowledge, and the generation of new knowledge raised from the best applied practices.
(e) Inter- and transdisciplinary actions have positive impacts in recognizing the role of the academy, government, and society as a management entity; in their responsibility to articulate and promote processes to achieve sustainable development; and in the analysis of problems and management options based on local worldview and lifestyles.

Fig. 35.4 Best practices (BP) of sustainable production: A cattle raising and B reforestation with maguey and demonstration trails; BP for sustainable rural housing: C bio-intensive gardens and medicinal and native plants and D backyard productive units; BP for micro-watershed rehabilitation: E using plant fascines, stone, and geotextile walls and F local biodiversity conservation. (Author: Clara Tinoco https://fcn.uaq.mx/fcn.uaq.mx › crcc)

35.5.2 WWF Model of Integrated Water Management in River Basins

The integrated water management model that the WWF-Gonzalo Río Arronte Foundation (FGRA) Alliance developed in a basin scale corresponding to WWF's freshwater ecoregions: the Conchos River in the Chihuahuan Desert Program; the Copalita-Zimatán-Huatulco rivers in the Oaxaca Coastal Sierra, within the Mexican

Table 35.2 Integrated management of micro-watersheds as a strategy for adaptation to climate change (own elaboration)

Climate change focus areas	Watershed management best practices
Decrease in freshwater reserves	Water use and management in the micro-watershed and local organization Eco-technologies for water supply, collection, storage and treatment, restoration of watercourses, etc. Decrease runoff, increase infiltration, promote moisture conservation, sediment control to preserve water reservoirs, revegetation of slopes, and spring maintenance
Desertification and soil degradation	Soil rehabilitation through water, soil, and biodiversity conservation practices, productive diversification, agroecological management, organic fertilizers, livestock management, etc. Food sovereignty through productive diversification schemes, bio-intensive, communitarian, and family gardens Integrated water-soil-biodiversity management, with SWBIOU (soil-water conservation and biodiversity) units Establishment of biodiversity protection areas and exclusion of livestock activities
Loss of biodiversity and ecosystem services	Biodiversity management, recovery of structure and function Maintenance of the vegetal cover and its recovery by revegetation, reforestation and management of the succession, propagation in situ of native species with greater resistance to the climatic variability, use of the biodiversity to favor the management of grounds, management of pollinators, and habitat management
Greenhouse gas mitigation	Reduction of dependence on external energy systems, use of clean technologies, conservation and management of carbon sinks, etc.

Forests Program; and the San Pedro Mezquital River, in the Gulf of California Program. The model implies a change of paradigm in water management from a vision where the environment is considered as a user of the resource to another where ecosystems are considered the source of services that guarantee social and economic development. The following are identified as the main causes of the deterioration of the country's watersheds: the lack of ecosystem criteria in water and land management, the incipient development of water governance in the watershed, the lack or limited valuation of the environmental services of the forest, the weakness of the municipalities to provide sustainable water services, the lack of awareness by society of belonging to a watershed, and the importance of ecosystems as service providers (González et al. 2009; WWF-FGRA 2006). This model integrates three initiatives considering forests, freshwater ecosystems, and coast-river ecosystems as conservation objects.

(a) Ecological Flow

The model responds to five strategies in which social participation is a transversal activity (Fig. 35.5). The aim is to consolidate the processes of water governance in the basin, oriented toward the follow-up of agreements to achieve sustainable extraction and balance (ecological flow). The Building Block Methodology (BBM) is used, by blocks or phases, and is oriented to the health of all the components of

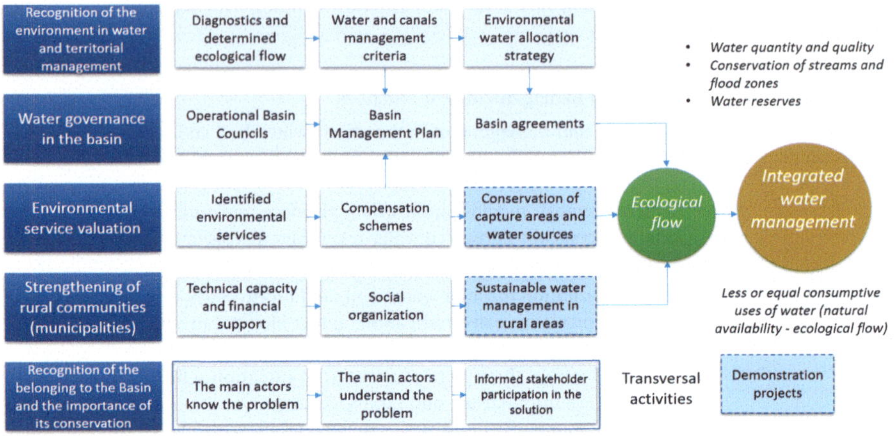

Fig. 35.5 WWF-FGRA partnership model of water management in river basins. (WWF 2019)

the riparian system, considering its structure and functionality (WWF 2019). Three relevant aspects related to social participation and protected areas are as follows:

- Participatory. Academic institutions and organizations, government entities, water users, and rural communities were involved in the process (WWF 2019).
- In support of conservation. The adoption of ecological flows allows the maintenance of the water and sediment regime necessary to maintain the hydrological functionality of associated natural protected areas (Santa Elena-Ocampo-Maderas del Carmen Canyon, Huatulco and Marismas Nacionales) (WWF 2019).
- Strategy for adaptation to climate change. Through its application, society and ecosystem resilience is improved to withstand extreme events (floods or drought) (WWF 2019).

(b) National Water Reserve Program

The knowledge associated to ecological flow experiences, described above, allowed the identification of the basins with a large water availability and that, due to their biological richness, ecological importance, and scarce water human pressure, were considered as national water reserves to guarantee flows for ecological protection. The reserve is a volume of water in a basin destined exclusively for nature protection and human consumption, determining the amount of water required by the flora and fauna and the growth of the population in the next 50 years, respecting water uses for current productive activities and those that will be developed in the future in a sustainable manner (WWF 2018). This coincides with society's need to regain its natural heritage through management that guarantees balanced basins, which is a strategic objective of the Water Agenda 2030 in Mexico (WWF 2019). The benefits of this initiative are as follows:

– Complements the conservation strategy of the country's most important ecosystems and their environmental services: 97 natural protected areas, 55 Ramsar sites, and 78,500 km^2 of additional basins
– Increases the resilience of ecosystems and society to scarcity situations, being an adaptation strategy
– Ensures connectivity across the entire basin and support multiple environmental services provided by ecosystem conservation to water and society management (WWF 2019)

In addition, in the WWF water program, Mexico has a social and technical process to transfer and exchange knowledge through sustainable water use (SWU) models (Fig. 35.6). SWU models are not pre-established but are designed together with the communities or beneficiary groups for different areas of social participation: family, groups of families, schools, and productive municipalities, establishing strong mechanisms of binding social participation. These mechanisms include the training of communities' habitants as process promoters, who help the community and the sectors to increase their capacity for participation and management to solve problems within the framework of environmental sustainability (Tinoco and Gonzalez 2015; WWF-FGRA 2013).

In terms of governance and social participation mechanisms, it includes:

(a) Communication

- Integrated information system for the conservation of the Río Conchos (www.rioconchos.org.mx)
- Monitoring system of social perception and participation (SISMOC) of the Copalita-Zimatán basin (www.comunitasac.org)
- Communication campaign on the natural richness of the San Pedro Mezquital basin (wwf.org.mx/sanpedromezquital)
- Diffusion materials

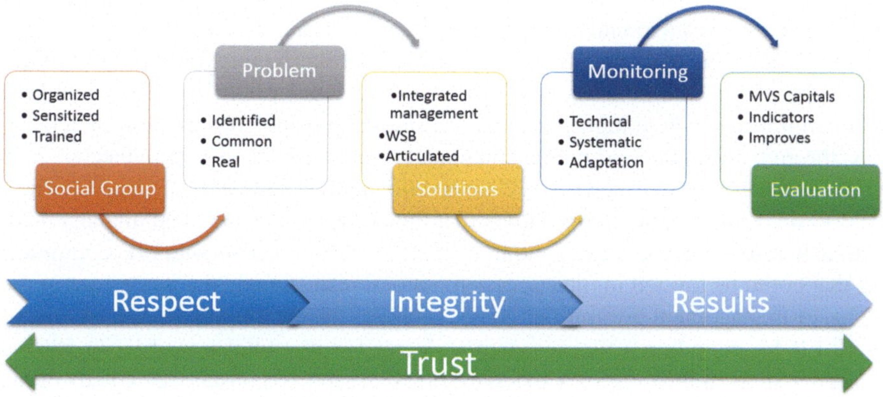

Fig. 35.6 SWU model. (WWF-FGRA 2013; WWF 2019)

(b) Deliberative spaces and capacities for participation in water management

- Creation of committees (community, basin, sub-basin).
- Participatory workshops for climate change water management adaptation plans, integrating management plans

(c) Participating entities

- Generation of inter-institutional working groups and operational management of basin councils

This is part of a "bottom-up" process, which incorporates the participation of users and local actors and the needs of the environment, in collaboration with basin councils, scientists, and experts. They seek to promote improvement and change in water management, translate experiences, define proposals into public policies at the national level, and offer an example at the international level (WWF 2019).

35.5.3 Model of Project C6: Integrating the Watershed Approach and Natural Protected Areas

According to Pineda et al. (2015), the limited territorial coverage of the NPAs in Mexico and the unequal proportion they have over the functional areas of the watersheds make their contribution in terms of natural heritage conservation less viable, especially for those NPAs located in lower functional areas of the watersheds. These authors consider that the approach of analysis and territorial intervention in the sub-basin offers tools to promote the conservation of the natural heritage. At the same time, they recognize other diverse problems related to production, climate change, and attention to risks in the face of natural hazards, among others. These, being based on the participation of the inhabitants, promotes agreements that encourage the conservation of the natural heritage as the basis for socioeconomic development and where the current NPAs can be considered according to their management plans, in an integrated manner within the ordinance of the sub-basin to guarantee the conservation of the natural capital from the local level. The integration of the watershed approach is an urgent need for the recovery of ecosystem services by and for the communities, where their participation strengthens the conservation of ecosystems through adequate land management and improvements in governance.

The project "Coastal Watershed Conservation in the Context of Climate Change" (C6) is an example of an initiative that integrates the watershed management approach with the conservation and management of NPAs. This supports communities that could be limited in their access to natural resources by the guidelines of the decree or management program of NPAs. It covers 6 watersheds with a total area of 1,842,176 ha, including 9 NPAs (two to be decreed) with a total area of 1,134,586 ha, in coastal areas of the Gulf of Mexico (Table 35.3) and 10 coastal watersheds of the southern Gulf of California (Table 35.4) with a total area of 1,968,649 ha, including 9 NPAs (four to be decreed) (CONANP 2013; C6 2019).

Table 35.3 Basins C6 Gulf of Mexico

Watershed (state)	Natural protected area	Factors that require attention
Tuxpan (Veracruz) 671,790 ha	Mesophile forest (to be declared) 340,980 ha	Uncontrolled slash and burn actions Extraction of firewood and irregular extraction of medicinal plants Poaching and smuggling of birds Pests and introduction of exotic species that affect the forest Loss of identity, poverty, and disorganization of indigenous communities Important mining activity in the area of influence
	Lobos-Tuxpan Reef System 30,571 ha	Port of Tuxpan and fishing activities in the Tamiahua lagoon
Antigua (Veracruz) 219,600 ha	Cofre de Perote 11,700 ha	Fires Irregular grazing Illegal logging and extraction of wood Contamination of water bodies
Jamapa (Veracruz) 406,184 ha	Pico de Orizaba	Fires Grazing inside the reserve Irregular grazing Illegal logging and extraction of wood
	Veracruz Reef System	Areas of influence: conurbation areas of Puerto de Veracruz and Boca del Río-Antón Lizardo
Coatzacoalcos-Huazuntlán River (Veracruz) 31,063 ha	Los Tuxtlas Biosphere Reserve 155,122 ha	Cultivation of corn for self-consumption, coffee, tobacco, vegetables Extensive cattle raising predominates Affections to mangrove and lagoon areas Seasonal tourism Community reserves for the protection of springs that require support Oil activity that affects coastal fishing. Native communities within traditional use zones
Temoloapa River (Veracruz) 25,161 ha		
	Los Tuxtlas Reef System (to be declared) 175,389 ha	Pemex's activities affect the NPA and fishing activities
Grijalva-Usumacinta (Tabasco) 488,378 ha	Centla swamps 302,707 ha	Unregulated fishing activity Subsistence agriculture Extensive livestock farming Poaching and trafficking of valuable species Unregulated tourism Gas exploitation
	Usumacinta canyon 46,128 ha	

Source: CONANP (2013)

These basins are highly vulnerable to the effects of climate change and are considered a priority for the conservation of their ecosystems because of their global importance for biodiversity (CONANP 2013). The objective of the C6 project is to promote the integrated management of cost-risk basins as a means of conserving biodiversity, contributing to the mitigation of climate change and increasing the

Table 35.4 Basins C6 Gulf of California and Pacific

Watershed (state)	Natural protected area	Factors that require attention
Piaxtla River (Sinaloa) 696,420 ha	Cacaxtla Plateau 50,862 ha	No information
Presidio River (Sinaloa) 289,361 ha	Mount Mojino (to be declared) 203,467 ha	
Baluarte River (Sinaloa) 151,946 ha		
	National marshes Sinaloa (to be declared)	
Acaponeta River (Nayarit) 161,884 ha	Isabel Island 194 ha	
	Marías Islands 641,542 ha	
	National marshes Nayarit 133,854 ha	
San Pedro River (Nayarit) 249,476 ha		
	Marías Islands	
Cuale River (Nayarit y Jalisco) 26,674 ha Las Juntas River (Jalisco) 32,773 ha Pitillal River (Jalisco) 43,207 ha El Tuito River (Jalisco) 44,492 ha	Marías Islands	Impacts of tourist activities Inadequate management of traditional agricultural and forestry activities Decrease of water in wells and galleries Real estate speculation and new roads Extensive agricultural and livestock practices Parking in the common area Sale of land
	El Cuale (to be declared) 121,651 ha	
	Islands Marietas 1383 ha	
Ameca River (Jalisco) 272,416 ha	El Cuale (to be declared)	
	Vallejo-Ameca Mountains (to be declared) 277,448 ha	

Source: CONANP (2013)

sustainability of land use, with emphasis on generating improvements in governance and social participation. The aim is to mitigate the social impacts of restrictions on access to natural resources in federal natural protected areas (NPAs), in order to generate positive effects on the conservation of the basins and their resources, while contributing to regional development and the well-being of the inhabitants. In this sense, its conservation strategy is based on voluntary participation in various incentive programs to promote adequate resource management (CONANP 2013).

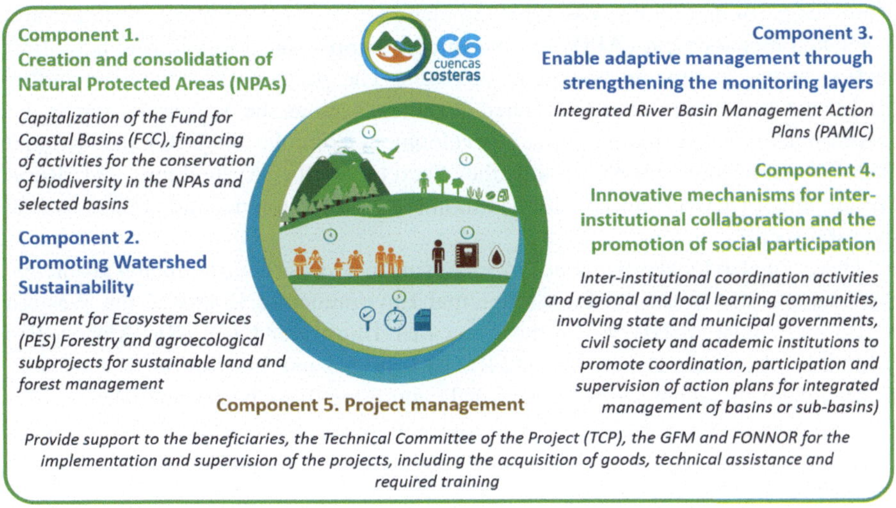

Fig. 35.7 Components of the C6 approach. (Source: CONANP 2013, C6 2019, http://www.c6.org.mx/)

It includes five components with four institutions with extensive experience in their field of action; these have been carrying out key activities for conservation in Mexico and to address the impacts of climate change and are articulated to generate an integrated strategy (Fig. 35.7). The National Commission of Natural Protected Areas (CONANP), the Mexican Fund for the Conservation of Nature (FMCN), the National Forestry Commission (CONAFOR), and the National Institute of Ecology and Climate Change (INECC) are all involved.

The mechanisms for social participation at the basin or sub-basin level promoted by the C6 project are:

(a) Coordination and consensus building among different stakeholders for better management of each basin and its governance
(b) Promotion of social participation and institutional transparency in decision-making
(c) Integration of action plans for the integrated management of each river basin (PAMIC) with the participation of the different stakeholders
(d) Supervision of the execution of the financed activities and coordination with other activities related to the PAMIC
(e) Supervision of the actions of C6 and coordination of other activities related to the GEMAP
(f) Exchange of experiences between basins and sub-basins
(g) Identification of capacity-building needs at the basin level and search for resources for financing (CONANP 2013)

The advisory councils and sub-councils of the NPAs (CNPA) are key forums for stakeholder mobilization required for work in the basins, including indigenous participation. The National Institute of Ecology and Climate Change (INECC) promotes participation processes for the elaboration of action plans for integrated watershed management (APIWM) and in community-based monitoring activities, such as water quality measurement in rivers. The project's technical committee establishes links with other C6-related agencies, such as the Secretariat for Social Development (SSD), the Indigenous Development Commission (IDC), the National Water Commission (NWC), and the Secretariat for Agriculture and Rural Development (SARD), to enable the alignment of investments and cooperation in the rural sector.

There are also local governance mechanisms to address environmental problems. An example of this is the Inter-municipal Environmental Board of the Western Sierra and Coast (IEBWSC) in the Ameca-El Tuito river basin (Rio Ameca, Rio Pitillal, Rio El Cuale, Las Juntas, and Rio El Tuito basins). In addition, as a measure to strengthen social participation, regional committees have been established to provide advisory services, linking C6 activities with other initiatives active in the region that may be complementary and support management among key stakeholders in the basins.

In several basins in Mexico, the inter-municipal boards allow rural municipalities to interact in a more balanced way with the cities around them and to add resources for programs that affect the integrated management of basins (solid waste, deforestation, etc.), for example, flexibility to initiate new emerging programs and to discontinue those that have already fulfilled their function; opportunities to trigger multiple interactions and collaborative networks among actors; and represent new spaces, platforms, or mechanisms to achieve timely and rapid application of scientific research results for the management of territory and natural resources to promote informed participation by inhabitants and institutional decision-makers (Graff 2018).

Part of the main problems in NPAs is the disorientation and rapid growth of populations and their extractive activities. The lack of knowledge of all sectors (governmental, social, productive) is related to the lack of valuation of the ecosystemic services, generating deficiencies in the design and implementation of public policies, that decreases NPAs in a fragmented way with respect to the territory and the diversity of ecosystem themes whose management plans are not appropriate to the context (Maldonado et al. 2018).

The complexity of the territorial socio-environmental problems requires an updating of environmental policies and inter-institutional teams, to effectively incorporate the participation of the inhabitants, and the generation of an integral vision that allows the articulation of the already made decisions, with innovative initiatives that support and promote self-organized and self-managed processes responding to the particular conditions of the territory. Some important components of social innovation, in order to manage protected areas more efficiently, are the satisfaction of social needs, value for society, social cohesion, the promotion of collective talent, generation of new relationships, social empowerment and participation, and integrated management strategies that value the connection between ecological and social systems. As a result, the watershed approach has the potential to support and improve the NPA management of protected areas (Maldonado et al. 2018).

35.6 Conclusions

The NPA territorial approach has its greatest strengths in the legal framework established for their creation and operation, as well as the knowledge that people have of their existence and operation. However, with exceptions, there are portions of natural or political-administrative territories with a limited vision of their natural context. And even if their creation and operation are intended to have important social participation components, this partially happens through the committees representing the community. To ensure their conservation, it will require significant efforts in the future to guarantee their ecological connectivity, so that they are not maintained as natural islands.

The IWM approach, following the three models described above, proves that land-based intervention at the sub-basin and micro-basin level promotes the conservation of natural heritage in an integrated manner with other problems and supports communities' participation to establish agreements that promote the conservation of natural heritage as the basis for socioeconomic development (Pineda et al. 2015). The watershed approach becomes an effective territorial management strategy for ecosystems' conservation, management, and adaptation to climate change. This is due to the importance given by social participation and the differentiated responsibilities of the actors involved in the watershed (which is a structural factor for the adequate management of any territory, since it is necessary to generate the conditions to promote processes of local knowledge and skills in the different actors involved). This appropriation will allow the agreements and actions derived from management programs to be maintained in the long term. Some advantages of the integration of this approach are:

(a) An integral systemic vision of the territory that favors the prevention and attention of risks related to climate change. The conservation and restoration of the ecosystems generate services for socioeconomic activities and prevent natural disasters as key adaptation elements. From the integrated watershed management approach, processes that can strengthen the capacity of ecosystems and their biodiversity to adapt to the impacts of climate change are established.
(b) The participatory nature of planning supports the understanding of the phenomenon, vulnerabilities, possible impacts, and risks. It also involves the society in the design and implementation of adaptation actions, considering the adjustment of the human and economic system according to the sustainable use of the basin territory.
(c) Planning allows the use of natural resources and conservation strategies that take into account the possible future scenarios. These will reduce impacts and risks, reducing the cumulative effects of inadequate management and integrating scientific, technical, and community knowledge.
(d) It allows the coordinated intervention of different watershed stakeholders, such as the public, private, and civil society sectors, in order to seek strategies that ensure the sustainability of the resources in the current scarcity and deterioration scenarios.

(e) It facilitates the generation of agreements, among the actors involved, for organized production, generation of strategies for the solutions with differentiated responsibilities, and strengthening local forms of governance, such as intermunicipal boards, committees, etc.

However, IWM faces the same approach problems with NPA: disarticulation among institutional actors, poorly functioning and contradictory public policies, lack of continuity in programs, low levels of concurrent funding, and lack of local capacities for informed participation in decision-making. It is necessary to establish mechanisms that allow the effective operation of public policies, where conservation and management of natural resources consider real (functional) participation and a strengthening of governance from the local level, and then strengthen the inhabitants' capacities and reduce the control of external institutions for economic, administrative, political, or technical knowledge issues. In addition, legal framework for IWM approach is weak, despite their recognition in various main laws as an ideal planning and intervention territory (water, forest, rural development, and climate change). Therefore, its growing interest along the country is due to two factors: the increasing number of successful watershed management cases and the learning of decision-makers and operators from academy, government, and NGOs.

35.6.1 Lessons Learned

In the current context, the answer to the initial question of this study is that, although the approaches to territorial management considered have advantages and disadvantages, the most appropriate one to favor participation and the possible appropriation of conservation-production strategies is the integrated watershed management. This, with the supporting analysis and alternative solutions of socio-ecosystemic nature, enhances the advantages of both and develops a better transition toward sustainability, like the C6 project which provides another vision that is the hybridization of both approaches.

The main challenges are the lack of coordination and agreement between institutions, which generate risks of negative socio-environmental impacts for the basin due to the incidence of disjointed programs and policies. Also the lack of compatibility between conservation policies and sustainable uses in management actions. With regard to social participation, there is still a lack of innovative mechanisms that promote the integration of visions and the joint construction of alternatives for conservation and sustainable use. The inhabitants recognize the conservation and restoration of ecosystems as an indispensable aspect for the maintenance of ecosystem services for the adequate use of natural resources and the satisfaction of local needs.

Participation should not be restricted to consultation and approval of programs but should be extended to a more active role in the decision-making. It is necessary to establish criteria or mechanisms to ensure balanced representation of the various water users and sectors and the territory, along with comprehensive educational communication and awareness-raising strategies in institutional, productive, and social sectors that have an impact on both urban and rural areas. Also, it is important to implement citizen bodies that not only promote participation but also have a greater influence on the decision-making, greater powers, training, and financing, so that effective communication channels and functional participation by the population are established, recognizing that power is not taken away from the institutions, rather than generating more equitable, transversal, and locally appropriate decision-making processes, which strengthens governance for ecosystems' management.

Acknowledgments To the Water Resources Center for Agriculture and Mining (CRHIAM) Project ANID/FONDAP/15130015.

References

Alcorn JB (2005) Dances around the fire: conservation organizations and community-based natural resource management. In: Communities and conservation: histories and politics of community-based natural resource management. Altamira Press, Walnut Creek, pp 37–68

Aldunce P, Beilin R, Handmer J, Howden M (2014) Framing disaster resilience: the implications of the diverse conceptualizations of "bouncing back". Disaster Prev Manag 23(3):252–270. https://doi.org/10.1108/DPM-07-2013-0130

Arriagada R, Aldunce P, Blanco G, Ibarra C, Moraga P, Nahuelhual L et al (2018) Climate change governance in the Anthropocene: emergence of polycentrism in Chile. Elem Sci Anth 6(1)

Bautista L (2007) Las vedas forestales en el México post-revolucionario. Thesis for the Master's degree in Regional Studies. Mexico. José María Luis Mora Research Institute, 186 pp

Brenner L, San German S (2012) Local governance for 'ecotourism' in the monarch butterfly biosphere reserve, Mexico. Alteridades 22(44):131–146

Berggruen N, Gardels N (2013) Intelligent governance for the 21st century: a middle way between west and east. Wiley

Botana A, Zugasti C, Ortega L, Miranda C (2019) Producción de espacios de conservación en Veracruz, México/Production of conservation spaces in Veracruz. Mexico Brazilian Journal of Development 5(6):5411–5433

Campos M, Velázquez A, Verdinelli G, Skutsch M, Juncà M, Priego-Santander A (2012) An interdisciplinary approach to depict landscape change drivers: a case study of the Ticuiz agrarian community in Michoacan Mexico. Applied Geography 32(2):409-419. https://doi.org/10.1016/j.apgeog.2011.06.004

C6 (2019) Coastal Basins and Climate Change Project. http://www.c6.org.mx/

Carvajal A, Gilio M (2015) Teaching and learning the basin: a multidisciplinary pedagogical model In: Burgos A, Bocco G, Sosa Ramírez J (Coordinators). Social dimensions in basin management. CIGA-UNAM, Mexico, pp 75–90

CECADESU, Education and Training Center for Sustainable Development (2015). Ministry of the Environment and Natural Resources

El Colegio de México (2012) Towards good governance for integrated water resource management. Regional process of the Americas. VI World Water Forum Marseille, France. Retrieved from http://www.unesco.org.uy/phi/fileadmin/phi/infocus/World_Water_Forum_2012/5_TOWARDS_GOOD_GOVERNANCE_FOR_INTEGRATED_WATER_RESOURCES_MANAGEMENT.pdf

CONAGUA, National Water Commission (2017) Water Atlas in Mexico. Access in 06 Junio 2019

CONAGUA, National Water Commission (2018a) National Water Commission. http://conaguagobmx Access on 10 Nov 2018

CONAGUA, National Water Commission (2018b) National Water Information System. http://sinaconaguagobmx Access on 23 August 2019
CONAGUA, National Water Commission (2018c) Basin Councils. https://www.gob.mx/conagua/documentos/consejos-de-cuenca
CONANP, National Commission of Natural Protected Areas (2013) Procedural framework for coastal watershed conservation project in the context of climate change (C6) IPP636 V5.. National Commission of Natural Protected Areas
Cotler H (2010) The watersheds of Mexico. Diagnosis and Prioritization. Pluralia Ediciones e Impresiones SA de CV, Mexico
Cotler H, Pineda R (2008) Manejo integral de cuencas en México ¿hacia dónde vamos? Boletín del archivo Histórico del agua 39(2008):16–21. Retrieved from http://132.248.9.34/hevila/Boletindelarchivohistoricodelagua/2008/vol13/no39/2.pdf
CR2 (2019) The Anthropocene in Chile: evidence and ways forward. Report to the Nation. Center for Climate and Resilience Research, Chile
Cusva A (2014) Updating and analysis of municipal socio-economic information, design and development of indicators for descriptive relational analysis with available mapping information of wetlands at scale 1: 100,000
de la Cruz-Hernández J, Ávila-Akerberg V, Herrejón M, Bordi I (2016) The disincorporation, reduction of category or surface area in protected areas (PADDD phenomenon) and its effect on forest resource management in an Ejido of Nevado de Toluca, Mexico. Teoría y Praxis 19:95–118
DOF, Diario Oficial de la Federación (2014) Ley de Aguas Nacionales, Cámara de Diputados. In: Congreso de los Estados Unidos Mexicanos
DOF, Diario Oficial de la Federación (2018) Law of Ecological Equilibrium and Environmental Protection Chamber of Deputies. In: Congress of the United Mexican States
DOF, Diario Oficial de la Federación (2019) Ley de Desarrollo Rural Sustentable (Última Reforma DOF 12-04-2019) Cámara de Diputados. In: Congreso de los Estados Unidos Mexicanos. http://www.diputados.gob.mx/LeyesBiblio/pdf/235_120419.pdf. Access 13 August 2019
Fernández A, Cruz F, del Razo I, de Jesús B, Márquez A (2002) Evaluation of the advance of the urban stain on the protected natural area of Cañada de Los Dinamos. Gaceta ecológica 62:56–67
Graf S (2018) Governance mechanisms for integrated river basin management in the context of climate change. VI Water Forum. University of Guadalajara
González ID, de la Lanza G, Sánchez R (2009) Workshop report: proposal of ecological flow in the Copalita-Zimatán-Huatulco basin. Alliance WWF-Gonzalo Río Arronte Foundation IAP Mexico. 19p
Halffter G (2011) Biosphere reserves: challenges and opportunities in Mexico. Mexican Zoological Act 27(1):177–189
LAN, Ley de Aguas Nacionales (1992) http://www.diputados.gob.mx/LeyesBiblio/ref/lan/LAN_orig_01dic92_ima.pdf. Access 13 August 2019
LAN, Ley de Aguas Nacionales (last reform 2016) http://www.diputados.gob.mx/LeyesBiblio/pdf/16_240316.pdf Access 13 August 2019
Luna E (2014) Social evaluation of the forest restoration program in priority watersheds with Mazahua localities of the Cutzamala priority micro-watershed system. Doctoral dissertation, Master's thesis, Universidad Autónoma de Querétaro, Mexico
Magrin GO, Marengo JA, Boulanger JP, Buckeridge MS, Castellanos E et al (2014) Central and South America. In: Barros VR, Field CB, Dokken DJ, Mastrandrea MD et al (eds) Climate change 2014: impacts, adaptation, and vulnerability. Part B: regional aspects. Contribution of working group II to the fifth assessment report of the intergovernmental panel on climate change. Cambridge University Press, Cambridge/New York, pp 1499–1566
Maldonado E, Torres H, Rojas H (2018) Environmental policy design and social innovation theoretical and methodological contributions to the sustainable management of natural protected areas. Econ Soc 38:111–129

Mascia MB, Pailler S, Krithivasan R, Roshchanka V, Burns D, Mlotha MJ, Peng N (2014) Protected area downgrading, downsizing, and degazettement (PADDD) in Africa, Asia, and Latin America and the Caribbean, 1900–2010. Biol Conserv 169:355–361

McGinnis M, Ostrom E (2014) Social-ecological system framework: initial changes and continuing challenges. Ecol Soc 19(2). https://doi.org/10.5751/ES-06387-190230

Merino L, Segura G (2007) Las políticas forestales y de conservación y sus impactos en las comunidades forestales en México. Community forests in Mexico Sustainable management of forest landscapes, pp 21–49

Miranda A, Altamirano A, Cayuela L, Lara A, González M (2017) Native forest loss in the Chilean biodiversity hotspot: revealing the evidence. Reg Environ Chang 17(1):285–297

Monroy A, Sánchez V, Briones M, Lira R, Maass J (2015) Representativeness of vegetation types in different conservation initiatives in Oaxaca, Mexico. Bosque (Valdivia) 36(2):199–210

Monterrubio-Solís C (2019) Formalization of areas intended for voluntary conservation in community and indigenous territories, advances and setbacks. EntreDiversidades 6(1 (12)):79–110

Nardi A, Mahmoud Y (2017) The importance of territorial scale in biodiversity conservation and natural resource exploitation analyses: a discussion from the Korup National Park (Cameroon). In: XII RAM Mercosur Anthropology Meeting

OECD, Organization for Economic Cooperation and Development (2012) Water governance in Latin America and the Caribbean: a multilevel approach. OECD Publishing, Paris. https://doi.org/10.1787/9789264079779-es

Ortiz S, Romo M (2016) Socioenvironmental impacts of water management in the protected natural area of Cuatro Ciénegas, Coahuila. Region Soc 28(66):195–230

Pineda R, González L, Tinoco C, Bustos D, Hernández V (2014) Creación del Centro Regional de Capacitación en Cuencas. Fundación Gonzalo Rio Arronte-Universidad Autónoma de Querétaro, México

Pineda F, Ríos E (2013) Watersheds. Foundations and perspectives for its management

Pineda R., Hernández J, Tobar R (2015) Conservation challenges for natural heritage from an integrated watershed management approach. DIGITAL CIENCIA@UAQRO

Pineda, R, Segura O. and González L (2016a) Learning from watershed management for rural area management to increase socio-environmental resilience: the case of La Joya micro basin. In: Hremesse, J., Servais, O., Schmitz and A. Tobar (COMP). Land use planning in disaster prevention: case studies in Quetzaltenango (Guatemala), Mexico and Canada. University of Louvain Press, Belgium, pp 201–213

Pineda R, Tinoco C, Barrera G (2016b) A proposal for strengthening local capacities in integrated watershed management from a climate change perspective. In: Espinoza M, Vidriales G, Barrera C, Escobedo M (eds) Memory of the IV National Congress on watershed management. Universidad Veracruzana, Institute of Ecology, SENDAS A.C, Xalapa Veracruz

Porter-Bolland L, Ellis EA, Guariguata MR, Ruiz I, Negrete S, Reyes V (2012) Community managed forests and forest protected areas: an assessment of their conservation effective-ness across the tropics. For Ecol Manag 268:6–17

Tinoco C and González I (2015) Diploma in watershed management. CECADESU – SEMARNAT, education and training Center for Sustainable Development of the Ministry of the Environment and natural resources

Tinoco C (2017) Systematization and analysis of qualitative methodologies to evaluate the social perception of adaptation to climate change. Project "construction of monitoring and evaluation schemes for adaptation in Mexico for the formulation of public policies based on evidence". INECC, National Institute of ecology and climate change, CONACYT, National Council for science and technology

Tinoco C, Barrera G, Dorantes C, González L, Pineda R (2017) La formación de promotores comunitarios. In Ministerio de Medio Ambiente (comp). Results report of the Chile – Mexico project sustainable use of ecosystem Services of Water Resources Department of local environmental management, environmental education and citizen participation Division. AGCID and AMEXCID. pp 19–25

Tlapa M, González Á, López S, Valverde B, Gutiérrez V, Bello G (2020) Deterioration factors of periurban protected natural areas in the Puebla Valley, Mexico. Estudios Demográficos y Urbanos, 35(1(103):51–82

UNEP, United Nations Environment Programme (2012) GEO 5 Global Environment Outlook

Wilder M (2010) Water governance in Mexico: political and economic apertures and a shifting state-citizen relationship. Ecol Soc 15(2)

WWF, World Wildlife Fund (2018) Press release on the water decrees. https://ceiba.org.mx/wwf-boletin-prensadecretos-reservas-agua/. Accessed on 03 December 2019

WWF, World Wildlife Fund (2019) http://www.wwf.org. Accessed on 03 December 2019

WWF-FGRA, World Wildlife Fund – Gonzalo Río Arronte Foundation (2006) General Strategy of the Watershed Management Program Development of New Models in Mexico. Alliance WWF-Gonzalo Río Arronte Foundation, I.A.P. (FGRA)

WWF-FGRA, World Wildlife Fund – Gonzalo Río Arronte Foundation (2013) Sustainable water use models as a strategy to reduce vulnerability and poverty. Climate change adaptation measures: water management in rural communities. Water Management in Watersheds Project: Development of New Models in Mexico Phase II. WWF-Gonzalo Río Arronte Foundation Alliance

Chapter 36
The Use of Geographical Environmental Perception in the Detection of Contaminated Urban Streams: Toward the Proposal of Environmental Policies in Chiapas, México

Sergio Hernández-Solorzano, Maria Eugenia González-Ávila, Elizabeth Olmos-Martínez, and Juana Isabel Vera López

Abstract This chapter identified and analyzed spatial, temporal, and occupational factors, among others, by assessing the perception of the communities located along El Riito micro-basin in the municipality of Tonalá, Chiapas, since pollution problems had been observed in the urban streams caused by human activities and the lack of management from local public institutions faced with such problem. For this purpose, a mixed method was used, which included, in the qualitative phase, the environmental perception with a geographic approach (PAEG, for its acronym in Spanish). This phase was assessed by interviews and surveys that allowed using statistical and spatial analyses and helped to detect the existing pollution in the streams. Additionally, their level of pollution had been related with the external factors previously mentioned influencing the joint responsibility of the problem, which suggested some actions of public policy with the participation of the communities. Finally, despite the factors previously detected, the communities have the willingness and disposition to participate in any program or activity to solve such problem. In the case of the authorities, although they have performed actions, they have been

S. Hernández-Solorzano (✉)
Independent Consultor, Monterrey, Nuevo León, Mexico

M. E. González-Ávila
Department of Urban Development and Environment, The North Border College, Tijuana, Mexico
e-mail: megamar@colef.mx

E. Olmos-Martínez
Department of Economic-Administrative Sciences, Autonomous University of the West, Cali, Mexico

J. I. Vera López
Department of Public Administration Studies, The North Border College, Tijuana, Mexico
e-mail: jvera@colef.mx

A. Ortega-Rubio (ed.), *Socio-ecological Studies in Natural Protected Areas*,
https://doi.org/10.1007/978-3-030-47264-1_36

brief and rather ineffective to decrease hydric pollution in favor of human health, hydric quality, and sustainability of the basin.

Keywords Perception · Water · Contamination · Urban streams

36.1 Introduction

The relationship of human beings with their environment has always been contradictory, on the one hand, destroying to survive, and on the other hand, guaranteeing reproduction of living organisms with the purpose of solving their food and other needs to improve their lives (Tommasino et al. 2000). This complex relationship has been evident after the environmental crisis caused by the degradation of environmental quality; this last one is understood as a series of environmental, social, cultural, and economic characteristics that qualify the status, which includes availability and access to nature components, as well as the possible alteration in the regional natural environment (Fundesnap 2008). Lovelock (1992) denotes such fact in his theory GAIA (name of ancient Greek goddess of the Earth), which refers to nature components having an effect on planet Earth in its totality, including living organisms, oceans, rocks, and atmosphere because they function as a superorganism that modifies actively its internal composition to assure its survival and where human beings shall also be affected.

One of the best examples of the relationship man:natural environment is given by the most simple and valuable resource in the planet, water; although a simple compound, it is essential for all living forms and socio-environmental and economic processes of human beings. With respect to the relationship, Lezama and Graizbord (2010) mentioned the need for human beings to have an integrating perspective in water management that allowed directing their behavior and participation on their natural environment as part of the water management process.

Under the previous premises, an obligated question is, *why study the perception of a natural resource, such as water?* One of the reasons is precisely to identify the relationship between human beings and a resource to know whether to appreciate it or if an element exists that potentially leads to value it (Arizpe et al. 1993; Padilla-Sotelo and Luna 2003).

The information previously mentioned took us to define three basic concepts: perception, environmental perception, and geographic environmental perception (PEAG, for its acronym in Spanish); this last one is the approach used in this study. According to Sescovich (2016), perception is the product received by the person while relating it with their environment, channeled by their sensorial organs, in such a way that when the individuals give it a meaning, at this moment the sensations have transformed into perception. This perception emerges from the individuals' knowledge of the environment and should allow later to integrate management with a fundamental role within the process, management, and use of the resources, such as water or another one understood individually but constructed socially because of the distinct impressions (Aguilar and Brenes 2013). Whereas the environmental perception defined by Moyano (2009) was the existing relationship between the human being and environment, which is understood as the way the individuals appreciate and value their natural environment, providing elements that potentially contribute to its conservation. While perception from the geographical point of view, in this study called environmental perception with a geographic approach (PAEG), is the act of decision on the environment based on psychological and anthropological knowledge (Capel 1973).

In the case of Mexico, environmental perception studies or perception of the sustainable use of the natural resources is scarce, which is likely because it is a method involving different disciplines, such as psychology, anthropology, and human geography, that may be complex to interpret or only represent populations with specific characteristics, making it difficult in some cases to generalize the perception toward similar communities. However, in the case of the geographic perception (Gutiérrez and Peña 1996) used in this study, it proposes integrating qualitative and quantitative variables that allow integrating the process of decision-making according to Yi Fu-Tuan (1975); in other words, it would adapt to environmental topics spatially and temporally, integrating qualitative factors (participation and perception within itself) for performance or proposals (Fernández 2008).

Thus, the objectives of this research study were to identify and assess pollution of El Riito urban stream located southeast of the state of Chiapas, Mexico; the methodological contribution was made by an environmental perception method with a geographical approach that allowed integrating qualitative and quantitative aspects of two communities located in the higher part (Tonalá) and the other one (Paredón) in the lower part of the basin.

The chapter discusses the research performed in six sections: (1) sets out the theoretical framework, (2) describes the study area, (3) explains the methodology, (4) shows the results, (5) develops the analysis and discusses the result, and finally (6) concludes on how the geographic and temporal factors influence pollution perception of the urban streams for both communities in this study; it is worth to note that in the measure that the inhabitants are located close to the stream, they tend to detect the pollution problem more easily compared to those who live farther. Something similar happens with time; at a greater time of residence and knowledge of the environment, changes are more easily detected by the population and with it, pollution. The previous information allowed us to consider that perception with a geographical approach is a good tool that allows knowing part of the feelings, perceptions, and knowledge that the population has of its natural environment, in this case pollution of the urban stream and how much they feel jointly responsible of the problem and willing to participate to solve it. This same section includes some proposals to be considered by the competent authorities and social actors in favor of a hydraulically sustainable basin.

36.2 Theoretical Framework

Conceptually, perception is the knowledge that the individuals get when they relate with the natural environment, channeled by their sensorial organs, in such a way that when they give it a meaning, at that moment, the sensations have been transformed into perception (Sescovich 2016). Nonetheless, the concept of perception has evolved because of the diverse disciplinary contributions, which have originated multiple definitions. One of the main disciplines is psychology, and in terms traditionally used in this field, it has been defined as a cognitive process of the conscience that achieves acknowledgment, interpretation, and meaning to elaborate judgment surrounding the sensations obtained from the physical and social environment (Vargas 1995). Other authors, such as Conroy (2005) and Stea and Downs (2003) who also dealt with the concept of psychological perception, center on the relationship with the cognitive response of the individuals to their natural environment.

However, the perception with an anthropological approach is defined through the experience, attributing qualitative characteristics to the environmental circumstances by points of reference based on specific cultural and ideological systems and reconstructed by the social group. These points of reference generate evidence of the reality, in this case, the environment (Vargas 1995). This point of view is retaken by Lazos (2010), who subsequently with Paré included information on the factors that influence these perceptive processes, as population density and size besides external variables, such as the system, means of communication, forms of education systems, consumption patterns, and incidence of religions and churches, among many other factors (Lazos and Paré 2000).

The previous information was supported by authors as Moyano et al. (2009) who referred to the existing relationship between human beings and their environment as a reflection of their environmental perceptions that may contribute to conservation. Olmos et al. (2016) provided an example of this type of perception when they mentioned that the way of perceiving and measuring the subjective validity, which the persons, groups, and human communities exert on the sense of change in the social, economic, and environmental settings, may produce changes in human behavior on the use and appropriation of natural resources. While in the study of a basin in Chiapas, Benez et al. (2010) detected that water quality perception of the communities was heterogeneous, which was interpreted as interconnection and complexity between the dimensions of the setting and the culture cosmovision aspects on water resource and the location of such populations in the basin.

In the case of the geographic perception dealt with in this study, it is characterized by the act of deciding on the environment, based on psychological and anthropological knowledge (Capel 1973). Whereas Bonnes et al. (2004) used this type of perception on management planning of natural resources, which arrived to the conclusion of the need of environmental education leading the population to raise awareness on the importance of the areas they inhabited and the existing natural resources. Additionally, the geographic perception contributes to detecting collective knowledge in relation to the past, traditions, competencies, productive structure, cultural patrimony, material resources, and the future (Millán 2002) which allows planning urban spaces, such as Bailly (1984) dealt with it considering the union between the subject and object, that is, including the representations that individuals have on the objects, spaces, and time and assessing the independent conception of the groups with certain similarities.

It is worth to highlight that perception studies with the geographic perception are characterized placing emphasis on analyzing the environment from a holistic view, trying to analyze the processes that allow the individuals to capture their surroundings, and introducing themselves within the process of defining their space (Valera 2002). Furthermore, this type of perception has important theoretical contributions of environmental perception, based on Brunswick's (1956) functionalism theory, Gibson's (1980) ecological perception theory, and Ames' (1951) transactional perception approach, seen as a transaction between the individuals and their environment.

Under the previous premise, the study sets out a methodology with an environmental perception with a geographic approach (PAEG) that integrates the absolute and relative space of environmental reality, which translates into a methodology both quantitative and qualitative. According to Creswell (2008) in Pereira (2011), they emphasized the application of this methodology as an exploratory strategy.

It is important to mention that PAEG considers the approach proposed by Adelbert Ames (1951), quoted in Valera (2002), where perception was considered as a transaction between the person and the environment; it started from the senses given under a transaction among the principles assumed, the perceptions, and the environmental characteristics at the time they allowed knowing and analyzing the variables that influenced the individuals. It is worth to mention that under the PAEG proposal few studies have been performed specifically applied to water bodies, above all in rivers and streams, although they have been the basis of several productive regional activities and receptors of waste generated by different human activities (Mora et al. 2002).

Before proceeding, the concept of urban stream (arroyo) should be clarified, which is not defined in the Law of National Waters (Ley de Aguas Nacionales, DOF 2016) or in the General Law of Ecological Equilibrium and Environmental Protection (Ley General de Equilibrio Ecológico y Protección al Ambiente, DOF 2018), which in this case shall refer to the concept developed academically. Under this premise, Guerrero (2011) defined streams as natural water currents with scarce water volume; thus, this study shall define an urban stream "as a superficial intermittent water current type that goes through or surrounds an urban city, and whose affectation (pollution) is determined by the productive activities of the city (industries, businesses, etc.) and population concentration (household discharge and solid waste, etc.), transforming them into pollution sources and possible sources of diseases."

36.3 Area of Study

The study area is located at 17° 59′ N, 14° 32′ S latitude and 90° 22′ E, 94° 14′ W longitude to the southeast of Mexico in the state of Chiapas (https://www.chiapas.gob.mx/ubicacion/), which borders to the north with the state of Tabasco, to the west with Veracruz and Oaxaca, to the south with the Pacific Ocean, and to the east with the Republic of Guatemala (Ihne 2004) (Fig. 36.1).

The micro-basin is located in the next central coordinate: 16° 5′ 12.33″ N latitude and 93° 45′ 48.96″ W longitude, specifically in the southwest of the state of Chiapas, taking as reference the municipality of Tonalá located in the isthmus coast of the state with a territorial extension of 1867.72 km^2 (Inafed 2009),with a surface area of 76.45 km^2, representing approximately 4.32% of 23.7 km long; it originates in the slopes of Sierra Madre de Chiapas at 622 m a.s.l. and flows into the wetland and lagoon of Mar Muerto in El Paredón (Arellano and Ruiz 2016) (Fig. 36.2).

According to data of the Public Registry of Water Rights (Registro Público de Derechos de Agua, REPDA 2015) in the micro-basin, 54 permits of superficial water use, 53 of underground water use, and 4 for wastewater discharge on the stream El Riito have been granted with an average value of an annual discharge of 450 934.18 m^3/year, in which the discharged water quality is unknown. Furthermore,

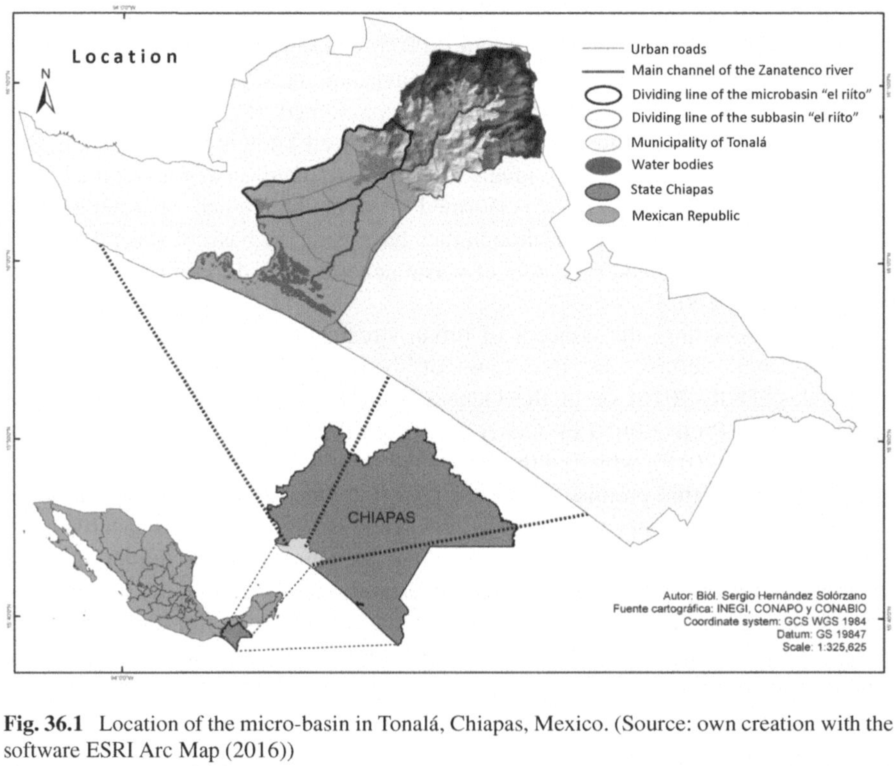

Fig. 36.1 Location of the micro-basin in Tonalá, Chiapas, Mexico. (Source: own creation with the software ESRI Arc Map (2016))

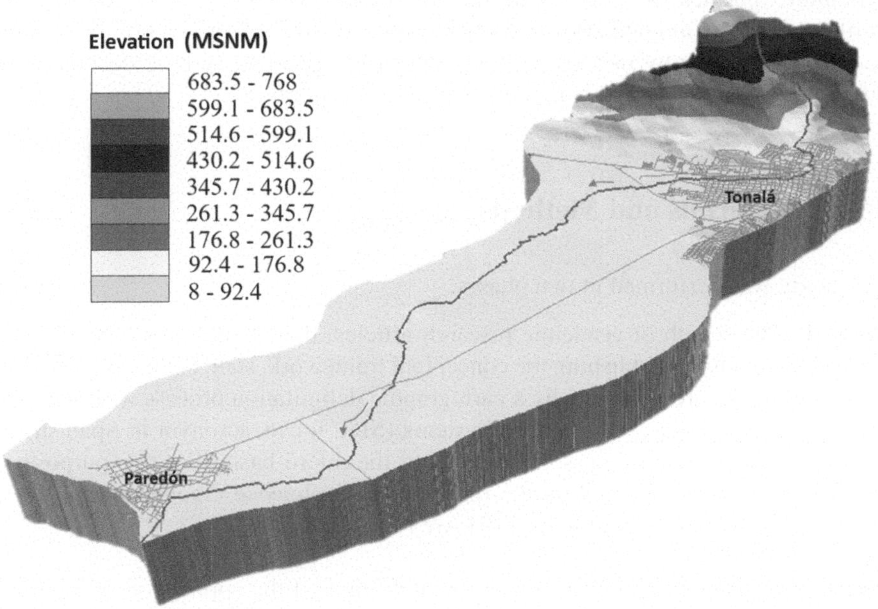

Fig. 36.2 Relief of the micro-basin and principal riverbed of the stream El Riíto, Tonalá, Chiapas, Mexico. (Source: own creation with the software ESRI Arc Map (2016))

on the use of soil and vegetation, only 7.65% of the total surface was found under a natural vegetal cover, while 92.35% was used for anthropic activities: urban settlements, cultivated grazing lands, and temporal agriculture with maize, bean, and fruit cultivations (INEGI 2012).

The comparison of PAEG was performed in the communities of Tonalá and El Paredón. On such communities at social level, Population and Housing Census of the years 1990, 2000, and 2010 in the National Institute of Statistics, Geography, and Informatics (www.inegi.org.mx) website estimated an increase in population growth rate of 16% from 1990 to 2000 and 13% from 2000 to 2010 in the community of Tonalá and an increase of 9% from 1990 to 2000 and 4% from 2000 to 2010 in the community of El Paredón. It is worth to mention that the municipality of Tonalá is catalogued with a degree of medium margination type, according to data from the Ministry of Social Development (Secretaria de Desarrollo Social, SEDESOL 2013).

Finally, on the economic characteristics and according to the 2011–2012 Municipal Development Plan (Plan de Desarrollo Municipal 2011) from the municipal government, Tonalá distinguishes itself by an important economic activity at state level, making it the third most important city where livestock stands out (89.90%) followed by agriculture (9.31%) and fisheries (0.80%) (H. Ayuntamiento de Tonalá 2011). The referred plan indicates that bovine livestock occupies the first position in production rate, while mango and maize production stands out in agriculture besides tamarind, watermelon, orange, and other fruit cultivations. Furthermore, the municipality is known nationally by the production of mango Ataulfo and finally by fishery, which is a locally important activity developed in different coastal lagoons although it is for self-consumption.

36.4 Materials and Methods

This study was performed in two phases.

Phase 1 The search of academic research articles, theses, and press articles was performed to allow establishing the conceptual framework, state of the art, and characteristics of the area of study by a cartographic delimitation process with the support of the Geographic Information Systems (SIG, for its acronym in Spanish) to perform a subsequent process characterizing the micro-basin. For this purpose, a delimitation was used based on the relief and hydrology in Google Earth (2017) software and its representation in ESRI Arc Map (2016).

Phase 2 Field work and observation research implied the application of a closed and structured survey directed to the urban population of the communities El Paredón and Tonalá. The survey was composed of 18 questions distributed in 4 sections, which were (1) perception of the problem, (2) origin, (3) effects, and (4) solutions facing hydric pollution, according to the methodological criteria described by Benez et al. (2010), Bustamante et al. (2016), and Valera (2002). Additionally, a representative sample of the urban population was determined, using a simple sampling that considered the population of both localities as the universe starting from the Population and Housing Census, indicating El Paredón had a total population of 6 126 inhabitants and Tonalá 35 322 (INEGI 2010).

Likewise, a sectorization of the area of study was performed using the Basic Geostatistics Areas (AGEBs for its acronym in Spanish); the micro-basin was formed by 24 AGEBs normalized for a better identification in zones from 1 to 24, considering zones 1–20 inhabitants for Tonalá and zones 21–24 for inhabitants of El Paredón; from the total, 13 AGEBs were characterized for having a population ≥1500 inhabitants. The analysis technique of the gathered data was based on descriptive statistics using the nonparametric test of qualitative variable of chi-square hypothesis by the software IBM SPSS (2014). The hypothesis setup was divided by:

- The environmental perception vs individual location

  $P_s > 0.05 = H_{0\ (null)}$ "The perception is independent of the location."
  $P_s < 0.05 = H_{1\ (alternative)}$ "The perception depends on the location."

- The environmental perception vs residence time

  $P_s > 0.05 = H_{0\ (null)}$ "The perception is independent of residence time."

$P_s < 0.05 = H_{1\ (\mathrm{alternative})}$ "The perception depends on residence time."

Additionally, this phase was supplemented with the collection of information with open interviews, which were structured with questions predesigned and applied by discretional sampling addressed to directors of governmental environmental institutions with social representation in the micro-basin.

36.5 Results Obtained

Data shown were obtained starting from a statistical analysis where sociodemographic variables stood out with the purpose of knowing the social environment where the research was developed, observing that the level of education of the respondents was basic and in their majority female population (see Table 36.1).

Table 36.1 Summary of social and environmental impacts by contaminants in El Riito stream, Chiapas

Author	Environmental impact and health	Observations
CEAS (2007)	Low quality in terms of the potability that is caused by the discharge of wastewater from domestic, recreational, and agricultural use	In the direct discharge of wastewater without treatment, its potability has decreased, so direct consumption for domestic use is no longer feasible.
	Low level of drinking water quality that makes aquatic life less viable	A large part of the population has access to health services.
	Likelihood of contamination of the El Riito and Borbollón streams, thereby contracting diseases of sanitary origin	
Arellano (2018)	Beating of the aquifer to provide the resource to the municipal capital Tonalá	
Tovilla (2004)	Serious erosive processes in the upper and middle basin of the municipality of Tonalá	There are diagnoses on the problem of water pollution, and there were general recommendations to address them.
	Depletion of the water tables of the Zanatenco River caused by the use of water and deforestation, which affected the cities of Tonalá	
	Lack of monitoring and sanitation of water from streams and rivers, being that wastewater and solid waste discharges are significant	
Graniel and Carrillo (2006)	Severe effects on the quality of water that reach the Riito are reported and a potential danger to the health of human populations, especially gastrointestinal diseases, is detected	
Machuca (2014)	Pollution of rivers and streams caused by activities that occur in Tonalá that severely affects the coastal area	The productive activities have contributed to the deterioration of water bodies and are located in the municipality of Tonalá and therefore affect the coastal area of the micro-basin.

Source: Own production with information gathered

36.5.1 Pollution Perception by Distance

The first category of environmental perception vs individual location determined that with Pearson chi-square of 36.01[a], the alternative hypothesis of dependence between the variables analyzed was accepted. While examining the relationship, the population sampling of those who live in the AGEBs 1, 2, 4, 6, and 7 were identified as tending to perceive pollution in its totality because they lived closer to the urban stream. On the contrary, with the AGEBs 14 and 21, whose population showed a greater number of cases of not knowing the environmental situation, they represented the AGEBs farther from the stream (Fig. 36.3).

36.5.2 Perception with Respect to Residence Time

The environmental perception vs residence time determined that with Pearson chi-square of 35.88[a], the alternative hypothesis of dependence between the variables analyzed was accepted. When the relationship was analyzed, the group from 13 to 23 years of residence (25.30%) was identified as having greater cases of hydric pollution perception. Thus, residence time represented a characteristic that allowed knowing that the inhabitants having 13 or more years of residence were those who tended to have greater experience and knowledge on how their environment has changed gradually, in this case, related to hydric pollution.

It is important to mention that in the rest of the markers that conformed the survey (sex, age, education, and occupation) significantly higher asymptotic values were observed to those shown previously (0.64, 0.81, 0.95, and 0.26), which statistically indicated a rejection of the alternative hypothesis and acceptance of the null hypothesis of independence between variables.

36.5.3 Problem Perception

In this criterion, 82.70% of the sampling admitted the existence of the stream pollution problem; from the sampling percentage that accepted the problem, 29.03% considered as pollution the fact that the stream had fetid smell, and 47.58% of the sampling assessed stream pollution as high degree (see Table 36.2).

36.5.4 Cause Perception

This criterion identified that 55.65% had the opinion that pollution was due to wastewater from the municipal sewage; on the other hand, from the perspective of the self-assessment as agents of change, 50.81% did not consider their daily activities caused the stream pollution (see Table 36.3).

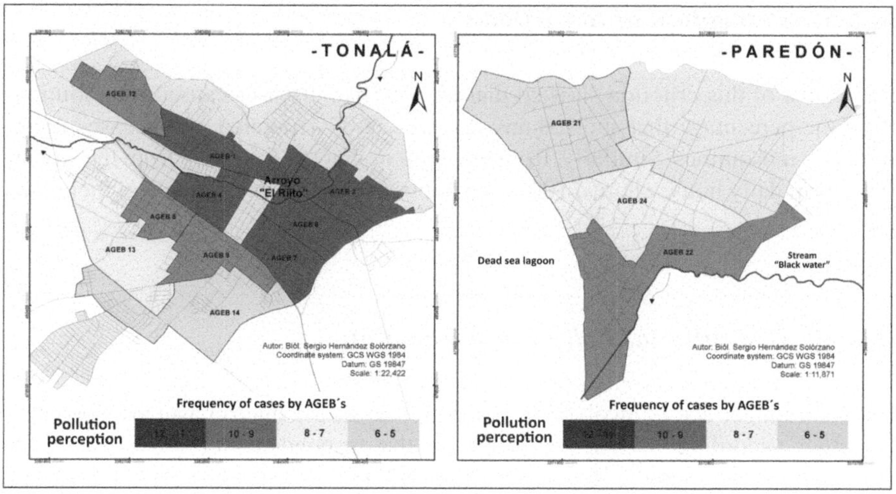

Fig. 36.3 Distribution of cases in communities that perceive pollution by environmental perception with a geographic approach (AGEB). (Source: own creation with the software ESRI Arc Map (2016))

Table 36.2 Evolution of the total population in the urban communities of Tonalá and El Paredón of the micro-basin El Riito, Chiapas, Mexico

Year	CLAVE	Location	Population (inhabitants)	Growth rate (1990–2000)	Growth rate (2000–2010)
1990	0001	Tonalá	26,919		
2000	0001	Tonalá	31,212	15.95%	13.17%
2010	0001	Tonalá	35,322		
1990	0040	Paredón	5319		
2000	0040	Paredón	5846	9.91%	4.79%
2010	0040	Paredón	6126		

Source: Compiled from data from the INEGI (1990, 2000 and 2010)

Table 36.3 Characteristics of residual water discharge in the micro-basin El Riito, Chiapas, Mexico

Invoice	Location	Volume/download (m^3/year)	Type of download	Download procedure	Receipt of download
11chs154980/23hada15	Paredón	157,680	Urban public	Municipal sewer	Riíto stream
11chs129993/23haoc08	Tonalá	1,634,826.40	Municipal	Municipal sewer	Riíto stream
11chs154570/23hsda14	Tonalá	3784.32	Urban public	Municipal sewer	Riíto stream
11chs132364/23esoc08	Tonalá	7446	Services	Sanitary wash, trail, and slaughter	Riíto stream

Source: Elaboration based on data recorded in REPDA (2015)

36.5.5 Perception of the Effects

The results of this criterion showed that 61.30% had diseases caused by pollution. From the percentage that showed any disease, 72.37% pointed out diseases transmitted by mosquitoes. Similarly, from the point of view of water use from the stream to perform daily house chores, 100% did not use stream water to perform them (see Table 36.4).

36.5.6 Institutional Pollution Perception

The institutional actors involved in hydric topics in the micro-basin are shown in Fig. 36.4 according to that indicated by the respondents of the Ministry of Environment and Natural Resources (SEMARNAT, for its acronym in Spanish) Region IX Istmo Costa, the Consejo de Cuenca de la Costa de Chiapas, and the Comité de Cuenca del Río Zanatenco. All of them knew and agreed that alteration in water quality existed in the stream El Riito, highlighting that the pollution affected both inhabitants of Tonalá and El Paredón and other rural communities settled south of the municipality. Additionally the representative of the Consejo de Cuenca de la Costa de Chiapas added that serious problems also existed in the municipalities of Mazatán and Huixtla located in the southern coast of Chiapas (interview performed on 12 January 2018).

The pollution problem referred to in this study has been evident when social demonstrations have taken place in front of the offices of SEMARNAT and the Comité de Cuenca del Río Zanatenco, which is why actions have been performed by the Temporal Employment Program (Programa de Empleo Temporal, PET, for its acronym in Spanish), such as recollection actions of solid waste on streets and avenues of the municipality of Tonalá.

On the other hand, the respondent of the Comité de Cuenca del Río Zanatenco mentioned that during the activities of the reforestation and cleaning projects, the inhabitants of the communities expressed their unrest due to the fetid odors and the presence of waste in the stream; this situation has made the inhabitants of El Paredón block the road that leads to the fishery bay and access to the municipal city hall of Tonalá in repeated occasions, so their demands may be heard, among them, the petition of solving the stream pollution. It is worth to point out that in general the institutional actors agreed that the main activities that modify the stream water quality are dumping waste (bags, cans, tires, etc.) and wastewater discharge from the municipal sewage system; additionally, they mentioned wastewater discharge of the cheese and meat enterprises.

An unfavorable expressed situation considered that a solution to the problem has not been provided due to the dissolution of SEMARNAT Region IX Istmo Costa, which at the moment is not performing projects to solve the pollution problem. This responsibility was turned to the competence of the central offices in Tuxtla Gutiérrez,

Table 36.4 Coverage of public or private health services provided to the population with rights in the sample of AGEBs of Tonalá and Paredón

Communities	AGEB	Total population	Population without health services	Population with health services	IMSS rightful population	ISSSTE rightful population	State ISSSTE rightful population	Population entitled to popular insurance
Tonalá	027A	2255	663	1582	561	308	108	571
	0284	4694	1397	3282	983	470	81	1672
	0301	2151	591	1555	569	221	144	598
	0320	2904	931	1967	778	314	105	759
	0335	3604	1435	2153	671	304	79	1040
	034A	1944	621	1322	537	234	119	440
	035A	2809	787	2016	726	250	146	893
	0405	1815	502	1309	684	161	93	352
	0513	2584	735	1832	508	388	36	894
	0528	2673	775	1886	715	220	16	867
Paredón	0439	1504	523	962	48	48	9	863
	0443	1751	864	883	70	40	0	767
	0636	1686	695	990	96	88	48	745

Source: Population and Housing Census 2010 in INEGI (2013)

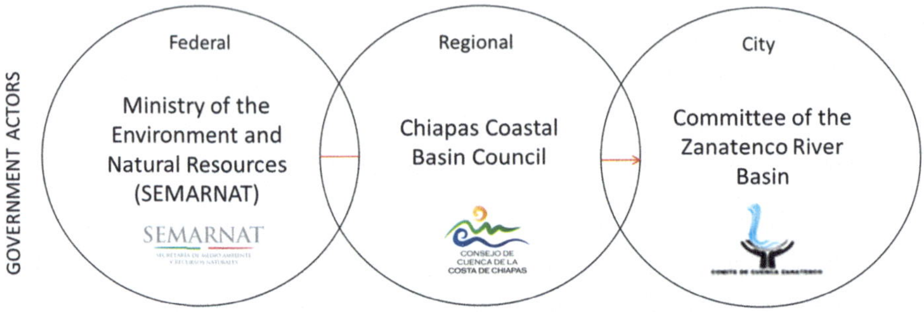

Fig. 36.4 Institutional actors interviewed. (Source: own production with field data)

deriving from an inadequate administration. Although the Comité de Cuenca del Río Zanatenco has tried to deal with the problem, the projects and actions performed, such as waste collection in the river bed and campaigns of environmental culture, have resulted insufficient.

To sum up, the institutional and social perceptions agree that a serious pollution problem exists in El Riito although the institutions indicated a series of actions and collaboration projects that were made to solve the problem. Apparently this situation has not been detected by the population, who considered that the institutions have not performed any action and including them as part of the problem for the waste discharge in the stream El Riito.

36.5.7 Solution Perception

Finally, this criterion identified that 46.77% had the opinion that the city hall of Tonalá should provide the solution to the pollution problem, followed by 23.39% who pointed out to the governmental environmental institutions; 16.94% pointed to the inhabitants of the municipality, and in a lesser degree, 12.90% expressed the solution should be in charge of the state government. Additionally, 77.40% would be willing to participate in any program to diminish pollution, while 22.60% were not willing to contribute to the solution to the problem. From the economic perspective, 55.60% would be willing to collaborate economically to create a fund for the solution and mitigate pollution, while 44.40% would not be open to contributing economically.

36.6 Analysis and Discussion of the Results

The perception assessed for the two communities was similar to those reported by Bush et al. (2000) and Martínez et al. (2007), who identified that the inhabitants of metropolitan areas perceived air pollution more than those who inhabited peripheral areas. Likewise, authors, such as Plata and Ibarra (2015) and Gómez (2017), referred that residents in adjacent areas to the polluted rivers were those who perceived the problem more. However, living close to the problem caused a constant interaction in such a way that they learned to live in that space with a condition that affected them and they got adapted to it.

Similarly, the behavior identified during the time of residence from the respondents was similar to that reported by Orzanco (1999), whose research indicated that population with transitory residence time tended to not have deep environmental knowledge on their surroundings. Whereas the permanent residents detected to a greater detail the events and changes in the environment. Something similar was reported by Giraldo (2013), indicating that longer residence periods in an area provided greater popular community knowledge.

In the perception of the pollution origin, the results obtained were equivalent to those reported by Crona et al. (2009), who argued that urban wastewater discharges were perceived as the main pollution cause in rivers. Furthermore, while analyzing the degree of pollution, the results showed that it was high in Tonalá and very high in El Paredón, similar to those reported by Plata and Ibarra (2015), who determined that wastewater discharge was the main problem perceived by a community located in the lower part of a basin and who assessed a very high degree of pollution.

In the perception of the stream pollution effects, the results agreed with data reported by the Health Ministry in Chiapas (Secretaría de Salud 2016), which has intensified prevention and control actions against diseases transmitted by vectors. On the other hand, in the use of stream water in house chores, the results indicated a complete negative, which contrasted with those of Isupov (2014), who considered that a clean stream that flows close to home should add possibilities of esthetic value using it at home and promoting recreational activities.

At the level of institutional perception, the three institutional actors have detected the problem and the social actors causing the problem, including wastewater discharges from the municipality; however, the actions performed by SEMARNAT, Consejo de Cuenca, and city hall (H. Ayuntamiento) have not been able to provide a definite solution to the problem. Inhabitants have even performed actions, such as public demonstrations and closing booths, ways in which the population makes their discontent evident or demands attention to the problem. For this reason, a disassociation exists between society and the institutions in charge of mitigating El Riito stream pollution.

Therefore, several factors were detected influencing this disassociation between society and environmental institutions; one of them is the participation that Aguiluz et al. (2001) reported, who mentioned that participation is the individual way and conceptualization of how to participate; on the other hand, the competencies and obligations of each one of the persons involved are unknown; moreover, a limited communication exists between the two parts. Additionally, the internal institutional problems limit collaboration and knowledge both of the problem and the solution to prevent and mitigate pollution in water bodies, such as that of El Riito stream.

As to the perception of the pollution solutions, the results identified that the municipal government should be the one to solve the problem. Pagaza (2009) argued that the municipalities had a responsibility in direct, immediate, and unavoidable terms toward the environment since, more than a legal obligation, it represented a social commitment between the government and society. On the other hand, a high citizen disposition to participate in environmental problems was identified in both localities, same as in the study of Runfola et al. (2008). Likewise, a considerable disposition was identified to carry out economic contributions to a fund for the solution of the problem. At the same time, Saldón (2012) indicated that the disposition to perform economic contributions represented a way of economically valuating an improvement in environmental quality.

Finally, another important aspect found in this research refers to the dominium of a population with a low education level and probably related to the high degrees of urban margination. This aspect was analyzed by Macías et al. (2010), who considered that when inhabitants economically weak do not have a basic academic degree, they mainly develop work where pay is not sufficient. Furthermore, the concept of the urban stream has not been legally typified, which is why it is unknown what institutions should take part in face of this socio-environmental situation. In this matter, regulations exist on wastewater discharge, such as the permit application issued by the National Water Commission (Comisión Nacional del Agua, CONAGUA-01-001-2017), in which the quality of discharge is set out, according to the Mexican official norms (NOM-001-SEMARNAT-1996, NOM-002-SEMARNAT-1996, and NOM-003-SEMARNAT-1997, Conagua 2000), as well as the repair or compensation for the environmental damage when waste discharge affects public health.

36.7 Conclusions

The perceptions of social actors of the communities analyzed and the governmental institutions (federal, state, and municipal) agreed in relation to water pollution of the urban stream called El Riito. Nevertheless, a disassociation exists between the activities performed by the institutions to diminish such pollution although the social actors of both communities identified and recognized themselves as part of the problem and had the purpose of participating collectively in favor of their environment. The communities considered that the local institutions were the main responsible entities that did not carry out the activities and actions of their legal competence. A similar response was reported when time of residence in site was assessed; in other words, those who had more time living in the locality perceived pollution more because they had more knowledge on how their environment has been gradually transformed; their perception sharpens same as the actions of the federal, state, and municipal institutions with respect to the problem and time of acting.

36.7.1 Lessons Learned

The proposal for the geographic perception provides, as learning, not only the knowledge generated by the social entities involved in the pollution problem but also information on the existing complexity in the socio-environmental systems in space and time. These means have opened the need for searching adequate strategies for environmental planning of urban spaces to solve, mitigate, and prevent the social problems in the framework of an integrated management of natural resources.

Even though this analysis proposal is the first approach to a complex problem, it allowed suggesting some recommendations and actions directed toward decision-makers at the level of the basin and municipality under a paradigm of an inclusive policy that implies a greater co-responsibility between social actors. Such recommendations are, for example, campaigns of water culture and care of natural resources, community meetings as a mechanism of participation, a community cleaning project, and promotion to the compliance of the environmental regulation in force, among others. The knowledge generated through this study should allow contributing to the design of truly public environmental policies directed toward the general interest under the scheme of a new hydric sustainability model in this basin and others in the country or elsewhere.

Furthermore, the contribution of this study with an environmental perception with geographic approach is a contribution on specific knowledge of the behavior of the micro-basin inhabitants facing and suffering alterations in their natural environment and how these spatial-temporal aspects, age, and interaction between social entities allow providing guidelines to specific or group actions, as the case may be.

References

Aguilar FM, Brenes VG (2013) La percepción como herramienta para la gestión del riesgo-aporte para la co-gestión comunitaria. Comisión Nacional de Prevención de Riesgos y Atención de Emergencias (CNE). San José, República de Costa Rica, Costa Rica. http://www.relaciger.org/revista/pdf/spa/doc1101/doc1101-contenido.pdf. Accessed on 6 Feb 2017

Aguiluz CA et al (2001) Planeación ambiental participativa: de la teoría a la práctica en San Cristóbal de Las Casas, Chiapas Estudios Demográficos y Urbanos, núm. 47, May Aug, pp 321–349

Ames AJ (1951) Visual perception and the rotating trapezoidal window. Psychological monographs: general and applied, 65(7). Pion Publication. Great Britain, Europe, pp 277–282. http://journals.sagepub.com/doi/pdf/10.1068/p160277. Accessed on Apr 2017

Arellano ML, Ruiz MLE (2016) Caracterización climatológica, variabilidad climática y eventos hidrometeorológicos extremos. Instituto de ciencias biológicas centro de investigaciones costeras, Tuxtla Gutiérrez, Chiapas, México, pp 1–103

Arellano Monterrosas, JL, Ruiz Meza, YL (2018) Evaluación y tendencias de los servicios ecosistémicos hidrológicos de la cuenca del río Zanatenco, Chiapas. Investigaciones Geográficas, Instituto de Geografía. UNAM issn (digital): 2448–7279. https://doi.org/10.14350/rig.59467

Arizpe L, Paz F, Velázquez M (1993) Cultura y cambio global: percepciones sociales sobre la deforestación en la selva lacandona. Centro Regional de Investigaciones Multidisciplinarias-Porrúa, México

Ayuntamiento MH (2011). Plan municipal de desarrollo 2011-2012. Instituto de Administración Pública del Estado de Chiapas. pp. 1–94

Bailly A (1984) Practique et perception de lé space. Les principaus corants de recherché dans le monde. Acta de table Ronde Iternationale. Universite de Pa et des Pays de LáDOUR

Benez MC, Edith F. Kauffer Michel, Álvarez GGC (2010).Percepciones ambientales de la calidad del agua superficial en la microcuenca del río Fogótico, Chiapas. Revista Frontera Norte 22(43), Tijuana, Baja California, México, pp 129–158

Bonnes MGC et al (2004) Inhabitant environmental perception in City of Rome within the framework for urban biosphere reserves of the UNESCO programme on man and biosphere. Ann N Y Acad Sci 1023(1), New York City, United States, pp 173–184

Brunswik Egon (1956). Systematic and Representative Design of Psychological Experiments with Results in Physical and Social Perception. University of California Press, Berkeley, California, United States, p.p. 143–202

Bush J et al (2000) Even the birds round here cough: stigma, air pollution and health in Teesside. Health Place 7(1), University of Newcastle, Newcastle Upon Tyne, England, pp 47–56. https://www.sciencedirect.com/science/article/abs/pii/S135382920000037X. Accessed on 31 2017

Bustamante-González Á, Jesús GG-D,Jaramillo-Villanueva JL, Vargas-López S (2016) Percepción de la contaminación del río Tlapaneco por la población ribereña.. Agricultura Sociedad y Desarrollo 13(1):47–62 Recuperado de: http://www.revistas-conacyt.unam.mx/asyd/index.php/asyd/article/view/27

Capel SH (1973) Percepción del medio y comportamiento geográfico. Revista de Geografía 7(1), pp 58–150. Universidad de La Rioja, Barcelona. https://core.ac.uk/download/pdf/39036479.pdf. Accessed on Dic 22, 2016

CEAS (Comisión Estatal del Agua y Saneamiento) (2007) Construcción de la planta de tratamiento de aguas residuales de la localidad de tonalá, Chiapas. Resumen ejecutivo. 1–45 pp. http://sinat.semarnat.gob.mx/dgiraDocs/documentos/chis/resumenes/2008/07CH2008HD004.pdf. Accessed on 20 Jan 2018

Conroy D R (2005) Space syntax and spatial cognition. World architecture: space syntax monograph. 11(185), Cambridge, United Kingdom, pp 41–45. ISSN 2044. Accessed on 29 Dic 2017

Creswell J (2008). Mixed methods research: state of the art [power point presentation]. University of Michigan. Recuperado de sitemaker.umich.edu/creswell.workshop/files/creswell_lecture_slides.pptConagua. (Comisión Nacional del Agua) (2006). El agua en México. Estado de México, México, pp 1–37. http://www.conagua.gob.mx/CONAGUA07/Publicaciones/Publicaciones/EL-AGUA-EN-MEXICO.pdf. Accessed on 29 Dic 2019

Crona B et al (2009) Murky water: analyzing risk perception and stakeholder vulnerability related to sewage impacts in mangroves of East Africa. Glob Environ Change, Arizona State University, United States, 19(2), pp 227–239. https://www.sciencedirect.com/science/article/pii/S0959378009000132. Accessed on 23 Apr 2018

DOF (Diario Oficial de la Federación) (2016) Ley de Aguas Nacionales, Última reforma publicada DOF 24-03, Cámara de diputados del h. Congreso de la unión, Estado de México, México, pp 1–107. http://www.diputados.gob.mx/LeyesBiblio/pdf/16_240316.pdf. Accessed on 24 Dic 2016

DOF (Diario Oficial de la Federación) (2018) Ley General de Equilibrio Ecológico y la Protección al Ambiente, Última reforma publicada DOF 05-06, Cámara de diputados del h. Congreso de la unión, México, D.F., pp 72–83. http://www.diputados.gob.mx/LeyesBiblio/pdf/148_050618.pdf. Accessed on 10 June 2019

Esri (Environmental Systems Research Institute) (2016) *ArcMap Software*. Version 2016 – desktop for Windows 10 1, Redlands, California, United States. En línea: <https://www.esri.com> [Consultado en Diciembre 1, 2016]

Fernández MY (2008) ¿Por qué estudiar las percepciones ambientales? Espiral, Estudios sobre Estado y Sociedad XV(43), Guadalajara, Jalisco, México, pp 179–202. http://www.scielo.org.mx/scielo.php?script=sci_abstract&pid=S1665-05652008000100006&lng=es&nrm=iso. Accessed on 23 2017

Fundesnap, (Fundación para el Desarrollo del Sistema Nacional de Áreas Protegidas) (2008) ¿Qué es Calidad Ambiental? Sustentabilidad corresponsable, La Paz, Sucre. http://fundesnap.org/files/comites_locales_cepf.pdf. Accessed on 21 2017

Gibson JJ (1980). Elementos básicos de psicología ambiental. Universidad de Barcelona, España, En línea: http://www.ub.edu/psicologia_ambiental/. Accessed on 23 Dic 2016

Giraldo VC (2013) Percepción y adaptación en poblaciones rurales de Manizales y aledaños: Dinámicas entorno a la construcción y vivencia del Cambio Climático y Variabilidad Climática. Universidad nacional de San Martín, Buenos Aires, Argentina, pp 1–182. http://www.academia.edu/20037870/Percepci%C3%B3n_y_adaptaci%C3%B3n_en_poblaciones_rurales_de_Manizales_y_aleda%C3%B1os_Din%C3%A1micas_entorno_a_la_construcci%C3%B3n_y_vivencia_del_Cambio_Clim%C3%A1tico_y_Variabilidad_Clim%C3%A1tica. Accessed on 13 May 2018

Gómez LPM (2017) Prácticas sociales en el barrio el guabo por la contaminación del río Machángara. Universidad Politécnica Salesiana. Quito. https://dspace.ups.edu.ec/bitstream/123456789/13537/1/UPS-QT11307.pdf. Accessed on 28 Mar 2018

Google (2017) Google Earth Software. Version 7.1.7.2606 for Windows. Edición Monterrey, Nuevo León, México. https://earth.google.com/web/. Accessed on 14 June 2017

Graniel CE, Carrillo CME (2006) Calidad del agua del río Zanatenco en el estado de Chiapas. Ingeniería 10(3):35–42. http://www.revista.ingenieria.uady.mx/volumen10/calidad.pdf. Accessed on 28 May 2018

Guerrero P (2011) La guía de Geografía. Geografía. Estado de México. México. http://geografia.laguia2000.com/general/arroyo. Accessed on 9 June 2017

Gutiérrez C and Peña S (1996) La percepción geográfica como factor en el desarrollo local. Centro de Estudios Sociales Rev. Última Década, núm. 4, 1996, pp 1–http://www.redalyc.org/pdf/195/19500412.pdf. Accessed on 20 May 2019

Ihne (Instituto de Historia Natural y Ecología) (2004) Programa de Ordenamiento Ecológico Territorial de la Subcuenca del Río Zanatenco. Tuxtla Gutiérrez, Chiapas, pp 2–9

INAFED (2009) Enciclopedia de los municipio y delegaciones de México. Retrieved December 18, 2016, from http://www.inafed.gob.mx/work/enciclopedia/EMM07chiapas/municipios/07097a.html

INEGI (Instituto Nacional de Estadística, Geografía e Informática) (2010) Censo de Población y Vivienda 2010. Datos. Ciudad de México. http://www.beta.inegi.org.mx/proyectos/ccpv/2010/. Accessed on 8 Oct 2018

INEGI (Instituto Nacional de Estadística, Geografía e Informática) (2012) Productos y servicios. Inventario de productos geográficos digitales. Datos. Ciudad de México, México. http://www.inegi.org.mx/geo/contenidos/recnat/usosuelo/. Accessed on 21 Apr 2017

Instituto Nacional de Estadística y Geografía. (2013). Censo de Población y Vivienda 2010. Retrieved April 25, 2017, from http://www.inegi.org.mx/sistemas/consulta_resultados/ageb_urb2010.aspx?c=28111

International Business Machines Corporation (IBM). Released. [Software de computación] (2014). IBM SPSS Statistics for Windows, Version 23.0. Armonk, NY: IBM Corp.

Isupov D (2014) Guía para el Mantenimiento de Arroyos Urbanos. Russian River Watershed Association, Ukiah, California, United State, pp 1–15. http://www.rrwatershed.org/wp-content/uploads/2016/05/CCGSpanishForWebsite.pdf. Accessed on 23 Abr 2018

Lazos EC (2010) Percepciones y responsabilidades sobre el deterioro ecológico en el sur de Veracruz. Población y medio ambiente, descifrando el rompecabezas. El Colegio Mexiquense. Sociedad Mexicana de Demografía. Toluca estado de México, México, pp 235–272

Lazos E, Paré L (2000) Miradas indígenas sobre una naturaleza entristecida: percepciones del deterioro ambiental entre nahuas del sur de Veracruz, México. IIS, UNAM-Plaza y Valdés Editores, Estado de México, México, pp 23–34

Lezama JL, Graizbord B (2010) Los grandes problemas de México, 1st ed. El colegio de México, Estado de México, México. http://2010.colmex.mx/16tomos/IV.pdf. Accessed 1 Jan 2017

Lovelock J (1992) The Evolving Gaia Theory. Tokio, Japan: Paper presentedat the United Nations University. 1992, http://archive.unu.edu/unupress/lecture1.html. Accessed 20 Mat 2019

Machuca JMA (2014) Diagnóstico del estado actual del uso y manejo de la zona costera en el Estado de Chiapas. Tesis de Maestría. El Colegio de la Frontera Sur, Tapachula, Chipas, México, 194 pp

Macías HM, del C et al (2010) Distribución territorial de los índices de marginación en las localidades urbanas del municipio Tlaquepaque, Jalisco. Sincronía, N°. 4, Universidad de Guadalajara, Jalisco, México. ISSN-e: 1562-384X, pp 1–7. http://sincronia.cucsh.udg.mx/maciashuertaetalwinter2010.htm. Accessed on 21 Apr 2018

Martínez LEM et al (2007) Contaminación atmosférica y efectos sobre la salud de la población, Universidad de Antioquia, Medellín, Colombia, pp 1–109. https://www.medellin.gov.co/irj/go/km/docs/wpccontent/Sites/Subportal%20del%20Ciudadano/Salud/Secciones/Publicaciones/Documentos/2012/Investigaciones/Contaminaci%C3%B3n%20atmosf%C3%A9rica%20y%20efectos%20sobre%20la%20salud%20de%20la%20poblaci%C3%B3n.pdf. Accessed on 31 May 2017

Millan EM (2002) La geografía de la percepción una metodología de análisis para el desarrollo rural. Papeles de Geografía, 133–149 p

Mora D, Carlos Felipe Portuguez, Brenes G (2002) Evaluación de la contaminación fecal de la cuenca del río Tempisque 1997–2000. Revista Costarricense de Salud Pública 11(20): San José, Costa Rica, pp 5–17

Moyano EPÁ, Lafuente R (2009) Políticas ambientales, cambio climático y opinión pública en escenarios regionales: el caso de Andalucía. Revista Internacional de Sociología 67 (3), Instituto de Estudios Sociales Avanzados, Córdoba, España, p.p. 681–699

Olmos ME et al (2016) Opinión pública y percepción sobre la conservación de la Reserva ecológica estatal Estero San José del Cabo y su zona de influencia, Revista de comunicación Vivat academia, Núm.135, Universidad Complutense de Madrid, España, pp. 24–40. https://doi.org/10.15178/va.2016.135.24-40. Accessed on 12 Dic 2017

Orzanco MG (1999) Problemas ambientales detectados por la población de Ushuaia (Tierra del Fuego, Argentina). Investigaciones Geográficas, Núm. 40, Universidad Nacional de La Plata, Buenos Aires, Argentina, pp 85–98. http://www.scielo.org.mx/scielo.php?script=sci_arttext&pid=S0188-46111999000300007. Accessed on 12 Feb 2017

Padilla Sotelo SL, Luna Moliner MA (2003) Percepción y conocimiento ambiental en la costa de Quintana Roo: una caracterización a través de encuestas. Perception and environmental knowledge across Quintana Roo's coast: a poll-based characterization, 52, 99–116. Retrieved from http://www.scielo.org.mx/pdf/igeo/n52/n52a7.pdf

Pagaza PI (2009) Responsabilidades municipales en materia ambiental. Convergencia 16(49), Universidad Autónoma del Estado de México, México, pp 291–308. http://www.scielo.org.mx/scielo.php?script=sci_arttext&pid=S1405-14352009000100012. Accessed on 23 April 2018

Pereira PZ (2011). Los diseños de método mixto en la investigación en educación: Una experiencia concreta Mixed Mthod Designs in Education Research: a Particular Experience. Revista Electronica Educare, Núm.1, Universidad Nacional Heredia, Costa Rica, pp 15–29. http://www.redalyc.org/pdf/1941/194118804003.pdf. Accessed on 7 Mar 2017

Plata ÁM, Ibarra VD (2015) Percepción local del estado ambiental en la Cuenca Baja del Río Manzanares. Luna Azul Núm. 42, Universidad Sergio Arboleda, Bogotá, Colombia, pp 1–21. http://www.scielo.org.co/pdf/luaz/n42/n42a15.pdf. Accessed on 13 June 2018

Repda (Registro Público de Derechos de Agua) (2015) Datos Estado de México, México, En línea: https://www.gob.mx/conagua/acciones-y-programas/registro-publico-de-derechos-de-agua-repda-55190. Accessed on 2 Aug 2017

Runfola J et al (2008) Diagnóstico de los desechos sólidos generados en las comunidades del programa agenda plátano, para el diseño de un sistema piloto de manejo integral en pequeñas comunidades, Universidad de Los Andes, Mérida, Venezuela. http://www.redisa.net/doc/artSim2008/gestion/A32.pdf. Accessed on 15 May 2018

Saldón M (2012) Valoración ambiental del reciclado de residuos : El caso de Quilmes, Argentina. Economia 34(34), Centro de Estudios de Estado y Sociedad, Buenos Aires, Argentina, pp 33–53. http://www.redalyc.org/html/1956/195631019003/. Accessed on 15 May 2018

Secretaría de Salud (2016) Jurisdicción Sanitaria de Tonalá y ayuntamientos de la región Istmo-Costa, unidos contra dengue, chikungunya y zika. Gobierno del estado, Tuxtla Gutiérrez, Chiapas, México, En línea: http://saludchiapas.gob.mx/?p=7749 [Consultado el April 23, 2018]

SEDESOL (Secretaria de Desarrollo Social) (2013) Resumen municipal. Sistema de Apoyo para la Planeación del PDZP. Dirección general adjunta de planeación microrregional, Estado de Chiapas, México, En línea: http://www.microrregiones.gob.mx/catloc/LocdeMun.aspx?tipo=clave&campo=loc&ent=07&mun=097 [Consultado el May 8, 2018]

Sescovich SR (2016) La naturaleza de las percepciones. Conducta Humana. Buenos Aires, Argentina. https://cmapspublic2.ihmc.us/rid=1NZTZ3Q1M-132T4RR-2GXB/PROCESO%20DE%20SOCIALIZACI%C3%93N.htm. Accessed on 22 2016

Stea D, Downs R (2003) Cognitive Mapping and spatial behavior, Transaction Publishers, New Brunswick, United State. En línea: https://books.google.com.mx/books?hl=es&lr=&id=dnwkmQ1fuUAC&oi=fnd&pg=PR8&dq=Environmental+Geographic+Perception+and+Cognition&ots=xp_72wXZzP&sig=KRNy-Im96fci66JG6obxMurDSGI#v=onepage&q=Environmental%20Geographic%20Perception%20and%20Cognition&f=false [Consultado el January 30, 2017]

Tommasino H Foladori G, Taks J (2000) La crisis ambiental contemporánea. Trabajo y capital, Montevideo, Uruguay, pp 9–26. En línea: http://tallerdesustentabilidad.ced.cl/wp/wp-content/uploads/2015/04/La-Crisis-Ambiental-ContemporáneaTommasino-Foladori-Taks.pdf. Accedy Jan 1 2017]

Tovilla HC (2004) La dimensión de la crisis ambiental en la costa del Chiapas y la necesidad de un programa de ordenamiento de las actividades, Instituto de investigaciones científicas, Estado de México, Mexico. En línea: https://archivos.juridicas.unam.mx/www/bjv/libros/6/2837/5.pdf [Consultado el May 30, 2017]

Valera S (2002) Elementos básicos de psicología ambiental. Departamento de psicología social y psicología cuantitativa, Universidad de Barcelona, España. En línea: http://www.ub.edu/psicologia_ambiental/ [Consultado el December 23, 2016]

Vargas MLM (1995) Los colores lacandones: un estudio sobre percepción visual, Tesis doctoral presentada a la Escuela Nacional de Antropología e Historia, Ciudad de México, México, pp 23–26

Yi-Fu Tuan (1975) Images and mental maps. Ann Assoc Am Geogr 65(2):205–213. https://onlinelibrary.wiley.com/doi/abs/10.1111/j.1467-8306.1975.tb01031.x. Accessed on 12 June 1997

Part VII
Concluding Remarks

Chapter 37
Concluding Remarks

Alfredo Ortega-Rubio and María Carmen Blázquez

Abstract Around the world, hot spots of biodiversity are often coincident with hot spots of cultural and even linguistic diversity. Many of those hot spots have become Natural Protected Areas. Therefore, Natural Protected Areas are the ideal starting point to address socio-ecological studies. The aim of this book is to explore the current trend of economic and sociological studies in the Natural Protected Areas of Mexico and other countries. In the first section of the book, authors review the foundations of the socio-ecological relationship, i.e., the evolution of the human-natural systems. This first section is dedicated to theoretical aspects linking Community Development and Conservation where a new paradigm emerges, one that takes account of socio-ecological and cultural diversity as well as environmental ethics. Politics of protection without the human component are no longer viable for Natural Protected Areas. In the second section, the authors describe methodologies and explain that an inter-disciplinary approach is necessary to incorporate the social, the economic and the ecological aspects in a socio-ecological matrix. Some chapters analyze tools for the integration of "human" and "natural" dimensions in the management of the environmental matrix. Other chapters describe research into the benefits of ecosystem services, long-term socio-ecological research, and the local governance of common resources. In the third section of the book, there are some study cases explaining the interactions among the local inhabitants and the environmental systems in the Protected Natural Areas. The book aims to help understand how social groups contextualize ecological knowledge, how anthropic activities contribute to modify the environmental matrix, how cultural and economic aspects influence the use, management and conservation of their ecological environment, and how to understand social phenomena in the light of ecological knowledge.

A. Ortega-Rubio
Red Nacional Áreas Naturales Protegidas RENANP-CONACYT, Centro de Investigaciones Biológicas del Noroeste CIBNOR-CONACYT, La Paz, Baja California Sur, Mexico
e-mail: aortega@cibnor.mx

M. C. Blázquez (✉)
Behavioural Ecology Lab. Environmental Planning and Conservation Department, Centro Investigaciones Biológicas Noroeste (CIBNOR), La Paz, B.C.S., México
e-mail: blazquez@cibnor.mx

A. Ortega-Rubio (ed.), *Socio-ecological Studies in Natural Protected Areas*,
https://doi.org/10.1007/978-3-030-47264-1_37

Keywords Biodiversity · Conservation · Management · Culture · Socioeconomic and socioecological methodology

The conservation of biodiversity, the use of natural resources, and the community's ancestral knowledge have become a major issue to our generation. The concept of conservation, as just the preservation of pristine ecosystems, has long ago been proven to be unrealistic. Biological conservation science now recognizes that the economic value of the ecosystems services is the most powerful tool to encourage local people in the conservation of their land. It is also a new way to generate income for the local economy. The aim of this book is to explore the current trend of economic and sociological studies in the Natural Protected Areas of Mexico and other countries.

The book is organized in VI sections, and here we comment on some of the main conclusions of each of them.

The first part is dedicated to theoretical aspects linking Community Development and Conservation. In this section the authors reviewed the sociobiological approach that allows the authorities to deal holistically with the social and ecological systems encouraging sustainability.

They also revised the importance of ethnology to manage natural areas through consumptive and non-consumptive uses of their environmental services. The understanding of how the collective action emerges, the importance of a new conservation scheme, and the need of solving common alternatives should allow the communities and institutions (governmental and social) to be better prepared for a joint land and ecosystem management.

In this theoretical section, the authors explain the complexities and difficulties of modeling the sustainability of the ecosystems, because they have a large number of elements, and they are rich in non-linear interactions, with interacting feedback loops that are recurrent open and operate in conditions that tend to unbalance the systems as a whole. The author's opinion is that effective ecosystems sustainability research, requires an interdisciplinary and holistic approach, and cannot be simply modeled or solved by using only one narrow discipline of science. The complexity of sustainability is such that it needs to involve multiple scientific areas of study.

As part of the theoretical frame, a new paradigm emerges that takes account of socio-ecological and cultural diversity as well as environmental ethics. Politics of protection without the human component are no longer viable for Natural Protected Areas. One chapter highlights the consequences of applying regulations and norms in a Natural Protected Area excluding the social factor of the indigenous communities living in it. If a series of restrictions (i.e., on hunting and logging) are imposed without the accompaniment of sufficient financial compensation, local communities are inclined to break any bans, precluding the accomplishment of any of the original conservation goals.

In the second section of the book, the authors described some methodological models to achieve the changes proposed in the first section. As an example, an integrated (socioeconomic, governance, and environmental) model shows that it is possible to generate plausible zoning scenarios for beach management. This model takes into account the dynamic character of the beach geomorphology, the legal aspects related to natural protected areas, the biodiversity status, and the connection with surrounding communities. The most complex element within the model was to define the boundaries between management categories in the 2DH dimension. The authors propose as a future line of study the definition of transition areas (buffers) between zones.

In this section, other authors present a systemic methodology based on socio-ecological modeling and sustainability indicators, for the Management of Galapagos National Park. Very interestingly, the model results show that the conservation of the protected sites depends strongly on the dynamics outside them. Therefore, in this case the demands of the population and tourism outside the boundaries of the park pose a serious threat to the fauna and flora, because of the increasing risk of introduction of foreign species into the National Park. Additionally, in the model for water indicators system, the authors show a deep understanding of the problem is achieved only by integrating diverse indicators, along with the their causes and the expected benefits. They also highlight the limitations, and the un-intended effects of different policies and management options. Participatory approaches are crucial to support decisions and to engage all parties on the measures required in order to achieve any proposed goals.

Another chapter is about methodological models that include the application of the Integrated Coastal Zone Management (ICZM) Strategy for the Socio-ecological System in a marine estuary, in SE Spain. The conceptual frameworks followed for the model of public management of the lagoon and its surroundings were DPSIWR (drivers, pressures, state, impact, and response model of intervention) and the Decalogue of ICZM. It had a technical- scientific basis, but an important part of its development took place in the political and social arena. It was of great importance that the technical team (officials and consultants) knew how to interpret the new conceptual frameworks, as well as know how to develop them within political-administrative scenarios. The influence of a sound participatory process was another key element of the strategy.

The methodological aspects in this section include one chapter that analyzes the politics applied in a coastal NPA, in Cuba. There, the tools used by the government to protect the biodiversity are mechanisms of environmental education and training in responsible aquaculture techniques, along with good management practices and follow-up. This model increased the social wellness of the inhabitants of the area at the same time.

This section also delves into how the socio-historic and cultural knowledge of the local inhabitants of natural protected areas reflect the relationship among local indigenous people and nature in an integrating theoretical framework with a cosmic and not an anthropocentric approach.

The third section of the book is devoted to the social groups and their ecological knowledge, with some examples explained in depth. One of them explains the social and cultural significance of the hunt, represented by the types of hunting that can be practiced, the age to start hunting, the links that can be established between hunters, as well as the hunting rituals that perform both pre- and post-hunting, and legends and myths around the activity.

In this section, other authors analyzed the Nagoya Protocol policy instruments in the context of the public administration. Its scope and the instruments developed to date to protect the intellectual property and to improve the Biodiversity Conservation. The results of the analysis show that a large number of patents and plant breeder rights on genetic resources and plant varieties refer to ownership by countries, companies, and institutions that are not part of the Nagoya Protocol. This situation creates uncertainty about the destination of the profits from the exploitation of varieties and genetic materials from the countries adhered to the Nagoya Protocol.

The socio-environmental effect of coffee production in tributaries of a river in a NPA in Mexico is discussed in another chapter. Coffee production in the high areas of the basin has a great socioeconomic and environmental importance, but during the washing process of coffee beans, great volumes of water (mead) and pulp in the humid benefit stage (HBS) are discharged without treatment. Both products cause environmental and social impacts. The study detected that both coffee growers and representatives of institutions and organizations have known, or detected, the problem of water contamination and were willing to participate in finding a solution. Nonetheless, coffee producers considered themselves part of the problem because they know the importance of maintaining a basin with good water quality required for consumption of the population and development of coffee cultivation.

The IV section of the book describes cases of biodiversity conservation success from socio-ecological approach. One of them is the main methods of resistance used by local communities against open-pit mining projects. Unfortunately, a very high proportion of concessions for projects overlap with Protected Natural Areas, but the law favors the mining companies over the interest of the local communities, making it difficult to preserve their environment and their health. This chapter demonstrates that open-pit mining is not a synonym of development or progress; it does not generate well-being. By specializing in gold extraction, the consequences of its high degree ecotoxicity and pollution represent a threat to life, traditions, and peacefulness of rural populations where mining enterprises are located. That is why local communities, faced with the difficulties posed by these predatory companies, organize themselves to face these social and ecological tragedies. This has encouraged alliances among local groups and organizations, which has succeeded in increasing the knowledge about these projects and giving them visibility by contributing to new strategies such as legal judgments, scientific collaborations, and public consultations to reject mining projects. This confrontation often results in deep and alarming crisis of human rights in the field of environmental activism, sometimes with violent scenarios. There are even murders and extrajudicial executions against territory defenders. Ecological, cultural, subsistence, and economic values of the populations are highlighted in these environmental conflicts. They are values expressed in different scales, and they are not measurable. An urgent reform of current mining legislation is needed so that it complies with the constitutional principles and guarantees fundamental rights of indigenous people and communities. A new model is necessary where the inhabitants in areas of natural resources may teach and continue the form of exploitation and sustainability that they have performed for years and centuries.

Another chapter describes the success of a Unit of Environmental Management to protect an endemic species of crocodile in Mexico. The local people working as a cooperative combine the use of the mangroves as an intensive crocodile farm and as an ecotourist center. Its main objective is to conserve the crocodiles and the mangroves through sustainable use. This allows the generation of a broad platform of proposals for scientific and technological research that can consolidate a model of sustainable use of the mangrove and biodiversity conservation. The dynamism and the short time in which this project has become successful are the results of adopting the instruments and interdisciplinary analysis of the science of sustainability in natural protected areas.

Two other cases of success of socio-ecosystem institutional management in Natural Protected Areas in Mexico are analyzed. One case is devoted to the success of the conservation of relict forests of pine and fir trees. The work performed by the staff of the federal management office and guards is relevant, taking into consideration the low budget assigned to them. The park staff should continue with outreach programs of environmental education activities since the inhabitants have to be the promoters for the protection of their forests. For the other case of socio-ecosystem institutional management, the staff of this reserve have significantly increased the level of consciousness in local people about the importance of conservation, in less than 20 years since its creation. The local communities have now become the protectors of their own resources. The support of the reserve staff for installing UMAs with cactus production have helped locals to develop a strategy of sustainable use of the resources, which have also slowed down emigration in the communities, at least on a small scale.

The section V is about community well-being and improvement derived from ecological conservation. Local people from a very small coastal village in Baja California peninsula changed its main activity from traditional fishing to offer tourist facilities for observation of marine life, specifically to dive with Bull sharks in the wild. This happened after a reef and the marine area around it was declared a National Park 30 years ago. As a result, the economic benefits are now greater than ever before. The authors estimated the specific economic value of diving with Bull sharks by using a method of revealed preferences (Travel cost) and calculating the consumer surplus (CS) of diving with Bull sharks in the National Park. The economic analysis model revealed a CS result of $694 USD per person per day with an average of a 4-day stay, and $7 595 097 USD annually, as economic benefits. This information was useful to promote improvements in management and conservation of Bull sharks at the Park. Just as other shark species, Bull sharks and their habitat represent an important form of natural capital with high potential to produce economic value. Although the reef is the main attraction of the Park, this result is relevant because diving with Bull sharks in natural conditions has become an attraction that will increase in the next years, if proper management prevails.

Another example focuses on successful collaborative socio-ecological work in a Natural Protected Area. It is the case of a cooperative formed by community members of a small coastal village on the Pacific coast of Mexico to develop ecotourism, under the sponsorship and advisement of government officials. The result of this collaboration has generated a network of collective actions for the benefit of the community and conservation of the Protected Natural Area, and management of the coastal wetland and endangered species, such as the Olive Ridley sea turtle. The cooperative has had a positive impact on the well-being of the community, mainly due to the integration of inhabitants that are not members of the fishing cooperative. Incorporating previously non-economically productive community members (at least not directly), such as older adults, youth, and housewives into the workforce, contributes to the development of the locality. Another strategy identified as a success is responsibility and collaboration which highlights activities such as turtle release, cleaning, and maintenance of beaches, among others. The authors identified also a gender equality, since women in the cooperative have important positions and responsibilities at the same level as men. A broad respect for women's decisions emerged, where women are perceived as important and relevant for the association. This research shows the positive aspects of ecotourism, among which are the conservation of nesting sites and endemic species, surveillance, landscape protection, solid waste management, as well as control of the carrying capacity for site visits. One of the most relevant actions related to the conservation of this area is the Clean Beach certification obtained in 2012, which has been endorsed five times and includes the year 2019.

A third example of success in this section discusses the integration of rural/coastal groups in Protected Areas management, through productive social micro-entrepreneurships (SMEs) with small-scale mariculture and coastal fishermen in Baja California. The authors established that mariculture comes in many modalities according to the type of technologies used, the degree of intensification, and the species employed. Between the industrial production of large private enterprises and the small-scale, rustic, community-based, there exists a wide range of practices, types, and scales that could provide opportunities for greater synergies within multiple-use MPAs. Multiple-use designs could also offer alternative livelihoods to develop sustainable financing opportunities for the MPA management. Using the narrative visualization approach, the authors review the extension (capacity-building) programs in which the role of technologies and knowledge-holders signify decisive influences on the configuration of SMEs into or around protected areas.

The final section of the book is devoted to the analysis of governance changes from socio-ecological approach. One chapter highlights the importance of the future management and protection of the fragile and conflicted Islas Marias Archipelago. The Archipelago has a variety of parties all with their own and different approach to the use of the local resources. There are those of who desire to protect and conserve the natural resources and those who are intent on exploiting them to obtain economic benefits. The author suggests an urgent need to devise better governance schemes that allow managing conflicts, building consensus, and harmonizing activities toward the ultimate goal of sustainability.

Another chapter analyzes necessary mechanisms and socioeconomic assessment for creating successful resource management policies for a new Protected Area in

SE Mexico. First, a municipal-level policy roadmap begins with the local government council, which must first approve specific proposals. After that, this decision-making entity authorizes the mayor of the village to declare the reserve. However, achieving a coherently derived strategy to produce such an outcome requires considerable measures and additional actions for guaranteeing such an endeavor's success. A localized management strategy such as one proposed in the study would have to develop significant policies. First to educate local populations and second to cultivate and stimulate political inclusion strategies that could substantiate the inherent internal inter-community cohesion feedback factors necessary for proposing this area as a candidate for Presidential decree.

Another case focuses on a new project designed to promote ecotourism on a NPA and its surroundings in SE Mexico. The idea is that the ecotourism will spread the NPA success in the region, reducing pressure on the reserve while benefiting local communities, showing that there are profitable low-key tourism models that are environmentally and socially viable.

The last chapter describes a successful case in effective management of a National Park in the Gulf of California. The success was attained through the governance, planning, and design of an integral strategy for a rookery that is the only California sea lion breeding colony in the Gulf of California with a healthy growing population. Under favorable conditions, the rookery might become a core population from which historic and current breeding colonies in the Gulf of California that are at serious risk of disappearing in the next 20 years might be recolonized. In recognition of the proper management of the Park, the marine national park was included in the Green List of Protected and Conserved Areas of the IUCN in 2018. The first one NPA listed for the 185 Federal NPAs from Mexico.

All around the world hot spots of biodiversity are coincident with hot spots of cultural and even linguistic diversity. Most of the biodiversity hot spots are decreed as Natural Protected Areas. Therefore, Natural Protected Areas must be the focus point to address socio-ecological studies. This new book focuses on describing the interactions among the local inhabitants and the environmental systems in the Protected Natural Areas of Mexico. It aims to help understand how social groups contextualize ecological knowledge, how anthropic activities contribute to modify the environmental matrix, how cultural and economic aspects influence the use, management and conservation of their ecological environment and how to understand social phenomena in the light of ecological knowledge.

This book intends to review the foundations, from the epistemological and historical bases of the conception of the socio-ecological relationship to the description of the evolution of the human-natural systems. In the methodological aspects, it aims to analyze the tools required for the integration of "human" and "natural" dimensions in the management of the environmental matrix, to describe the research with the approach of ecosystem services, long-term socio-ecological research and the local governance of common resources. In the case studies section, this book reviews valuable and recent experiences on the retro-interactions of local inhabitants with their environmental matrix. Such publication would have a dramatic impact on researchers and professionals all over the world especially in Latin-American countries.

Undoubtedly, this is the very first book in English incorporating the Natural Protected Areas of Mexico and the region, the social and economic aspects of the local inhabitants with the environmental management and sustainability. The inter-disciplinary approach required to incorporate the social, the economic, and the ecological aspects in the socio-ecological matrix, is addressed by a wide array of author´s from all these disciplines, resulting in a truly multidisciplinary book. All authors are academic authorities, with a solid international prestige in socio-ecology.

As part of a developing science, this book provides both theoretical basis and methodological tools and key case studies, which will contribute for the advancement of the socio-ecological perspectives in Natural Protected Areas for Mexico, which can be easily extrapolated to other Latin-American countries.

Index

A. Ortega-Rubio (ed.), *Socio-ecological Studies in Natural Protected Areas*,
https://doi.org/10.1007/978-3-030-47264-1

L

M

Q

T

The manufacturer's authorised representative in the EU is Springer Nature Customer Service Centre GmbH, Europaplatz 3, 69115 Heidelberg, Germany. If you have any concerns regarding our products, please contact ProductSafety@springernature.com

Printed and bound by CPI Group (UK) Ltd, Croydon, CR0 4YY
30/09/2026
02225699-0017